Algebra

for JEE Main & Advanced

2000+
Problems with Solutions...

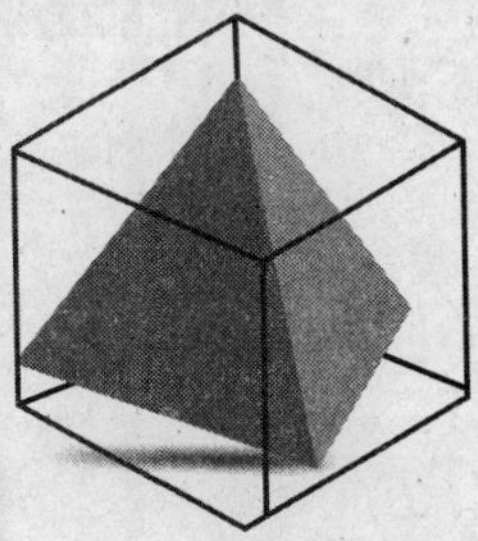

Unprobleм JEE

Algebra

for JEE Main & Advanced

2000+

Problems with Solutions...

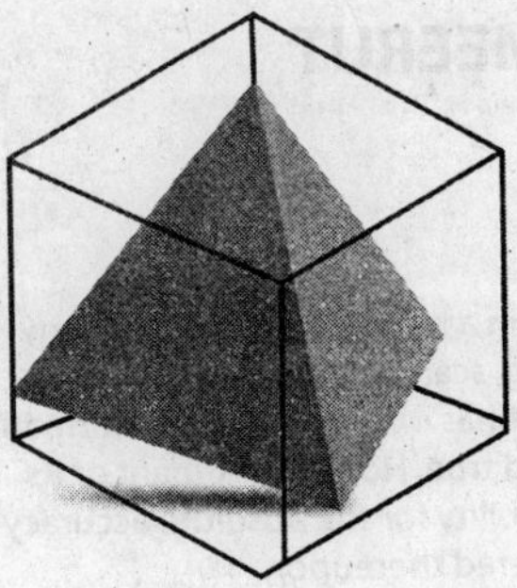

Dr. Amit Kumar Singh

MSc.(BHU), Ph D (IIT Delhi)

arihant

ARIHANT PRAKASHAN (Series), MEERUT

arihant

ARIHANT PRAKASHAN (Series), MEERUT

卐 **Administrative & Production Offices**

Regd. Office

'Ramchhaya' 4577/15, Agarwal Road, Darya Ganj, New Delhi -110002
Tele: 011- 47630600, 43518550

卐 **Head Office**

Kalindi, TP Nagar, Meerut (UP) - 250002
Tel: 0121-7156203, 7156204

卐 **Sales & Support Offices**

Agra, Ahmedabad, Bengaluru, Bareilly, Chennai, Delhi, Guwahati, Hyderabad, Jaipur, Jhansi, Kolkata, Lucknow, Nagpur & Pune.

卐 **ISBN** : 978-93-26199-94-0

卐 **PRICE** : ₹ 425.00

PO No. : TXT-59-T050278-11-22

Published by Arihant Publications (India) Ltd.

For further information about the books published by Arihant, log on to www.arihantbooks.com or e-mail at info@arihantbooks.com

Follow us on

Preface

Unproblem JEE Series is an attempt to enrich and enliven the Mathematical Problem Skills in JEE & other engineering entrances aspirants. Most of the Problem Books available in the segment are not substantially capable to serve the purpose for which an engineering aspirant goes for the problem books. They are just a compilation of problems packed in different chapters without any logical arrangement & purposeful grading. Thus, they fail to provide the readers with the approach required to solve complex & structured problems generally asked in JEE.

The First book of Unproblem JEE Series, **'Problems in Algebra'** will help students to develop the insight and intuition necessary to master the Algebraic Techniques & Manipulative Skills required to solve different types of JEE Level Problems.

Problems in each chapter are arranged to promote problem-solving ability and make the students understand that concepts & techniques are logically connected. In this way, students will become comfortable with concepts & will know how and when to use them while solving a problem.

The chapters in the book are divided into three levels and each level is a complete treatment of the chapter but at different levels of difficulty.

- Each chapter is divided into three major heads; **Level 1** 'Starter Level', **Level 2** 'JEE Main Level' & **Level 3** 'JEE Advanced Level'.
- **Level 1 'Starter Level'** consists of preliminary questions that test basic concepts. After completing this level student will have a solid understanding of fundamentals of the chapter necessary to solve the problems.
- **Level 2 'JEE Main Level'** has single choice correct questions and numerical value questions aimed especially for JEE Main.
- **Level 3 'JEE Advanced Level'** consists of various types of questions; MCQ with one or more than One Correct Answer, Comprehension Based Questions, Match the Columns and Numerical Value problems as per JEE Advanced Pattern.
- The problems given in JEE Advanced Level are made to be conceptually challenging resulting in practical and relevant practice for JEE Advanced.
- Detailed solutions to all the problems have been made available online just to help the students in case they stuck somewhere while solving some questions.

At the end, I can say confidently that if you rigorously practice the problems given in this book then surely you will get an edge in Mathematics and will increase your chance of getting good JEE Score.

However, I have taken the utmost care while writing this book but if any error or whatsoever has skipped my attention then please bring it to my notice so that it can be rectified. I would also like to appeal all my fellow Mathematics Teachers to have a look at the book and share their valuable suggestions for its further improvement.

Dr. Amit Kumar Singh
MSc (BHU), Ph D (IIT Delhi)

Contents

CHAPTER

01

SETS, RELATIONS AND FUNCTIONS

LEVEL 01 \\ Starter Level

1.1 Set and Its Properties

1 Write the solution set of the following equation in roster form

(i) $x + \frac{1}{2} = 0$

(ii) $x^2 + x - 2 = 0$

(iii) $x^2 + 4x + 4 = 0$

2 Write the following sets in the roster form

(i) $A = \{x : x$ is a positive integer and $x^2 < 49\}$

(ii) $B = \{x : x$ is a whole number less than 10 and $2^x - 1$ is a multiple of 3$\}$

(iii) $C = \{x : x$ is a prime number which is a divisor of 60$\}$

3 Write the following sets in the set-builder form

(i) $A = \left\{\frac{1}{2}, \frac{4}{3}, \frac{9}{4}, \frac{16}{5}, \frac{25}{6}\right\}$

(ii) $B = \{A, G, L, X, Y\}$

(iii) $C = \{1, 27, 125, 343, 729\}$

4 State which of the following sets are finite or infinite

(i) $\{x : x \in N$ and $(x - 1)(x - 2) = 0\}$

(ii) $\{x : x \in N$ and $2x - 1 = 0\}$

(iii) $\{x : x \in N$ and x is prime$\}$

(iv) $\{x : x \in N$ and x is odd$\}$

5 Are the following pairs of equal sets?

(i) $A = \{x : x \text{ is a prime factor of } 6\}$
$B = \{x : x \text{ is a solution of } x^2 - 5x + 6 = 0\}$

(ii) $A = \{x : x \text{ is a natural number}, x > 1\}$
$B = \{x : x \text{ is a natural number}, x \geq 1\}$

(iii) $A = \{x : x \text{ is a letter of word PRIME}\}$
$B = \{y : y \text{ is a letter in the word EMPIRE}\}$

6 If $x \in A$ and $A \not\subset B$, then $x \in B$. Is this statement true?

7 Let A, B and C be three sets. If $A \in B$ and $B \subset C$, is it true that $A \subset C$? If not, give an example.

8 Find the power set of $A = \{1, 2, 3\}$ and a total number of elements in the power set.

9 How many elements are there for the power set of an empty set?

10 Consider the following sets

A = Set of all rectangles in the plane

B = Set of all squares in the same plane

C = Set of all parallelogram in the same plane

Find the following sets.

(i) $A \cap C$ (ii) $B \cap C$

(iii) $B \cup C$ (iv) $A \cap B \cap C$

(v) $A - B$ (vi) $C - A$

11 Draw the Venn-diagrams to illustrate the following relationship among sets E, M and U, where E is the set of students studying English in a school, M is the set of students studying Mathematics in a same school, U is the set of all student in that school.

(i) All the students who study Mathematics study English, but some students who study English do not study Mathematics.

(ii) There is no student who studies both subject.

(iii) Some of the students study Mathematics but do not study English, some study English but do not study Mathematics, and some study both.

(iv) Not all students study Mathematics, but every students studying English studies Mathematics.

12 For all sets A, B and C

Is $(A - B) \cap (C - B) = (A \cap C) - B$?

Justify your answer.

13 If A and B be two sets containing 6 and 3 elements, respectively. What can be the minimum number of elements in $A \cup B$? Also, find the maximum number of elements in $A \cup B$.

14 In a group of 1000 people, there are 750 who can speak Hindi and 400 who can speak English.

(i) How many can speak Hindi only?

(ii) How many can speak English only?

(iii) How many can speak both Hindi and English?

15 In a group of 500 people, 350 speak Hindi and 300 speak English. It is given that each person speaks atleast one languages.

(i) How many people can speak both Hindi and English?

(ii) How many people can speak Hindi only?

(iii) How many people can speak English only?

16 An investigator interviewed 100 students to determine their preference for three drinks : Milk (M), Coffee (C), and Tea (T). He reported the following

10 students had all the three drinks M, C and T, 20 had M and C, 30 had C and T, 25 had M and T, 12 had M only, 5 had C only and 8 had T only. Using a Venn-diagram, find how many did not take any of three drinks.

17 In a group of 50 students, the number of students learning French, English, Sanskrit were found to be as follows : French = 17, English = 13, Sanskrit = 15, French and English = 9, English and Sanskrit = 4, French and Sanskrit = 5 and English, French and Sanskrit = 3.

Find the number of students who are learning

(i) French only
(ii) English only
(iii) Sanskrit only
(iv) English and Sanskrit but not French
(v) French and Sanskrit but not English
(vi) French and English but not Sanskrit
(vii) atleast one of three languages
(viii) none of three languages
(ix) exactly one language
(x) exactly two language.

1.2 Relation and Its Types

18 Let $A = \{1, 2, 3, 4\}$ and $B = \{5, 7, 9\}$

Determine

(i) $A \times B$
(ii) $B \times A$
(iii) Is $A \times B = B \times A$?
(iv) Is $n(A \times B) = n(B \times A)$?

19 The cartesian product of $A \times A$ has 9 elements among which are found (1, 0) and (0, 2). Find the set A and the remaining elements of $A \times A$.

20 If $A = \{1, 2\}$, then find the set $A \times A \times A$.

21 Find the domain and range of the relation R given by

$R = \{(x, y) : y = x + \frac{6}{x}$; where $x, y \in N, x < 6\}$

22 Let $A = \{1, 2, 3\}$ and $R = \{(a, b) : a, b \in A$, a divides b and b divides $a\}$. Is R is an identity relation on A?

23 Discuss the symmetric and reflexive relation on R defined on the set of natural numbers by $x R y \Leftrightarrow 2x^2 - 3xy + y^2 = 0$

24 Justify that R in the set $\{1, 2, 3\}$ given by $R = \{(1, 1), (2, 2), (3, 3), (1, 2), (2, 3)\}$ is reflexive but neither symmetric nor transitive.

25 Let R be the relation in the set $\{1, 2, 3, 4\}$ given by $R = \{(1, 2), (2, 2), (1, 1), (4, 4), (1, 3), (3, 3), (3, 2)\}$. Discuss the equivalence relation on R.

26 Let T be the set of all triangles in a plane with R a relation in T given by $R = \{(T_1, T_2) : T_1$ is congruent to $T_2\}$. Is R an equivalence relation? Justify your answer.

27 Examine the equivalence relation of the following

(i) Let N be the set of natural numbers. The relation R is defined on $N \times N$ as follows $(a, b) R (c, d) \Leftrightarrow a + d = b + c$

(ii) A relation R on the set of complex numbers is defined by $z_1 R z_2 \Leftrightarrow \frac{z_1 - z_2}{z_1 + z_2}$ is real.

(iii) $R = \{(x, y) : x = y$ or $x - y = 1, x; y \in N\}$, where N is the set of natural numbers.

28 If R_1 and R_2 are equivalence relations in a set A, then discuss the equivalence for $R_1 \cap R_2$.

29 Let $I = \{0, \pm 1, \pm 2, \pm 3, \ldots\}$ and $R = \left\{(a, b) : \frac{a-b}{4} = k, k \in I\right\}$ is an equivalence relation. Find the equivalence class of [0] and [1].

30 Let $A = \{1, 2, 3, 4\}$. Let R be the equivalence relation on $A \times A$ defined by $(a, b) R (c, d)$, iff $a + d = b + c$. Find the equivalence class $[(1, 3)]$.

31 Let R be a relation such that $R = \{(1, 4), (3, 7), (4, 5), (4, 6), (7, 6)\}$. Then, find

(i) RoR

(ii) $R^{-1} o R^{-1}$

(iii) $(R^{-1} o R)^{-1}$

1.3 Real Number System and Functions

32 Find the value of x^2 for the given values of x.

(i) $x < 2$

(ii) $x > -1$

(iii) $x \geq 2$

(iv) $x < -1$

33 Solve $(x^2 - 4)\sqrt{x^2 - 1} < 0$

34 Find the value of $\frac{1}{x}$ for the given values of x.

(i) $x > 3$

(ii) $x < -2$

(iii) $x \in (-1, 3) - \{0\}$

35 Solve $\left|1 + \frac{3}{x}\right| > 2$

36 Solve $\frac{x}{x+2} \leq \frac{1}{|x|}$

37 Solve $x^2 - x - 1 < 0$

38 Solve $(x - 1)(x - 2)(1 - 2x) > 0$

39 For $x \in R$, find all possible values of

(i) $|x - 3| - 2$

(ii) $4 - |2x + 3|$

40 Is the following relation a function? Justify your answer.

(i) $R_1 = \{(2, 3), \left(\frac{1}{2}, 0\right), (2, 7), (-4, 6)\}$

(ii) $R_2 = \{(x, |x|)\}$, x is a real number?

41 Find the domain for which the functions $f(x) = 2x^2 - 1$ and $g(x) = 1 - 3x$ are equal.

42 Let N be the set of all natural numbers. Consider $f : N \to N : f(x) = 2x, \forall x \in N$. Identify that f is one-one and into.

43 Verify that the mapping $f : R \to R : f(x) = \cos x, \forall x \in R$ is neither one-one nor onto.

44 If $f : R \to R$ and $g : R \to R$ be two mapping such that $f(x) = \sin x$ and $g(x) = x^2$. Then, find

(i) $fog(x)$

(ii) $gof(x)$

(iii) $fog\left(\frac{\sqrt{\pi}}{2}\right)$

(iv) $gof\left(\frac{\pi}{3}\right)$

45 If the mapping f and g are given by

$$f = \{(1, 2), (3, 5), (4, 1)\} \text{ and } g = \{(2, 3), (5, 1), (1, 3)\}$$

Write down the pairs in the mapping fog and gof.

46 Let $f : R \to R$ be defined by

$f(x) = \cos(5x + 2)$. Is f invertible? Justify your answer.

47 If the functions f and g defined from the set of real numbers R to R such that $f(x) = e^x$ and $g(x) = 3x - 2$, then find functions fog and gof. Also, find the domain of the functions $(fog)^{-1}$ and $(gof)^{-1}$.

48 If the functions f, g and h are defined from the set of real numbers R to R such that

$$f(x) = x^2 - 1,\ g(x) = \sqrt{x^2 + 1} \text{ and } h(x) = \begin{cases} 0, & \text{if } x \le 0 \\ x, & \text{if } x \ge 0 \end{cases}$$

Then, find the composite function $hofog$ and determine the function fog is invertible? Is h is invertible? Justify the answer.

LEVEL 02 \\ JEE Main Level

Single Choice Correct Questions

1.1 Sets and Its Properties

1 Let $S = \{x \in R : x \ge 0\}$ and $2|\sqrt{x} - 3| + \sqrt{x}(\sqrt{x} - 6) + 6 = 0$. Then, S **[JEE Main 2018]**

(a) contains exactly one element (b) contains exactly two elements
(c) contains exactly four elements (d) is an empty set

2 Two finite sets have m and n $(m > n)$ elements. The number of subsets of the first set is 112 more than that of the second set. The value of mn is

(a) 18 (b) 28 (c) 32 (d) 36

3 Let $S = \{1, 2, 3, \ldots, 100\}$. The number of non-empty subsets A of S such that the product of elements in A is even, is **[JEE Main 2019]**

(a) $2^{50}(2^{50} - 1)$ (b) $2^{100} - 1$ (c) $2^{50} - 1$ (d) $2^{50} + 1$

4 Let $S = \{1, 2, 3, 4\}$. The total number of unordered pairs of disjoint subsets of S is equal to

(a) 25 (b) 34 (c) 42 (d) 41

5 If $n(A) = 3$, $n(B) = 6$ and $A \subseteq B$, then the number of elements in $A \cup B$ is equal to

(a) 3 (b) 9 (c) 6 (d) None of these

6 If the sets A and B are defined as

$$A = \left\{(x, y) \mid y = \frac{1}{x}, x \ne 0, x \in R\right\} \text{ and } B = \{(x, y) \mid y = -x, x \in R\}, \text{ then}$$

(a) $A \cap B = A$ (b) $A \cap B = B$ (c) $A \cap B = \phi$ (d) $A \cup B = A$

7 If $X = \{4^n - 3n - 1 : n \in N\}$ and $Y = \{9(n - 1) : n \in N\}$, where N is the set of natural numbers, then $X \cup Y$ is equal to

(a) N (b) $Y - X$ (c) X (d) Y

8 If $aN = \{ax : x \in N\}$, then the set $4N \cap 6N$ is

(a) $8N$ (b) $10N$ (c) $12N$ (d) None of these

9 Let $\bigcup_{i=1}^{50} X_i = \bigcup_{i=1}^{n} Y_i = T$, where each X_i contains 10 elements and each Y_i contains 5 elements. If each element of the set T is an element of exactly 20 of sets X_i's and exactly 6 of sets Y_i's, then n is equal to **[JEE Main 2020]**

(a) 15 (b) 50 (c) 45 (d) 30

10 Shaded region for the figure is

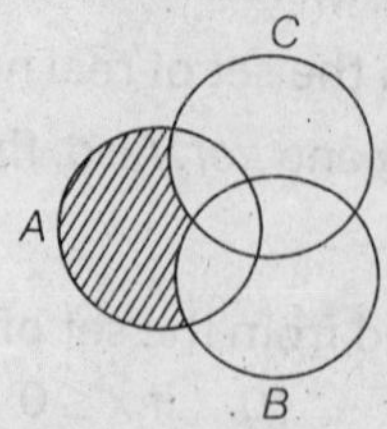

(a) $A \cap (B \cup C)$ (b) $A \cup (B \cap C)$ (c) $A \cap (B - C)$ (d) $A - (B \cup C)$

11 In a school, there are three types of games to be played. Some of the students play two types of games, but none play all the three games. Which Venn-diagram can justify the above statement? **[JEE Main 2021]**

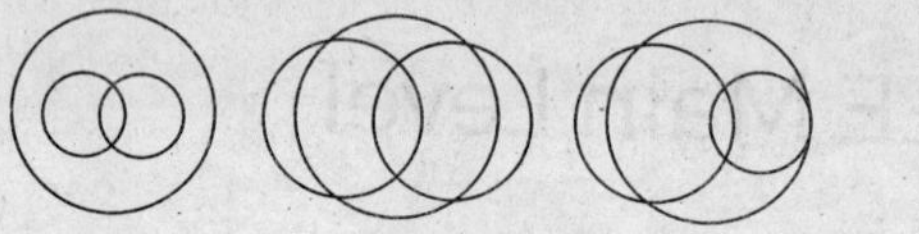

(P) (Q) (R)

(a) P and Q (b) P and R (c) Q and R (d) None of these

12 If A, B and C are three sets such that $A \cap B = A \cap C$ and $A \cup B = A \cup C$, then

(a) $A = C$ (b) $B = C$ (c) $A \cap B = \phi$ (d) $A = B$

13 Let A, B and C be sets such that $\phi \neq A \cap B \subseteq C$. Then, which of the following statement is not true?

(a) $B \cup C \neq \phi$ (b) If $(A - B) \subseteq C$, then $A \subseteq C$

(c) $(C \cup A) \cap (C \cup B) = C$ (d) If $(A - C) \subseteq B$, then $A \subseteq B$

14 Let $P = \{\theta : \sin\theta - \cos\theta = \sqrt{2}\cos\theta\}$ and $Q = \{\theta : \sin\theta + \cos\theta = \sqrt{2}\sin\theta\}$ be two sets. Then,

(a) $P \subset Q$ and $Q - P \neq \phi$ (b) $Q \not\subset P$

(c) $P \not\subset Q$ (d) $P = Q$

15 If X and Y are two sets, then $X \cap (X \cup Y)^c$ equal to

(a) X (b) Y (c) ϕ (d) None of these

16 The set $(A \cap B')' \cup (B \cap C)$ is equal to

(a) $A' \cup B \cup C$ (b) $A' \cup B$ (c) $A' \cup C'$ (d) $A' \cap B$

17 For sets A and B, $(A \cup B)' \cup (A' \cap B)$ equals

(a) A' (b) B' (c) A (d) None of these

18 Which is the simplified representation of $(A' \cap B' \cap C) \cup (B \cap C) \cup (A \cap C)$, where A, B and C are subsets of set X?

(a) A (b) B (c) C (d) $X \cap (A \cup B \cup C)$

19 Two sets A and B are as under

$A = \{(a, b) \in R \times R : |a - 5| < 1 \text{ and } |b - 5| < 1\}$ and $B = \{(a, b) \in R \times R : 4(a - 6)^2 + 9(b - 5)^2 \leq 36\}$

Then, **[JEE Main 2018]**

(a) $B \subset A$ (b) $A \subset B$

(c) $A \cap B = \phi$ (an empty set) (d) neither $A \subset B$ nor $B \subset A$.

20 Let F_1 be the set of parallelograms, F_2 be the set of rectangles, F_3 be the set of rhombuses, F_4 be the set of squares and F_5 be the set of trapeziums in a plane. Then, F_1 may be equal to

(a) $F_2 \cap F_3$ (b) $F_3 \cap F_4$ (c) $F_2 \cup F_5$ (d) $F_2 \cup F_3 \cup F_4 \cup F_1$

21 A survey shows that 73% of the persons working in a office like coffee whereas 65% like tea. If x denotes the percentage of them, who like both coffee and tea, then x cannot be

(a) 63 (b) 36 (c) 54 (d) 38 **[JEE Main 2020]**

22 A survey shows that 63% of the people watch a news channel whereas, 76% watch on entertainment channel at particular time. If x% of the people watch both types of channels, then

(a) $x = 25$ (b) $x \geq 63$ (c) $39 \leq x \leq 63$ (d) $x = 39$

23 Out of all the patients in a hospital 89% are found to be suffering from heart ailment and 98% are suffering from lungs infection. If K% of them are suffering from both ailments, then K can not belong to the set **[JEE Main 2021]**

(a) {80, 83, 86, 89} (b) {84, 86, 88, 90}
(c) {79, 81, 83, 85} (d) {84, 87, 90, 93}

24 In a town of 10000 families, it was found that 40% families buy newspaper A, 20% buy newspaper B and 10% buy newspaper C. Also, 5% families buy newspapers A and B, 3% buy newspapers C and B and 4% buy newspapers A and C. If 2% families buy all the three newspapers, then number of families which buy newspaper A only, is

(a) 3100 (b) 3300 (c) 2900 (d) 1400

25 Let Z be the set of integers. If $A = \{x \in Z : 2^{(x+2)(x^2-5x+6)} = 1\}$ and $B = \{x \in Z : -3 < 2x - 1 < 9\}$, then the number of subsets of the set $A \times B$ is **[JEE Main 2019]**

(a) 2^{12} (b) 2^{18} (c) 2^{15} (d) 2^{10}

1.2 Relation and Its Types

26 Let N be the set of natural numbers and a relation R on N be defined by $R = \{(x, y) \in N \times N : x^3 - 3x^2y - xy^2 + 3y^3 = 0\}$. Then the relation R is **[JEE Main 2021]**

(a) symmetric but neither reflexive nor transitive
(b) reflexive but neither symmetric nor transitive
(c) reflexive and symmetric, but not transitive
(d) an equivalence relation

27 Let R be the relation on the set R of all real numbers defined by aRb, iff $|a - b| \leq 1$. Then, R is

(a) reflexive and symmetric (b) symmetric only
(c) transitive (d) anti-symmetric

28 Let R_1 and R_2 be two relations defined as follows **[JEE Main 2020]**

$R_1 = \{(a,b) \in R^2 : a^2 + b^2 \in Q\}$ and

$R_2 = \{(a, b) \in R^2 : a^2 + b^2 \notin Q\}$, where Q is the set of all rational number. Then,

(a) neither R_1 nor R_2 is transitive (b) R_2 is transitive but R_1 is not transitive
(c) R_1 is transitive but R_2 is not transitive (d) R_1 and R_2 both transitive

29 Which of the following is not correct for relation R on the set of real numbers?

(a) $(x, y) \in R \Leftrightarrow 0 < |x| - |y| \leq 1$ is neither transitive nor symmetric
(b) $(x, y) \in R \Leftrightarrow 0 < |x - y| \leq 1$ is symmetric and transitive
(c) $(x, y) \in R \Leftrightarrow |x| - |y| \leq 1$ is reflexive but not symmetric
(d) $(x, y) \in R \Leftrightarrow |x - y| \leq 1$ is reflexive and symmetric

30 Define a relation R over a class of $n \times n$ real matrices A and B as "ARB, if there exists a non-singular matrix P such that $PAP^{-1} = B$". Then which of the following is true ? [JEE Main 2021]

(a) R is symmetric, transitive but not reflexive
(b) R is reflexive, symmetric but not transitive
(c) R is an equivalence relation
(d) R is reflexive, transitive but not symmetric

31 Let R be the real line. Consider the following subsets of the plane $R \times R$
$S = \{(x, y) : x + 1 = y \text{ and } 0 < x < 2\}$
$T = \{(x, y) : x - y \text{ is an integer}\}$
Which one of the following is true?
(a) T is an equivalence relation on R but S is not
(b) Neither S nor T is an equivalence relation on R
(c) Both S and T are equivalence relation on R
(d) S is an equivalence relation on R but T is not

32 If $R = \{(x, y) : x, y \in Z, x^2 + 3y^2 \le 8\}$ is a relation on the set of integers Z, then the domain of R^{-1} is [JEE Main 2020]
(a) $\{-2, -1, 1, 2\}$ (b) $\{0, 1\}$ (c) $\{-2, -1, 0, 1, 2\}$ (d) $\{-1, 0, 1\}$

33 Let $R = \{(P, Q) \mid$, P and Q are at the same distance from the origin$\}$ be a relation, then the equivalence class of $(1, -1)$ is the set [JEE Main 2021]
(a) $S = \{(x, y) \mid x^2 + y^2 = 4\}$ (b) $S = \{(x, y) \mid x^2 + y^2 = 1\}$
(c) $S = \{(x, y) \mid x^2 + y^2 = \sqrt{2}\}$ (d) $S = \{(x, y) \mid x^2 + y^2 = 2\}$

1.3 Real Number System and Functions

34 Complete solution set of inequality $\dfrac{(x+2)(x+3)}{(x-2)(x-3)} \le 1$ is
(a) $(-\infty, 0)$ (b) $(-\infty, 0] \cup (2, 3)$
(c) $(2, 3)$ (d) $(-\infty, 2) \cup (3, \infty)$

35 The number of integral values of x, if $5x - 1 < (x+1)^2 < 7x - 3$ is
(a) 0 (b) 1 (c) 2 (d) 3

36 The solution of the inequality $\dfrac{x+7}{x-5} + \dfrac{3x+1}{2} \ge 0$ is
(a) $[1, 3] \cup (5, \infty)$ (b) $(1, 3) \cup (5, \infty)$ (c) $(-\infty, 1) \cup (5, \infty)$ (d) None of these

37 If set $A = \left\{x \in R \mid \dfrac{x^2(5-x)(1-2x)}{(5x+1)(x+2)} < 0\right\}$ and set $B = \left\{x \in R \mid \dfrac{3x+1}{6x^3 + x^2 - x} > 0\right\}$, then $A \cap B$ does not contain
(a) $(1, 4)$ (b) $(5, 11)$ (c) $\left(-\dfrac{3}{2}, -\dfrac{1}{2}\right)$ (d) None of these

38 Number of integers satisfying the inequality, $x^4 - 29x^2 + 100 \le 0$ is
(a) 2 (b) 4 (c) 6 (d) 8

39 If $n > 0$ and exactly 15 integers satisfy $(x+6)(x-4)(x-5)(2x-n) \le 0$, then sum of digits of the least possible value of n is
(a) 10 (b) 12 (c) 14 (d) 16

40 The complete solution set of inequality $\frac{(x-5)^{1005}(x+8)^{1008}(x-1)}{x^{1006}(x-2)^3(x-3)^5(x-6)(x+9)^{1010}} \le 0$ is

(a) $(-\infty, -9) \cup (-8, 0) \cup (0, 1) \cup (2, 3) \cup [5, 6)$
(b) $(-\infty, -9) \cup (-9, 0) \cup (0, 1) \cup (2, 3) \cup (5, 6)$
(c) $(-\infty, -9) \cup (-9, 0) \cup (0, 1] \cup (2, 3) \cup [5, 6)$
(d) $(-\infty, 0) \cup (0, 1] \cup (2, 3) \cup [5, 6]$

41 Sum of the solutions of the equation $|x|^3 - 4|x|^2 + 3|x| = 0$ is

(a) 4 (b) 3 (c) 0 (d) 1

42 Number of integral roots of $|x-1||x^2-2| = 2$ is

(a) 0 (b) 1 (c) 2 (d) 3

43 The solution set of the inequality $\frac{|x+2|-x}{x} < 2$ is

(a) $(0, 1)$ (b) $[0, 2)$ (c) $(-\infty, 0) \cup (1, \infty)$ (d) None of these

44 Number of solutions of the equation $|2 - |x|| = x + 4$ is

(a) 0 (b) 1 (c) 2 (d) infinite

45 Number of integral values of x satisfying the inequality $\frac{x^2+6x-7}{|x+2||x+3|} < 0$ is

(a) 5 (b) 6 (c) 7 (d) 8

46 If $-4 \le x < 2$, then $||x+2|-3|$ lies in interval

(a) $(1, 3)$ (b) $[1, 3)$ (c) $[0, 3]$ (d) $[0, \infty)$

47 Let $f: R \to R$ be defined by $f(x) = \frac{x}{1+x^2}$, $x \in R$. Then, the range of f is

(a) $\left[-\frac{1}{2}, \frac{1}{2}\right]$ (b) $R - [-1, 1]$ (c) $R - \left[-\frac{1}{2}, \frac{1}{2}\right]$ (d) $(-1, 1) - \phi)$

48 Let $f(x) = a^x$ $(a > 0)$ be written as $f(x) = f_1(x) + f_2(x)$, where $f_1(x)$ is an even function and $f_2(x)$ is an odd. Then, $f_1(x+y) + f_1(x-y)$ equals [JEE Main 2019]

(a) $2f_1(x) f_1(y)$ (b) $2f_1(x+y) f_1(x-y)$
(c) $2f_1(x) f_2(y)$ (d) $2f_1(x+y) f_2(x-y)$

49 Let $f(n) = \left[\frac{1}{3} + \frac{3n}{100}\right] n$, where $[n]$ denotes the greatest integer less than or equal to n. Then, $\sum_{n=1}^{56} f(n)$ is

(a) 56 (b) 689 (c) 1287 (d) 1399

50 If $[x]$ be the greatest integer less than or equal to x, then $\sum_{n=8}^{100} \left[\frac{(-1)^n n}{2}\right]$ is equal to [JEE Main 2021]

(a) 0 (b) 4 (c) −2 (d) 2

51 If $f(x) = \log_e \left(\frac{1-x}{1+x}\right)$, $|x| < 1$, then $f\left(\frac{2x}{1+x^2}\right)$ is equal to

(a) $2f(x)$ (b) $2f(x^2)$ (c) $[f(x)]^2$ (d) $-2f(x)$

52 If the function $f: R \to \{1, -1\} \to A$ defined by $f(x) = \dfrac{x^2}{1-x^2}$ is surjective, then A is equal to **[JEE Main 2019]**

(a) $R - \{-1\}$ (b) $[0, \infty)$ (c) $R - [-1, 0)$ (d) $R - (-1, 0)$

53 The domain of the function $f(x) = \sin^{-1}\left(\dfrac{|x|+5}{x^2+1}\right)$ is $[-\infty, -a] \cup [a, \infty]$, then a is equal to

(a) $\dfrac{\sqrt{17}}{2}$ (b) $\dfrac{\sqrt{17}-1}{2}$ (c) $\dfrac{1+\sqrt{17}}{2}$ (d) $\dfrac{\sqrt{17}}{2}+1$

54 If $f(x) = \dfrac{1}{1-x}$, $g(x) = f\{f(x)\}$ and $h(x) = f[f\{f(x)\}]$, then the value of $f(x) \cdot g(x) \cdot h(x)$ is

(a) 6 (b) -1 (c) 1 (d) 2

55 Let $f: N \to N$ be a function such that $f(m+n) = f(m) + f(n)$ for every $m, n \in N$. If $f(6) = 18$, then $f(2) \cdot f(3)$ is equal to **[JEE Main 2021]**

(a) 6 (b) 54 (c) 18 (d) 36

Numerical Value Type Questions

56 In a certain city, only two newspapers A and B are published. It is known that 25% of the city population read A and 20% read B, while 8% read both. It is also known that 30% of those who read A but not B, look into advertisements and 40% of those who read B not A, look into advertisements while 50% of those who reads both A and B look into the advertisements. If x% of the population read an advertisements. Then, the value of x is ....

57 Let $A = \{n \in N \mid n^2 \le n + 10000\}$, $B = \{3k+1 \mid k \in N\}$ and $C = \{2k \mid k \in N\}$, then the sum of all the elements of the set $A \cap (B - C)$ is equal to .... **[JEE Main 2021]**

58 Given the relation $R = \{(1, 2), (2, 3)\}$ on the set $A = \{1, 2, 3\}$, the minimum number of ordered pairs which can be added to R to make it an equivalence relation, is ....

59 Let $a, b, c \in R$. If $f(x) = ax^2 + bx + c$ be such that $a + b + c = 3$ and $f(x+y) = f(x) + f(y) + xy$, $\forall\, x, y \in R$, then the value of $\sum_{n=1}^{10} f(n)$ is .... **[JEE Main 2020]**

60 Let $f: R \to R$ is a function defined by $f(x) = 10x - 7$. If $g = f^{-1}$ and $g(x) = \dfrac{Ax+B}{C}$, then the value of $A + B + C$ is ....

61 If $A = \{x \in R: |x-2| > 1\}$, $B = \{x \in R: \sqrt{x^2-3} > 1\}$ and $C = \{x \in R: |x-4| \ge 2\}$ and Z is the set of all integers, then the number of subsets of the set $(A \cap B \cap C)^C \cap Z$ is ....

62 If $f: R^+ \to A$, where $A = \{x: -5 < x < \infty\}$ is defined by $f(x) = x^2 - 5$ and if $f^{-1}(13) = \{-\lambda\sqrt{\lambda-1}, \lambda\sqrt{\lambda-1}\}$, then value of λ is ....

63 Let $\{S = 1, 2, 3, 4, 5, 6, 7\}$. Then the number of possible functions $f: S \to S$ such that $f(m \cdot n) = f(m) \cdot f(n)$ for every $m, n \in S$ and $m \cdot n \in S$ is equal to .... **[JEE Main 2021]**

LEVEL 03 \\ JEE Advanced Level

Single Choice Correct Questions

1 Suppose $A_1, A_2, \ldots A_{30}$ are thirty sets each with five elements and $B_1, B_2, \ldots B_n$ are sets each with three elements. Let $\bigcup_{i=1}^{30} A_i = \bigcup_{j=1}^{n} B_j = S$. Assume that each element of S belongs to exactly ten of the A_i's and to exactly nine of the B_j's. Then, the value of n is

(a) 45 (b) 55 (c) 65 (d) 90

2 Let U be the universal set and $A \cup B \cup C = U$, then $[(A - B) \cup (B - C) \cup (C - A)]'$ equals

(a) $A \cup B \cup C$ (b) $A \cap B \cap C$ (c) $A \cup (B \cap C)$ (d) $A \cap (B \cup C)$

3 In a certain town, 25% of the families own a phone and 15% own a car, 65% families own neither a phone nor a car and 2000 families own both a car and a phone. Consider the following three statements

(A) 5% families own both a car and a phone

(B) 35% families own either a car or a phone

(C) 40000 families live in the town

Then,

(a) only (A) and (C) are correct (b) only (B) and (C) are correct

(c) only (A) and (B) are correct (d) All (A), (B) and (C) are correct

4 In a class of 140 students numbered 1 to 140, all even numbered students opted Mathematics course, those whose number is divisible by 3 opted Physics course and those whose number is divisible by 5 opted Chemistry course. Then, the number of students who did not opt for any of the three courses is

(a) 102 (b) 42 (c) 1 (d) 38

5 In statistical survey of 1003 families of Kolkata it was found that 63 families has neither a radio nor a TV, 794 families has a radio and 187 has TV. The number of families in that group having both a radio and TV is

(a) 36 (b) 41 (c) 32 (d) None of these

6 Let R and S be two non-void relations on a set A. Which of the following statement is false?

(a) R and S are transitive $\Rightarrow R \cup S$ is transitive

(b) R and S are transitive $\Rightarrow R \cap S$ is symmetric

(c) R and S are symmetric $\Rightarrow R \cup S$ is symmetric

(d) R and S are reflexive $\Rightarrow R \cap S$ is reflexive

7 $\frac{n}{m}$ means that n is a factor of m, then the relation f is

(a) reflexive and symmetric (b) transitive and symmetric

(c) reflexive, transitive and symmetric (d) reflexive, transitive but not symmetric

8 The complete solution set of the equation $|x^2 - 5x + 6| + |x^2 + 12x + 27| = |17x + 21|$ is

(a) $x \in [-9, 3]$ (b) $x \in [-3, 2) \cup (2, 3]$ (c) $x \in [-9, -3] \cup [2, 3]$ (d) $x \in (-2, 3)$

9 If $|x^2 - 2x + 2| - |2x^2 - 5x + 2| = |x^2 - 3x|$, then the set of values of x is

(a) $(-\infty, 0] \cup [3, \infty)$ (b) $\left[0, \frac{1}{2}\right] \cup [2, 3]$

(c) $(-\infty, 0] \cup \left[\frac{1}{2}, 2\right] \cup [3, \infty)$ (d) $[0, 2] \cup [3, \infty)$

10 Complete set of values of x satisfying inequality $||x-1|-5|<2x-5$ is

(a) $\left(\frac{5}{2}, \infty\right)$ (b) $\left(\frac{11}{3}, \infty\right)$ (c) $(-1, \infty)$ (d) $\left(-\infty, \frac{1}{3}\right)$

11 Domain of the function, $f(x)=\frac{3}{4-x^2}+\log_{10}(x^3-x)$ is

(a) $(1, 2)$ (b) $(-1, 0) \cup (1, 2)$
(c) $(1, 2) \cup (2, \infty)$ (d) $(-1, 0) \cup (1, 2) \cup (2, \infty)$

12 If $f(x)+2f\left(\frac{1}{x}\right)=3x, x \neq 0$ and $S=\{x \in R : f(x)=f(-x)\}$, then S

(a) is an empty set (b) contains exactly one element
(c) contains exactly two element (d) contains more than two elements

13 If $f: R \to R$ satisfies $f(x+y)=f(x)+f(y)$, for all $x, y \in R$ and $f(1)=7$, then $\sum_{r=1}^{n} f(r)$ is

(a) $\frac{7n}{2}$ (b) $\frac{7(n+1)}{2}$ (c) $7n(n+1)$ (d) $\frac{7n(n+1)}{2}$

More than One Correct

14 If $A=\{x \mid \cos x>-\frac{1}{2}$ and $0 \leq x \leq \pi\}$ and $B=\{x \mid \sin x>\frac{1}{2}$ and $\frac{\pi}{3} \leq x \leq \pi\}$ and, if $\pi\lambda \leq A \cap B<\pi\mu$, then

(a) $\lambda=\frac{1}{3}$ (b) $\lambda=\frac{2}{3}$ (c) $\mu=\frac{1}{3}$ (d) $\mu=\frac{2}{3}$

15 Let $X=\{1, 2, 3, 4, 5\}$ and $Y=\{1, 3, 5, 7, 9\}$. Which of following is/are relations from X to Y?

(a) $R_1=\{(x, y): y=2+x, x \in X, y \in Y\}$ (b) $R_2=\{(1, 1), (2, 1), (3, 3), (4, 3), (5, 5)\}$
(c) $R_3=\{(1, 1), (1, 3), (3, 5), (3, 7), (5, 7)\}$ (d) $R_4=\{(1, 3), (2, 5), (2, 4), (7, 9)\}$

16 Let L be the set of all straight lines in the euclidean plane. Two lines l_1 and l_2 are said to be related by the relation R, iff l_1 is parallel to l_2. Then, the relation R is

(a) reflexive (b) symmetric (c) transitive (d) equivalence

17 If Z is the set of integers and, if the relation R is defined over Z by aRb, iff $a-b$ is an even integer, $a, b \in Z$, then relation R is

(a) reflexive (b) anti-symmetric
(c) symmetric (d) equivalence

18 Let $f: R \to R, g: R \to R$ be two functions given by $f(x)=2x-3$ and $g(x)=x^3+5$. Then,

(a) $fog(x)=2x^3+8$ (b) $fog(x)=2x^3+7$
(c) $(fog)^{-1}(x)=\left(\frac{x-8}{2}\right)^{1/3}$ (d) $(fog)^{-1}(x)=\left(\frac{x-7}{2}\right)^{1/3}$

19 Let $f:[0, 1] \to [0, 1]$ be defined by $f(x)=\frac{1-x}{1+x}, 0 \leq x \leq 1$ and $g:[0, 1] \to [0, 1]$ be defined by $g(x)=4x(1-x), 0 \leq x \leq 1$. Then,

(a) $fog(x)=\frac{(2x+1)^2}{1+4x-4x^2}$ (b) $fog(x)=\frac{(2x-1)^2}{1+4x-4x^2}$
(c) $gof(x)=\frac{8x(1+x)}{(1-x)^2}$ (d) $gof(x)=\frac{8x(1-x)}{(1+x)^2}$

20 If $f(x)=\dfrac{a-x}{a+x}$, the domain of $f^{-1}(x)$ contains

(a) $(-\infty, \infty)$ (b) $(-\infty, -1)$ (c) $(-1, \infty)$ (d) $(0, \infty)$

21 If $f(x)=\dfrac{\sin([x]\pi)}{x^2+x+1}$, where [.] denotes the greatest integer function, then

(a) f is one-one
(b) f is not one-one and non-constant
(c) f is constant function
(d) f is zero function

22 Let the function $f: R-\{-b\} \to R-\{1\}$ be defined by $f(x)=\dfrac{x+a}{x+b}$ $(a \neq b)$, then

(a) f is one-one but not onto
(b) f is onto but not one-one
(c) f is both one-one and onto
(d) $f^{-1}(2)=a-2b$

Linked Comprehension Type

Paragraph I Directions (Q.Nos. 23-25)

R_1 on Z defined by $(a, b) \in R_1$, iff $|a-b| \leq 7$, R_2 on Q defined by $(a, b) \in R_2$, iff $ab=4$ and R_3 on R defined by $(a, b) \in R_3$, iff $a^2-4ab+3b^2=0$.

23 Relation R_1 is

(a) reflexive and symmetric
(b) symmetric and transitive
(c) reflexive and transitive
(d) equivalence

24 Relation R_2 is

(a) reflexive (b) symmetric (c) transitive (d) equivalence

25 Relation R_3 is

(a) reflexive (b) symmetric (c) transitive (d) equivalence

Matrix Match Type Questions

26 The function defined have domain R

	Column-I		Column-II
(i)	$7x+1$	A.	onto on $[-1, 1]$ but not one-one on $[0, \pi]$
(ii)	$\cos x$	B.	one-one on $[0, \pi]$ but not onto on R
(iii)	$\sin x$	C.	one-one and onto on R
(iv)	$1+\ln x$	D.	one-one on $(0, \infty)$

27

	Column-I		Column-II
(i)	The set of value(s) of x satisfying the equations $7^{\frac{2x^2-5x-9}{2}}=(\sqrt{2})^{3\log_2 7}$ lies in the interval	A.	$[1,2]$
(ii)	The set of value(s) of x satisfying the equation $4^{x+\sqrt{x^2-2}}-5\cdot 2^{x-1+\sqrt{x^2-2}}=6$ lies in the interval	B.	$[-2,3]$
(iii)	The set of value(s) of x satisfying the equation $\sqrt{\lvert x-3\rvert^{x-1}}=\sqrt[4]{\lvert x-3\rvert^{x+1}}$ lies in the interval	C.	$[-2,4]$
		D.	$[1,5]$

Numerical Value Type Questions

Question Stem (Q.Nos. 28-30) The answer to each question is a **NUMERICAL VALUE.**

If $A = \{x : |x| < 2\}$, $B = \{x : |x - 5| \leq 2\}$, $C = \{x : |x| > x\}$ and $D = \{x : |x| < x\}$

28 The number of integral values in $A \cup B$ is ....

29 The number of integral values in $A \cap C$ is ....

30 The number of integral values in $A \cap D$ is ....

Question Stem (Q.Nos. 31-33) The answer to each question is a **NUMERICAL VALUE.**

Let f and g be real valued functions defined as

$$f(x) = \begin{cases} 7x^2 + x - 8; & x \leq 1 \\ 4x + 5; & 1 < x \leq 7 \\ 8x + 3; & x > 7 \end{cases}$$

and $$g(x) = \begin{cases} |x|, & x < -3 \\ 0, & -3 \leq x < 2 \\ x^2 + 4, & x \geq 2 \end{cases}$$

31 The value of $gof(0) + fog(-3)$ is ....

32 The value of $2\,fog(7) - gof(6)$ is ....

33 The value of $4\ gof\ (2) - fog\ (9)$ is ....

34 Two finite sets have m and n elements. The total number of subsets of the first set is 48 more than the total number of subsets of the second set. The value of $(m - n)$ is ....

35 If $S = R$, $A = \{x : -3 \leq x < 7\}$ and $B = \{x : 0 < x < 10\}$ the number of positive integers in $A\ \Delta\ B$ is .... (where Δ is symmetric difference of two sets)

36 In a group of 45 students, 22 can speak Hindi only and 12 can speak English only. If $(2\lambda + 1)$ student can speak both Hindi and English, the value of λ is ....

37 If two sets A and B are having 99 elements in common the number of elements to each of the sets $A \times B$ and $B \times A$ are $121\,\lambda^2$, the value of λ is ....

38 Let the function $f : [0, 1] \to R$ be defined by $f(x) = \dfrac{4^x}{4^x + 2}$, then the value of

$f\left(\frac{1}{40}\right) + f\left(\frac{2}{40}\right) + f\left(\frac{3}{40}\right) + \ldots f\left(\frac{39}{40}\right) - f\left(\frac{1}{2}\right)$ is .... **[JEE Advanced 2020]**

ANSWER KEY

Level 1

1. (i) $\left\{-\frac{1}{2}\right\}$ (ii) $\{1, -2\}$ (iii) $\{-2\}$

2. (i) $\{1, 2, 3, 4, 5, 6\}$ (ii) $\{0, 2, 4, 6, 8\}$ (iii) $\{2, 3, 5\}$

3. (i) $\left\{x : x = \frac{n^2}{n+1}, n \in N, 1 \le n \le 5\right\}$ (ii) $\{x : x$ is the letter of the word GALAXY$\}$

(iii) $\{x : x = (2n-1)^3, n \in N, 1 \le n \le 5\}$

4. (i) Finite (ii) Finite (iii) Infinite (iv) Infinite

5. (i) Equal sets (ii) Unequal sets (iii) Equal sets

6. No **7.** No **8.** $\{\phi, \{1\}, \{2\}, \{3\}, \{1, 2\}, \{2, 3\}, \{1, 3\}, \{1, 2, 3,\}\}$ and 8 **9.** 1

10. (i) Set of all rectangles

(ii) Set of all squares

(iii) Set of all parallelogram

(iv) Set of all squares

(v) Set of all rectangles which are not squares

(vi) Set of all parallelgoram which are not rectangles

12. Yes **13.** 6 and 9

14. (i) 600 (ii) 250 (iii) 150

15. (i) 150 (ii) 200 (iii) 150

16. 20

17. (i) 6 (ii) 3 (iii) 9 (iv) 1 (v) 2 (vi) 6

(vii) 30 (viii) 20 (ix) 18 (x) 9

18. (i) $\{(1, 5), (1, 7), (1, 9), (2, 5), (2, 7), (2, 9), (3, 5), (3, 7), (3, 9), (4, 5), (4, 7), (4, 9)\}$

(ii) $\{(5, 1), (5, 2), (5, 3), (5, 4), (7, 1), (7, 2), (7, 3), (7, 4), (9, 1), (9, 2), (9, 3), (9, 4)\}$

(iii) No

(iv) Yes

19. $\{0, 1, 2\}$ and 7 **20.** $\{(1, 1, 1), (1, 1, 2), (1, 2, 1), (1, 2, 2), (2, 1, 1), (2, 1, 2), (2, 2, 1), (2, 2, 2)\}$

21. $\{1, 2, 3\}$ and $\{7, 5\}$ **22.** Yes **26.** Yes

27. (i) Yes (ii) Yes (iii) No

29. $\{0, \pm 4, \pm 8, \pm 12\}$ and $\{\ldots\ldots, -11, -7, -3, 1, 5, 9, \ldots\ldots\}$

30. $\{(1, 3), (2, 4)\}$

31. (i) $\{(1, 5), (1, 6), (3, 6)\}$ (ii) $\{(5, 1), (6, 1), (6, 3)\}$

(iii) $\{(1, 1), (3, 3), (4, 4), (4, 7), (7, 4), (7, 7)\}$

32. (i) $[0, \infty)$ (ii) $[0, \infty)$ (iii) $[4, \infty)$ (iv) $(1, \infty)$

33. $(-2, -1) \cup (1, 2)$

34. (i) $\left(0, \frac{1}{3}\right)$ (ii) $\left(-\frac{1}{2}, 0\right)$ (iii) $(-\infty, -1) \cup \left(\frac{1}{3}, \infty\right)$

35. $(-1, 0) \cup (0, 3)$ **36.** $(-2, 2]$ **37.** $\left(\frac{1-\sqrt{5}}{2}, \frac{1+\sqrt{5}}{2}\right)$ **38.** $\left(-\infty, \frac{1}{2}\right) \cup (1, 2)$

39. (i) $[-2, \infty)$ (ii) $(-\infty, 4]$

40. (i) No (ii) Yes **41.** $\left\{\frac{1}{2}, -2\right\}$

44. (i) $\sin x^2$ (ii) $\sin^2 x$ (iii) $\frac{1}{\sqrt{2}}$ (iv) $\frac{3}{4}$

45. $\{(2, 5), (5, 2), (1, 5)\}$ and $\{(1, 3), (3, 1), (4, 3)\}$

46. No **47.** e^{3x-2} and $3e^x - 2$; $(0, \infty)$ and $(-2, \infty)$ **48.** x^2, not invertible

Level 2

1. (b)	**2.** (b)	**3.** (a)	**4.** (d)	**5.** (c)	**6.** (c)	**7.** (d)
8. (c)	**9.** (d)	**10.** (d)	**11.** (d)	**12.** (b)	**13.** (c)	**14.** (d)
15. (c)	**16.** (b)	**17.** (a)	**18.** (c)	**19.** (b)	**20.** (d)	**21.** (b)
22. (c)	**23.** (c)	**24.** (b)	**25.** (c)	**26.** (b)	**27.** (a)	**28.** (a)
29. (b)	**30.** (c)	**31.** (a)	**32.** (d)	**33.** (d)	**34.** (b)	**35.** (b)
36. (a)	**37.** (b)	**38.** (d)	**39.** (d)	**40.** (c)	**41.** (c)	**42.** (d)
43. (c)	**44.** (b)	**45.** (a)	**46.** (c)	**47.** (a)	**48.** (c)	**49.** (d)
50. (b)	**51.** (a)	**52.** (c)	**53.** (c)	**54.** (b)	**55.** (b)	**56.** (13.9)
57. (832)	**58.** (7)	**59.** (330)	**60.** (18)	**61.** (256)	**62.** (3)	**63.** (490)

Level 3

1. (a)	**2.** (b)	**3.** (c)	**4.** (d)	**5.** (b)	**6.** (a)	**7.** (d)
8. (c)	**9.** (b)	**10.** (b)	**11.** (d)	**12.** (c)	**13.** (d)	**14.** (a, d)
15. (b, c)	**16.** (a, b, c, d)	**17.** (a, c, d)	**18.** (b, d)	**19.** (b, d)	**20.** (b, c, d)	**21.** (c, d)
22. (c, d)	**23.** (a)	**24.** (b)	**25.** (a)			

26. (i)–C, (ii)–B, (iii)–A, (iv)–D

27. (i)–C, (ii)–A, B, C, D, (iii)–B, C, D

28. (8)	**29.** (1)	**30.** (0)	**31.** (0)	**32.** (9)	**33.** (9)	**34.** (2)
35. (3)	**36.** (5)	**37.** (9)	**38.** (19)			

SOLUTIONS

Level 1

1. (i) The given equation can be written as

$$x+\frac{1}{2}=0 \Rightarrow x=-\frac{1}{2}$$

$\therefore$ The solution set $=\left\{-\frac{1}{2}\right\}$

(ii) The given equation can be written as
$(x-1)(x+2)=0$, i.e. $x=1,-2$
$\therefore$ The solution set $=\{1,-2\}$

(iii) The given equation can be written as
$(x+2)^2=0; x=-2,-2$
$\therefore$ The solution set $=\{-2,-2\}=\{-2\}$
It is noted that while writing the set in Roster form an element is not generally repeated.

2. (i) $A=\{1,2,3,4,5,6\}$
(ii) $B=\{0,2,4,6,8\}$
(iii) $C=\{2,3,5\}$

3. (i) $A=\left\{x: x=\frac{n^2}{n+1}, n\in N, 1\leq n\leq 5\right\}$
(ii) $B=\{x: x$ is the letter of the word GALAXY$\}$
(iii) $C=\{x: x=(2n-1)^3, n\in N, 1\leq n\leq 5\}$

4. (i) Given set $=\{1,2\}$, the set consist definite number of elements, then it is a finite set.
(ii) Given set $=\phi$, it is a finite set.
(iii) The given set is the set of all prime numbers and since set of prime number is infinite.
Hence, the given set is infinite.
(iv) Since, there are infinite number of odd numbers.
Hence, the given set is infinite.

5. (i) $A=\{2,3\}$
$B=\{x: x$ is a solution of $x^2-5x+6=0\}$

$$x^2-5x+6=0$$
$$\Rightarrow (x-2)(x-3)=0$$
$$\Rightarrow x=2 \text{ or } x=3$$

$\therefore B=\{2,3\}$
$\therefore A=B$ (equal set)

(ii) $A=\{2,3,4,\ldots\}$
$B=\{1,2,3,4,\ldots\}$
We observe that $1\in B$ but $1\notin A$.
$\therefore A\neq B$

(iii) $A=\{x: x$ is a letter in the word PRIME$\}$
$A=\{P,R,I,M,E\}$
$B=\{E,M,I,R,P\}=\{P,R,I,M,E\}$
$\therefore A=B$

6. Let $A=\{3,5,7\}$ and $B=\{3,4,6\}$
Now, $5\in A$ and $A\not\subset B$
However, $5\notin B$
So, the given statement is false.

7. No,
Let $A=\{1\}, B=\{\{1\},2\}$ and $C=\{\{1\},2,3\}$
Here, $A\in B$ as $A=\{1\}$ and $B\subset C$.
But $A\not\subset C$ as $1\in A$ and $1\notin C$.

8. Given, $A=\{1,2,3\}$
Total number of elements in power set of n elements $=2^n$.
Here, $n=3$, so $2^3=8$, which shows that there are eight elements in the power set of A.
$\therefore P(A)=\{\phi,\{1\},\{2\},\{3\},\{1,2\},\{2,3\},\{1,3\},\{1,2,3\}\}$

9. An empty set has zero elements.
$\therefore$ Number of elements of power set $=2^0=1$.
Hence, there is only one element in the power set which is empty set itself.
If $A=\phi$, then $P(A)=\{\phi\}$

10. $\because$ All squares are rectangle, $B\subset A$
Also, all squares are rectangles are parallelogram.
So, $A,B\subset C$
$\therefore B\subset A\subset C$
(i) $A\cap C=$ Set of all rectangles $=A$
(ii) $B\cap C=$ Set of all squares $=B$
(iii) $B\cup C=$ Set of all parallelogram $=C$
(iv) $A\cap B\cap C=$ Set of all squares $=B$
(v) $A-B=$ Set of all rectangles which are not squares
(vi) $C-A=$ Set of all parallelogram which are not rectangles.

11. (i) $M\subset E\subset U$

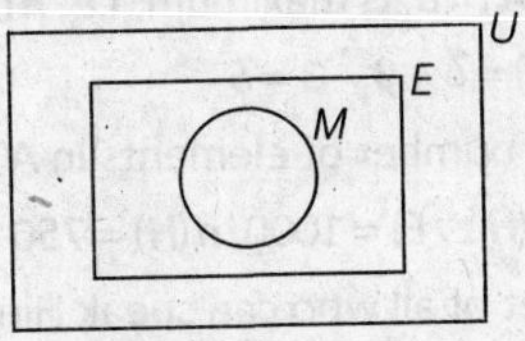

(ii) $E\cap M=\phi$

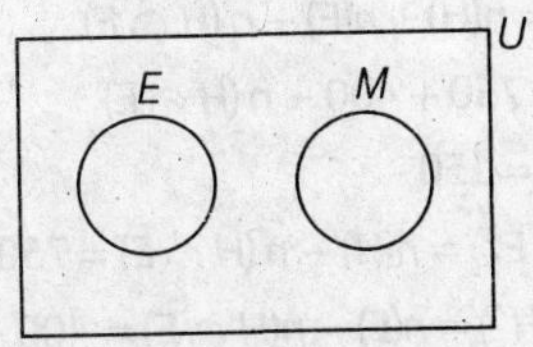

(iii)

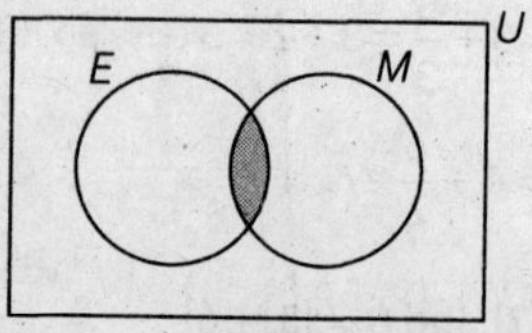

(iv) $E \subset M \subset U$

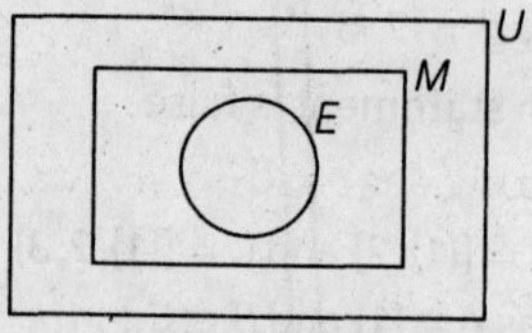

12. 'Yes'

Let $x \in (A-B) \cap (C-B)$

$\Rightarrow x \in (A-B)$ and $x \in (C-B)$

$\Rightarrow (x \in A$ and $x \notin B)$ and $(x \in C$ and $x \notin B)$

$\Rightarrow (x \in A$ and $x \in C)$ and $x \notin B$

$\Rightarrow x \in (A \cap C)$ and $x \notin B$

$\Rightarrow x \in (A \cap C) - B$

Thus, $(A-B) \cap (C-B) \subset (A \cap C) - B$ ...(i)

Now conversely,

Let $y \in (A \cap C) - B$

$\Rightarrow y \in (A \cap C)$ and $y \notin B$

$\Rightarrow (y \in A$ and $y \notin B)$ and $(y \in C$ and $y \notin B)$

$\Rightarrow y \in (A-B)$ and $y \in (C-B)$

$\Rightarrow y \in (A-B) \cap (C-B)$

Thus, $(A \cap C) - B \subset (A-B) \cap (C-B)$ ...(ii)

From Eqs. (i) and (ii),

$(A-B) \cap (C-B) = (A \cap C) - B$

13. We have,

$n(A \cup B) = n(A) + n(B) - n(A \cap B)$

Case I If $n(A \cap B)$ is minimum, i.e. $n(A \cap B) = 0$

$\therefore n(A \cup B) = n(A) + n(B) = 6 + 3 = 9$

$\therefore$ Maximum number of elements in $A \cup B = 9$

Case II If $n(A \cap B)$ is maximum, i.e. $n(A \cap B) = 3$

$\therefore \quad n(A \cup B) = 6 + 3 - 3 = 6$

$\therefore$ Minimum number of elements in $A \cup B$ is 6.

14. We have, $n(H \cup E) = 1000$, $n(H) = 750$ and $n(E) = 400$,

where H : Set of all who can speak Hindi

E : Set of all who can speak English

$\because n(H \cup E) = n(H) + n(E) - n(H \cap E)$

$1000 = 750 + 400 - n(H \cap E)$

$\Rightarrow n(H \cap E) = 150$

Also, $n(H \cap E') = n(H) - n(H \cap E) = 750 - 150 = 600$

and $n(E \cap H') = n(E) - n(H \cap E) = 400 - 150 = 250$

Thus,

(i) 600 people speak Hindi only

(ii) 250 people speak English only

(iii) 150 people speak both Hindi and English

15. Let H be the set of people who speak Hindi and E be the set of people who speak English.

$\therefore n(H \cup E) = 500, n(H) = 350$ and $n(E) = 300$

Now, $n(H \cup E) = n(H) + n(E) - n(H \cap E)$

$500 = 350 + 300 - n(H \cap E)$

$\Rightarrow \quad n(H \cap E) = 150$

(i) Thus, 150 people can speak both Hindi and English

(ii) Number of people who speak Hindi only
$= n(H) - n(H \cap E) = 350 - 150 = 200$

(iii) Number of people who speak English only
$= n(E) - n(H \cap E) = 300 - 150 = 150$

16. $n(U) = 100$ (universal set)

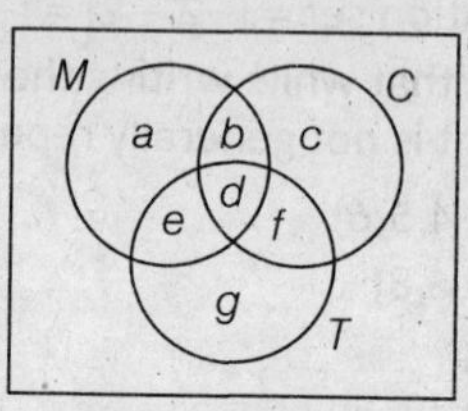

$n(M \cap C \cap T) = d = 10$

$n(M \cap C) = b + d = 20$

$\Rightarrow \quad b = 20 - d = 20 - 10$

$\Rightarrow \quad b = 10$

$n(C \cap T) = d + f = 30$

$\Rightarrow \quad f = 20$

and $\quad n(M \cap T) = d + e = 25$

$\Rightarrow \quad e = 15$

$\therefore n(\text{only } M) = a = 12$; $n(\text{only } C) = 5 = c$

and $n(\text{only } T) = 8 = g$

$\therefore n(M \cup C \cup T) = a + b + c + d + e + f + g$

$= 12 + 10 + 5 + 10 + 15 + 20 + 8 = 80$

$\therefore n(M \cup C \cup T)' = n(U) - n(M \cup C \cup T)$

$= 100 - 80 = 20$

17. Let F, E and S be the sets of student who learn French, English and Sanskrit, respectively.

Then, $n(U) = 50, n(F) = 17, n(E) = 13, n(S) = 15, n(F \cap E) = 9$, $n(E \cap S) = 4, n(F \cap S) = 5$, and $n(F \cap E \cap S) = 3$.

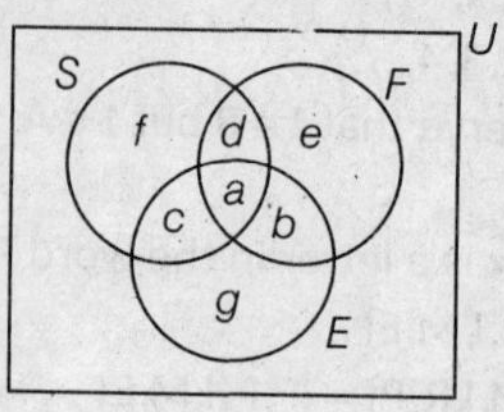

From the figure, we have

$a = n(E \cap F \cap S) = 3$ and $a + d = n(F \cap S) = 5$

$\Rightarrow \quad d = 2$

$\Rightarrow \quad a + b = n(F \cap E) = 9$

$\Rightarrow \quad b = 6$

$a + c = n(E \cap S) = 4$

$\Rightarrow \quad c = 1$

$a + b + d + e = n(F)$

$3 + 6 + 2 + e = 17$

$\Rightarrow \quad e = 6$

$a + b + c + g = n(E) = 13$

$\Rightarrow \quad 3 + 6 + 1 + g = 13 \Rightarrow g = 3$

$a + c + d + f = n(S) = 15$

$\Rightarrow \quad 3 + 1 + 2 + f = 15 \Rightarrow f = 9$

(i) $e = 6$

(ii) $g = 3$

(iii) $f = 9$

(iv) $c = 1$

(v) $d = 2$

(vi) $b = 6$

(vii) atleast one language $= a + b + c + d + e + f + g = 30$

(viii) none of three languages $= 50 - 30 = 20$

(ix) exactly one language $= e + f + g = 6 + 3 + 9 = 18$

(x) exactly two languages $= b + c + d = 6 + 2 + 1 = 9$

18. $\because A = \{1, 2, 3, 4\}$ and $B = \{5, 7, 9\}$.

Therefore,

(i) $A \times B = \{(1, 5), (1, 7), (1, 9), (2, 5), (2, 7), (2, 9), (3, 5), (3, 7), (3, 9), (4, 5), (4, 7), (4, 9)\}$

(ii) $B \times A = \{(5, 1), (5, 2), (5, 3), (5, 4), (7, 1), (7, 2), (7, 3), (7, 4), (9, 1), (9, 2), (9, 3), (9, 4)\}$.

(iii) No, $A \times B \neq B \times A$. Since, $A \times B$ and $B \times A$ do not have exactly the same ordered pairs.

(iv) $n(A \times B) = n(A) \times n(B) = 4 \times 3 = 12$

and $n(B \times A) = n(B) \times n(A) = 3 \times 4 = 12$

Hence, $n(A \times B) = n(B \times A)$

19. $A = \{0, 1, 2\}$ [as (1, 0) and (0, 2) are the elements of $A \times A$]

Now, $A \times A = \{(0, 0), (0, 1), (0, 2), (1, 0), (1, 1), (1, 2), (2, 0), (2, 1), (2, 2)\}$

20. $\because A = \{1, 2\}$

$\therefore A \times A = \{1, 2\} \times \{1, 2\}$

$= \{(1, 1), (1, 2), (2, 1), (2, 2)\}$

Now, $A \times A \times A = \{(1, 1), (1, 2), (2, 1)\ (2, 2)\} \times \{1, 2\}$

$= \{(1, 1, 1), (1, 1, 2), (1, 2, 1), (1, 2, 2), (2, 1, 1), (2, 1, 2), (2, 2, 1), (2, 2, 2)\}$

21. When $x = 1, y = 7$, so $(1, 7) \in R$.

Again for

$x = 2, y = 2 + \frac{6}{2} = 2 + 3 = 5 \in N$, so $(2, 5) \in R$

For $x = 3, y = 3 + \frac{6}{3} = 3 + 2 = 5 \in N$, so, $(3, 5) \in R$

For $x = 4, y = 4 + \frac{6}{4} \notin N$ and for $x = 5, y = 5 + \frac{6}{5} \notin N$

Then, $R = \{(1, 7), (2, 5), (3, 5)\}$

where Domain of $R = \{1, 2, 3\}$

and Range of $R = \{7, 5\}$

22. Given, $A = \{1, 2, 3\}$

$a \in A, b \in A$, a divides b and b divides a.

$\Rightarrow \quad a = b$

$\therefore R = \{(a, a), a \in A\} = \{(1, 1), (2, 2), (3, 3)\}$

Thus, R is an identity relation on A.

23. Given that $x\,R\,y \Rightarrow 2x^2 - 3xy + y^2 = 0, \forall\ x \in N$

$\therefore x\,R\,x$ as $2x^2 - 3x^2 + x^2 = 0$

$\therefore R$ is reflexive.

For $x = 1$ and $y = 2$, $2x^2 - 3xy + y^2 = 0$

$\therefore 1\,R\,2$

But $2 \not R\, 1$ as $2 \cdot 2^2 - 3(2)(1) + (1)^2 \neq 0$

$\therefore$ R is not symmetric.

24. R is reflexive, since (1, 1), (2, 2) and (3, 3) lie in R. Also, R is not symmetric as $(1, 2) \in R$ but $(2, 1) \notin R$. Similarly, R is not transitive as $(1, 2) \in R$ and $(2, 3) \in R$ but $(1, 3) \notin R$.

25. $\because R = \{(1, 2), (2, 2), (1, 1), (4, 4), (1, 3), (3, 3), (3, 2)\}$

Since, $\{(1, 1), (2, 2), (3, 3), (4, 4)\} \in R$.

$\therefore R$ is reflexive.

Here, $(1, 2) \in R$, but $(2, 1) \notin R$

$\therefore R$ is not symmetric.

Here, $(1, 3), (3, 2) \in R$ and $(1, 2) \in R$

$\therefore R$ is transitive.

Hence, R is reflexive and transitive but not symmetric. Therefore, R is not an equivalence relation.

26. R is reflexive, since every triangle is congruent to itself. Further $(T_1, T_2) \in R \Rightarrow T_1$ is congruent to $T_2 \Rightarrow T_2$ is congruent to T_1

$\Rightarrow (T_2, T_1) \in R$. Hence, R is symmetric.

Moreover, $(T_1, T_2), (T_2, T_3) \in R \Rightarrow T_1$ is congruent to T_2 and T_2 is congruent to $T_3 \Rightarrow T_1$ is congruent to T_3 $\Rightarrow (T_1, T_3) \in R$.

$\therefore R$ is transitive.

Hence, R is an equivalence relation.

27. (i) Let $a, b \in N$

$(a, b)\,R\,(a, b) \Rightarrow a + b = b + a$

$\therefore R$ is reflexive.

$(a, b)\,R\,(c, d) \Rightarrow a + d = b + c$

$\Rightarrow c + b = d + a \Rightarrow (c, d)\,R\,(a, b)$

$\therefore$ R is symmetric

Let $(a,b)\,R\,(c,d)$ and $(c,d)\,R\,(e,f)$

$\Rightarrow a+d=b+c$ and $c+f=d+e$

$\Rightarrow a+d+c+f=b+c+d+e$

$\Rightarrow (a+f)=(b+e) \Rightarrow (a,b)\,R\,(e,f)$

$\therefore R$ is transitive.

Thus, R is an equivalence relation.

(ii) $z_1 R z_1 \Rightarrow \dfrac{z_1-z_1}{z_1+z_1}, \forall z_1 \in C \Rightarrow 0$ is real.

$\therefore R$ is reflexive.

Let $z_1 R z_2 \Rightarrow \dfrac{z_1-z_2}{z_1+z_2}$ is real; $z_1, z_2 \in C$

$\Rightarrow -\left(\dfrac{z_2-z_1}{z_1+z_2}\right)$ is also real.

$\Rightarrow \left(\dfrac{z_2-z_1}{z_1+z_2}\right)$ is real.

$\Rightarrow z_2 R z_1, \forall z_1, z_2 \in C$

$\therefore R$ is symmetric.

Let $z_1 R z_2$ and $z_2 R z_3$

$\because z_1 R z_2 \Rightarrow \left(\dfrac{z_1-z_2}{z_1+z_2}\right)$ is real.

$$\Rightarrow \overline{\left(\frac{z_1-z_2}{z_1+z_2}\right)} = -\left(\frac{z_1-z_2}{z_1+z_2}\right)$$

$$\Rightarrow \frac{\bar{z}_1-\bar{z}_2}{\bar{z}_1+\bar{z}_2} + \left(\frac{z_1-z_2}{z_1+z_2}\right) = 0$$

$$\Rightarrow \quad 2(z_1\bar{z}_1 - z_2\bar{z}_2) = 0$$

$$\Rightarrow \quad |z_1|^2 = |z_2|^2$$

Similarly, if $z_2 R z_3 \Rightarrow |z_2|^2 = |z_3|^2$

$\therefore$ From above, we can write $|z_1|^2 = |z_3|^2$

$\Rightarrow z_1 R z_3$.

$\therefore R$ is transitive.

Thus, R is an equivalence relation.

(iii) $R = \{(x,y) : x = y \text{ or } x - y = 1;\ x, y \in N\}$

$\because \quad x = x$

So, $(x,x) \in R, \forall x \in N$

$\therefore$ R is reflexive.

Let $(3,2) \in R$ and $3-2=1$

But $(2,3) \notin R$ as $2-3=-1 \neq 1$

$\therefore$ R is not symmetric.

Now, let $(3,2) \in R$ and $(2,1) \in R$

$\therefore$ $3-2=1$ and $2-1=1$

But $(3,1) \notin R$ as $3-1=2 \neq 1$

$\therefore$ R is not transitive.

Thus, R is not an equivalence relation.

28. $\because R_1$ and R_2 are equivalence relations,

$(a,a) \in R_1$ and $(a,a) \in R_2, \forall a \in A$.

$\Rightarrow (a,a) \in R_1 \cap R_2, \forall a$, showing $R_1 \cap R_2$ is reflexive.

Further, $(a,b) \in R_1 \cap R_2 \Rightarrow (a,b) \in R_1$ and $(a,b) \in R_2$

$\Rightarrow (b,a) \in R_1$ and $(b,a) \in R_2$

$\Rightarrow (b,a) \in R_1 \cap R_2$. Hence, $R_1 \cap R_2$ is symmetric.

Similarly, $(a,b) \in R_1 \cap R_2$ and $(b,c) \in R_1 \cap R_2$

$\Rightarrow (a,c) \in R_1$ and $(a,c) \in R_2$

$\Rightarrow (a,c) \in R_1 \cap R_2$. Thus, shows that $R_1 \cap R_2$ is transitive. Thus, $R_1 \cap R_2$ is an equivalence relation.

29. Given, $\dfrac{a-b}{4} = k$

$\Rightarrow a = 4k + b$, where $0 \le b < 4$

It is clear that b has only value in 0, 1, 2, 3.

The equivalence class of $[0] = \{x : x - 0 = 4k\}$

$= \{0, \pm 4, \pm 8, \pm 12, \ldots\}$, where $k = 0, \pm 1, \pm 2, \ldots$

Similarly, the equivalence class of

$[1] = \{x : x - 1 = 4k\} = \{x : x = 4k + 1\}$

$[1] = \{\ldots, -11, -7, -3, 1, 5, 9, \ldots\}$

30. Given, $A = \{1, 2, 3, 4\}$

R is defined as

$(a,b)\,R\,(c,d)$, iff $a+d=b+c$

We need to find $[(1,3)]$

Taking $a=1$ and $b=3$

$\Rightarrow 1+d=3+c \Rightarrow d-c=2$ ...(i)

So, in our relation $[(1,3)]$, we need to find the values of c and d which satisfy $d-c=2$, where $c, d \in A$

Numbers	$d-c$	**(c, d) as per condition**
$d=2, c=1$	1	not possible
$d=3, c=1$	2	possible
$d=3, c=2$	1	not possible
$d=4, c=1$	3	not possible
$d=4, c=2$	2	possible
$d=4, c=3$	1	not possible

Thus, $(c,d)=(1,3)$ and $(c,d)=(2,4)$ is possible as per given statement of the condition (i)

$\therefore$ The equivalence class

$[(1,3)] = \{(1,3), (2,4)\}$

31. $\because$ Domain of $R = \{1, 3, 4, 7\}$

Range of $R = \{4, 7, 5, 6\}$

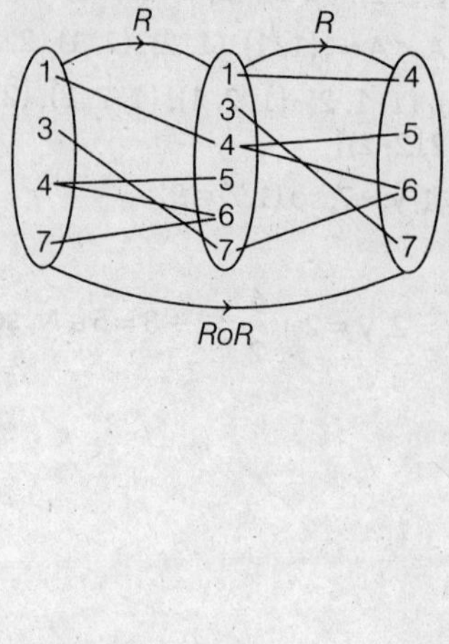

We see that

$1 \to 4 \to 5 \Rightarrow (1,5) \in RoR$

$1 \to 4 \to 6 \Rightarrow (1,6) \in RoR$

$3 \to 7 \to 6 \Rightarrow (3,6) \in RoR$

(i) $RoR = \{(1,5),(1,6),(3,6)\}$

(ii) $\because R^{-1}oR^{-1} = (RoR)^{-1} = \{(5,1),(6,1),(6,3)\}$

(iii) We know that

$(R^{-1}oR)^{-1} = R^{-1}o(R^{-1})^{-1} = R^{-1}oR$

$\because$ Domain of $R = \{1,3,4,7\}$

and Range of $R = \{4,5,6,7\}$

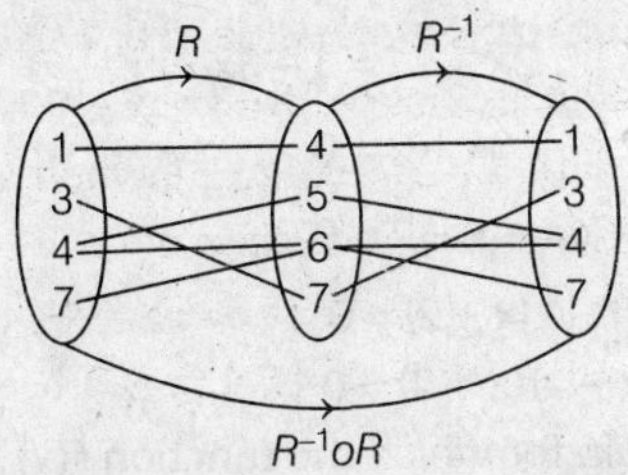

Domain of $R^{-1} = \{4,5,6,7\}$ and Range of $R^{-1} = \{1,3,4,7\}$

$1 \xrightarrow{R} 4 \xrightarrow{R^{-1}} 1 \Rightarrow (1,1) \in R^{-1}oR$

$3 \xrightarrow{R} 7 \xrightarrow{R^{-1}} 3 \Rightarrow (3,3) \in R^{-1}oR$

$4 \xrightarrow{R} 5 \xrightarrow{R^{-1}} 4 \Rightarrow (4,4) \in R^{-1}oR$

$4 \xrightarrow{R} 6 \xrightarrow{R^{-1}} 4 \Rightarrow (4,4) \in R^{-1}oR$

$4 \xrightarrow{R} 6 \xrightarrow{R^{-1}} 7 \Rightarrow (4,7) \in R^{-1}oR$

$7 \xrightarrow{R} 6 \xrightarrow{R^{-1}} 4 \Rightarrow (7,4) \in R^{-1}oR$

$7 \xrightarrow{R} 6 \xrightarrow{R^{-1}} 7 \Rightarrow (7,7) \in R^{-1}oR$

$\therefore (R^{-1}oR)^{-1} = R^{-1}oR = \{(1,1), (3,3), (4,4), (4,7), (7,4), (7,7)\}$

32. (i) $x < 2 \Rightarrow x \in (-\infty, 0) \cup [0, 2)$

For $x \in [0,2), x^2 \in [0,4)$

For $x \in (-\infty, 0), x^2 \in (0, \infty)$

$\Rightarrow$ For $x < 2, x^2 \in [0,4) \cup (0,\infty)$

$\Rightarrow \quad x^2 \in [0, \infty)$

(ii) When $x > -1 \Rightarrow x \in (-1, 0) \cup [0, \infty)$

For $x \in (-1, 0), x^2 \in (0,1)$

For $x \in [0, \infty), x^2 \in [0, \infty)$

$\Rightarrow$ For $x = -1, x^2 \in (0,1) \cup [0, \infty)$

$\Rightarrow \quad x^2 \in [0, \infty)$

(iii) For $x \geq 2, x \in [2, \infty); x^2 \in [4, \infty)$

(iv) For $x < -1, x \in (-\infty, -1) \Rightarrow x^2 \in (1, \infty)$

33. $(x^2 - 4)\sqrt{x^2 - 1} < 0$

We must have,

$x^2 - 1 \geq 0$ or $x^2 \geq 1$

or $\quad x \leq -1$ or $x \geq 1$ ...(i)

Also, $\quad (x^2 - 4)\sqrt{x^2 - 1} < 0$

or $x^2 - 4 < 0$ or $x^2 < 4$ or $-2 < x < 2$ and $x \neq 1$ ...(ii)

From Eqs. (i) and (ii), we have

$x \in (-2, -1) \cup (1, 2)$.

34. (i) $x > 3$, where $3 < x < \infty$

$\Rightarrow \quad \frac{1}{3} > \frac{1}{x} > \frac{1}{\to \infty}$ ($\to \infty$ means tends to infinity)

$\Rightarrow \quad 0 < \frac{1}{x} < \frac{1}{3}$

(ii) $x < -2$, where $-\infty < x < -2$

$\Rightarrow \frac{1}{\to -\infty} > \frac{1}{x} > \frac{1}{-2}$

$\Rightarrow \quad 0 > \frac{1}{x} > \frac{-1}{2} \Rightarrow -\frac{1}{2} < \frac{1}{x} < 0$

(iii) $x \in (-1, 3) - \{0\}$

$\Rightarrow x \in (-1, 0) \cup (0, 3)$

For $x \in (-1, 0)$, we have

$\Rightarrow \frac{1}{-1} > \frac{1}{x} > \frac{1}{\to 0^-}$ ($\to 0^-$ means the value of x approaches to 0 from left hand side or negative side)

$\Rightarrow \quad -\infty < \frac{1}{x} < -1$ ...(a)

For $x \in (0, 3)$, we have

$\frac{1}{\to 0^+} > \frac{1}{x} > \frac{1}{3}$

$\Rightarrow \quad \infty > \frac{1}{x} > \frac{1}{3}$

$\Rightarrow \quad \frac{1}{3} < \frac{1}{x} < \infty$ ...(b)

From Eqs. (a) and (b),

$\frac{1}{x} \in (-\infty, -1) \cup \left(\frac{1}{3}, \infty\right)$

35. $1 + \frac{3}{x} > 2$ or $1 + \frac{3}{x} < -2$

$\Rightarrow 1 + \frac{3}{x} - 2 > 0$ or $1 + \frac{3}{x} + 2 < 0$

$\Rightarrow \quad \frac{3 - x}{x} > 0$ or $\frac{3(x+1)}{x} < 0$

$\Rightarrow \quad \frac{x - 3}{x} < 0$ or $\frac{3(x+1)}{x} < 0$

$\Rightarrow \quad x \in (0, 3)$ or $x \in (-1, 0)$

$\therefore \quad x \in (-1, 0) \cup (0, 3)$.

36. $\frac{x}{x+2} \le \frac{1}{|x|}$

or $\frac{x|x|}{x+2} \le 1, x \ne 0$

or $\frac{x|x|}{x+2} - 1 \le 0, x \ne 0$

or $\frac{x|x| - x - 2}{x+2} \le 0, x \ne 0$

Case I $x > 0, \frac{x^2 - x - 2}{x+2} \le 0$

or $\frac{(x-2)(x+1)}{x+2} \le 0$

or $0 < x \le 2$ as $x > 0$ ...(i)

Case II $x < 0, \frac{-x^2 - x - 2}{x+2} \le 0$

or $\frac{x^2 + x + 2}{x+2} \ge 0 \Rightarrow -2 < x < 0$ ...(ii)

(as $x^2 + x + 2 > 0, \forall x \in R$ and $x < 0$)

From Eqs. (i) and (ii), we have

$x \in (-2, 2] - \{0\}$

37. Let us factorize $x^2 - x - 1$

$$x = \frac{1 \pm \sqrt{5}}{2}$$

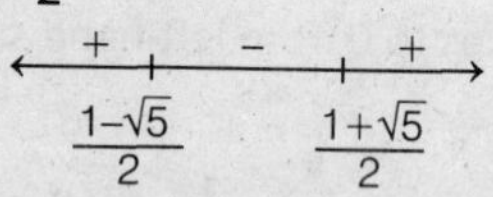

From the sign scheme of $x^2 - x - 1$, as shown in figure, we have

$$x^2 - x - 1 < 0 \Rightarrow x \in \left(\frac{1-\sqrt{5}}{2}, \frac{1+\sqrt{5}}{2}\right)$$

38. We have, $(x-1)(x-2)(1-2x) > 0$

or $-(x-1)(x-2)(2x-1) > 0$

or $(x-1)(x-2)(2x-1) < 0$

On the number line mark $x = \frac{1}{2}, 1, 2$

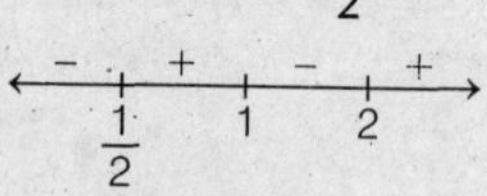

Hence, from the sign scheme,

$$x \in \left(-\infty, \frac{1}{2}\right) \cup (1, 2)$$

39. (i) $\because |x-3| \ge 0, \forall x \in R$

$\Rightarrow |x-3| - 2 \ge -2$

$\Rightarrow |x-3| - 2 \in [-2, \infty)$

(ii) $\because |2x+3| \ge 0, \forall x \in R$

$\Rightarrow -|2x+3| \le 0$

$\Rightarrow 4 - |2x+3| \le 4$

$\Rightarrow 4 - |2x+3| \in (-\infty, 4]$

40. (i) Since, $(2, 3)$ and $(2, 7) \in R_1$

$\Rightarrow R_1(2) = 3$ and $R_1(2) = 7$

So, $R_1(2)$ does not have an unique image.

Thus, R_1 is not a function.

(ii) $R_2 = \{(x, |x|) \mid x \in R\}$

For every $x \in R$, there will be unique image as $|x| \in R$.

$\therefore R_2$ is a function

41. For $f(x) = g(x)$

$\Rightarrow 2x^2 - 1 = 1 - 3x$

$\Rightarrow 2x^2 + 3x - 2 = 0$

$\Rightarrow 2x^2 + 4x - x - 2 = 0$

$\Rightarrow 2x(x+2) - 1(x+2) = 0$

$\Rightarrow (2x-1)(x+2) = 0$

Thus, domain for which the function $f(x) = g(x)$ is $\left\{\frac{1}{2}, -2\right\}$.

42. Let $x_1, x_2 \in N$, then

$f(x_1) = f(x_2) \Rightarrow 2x_1 = 2x_2$

$\Rightarrow x_1 = x_2$

$\therefore f$ is one-one.

Let $y = 2x$, then $x = \frac{y}{2}$

Now, if we put $y = 3$, then $x = \frac{3}{2} \notin N$

This show that $3 \in N$ has no pre-image in N. So, f is into.

Hence, f is one-one and into.

43. Let $x_1, x_2 \in R$

Then $f(x_1) = f(x_2)$

$\Rightarrow \cos x_1 = \cos x_2$

$\Rightarrow x_1 = 2n\pi \pm x_2$

$\Rightarrow x_1 \ne x_2$

$\therefore f$ is not one-one.

Let $y = \cos x$ but $-1 \le \cos x \le 1$.

$\therefore y \in [-1, 1]$

$\Rightarrow [-1, 1] \subset R$

So, f is into (not onto)

Hence, f is neither one-one nor onto.

44. Let $x \in R$

(i) $\therefore fog(x) = f(g(x)) = f(x^2) = \sin x^2$

(ii) $gof(x) = g(f(x)) = g(\sin x) = (\sin x)^2 = \sin^2 x$

(iii) $fog\left(\frac{\sqrt{\pi}}{2}\right) = \sin\left(\frac{\sqrt{\pi}}{2}\right)^2 = \sin\frac{\pi}{4} = \frac{1}{\sqrt{2}}$

(iv) $gof\left(\frac{\pi}{3}\right) = \sin^2\frac{\pi}{3} = \left(\frac{\sqrt{3}}{2}\right)^2 = \frac{3}{4}$

45. Domain of $f = \{1, 3, 4\}$, Range of $f = \{2, 5, 1\}$ and Domain of $g = \{2, 5, 1\}$, Range of $g = \{3, 1\}$

$\because$ Range of f = Domain of g

$\therefore$ gof is defined

$gof(1) = g(f(1)) = g(2) = 3;$

$gof(4) = g(f(4)) = g(1) = 3;$

$gof(3) = g(f(3)) = g(5) = 1$

Hence, $gof = \{(1,3),(3,1),(4,3)\}$

Now, since range of $g \subset$ domain of f

$\therefore$ fog is defined.

$fog(2) = f(g(2)) = f(3) = 5;$

$fog(5) = f(g(5)) = f(1) = 2;$

$fog(1) = f(g(1)) = f(3) = 5;$

Hence, $fog = \{(2,5),(5,2),(1,5)\}$

46. For invertible of f, f must be bijective (one-one and onto).

If $x_1, x_2 \in R$

$f(x_1) = f(x_2)$

$\Rightarrow \cos(5x_1 + 2) = \cos(5x_2 + 2)$

$\Rightarrow \quad 5x_1 + 2 = 2n\pi \pm (5x_2 + 2)$

$\Rightarrow \quad x_1 \neq x_2$

$\therefore$ f is not one-one.

But $-1 \leq \cos(5x+2) \leq 1$

$-1 \leq f(x) \leq 1$

Range $= [-1, 1] \subset R$

$\therefore$ f is into (not onto) mapping.

Hence, the function $f(x)$ is not bijective and so, it is not invertible.

47. $fog(x) = f(g(x)) = f(3x-2) = e^{3x-2}$

and $gof(x) = g(f(x)) = g(e^x) = 3e^x - 2$

Let $(fog)\,x = y \Rightarrow e^{3x-2} = y$

$\Rightarrow \quad 3x - 2 = \log y$

$\Rightarrow \quad x = \dfrac{2 + \log y}{3}$

$\Rightarrow \quad (fog)^{-1} y = \dfrac{2 + \log y}{3}$

$\Rightarrow y > 0$, so domain of $(fog)^{-1}$ is $(0, \infty)$

Now, again let $(gof)x = y$

$\Rightarrow y = 3e^x - 2 \Rightarrow e^x = \dfrac{y+2}{3}$

$\therefore \quad x = \log_e\left(\dfrac{y+2}{3}\right)$

$\Rightarrow (gof)^{-1}(y) = \log_e\left(\dfrac{y+2}{3}\right)$

Clearly, $y + 2 > 0$

$\Rightarrow \quad y > -2$

$\therefore$ Domain of $(gof)^{-1}$ is $(-2, \infty)$.

48. $(hofog)\,x = hof\,(g\,(x)) = hof\,(\sqrt{x^2+1})$

$= h\,\{f(\sqrt{x^2+1})\}$

$= h\,(x^2 + 1 - 1) = h(x^2) = x^2$

$\because \quad x^2 \geq 0$

and $(fog)\,x = f(g(x)) = f(\sqrt{x^2+1})$

$= (\sqrt{x^2+1})^2 - 1 = x^2 + 1 - 1 = x^2$

Let $y = (fog)x = x^2, \forall\, x \in R$

If $x = 1$, then $y = 1$

and if $x = -1$, then $y = 1$

So, fog is not one-one, so it is not invertible.

Here, $h(x) = \begin{cases} 0, & x \leq 0 \\ x, & x \geq 0 \end{cases}$

For $x = -1$, $h(-1) = 0$

For $x = 1$, $h(1) = 1$

For $x = -2$, $h(-2) = 0$,

For $x = 2$, $h(2) = 2$

$\therefore$ h is not identity function.

Level 2

1. (b) **Case I** $x \in [0, 9]$

$2(3 - \sqrt{x}) + x - 6\sqrt{x} + 6 = 0$

$\Rightarrow \quad x - 8\sqrt{x} + 12 = 0$

$\Rightarrow \quad \sqrt{x} = 4, 2$

$\Rightarrow \quad x = 16, 4$

Case II $x \in [9, \infty)$

$2(\sqrt{x} - 3) + x - 6\sqrt{x} + 6 = 0$

$\Rightarrow \quad x - 4\sqrt{x} = 0$

$\Rightarrow \quad x = 16, 0$

$\Rightarrow \quad x = 16 \quad [\because x \in [9, \infty)]$

Hence, $x = 4$ and 16.

2. (b) According to the question,

$2^m - 2^n = 112$

$\Rightarrow 2^n(2^{m-n} - 1) = 2^4 \times 7$

Comparing exponents on both sides, we get

$2^n = 2^4$ and $2^{m-n} - 1 = 7$

Since, $\quad 2^n = 2^4 \Rightarrow n = 4$

Now, $2^{m-n} = 8$

$\Rightarrow \quad 2^{m-n} = 2^3$

$\Rightarrow \quad m - n = 3$

$\Rightarrow \quad m - 4 = 3$

$\therefore \quad m = 7$

$\therefore \quad mn = 7 \times 4 = 28$

3. (a) $\because$ Product of two even numbers is always even and product of two odd numbers is always odd.

$\therefore$ Number of required subsets

$= $ Total number of subsets

$-$ Total number of subsets having only odd

$= 2^{100} - 2^{50}$

$= 2^{50}(2^{50} - 1)$

4. (d) $S = \{1, 2, 3, 4\}$

Let A and B be two disjoint subsets of S

Now, for any element $a \in S$, has got three possibilities either, it is in A or B or none.

For every element out of 4 elements there are three choices.

$\therefore$ Total options $= 3^4 = 81$

Here, $A \neq B$, except when

$A = B = \phi$

$\therefore 81 - 1 = 80$ (order pairs) are there for which $A \neq B$.

Hence, total number of unordered pairs of disjoint subsets $= \frac{80}{2} + 1 = 41$

5. (c) Since, $A \subseteq B, A \cup B = B$

So, $n(A \cup B) = n(B) = 6$

6. (c) For $A \cap B$, we have

$$-x = \frac{1}{x}$$

$\Rightarrow \quad -x^2 = 1$

$\Rightarrow \quad x^2 = -1,$

which is not possible

$\therefore \quad A \cap B = \phi$

7. (d) $\because X = \{4^n - 3n - 1 : n \in N\}$

$\Rightarrow \quad X = \{0, 9, 54, 243, \ldots\}$

and $Y = \{9(n-1) : n \in N\}$

$\Rightarrow \quad Y = \{0, 9, 18, 27, \ldots\}$

It is clear that $X \subset Y$

$\therefore \quad X \cup Y = Y$

8. (c) $4N = \{4, 8, 12, 16, 20, 24, \ldots\}$ and $6N = \{6, 12, 18, 24, \ldots\}$

Hence, $4N \cap 6N = \{12, 24, 36, \ldots\} = 12N$

9. (d) $\bigcup_{i=1}^{50} X_i = \bigcup_{i=1}^{n} Y_i = T$

$\because n(X_i) = 10$ and $n(Y_i) = 5$

So, $\bigcup_{i=1}^{50} X_i = 500$ and $\bigcup_{i=1}^{n} Y_i = 5n$

$\Rightarrow \quad \frac{500}{20} = \frac{5n}{6}$

$\Rightarrow \quad n = 30.$

10. (d) Shaded region in the figure is clearly $A - (B \cup C)$.

11. (d)

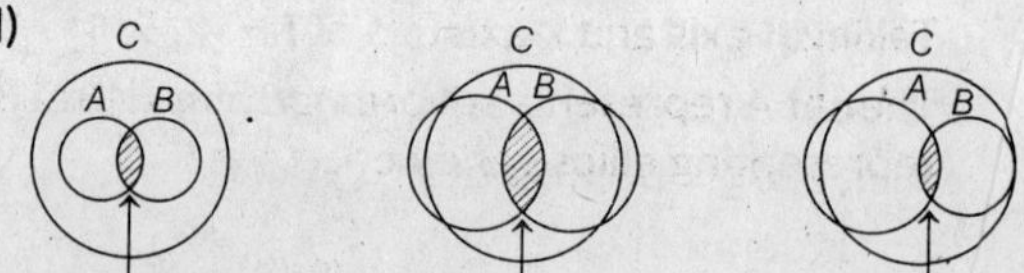

The shaded region of this Venn-diagram represents the students who play all three types of games. | The shaded region of this Venn-diagram represents the students who play all three type of games. | The shaded region of this Venn-diagram represents the students who play all three type of games.

As now, play all three games, the intersection of all three circles must be zero.

12. (b) Given, $A \cap B = A \cap C$

and $\quad A \cup B = A \cup C$

$\therefore \quad B = C$

13. (c) Options (a), (b) and (d) are always correct

In option (c), if $A = C$, then $A - C = \phi$

Clearly $\phi \subseteq B$ but $A \subseteq B$ is not always true.

14. (d) $\because P = \{\theta : \sin\theta - \cos\theta = \sqrt{2}\cos\theta\}$

$\Rightarrow \quad \sin\theta = (\sqrt{2} + 1)\cos\theta$

$\Rightarrow \quad \tan\theta = \sqrt{2} + 1 \quad$ ...(i)

and $\quad Q = \{\theta : \sin\theta + \cos\theta = \sqrt{2}\sin\theta\}$

$\Rightarrow \cos\theta = (\sqrt{2} - 1)\sin\theta$

$\Rightarrow \tan\theta = \sqrt{2} + 1 \quad$ ...(ii)

From Eqs. (i) and (ii),

$P = Q$

15. (c) $X \cap (X \cup Y)^c = X \cap (X^c \cap Y^c)$

$= (X \cap X^c) \cap Y^c$

$= \phi \cap Y^c = \phi$

16. (b) $(A \cap B')' \cup (B \cap C) = ((A' \cup (B')') \cup (B \cap C)$

$= (A' \cup B) \cup (C \cap B)$

$= A' \cup (B \cup (C \cap B))$

$= A' \cup B$

17. (a) $(A \cup B)' \cup (A' \cap B) = (A' \cap B') \cup (A' \cap B)$

$= (A' \cup A') \cap (A' \cup B) \cap (B' \cup A') \cap (B' \cup B)$

$= A' \cap \{A' \cup (B \cap B')\} \cap U$

$= A' \cap \{A' \cup \phi\} \cap U$

$= A' \cap A' \cap U$

$= A' \cap U = A'$

18. (c) $[(A' \cap B' \cap C)] \cup [(B \cap C) \cup (A \cap C)]$

$= [(A \cup B)' \cap C] \cup [(B \cup A) \cap C]$

$= [(A \cup B)' \cup (B \cup A)] \cap C$

$= X \cap C = C$

19. (b) We have, $|a - 5| < 1$ and $|b - 5| < 1$

$\therefore -1 < a - 5 < 1$ and $-1 < b - 5 < 1$

$\Rightarrow 4 < a < 6$ and $4 < b < 6$

Now, $4(a-6)^2 + 9(b-5)^2 \leq 36$

$\Rightarrow \quad \frac{(a-6)^2}{9} + \frac{(b-5)^2}{4} \leq 1$

Taking X-axis and Y-axis

The set A represents square $PQRS$ inside set B representing ellipse and hence $A \subset B$.

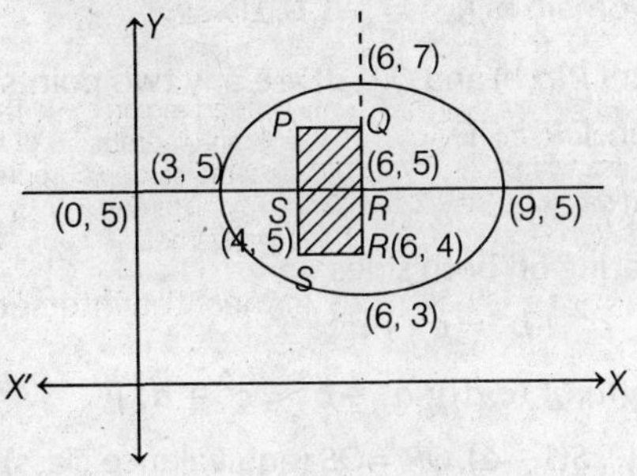

20. (d) Every rectangle, rhombus, square in a plane is a parallelogram but every trapezium is not a parallelogram.

$\therefore \quad F_1 = F_2 \cup F_3 \cup F_4 \cup F_1$

21. (b) Given that $n(C) = 73$ and $n(T) = 65$

Let $n(C \cap T) = x$

$\therefore \quad 65 \geq n(C \cap T) \geq 65 + 73 - 100$

$\Rightarrow \quad 65 \geq x \geq 38$

$\Rightarrow \quad x \neq 36$

22. (c) Let A be the set of those people who watch news channel and B be the set of those people who watch entertainment channel, then

$$n(A) = 63, n(B) = 76 \text{ and } n(A \cap B) = x$$

$\because \quad n(A \cup B) \leq 100$

$\Rightarrow \quad n(A) + n(B) - n(A \cap B) \leq 100$

$\Rightarrow \quad 63 + 76 - x \leq 100$

$\quad x \geq 39$

Also, $n(A \cap B) \leq \min\{n(A), n(B)\}$

$\Rightarrow \quad x \leq 63$

$\therefore \quad 39 \leq x \leq 63$

23. (c) Let A = Patient suffering from heart ailment and B = Set of patient suffering from lungs infection

Given, $n(A) = 89\%$ and $n(B) = 98\%$

$n(A \cup B) \geq n(A) + n(B) - n(A \cap B)$

$\Rightarrow \quad 100 \geq 89 + 98 - n(A \cap B)$

$\Rightarrow \quad 87 \leq n(A \cap B)$

Also, $n(A \cap B) = \min\{n(A), n(B)\}$

$\Rightarrow \quad n(A \cap B) \leq 89$

$\therefore \quad 87 \leq n(A \cap B) \leq 89$

So, $n(A \cap B) \notin \{79, 81, 83, 85\}$.

24. (b) $n(A) = 40\%$ of $10000 = 4000$

$n(B) = 20\%$ of $10000 = 2000$

$n(C) = 10\%$ of $10000 = 1000$

$n(A \cap B) = 5\%$ of $10000 = 500$

$n(C \cap A) = 4\%$ of $10000 = 400$

$n(A \cap B \cap C) = 2\%$ of $10000 = 200$

Number of families which buy newspaper A only

$= n(A \cap B^c \cap C^c)$

$= n[A \cap (B \cup C)^c]$

$= n(A) - n[A \cap (B \cup C)]$

$= n(A) - n[(A \cap B) \cup (A \cap C)]$

$= n(A) - [n(A \cap B) + n(A \cap C) - n(A \cap B \cap C)]$

$= 4000 - [500 + 400 - 200]$

$= 4000 - 700$

$= 3300$

25. (c) Given, set $A = \{x \in Z : 2^{(x+2)(x^2 - 5x + 6)} = 1\}$

Consider, $2^{(x+2)(x^2 - 5x + 6)} = 1 = 2^0$

$\Rightarrow \quad (x + 2)(x^2 - 5x + 6) = 0$

$\Rightarrow \quad (x + 2)(x - 3)(x - 2) = 0$

$\Rightarrow \quad x = -2, 2, 3$

$\Rightarrow \quad A = \{-2, 2, 3\}$

Also, $B = \{x \in Z : -3 < 2x - 1 < 9\}$

Consider, $-3 < 2x - 1 < 9, x \in Z$

$\Rightarrow \quad -2 < 2x < 10, x \in Z$

$\Rightarrow \quad -1 < x < 5, x \in Z$

$\Rightarrow \quad B = \{0, 1, 2, 3, 4\}$

So, $A \times B$ has 15 elements.

$\therefore$ Number of subsets of $A \times B = 2^{15}$.

26. (b) Given, relation R on N is defined by

$R = \{(x, y) \in N \times N : x^3 - 3x^2 - xy^2 + 3y^3 = 0\}$

$x^3 - 3x^2y - xy^2 + 3y^3 = 0$

$\Rightarrow \quad x^3 - xy^2 - 3x^2y + 3y^3 = 0$

$\Rightarrow \quad x(x^2 - y^2) - 3y(x^2 - y^2) = 0$

$\Rightarrow \quad (x - 3y)(x^2 - y^2) = 0$

$\Rightarrow \quad (x - 3y)(x - y)(x + y) = 0$

Now, $\quad x - x = 0$

$\Rightarrow \quad x = x, \forall (x, x) \in N \times N$

So, R is a reflexive relation.

But not symmetric and transitive relation because, (3, 1) satisfies but (1, 3) does not. Also, (3, 1) and (1, −1) satisfies but (3, −1) does not.

Hence, relation R is reflexive but neither symmetric nor transitive.

27. (a) $\because |a - a| = 0 < 1, aRa, \forall a \in R$ (reflexive)

Again, $aRb \Rightarrow |a - b| \leq 1$

$\Rightarrow \quad |b - a| \leq 1$

$\Rightarrow \quad bRa$

$\therefore R$ is symmetric

Again, $1R2$ and $2R1$ but $2 \neq 1$

$\therefore R$ is not anti-symmetric.

Further, $1R2$ and $2R3$ but $1 \not R 3$. $[|1 - 3| = 2 > 1|]$

$\therefore R$ is not transitive.

28. (a) For R_1, let $a=1+\sqrt{2}, b=1-\sqrt{2}$ and $c=8^{\frac{1}{4}}$

$aR_1 b = a^2+b^2=(1+\sqrt{2})^2+(1-\sqrt{2})^2=6\in Q$

$bR_1 c = b^2+c^2=(1-\sqrt{2})^2+(8^{1/4})^2=3\in Q$

$aR_1 c = a^2+c^2=(1+\sqrt{2})^2+(8^{1/4})^2=3+2\sqrt{2}\notin Q$

$\therefore$ R_1 is not transitive

For R_2, let $a=1+\sqrt{2}, b=\sqrt{2}$ and $c=1-\sqrt{2}$

$aR_2 b = a^2+b^2=5+2\sqrt{2}\notin Q$

$bR_2 c = b^2+c^2=5-2\sqrt{2}\notin Q$

$aR_2 c = a^2+c^2=6\in Q$

$\therefore$ R_2 is not transitive.

29. (b) According to the question, let's consider option (b) (2, 3) and (3, 4) satisfy $0<|x-y|\le 1$ but (2, 4) does not satisfy it.

30. (c) For reflexive relation, $\forall (A,A)\in R$ for matrix P.

$\Rightarrow \quad A=PAP^{-1}$ is true for $P=1$

So, R is reflexive relation.

For symmetric relation,

Let $(A,B)\in R$ for matrix P.

$\Rightarrow \quad A=PBP^{-1}$

After pre-multiply by P^{-1} and post-multiply by P, we get

$P^{-1}AP=B$

So, $(B,A)\in R$ for matrix P^{-1}.

So, R is a symmetric relation.

For transitive relation,

Let ARB and BRC

So, $\quad A=PBP^{-1}$ and $B=PCP^{-1}$

Now, $\quad A=P(PCP^{-1})P^{-1}$

$\Rightarrow \quad A=(P)^2C(P^{-1})^2$

$\Rightarrow \quad A=(P)^2C(P^2)^{-1}$

$\therefore (A,C)\in R$ for matrix P^2.

$\therefore R$ is transitive relation.

Hence, R is an equivalence relation.

31. (a) $\because (1,2)\in S$ but $(2,1)\notin S$, so S is not symmetric

Hence, S is not equivalence relation.

Given, $T=\{(x,y): x-y$ is an integer$\}$

Now, $x-x=0\in I$, reflexive

Again, now $(x-y)\in I$

$\Rightarrow (y-x)\in I$ (symmetric)

Let $x-y={}_1, y-z={}_2$, where ${}_1, {}_2\in$

[1: set of integers]

Now, $x-z=(x-y)+(y-z)={}_1+{}_2\in$

So, T is transitive. Hence, T is an equivalence relation.

32. (d) $\because R=\{(x,y): x,y\in Z, x^2+3y^2\le 8\}$

$\therefore \quad R=\{(1,1),(2,1),(1,-1),(0,1),(1,0)\}$

$\Rightarrow$ Domain of $R^{-1}=\{-1,0,1\}$

33. (d) Let $P(a,b)$ and $Q(c,d)$ are any two points.

Given, $OP=OQ$

i.e. $\sqrt{a^2+b^2}=\sqrt{c^2+d^2}$

Squaring on both sides,

$a^2+b^2=c^2+d^2 \quad$ ... (i)

$R=\{((a,b),(c,d)): a^2+b^2=c^2+d^2\}$

$R(x,y)$, $S(1,-1)$, $OR=OS$ (equivalence class)

This gives $OR=\sqrt{x^2+y^2}$ and $OS=\sqrt{2}$

$\Rightarrow \quad \sqrt{x^2+y^2}=\sqrt{2}$

$\Rightarrow \quad x^2+y^2=2$ (squaring on both sides)

$\therefore \quad S=\{(x,y): x^2+y^2=2\}$

34. (b) $\dfrac{(x+2)(x+3)}{(x-2)(x-3)}\le 1$

$\Rightarrow \dfrac{x^2+5x+6}{x^2-5x+6}-1\le 0 \Rightarrow \dfrac{10x}{(x-2)(x-3)}\le 0$

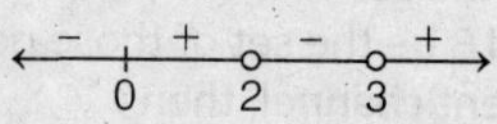

$x\in(-\infty,0]\cup(2,3)$

35. (b) We have, $(x+1)^2>5x-1$

$\Rightarrow \quad x^2-3x+2>0$

$\Rightarrow \quad (x-1)(x-2)>0$

$\Rightarrow \quad x<1$ or $x>2 \quad$ ...(i)

and $\quad (x+1)^2<7x-3$

$\Rightarrow \quad x^2-5x+4<0$

$\Rightarrow \quad (x-1)(x-4)<0$

$\Rightarrow \quad 1<x<4 \quad$ ...(ii)

From Eqs. (i) and (ii) common values are $x\in(2,4)$, therefore there is only one integral value of x which is 3.

36. (a) We have, $\dfrac{x+7}{x-5}+\dfrac{3x+1}{2}\ge 0$

$\Rightarrow \dfrac{2x+14+3x^2-14x-5}{2(x-5)}\ge 0$

$\Rightarrow \dfrac{3x^2-12x+9}{2(x-5)}\ge 0$

$\Rightarrow \dfrac{(x-1)(x-3)}{(x-5)}\ge 0$

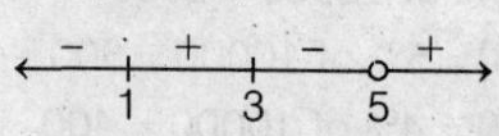

From the sign scheme $x\in[1,3]\cup(5,\infty)$

37. (b) We have, $\dfrac{x^2(x-5)(2x-1)}{(5x+1)(x+2)}<0$

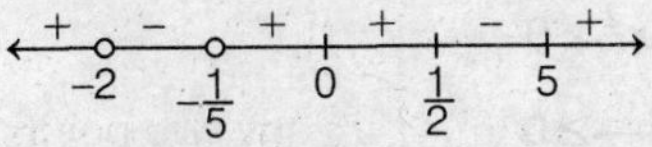

From the sign scheme solution is

$$\left(-2,-\frac{1}{5}\right)\cup\left(\frac{1}{2},5\right) \quad \text{...(i)}$$

Now, $\dfrac{3x+1}{x(6x^2+x-1)}>0$

$\Rightarrow \dfrac{3x+1}{x(3x-1)(2x+1)}>0$

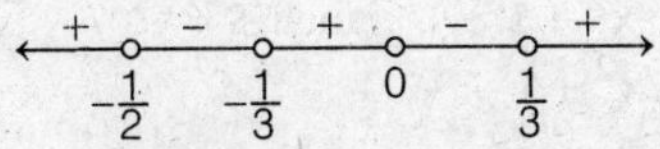

From the sign scheme solution is

$$\left(-\infty,-\frac{1}{2}\right)\cup\left(-\frac{1}{3},0\right)\cup\left(\frac{1}{3},\infty\right) \quad \text{...(ii)}$$

Clearly from Eqs. (i) and (ii), options (a) and (c) are correct.

38. (d) We have, $x^4-29x^2+100\le 0$

$\Rightarrow \quad (x^2-4)(x^2-25)\le 0$

$\Rightarrow \quad 4\le x^2\le 25$

$\Rightarrow \quad x\in[-5,-2]\cup[2,5]$

Integers satisfying above are $\pm 5,\pm 4,\pm 3,\pm 2$, so there are 8 integers.

39. (d) Hence, $x\in[-6,4]\cup\left[5,\dfrac{n}{2}\right]$

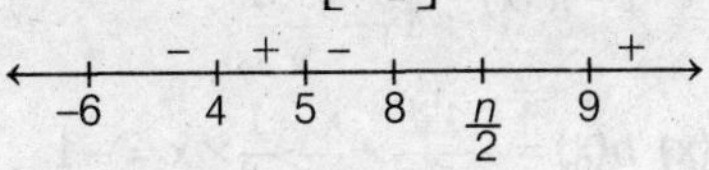

Now, there are 11 integers in $[-6,4]$

So, there must be 4 integers in $\left[5,\dfrac{n}{2}\right]$

So, $\dfrac{n}{2}\in(8,9) \Rightarrow n\in(16,18)$

40. (c) We have,

$$\frac{(x-5)^{1005}(x+8)^{1008}(x-1)}{x^{1006}(x-2)^3(x-3)^5(x-6)(x+9)^{1010}}\le 0$$

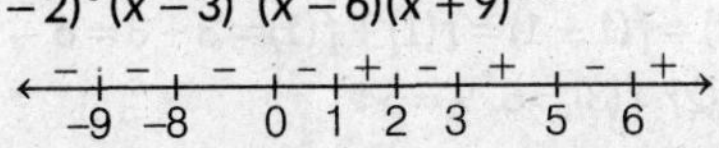

From the sign scheme and as per the inequality,

$$x\in(-\infty,-9)\cup(-9,0)\cup(0,1]\cup(2,3)\cup[5,6)$$

41. (c) We have, $|x|^3-4|x|^2+3|x|=0$

$\Rightarrow \quad |x|(|x|^2-4|x|+3)=0$

$\Rightarrow \quad |x|(|x|-1)(|x|-3)=0$

$\Rightarrow \quad |x|=0,1,3$

$\Rightarrow \quad x=0,1,3,-1,-3$

42. (d) We have, $|x-1||x^2-2|=2$

Since, x is an integer, we have,

$|x-1|=1$ and $|x^2-2|=2$

or $\quad |x-1|=2$ and $|x^2-2|=1$

$\therefore \quad x=0,2$ and $x=0,\pm 2$

or $\quad x=-1,3$ and $x=\pm 1$

Therefore, solutions are $-1,0,2$.

43. (c) $\dfrac{|x+2|-x}{x}<2$

$\Rightarrow \dfrac{|x+2|-3x}{x}<0$

Let $|x+2|-3x=0$

$\Rightarrow \quad x+2=\pm 3x \Rightarrow x=1$

Sign scheme of the expression $\dfrac{|x+2|-3x}{x}$ is as shown in the following figure

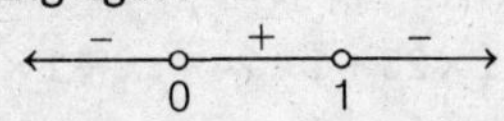

From the sign scheme, $x\in(-\infty,0)\cup(1,\infty)$.

44. (b) $|2-|x||=x+4$ ...(i)

$\Rightarrow \; 2-|x|=x+4$ ...(ii)

or $\; |x|-2=x+4$ ...(iii)

From Eq. (i), $2-x=x+4$ or $x=-1$

But this value does not satisfy Eq. (i)

From Eq. (iii), $-x-2=x+4$ or $x=-3$, which satisfies Eq. (i)

So, equation has only one solution.

45. (a) $\dfrac{x^2+6x-7}{|x+2||x+3|}<0$

Clearly, $x\ne -2,-3$

Now, $x^2+6x-7<0$

$\Rightarrow \; (x+7)(x-1)<0 \Rightarrow x\in(-7,1)-\{-3,-2\}$

So, integers are $\{-6,-5,-4,-1,0\}$

46. (c) We have, $-4\le x<2$

$\Rightarrow \quad -2\le x+2<4$

$\Rightarrow \quad 0\le|x+2|<4$

$\Rightarrow \quad -3\le|x+2|-3<1$

$\Rightarrow \quad 0\le||x+2|-3|\le 3$

47. (a) $f(x)=\dfrac{x}{1+x^2}, x\in R$

Let $y=\dfrac{x}{1+x^2}$

$\Rightarrow \quad yx^2-x+y=0 \Rightarrow x=\dfrac{1\pm\sqrt{1-4y^2}}{2}$

$\Rightarrow \quad 1-4y^2\ge 0 \Rightarrow 1\ge 4y^2$

$\Rightarrow \quad |y|\le\dfrac{1}{2} \Rightarrow -\dfrac{1}{2}\le y\le\dfrac{1}{2}$

$\therefore$ The range of f is $\left[-\dfrac{1}{2},\dfrac{1}{2}\right]$

48. (c) $\because f(x)=a^x=\underset{(f_1(x)=\text{even})}{\dfrac{a^x+a^{-x}}{2}}+\underset{(f_2(x)=\text{odd})}{\dfrac{a^x-a^{-x}}{2}}$

$$\therefore f_1(x+y)+f_1(x-y)=\frac{a^{x+y}+a^{-x-y}}{2}+\frac{a^{x-y}-a^{-x+y}}{2}$$

$$=\frac{1}{2}[a^x(a^y+a^{-y})]+a^{-x}(a^y+a^{-y})]$$

$$=\frac{(a^x+a^{-x})(a^y+a^{-y})}{2}$$

$$=2f_1(x)\,f_2(y)$$

49. (d) $\because f(n)=\left[\dfrac{1}{3}+\dfrac{3n}{100}\right]n$

For $n=1,2,\ldots,22$, $f(n)=0$

For $n=23,24,\ldots,55$, $f(n)=1\times n$

and for $n=56$, $f(n)=2\times n$

So, $\sum_{n=1}^{56} f(n)=1\times 23+1\times 24+\ldots+1\times 55+2\times 56$

$$=23+24+\ldots+55+112$$

Since, 23, 24, ..., 55 are in AP thus, using the formulation of sum of an AP, we get

$$=\frac{33}{2}[46+32]+112=\frac{33}{2}\times 78+112=1399$$

50. (b) We have,

$$\sum_{n=8}^{100}\left[\frac{(-1)^n n}{2}\right]\quad[\because [x] \text{ is the greatest integer function}]$$

Substitute the values of n

$$=[4]+[-4.5]+[5]+[-5.5]+\ldots+[-49.5]+[50]$$

$$=4-5+5-6+\ldots-50+50=4$$

51. (a) $\because f(x)=\log_e\left(\dfrac{1-x}{1+x}\right), |x|<1$

$$\therefore f\left(\frac{2x}{1+x^2}\right)=\log_e\left(\frac{1-\dfrac{2x}{1+x^2}}{1+\dfrac{2x}{1+x^2}}\right)$$

$$=\log_e\left(\frac{1+x^2-2x}{1+x^2+2x}\right)$$

$$=\log_e\left(\frac{1-x}{1+x}\right)^2$$

$$=2\log_e\left(\frac{1-x}{1+x}\right)$$

$$=2f(x)$$

52. (c) Given that $f: R-\{1,-1\}\to A$ defined as

$$f(x)=\frac{x^2}{1-x^2}=y \text{ (let)}$$

$\Rightarrow \quad x^2=y(1-x^2)\Rightarrow x^2(1+y)=y$

$\Rightarrow \quad x^2=\dfrac{y}{1+y};\ y\neq -1$

$\because \quad x^2\geq 0$

$$\frac{y}{1+y}\geq 0$$

$\Rightarrow y\in(-\infty,-1)\cup[0,\infty)$

$\because$ For surjective, range of f = codomain

$\therefore$ Set A should be $R-[-1,0)$

53. (c) $\because f(x)=\sin^{-1}\left(\dfrac{|x|+5}{x^2+1}\right)$

$$-1\leq\frac{|x|+5}{x^2+1}\leq 1$$

$\Rightarrow |x|+5\leq x^2+1 \qquad [\because x^2+1\neq 0]$

$\Rightarrow x^2-|x|-4\geq 0$

$$\Rightarrow\left(|x|-\frac{1+\sqrt{17}}{2}\right)\left(|x|-\frac{1-\sqrt{17}}{2}\right)\geq 0$$

$$\Rightarrow x\in\left(-\infty,-\frac{1+\sqrt{17}}{2}\right]\cup\left[\frac{1+\sqrt{17}}{2},\infty\right)$$

$$\therefore\ a=\frac{1+\sqrt{17}}{2}$$

54. (b) $\because g(x)=f[f(x)]=f\left(\dfrac{1}{1-x}\right)=\dfrac{1}{1-\dfrac{1}{1-x}}=\dfrac{x-1}{x}$

$$h(x)=f[f\{f(x)\}]=f(g(x))$$

$$=\frac{1}{1-g(x)}=\frac{1}{1-\dfrac{x-1}{x}}=x$$

$$\therefore f(x)\cdot g(x)\cdot h(x)=\frac{1}{1-x}\times\frac{x-1}{x}\times x=-1$$

55. (b) $f(m+n)=f(m)+f(n), m,n\in N$

$\therefore \quad f(3+3)=f(3)+f(3)$

$\Rightarrow \quad f(6)=2f(3)=18 \qquad [\because f(6)=18]$

Also, $f(3)=f(2+1)=f(2)+f(1)$

$=f(1+1)+f(1)$

$f(3)=f(1)+f(1)+f(1)$

$\Rightarrow \quad 9=3f(1)\Rightarrow f(1)=3$

$\therefore \quad f(2)=f(1+1)=f(1)+f(1)=3+3=6$

Hence, $f(2)\cdot f(3)=6\cdot 9=54$

56. (13.9) Let C = Set of people who read newspaper A and D = Set of people who read newspaper B

Given, $n(C)=25$, $n(D)=20$ and $n(C\cap D)=8$

$\therefore n(C\cap D')=n(C)-n(C\cap D)=25-8=17$

But total number of people who read A but not $B=30\%$

$\therefore$ % of people reading newspaper A but not $B=30\%$ of 17

$$=\frac{30\times 17}{100}=\frac{51}{10}$$

and $n(C' \cap D) = n(D) - n(C \cap D) = 20 - 8 = 12$

Also, total number of people who read newspaper B but not $A = 40\%$.

$\therefore$ % of people reading newspaper B but not $A = 40\%$ of 12

$$= \frac{40 \times 12}{100} = \frac{24}{5}$$

and given total people who read newspaper A and $B = 50\%$

$\therefore$ % of people who read newspaper A and $B = 50\%$ of $8 = \frac{50 \times 8}{100} = 4$

$\therefore$ Percentage of people reading an advertisement

$$= \left(\frac{51}{10} + \frac{24}{5} + 4\right)\% = 13.9\%$$

$\therefore \quad x = 13.9$

57. (832) Let $A = \{n \in N \mid n^2 \le n + 10000\}$

$$n^2 \le n + 10000$$

$$n^2 - n \le 10000$$

$\Rightarrow \quad n(n-1) \le 100 \times 100$

$\Rightarrow \quad A = \{1, 2, 3, \ldots\ldots, 100\}$

Now, $\quad B = \{3k + 1 \mid k \in N\}$

$B = \{4, 7, 10, 13, \ldots\}$

and $\quad C = \{2k \mid k \in N\}$

$C = \{2, 4, 6, 8, \ldots\}$

So, $\quad B - C = \{7, 13, 19, \ldots\ldots, 97, \ldots\}$

So, $A \cap (B - C) = \{7, 13, 19, \ldots\ldots, 97\}$

This form an AP with common difference $(d = 6)$

$\Rightarrow \quad 97 = 7 + (n-1)6$

$$n = \frac{97 - 7}{6} + 1 = 16 \qquad [\because a_n = a + (n-1)d]$$

Hence, sum $= \frac{16}{2}[7 + 97]$

$$= 832 \qquad \left\{\because S_n = \frac{n}{2}(a + l)\right\}$$

58. (7) $\because$ R is reflexive, if it contains (1, 1), (2, 2), (3, 3)

$\because (1, 2) \in R$ and $(2, 3) \in R$

$\therefore R$ is symmetric, if $(2, 1), (3, 2) \in R$

Now, $R =$ {(1, 1), (2, 2), (3, 3), (2, 1), (3, 2), (2, 3), (1, 2)}

R will be transitive, if $(3, 1), (1, 3) \in R$

Thus, R becomes an equivalence relation by adding (1, 1), (2, 2), (3, 3), (2, 1), (3, 2), (1, 3), (3, 1)

Hence, the total number of ordered pair is 7.

59. (330) We have, $f(x) = ax^2 + bx + c$

Now, $f(x + y) = f(x) + f(y) + xy$

Put $y = 0$;

$$f(x) = f(x) + f(0)$$

$\Rightarrow \quad f(0) = 0 \Rightarrow c = 0$

Again, put $y = -x$

$$f(0) = f(x) + f(-x) - x^2$$

$$0 = ax^2 + bx + ax^2 - bx - x^2 \qquad [\text{as } c = 0]$$

$\Rightarrow \quad 2ax^2 - x^2 = 0$

$\Rightarrow \quad a = \frac{1}{2}$

Also, $a + b + c = 3$

$\Rightarrow \quad \frac{1}{2} + b + 0 = 3 \Rightarrow b = \frac{5}{2}$

$\therefore \quad f(x) = \frac{x^2 + 5x}{2}$

Now, $\sum_{n=1}^{10} f(n) = \frac{1}{2}\sum_{n=1}^{10} n^2 + \frac{5}{2}\sum_{n=1}^{10} n$

$$= \frac{1}{2} \times \frac{10 \times 11 \times 21}{6} + \frac{5}{2} \times \frac{10 \times 11}{2}$$

$$= \frac{385}{2} + \frac{275}{2}$$

$$= \frac{660}{2} = 330$$

60. (18) $f: R \to R, f(x) = 10x - 7$

It is clear that f is one-one and onto.

$\therefore$ Let $y = 10x - 7$

$\Rightarrow \quad x = \frac{y + 7}{10}$

$\Rightarrow \quad f^{-1}(y) = \frac{y + 7}{10}$

$\therefore \quad g(x) = \frac{x + 7}{10}$

$\Rightarrow A = 1, B = 7$ and $C = 10$

$\therefore \ A + B + C = 1 + 7 + 10 = 18$

61. (256) $A = \{x \in R : |x - 2| > 1\}$

$\Rightarrow \quad A = (-\infty, 1) \cup (3, \infty)$

$B = \{x \in R : \sqrt{x^2 - 3} > 1\}$

$\Rightarrow \quad B = (-\infty, -2) \cup (2, \infty)$

$C = \{x \in R : |x - 4| \ge 2\}$

$\Rightarrow \quad C = (-\infty, 2] \cup [6, \infty)$

$\Rightarrow \quad A \cap B \cap C = (-\infty, -2) \cup [6, \infty)$

$\Rightarrow \quad (A \cap B \cap C)^C = [-2, 6)$

$\therefore \quad (A \cap B \cap C)^C \cap Z = \{-2, -1, 0, 1, 2, 3, 4, 5\}$

Number of subsets of $(A \cap B \cap C)^C \cap Z = 2^8 = 256$

62. (3) $f^{-1}(13) = \{x : f(x) = 13\} = \{x : x^2 - 5 = 13\}$

$$= \{x : x^2 = 18\} = \{x : x = \pm 3\sqrt{2}\}$$

$$= \{-3\sqrt{2}, 3\sqrt{2}\}$$

$$= \{-\lambda\sqrt{\lambda - 1}, \lambda\sqrt{\lambda - 1}\}$$

$\therefore \quad \lambda = 3$

63. (490) $S=\{1, 2, 3, 4, 5, 6, 7\}$

$f: S \to S$

$f(m\cdot n)=f(m)\,f(n)$

$m, n \in S \Rightarrow m, n \in S$

If $mn \in S \Rightarrow mn \le 7$

So, $(1\cdot 1, 1\cdot 2, \ldots, 1\cdot 7) \le 7$

$(2\cdot 2, 2\cdot 3) \le 7$

When $m=1$, $f(n)=f(1)\cdot f(n) \Rightarrow f(1)=1$

When $m=n=2$,

$$f(4)=f(2)\,f(2)=\begin{cases} f(2)=1 \Rightarrow f(4)=1 \\ f(2)=2 \Rightarrow f(4)=4 \end{cases}$$

When $m=2, n=3$

$$f(6)=f(2)\,f(3)\begin{cases} \text{When, } f(2)=1, \\ f(3) \;=1 \text{ to } 7 \\ \text{When, } f(2)=2, \\ f(3)=1 \text{ or } 2 \text{ or } 3 \end{cases}$$

and $f(5)$, $f(7)$ can take any value (1-7)

$[\because f(5)=f(1)\cdot f(5) \le 7$ and $f(7)=f(1)\cdot f(7) \le 7]$

The possible combination is

$f(1)=1$	$f(1)=1$
$f(2)=1$	$f(2)=2$
$f(3)=(1-7)$	$f(3)=(1-3)$
$f(4)=1$	$f(4)=4$
$f(5)=(1-7)$	$f(5)=(1-7)$
$f(6)=f(3)$	$f(6)=f(3)$
$f(7)=(1-7)$	$f(7)=(1-7)$

So, total $=(1\times 1\times 7\times 1\times 7\times 1\times 7)$

$+(1\times 1\times 3\times 1\times 7\times 1\times 7)$

$=490$

Level 3

1. (a) Given that $\bigcup\limits_{i=1}^{30} A_i=\bigcup\limits_{j=1}^{n} B_j=S$

and each A_i's contain 5 elements.

So, total number of elements in $A_i=5\times 30=150$.

$\because$ Each element of S belongs to exactly 10 of the A_i's

$$\therefore \quad n(S)=n\left\{\bigcup_{i=1}^{30} A_i\right\}=\frac{150}{10}=15$$

Now, each B_j's contains 3 elements

So, total number of elements in $B_j=3\times n=3n$

$\because$ Each elements of S belongs to exactly nine of the B_j's

$$\therefore \quad n(S)=n\left\{\bigcup_{j=1}^{n} B_j\right\}=\frac{3n}{9}$$

$$\Rightarrow \quad \frac{3n}{9}=15 \Rightarrow n=45$$

2. (b) $A-B \to$ Regions 3 and 5

$B-C \to$ Regions 2 and 6

$C-A \to$ Regions 4 and 7

$\therefore (A-B)\cup(B-C)\cup(C-A) \to$ Regions 2, 3, 4, 5, 6, 7

$\therefore [(A-B)\cup(B-C)\cup(C-A)]' \to$ Regions 1, which is $A\cap B\cap C$.

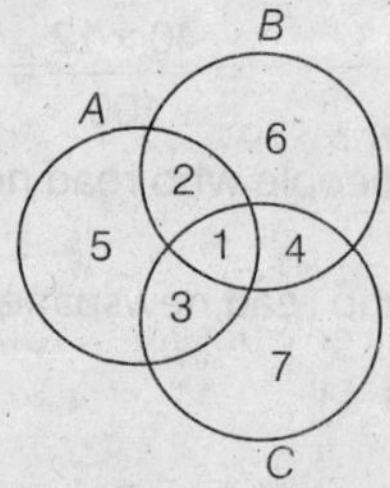

3. (c) $n(P)=25\%$, $n(C)=15\%$

and $n(P'\cup C')=65\%$

$\Rightarrow n(P\cup C)'=65\%$

$n(P\cup C)=35\%$

$\therefore n(P\cap C)=n(P)+n(C)-n(P\cup C)$

$=25+15-35=5$

$x\times 5\%=2000$

$\Rightarrow x=40000$

4. (d) $P=\{30, 60, 90, 120\}$

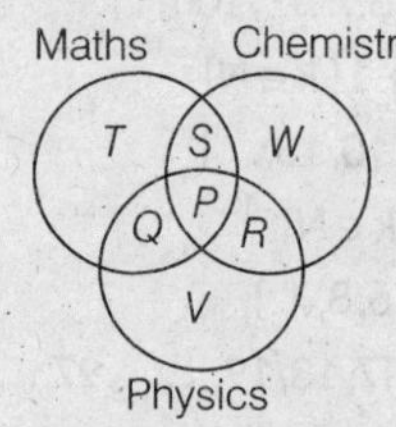

$\Rightarrow n(P)=4$

$Q=\{6n: n\in N, 1\le n\le 23\}-P$

$\Rightarrow n(Q)=19$

$R=\{15n: n\in N, 1\le n\le 9\}-P$

$\Rightarrow n(R)=5$

$S=\{10n: n\in N, 1\le n\le 14\}-P$

$\Rightarrow n(S)=10$

$n(T)=70-n(P)-n(Q)-n(S)=70-33=37$

$n(V)=46-n(P)-n(Q)-n(R)=46-28=18$

$n(W)=28-n(P)-n(R)-n(S)=28-19=9$

Numbered of required students

$=140-(4+19+5+10+37+18+9)$

$=140-102=38$

5. (b) Let R be the set of families having a radio and T be the set of families having a TV.

Then, $n(R\cup T)=$ Number of families having atleast one of the radio and TV $=1003-63=940$

$\therefore n(R)=794$ and $n(T)=187$

Let x be the number of families having both a radio and a TV, then number of families who have only radio $=794-x$, number of families who have only TV $=187-x$

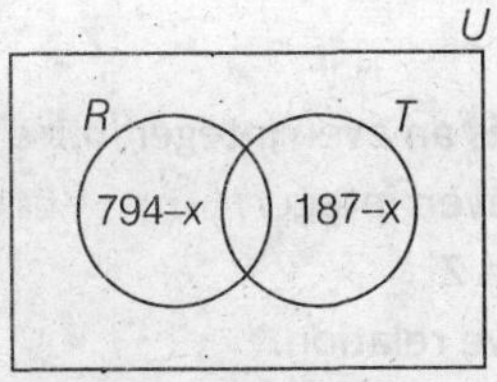

From venn-diagram, $794 - x + x + 187 - x = 940$

$\Rightarrow \quad 981 - x = 940$

$\therefore \quad x = 981 - 940 = 41$

Hence, the required number of families having both a radio and a TV is 41.

6. (a) Let $A = \{1, 2, 3\}$ and $R = \{(1, 1), (1, 2)\}$ and $S = \{(2, 2), (2, 3)\}$ be the transitive relation on A

Then, $R \cup S = \{(1, 1), (1, 2), (2, 2), (2, 3)\}$

$R \cup S$ is not transitive, because $(1, 2) \in R \cup S$ and $(2, 3) \in R \cup S$ but $(1, 3) \notin R \cup S$

7. (d) $\because \frac{n}{m}$ means that n is a factor of m. So, f is reflexive.

$\because$ A number is a factor of itself.

Now, if n is a factor of m, then m is not a factor of n.

$\therefore f$ is not symmetric.

Let n is a factor of m and m is a factor of p, then it is true that n is a factor of p.

$\therefore f$ is transitive.

8. (c) We have, $|x^2 - 5x + 6| + |x^2 + 12x + 27| = |17x + 21|$

$$= |(x^2 + 12x + 27) - (x^2 - 5x + 6)|$$

$\Rightarrow (x^2 - 5x + 6)(x^2 + 12x + 27) \le 0$

$\Rightarrow (x - 2)(x - 3)(x + 3)(x + 9) \le 0$

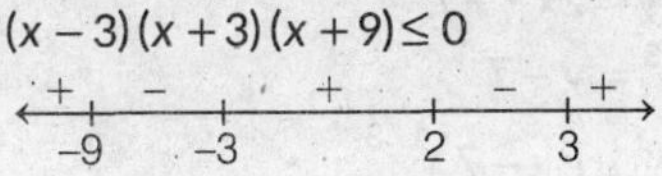

From the sign scheme, $x \in [-9, -3] \cup [2, 3]$

9. (b) We have, $|2x^2 - 5x + 2| + |x^2 - 3x| = |x^2 - 2x + 2|$

$$= |(2x^2 - 5x + 2) - (x^2 - 3x)|$$

$\therefore \quad (2x^2 - 5x + 2)(x^2 - 3x) \le 0$

$\Rightarrow \quad (2x - 1)(x - 2)\, x\, (x - 3) \le 0$

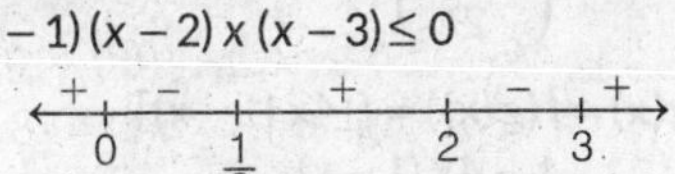

From the sign scheme, we have

$$x \in \left[0, \frac{1}{2}\right] \cup [2, 3]$$

10. (b) We have, $||x - 1| - 5| < 2x - 5$

$\therefore \quad 2x - 5 > 0$

$\Rightarrow \quad x > \frac{5}{2}$ ...(i)

Now, from Eq. (i),

$$-2x + 5 < |x - 1| - 5 < 2x - 5$$

$$-2x + 10 < |x - 1| < 2x$$

Case I $x \ge 1$

$\therefore \quad -2x + 10 < x - 1 < 2x$

$\Rightarrow \quad 3x > 11$ and $x > -1$

$\Rightarrow \quad x > \frac{11}{3}$ ...(ii)

Case II $x < 1$

$\therefore \quad -2x + 10 < 1 - x < 2x$

$\Rightarrow \quad x > 9$ and $x > \frac{1}{3}$

$\Rightarrow \quad x > 9$, not possible as $x < 1$

Therefore, $x \in \left(\frac{11}{3}, \infty\right)$

11. (d) $\because f(x) = \frac{3}{4 - x^2} + \log_{10}(x^3 - x)$

For domain of $f(x)$.

$$x^3 - x > 0$$

$\Rightarrow \quad x(x^2 - 1) > 0$

$\Rightarrow \quad x \in (-1, 0) \cup (1, \infty)$ and $4 - x^2 \ne 0$

$\Rightarrow \quad x \ne \pm 2$

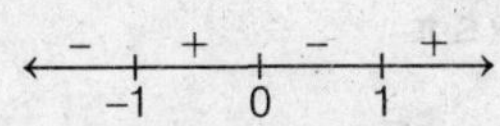

$\therefore$ The domain of the function is

$$(-1, 0) \cup (1, 2) \cup (2, \infty)$$

12. (c) We have,

$$f(x) + 2f\left(\frac{1}{x}\right) = 3x, x \ne 0 \quad \text{...(i)}$$

$\Rightarrow \quad f\left(\frac{1}{x}\right) + 2f(x) = \frac{3}{x}$ ...(ii)

From Eqs. (i) and (ii), we get

$$f(x) = \frac{2}{x} - x$$

Now, consider $f(x) = f(-x)$

$\Rightarrow \quad \frac{2}{x} - x = -\frac{2}{x} + x$

$\Rightarrow \quad \frac{4}{x} = 2x$

$\Rightarrow \quad 2x^2 = 4$

$\Rightarrow \quad x = \pm\sqrt{2}$

Hence, S contains exactly two elements.

13. (d) $\sum_{r=1}^{n} f(r) = f(1) + f(2) + f(3) + \ldots + f(n)$

$\because \quad f(1) = 7$

and $f(x + y) = f(x) + f(y)$

Putting $x = 1$ and $y = 1$, we get

$$f(1 + 1) = f(1) + f(1) = 2f(1)$$

$\Rightarrow \quad f(2) = 2f(1)$

Similarly, $f(3) = 3f(1)$

$$f(n) = nf(1)$$

$$\Rightarrow \sum_{r=1}^{n} f(r) = f(1) + 2f(1) + 3f(1) + \dots nf(1)$$

$$= f(1)[1 + 2 + 3 + \dots + n]$$

$$= 7 \times \frac{n(n+1)}{2} \quad [\because f(1) = 7]$$

14. (a,d) $\cos x > -\frac{1}{2}$ and $0 \le x \le \pi$ (given)

$$\Rightarrow -\frac{2\pi}{3} < x < \frac{2\pi}{3} \text{ and } 0 \le x \le \pi$$

$$\Rightarrow 0 \le x < \frac{2\pi}{3}$$

$$\therefore A = \left[0, \frac{2\pi}{3}\right)$$

Again, $\sin x > \frac{1}{2}$ and $\frac{\pi}{3} \le x \le \pi$

$$\Rightarrow \frac{\pi}{6} < x < \frac{5\pi}{6}$$

and $\frac{\pi}{3} \le x \le \pi$

$$\Rightarrow \frac{\pi}{3} \le x < \frac{5\pi}{6}$$

$$\Rightarrow B = \left[\frac{\pi}{3}, \frac{5\pi}{6}\right)$$

Now, $A \cap B = \left[\frac{\pi}{3}, \frac{2\pi}{3}\right)$

$$\Rightarrow \frac{\pi}{3} \le A \cap B < \frac{2\pi}{3}$$

Here, $\lambda = \frac{1}{3}$ and $\mu = \frac{2}{3}$.

15. (b, c) $X = \{1, 2, 3, 4, 5\}$ and $Y = \{1, 3, 5, 7, 9\}$

(a) $R_1 = \{(x, y) : y = 2 + x, x \in X, y \in Y\}$

$\Rightarrow R_1 = \{(1,3), (2,4), (3,5), (4,6), (5,7)\}$

$R_1 \not\subset x \times y$

(b) $R_2 \subseteq X \times Y$

(c) $R_3 \subseteq X \times Y$

(d) $R_4 \not\subseteq X \times Y$ $\quad [\because (2,4) \in R_4$ but $4 \notin Y]$

16. (a, b, c, d)

$\because$ A line is parallel to itself. So, R is reflexive.

If $l_1 \parallel l_2$, then $l_2 \parallel l_1$.

$\therefore$ R is the symmetric relation.

Let $l_1, l_2, l_3 \in L$, where L is the set of all straight lines.

$$l_1 R l_2 \Rightarrow l_1 \parallel l_2$$

$$l_2 R l_3 \Rightarrow l_2 \parallel l_3 \Rightarrow l_1 \parallel l_3$$

$\therefore$ $l_1 R l_3$.

Then, R is transitive relation.

So, R is an equivalence relation.

17. (a, c, d)

$aRb \Leftrightarrow a - b$ is an even integer, $a, b \in Z$

$\because a - a = 0$ (even integer)

$(a, a) \in R, \forall a \in Z$

$\therefore$ R is reflexive relation.

Let $(a, b) \in R \Rightarrow a - b$ is an even integer.

$\Rightarrow -(b - a)$ is an even integer.

$\Rightarrow (b - a)$ is an even integer

$\Rightarrow (b, a) \in R$

$\therefore R$ is symmetric relation.

Now, let $(a, b), (b, c) \in R$, then $(a - b)$ and $(b - c)$ is an even integer.

So, let $a - b = 2x_1, x_1 \in Z$

and $b - c = 2x_2, x_2 \in Z$

$\therefore (a - b) + (b - c) = 2(x_1 + x_2) = 2x_3$

[where, $x_3 = (x_1 + x_2) \in Z$]

$$\Rightarrow a - c = 2x_3, x_3 \in Z$$

$\therefore (a - c)$ is an even integer.

$\therefore aRb$ and $bRc \Rightarrow aRc$. So, R is transitive.

Hence, R is an equivalence relation.

A relation R on given set A is said to be anti-symmetric, iff $(a, b) \in R$ and $(b, a) \in R \Rightarrow a = b, \forall a, b \in A$.

$\therefore$ So, relation is not anti-symmetric.

18. (b, d) $\because f(x) = 2x - 3$ and $g(x) = x^3 + 5$

$$\Rightarrow fog(x) = f(g(x)) = f(x^3 + 5) = 2(x^3 + 5) - 3$$

$$\Rightarrow fog(x) = 2x^3 + 7$$

Now, let $y = 2x^3 + 7$

$$\Rightarrow 2x^3 = y - 7$$

$$\Rightarrow x = \left(\frac{y - 7}{2}\right)^{1/3}$$

$$\Rightarrow y = \left(\frac{x - 7}{2}\right)^{1/3}$$

$$\Rightarrow (fog)^{-1} x = \left(\frac{x - 7}{2}\right)^{1/3}$$

19. (b, d) $fog(x) = f(g(x)) = f[4x(1 - x)]$

$$= \frac{1 - 4x(1 - x)}{1 + 4x(1 - x)}$$

$$= \frac{1 - 4x + 4x^2}{1 + 4x - 4x^2} = \frac{(2x - 1)^2}{1 + 4x - 4x^2}$$

and $gof(x) = g[f(x)]$

$$= g\left[\frac{1 - x}{1 + x}\right] \left[\because f(x) = \frac{1 - x}{1 + x}\right]$$

$$= 4\left(\frac{1 - x}{1 + x}\right)\left(1 - \frac{1 - x}{1 + x}\right)$$

$$= 4\left(\frac{1 - x}{1 + x}\right)\left(\frac{2x}{1 + x}\right) = \frac{8x(1 - x)}{(1 + x)^2}$$

20. (b, c, d) Let $y = f(x) = \frac{a-x}{a+x}$

$\Rightarrow \quad ay + xy = a - x$

$\therefore \quad x = \frac{a(1-y)}{(1+y)} = f^{-1}(y)$

$\Rightarrow \quad f^{-1}(x) = \frac{a(1-x)}{(1+x)}$

$\therefore f^{-1}(x)$ is not defined for $x = -1$.

Domain of $f^{-1}(x) \in (-\infty, -1) \cup (-1, \infty)$

Now, for $a = 0$, the given function $f(x) = -1$, which is constant. Thus, $f^{-1}(x)$ is not defined.

Therefore, $a \neq 0$.

21. (c, d) $\because \sin([x]\pi) = 0$

$\therefore \quad f(x) = 0$

$\Rightarrow f(x)$ is a constant function and also $f(x)$ is a zero function.

22. (c, d) $f: R - \{-b\} \to R - \{1\}$

$$f(x) = \frac{x+a}{x+b} \qquad [a \neq b]$$

Let $x_1, x_2 \in D_f$

$f(x_1) = f(x_2)$

$\Rightarrow \quad \frac{x_1 + a}{x_1 + b} = \frac{x_2 + a}{x_2 + b}$

$\Rightarrow x_1x_2 + bx_1 + ax_2 + ab = x_1x_2 + ax_1 + bx_2 + ab$

$\Rightarrow \quad b(x_1 - x_2) = a(x_1 - x_2)$

$\Rightarrow \quad (x_1 - x_2)(b - a) = 0$

$\Rightarrow \quad x_1 = x_2 \qquad [\because a \neq b]$

Now, let $y = \frac{x+a}{x+b}$

$\Rightarrow \quad x = \frac{a - by}{y - 1}$

$\Rightarrow \quad f^{-1}(y) = \frac{a - by}{y - 1}; y \neq 1$

$\therefore x$ is defined, $\forall\, y \in R - \{1\}$

$\Rightarrow f$ is onto.

Also, $f^{-1}(2) = \frac{a - 2b}{2 - 1} = a - 2b$.

Sol. (Q.Nos. 23-25)

We have, $(a, b) \in R_1$, iff $|a - b| \leq 7$, where $a, b \in Z$.

23. (a) **Reflexivity** : Let $a \in Z, a - a = 0$

$\Rightarrow \quad |a - a| \leq 7 \Rightarrow 0 < 7$

$\Rightarrow \quad (a, a) \in R_1$ (reflexive)

Symmetry : $(a, b) \in R_1 \Rightarrow |a - b| \leq 7$

$\Rightarrow \quad |b - a| \leq 7 \Rightarrow (b, a) \in R_1$

$\therefore R_1$ is symmetric.

Transitivity : $(2, 6), (6, 10) \in R_1$ but $(2, 10) \notin R_1$ (not transitive)

We have $(a, b) \in R_2$, iff $ab = 4$, where $a, b \in Q$

24. (b) **Reflexivity** : $5 \in Q$ and $5 \cdot 5 = 25 \neq 4$ **(not reflexive)**

Symmetry : $(a, b) \in R_2$

$\Rightarrow \quad ab = 4$

$\Rightarrow \quad ba = 4$

$\Rightarrow \quad (b, a) \in R_2$

$\therefore R_2$ is symmetric.

Transitivity : $\left(8, \frac{1}{2}\right), \left(\frac{1}{2}, 8\right) \in R_2$ but $(8, 8) \notin R_2$.

$\therefore R_2$ is not transitive.

25. (a) We have, $(a, b) \in R_3$,

iff $a^2 - 4ab + 3b^2 = 0$, where $a, b \in R$

Reflexivity : $(a, a) \in R_3$

$\because a^2 - 4a \cdot a + 3a^2 = 0$

$\therefore R_3$ is reflexive.

Symmetry : Let $(a, b) \in R_3$

$\Rightarrow \quad a^2 - 4ab + 3b^2 = 0$,

we get $a = b$ and $a = 3b$ and if $(b, a) \in R_3$

$\Rightarrow b^2 - 4ab + 3a^2 = 0$, we get $b = a$ and $b = 3a$.

$\therefore \quad (a, b) \in R_3 \neq (b, a) \in R_3$

$\therefore R_3$ is not symmetry.

Transitivity : We have, $(3, 1)\left(1, \frac{1}{3}\right) \in R_3$

as $3^2 - 4(3)(1) + 3(1)^2 = 9 - 12 + 3 = 0$

and $(1)^2 - 4(1)\left(\frac{1}{3}\right) + 3\left(\frac{1}{3}\right)^2 = 1 - \frac{4}{3} + \frac{1}{3} = 0$

But $\left(3, \frac{1}{3}\right) \notin R_3$

$\because \quad (3)^2 - 4(3)\left(\frac{1}{3}\right) + 3\left(\frac{1}{3}\right)^2 = \frac{16}{3} \neq 0$

$\therefore R_3$ is not transitive.

26. (i) $\to$ C, (ii) $\to$ B, (iii) $\to$ A, (iv) $\to$ D

(i) Let $f(x) = 7x + 1$, linear function which is one-one and onto in R

(ii) $f(x) = \cos x$, for $x \in [0, \pi], f(x) \in [-1, 1]$

$\therefore f$ is one-one on $[0, \pi], \forall\, x \in R, y \in [-1, 1]$

Also, $f(x)$ is not onto on R.

(iii) $f(x) = \sin x$, for $x \in [0, \pi], f(x), \in [0, 1]$

$$f\left(\frac{\pi}{3}\right) = f\left(\frac{2\pi}{3}\right) = \frac{\sqrt{3}}{2}$$

$\therefore f$ is not one-one on $(0, \pi), \forall\, x \in R$ and $y \in [-1, 1]$.

$\therefore f$ is onto on $[-1, 1]$

(iv) $f(x) = 1 + \ln(x), f(x)$ is defined for $x \in (0, \infty)$

Let $x_1, x_2 \in D_f$

Then, $f(x_1) = f(x_2) \Rightarrow x_1 = x_2$

$\therefore f$ is one-one, $\forall\, x\, (0, \infty)$

27. (i) → C; (ii) → (A, B, C, D), (iii) → (B, C, D)

(i) $7^{\frac{2x^2-5x-9}{2}} = (2)^{\frac{3}{2}\log_2 7}$

$\Rightarrow \frac{2x^2-5x-9}{2} = \frac{3}{2}$

$\Rightarrow 2x^2-5x-12=0$

Hence, $x = 4, \frac{-3}{2}$

(ii) $2^{2(x+\sqrt{x^2-2})} = 5\cdot 2^{x-1+\sqrt{x^2-2}} + 6$

Let $\lambda = 2^{\sqrt{x^2-2}+x}$

$\Rightarrow \quad \lambda^2 - \frac{5\lambda}{2} = 6$

$\Rightarrow 2\lambda^2 - 5\lambda - 12 = 0$

$\Rightarrow \quad \lambda = 4, -\frac{3}{2}$

Hence, $2^{\sqrt{x^2-2}+x} = 2^2$

$\Rightarrow \quad x = \frac{3}{2}$

(iii) $\because \sqrt{|x-3|^{x-1}} = \sqrt[4]{|x-3|^{x+1}}$

$\Rightarrow |x-3|^{\frac{x-1}{2}} = |x-3|^{\frac{x+1}{4}}$

$\therefore \quad x-3=0 \Rightarrow x=3$

or $\quad x-3=1 \Rightarrow x=4$ or $x-3=-1$

$\Rightarrow \quad x=2$

Also, $\frac{x-1}{2} = \frac{x+1}{4}$

$\Rightarrow \quad 2(x-1) = x+1 \Rightarrow x=3$

Sol. (Q.Nos. 28-30)

28 → (8), 29 → (1), 30 → (0)

$\because A = \{x : |x| < 2\} = \{x : -2 < x < 2\} = (-2, 2)$

$B = \{x : |x-5| \le 2\} = \{x : -2 \le x-5 \le 2\} = [3, 7]$

$C = \{x : |x| > x\} = \{x : x < 0\} = (-\infty, 0)$

and $D = \{x : |x| < x\} = \phi$

$\therefore \quad A \cup B = (-2, 2) \cup [3, 7]$

$\therefore$ Integral values are

−1, 0, 1, 3, 4, 5, 6, 7

$\therefore$ Number of integral values is 8

Similarly, $A \cap C = (-2, 2) \cap (-\infty, 0) = (-2, 0)$.

$\therefore$ Number of integral value in $A \cap C$ is 1.

Further, $A \cap D = (-2, 2) \cap \phi = \phi$

$\therefore$ Number of integral values in $A \cap D$ is 0.

Sol. (Q.Nos. 31-33)

31 → (0), 32 → (9), 33 → (9)

$gof(0) = g(f(0))$

$= g[7(0)^2 + 0 - 8] = g(-8) = |-8| = 8$

and $fog(-3) = f(g(-3)) = f[0] = -8$

$\therefore gof(0) + fog(-3) = 8 - 8 = 0$

$fog(7) = f(g(7)) = f[7^2+4] = f(53) = 8(53) + 3 = 427$

$gof(6) = g(f(6)) = g(4 \times 6 + 5) = g(29) = 29^2 + 4 = 845$

$\therefore 2\, fog(7) - gof(6) = 2 \times 427 - 845 = 9$

$gof(2) = g(f(2)) = g(4 \times 2 + 5)$

$= g(13) = 13^2 + 4 = 173$

$fog(9) = f(g(9)) = f(9^2+4) = f(85) = 8 \times 85 + 3 = 683$

$\therefore \; 4\, gof(2) - fog(9) = 4 \times 173 - 683 = 9$

34. (2) As, $2^m - 2^n = 48 = 2^4 \times 3$

$\Rightarrow \quad 2^n(2^{m-n} - 1) = 2^4 \times (2^2 - 1)$

$\Rightarrow \quad n = 4$ and $m - n = 2$

$\Rightarrow \quad n = 4$ and $m = 6$

Now, $m - n = 6 - 4 = 2$

35. (3) Here, $A = [-3, 7), B = (0, 10)$

and $\quad S = (-\infty, \infty)$

$\therefore \quad A - B = [-3, 0]$ and $B - A = [7, 10)$

$\therefore \quad A \Delta B = (A - B) \cup (B - A) = [-3, 0] \cup [7, 10)$

$\therefore$ Positive integers are 7, 8, 9

$\therefore$ Number of positive integers is 3

36. (5) Given, $a = 22, c = 12$

and $a + b + c = 45$

$22 + 12 + b = 45$

$b = 11 = 2\lambda + 1$

$\Rightarrow \quad \lambda = 5$

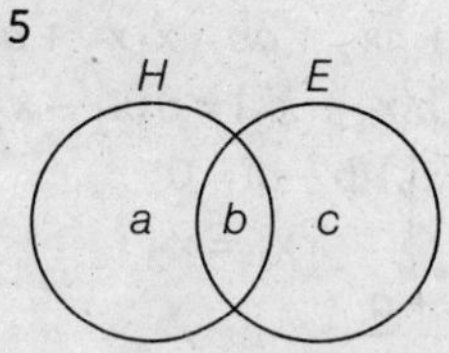

37. (9) $n[(A \times B) \cap (B \times A)] = n[(A \cap B) \times (B \cap A)]$

$= n(A \cap B) \times n(B \cap A)$

$= 99 \times 99$

$= 121 \times 9^2$

$\Rightarrow \quad \lambda = 9$

38. (19) $\because f(x) + f(1-x) = \frac{4^x}{4^x+2} + \frac{4^{1-x}}{4^{1-x}+2}$

$\Rightarrow \quad f(x) + f(1-x) = \frac{4^x}{4^x+2} + \frac{2}{2+4^x} = 1$

$\therefore f\left(\frac{1}{40}\right) + f\left(\frac{2}{40}\right) + \ldots + f\left(\frac{39}{40}\right) - f\left(\frac{1}{2}\right)$

$= \left[f\left(\frac{1}{40}\right) + f\left(\frac{39}{40}\right)\right] + \left[f\left(\frac{2}{40}\right) + f\left(\frac{38}{40}\right)\right]$

$+ \ldots \text{(19 times)} + f\left(\frac{20}{40}\right) - f\left(\frac{1}{2}\right)$

$= 1 + 1 + 1 + \ldots + \text{(19 times)} + f\left(\frac{1}{2}\right) - f\left(\frac{1}{2}\right)$

$= 19$

CHAPTER

02

THEORY OF EQUATIONS

LEVEL 01 \\ Starter Level

2.1 Quadratic Equation and Its Roots

1 If equation $(\lambda^2 - 5\lambda + 6)x^2 + (\lambda^2 - 3\lambda + 2)x + (\lambda^2 - 4) = 0$ is satisfied by more than two values of x, then find the parameter λ.

2 Find all values of parameter *a* for which the quadratic equation $(a+1)x^2 + 2(a+1)x + a - 2 = 0$ has
   (i) two distinct roots
   (ii) no real roots
   (iii) two equal roots

3 The equation $x^2 + 12x = 3kx^2 + 2$ has real roots. Then,
   (i) find the greatest value of *k* such that $k \in Z$, where *Z* is set of integers.
   (ii) find one rational value of *k* for which the above equation has rational roots.

4 If one root of the equation $x^2 - \sqrt{5}x - 19 = 0$ is $\dfrac{9+\sqrt{5}}{2}$, then find the other root.

5 Find *a*, if $ax^2 - 4x + 9 = 0$ has integral roots.

6 If α and β are roots of the equation $x^2 + \sqrt{2}x + 3 = 0$, then find the monic quadratic equation having zeroes $\alpha^2 + \beta^2$ and $2\alpha\beta$.

7 Let α and β be the roots of the quadratic equation

$$(x-2)(x-3)+(x-3)(x+1)+(x+1)(x-2)=0.$$

Find the value of $\frac{1}{(\alpha+1)(\beta+1)}+\frac{1}{(\alpha-2)(\beta-2)}+\frac{1}{(\alpha-3)(\beta-3)}$.

8 If α and β are roots of $x^2-3x+1=0$, then find the value of $\alpha^4+\beta^4$.

2.2 Transformation of Quadratic Equations and Common Roots

9 If α and β are the roots of $ax^2+bx+c=0$, then find the roots of the equation $ax^2-bx(x-1)+c(x-1)^2=0$.

10 Find the value of λ, so that the equation $x^2-x-12=0$ and $\lambda x^2+10x+3=0$ may have one root in common. Also, find the common root.

2.3 Quadratic Expressions and Location of Roots

11 If $f(x)=x^2+4x+4$, then solve the following inequalities :

(i) $f(x)\geq 0$ (ii) $f(x)\leq 0$
(iii) $f(a)>0$ (iv) $f(x)<0$

12 Find all values of p, so that 6 lies between roots of the equation $x^2+2(p-3)x+9=0$.

13 Let $f(x)=\frac{(x-3)(x+4)(x+6)}{(x+3)(x-5)}$. Find intervals, when $f(x)$ is positive or negative.

14 Solve the equation $\left|\frac{x^2-8x+12}{x^2-10x+21}\right|=-\frac{x^2-8x+12}{x^2-10x+21}$.

15 Find the values of m, for which both roots of equation $x^2-mx+1=0$ are less than unity.

16 Let $x^2-(m-3)x+m=0$ $(m\in R)$ be quadratic equation. Find the value of m for which

(i) both the roots are smaller than 2.
(ii) both the roots are greater than 2.
(iii) one root is smaller than 2 and other root is greater than 2.
(iv) exactly one root lies in the interval (1, 2).

17 If $c<0$ and the equation $ax^2+bx+c=0$ does not have any real root, then verify that

(i) $a-b+c<0$ (ii) $9a+3b+c<0$

18 If $a<b<c<d$, then discuss the nature of roots of the equation $(x-a)(x-c)+2(x-b)(x-d)=0$.

2.4 Equations of Higher Degree

19 If α, β and γ are the roots of the equation $x^3-px^2+qx-r=0$, find

(i) $\Sigma\alpha^2$ (ii) $\Sigma\alpha^2\beta$

20 Find the conditions, if roots of the equation $x^3-px^2+qx-r=0$ are in

(i) AP (ii) GP

2.5 Irrational Equations/Inequations, Exponential Equations /Inequations and Logarithmic Equations/Inequations

21 Solve the equation $\sqrt{x^2 - 3x} = 4x^2 - 12x - 3$.

22 Solve the inequation $\sqrt[5]{\left[\frac{3}{x+1} + \frac{7}{x+2}\right]} < \sqrt[5]{\frac{6}{x-1}}$.

23 Solve the equation $\sqrt{(6-x)}\,(3^{x^2 - 7.2x + 3.9} - 9\sqrt{3}) = 0$.

24 Solve the equation $2^{x^2 + 7x + 10} = 1$.

25 Solve the inequation $3^{x+2} > \left(\frac{1}{9}\right)^{1/x}$.

26 Solve the inequation $(\log_5 x)^2 + \log_5 x < 2$.

27 Solve the equation $2^{2^x} - 18 \cdot 2^x + 32 = 0$.

LEVEL 02 \\ JEE Main Level

Single Choice Correct Questions

2.1 Quadratic Equation and Its Roots

1 The number of real solutions of the equation $x^2 - 3|x| + 2 = 0$ is

(a) 3 (b) 2 (c) 4 (d) 1

2 If the roots of the equation $x^2 - bx + c = 0$ be two consecutive integers, then $b^2 - 4ac$ is equal to

(a) -2 (b) 3 (c) 2 (d) 1

3 The real roots of the equation $x^2 + 5|x| + 4 = 0$ are

(a) $\{-1, -4\}$ (b) $\{1, 4\}$ (c) $\{-4, 4\}$ (d) None of these

4 The number of real solutions of $|x| + 2\sqrt{5 - 4x - x^2} = 16$ is/are

(a) 6 (b) 1 (c) 0 (d) 4

5 The number of integral values of a for which the quadratic equation $(x+a)(x+1991)+1=0$ has integral roots, are

(a) 3 (b) 0 (c) 1 (d) 2

6 Consider the equation $x^2 + 2x - n = 0$, where $n \in N$ and $n \in [5, 100]$. The total number of different values of n, so that the given equation has integral roots, is

(a) 8 (b) 3 (c) 6 (d) 4

7 If c and d are the roots of $(x-a)(x-b) = k$, then the roots of $(x-c)(x-d) + k = 0$ are

(a) a and b (b) $-a$ and $-b$ (c) ac and bd (d) None of these

8 The sum of all real values of x satisfying the equation $(x^2 - 5x + 5)^{x^2 + 4x - 60} = 1$ is **[JEE Main 2016]**

(a) -4 (b) 6 (c) 5 (d) 3

9 If $a > b > 0$, then the value of $\sqrt{ab + (a-b)\sqrt{ab + (a-b)\sqrt{ab + (a-b)\sqrt{ab + \dots}}}}$ is

(a) independent of only b (b) independent of only a
(c) independent of both a and b (d) dependent of both a and b

10 The value of $3 + \cfrac{1}{4 + \cfrac{1}{3 + \cfrac{1}{4 + \cfrac{1}{3 + \dots \infty}}}}$ is equal to **[JEE Main 2021]**

(a) $1.5 + \sqrt{3}$ (b) $2 + \sqrt{3}$ (c) $3 + 2\sqrt{3}$ (d) $4 + \sqrt{3}$

11 The equation $2\cos^2 \frac{x}{2} \sin^2 x = x^2 + \frac{1}{x^2}, 0 \le x \le \frac{\pi}{2}$ has

(a) one real solution (b) no real solution
(c) more than one real solution (d) None of these

12 Both the roots of the equation $(x-b)(x-c) + (x-a)(x-c) + (x-a)(x-b) = 0$ are always

(a) positive (b) real (c) negative (d) None of these

13 If for a positive integer n, the quadratic equation
$x(x+1) + (x+1)(x+2) + \dots (x + \overline{n-1})(x+n) = 10n$ has two consecutive integral solutions, then n is equal to **[JEE Main 2017]**

(a) 11 (b) 12 (c) 9 (d) 10

14 Let $p, q \in \{1, 2, 3, 4\}$, the number of equations of the form $px^2 + qx + 1 = 0$ having real roots, is

(a) 15 (b) 9 (c) 7 (d) 8

15 If p, q and r positive are in AP, the roots of the quadratic equation $px^2 + qx + r = 0$ are all real for

(a) $\left|\frac{r}{p} - 7\right| \ge 4\sqrt{3}$ (b) $\left|\frac{p}{r} - 7\right| \ge 4\sqrt{3}$ (c) all p and r (d) no p and r

16 Let two numbers have arithmetic mean 9 and geometric mean 4, then these numbers are the roots of the quadratic equation

(a) $x^2 - 18x - 16 = 0$ (b) $x^2 - 18x + 16 = 0$
(c) $x^2 - 18x - 16 = 0$ (d) $x^2 + 18x + 16 = 0$

17 If α and β are the distinct roots of the equation $x^2 + (3)^{1/4} x + 3^{1/2} = 0$, then the value of $\alpha^{96}(\alpha^{12} - 1) + \beta^{96}(\beta^{12} - 1)$ is equal to **[JEE Main 2021]**

(a) 56×3^{25} (b) 56×3^{24} (c) 52×3^{24} (d) 28×3^{25}

18 Let α and β be the roots of the equation $5x^2 + 6x - 2 = 0$. If $S_n = \alpha^2 + \beta^2, n = 1, 2, 3, \dots$, then **[JEE Main 2020]**

(a) $6S_6 + 5S_5 = 2S_4$ (b) $6S_6 + 5S_5 + 2S_4 = 0$
(c) $5S_6 + 6S_5 = 2S_4$ (d) $5S_6 + 6S_5 + 2S_4 = 0$

19 If $\alpha \ne \beta$ but $\alpha^2 = 5\alpha - 3$ and $\beta^2 = 5\beta - 3$, then the equation having $\frac{\alpha}{\beta}$ and $\frac{\beta}{\alpha}$ as its roots is

(a) $3x^2 - 19x + 3 = 0$ (b) $3x^2 + 19x - 3 = 0$
(c) $3x^2 - 19x - 3 = 0$ (d) $x^2 - 5x + 3 = 0$

20 Difference between the corresponding roots of $x^2 + ax + b = 0$ and $x^2 + bx + a = 0$ is same and $a \ne b$, then

(a) $a + b + 4 = 0$ (b) $a + b - 4 = 0$ (c) $a - b - 4 = 0$ (d) $a - b + 4 = 0$

21 The value of a for which one root of the quadratic equation $(a^2 - 5a + 3)x^2 + (3a - 1)x + 2 = 0$ is twice as large as the other, is

(a) $-\frac{1}{3}$ (b) $\frac{2}{3}$ (c) $-\frac{2}{3}$ (d) $\frac{1}{3}$

22 If α and β are the roots of the equation $x^2 + px + 2 = 0$ and $\frac{1}{\alpha}$ and $\frac{1}{\beta}$ are the roots of the equation $2x^2 + 2qx + 1 = 0$, then $\left(\alpha - \frac{1}{\alpha}\right)\left(\beta - \frac{1}{\beta}\right)\left(\alpha + \frac{1}{\beta}\right)\left(\beta + \frac{1}{\alpha}\right)$ is equal to **[JEE Main 2020]**

(a) $\frac{9}{4}(9 + q^2)$ (b) $\frac{9}{4}(9 - q^2)$ (c) $\frac{9}{4}(9 + p^2)$ (d) $\frac{9}{4}(9 - p^2)$

23 Let $\lambda \neq 0$ be in R. If α and β are roots of the equation $x^2 - x + 2\lambda = 0$ and α and γ are the roots of the equation $3x^2 - 10x + 27\lambda = 0$, then $\frac{\beta\gamma}{\lambda}$ is equal to **[JEE Main 2020]**

(a) 27 (b) 18 (c) 9 (d) 36

24 The value of a for which the sum of the squares of the roots of the equation $x^2 - (a - 2)x - a - 1 = 0$ assume the least value, is

(a) 1 (b) 0 (c) 3 (d) 2

25 In a ΔPQR, $\angle R = \frac{\pi}{2}$, if $\tan\left(\frac{P}{2}\right)$ and $\tan\left(\frac{Q}{2}\right)$ are the roots of $ax^2 + bx + c = 0$, $a \neq 0$, then

(a) $a = b + c$ (b) $c = a + b$ (c) $b = c$ (d) $b = a + c$

26 If a, b and c are in GP, then the ratio of the roots of the equation $ax^2 + bx + c = 0$ is

(a) $\frac{1}{2}$ (b) $\frac{-1 \pm i\sqrt{3}}{2}$ (c) $\frac{1 \pm i}{2}$ (d) $\pm\sqrt{2}$

27 If the roots of $ax^2 + bx + c = 0$ are the reciprocals of those of $lx^2 + mx + n = 0$, then $a : b : c$ is equal to

(a) $n : m : l$ (b) $l : m : n$ (c) $m : n : l$ (d) $n : l : m$

28 In writing an equation of the form $ax^2 + bx + c = 0$, the coefficient of x is written incorrectly and roots are found to be equal. Again in writing the same equation the constant term is written incorrectly and it is found that one root is equal to those of the previous wrong equation while the other is double of it. If α and β are the roots of correct equation, then $(\alpha - \beta)^2$ is equal to

(a) 5 (b) $5\alpha\beta$ (c) $-4\alpha\beta$ (d) -4

29 The roots α and β of the quadratic equation $px^2 + qx + r = 0$ are real and of opposite signs, the roots of $\alpha(x - \beta)^2 + \beta(x - \alpha)^2 = 0$ are

(a) positive (b) negative (c) of opposite signs (d) non-real

30 If α and β are real and α^2, β^2 are the roots of the equation $a^2x^2 - x + 1 - a^2 = 0 \left(\frac{1}{\sqrt{2}} < a < 1\right)$ and $\beta^2 \neq 1$, then β^2 is equal to

(a) a^2 (b) $\frac{1 - a^2}{a^2}$ (c) $1 - a^2$ (d) $1 + a^2$

31 If each of the roots of $ax^2 + bx + c = 0$ increases by the same quantity, then $\frac{b^2 - 4ac}{a^2}$ is

(a) a function of a (b) a function of b (c) a function of c (d) invariant

32 Let α and β be the roots of equation $px^2 + qx + r = 0, p \neq 0$. If p, q and r are in AP and $\frac{1}{\alpha} + \frac{1}{\beta} = 4$, then the value of $|\alpha - \beta|$ is

(a) $\frac{\sqrt{61}}{9}$ (b) $\frac{2\sqrt{17}}{9}$ (c) $\frac{\sqrt{34}}{9}$ (d) $\frac{2\sqrt{13}}{9}$

33 If one root is square of the other root of the equation $x^2 + px + q = 0$, then the relation between p and q is

(a) $p^3 - q(3p - 1) + q^2 = 0$ (b) $p^3 - q(3p + 1) + q^2 = 0$
(c) $p^3 + q(3p - 1) + q^2 = 0$ (d) $p^3 + q(3p + 1) + q^2 = 0$

34 Let α and β be the roots of $x^2 - 6x - 2 = 0$ with $\alpha > \beta$. If $a_n = \alpha^n - \beta^n$ for $n \geq 1$, then the value of $\frac{a_{10} - 2a_8}{2a_9}$ is

(a) 1 (b) 2 (c) 3 (d) 4

35 The set of all values of $k > -1$, for which the equation $(3x^2 + 4x + 3)^2 - (k + 1)(3x^2 + 4x + 3)(3x^2 + 4x + 2) + k(3x^2 + 4x + 2)^2 = 0$ has real roots, is **[JEE Main 2021]**

(a) $\left(1, \frac{5}{2}\right]$ (b) $[2, 3)$ (c) $\left[-\frac{1}{2}, 1\right)$ (d) $\left(\frac{1}{2}, \frac{3}{2}\right] - \{1\}$

36 If the difference between the roots of the equation $x^2 + ax + 1 = 0$ is less than $\sqrt{5}$, then the set of possible value of a is

(a) $(3, \infty)$ (b) $(-\infty, -3)$ (c) $(-3, 3)$ (d) $(-3, \infty)$

2.2 Transformation of Quadratic Equations and Common Roots

37 Let p and q be real numbers such that $p \neq 0, p^3 \neq q$ and $p^3 \neq -q$. If α and β are non-zero complex numbers satisfying $\alpha + \beta = -p$ and $\alpha^3 + \beta^3 = q$, then a quadratic equation having $\frac{\alpha}{\beta}$ and $\frac{\beta}{\alpha}$ as its roots is

(a) $(p^3 + q)x^2 - (p^3 + 2q)x + (p^3 + q) = 0$ (b) $(p^3 + q)x^2 - (p^3 - 2q)x + (p^3 + q) = 0$
(c) $(p^3 - q)x^2 - (5p^3 - 2q)x + (p^3 - q) = 0$ (d) $(p^3 - q)x^2 - (5p^3 + 2q)x + (p^3 - q) = 0$

38 If α and β are the roots of $x^2 + px + q = 0$ and α^4, β^4 are the roots of $x^2 - rx + s = 0$, then the equation $x^2 - 4qx + 2q^2 - r = 0$ has always

(a) two real roots (b) two positive roots
(c) two negative roots (d) one positive and one negative root

39 Let p and q be two positive numbers, such that $p + q = 2$ and $p^4 + q^4 = 272$. Then, p and q are roots of the equation **[JEE Main 2021]**

(a) $x^2 - 2x + 136 = 0$ (b) $x^2 - 2x + 16 = 0$ (c) $x^2 - 2x + 8 = 0$ (d) $x^2 - 2x + 2 = 0$

40 If p, q and r are non-zero real numbers, then two equations $2a^2x^2 - 2abx + b^2 = 0$ and $p^2x^2 + 3pqx + q^2 = 0$ have

(a) no common root (b) one common root if $2a^2 + b^2 = p^2 + q^2$
(c) two common roots if $3pq = 2ab$ (d) two common roots if $3qb = 2ap$

41 If the quadratic equations $ax^2 + 2cx + b = 0$ and $ax^2 + 2bx + c = 0 (b \neq c)$ have a common root, then $a + 4b + 4c$ is equal to

(a) -2 (b) -1 (c) 0 (d) 1

42 If the equations $x^2 + 2x + 3 = 0$ and $ax^2 + bx + c = 0$, $a, b, c \in R$ have a common root, then $a : b : c$ is

(a) 1 : 2 : 3 (b) 3 : 2 : 1 (c) 1 : 3 : 2 (d) 3 : 1 : 2

43 If every pair from among the equations $x^2 + px + qr = 0$, $x^2 + qx + rp = 0$ and $x^2 + rx + pq = 0$, where p, q and r are unequal non-zero numbers, have a common root, then the value of $\left(\frac{\text{sum of common roots}}{\text{product of common roots}}\right)$ is

(a) $\frac{\Sigma p}{pqr}$ (b) $(\Sigma p)^2$ (c) $\Sigma \frac{1}{p}$ (d) 0

44 If the equation $ax^2 + 2bx + 3c = 0$ and $3x^2 + 8x + 15 = 0$ have a common root, where a, b and c are the lengths of the sides of a ΔABC, then $\sin^2 A + \sin^2 B + \sin^2 C$ is equal to

(a) 1 (b) $\frac{3}{2}$ (c) $\sqrt{2}$ (d) 2

45 If the equations $k(6x^2 + 3) + rx + 2x^2 - 1 = 0$ and $6k(2x^2 - 1) + px + 4x^2 - 2 = 0$ have both the roots common, then the value of $2r - p$ is

(a) 0 (b) 1 (c) 2 (d) None of these

46 A value of b for which the equation $x^2 + bx - 1 = 0$ and $x^2 + x + b = 0$ have one root in common, is

(a) $-\sqrt{2}$ (b) $-i\sqrt{3}$ (c) $i\sqrt{5}$ (d) $\sqrt{2}$

47 The quadratic equations $x^2 - 6x + a = 0$ and $x^2 - cx + 6 = 0$ have one root in common, the other roots of the first and second equations are integers in the ratio 4 : 3, then the common root is

(a) 1 (b) 4 (c) 3 (d) 2

2.3 Quadratic Expressions and Location of Roots

48 If the roots of the equation $bx^2 + cx + a = 0$ are imaginary, then for all real values of x, the expression $3b^2x^2 + 6bcx + 2c^2$ is

(a) greater than $4ab$ (b) less than $4ab$

(c) greater than $-4ab$ (d) less than $-4ab$

49 If x is real, then maximum value of $\frac{3x^2 + 9x + 17}{3x^2 + 9x + 7}$ is

(a) $\frac{1}{4}$ (b) 41 (c) 1 (d) $\frac{17}{7}$

50 The number of points of intersection of two curves $y = 2\sin x$ and $y = 5x^2 + 2x + 3$ is

(a) 0 (b) 1 (c) 2 (d) ∞

51 If $x \in R$, then the maximum and minimum values of $\frac{x^2 + 14x + 9}{x^2 + 2x + 3}$ are

(a) 3, 1 (b) 0, ∞ (c) 4, -5 (d) ∞, $-\infty$

52 The set of all real numbers x for which $x^2 - |x + 2| + x > 0$, is

(a) $(-\infty, -2) \cup (2, \infty)$ (b) $(-\infty, -\sqrt{2}) \cup (\sqrt{2}, \infty)$

(c) $(-\infty, -1) \cup (1, \infty)$ (d) $(\sqrt{2}, \infty)$

53 For all $x \in R$, if $mx^2 - 9mx + 5m + 1 > 0$, then m lies in the

(a) $\left(-\frac{4}{61}, 0\right)$ (b) $\left(0, -\frac{4}{61}\right)$

(c) $\left(0, \frac{4}{61}\right)$ (d) None of these

54 Let a, b and c be the sides of a triangle, where $a \neq b \neq c$ and $\lambda \in R$. If the roots of the equation $x^2 + 2(a+b+c)x + 3\lambda(ab+bc+ca) = 0$ are real, then

(a) $\lambda < \frac{4}{3}$ (b) $\lambda > \frac{5}{3}$ (c) $\lambda \in \left(\frac{1}{3}, \frac{5}{3}\right)$ (d) $\lambda \in \left(\frac{4}{3}, \frac{5}{3}\right)$

55 All the values of m for which both roots of the equation $x^2 - 2mx + m^2 - 1 = 0$ are greater than -2 but less than -4 lie in the interval, are

(a) $-2 < m < 0$ (b) $m > 3$ (c) $-1 < m < 3$ (d) $1 < m < 4$

56 If α and $\beta (\alpha < \beta)$ are the roots of the equation $x^2 + bx + c = 0$, where $c < 0 < b$, then

(a) $0 < \alpha < \beta$ (b) $\alpha < 0 < \beta < |\alpha|$ (c) $\alpha < \beta < 0$ (d) $\alpha < 0 < |\alpha| < \beta$

57 If the roots of $x^2 + x + a = 0$ exceed a, then

(a) $2 < a < 3$ (b) $a > 3$ (c) $-3 < a < 3$ (d) $a < -2$

58 If x is real and the expression $\frac{x^2 + 2x - 11}{x - 3}$ takes all real values except those which lie between a and b, then a and b are

(a) $-12, -4$ (b) $-12, 12$ (c) $4, 12$ (d) $-4, 4$

59 Set of all real values of a such that $f(x) = \frac{(2a-1)x^2 + 2(a+1)x + (2a-1)}{x^2 - 2x + 40}$ is always negative, is

(a) $(-\infty, 0)$ (b) $(0, \infty)$ (c) $(-\infty, 1/2)$ (d) None of these

60 The graph of a quadratic polynomial $y = ax^2 + bx + c$; $a, b, c \in R$ is as shown

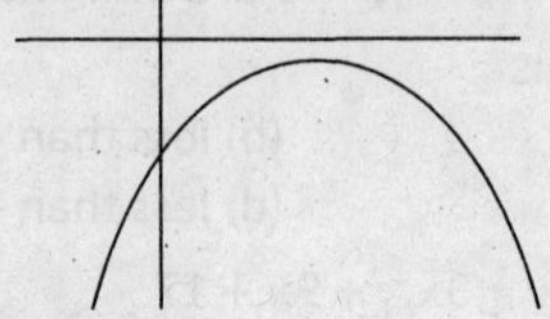

Which one of the following is not correct?

(a) $b^2 - 4ac < 0$ (b) $\frac{c}{a} < 0$

(c) c is negative (d) Abscissa corresponding to the vertex is $\left(\frac{-b}{2a}\right)$

61 The set of all real values of λ for which the quadratic equation, $(\lambda^2 + 1)x^2 - 4\lambda x + 2 = 0$ always have exactly one root in the interval $(0, 1)$, is **[JEE Main 2020]**

(a) $(0, 2)$ (b) $(2, 4)$ (c) $(1, 3]$ (d) $(-3, -1)$

62 Let a, b and c be real numbers, $a \neq 0$. If α is a root of $a^2x^2 + bx + c = 0$, β is the root of $a^2x^2 - bx - c = 0$ and $0 < \alpha < \beta$, then equation $a^2x^2 + 2bx + 2c = 0$ has a root γ, then always satisfies

(a) $\gamma = \frac{\alpha + \beta}{2}$ (b) $\gamma = \alpha + \frac{\beta}{2}$ (c) $\gamma = \alpha$ (d) $\alpha < \gamma < \beta$

2.4 Equations of Higher Degree

63 If roots of an equation $x^n - 1 = 0$ are $1, a_1, a_2, \ldots a_{n-1}$, then the value of $(1-a_1)(1-a_2)(1-a_3)\ldots(1-a_{n-1})$ will be

(a) n (b) n^2 (c) n^n (d) 0

64 If α, β, γ and δ are the roots of the equation $x^4 + 4x^3 - 6x^2 + 7x - 9 = 0$, then the value of $(1+\alpha^2)(1+\beta^2)(1+\gamma^2)(1+\delta^2)$ is

(a) 9 (b) 11 (c) 13 (d) 5

65 Let $P(x) = 0$ be a polynomial equation of the least possible degree with rational coefficient, having $\sqrt[3]{7} + \sqrt[3]{49}$ as one of its roots, then the product of all the roots of $P(x) = 0$ is

(a) 56 (b) 63 (c) 7 (d) 49

66 $P(x)$ is a polynomial with integral coefficient such that for four distinct integers a, b, c and d, $P(a) = P(b) = P(c) = P(d) = 3$. If $P(e) = 5$ (e is an integer), then

(a) $e = 1$ (b) $e = 3$
(c) $e = 4$ (d) no real value of e

67 If $(x^2 + px + 1)$ is a factor of $(ax^3 + bx + c)$, then

(a) $a^2 + c^2 = -ab$ (b) $a^2 - c^2 = -ab$
(c) $a^2 - c^2 = ab$ (d) None of these

68 If α and β are the roots of $x^2 + px + q = 0$ and $x^{2n} + p^n x^n + q^n = 0$ and if $\left(\frac{\alpha}{\beta}\right), \left(\frac{\beta}{\alpha}\right)$ are the roots of $x^n + 1 + (x+1)^n = 0$. Then, $n (\in N)$

(a) must be an odd integer (b) may be any integer
(c) must be an even integer (d) cannot say anything

2.5 Irrational Equations/Inequations, Exponential Equations /Inequations and Logarithmic Equations/Inequations

69 The sum of the solutions of the equation $|\sqrt{x} - 2| + \sqrt{x}(\sqrt{x} - 4) + 2 = 0 \; (x > 0)$ is equal to **[JEE Main 2019]**

(a) 9 (b) 12 (c) 4 (d) 10

70 Let $S = \{x \in R : x \geq 0$ and $2|\sqrt{x} - 3| + \sqrt{x}(\sqrt{x} - 6) + 6 = 0\}$, then S **[JEE Main 2018]**

(a) contains exactly four elements (b) is an empty set
(c) contains exactly one element (d) contains exactly two elements

71 The sum of the roots of the equation $x + 1 - 2\log_2(3 + 2^x) + 2\log_4(10 - 2^{-x}) = 0$ is **[JEE Main 2021]**

(a) $\log_2 14$ (b) $\log_2 11$ (c) $\log_2 12$ (d) $\log_2 13$

72 Number of real roots of the equation $e^{\sin x} - e^{-\sin x} - 4 = 0$ is

(a) 1 (b) 0 (c) 2 (d) None of these

73 The number of real roots of the equation $e^{4x} + e^{3x} - 4e^{2x} + e^x + 1 = 0$ is **[JEE Main 2020]**

(a) 1 (b) 3 (c) 2 (d) 4

74 If all the solutions of $a^{\cos x} + a^{-\cos x} = 6 (a > 1)$ are real, then the set of values of a is

(a) $[3 + 2\sqrt{2}, +\infty)$ (b) $(6, 12)$
(c) $[1, 3 + 2\sqrt{2})$ (d) None of these

75 The number of real solutions of the equation $\left(\frac{9}{10}\right)^x = -3 + x - x^2$ is

(a) 2 (b) 0 (c) 1 (d) None of these

76 The number of solutions of the equation $\sin(e)^x = 5^x + 5^{-x}$ is

(a) 0 (b) 1 (c) 2 (d) infinitely many

77 Let $[x]$ denote the greatest integer less than or equal to x. Then, the values of $x \in R$ satisfying the equation $[e^x]^2 + [e^x + 1] - 3 = 0$ lie in the interval **[JEE Main 2021]**

(a) $\left[0, \frac{1}{e}\right)$ (b) $[\log_e 2, \log_e 3)$

(c) $[1, e)$ (d) $[0, \log_e 2)$

78 The value of b for which the equation, $2\log_{1/25}(bx + 28) = -\log_5(12 - 4x - x^2)$ has coincident roots, is/are

(a) $b = -12$ (b) $b = 4$

(c) $b = 4$ or $b = -12$ (d) $b = -4$ or $b = 12$

Numerical Value Type Questions

79 If the roots of the equation $x^2 - bx + c = 0$ is two consecutive integers, then the value of $\frac{b^2 - 4c}{2}$ is ....

80 Let α and β be two real numbers, such that $\alpha + \beta = 1$ and $\alpha\beta = -1$. Let $p_n = (\alpha)^n + (\beta)^n$, $p_{n-1} = 11$ and $p_{n+1} = 29$, for some integer $n \geq 1$. Then, the value of p_n^2 is ______. **[JEE Main 2021]**

81 Let α and β be the solutions of the quadratic equation $x^2 - 1154x + 1 = 0$, then the value of $\alpha^{\frac{1}{4}} + \beta^{\frac{1}{4}}$ is equal to ....

82 If $b^2 - 4ac \leq 0$ (where $a \neq 0$ and $a, b, c, x, y \in R$) satisfies the systems $ax^2 + x(b - 3) + c + y = 0$ and $ay^2 + y(b + 1) + c + 3x = 0$, then the value of $\frac{y}{x}$ is .....

83 If the equation $x^2 + 2(\lambda + 1)x + \lambda^2 + \lambda + 7 = 0$ has only negative roots, then the least value of λ is ....

84 The value of a for which the sum of the squares of the roots of the equation $x^2 - (a - 2)x - (a + 1) = 0$ assume the least value, is ....

85 The sum of all integral values of k $(k \neq 0)$ for which the equation $\frac{2}{x-1} - \frac{1}{x-2} = \frac{2}{k}$ in x has no real roots, is **[JEE Main 2021]**

86 If $x^2 + 3x + 5 = 0$ and $ax^2 + bx + c = 0$ have a common root and $a, b, c \in N$, then the minimum value of $a + b + c$ is ....

87 If $ax^2 + bx + c = 0$ and $bx^2 + cx + a = 0$ have a common root and a, b and c are non-zero real numbers, then find the value of $\frac{a^3 + b^3 + c^3}{5abc}$ is ....

88 If equations $x^2+ax+12=0$, $x^2+bx+15=0$ and $x^2+(a+b)x+36=0$ have a common positive root, then find the value of $\frac{|a+b|}{2}$ is ....

89 If $x^2+ax+10=0$ and $x^2+bx-10=0$ have a common root, then the value of $\frac{a^2-b^2}{3}$ is ....

90 Let $\lambda \neq 0$ be in R. If α and β are the roots of the equation $x^2-x+2\lambda=0$ and α and γ are the roots of equation $3x^2-10x+27\lambda=0$, then $\frac{\beta\gamma}{\lambda}$ is equal to **[JEE Main 2021]**

91 If a root of the equation $n^2\sin x^2-2\sin x-(2n+1)=0$ lies in $\left[0, \frac{\pi}{2}\right]$, then the minimum positive integer value of n is ......

92 Let $P(x)=\frac{5}{3}-6x-9x^2$ and $Q(y)=-4y^2+4y+\frac{13}{2}$. If there exist a unique pair of real numbers (x, y) such that $P(x)\cdot Q(y)=20$, then the value of $(6x+10y)$ is ......

93 $f: R \to R$, $f(x)=\frac{3x^2+mx+n}{x^2+1}$. If the range of this function is $[-4, 3]$, then the value of $|m+n|$ is ....

94 If a, b and c are non-zero real numbers, then the minimum value of $\left\{\frac{(a^4+3a^2+1)(b^4+5b^2+1)(c^4+7c^2+1)}{a^2b^2c^2}\right\}$ is ....

95 Let $x^2+y^2+xy+1 \geq a(x+y)\ \forall x, y \in R$, then the number of possible integer(s) in the range of a is ....

96 If $(a^2-4a+13)x^2+(a+2)x-2=0$ doesn't have two distinct real roots, then the maximum value of a^2-15a is A, then the value of $|A|$ is ....

97 If the cubic $2x^3-9x^2+12x+k=0$ has two equal roots, then maximum value of $|k|$ is ....

98 If the roots of the equation $10x^3-cx^2-54x-27=0$ are in HP, then the value of c is ....

99 Suppose, $a, b, c \in I$ such that the greatest common divisor of x^2+ax+b and x^2+bx+c is $(x-1)$ and the least common multiple of x^2+ax+b and x^2+bx+c is (x^3-4x^2+x+6). Then, the value of $|a+b+c|$ is ....

100 The number of real roots of the equation $e^{4x}-e^{3x}-4e^{2x}-e^{x}+1=0$ is equal to .... **[JEE Main 2021]**

101 Integral part of the product of non-real roots of the equation $x^4-4x^3+6x^2-4x=69$ is ....

102 The function $f(x)=ax^3+bx^2+cx+d$ has three positive zeroes. If the sum of the zeroes of $f(x)$ is 4, then the largest possible integral values of $\frac{c}{a}$ is ......

103 The sum of the roots of the equation $2^{33x-2}+2^{11x+2}=2^{22x+1}$ is ....

LEVEL 03 \\ JEE Advanced Level

Single Choice Correct Questions

1 If $x = 1 + \cfrac{1}{3 + \cfrac{1}{2 + \cfrac{1}{3 + \cfrac{1}{2 \ldots \infty}}}}$, then the value of x is

(a) $\sqrt{\frac{5}{2}}$ (b) $\sqrt{\frac{3}{2}}$ (c) $\sqrt{\frac{7}{2}}$ (d) $\sqrt{\frac{5}{3}}$

2 The sum of the non-real roots of $(x^2 + x - 2)(x^2 + x - 3) = 12$ is

(a) -1 (b) 1 (c) -6 (d) 6

3 The number of irrational roots of the equation $\frac{4x}{x^2 + x + 3} + \frac{5x}{x^2 - 5x + 3} = -\frac{3}{2}$ is

(a) 4 (b) 0 (c) 1 (d) 2

4 Let a, b and c be real numbers such that $4a + 2b + c = 0$ and $ab > 0$, then the equation $ax^2 + bx + c = 0$ has

(a) complex roots (b) exactly one root
(c) real roots (d) None of these

5 If $a \in (-1, 1)$, then roots of the quadratic equation $(a - 1)x^2 + ax + \sqrt{1 - a^2} = 0$ are

(a) real (b) imaginary
(c) both equal (d) None of these

6 If a, b and c are three distinct positive real numbers, then the real number of real roots of $ax^2 + 2b|x| - c = 0$ is

(a) 0 (b) 4 (c) 2 (d) None of these

7 If $(ax^2 + c)y + (a'x^2 + c') = 0$ and x is a rational function of y and ac is negative, then

(a) $ac' + a'c = 0$ (b) $\frac{a}{a'} = \frac{c}{c'}$
(c) $a^2 + c^2 = a'^2 + c'^2$ (d) $aa' + cc' = 1$

8 If α and β, α and γ, α and δ are the roots of the equations $ax^2 + 2bx + c = 0$, $2bx^2 + cx + a = 0$ and $cx^2 + ax + 2b = 0$, respectively, where a, b and c are positive real numbers, then $\alpha + \alpha^2$ is equal to

(a) abc (b) $a + 2b + c$ (c) -1 (d) 0

9 If $x^2 + ax - 3x - (a + 2) = 0$ has real and distinct roots, then the minimum value of $(a^2 + 1)(a^2 + 2)$ is

(a) 1 (b) 0 (c) $\frac{1}{2}$ (d) $\frac{1}{4}$

10 If α and β are two roots of the equation $x^2 - 64x + 256 = 0$, then the value of

$$\left(\frac{\alpha^3}{\beta^5}\right)^{\frac{1}{8}} + \left(\frac{\beta^3}{\alpha^5}\right)^{\frac{1}{8}}$$ is

(a) 2 (b) 3 (c) 1 (d) 4

11 If α and β are the roots of the equation $\frac{x^2 + px - 1}{(2p)^2}$, where $p \in R$, then the minimum value of $\alpha^4 + \beta^4$ is

(a) $2\sqrt{2}$ (b) $2 - \sqrt{2}$ (c) 2 (d) $2 + \sqrt{2}$

12 If the roots of the equation $ax^2 + bx + c = 0$ are of the form $(k+1)/k$ and $(k+2)/(k+1)$, then $(a+b+c)^2$ is equal to

(a) $2b^2 - ac$ (b) a^2
(c) $b^2 - 4ac$ (d) None of these

13 A quadratic equation with integral coefficients has two different prime numbers as its roots. If the sum of the coefficient of the equation is prime, then the sum of the roots is

(a) 1 (b) 5 (c) 7 (d) 11

14 If α and β are the roots of the equation $2x(2x+1) = 1$, then β is equal to

(a) $2\alpha(\alpha + 1)$ (b) $-2\alpha(\alpha + 1)$ (c) $2\alpha(\alpha - 1)$ (d) $2\alpha^2$

15 If α and β are the roots of the equation $ax^2 + bx + c = 0$, then the value of $\frac{a\alpha^2 + c}{a\alpha + b} + \frac{a\beta^2 + c}{a\beta + b}$ is

(a) $\frac{b(b^2 - 2ac)}{4a}$ (b) $\frac{b^2 - 4ac}{2a}$ (c) $\frac{b(b^2 - 2ac)}{a^2c}$ (d) None of these

16 Suppose a and b denote the distinct real roots of the quadratic polynomial $x^2 + 20x - 2020$ and suppose c and d denote the distinct complex roots of the quadratic polynomial $x^2 - 20x + 2020$. Then, the value of $ac(a-c) + ad(a-d) + bc(b-c) + bd(b-d)$ is **[JEE Adv. 2020]**

(a) 0 (b) 8000 (c) 8080 (d) 16000

17 Let $\frac{-\pi}{6} < \theta < -\frac{\pi}{12}$. Suppose α_1 and β_1 are the roots of the equation $x^2 - 2x\sec\theta + 1 = 0$ and α_2 and β_2 are the roots of the equation $x^2 + 2x\tan\theta - 1 = 0$. If $\alpha_1 > \beta_1$ and $\alpha_2 > \beta_2$, then $\alpha_1 + \beta_2$ is equal to

(a) $2(\sec\theta - \tan\theta)$ (b) $2\sec\theta$ (c) $-2\tan\theta$ (d) 0

18 If one root of $x^2 - x - k = 0$ is square of the other, then k is equal to

(a) $2 \pm \sqrt{5}$ (b) $2 \pm \sqrt{3}$ (c) $3 \pm \sqrt{2}$ (d) $5 \pm \sqrt{2}$

19 If α and β are the roots of $ax^2 + c = bx$, then the equation $(a + cy)^2 = b^2y$ in y has the roots

(a) $\alpha\beta^{-1}$ and $\alpha^{-1}\beta$ (b) α^{-2} and β^{-2} (c) α^{-1} and β^{-1} (d) α^2 and β^2

20 If $a(p+q)^2 + 2bpq + c = 0$ and $a(p+r)^2 + 2bpr + c = 0$ $(a \neq 0)$, then

(a) $qr = p^2$ (b) $qr = p^2 + \frac{c}{a}$ (c) $qr = -p^2$ (d) None of these

21 If α and β are the non-zero roots of $ax^2 + bx + c = 0$ and α^2, β^2 are the roots of $a^2x^2 + b^2x + c^2 = 0$, then a, b and c are in

(a) GP (b) HP (c) AP (d) None of these

22 If α and β are the roots of $x^2 + px + q = 0$ and γ, δ are the roots of $x^2 + px + r = 0$, then $\frac{(\alpha - \gamma)(\alpha - \delta)}{(\beta - \gamma)(\beta - \delta)}$ is equal to

(a) 1 (b) q (c) r (d) $q + r$

23 If α and β are roots of the equation $ax^2+bx+c=0$, then the roots of the equation $a(2x+1)^2-b(2x+1)(3-x)+c(3-x)^2=0$ are

(a) $\frac{2\alpha+1}{\alpha-3}$ and $\frac{2\beta+1}{\beta-3}$
(b) $\frac{3\alpha+1}{\alpha-2}$ and $\frac{3\beta+1}{\beta-2}$
(c) $\frac{2\alpha-1}{\alpha-2}$ and $\frac{2\beta+1}{\beta-2}$
(d) None of these

24 If α and β are the roots of the equation $x^2-2x+3=0$, then the equation whose roots are $P=\alpha^3-3\alpha^2+5\alpha-2$ and $Q=\beta^3-\beta^2+\beta+5$ is

(a) $x^2+3x+2=0$
(b) $x^2-3x-2=0$
(c) $x^2-3x+2=0$
(d) None of these

25 If the roots of the equation $x^2+2ax+b=0$ are real and distinct and they differ by atmost $2m$, then b lies in the interval

(a) (a^2, a^2+m^2)
(b) (a^2-m^2, a^2)
(c) $[a^2-m^2, a^2)$
(d) None of these

26 If x is real, then $\frac{x}{(x^2-5x+9)}$ lies between

(a) -1 and $-\frac{1}{11}$
(b) 1 and $-\frac{1}{11}$
(c) 1 and $\frac{1}{11}$
(d) None of these

27 If the equation $x^2-3px+2q=0$ and $x^2-3ax+2b=0$ have a common root and the other root of the second equation is the reciprocal of the other root of the first, then $(2q-2b)^2$ is equal to

(a) $36pa$
(b) $18pa(q-b)^2$
(c) $36bq(p-a)^2$
(d) $18bq(p-a)^2$

28 If $b_1b_2=2(c_1+c_2)$, then atleast one of the equation $x^2+b_1x+c_1=0$ and $x^2+b_2x+c_2=0$ has

(a) imaginary roots
(b) real roots
(c) purely imaginary roots
(d) None of these

29 If the roots of the equation $(a-1)(x^2+x+1)^2=(a+1)(x^4+x^2+1)$ are real and distinct, then the value of $a\in$

(a) $(-\infty, 3)$
(b) $(-\infty, -2)\cup(2,\infty)$
(c) $(-2, 2)$
(d) $(-3, \infty)$

30 If $a, b, c, d \in R$, then the equation $(x^2+ax-3b)(x^2-cx+b)(x^2-dx+2b)=0$ has

(a) 6 real roots
(b) atleast 2 real roots
(c) 4 real roots
(d) 3 real roots

31 One of the roots of $ax^2+bx+c=0$ is greater than 2 and the other is less than -1. If the roots of $cx^2+bx+a=0$ are α and β, then

(a) $0<\alpha<\frac{1}{2}$ and $-1<\beta<0$
(b) $\alpha<\frac{1}{2}$ and $\beta<-1$
(c) $\alpha>\frac{1}{2}$ and $\beta>-1$
(d) $\alpha<2$ and $\beta>-1$

32 The equation $(x^2+x+1)^2+1=(x^2+x+1)(x^2-x-5)$ for $x\in(-2, 3)$ will have number of solutions

(a) 1
(b) 2
(c) 3
(d) 0

33 The quadratic equation $p(x)=0$ with real coefficients has purely imaginary roots. Then, the equation $p\{p(x)\}=0$ has

(a) one purely imaginary root
(b) all real roots
(c) two real and two purely imaginary roots
(d) neither real nor purely imaginary roots

34 The value of the expression $x^4-8x^3+18x^2-8x+2$, when $x=2+\sqrt{3}$, is

(a) 2 (b) 1 (c) 0 (d) 3

35 The sum of values of x satisfying the equation $(31+8\sqrt{15})^{x^2-3}+1=(32+8\sqrt{15})^{x^2-3}$ is

(a) 3 (b) 0 (c) 2 (d) None of these

More than One Correct

36 If a, b and c are in GP, then the roots of the equation $ax^2+bx+c=0$ are in the ratio

(a) $\frac{1}{2}(-1+i\sqrt{3})$ (b) $\frac{1}{2}(1-i\sqrt{3})$ (c) $\frac{1}{2}(-1-i\sqrt{3})$ (d) $\frac{1}{2}(1+i\sqrt{3})$

37 If the quadratic equation $ax^2+bx+c=0\,(a>0)$ has $\sec^2\theta$ and $\operatorname{cosec}^2\theta$ as its roots, then which of the following must hold good?

(a) $b+c=0$ (b) $b^2-4ac\geq 0$
(c) $c\geq 4a$ (d) $4a+b\geq 0$

38 If one root of the quadratic equation $px^2+qx+r=0\,(p\neq 0)$ is $\frac{\sqrt{a}}{\sqrt{a}+\sqrt{a-b}}$, where $p, q, r\in Q$, and a, b are positive integers which are not perfect square of any integer, then the other root is

(a) $\frac{\sqrt{a}}{\sqrt{a}-\sqrt{a-b}}$ (b) $\frac{\sqrt{a}-\sqrt{a-b}}{\sqrt{b}}$ (c) $a+\frac{\sqrt{a(a-b)}}{b}$ (d) $\frac{a+\sqrt{a(a-b)}}{b}$

39 If $c\neq 0$ and the equation $\frac{p}{2x}=\frac{a}{x+c}+\frac{b}{x-c}$ has two equal roots, then p can be

(a) $(\sqrt{a}-\sqrt{b})^2$ (b) $(\sqrt{a}+\sqrt{b})^2$ (c) $a+b$ (d) $a-b$

40 Given that α and γ are roots of the equation $x^2-4x+A=0$ and β, δ are the roots of the equation $x^2-6x+B=0$, such that α, β, γ and δ are in AP, then

(a) $A=3$ (b) $A=4$ (c) $B=2$ (d) $B=8$

41 If $a, b, c\in R$ and $abc<0$, then the equation $bcx^2+2(b+c-a)x+a=0$ has

(a) both positive roots (b) both negative roots
(c) real roots (d) one positive and one negative roots

42 If the roots of the equation $x^2+ax+b=0$ are c and d, then roots of the equation $x^2+(2c+a)x+c^2+ac+b=0$ are

(a) c (b) $d-c$ (c) $2c$ (d) 0

43 Let S be the set of all non-zero real numbers α such that the quadratic equation $\alpha x^2-x+\alpha=0$ has two distinct real roots x_1 and x_2 satisfying the inequality $|x_1-x_2|<1$. Which of the following is (are) a subset(s) of S?

(a) $\left(-\frac{1}{2},-\frac{1}{\sqrt{5}}\right)$ (b) $\left(-\frac{1}{\sqrt{5}},0\right)$ (c) $\left(0,\frac{1}{\sqrt{5}}\right)$ (d) $\left(\frac{1}{\sqrt{5}},\frac{1}{2}\right)$

44 If $f(x) = ax^2 + bx + c$, where $a \neq 0, b, c \in R$, then which of the following conditions implies that $f(x)$ has real roots?

(a) $a + b + c = 0$ (b) a and c are of opposite sign
(c) $4ac - b^2 < 0$ (d) a and b are of opposite sign

45 If $\cos^4 \theta + \alpha$ and $\sin^4 \theta + \alpha$ are the roots of the equation $x^2 + 2bx + b = 0$ and $\cos^2 \theta + \beta$, $\sin^2 \theta + \beta$ are the roots of the equation $x^2 + 4x + 2 = 0$, then values of b are

(a) 2 (b) -1 (c) -2 (d) 1

46 If the equation $ax^2 + bx + c = 0$, $a, b, c \in R$ have not real roots, then

(a) $c(a + b + c) > 0$ (b) $c(a - b + c) > 0$
(c) $c(4a - 2b + c) > 0$ (d) $c(a - b - c) > 0$

47 If $x, y \in R$ and $2x^2 + 6xy + 5y^2 = 1$, then

(a) $|x| \leq \sqrt{5}$ (b) $|x| \geq \sqrt{5}$ (c) $y^2 \leq 2$ (d) $y^2 \leq 4$

48 If the equations $4x^2 - x - 1 = 0$ and $3x^2 + (\lambda + \mu)x + \lambda - \mu = 0$ have a common root, then the rational values of λ and μ are

(a) $\lambda = \frac{-3}{4}$ (b) $\lambda = 0$ (c) $\mu = \frac{3}{4}$ (d) $\mu = 0$

49 If the quadratic equations $x^2 + bx + c = 0$ and $bx^2 + cx + 1 = 0$ have a common root, then

(a) $b + c + 1 = 0$ (b) $b^2 + c^2 - 1 = bc - b - c$
(c) $b + c - 1 = 0$ (d) $b^2 + c^2 + 1 = bc + b + c$

50 Let $a, b, c \in Q^+$ satisfying $a > b > c$. Which of the following statement(s) hold true for the quadratic polynomial $f(x) = (a + b - 2c)x^2 + (b + c - 2a)x + (c + a - 2b)$?

(a) The mouth of the parabola $y = f(x)$ opens upward
(b) Both roots of the equation $f(x) = 0$ are rational
(c) The x-coordinate of vertex of the graph is positive
(d) The product of the roots is always negative

51 If $ax^2 + (b - c)x + a - b - c = 0$ has unequal real roots for all $c \in R$, then

(a) $b < 0 < a$ (b) $a < 0 < b$ (c) $b < a < 0$ (d) $b > a > 0$

52 If the range of function $f(x) = \frac{x + 1}{k + x^2}$ contains the interval $[0, 1]$, then the value of k can be equal to

(a) 0 (b) 0.5 (c) 1.25 (d) 1.5

53 If $\frac{(x^2 + ax + 3)}{(x^2 + x + a)}$ takes all real values for possible real values of x, then possible value/values of a is/are

(a) -3 (b) 2 (c) -1 (d) -4

54 If $x^3 + 3x^2 - 9x + c$ is of the form $(x - \alpha)^2 (x - \beta)$, then c is equal to

(a) 27 (b) -27 (c) 5 (d) -5

55 If the equation whose roots are the squares of the roots of the cubic $x^3 - ax^2 + bx - 1 = 0$ is identical with the given cubic equation, then

(a) $a = 0$ and $b = 2$ (b) $a = b = 0$
(c) $a = b = 3$ (d) a and b are roots of $x^2 + x + 2 = 0$

56 If the roots of the equation $x^3 + px^2 + qx - 1 = 0$ form an increasing GP, where p and q are real, then
(a) $p + q = 0$
(b) $p \in (-3, \infty)$
(c) one of the roots is unity
(d) one root is smaller than 1 and one root is greater than 1

57 The equation $2^{2x} + (a-1)2^{x+1} + a = 0$ has roots of opposite sign, then exhaustive set of values of a is
(a) $a \in (-1, 0)$ (b) $a < 0$ (c) $a \in \left(-\infty, \frac{1}{3}\right)$ (d) $a \in \left(0, \frac{1}{3}\right)$

Linked Comprehension Type

Paragraph I Directions (Q.Nos. 58-60)

Let $f(x) = x^2 + b_1x + c_1$ and $g(x) = x^2 + b_2x + c_2$. Let the real roots of $f(x) = 0$ be α, β and real roots of $g(x) = 0$ be $\alpha + h$ and $\beta + h$. The least value of $f(x)$ is $-\frac{1}{4}$. The least value of $g(x)$ occurs at $x = \frac{7}{2}$.

58 The least value of $g(x)$ is
(a) $-\frac{1}{4}$ (b) -1 (c) $-\frac{1}{3}$ (d) $-\frac{1}{2}$

59 The value of b_2 is
(a) -5 (b) 9 (c) -8 (d) -7

60 The roots of $f(x) = 0$ are
(a) $3, -4$ (b) $-3, 4$ (c) $3, 4$ (d) $-3, -4$

Paragraph II Directions (Q.Nos. 61-63)

In the given figure, vertices of ΔABC lie on $y = f(x) = ax^2 + bx + c$. The ΔABC is right angled isosceles triangle whose hypotenuse $AC = 4\sqrt{2}$ units.

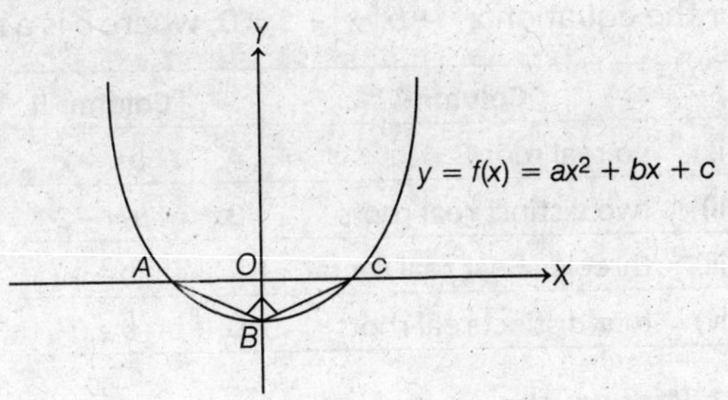

61 $y = f(x)$ is given by
(a) $y = x^2 - 2\sqrt{2}$ (b) $y = x^2 - 12$ (c) $y = \frac{x^2}{2} - 2$ (d) $y = \frac{x^2}{2\sqrt{2}} - 2\sqrt{2}$

62 Minimum value of $y = f(x)$ is
(a) -4 (b) -2
(c) $-2\sqrt{2}$ (d) None of these

63 Number of integral values of k for which one root of $f(x)=0$ is more than k and other less than k

(a) 6 (b) 4 (c) 5 (d) 7

Paragraph III Directions (Q.Nos. 64-66)

Consider an unknown polynomial which when divided by $(x-3)$ and $(x-4)$ leaves remainders 2 and 1, respectively. Let $R(x)$ be the remainder, when this polynomial is divided by $(x-3)(x-4)$.

64 If equation $R(x)=x^2+ax+1$ has two distinct real roots, then exhaustive values of a are

(a) $(-2, 2)$ (b) $(-\infty, -2)\cup(2, \infty)$

(c) $(-2, \infty)$ (d) all real numbers

65 If $R(x)=px^2+(q-1)x+6$ has no distinct real roots and $p>0$, then the least value of $3p+q$ is

(a) -2 (b) $\frac{2}{2}$

(c) $-\frac{1}{3}$ (d) None of these

66 Range of $f(x)=\dfrac{R(x)}{(x^2-3x+2)}$ is

(a) $(-2, 2)$ (b) $(-\infty, -2-\sqrt{3})\cup(-2+\sqrt{3}, \infty)$

(c) $(-\infty, -7-4\sqrt{3})\cup(-7+4\sqrt{3}, \infty)$ (d) None of these

Matrix Match Type Questions

67 Let $a, b, c \in R$ be such that $b^2-ac\geq 0$ the equation $ax^2+2bx+c=0$ has

Column-I	Column-II
(i) two positive roots	A. $a, -c$ are of the same sign
(ii) two negative roots	B. $a, -b, c$ are of the same sign
(iii) one positive and one negative roots	C. a, b, c are of the same sign
(iv) equal roots	D. a, b, c are in GP

68 Match the following for the equation $x^2+b|x|+1=0$, where b is a real parameter

Column-I	Column-II
(i) no real root	A. $b<-2$
(ii) two distinct real roots	B. $b=-2$
(iii) three distinct real roots	C. $b\in\phi$
(iv) four distinct real roots	D. $b\geq 0$

69 The set of value(s) of $k\in R$ for which

Column-I	Column-II
(i) $kx^2-(k+1)x+2k-1=0$ has no real roots	A. $\{1, -2\}$
(ii) $x^2-2(4k-1)x+15k^2-2k-7>0$ for each x	B. $\left\{-\infty, -\frac{1}{7}\right\}\cup(1, \infty)$
(iii) sum of roots of $x^2+(2-k-k^2)x-k^2=0$ is zero	C. $\{-4\}$
(iv) the roots of $x^2+(2k-1)x+k^2+2=0$ are in the ratio $1:2$	D. $\{2, 4\}$

70 Match equations on left with properties on right.

Column-I		Column-II	
(i)	$a<b<c<d$ and equation is $(x-a)(x-c)+\pi(x-b)(x-d)=0$	A.	real roots
(ii)	$a>0$, $a+b+c<0$ and equation is $ax^2+bx+c=0$	B.	distinct real roots
(iii)	$b, c \in I$ and the equation $x^2+bx+c=0$ has rational roots	C.	integer roots
(iv)	$a, b, c, d \in R$ are in GP and equation is $(a^2+b^2+c^2)x^2+2(ab+bc+cd)x+(b^2+c^2+d^2)=0$	D.	discriminant ≥ 0

71 Given the equation $x^4+x^2(1-2k)+k^2-1=0$, then the values of k for which given equation has

Column-I		Column-II	
(i)	no solution	A.	2
(ii)	one solution	B.	-1
(iii)	two solutions	C.	$(-\infty,-1)\cup\left(\frac{5}{4},\infty\right)$
(iv)	three solutions	D.	$(-1,1)\cup\left\{\frac{5}{4}\right\}$

Numerical Value Type Questions

Question Step Directions (Q.Nos. 72-73) The answer to each question is a **NUMERICAL VALUE.**

One root of the equation $x^4-5x^3+ax^2+bx+c=0$ is $3+\sqrt{2}$. If all the roots of the equations are real given that a, b, c are rational parameters, then

72 The greatest value of 'a' is ....

73 The least value of 'b' is ....

Question Step Directions (Q.Nos. 74-75) The answer to each question is a **NUMERICAL VALUE.**

Let m be a real number, such that the roots α and β of the equation $x^2+(m-4)x+m^2-3m+3=0$ are real numbers, then

74 The value of m for which $\alpha^2+\beta^2=6$ is ....

75 The maximum value of $\frac{m\alpha^2}{1-\alpha}+\frac{m\beta^2}{1-\beta}+8$ is ....

76 If a and b are positive numbers and each of the equations $x^2+ax+2b=0$ and $x^2+2bx+a=0$ has real roots, then the smallest possible value of $(a+b)$ is ......

77 Sum of the values of x satisfying the equation $\sqrt{2x+\sqrt{2x+4}}=4$ is ......

78 Let $a, b \in R$ and $ab \neq 1$. If $6a^2+20a+15=0$ and $15b^2+20b+6=0$, then the value of $\frac{4030b^3}{ab^2-9(ab+1)^3}$ is ....

79 If $a^2-4a+1=4$, then value of $\frac{a^3-a^2+a-1}{a^2-1}(a^2\neq 1)$ is equal to ....

80 Given α and β are the roots of the quadratic equation $x^2 - 4x + k = 0\ (k \neq 0)$. If $\alpha\beta, \alpha\beta^2 \times \alpha^2\beta$ and $\alpha^3 + \beta^3$ are in GP. Then, the value of $7k$ is ......

81 If α and β are the roots of the equation $x^2 - 6x + 12 = 0$ and the value of $(\alpha - 2)^{24} - \frac{(\beta - 6)^8}{\alpha^8} + 1$ is 4^a, then the value of a is ....

82 Let α_1, β_1 be the roots of $x^2 - 6x + p = 0$ and α_2, β_2 be the roots of $x^2 - 54x + q = 0$. If $\alpha_1, \beta_1, \alpha_2$ and β_2 form an increasing GP, then the value of $(q - p)$ is ....

83 The quadratic equation $x^2 + mx + n = 0$ has roots which are twice those of $x^2 + px + m = 0$ and m, n and $p \neq 0$, the value of $\frac{n}{p}$ is .....

84 a, b and c are all different and non-zero real numbers in arithmetic progression. If the roots of quadratic equation $ax^2 + bx + c = 0$ are α and β such that $\frac{1}{\alpha} + \frac{1}{\beta}, \alpha + \beta$ and $\alpha^2 + \beta^2$ are in GP, then the value of $\frac{a}{c}$ will be ......

85 If $\sin^2 \alpha, \cos^2 \alpha$ and $-\operatorname{cosec}^2 \alpha$ are the zeroes of $P(x) = x^3 + x^2 + ax + b,\ (a, b \in R)$, then $P(2)$ is equal to ....

86 If there exists atleast one real x which satisfies both the equations $x^2 + 2x \sin y + 1 = 0$, where $y \in \left(0, \frac{\pi}{2}\right)$ and $ax^2 + x + 1 = 0$, then the value of $(a + \sin y)$ is ....

87 If the equation $2x^2 + 4xy + 7y^2 - 12x - 2y + t = 0$, where t is a parameter has exactly one real solution of the form (x, y), then $(x + y)$ is equal to ......

88 Suppose a, b and c are the roots of the cubic $x^3 - x^2 - 2 = 0$, then the value of $(a^3 + b^3 + c^3)$ is ....

89 Let $P(x) = x^3 - 8x^2 + cx - d$ be a polynomial with real coefficients and with all its roots being distinct positive integers. Then, the sum of all possible values c is ....

90 Given that $x^2 - 3x + 1 = 0$, then the value of the expression $y = x^9 + x^7 + x^{-9} + x^{-7}$ is ....

91 If equation $x^4 - (3m + 2)x^2 - m^2 = 0\ (m > 0)$ has four real solutions which are in AP, then the value of m is ....

92 The value of x in the given equation $4^x - 3^{x - \frac{1}{2}} = 3^{x + \frac{1}{2}} - 2^{2x - 1}$ is $\frac{a}{b}$, where a and b don't have any factor. Then, the value of $\frac{a + b}{3}$ is ....

93 The sum of the roots of the equation $2^{x+2} \cdot 27^{\frac{x}{x-1}} = 9$ are given by $-a - \log_b c$. Then, the value of $\frac{ac}{b}$ is ....

ANSWER KEY

Level 1

1. $\lambda = 2$

2. (i) $a > -1, D > 0$ (ii) $a < -1, D < 0$ (iii) $a = -1, D = 0$

3. (i) $k = 6$ (ii) $k = \frac{19}{3}$

4. $\alpha = \frac{-9+\sqrt{5}}{2}$ **5.** $a = \frac{1}{3}$ **6.** $x^2 - 2x - 24 = 0$ **7.** 0

8. 47 **9.** $\frac{\alpha}{\alpha+1}$ and $\frac{\beta}{\beta+1}$

10. $\lambda = \frac{-43}{16}$, common root is $x = 4$, and for $\lambda = 3$, common root is $x = -3$

11. (i) $x \in R$ (ii) $x \in \{-2\}$ (iii) $x \in R - \{-2\}$ (iv) $x \in \phi$

12. $p \in \left(-\infty, -\frac{3}{4}\right)$

13. For $f(x) > 0, \forall x \in (-6, -4) \cup (-3, 3) \cup (5, \infty)$
For $f(x) < 0, \forall x \in (-\infty, -6) \cup (-4, -3) \cup (3, 5)$ **14.** $x \in [2, 3) \cup [6, 7)$

15. (i) $m \in (-\infty, -2] \cup [2, \infty)$ (ii) $m \in (-\infty, 2)$ (iii) $m \in (-\infty, -2]$

16. (i) $m \in (-\infty, 1]$ (ii) $m \in [9, 10]$ (iii) $m \in (10, \infty)$ (iv) $m \in (10, \infty)$

18. real or unequal

19. (i) $p^2 - 2q$ (ii) $pq - 3r$

20. (i) $2p^3 - 9pq + 27r = 0$ (ii) $p^3 r = q^3$

21. $x = \frac{3 \pm \sqrt{13}}{2}$

22. $x \in (-\infty, -2) \cup \left(\frac{-5}{4}, 1\right) \cup (1, 5)$ **23.** $x_1 = 6$ and $x_2 = \frac{1}{5}$

24. $x = -5, -2$ **25.** $x \in (0, \infty)$

26. $x \in \left(\frac{1}{25}, 5\right)$ **27.** $x = 1$ or $x = 4$

Level 2

1. (c)	**2.** (d)	**3.** (d)	**4.** (c)	**5.** (d)	**6.** (a)	**7.** (a)
8. (d)	**9.** (a)	**10.** (a)	**11.** (b)	**12.** (b)	**13.** (a)	**14.** (c)
15. (b)	**16.** (b)	**17.** (c)	**18.** (c)	**19.** (a)	**20.** (a)	**21.** (b)
22. (d)	**23.** (b)	**24.** (a)	**25.** (b)	**26.** (b)	**27.** (a)	**28.** (b)
29. (c)	**30.** (b)	**31.** (d)	**32.** (d)	**33.** (a)	**34.** (c)	**35.** (a)
36. (c)	**37.** (b)	**38.** (d)	**39.** (b)	**40.** (a)	**41.** (c)	**42.** (a)
43. (a)	**44.** (d)	**45.** (a)	**46.** (b)	**47.** (d)	**48.** (c)	**49.** (b)
50. (a)	**51.** (c)	**52.** (b)	**53.** (c)	**54.** (a)	**55.** (c)	**56.** (b)
57. (d)	**58.** (c)	**59.** (a)	**60.** (b)	**61.** (c)	**62.** (d)	**63.** (a)
64. (c)	**65.** (a)	**66.** (d)	**67.** (c)	**68.** (c)	**69.** (d)	**70.** (d)
71. (b)	**72.** (b)	**73.** (a)	**74.** (a)	**75.** (b)	**76.** (a)	**77.** (d)
78. (b)	**79.** (0.5)	**80.** (324)	**81.** (6)	**82.** (3)	**83.** (6)	**84.** (1)
85. (66)	**86.** (9)	**87.** (0.6)	**88.** (7.5)	**89.** (13.334)	**90.** (18)	**91.** (3)
92. (3)	**93.** (4)	**94.** (315)	**95.** (3)	**96.** (9)	**97.** (5)	**98.** (9)
99. (6)	**100.** (2)	**101.** (9)	**102.** (5)	**103.** (0.181)		

Level 3

1. (d)	**2.** (a)	**3.** (d)	**4.** (c)	**5.** (a)	**6.** (c)	**7.** (b)
8. (c)	**9.** (c)	**10.** (a)	**11.** (d)	**12.** (c)	**13.** (b)	**14.** (b)
15. (c)	**16.** (d)	**17.** (c)	**18.** (a)	**19.** (b)	**20.** (b)	**21.** (a)
22. (a)	**23.** (b)	**24.** (c)	**25.** (c)	**26.** (b)	**27.** (c)	**28.** (b)
29. (b)	**30.** (b)	**31.** (a)	**32.** (d)	**33.** (d)	**34.** (b)	**35.** (b)
36. (a, c)	**37.** (a, b, c)	**38.** (a, d)	**39.** (a, b)	**40.** (a, d)	**41.** (c, d)	**42.** (b, d)
43. (a, d)	**44.** (a, b, c)	**45.** (a, b)	**46.** (a, b, c)	**47.** (a, c)	**48.** (a, d)	**49.** (a, d)
50. (a, b, c)	**51.** (c, d)	**52.** (a, b, c)	**53.** (a, d)	**54.** (b, c)	**55.** (b, c, d)	**56.** (a, c, d)
57. (c)	**58.** (a)	**59.** (d)	**60.** (c)	**61.** (d)	**62.** (c)	**63.** (c)
64. (d)	**65.** (c)	**66.** (c)				

67. (i)–B, (ii)–C, (iii)–A, (iv)–D

68. (i)–D, (ii)–B, (iii)–C, (iv)–A

69. (i)–B, (ii)–D, (iii)–A, (iv)–C

70. (i)–B, (ii)–A, (iii)–C, (iv)–D

71. (i)–C, (ii)–B, (iii)–D, (iv)–A

72. (1.25)	**73.** (5.5)	**74.** (1.236)	**75.** (13.44)	**76.** (6)	**77.** (6)	**78.** (12)
79. (4)	**80.** (16)	**81.** (12)	**82.** (540)	**83.** (8)	**84.** (3)	**85.** (9)
86. (1.5)	**87.** (3)	**88.** (7)	**89.** (36)	**90.** (6621)	**91.** (6)	**92.** (1.667)
93. (1.5)						

SOLUTIONS

Level 1

1. If an equation of degree two is satisfied by more than two values of unknown, then it must be identity.
Then, we must have
$$\lambda^2-5\lambda+6=0, \lambda^2-3\lambda+2=0 \text{ and } \lambda^2-4=0$$
$$\Rightarrow \lambda=2,3 \text{ and } \lambda=2,1 \text{ and } \lambda=2,-2$$
Common value of λ which satisfies each condition is
$$\lambda=2.$$

2. For quadratic equation $a\neq-1$ and the discriminant
$$D=4(a+1)^2-4(a+1)(a-2)$$
$$=4(a+1)(a+1-a+2)$$
$$=12(a+1)$$
(i) If $a>-1$, then $D>0$, this equation has distinct roots.
(ii) If $a<-1$, then $D<0$, this equation has no real roots.
(iii) This equation can not have two equal roots, since $D=0$ only for $a=-1$ and this is not acceptable.

3. (i) Given, $x^2+12x=3kx^2+2$
$$\Rightarrow x^2(3k-1)-12x+2=0$$
Given, $D\geq 0$ [$\because$ roots are real]
$$(-12)^2-4(3k-1)(2)\geq 0$$
$$\Rightarrow 144-24k+8\geq 0$$
$$\Rightarrow 152\geq 24k$$
$$\therefore k\leq\frac{19}{3}=6\frac{1}{3}$$
$\because$ k needs to be an integer (Z), then greatest value of k is 6.
(ii) For rational roots, $D=0$
As above, we get
$$152-24k=0$$
$$\Rightarrow k=\frac{19}{3}$$

4. Let other root be α.
Sum of roots $=\sqrt{5}$
$$\Rightarrow \frac{9+\sqrt{5}}{2}+\alpha=\sqrt{5}$$
$$\Rightarrow \alpha=\frac{-9+\sqrt{5}}{2}$$

5. Let $a=\frac{1}{b}$, so that the given equation becomes
$$x^2-4bx+9b=0$$
The equation has integral roots if b is an integer and $16b^2-36b$ is a perfect square.
Let $b(4b-9)=k^2$
$$\Rightarrow 4b^2-9b-k^2=0$$
$$\Rightarrow \left(2b-\frac{9}{4}\right)^2-k^2=\frac{81}{16}$$
$$\Rightarrow (8b-9)^2-16k^2=81$$
$$\Rightarrow (8b-9-4k)(8b-9+4k)=3\times 27$$
$\because$ b and k are integers.
$$\therefore 8b-9-4k=3 \text{ and } 8b-9+4k=27$$
$$\Rightarrow b=3$$
Hence, $a=\frac{1}{3}$

6. The monic quadratic equation having $\alpha^2+\beta^2$ and $2\alpha\beta$ as its roots is given by
$$x^2-[\alpha^2+\beta^2+2\alpha\beta]x+2\alpha\beta(\alpha^2+\beta^2)=0$$
$$\because \alpha+\beta=-\sqrt{2} \text{ and } \alpha\beta=3$$
$$\therefore \alpha^2+\beta^2=(\alpha+\beta)^2-2\alpha\beta$$
$$=2-6=-4$$
Also, $\alpha^2+\beta^2+2\alpha\beta=2$ and $2\alpha\beta(\alpha^2+\beta^2)=-24$
Hence, the required quadratic equation is
$$x^2-2x-24=0$$

7. Quadratic equation is
$$3x^2-8x+1=0 \quad \text{[on simplifying]}$$
Let α and β be the roots.
$$\therefore \alpha+\beta=\frac{8}{3} \text{ and } \alpha\beta=\frac{1}{3}$$
$$\Rightarrow \frac{1}{\alpha\beta+(\alpha+\beta)+1}+\frac{1}{\alpha\beta-2(\alpha+\beta)+4}+\frac{1}{\alpha\beta-3(\alpha+\beta)+9}$$
$$=\frac{1}{4}-1+\frac{3}{4}=\frac{-3}{4}+\frac{3}{4}=0$$

8. Given, $x^2-3x+1=0$
$$\therefore \alpha+\beta=3 \text{ and } \alpha\beta=1$$
$$\Rightarrow \alpha^2+\beta^2=(\alpha+\beta)^2-2\alpha\beta$$
$$=(3)^2-2(1)=7$$
$$\Rightarrow \alpha^4+\beta^4=(\alpha^2+\beta^2)^2-2(\alpha\beta)^2$$
$$=(7)^2-2(1)^2$$
$$=49-2=47$$

9. Given, $ax^2-bx(x-1)+c(x-1)^2=0$
$$\Rightarrow a\left(\frac{x}{x-1}\right)^2-b\left(\frac{x}{x-1}\right)+c=0$$
$$\Rightarrow a\left(\frac{x}{1-x}\right)^2+b\left(\frac{x}{1-x}\right)+c=0$$

Now, α and β are the roots of $ax^2+bx+c=0$.

Then, $\alpha=\dfrac{x}{1-x}$ and $\beta=\dfrac{x}{1-x}$

[as roots satisfies the equation]

$\Rightarrow \quad x=\dfrac{\alpha}{\alpha+1}$ and $x=\dfrac{\beta}{1+\beta}$

Hence, $\dfrac{\alpha}{\alpha+1}$ and $\dfrac{\beta}{\beta+1}$ are the roots of the equation.

10. Given, $x^2-x-12=0$

$\Rightarrow \quad (x-4)(x+3)=0$

$\therefore\ x=4,-3$

If $x=4$ is a common root, then

$\lambda(4)^2+10(4)+3=0$

$\therefore \qquad \lambda=\dfrac{-43}{16}$

and if $x=-3$ is a common root, then

$\lambda(-3)^2+10(-3)+3=0$

$\Rightarrow \qquad \lambda=3$

Hence, for $\lambda=\dfrac{-43}{16}$, common root is $x=4$

and for $\lambda=3$, common root is $x=-3$.

11. Given, $f(x)=x^2+4x+4$

$=(x+2)^2$

Let us find D

$\therefore\ b^2-4ac=(4)^2-4(1)(4)=0$

$\Rightarrow \qquad D=0$

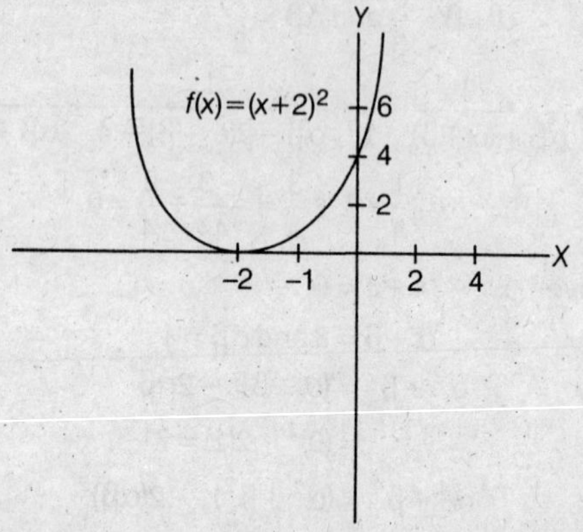

$\therefore$ Roots of the following equation are real and equal.

Also, observe that $a=1>0$

As, $D=0$ and $a>0$, we get

$f(x)\geq 0\ \forall\ x\in R$, as above, we get

(i) $f(x)\geq 0$ is true $\forall\ x\in R$

(ii) $f(x)\leq 0$ is true for $x\in\{-2\}$

(iii) $f(x)>0$ is true $\forall\ x\in R-\{-2\}$

(iv) $f(x)<0$ is true for no value of x i.e. $x\in\phi$.

12. Let $f(x)=x^2+2(p-3)x+9$, as 6 lies between the roots of $f(x)=0$, we can take $D>0$ and $a\cdot f(6)<0$, where a is the coefficient of x^2.

(i) $\qquad D>0$

$\Rightarrow [2(p-3)]^2-4\cdot1\cdot9>0$

$\Rightarrow \qquad (p-3)^2-9>0$

$\Rightarrow \qquad p(p-6)>0$

$\Rightarrow \qquad p\in(-\infty,0)\cup(6,\infty)$

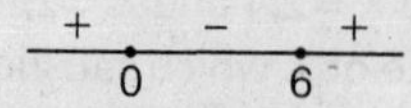

(ii) Consider, $\qquad a\cdot f(6)<0$

$1\cdot\{36+12(p-3)+9\}<0$

$\Rightarrow \qquad 12p+9<0$

$\Rightarrow \qquad p+\dfrac{3}{4}<0$

$\Rightarrow \qquad p\in\left(-\infty,\dfrac{-3}{4}\right)$

Hence, the values of p satisfying Eqs. (i) and (ii) at the same time, are $p\in\left(-\infty,\dfrac{-3}{4}\right)$.

13. We have, $f(x)=\dfrac{(x-3)(x+4)(x+6)}{(x+3)(x-5)}$

The critical points are $-6,-4,-3,3,5$

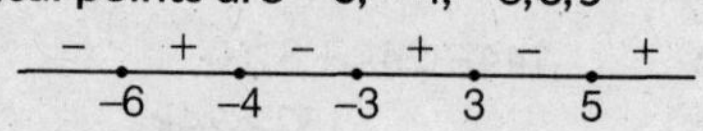

For $f(x)>0,\ \forall\ x\in(-6,-4)\cup(-3,3)\cup(5,\infty)$

For $f(x)<0,\ \forall\ x\in(-\infty,-6)\cup(-4,-3)\cup(3,5)$.

14. This equation has the form $|f(x)|=-f(x)$

When $f(x)=\dfrac{x^2-8x+12}{x^2-10x+21}$

such an equation is equivalent to the collection of systems

$f(x)=-f(x)$, if $f(x)\geq 0$

and $\quad f(x)=f(x)$, if $f(x)<0$

The first system is equivalent to $f(x)=0$

and the second system is equivalent to $f(x)<0$.

On combining, we get

$$f(x)\leq 0$$

$$\frac{x^2-8x+12}{x^2-10x+21}\leq 0$$

$$\Rightarrow \quad \frac{(x-2)(x-6)}{(x-3)(x-7)}\leq 0$$

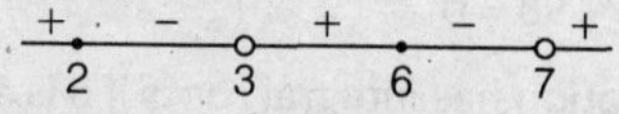

$\therefore$ By Wavy curve method,

$x\in[2,3)\cup[6,7)$.

15. Let $f(x) = x^2 - mx + 1$, as both roots of $f(x) = 0$ are less than 1, we can take $D \geq 0$, $af(1) > 0$ and $-\frac{b}{2a} < 1$ where a and b represent the coefficient of x^2 and x for the given function $f(x)$.

(i) Consider, $D \geq 0 \Rightarrow (-m)^2 - 4 \cdot 1 \cdot 1 \geq 0$

$\Rightarrow \quad (m+2)(m-2) \geq 0$

$\Rightarrow \quad m \in (-\infty, -2] \cup [2, \infty)$ ...(i)

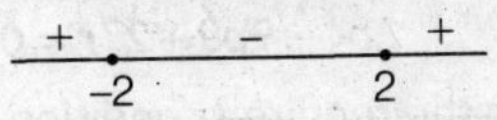

(ii) Consider, $af(1) > 0 \Rightarrow 1(1 - m + 1) > 0$

$\Rightarrow \quad m - 2 < 0$

$\Rightarrow \quad m < 2$

$\Rightarrow \quad m \in (-\infty, 2)$ ... (ii)

(iii) Consider, $\left(-\frac{b}{2a} < 1\right)$

$\Rightarrow \quad \frac{m}{2} < 1$

$\Rightarrow \quad m < 2$

$\Rightarrow \quad m \in (-\infty, 2)$ ... (iii)

Hence, the values of m satisfying Eqs. (i), (ii) and (iii) at the same time, are $m \in (-\infty, -2]$.

16. Let $f(x) = x^2 - (m-3)x + m$

Here, $a = 1, b = -(m-3)$ and $c = m$

$D = b^2 - 4ac = (m-3)^2 - 4m$

$= m^2 - 10m + 9 = (m-1)(m-9)$

and x-coordinate of vertex $= -\frac{b}{2a} = \frac{(m-3)}{2}$

(i) Both the roots are smaller than 2

$D \geq 0$

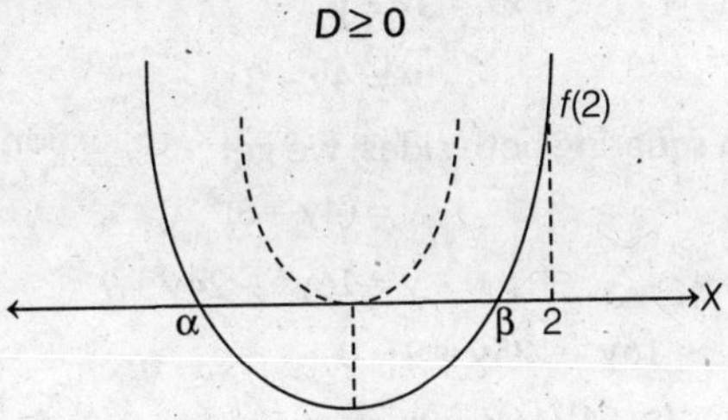

i.e. $(m-1)(m-9) \geq 0$

$\therefore \quad m \in (-\infty, 1] \cup [9, \infty)$ ...(i)

$f(2) > 0$

i.e. $4 - 2(m-3) + m > 0 \Rightarrow m < 10$

$\therefore \quad m \in (-\infty, 10)$ ... (ii)

and x-coordinate of vertex < 2

i.e. $\frac{(m-3)}{2} < 2 \Rightarrow m < 7$

$\therefore \quad m \in (-\infty, 7)$ ... (iii)

On combining Eqs. (i), (ii) and (iii), we get

$m \in (-\infty, 1]$

(ii) Both the roots are greater than 2

$D \geq 0$

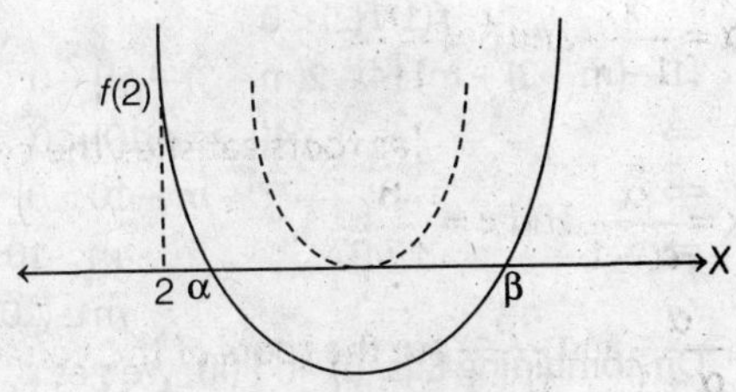

i.e. $(m-1)(m-9) \geq 0$

$\therefore \quad m \in (-\infty, 1] \cup [9, \infty)$ **... (i)**

$f(2) > 0$

i.e. $4 - 2(m-3) + m > 0$

$\Rightarrow \quad m < 10$

$\therefore \quad m \in (-\infty, 10)$ **... (ii)**

and x-coordinate of vertex > 2

i.e. $\frac{(m-3)}{2} > 2$

$\Rightarrow \quad m > 7$

$\therefore \quad m \in (7, \infty)$ **...(iii)**

On combining Eqs. (i), (ii) and (iii), we get

$m \in [9, 10)$

(iii) One root is smaller than 2 and the other root is greater than 2

$D > 0$

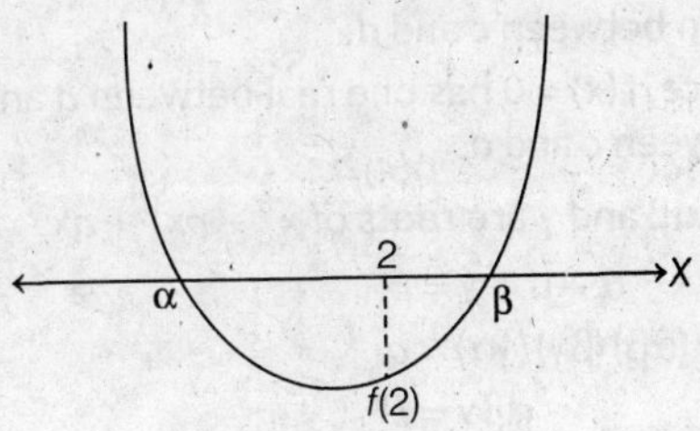

i.e. $(m-1)(m-9) > 0$

$\therefore \quad m \in (-\infty, 1) \cup (9, \infty)$ **... (i)**

$f(2) < 0$

i.e. $4 - 2(m-3) + m < 0$

$\therefore \quad m > 10$

$\therefore \quad m \in (10, \infty)$ **... (ii)**

On combining Eqs. (i) and (ii), we get

$m \in (10, \infty)$

(iv) Exactly one root lies in the interval (1, 2)

$D > 0$

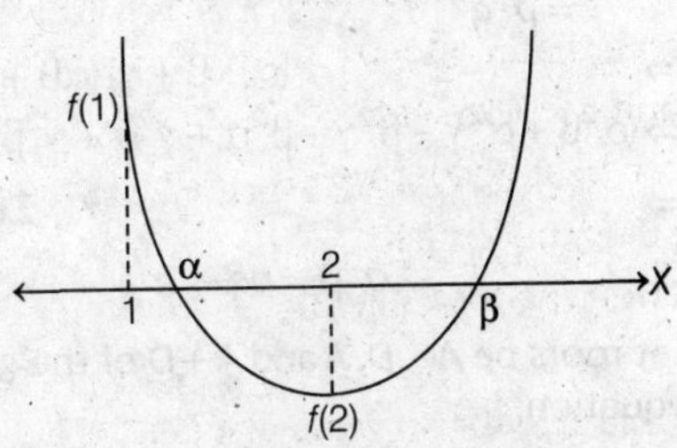

i.e. $(m-1)(m-9)>0$

$\therefore \quad m \in (-\infty, 1) \cup (9, \infty)$ ... (i)

$f(1)\, f(2) < 0$

$\{1-(m-3)+m\}\{4-2(m-3)+m\} < 0$

$\Rightarrow \quad 4(-m+10) < 0$

$\Rightarrow \quad m-10 > 0$

$\Rightarrow \quad m > 10$

$\therefore \quad m \in (10, \infty)$ ... (ii)

On combining Eqs. (i) and (ii), we get

$m \in (10, \infty)$

17. Let $f(x) = ax^2 + bx + c$

$f(0) = c < 0$

Since, $f(x)$ has no real root, the graph of $f(x)$ must open downward and should not cross the X-axis.

$\Rightarrow f(x) = ax^2 + bx + c < 0, \forall\, x \in R$

Hence, $f(-1) = a - b + c < 0$

and $\quad f(3) = 9a + 3b + c < 0$

18. Consider,

$f(x) = (x-a)(x-c) + 2(x-b)(x-d)$

$f(a) = 2(a-b)(a-d) > 0$

$f(b) = (b-a)(b-c) < 0$

$f(c) = 2(c-b)(c-d) < 0$

$f(d) = (d-a)(d-c) > 0$

We see that $f(x)$ changes sign between a and b and again between c and d.

Hence, $f(x) = 0$ has one real between a and b and other between c and d.

19. Let α, β and γ are roots of $x^3 - px^2 + qx - r = 0$

$\therefore \quad \alpha + \beta + \gamma = p$

$(\alpha\beta)(\beta\gamma)(\gamma\alpha) = q$

and $\quad \alpha\beta\gamma = r$

(i) $\Sigma\alpha^2 = \Sigma\alpha \cdot \Sigma\alpha$

$= p \cdot p = p^2$

$\Rightarrow \quad (\alpha+\beta+\gamma)(\alpha+\beta+\gamma) = p^2$

$\Rightarrow \alpha^2 + \beta^2 + \gamma^2 + 2(\alpha\beta + \beta\gamma + \alpha\gamma) = p^2$

$\Rightarrow \quad \alpha^2 + \beta^2 + \gamma^2 + 2(q) = p^2$

$\Rightarrow \quad \alpha^2 + \beta^2 + \gamma^2 = p^2 - 2q$

$\Rightarrow \quad \Sigma\alpha^2 = p^2 - 2q$

(ii) $\Sigma\alpha^2\beta = \Sigma\alpha \cdot \Sigma\alpha\beta$

$= p \cdot q$

$\Rightarrow \quad (\alpha+\beta+\gamma)(\alpha\beta+\beta\gamma+\alpha\gamma) = pq$

$\Rightarrow (\alpha^2\beta + \alpha^2\gamma + \beta^2\gamma + \beta^2\alpha + \gamma^2\alpha + \gamma^2\beta) + 3\alpha\beta\gamma = pq$

$\Rightarrow \quad \Sigma\alpha^2\beta + 3r = pq$

$\Sigma\alpha^2\beta = pq - 3r$

20. (i) Let roots be $A-D, A$ and $A+D$ of the given equation.

$\therefore \quad A - D + A + A + D = p$

$\Rightarrow \quad A = \dfrac{p}{3}$

Now, A is the root of the given equation.

Then, $\quad A^3 - pA^2 + qA - r = 0$

$\Rightarrow \left(\dfrac{p}{3}\right)^3 - p\left(\dfrac{p}{3}\right)^2 + q\left(\dfrac{p}{3}\right) - r = 0$

$\Rightarrow \quad p^3 - 3p^3 + 9qp - 27r = 0$

or $\quad 2p^3 - 9pq + 27r = 0$

which is the required condition.

(ii) Let roots of the given equation are $\dfrac{A}{R}$, A and AR, then

$\dfrac{A}{R} \cdot A \cdot AR = (-1)^3 \left(\dfrac{-r}{1}\right) = r$

$\Rightarrow \quad A^3 = r$

$\Rightarrow \quad A = (r)^{1/3}$

Now, A is the root of the given equation, then

$A^3 - pA^2 + qA - r = 0$

$\Rightarrow \quad r - pr^{2/3} + qr^{1/3} - r = 0$

$\Rightarrow \quad p(r)^{2/3} = q(r)^{1/3}$

$\Rightarrow \quad p^3r^2 = q^3r$

$\Rightarrow \quad p^3r = q^3$

which is required condition.

21. Given that

$\sqrt{x^2 - 3x} = 4x^2 - 12x - 3$

$\Rightarrow \quad \sqrt{x^2 - 3x} = 4(x^2 - 3x) - 3$

Let $\quad x^2 - 3x = y$ ...(i)

$\Rightarrow \quad \sqrt{y} = 4y - 3$ ...(ii)

On squaring both sides, we get

$y = (4y-3)^2$

$\Rightarrow \quad y = 16y^2 - 24y + 9$

$\Rightarrow \quad 16y^2 - 25y + 9 = 0$

$\Rightarrow \quad (y-1)(16y-9) = 0$

$\Rightarrow \quad y - 1 = 0$ or $16y - 9 = 0$

$y = 1$ or $y = \dfrac{9}{16}$

Only $y = 1$ satisfies the Eq. (ii), so neglecting $y = \dfrac{9}{16}$.

Putting $y = 1$ in Eq. (i), we get

$x^2 - 3x = 1$

$\Rightarrow \quad x^2 - 3x - 1 = 0$

$\Rightarrow \quad x = \dfrac{3 \pm \sqrt{13}}{2}$

22. The given inequation is equivalent to

$$\frac{3}{x+1}+\frac{7}{x+2}<\frac{6}{x-1}$$

$$\Rightarrow \quad \frac{4x^2-15x-25}{(x+1)(x+2)(x-1)}<0$$

$$\Rightarrow \quad \frac{\left(x+\frac{5}{4}\right)(x-5)}{(x+1)(x+2)(x-1)}<0$$

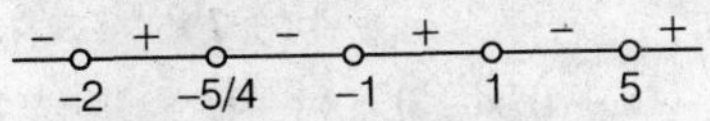

$$\Rightarrow \quad x \in (-\infty,-2) \cup \left(\frac{-5}{4},1\right) \cup (1,5)$$

23. We have, $\sqrt{(6-x)}\,(3^{x^2-7.2x+3.9}-9\sqrt{3})=0$

The equation is defined for $6-x \geq 0$ i.e. $x \leq 6$ ... (i)

The equation is equivalent to the collection of equations

$$\sqrt{6-x}=0 \text{ and } 3^{x^2-7.2x+3.9}-9\sqrt{3}=0$$

$$\therefore \quad x_1=6 \text{ and } 3^{x^2-7.2x+3.9}=3^{2.5}$$

$$x^2-7.2x+3.9=2.5$$

$$x^2-7.2x+1.4=0$$

We find that $x_2=\frac{1}{5}$ and $x_3=7$

Hence, solution of the original equation are (which satisfies Eq. (i))

$$x_1=6 \text{ and } x_2=\frac{1}{5}$$

24. The equation is equivalent to

$$2^{x^2+7x+10}=2^0$$

$$\Rightarrow \quad x^2+7x+10=0$$

$$\Rightarrow \quad (x+5)(x+2)=0$$

$$\Rightarrow \quad x=-5,-2$$

25. We have, $3^{x+2}>(3^{-2})^{1/x}$

$$\Rightarrow \quad 3^{x+2}>3^{-2/x}$$

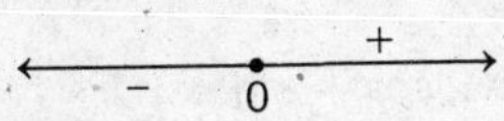

Here, base $3>1$

$$\Rightarrow \quad x+2>-\frac{2}{x}$$

$$\Rightarrow \quad \frac{x^2+2x+2}{x}>0 \Rightarrow \frac{(x+1)^2+1}{x}>0$$

$$\Rightarrow \quad x>0$$

$$\therefore \quad x \in (0,\infty)$$

26. Let $\log_5 x = y$

$$\Rightarrow \quad y^2+y-2<0$$

$$\Rightarrow \quad (y-1)(y+2)<0$$

$$\Rightarrow \quad -2<y<1$$

$$\Rightarrow \quad -2\log_5 5<\log_5 x<\log_5 5 \qquad [\because \log_a a=1]$$

$$\Rightarrow \quad \log_5(5^{-2})<\log_5 x<\log_5 5$$

$$\Rightarrow \quad \log_5\left(\frac{1}{25}\right)<\log_5 x<\log_5 5$$

$$\Rightarrow \quad \frac{1}{25}<x<5$$

$$\Rightarrow \quad x \in \left(\frac{1}{25},5\right)$$

27. Given that

$$2^{2x}-18\cdot 2^x+32=0$$

$$\Rightarrow \quad (2^x)^2-18\cdot 2^x+32=0$$

Let $2^7=y$

$$y^2-18y+32=0$$

$$\Rightarrow \quad (y-16)(y-2)=0$$

$$\Rightarrow \quad y=2 \text{ or } y=16$$

$$\Rightarrow \quad 2^x=2 \text{ or } 2^x=16=2^4$$

$$\Rightarrow \quad x=1 \text{ or } x=4$$

Hence, roots of the original equation are 1 and 4.

Level 2

1. (c) Given, $x^2-3|x|+2=0$

$$\Rightarrow \quad |x|^2-3|x|+2=0$$

$$\Rightarrow \quad (|x|-2)(|x|-1)=0$$

$$\Rightarrow \quad |x|=1,2 \text{ or } x=\pm 1,\pm 2$$

$\therefore$ Number of solutions = 4

2. (d) Let α and $\alpha+1$ be roots.

$$\Rightarrow \quad \alpha+\alpha+1=b$$

and $\quad \alpha(\alpha+1)=c$

$$\therefore \quad b^2-4ac=(2\alpha+1)^2-4\alpha(\alpha+1)$$

$$=1$$

3. (d) **Case I** $x \geq 0$

$\therefore$ The equation becomes $x^2+5x+4=0$

or $x=-1,-4$ but $x \geq 0$

$\therefore$ Both values are non-admissible.

Case II $x \leq 0$

The equation becomes $x^2-5x+4=0$ or $x=1,4$

Both values are non-admissible.

$\therefore$ No real roots

4. (c) We have, $2\sqrt{5-4x-x^2}=16-|x|$

$$\Rightarrow \quad 4(5-4x-x^2)=256-32|x|+x^2$$

$$\Rightarrow 5x^2-32|x|+16x+236=0$$

$\Rightarrow \quad 5x^2-16x+236=0$, when $x \geq 0$

and $5x^2+48x+236=0$, when $x<0$

In both cases, equation has non-real roots.

5. (d) We have, $(x+a)(x+1991)+1=0$

or $(x+a)(x+1991)=-1$

$\Rightarrow$ $(x+a)=1$ and $x+1991=-1$

$\Rightarrow$ $a=1993$

or $x+a=-1$ and $x+1991=1$

$\Rightarrow$ $a=1989$

6. (a) We have, $x^2+2x-n=0$

$\Rightarrow$ $(x+1)^2=n+1$

$\Rightarrow$ $x=-1\pm\sqrt{n+1}$

Thus, $(n+1)$ should be a perfect square.

Now, $n\in[5,100]$

$\Rightarrow n+1\in[6,101]$

Perfect square values of $(n+1)$ are

9, 16, 25, 36, 49, 64, 81 and 100

Hence, number of values is 8.

7. (a) If α and β are the roots of the equation

$$ax^2+bx+c=a(x-\alpha)(x-\beta)$$

$\therefore$ $(x-a)(x-b)-k=(x-c)(x-d)$

$\Rightarrow$ $(x-c)(x-d)+k=(x-a)(x-b)$

$\therefore$ The roots of $(x-c)(x-d)+k=0$ are a and b.

8. (d) We have,

$(x^2-5x+5)^{x^2+4x-60}=1$

Case I $x^2+4x-60=0$

$\Rightarrow$ $(x-6)(x+10)=0$

$\therefore$ $x=6,-10$

Case II $x^2-5x+5=1$

$\Rightarrow$ $x^2-5x+4=0$

$\Rightarrow$ $(x-1)(x-4)=0$

$\therefore$ $x=1,4$

Case III $x^2-5x+5=-1$ and $x^2-4x-60$ is even number.

$\therefore$ $x^2-5x+6=0$

$\Rightarrow$ $(x-2)(x-3)=0$

$\therefore$ $x=2,3$

But for $x=3$, $x^2-4x-60$ is odd.

Hence, $x=2$ only acceptable solution in this case.

Thus, solutions are $6,-10,1,4,2$.

$\therefore$ Sum of solutions is 3.

9. (a) Let $y=\sqrt{ab+(a-b)y}$

$\Rightarrow$ $y^2=ab+(a-b)y$

$\Rightarrow$ $y^2-(a-b)y-ab=0$

$\Rightarrow$ $y(y-a)+b(y-a)=0$

$\Rightarrow$ $(y-a)(y+b)=0$

$\Rightarrow$ $y=a$ or $y=-b$

As, $y>0\Rightarrow y=a$

10. (a) Let $x=3+\cfrac{1}{4+\cfrac{1}{3+\cfrac{1}{4+\cfrac{1}{3+\ldots\infty}}}}$

So, $x=3+\dfrac{1}{4+\dfrac{1}{x}}=3+\dfrac{1}{\dfrac{4x+1}{x}}=3+\dfrac{x}{4x+1}$

$\Rightarrow$ $(x-3)=\dfrac{x}{4x+1}$

$\Rightarrow$ $(4x+1)(x-3)=x$

$\Rightarrow$ $4x^2-12x+x-3=x$

$\Rightarrow$ $4x^2-12x-3=0$

$\Rightarrow$ $x=\dfrac{3\pm2\sqrt3}{2}$

$\Rightarrow$ $x=\dfrac{3}{2}\pm\sqrt3=1.5\pm\sqrt3$

But from above, $x>0$

$\therefore$ Only positive value of x is accepted.

$\therefore x=1.5+\sqrt3$

11. (b) We have for any real x,

$$\cos^2\frac{x}{2}\le1 \text{ and } \sin^2x\le1.$$

$\therefore$ $2\cos^2\dfrac{x}{2}\sin^2x\le2$

$\Rightarrow$ LHS ≤2

Again for any real x,

$$x^2+\frac{1}{x^2}=x^2+\frac{1}{x^2}-2+2=\left(x-\frac{1}{x}\right)^2+2\ge2$$

$\therefore$ RHS ≥2

Thus, given equation have solution only

LHS = RHS = 2

$\Rightarrow 2\cos^2\dfrac{x}{2}\sin^2x=x^2+\dfrac{1}{x^2}=2$

Now, $x^2+\dfrac{1}{x^2}=2\Rightarrow x=\pm1$

But for $x=\pm1$, $\cos^2\dfrac{x}{2}\sin^2x\ne1$

i.e. LHS and RHS cannot be simultaneously equal to 2 for any value of x.

$\therefore$ The equation has no real solution.

12. (b) The given equation is

$(x-b)(x-c)+(x-a)(x-c)+(x-a)(x-b)=0$

$\Rightarrow$ $3x^2-2(a+b+c)x+(ab+bc+ca)=0$

Discriminant $=4(a+b+c)^2-12(ab+bc+ca)$

$=4[a^2+b^2+c^2-ab-bc-ca]$

$=2[(a-b)^2+(b-c)^2+(c-a)^2]\ge0,\forall\, a,b,c$

$\therefore$ Roots of given equation are always real.

13. (a) We have,

$x(x+1)+(x+1)(x+2)+\ldots+(x+\overline{n-1})(x+n)=10n$

Roots of above equation are consecutive integers.

Let roots be α and $\alpha+1$

$\Rightarrow \quad \alpha(\alpha+1)+(\alpha+1)(\alpha+2)+\ldots+(\alpha+\overline{n-1})(\alpha+n)$
$= 10n \ldots$ (i)

and $(\alpha+1)(\alpha+2)+(\alpha+2)(\alpha+3)+\ldots+$
$(\alpha+n)(\alpha+n+1)=10n \ldots$(ii)

Subtracting Eq. (ii) from Eq. (i), we get

$$\alpha(\alpha+1)-(\alpha+n)(\alpha+n+1)=0$$

$$\Rightarrow \alpha^2+\alpha-\alpha^2-(2n+1)\alpha-n(n+1)=0$$

$$\Rightarrow \quad \alpha=-\frac{n+1}{2}$$

Putting this value in the original equation

i.e. $nx^2+x\sum_{r=1}^{n}(2r-1)+\sum_{r=1}^{n}(r-1)r=10n$, we get

$$\frac{n(n+1)^2}{4}-\frac{(n+1)}{2}n^2+\frac{n(n+1)(2n+1)}{6}-\frac{n(n+1)}{2}=10n$$

$$\Rightarrow \quad 3(n+1)^2-6n(n+1)+2(n+1)(2n+1)-6(n+1)=120$$

$$\Rightarrow \quad n^2=121$$

$$\Rightarrow \quad n=11$$

14. (c) For the equation $px^2+qx+1=0$ to have real roots

$$D\geq 0$$

$$\Rightarrow \quad q^2\geq 4p$$

If $p=1$, then $q^2\geq 4$

$\Rightarrow \quad q=2,3,4$

If $p=2$, then $q^2\geq 8$

$\Rightarrow \quad q=3,4$

If $p=3$, then $q^2\geq 12 \Rightarrow q=4$

If $p=4$, then $q^2\geq 16 \Rightarrow q=4$

$\therefore$ Number of required equations $=7$

15. (b) For real roots, $q^2-4pr\geq 0$

$\Rightarrow \quad \left(\frac{p+r}{2}\right)^2-4pr\geq 0 \quad$ [$\because p, q$ and r are in AP]

$$\Rightarrow \quad p^2+r^2-14pr\geq 0$$

$$\Rightarrow \quad \frac{p^2}{r^2}-14\frac{p}{r}+1\geq 0$$

$$\Rightarrow \quad \left(\frac{p}{r}-7\right)^2-48\geq 0$$

$$\Rightarrow \quad \left|\frac{p}{r}-7\right|\geq 4\sqrt{3}$$

16. (b) Let two numbers be a and b.

$\therefore$ Equation with roots a and b is

$$x^2-(a+b)x+ab=0$$

$$\Rightarrow \quad x^2-18x+16=0$$

17. (c) $x^2+3^{\frac{1}{4}}x+3^{\frac{1}{2}}=0$

$$\therefore \quad x=\frac{-3^{1/4}\pm\sqrt{3^{1/2}-4\cdot 3^{1/2}}}{2}$$

$$=\frac{3^{1/4}(-1\pm\sqrt{3}i)}{2}$$

$$=3^{1/4}\left(\frac{-1+\sqrt{3}i}{2}\right) \text{ or } 3^{1/4}\left(\frac{-1-\sqrt{3}i}{2}\right)$$

$$=3^{1/4}\omega, \text{ or } 3^{1/4}\omega^2$$

Now, $\alpha^{96}(\alpha^{12}-1)+\beta^{96}(\beta^{12}-1)$

$$=\alpha^{108}-\alpha^{96}+\beta^{108}-\beta^{96}$$

$$=(\alpha^{108}+\beta^{108})-(\alpha^{96}+\beta^{96})$$

$$=\{(3^{1/4}\omega)^{108}+(3^{1/4}\omega^2)^{108}\}$$
$$-\{(3^{1/4}\omega)^{96}+(3^{1/4}\omega^2)^{96}\}$$

$$=3^{27}(\omega^{108}+\omega^{216})-3^{24}(\omega^{96}+\omega^{192})$$

$$=3^{27}(2)-3^{24}(2)=3^{24}(54)-3^{24}(2)$$

$$=3^{24}(52)=52\times 3^{24}$$

18. (c) Since, α and β are the roots of the equation

$$5x^2+6x-2=0$$

Then, $5\alpha^2+6\alpha-2=0$ and $5\beta^2+6\beta-2=0$

$$\Rightarrow \quad 5\alpha^2+6\alpha=2 \text{ and } 5\beta^2+6\beta=2$$

$$5S_6+6S_5=5(\alpha^6+\beta^6)+6(\alpha^5+\beta^5)$$

$$=(5\alpha^6+6\alpha^5)+(5\beta^6+6\beta^5)$$

$$=\alpha^4(5\alpha^2+6\alpha)+\beta^4(5\beta^2+6\beta)$$

$$=2(\alpha^4+\beta^4)=2S_4$$

19. (a) We have,

$$\alpha^2=5\alpha-3 \text{ and } \beta^2=5\beta-3$$

$\Rightarrow$ α and β are roots of equation,

$$x^2=5x-3 \text{ or } x^2-5x+3=0$$

$\therefore \quad \alpha+\beta=5$ and $\alpha\beta=3$

Thus, the equation having $\frac{\alpha}{\beta}$ and $\frac{\beta}{\alpha}$ as its roots is

$$x^2-x\left(\frac{\alpha}{\beta}+\frac{\beta}{\alpha}\right)+\frac{\alpha\beta}{\alpha\beta}=0$$

$$\Rightarrow \quad x^2-x\left(\frac{\alpha^2+\beta^2}{\alpha\beta}\right)+1=0$$

or $\quad 3x^2-19x+3=0$

20. (a) Let α,β and γ,δ be the roots of the equations $x^2+ax+b=0$ and $x^2+bx+a=0$, respectively.

$\therefore \alpha+\beta=-a, \alpha\beta=b$ and $\gamma+\delta=-b, \gamma\delta=a$

Given, $\quad |\alpha-\beta|=|\gamma-\delta|$

$\Rightarrow \quad (\alpha-\beta)^2=(\gamma-\delta)^2$

$$\Rightarrow (\alpha+\beta)^2 - 4\alpha\beta = (\gamma+\delta)^2 - 4\gamma\delta$$

$$\Rightarrow \quad a^2 - 4b = b^2 - 4a \Rightarrow (a^2 - b^2) + 4(a-b) = 0$$

$$\Rightarrow \quad a + b + 4 = 0 \quad [\because a \neq b]$$

21. (b) $3\alpha = \dfrac{1-3a}{a^2-5a+3}$ and $2\alpha^2 = \dfrac{2}{a^2-5a+3}$

$$\Rightarrow \quad 2\left[\frac{1}{9}\frac{(1-3a)^2}{(a^2-5a+3)^2}\right] = \frac{2}{a^2-5a+3}$$

$$\Rightarrow \quad \frac{(1-3a)^2}{(a^2-5a+3)} = 9$$

or $\quad 9a^2 - 6a + 1 = 9a^2 - 45a + 27$

or $\quad 39a = 26$ or $a = \dfrac{2}{3}$

22. (d) $\alpha \cdot \beta = 2$ and $\alpha + \beta = -p$

Also, $\dfrac{1}{\alpha} + \dfrac{1}{\beta} = -q \Rightarrow p = 2q$

Now, $\left(\alpha - \dfrac{1}{\alpha}\right)\left(\beta - \dfrac{1}{\beta}\right)\left(\alpha + \dfrac{1}{\beta}\right)\left(\beta + \dfrac{1}{\alpha}\right)$

$$= \left[\alpha\beta + \frac{1}{\alpha\beta} - \frac{\alpha}{\beta} - \frac{\beta}{\alpha}\right]\left[\alpha\beta + \frac{1}{\alpha\beta} + 2\right]$$

$$= \frac{9}{2}\left[\frac{5}{2} - \frac{\alpha^2+\beta^2}{2}\right] = \frac{9}{4}[5 - (p^2 - 4)]$$

$$= \frac{9}{4}(9 - p^2) \quad [\because \alpha^2 + \beta^2 = (\alpha+\beta)^2 - 2\alpha\beta]$$

23. (b) Since, α is common root of $x^2 - x + 2\lambda = 0$ ...(i)

and $\quad 3x^2 - 10x + 27\lambda = 0$

$\therefore \quad 3\alpha^2 - 10\alpha + 27\lambda = 0$ ... (ii)

On multiplying Eq. (i) by 3 and replacing α in place of x,

$3\alpha^2 - 3\alpha + 6\lambda = 0$ ... (iii)

$\therefore$ On subtracting Eq. (iii) from Eq. (ii), we get

$$\alpha = 3\lambda$$

Now, $\alpha\beta = 2\lambda \Rightarrow 3\lambda \cdot \beta = 2\lambda$

$$\Rightarrow \quad \beta = \frac{2}{3}$$

$$\Rightarrow \quad \alpha + \beta = 1$$

$$\Rightarrow \quad 3\lambda + \frac{2}{3} = 1$$

$$\Rightarrow \quad \lambda = \frac{1}{9} \text{ and } \alpha\gamma = 9\lambda$$

$$\Rightarrow \quad 3\lambda \cdot \gamma = 9\lambda \Rightarrow \gamma = 3$$

$$\therefore \quad \frac{\beta\gamma}{\lambda} = 18$$

24. (a) We have, $x^2 - (a-2)x - a - 1 = 0$

$\Rightarrow \quad \alpha + \beta = a - 2$ and $\alpha\beta = -(a+1)$

$$\Rightarrow \quad \alpha^2 + \beta^2 = (\alpha+\beta)^2 - 2\alpha\beta$$

$$= a^2 - 2a + 6 = (a-1)^2 + 5$$

$$\Rightarrow \quad a = 1$$

25. (b) Given, $\tan\left(\dfrac{P}{2}\right)$ and $\tan\left(\dfrac{Q}{2}\right)$ are the roots of $ax^2 + bx + c = 0$

$\therefore \quad \tan\left(\dfrac{P}{2}\right) + \tan\left(\dfrac{Q}{2}\right) = \dfrac{-b}{a}$ and $\tan\left(\dfrac{P}{2}\right) \cdot \tan\left(\dfrac{Q}{2}\right) = \dfrac{c}{a}$

$$\frac{\tan\left(\frac{P}{2}\right) + \tan\left(\frac{Q}{2}\right)}{1 - \tan\left(\frac{P}{2}\right)\tan\left(\frac{Q}{2}\right)} = \tan\left(\frac{P}{2} + \frac{Q}{2}\right) = 1 \quad [\because P + Q = 90^\circ]$$

$$\Rightarrow \quad \frac{\frac{-b}{a}}{1 - \frac{c}{a}} = 1 \Rightarrow -\frac{b}{a} = \frac{a}{a} - \frac{c}{a}$$

$$\Rightarrow \quad -b = a - c \text{ or } c = a + b$$

26. (b) Here, $\alpha + \beta = \dfrac{-b}{a}, \alpha\beta = \dfrac{c}{a}$ and a, b, c are in GP.

$$\therefore \quad b^2 = ca \Rightarrow \left(\frac{b}{a}\right)^2 = \frac{c}{a}$$

or $\quad (\alpha+\beta)^2 = \alpha\beta$

$$\Rightarrow \quad \alpha^2 + \beta^2 + \alpha\beta = 0$$

$$\Rightarrow \left(\frac{\alpha}{\beta}\right)^2 + \left(\frac{\alpha}{\beta}\right) + 1 = 0$$

$$\therefore \quad \frac{\alpha}{\beta} = \frac{-1 \pm \sqrt{3}i}{2}$$

27. (a) If α and β are the roots, then

$$\alpha + \beta = \frac{-b}{a} \text{ and } \alpha\beta = \frac{c}{a}$$

Now, the roots of $lx^2 + mx + n = 0$ are $\dfrac{1}{\alpha}$ and $\dfrac{1}{\beta}$.

$$\therefore \quad \frac{1}{\alpha} + \frac{1}{\beta} = -\frac{m}{l} \text{ and } \frac{1}{\alpha} \cdot \frac{1}{\beta} = \frac{n}{l}$$

$$\Rightarrow \quad \frac{\alpha+\beta}{\alpha\beta} = -\frac{m}{l} \text{ and } \frac{a}{c} = \frac{n}{l}$$

$$\Rightarrow \quad -\frac{b}{c} = -\frac{m}{l} \text{ and } \frac{a}{c} = \frac{n}{l} \Rightarrow \frac{a}{n} = \frac{b}{m} = \frac{c}{l}$$

$$\therefore \quad a : b : c = n : m : l$$

28. (b) Let the correct equation is

$$ax^2 + bx + c = 0, \text{ then } \alpha + \beta = -\frac{b}{a} \text{ and } \alpha\beta = \frac{c}{a}$$

When b is written incorrectly, then the roots are equal say γ and γ

$$\therefore \quad \gamma + 2\gamma = \frac{-b}{a} \text{ and } \gamma^2 = \frac{c}{a} = \alpha\beta \quad \text{...(i)}$$

$$\Rightarrow \quad 3\gamma = \alpha + \beta$$

$$\Rightarrow \quad 9\gamma^2 = (\alpha+\beta)^2$$

$$\Rightarrow \quad 9\alpha\beta = (\alpha-\beta)^2 + 4\alpha\beta \quad \text{[from Eq. (i)]}$$

$$\therefore \quad (\alpha-\beta)^2 = 5\alpha\beta$$

29. (c) We have, $\alpha+\beta=\frac{-q}{p}$ and $\alpha\beta=\frac{r}{p}$

Now, the given equation

$$\alpha(x-\beta)^2+\beta(x-\alpha)^2=0$$

$$\Rightarrow (\alpha+\beta)x^2-4\alpha\beta x+\alpha\beta(\alpha+\beta)=0$$

$$\Rightarrow \left(-\frac{q}{p}\right)x^2-4\frac{r}{p}x+\frac{r}{p}\left(\frac{-q}{p}\right)=0$$

$$\Rightarrow \quad pqx^2+4prx+rq=0 \quad \text{... (i)}$$

Since, α and β have opposite signs, therefore p and r must have opposite signs.

$\Rightarrow pq$ and rq must have opposite signs.

$\therefore$ Roots of Eq. (i) have opposite signs.

30. (b) Given, $a^2x^2-x+1-a^2=0$

$$\alpha^2+\beta^2=\frac{1}{a^2} \text{ and } \alpha^2\beta^2=\frac{1-a^2}{a^2}$$

$$\Rightarrow \quad \alpha^2+\beta^2-1=\alpha^2\beta^2$$

$$\Rightarrow \quad (\alpha^2-1)(\beta^2-1)=0$$

$\because \ \beta^2\neq 1 \Rightarrow \alpha^2=1$

So, $\beta^2=\frac{1-a^2}{a^2}$

31. (d) If α and β are the roots, then

$$\alpha+\beta=\frac{-b}{a} \text{ and } \alpha\beta=\frac{c}{a}$$

$$\therefore \quad \frac{b^2-4ac}{a^2}=\left(\frac{b}{a}\right)^2-4\left(\frac{c}{a}\right)$$

$$=(\alpha+\beta)^2-4\alpha\beta=(\alpha-\beta)^2$$

If α and β increase by the same quantity, then $(\alpha-\beta)^2$ remain the same

i.e. $\frac{b^2-4ac}{a^2}$ is invariant.

32. (d) We have,

$$\frac{1}{\alpha}+\frac{1}{\beta}=4 \text{ and } 2q=p+r$$

$$\therefore \quad \frac{2q}{p}=1+\frac{r}{p}$$

$$\Rightarrow \quad -2(\alpha+\beta)=1+\alpha\beta$$

$$\Rightarrow \quad -2\left(\frac{1}{\alpha}+\frac{1}{\beta}\right)=\frac{1}{\alpha\beta}+1$$

$$\Rightarrow \quad \frac{1}{\alpha\beta}=-9$$

$$\therefore \quad \alpha\beta=-\frac{1}{9}$$

$$\therefore \quad \alpha+\beta=-\frac{4}{9}$$

Equation having roots α and β is $9x^2+4x-1=0$

$$\therefore \quad |\alpha-\beta|=\sqrt{(\alpha+\beta)^2-4\alpha\beta}$$

$$=\sqrt{\frac{16}{81}+\frac{4}{9}}=\frac{2\sqrt{13}}{9}$$

33. (a) We have, $x^2+px+q=0$

Let roots be α and α^2.

$$\Rightarrow \quad \alpha+\alpha^2=-p$$

and $\quad \alpha\alpha^2=q \Rightarrow \alpha=(q)^{1/3}$

$$\therefore \quad (q)^{1/3}+(q^{1/3})^2=-p$$

Taking cube of both sides, we get

$$q+q^2+3q(q^{1/3}+q^{2/3})=-p^3$$

$$\Rightarrow \quad q+q^2-3pq=-p^3$$

$$\Rightarrow \quad p^3+q^2-q(3p-1)=0$$

34. (c) $\because$ α and β are the roots of $x^2-6x-2=0$

$$\therefore \quad \alpha^2-6\alpha-2=0$$

$$\Rightarrow \quad \alpha^{10}-6\alpha^9-2\alpha^8=0$$

$$\Rightarrow \quad \alpha^{10}-2\alpha^8=6\alpha^9 \quad \text{... (i)}$$

Similarly, $\quad \beta^{10}-2\beta^8=6\beta^9 \quad$ ... (ii)

From Eqs. (i) and (ii),

$$\alpha^{10}-\beta^{10}-2(\alpha^8-\beta^8)=6(\alpha^9-\beta^9)$$

$$\Rightarrow \quad a_{10}-2a_8=6a_9$$

$$\Rightarrow \quad \frac{a_{10}-2a_8}{2a_9}=3$$

35. (a) Given, $(3x^2+4x+3)^2-(k+1)(3x^2+4x+3)$ $(3x^2+4x+2)+k(3x^2+4x+2)^2=0$

Let $\quad y=3x^2+4x+2$

Then, given equation becomes

$$(y+1)^2-(k+1)y(y+1)+ky^2=0$$

$$\Rightarrow \quad y^2+2y+1-ky^2-ky-y^2-y+ky^2=0$$

$$\Rightarrow \quad y+1-ky=0$$

$$\Rightarrow \quad y(1-k)=-1$$

$$\Rightarrow \quad y=\frac{1}{k-1}$$

$$\Rightarrow \quad 3x^2+4x+2-\frac{1}{k-1}=0$$

For real roots, $D\geq 0$

$$\Rightarrow \quad 16-4\cdot 3\cdot\left(2-\frac{1}{k-1}\right)\geq 0$$

$$\Rightarrow \quad -8+\frac{12}{k-1}\geq 0 \Rightarrow \frac{3}{k-1}\geq 2$$

$$\Rightarrow \quad \frac{3-2k+2}{k-1}\geq 0 \Rightarrow \frac{2k-5}{k-1}\leq 0$$

$$\Rightarrow \quad k\in\left(1,\frac{5}{2}\right] \quad [\because k\neq 1]$$

36. (c) Let α and β are roots of the equation $x^2 + ax + 1 = 0$

$\therefore \alpha + \beta = -a$ and $\alpha\beta = 1$

Given, $|\alpha - \beta| < \sqrt{5}$

$\Rightarrow \sqrt{(\alpha+\beta)^2 - 4\alpha\beta} < \sqrt{5}$ $[\because (\alpha-\beta)^2 = (\alpha+\beta)^2 - 4\alpha\beta]$

$\Rightarrow \sqrt{a^2 - 4} < \sqrt{5}$

$\Rightarrow a^2 - 4 < 5$

$\Rightarrow a^2 - 9 < 0$

$\Rightarrow a^2 < 9$

$\Rightarrow -3 < a < 3$

$\Rightarrow a \in (-3, 3)$

Also, $D \geq 0$

$\Rightarrow a^2 - 4 \geq 0$

$\Rightarrow a \leq -2$ or $a \geq 2$

$\therefore a \in (-3, 3)$

37. (b) Given that

$$\alpha + \beta = -p \text{ and } \alpha^3 + \beta^3 = q$$

3 (is missed to multiply in second terms)

$\Rightarrow (\alpha+\beta)^3 - 3\alpha\beta(\alpha+\beta) = q$

$\Rightarrow -p^3 - 3\alpha\beta(-p) = q$

$\Rightarrow \alpha\beta = \dfrac{p^3 + q}{3p}$

Now, for required quadratic equation,

$$\text{Sum of roots} = \frac{\alpha}{\beta} + \frac{\beta}{\alpha} = \frac{\alpha^2 + \beta^2}{\alpha\beta}$$

$$= \frac{(\alpha+\beta)^2 - 2\alpha\beta}{\alpha\beta}$$

$$= \frac{p^2 - 2\left(\dfrac{p^3+q}{3p}\right)}{\dfrac{p^3+q}{3p}}$$

$$= \frac{3p^3 - 2p^3 - 2q}{p^3 + q} = \frac{p^3 - 2q}{p^3 + q}$$

and product of roots $\dfrac{\alpha}{\beta} \cdot \dfrac{\beta}{\alpha} = 1$

$\therefore$ Required equation is

$$x^2 - \left(\frac{p^3 - 2q}{p^3 + q}\right)x + 1 = 0$$

or $(p^3 + q)x^2 - (p^3 - 2q)x + (p^3 + q) = 0$

38. (d) Given, α and β are the roots of $x^2 + px + q = 0$

$\therefore \alpha + \beta = -p$

and $\alpha\beta = q$ ...(i)

α^4 and β^4 are the roots of $x^2 - rx + s = 0$

$\Rightarrow \alpha^4 + \beta^4 = r$ and $\alpha^4\beta^4 = s$ ... (ii)

Now, $x^2 - 4qx + 2q^2 - r = 0$

$\Rightarrow (x - 2q)^2 - 4q^2 + 2q^2 - r = 0$

$\Rightarrow (x - 2q)^2 = 2q^2 + r$

$\Rightarrow (x - 2\alpha\beta)^2 = 2\alpha^2\beta^2 + \alpha^4 + \beta^4$

[from Eqs. (i) and (ii)]

$\Rightarrow (x - 2\alpha\beta)^2 = (\alpha^2 + \beta^2)^2$

$\Rightarrow x = 2\alpha\beta \pm (\alpha^2 + \beta^2)$

$\Rightarrow x = \alpha^2 + \beta^2 + 2\alpha\beta, -\alpha^2 - \beta^2 + 2\alpha\beta$

$\Rightarrow x = (\alpha+\beta)^2, -(\alpha-\beta)^2$

Hence, the equation has one positive and one negative root.

39. (b) Given, p, q are positive numbers.

$p + q = 2$ ...(i)

$p^4 + q^4 = 272$

$\Rightarrow (p^2 + q^2)^2 - 2p^2q^2 = 272$

$\Rightarrow [(p+q)^2 - 2pq]^2 - 2p^2q^2 = 272$

$\Rightarrow [(2)^2 - 2pq]^2 - 2p^2q^2 = 272$ [from Eq (i)]

$\Rightarrow (4 - 2pq)^2 - 2p^2q^2 = 272$

$\Rightarrow 16 + 4p^2q^2 - 16pq - 2p^2q^2 = 272$

$\Rightarrow 2p^2q^2 - 16pq - 256 = 0$

$\Rightarrow p^2q^2 - 8pq - 128 = 0$

$$\therefore pq = \frac{8 \pm \sqrt{64 + 4 \times 128}}{2 \times 1}$$

$$= \frac{8 \pm \sqrt{576}}{2} = \frac{8 \pm 24}{2}$$

$\therefore pq = 16, -8$

Here, $pq = -8$ is not possible as p and q are positive.

$\therefore pq = 16$

Now, the equation whose roots are p and q is

$x^2 - (p+q)x + pq = 0$

$\Rightarrow x^2 - 2x + 16 = 0$

40. (a) Discriminant of the equation $2a^2x^2 - 2abx + b^2 = 0$ is $-4a^2b^2 < 0$ and that of the equation $p^2x^2 + 3pqx + q^2 = 0$ is $5p^2q^2 > 0$

There cannot be any common root.

41. (c) Let α be the common root of given equations, then $a\alpha^2 + 2c\alpha + b = 0$ and $a\alpha^2 + 2b\alpha + c = 0$

Subtracting, we get

$2\alpha(c - b) + (b - c) = 0$

$\Rightarrow \alpha = \dfrac{1}{2}$ $[b \neq c]$

Putting $\alpha = \dfrac{1}{2}$ in $a\alpha^2 + 2c\alpha + b = 0$, we get

$a + 4b + 4c = 0$

42. (a) $x^2+2x+3=0$

$$D=2^2-4\cdot1\cdot3<0$$

i.e. both roots are complex.

Hence, both roots are common of $x^2+2x+3=0$ and $ax^2+bx+c=0$

$$\therefore \quad \frac{a}{1}=\frac{b}{2}=\frac{c}{3}$$

43. (a) Let the roots of three equations be (α,β), (β,γ) and (γ,δ), then substituting β in first two equations, we get

$$\beta^2+p\beta+qr=0 \text{ and } \beta^2+q\beta+rp=0$$

Subtracting, we get

$$\beta(p-q)+r(q-p)=0$$

$$\Rightarrow \quad \beta=r \text{ if } p\neq q$$

Similarly, we will get

$$\alpha=q \text{ and } \gamma=p$$

$$\therefore \quad \frac{\alpha+\beta+\gamma}{\alpha\beta\gamma}=\frac{p+q+r}{pqr}=\frac{\Sigma p}{pqr}$$

44. (d) Discriminant of equation

$$3x^2+8x+15=0$$

i.e. $D=64-180=-116<0$

So, its roots are imaginary and therefore conjugate of each other, therefore one common root means both the roots are common.

$$\therefore \quad \frac{a}{3}=\frac{2b}{8}=\frac{3c}{15}$$

$$\Rightarrow \quad \frac{a}{3}=\frac{b}{4}=\frac{c}{5}=k \text{ (say)}, k\neq0$$

$$\Rightarrow \quad a=3k, b=4k \text{ and } c=5k$$

Now, $a^2+b^2=c^2$

$\Rightarrow \Delta ABC$ is right angled.

$\therefore \ \sin^2A+\sin^2B=\sin^2C$

$\Rightarrow \ \sin^2A+\sin^2B+\sin^2C$

$$=2\sin^2C=2\sin^2 90°=2$$

45. (a) $(6k+2)x^2+4x+3k-1=0$

$(12k+4)x^2+px+6k-2=0$ have both roots common.

So, $\dfrac{6k+2}{12k+4}=\dfrac{r}{p}=\dfrac{3k-1}{6k-2}$

$$\Rightarrow \quad \frac{r}{p}=\frac{1}{2}$$

$\Rightarrow \quad 2r-p=0$

46. (b) Let α be the common root of given equations, then

$$\alpha^2+b\alpha-1=0 \quad \text{...(i)}$$

and

$$\alpha^2+\alpha+b=0 \quad \text{...(ii)}$$

Subtracting Eq. (ii) from Eq. (i), we get

$$(b-1)\alpha-(b+1)=0$$

or

$$\alpha=\frac{b+1}{b-1}$$

Substituting this value of α in Eq. (i), we get

$$\left(\frac{b+1}{b-1}\right)^2+b\left(\frac{b+1}{b-1}\right)-1=0$$

or

$$b^3+3b=0$$

$$\Rightarrow \quad b=0, i\sqrt3, -i\sqrt3$$

47. (d) Let the roots of the equation $x^2-6x+a=0$ be α and 4β and that of the equation $x^2-cx+6=0$ be α and 3β, then $\alpha+4\beta=6$, $4\alpha\beta=a$

and $\quad \alpha+3\beta=c, 3\alpha\beta=6$

$$\Rightarrow \quad a=8$$

$\therefore$ Equation becomes $x^2-6x+8=0$

$$\Rightarrow \quad (x-2)(x-4)=0$$

$\Rightarrow$ Roots are 2 and 4

$$\Rightarrow \quad \alpha=2 \text{ and } \beta=1$$

$\therefore$ Common root is 2.

48. (c) Given, $bx^2+cx+a=0$

Since, the roots are imaginary.

$$\therefore \quad c^2-4ab<0$$

$$\Rightarrow \quad c^2<4ab \Rightarrow -c^2>-4ab$$

Now consider, $3b^2x^2+6bcx+2c^2$

Since, $3b^2>0$, the given expression has a minimum value.

So, minimum value $=\dfrac{4(3b^2)(2c^2)-36b^2c^2}{4(3b^2)}$

$$=-\frac{12b^2c^2}{12b^2}=-c^2>-4ab$$

49. (b) Let $y=\dfrac{3x^2+9x+17}{3x^2+9x+7}$

$$3x^2(y-1)+9x(y-1)+7y-17=0$$

$$D\geq0 \quad [\because x \text{ is real}]$$

$$81(y-1)^2-4\times3(y-1)(7y-17)\geq0$$

$$\Rightarrow \quad (y-1)(y-41)\leq0$$

$$\Rightarrow \quad 1\leq y\leq41$$

50. (a) We have,

$$y=5x^2+2x+3=5\left(x+\frac{1}{5}\right)^2+\frac{14}{5}>2, \forall x\in R$$

While $y=2\sin x\leq2, \forall x\in R$

$\Rightarrow$ Two curves do not meet at all.

51. (c) Let $y=\dfrac{x^2+14x+9}{x^2+2x+3}$

$$\Rightarrow \quad (y-1)x^2+2(y-7)x+3(y-3)=0 \ x\in R$$

$$\Rightarrow \quad 4(y-7)^2-4\cdot3(y-1)(y-3)\geq0$$

$\Rightarrow \quad -2y^2 - 2y + 40 \geq 0$

$\Rightarrow \quad y^2 + y - 20 < 0$

$\therefore \quad -5 \leq y \leq 4$

52. (b) For $x < -2, |x+2| = -(x+2)$

and the inequality becomes

$$x^2 + x + 2 + x > 0$$

$\Rightarrow \quad (x+1)^2 + 1 > 0$

which is valid $\forall x \in R$ but $x < -2$

$\therefore \quad x \in (-\infty, -2)$ ...(i)

For $x \geq 2, |x+2| = x+2$

and the inequality becomes

$$x^2 - x - 2 + x > 0$$

$\Rightarrow \quad x^2 > 2$

$\Rightarrow \quad x > \sqrt{2}$ or $x < -\sqrt{2}$

i.e. $x \in (-\infty, -\sqrt{2}) \cup (\sqrt{2}, \infty)$

But $x \geq -2$

$\Rightarrow \quad x \in [(-2, -\sqrt{2}) \cup (\sqrt{2}, \infty)$ ... (ii)

From Eqs. (i) and (ii),

$$x \in (-\infty, -2) \cup (-2, -\sqrt{2}) \cup (\sqrt{2}, \infty)$$

$\Rightarrow \quad x \in (-\infty, -\sqrt{2}) \cup (\sqrt{2}, \infty)$

53. (c) If $m = 0$, then we get $1 > 0$, which is true if $m \neq 0$, then we have,

$$m > 0 \text{ and } (-9m)^2 - 4m(5m+1) < 0$$

$\Rightarrow \quad 0 < m < \frac{4}{61}$

$\therefore \quad m \in \left(0, \frac{4}{61}\right)$

54. (a) $\because a, b$ and c are sides of a triangle and $a \neq b \neq c$

$\therefore \quad |a-b| < |c|$

$\Rightarrow \quad a^2 + b^2 - 2ab < c^2$

Similarly, we have

$$b^2 + c^2 - 2bc < a^2, c^2 + a^2 - 2ca < b^2$$

On adding, we get

$$a^2 + b^2 + c^2 < 2(ab + bc + ca)$$

$\Rightarrow \quad \frac{a^2 + b^2 + c^2}{ab + bc + ca} < 2$ ...(i)

$\because$ Roots of the given equation are real.

$\therefore (a+b+c)^2 - 3\lambda(ab + bc + ca) \geq 0$

$\Rightarrow \quad \frac{a^2 + b^2 + c^2}{ab + bc + ca} \geq 3\lambda - 2$ ...(ii)

From Eqs. (i) and (ii), we get

$$3\lambda - 2 < 2$$

$\Rightarrow \quad \lambda < \frac{4}{3}$

55. (c) Given equation

$$x^2 - 2mx + m^2 - 1 = 0$$

$\Rightarrow \quad (x-m)^2 - 1 = 0$

$\Rightarrow \quad x = m-1, m+1$

$$m - 1 > -2 \text{ and } m + 1 < 4$$

$\Rightarrow \quad m > -1$ and $m < 3$

or $\quad -1 < m < 3$

56. (b) Given, $c < 0 < b$ and $\alpha + \beta = -b$ ...(i)

$$\alpha\beta = c$$ ...(ii)

From Eq. (ii), $c < 0 \Rightarrow \alpha\beta < 0$

Either α is negative or β is positive and second quantity is positive.

From Eq. (i), $b > 0$

$\Rightarrow \quad -b < 0 \Rightarrow \alpha + \beta < 0$

The sum is negative.

Modulus of negative quantity is > modulus of positive quantity but $\alpha < \beta$ is given, therefore it is clear that α is negative and β is positive and modulus of α is greater than modulus of β.

$\Rightarrow \quad \alpha < 0 < \beta < |\alpha|$

57. (d) Let $f(x) = x^2 + x + a$

$$D \geq 0$$

$\Rightarrow \quad 1 - 4a \geq 0 \Rightarrow a \leq \frac{1}{4}$ ...(i)

$$f(a) > 0$$

$\Rightarrow \quad a^2 + a + a > 0$

$\Rightarrow \quad a(a+2) > 0$

$\Rightarrow \quad a < -2$ or $a > 0$ ...(ii)

From Eqs. (i) and (ii), we get

$\Rightarrow \quad a < -2$

58. (c) Let $y = \frac{x^2 + 2x - 11}{x - 3}$

$\Rightarrow x^2 + (2-y)x + 3y - 11 = 0, x \in R$

So, $(2-y)^2 - 4(3y - 11) \geq 0$

$\therefore \quad y \leq 4$ or $y \geq 12$

59. (a) We have, $f(x) = \frac{(2a-1)x^2 + 2(a+1)x + (2a-1)}{x^2 - 2x + 40}$

Since, $x^2 - 2x + 40 > 0$ for all real x

$(2a-1)x^2 + 2(a+1)x + (2a-1) < 0 \, \forall x \in R$

$\therefore \quad 2a - 1 < 0$ or $a < \frac{1}{2}$ ...(i)

and $\quad D < 0$

$\Rightarrow 4(a+1)^2 - 4(2a-1)^2 < 0$

$\Rightarrow \quad a(a-2) > 0$

$\Rightarrow \quad a < 0$ or $a > 2$ ...(ii)

From Eqs. (i) and (ii), $a < 0$

60. (b) Here, $a<0$

Cut-off Y-axis, $x=0$

$\Rightarrow \quad y=c<0$ [from graph]

$\therefore \quad c<0$

x-coordinate of vertex >0

$\Rightarrow \quad \frac{-b}{2a}>0$

$\Rightarrow \quad \frac{b}{a}<0$

But $\quad a<0$

$\therefore \quad b>0$

and y-coordinate of vertex <0

$\Rightarrow \quad \frac{-D}{4a}<0$

$\Rightarrow \quad \frac{D}{4a}>0$

$\therefore \quad D<0$ $[\because a<0]$

i.e. $\quad b^2-4ac<0$

$\therefore \quad \frac{c}{a}>0$ $[\because c<0, a<0]$

61. (c) The given quadratic equation is

$$(\lambda^2+1)x^2-4\lambda x+2=0$$

$\because$ One root is in the interval (0, 1).

$\therefore \quad f(0)\,f(1)\le 0$

$\Rightarrow \quad 2(\lambda^2+1-4\lambda+2)\le 0$

$\Rightarrow \quad 2(\lambda^2-4\lambda+3)\le 0$

$(\lambda-1)(\lambda-3)\le 0$

$\Rightarrow \quad \lambda\in[1,3]$

But at $\lambda=1$, both roots are 1, so

$\lambda\ne 1$ [as both roots do not lie between (0, 1)]

Now, for $\lambda=3$, we get

$x=1, \frac{1}{5}$

$\therefore \quad \lambda\in(1,3]$

62. (d) If $f(\alpha)$ and $f(\beta)$ are of opposite sign, then there must lie a value γ between α and β such that $f(\gamma)=0$

a, b and c are real numbers and $a\ne 0$

As, α is a root of $a^2x^2+bx+c=0$

$\therefore \quad a^2\alpha^2+b\alpha+c=0$ ...(i)

Also, β is a root of $a^2x^2-bx-c=0$

$a^2\beta^2-b\beta-c=0$ ...(ii)

Now, let $f(x)=a^2x^2+2bx+2c$

Then, $f(\alpha)=a^2\alpha^2+2b\alpha+2c$

$=a^2\alpha^2+2(b\alpha+c)$

$=a^2\alpha^2+2(-a^2\alpha^2)$ [using Eq. (i)]

$=-a^2\alpha^2$

and $f(\beta)=a^2\beta^2+2b\beta+2c$

$=a^2\beta^2+2(b\beta+c)$

$=a^2\beta^2+2(a^2\beta^2)$ [using Eq. (ii)]

$=3a^2\beta^2>0$

Since, $f(\alpha)$ and $f(\beta)$ are of opposite signs and γ is a root of equation $f(x)=0$

$\therefore$ γ must lie between α and β.

63. (a) Clearly, $x^n-1=(x-1)(x-a_1)(x-a_2)\ldots(x-a_{n-1})$

$\Rightarrow \quad \frac{x^n-1}{x-1}=(x-a_1)(x-a_2)\ldots(x-a_{n-1})$

$\Rightarrow 1+x+x^2+\ldots x^{n-1}=(x-a_1)(x-a_2)\ldots(x-a_{n-1})$

$\Rightarrow \quad n=(1-a_1)(1-a_2)\ldots(1-a_{n-1})$

[putting $x=1$]

64. (c) Since, α, β, γ and δ are the roots of the given equation, we have

$$x^4+4x^3-6x^2+7x-9=(x-\alpha)(x-\beta)(x-\gamma)(x-\delta)$$

Putting $x=i$ and $x=-i$, we get

$$1-4i+6+7i-9=(i-\alpha)(i-\beta)(i-\gamma)(i-\delta)$$

and $1+4i+6-7i-9=(-i-\alpha)(-i-\beta)(-i-\gamma)(-i-\delta)$

Multiplying these two equations, we get

$$(-2+3i)(-2-3i)=(1+\alpha^2)(1+\beta^2)(1+\gamma^2)(1+\delta^2)$$

or $\quad 13=(1+\alpha^2)(1+\beta^2)(1+\gamma^2)(1+\delta^2)$

65. (a) $x=\sqrt[3]{7}+\sqrt[3]{49}$

$\Rightarrow x^3=7+49+3\sqrt[3]{7}\,\sqrt[3]{49}\,(\sqrt[3]{7}+\sqrt[3]{49})$

$=56+21x$

or $x^3-21x-56=0$

Therefore, the product of roots is 56.

66. (d) $P(a)=P(b)=P(c)=P(d)=3$

$\Rightarrow P(x)=3$ has a, b, c and d as its roots

$\Rightarrow P(x)-3=(x-a)(x-b)(x-c)(x-d)Q(x)$

[$\because$ $Q(x)$ has integral coefficient]

Given, $P(e)=5$, then

$$(e-a)(e-b)(e-c)(e-d)Q(e)=5-3=2$$

All factors are non-zero and five distinct factor of 2 is not possible. Thus, real value of e is not possible.

67. (c) As, (x^2+px+1) is a factor of ax^3+bx+c, we can assume that zeroes of x^2+px+1 are α, β and that of ax^3+bx+c be α, β, γ so that

$\alpha+\beta=-p$ ...(i)

$\alpha\beta=1$ ...(ii)

and $\quad \alpha+\beta+\gamma=0$ ...(iii)

$\alpha\beta+\beta\gamma+\gamma\alpha=\frac{b}{a}$ ...(iv)

$\alpha\beta\gamma=\frac{-c}{a}$ ...(v)

Solving Eqs. (ii) and (v), we get

$$\gamma = \frac{-c}{a}$$

Also, from Eqs. (i) and (iii), we get

$$\gamma = p$$

$$\therefore \quad p = \gamma = \frac{-c}{a}$$

Using Eqs. (i), (ii) and (iv), we get

$$1 + \gamma(-p) = \frac{b}{a}$$

$$\Rightarrow 1 + \left(-\frac{c}{a}\right)\left(\frac{c}{a}\right) = \frac{b}{a} \qquad \left[\text{using } \gamma = p = \frac{-c}{a}\right]$$

$$\Rightarrow \quad 1 - \frac{c^2}{a^2} = \frac{b}{a} \Rightarrow a^2 - c^2 = ab$$

68. (c) We have, $\alpha + \beta = -p$ and $\alpha\beta = q$ ...(i)

Also, since α and β are the roots of $x^{2n} + p^n x^n + q^n = 0$, we have,

$$\alpha^{2n} + p^n\alpha^n + q^n = 0 \text{ and } \beta^{2n} + p^n\beta^n + q^n = 0$$

Subtracting the above relations, we get

$$(\alpha^{2n} - \beta^{2n}) + p^n(\alpha^n - \beta^n) = 0$$

$$\therefore \quad \alpha^n + \beta^n = -p^n \qquad \text{...(ii)}$$

Given, α/β is a root of

$$x^n + 1 + (x+1)^n = 0$$

So, $(\alpha/\beta)^n + 1 + [(\alpha/\beta) + 1]^n = 0$

$$\Rightarrow \quad [\alpha^n + \beta^n] + (\alpha + \beta)^n = 0$$

$$\Rightarrow \quad -p^n + (-p)^n = 0 \quad \text{[using Eqs. (i) and (ii)]}$$

It is possible only when n is even.

69. (d) We have,

$$|\sqrt{x} - 2| + \sqrt{x}(\sqrt{x} - 4) + 2 = 0$$

$$\Rightarrow \quad |\sqrt{x} - 2| + x - 4\sqrt{x} + 2 + 2 = 2$$

$$\Rightarrow \quad |\sqrt{x} - 2| + (\sqrt{x} - 2)^2 = 2$$

$$\Rightarrow \quad y^2 + y - 2 = 0 \qquad [\text{let } y = |\sqrt{x} - 2|]$$

$$\Rightarrow \quad (y+2)(y-1) = 0$$

$$\Rightarrow \quad y = 1, -2$$

$$\therefore \quad y = 1 \qquad [\because y = |\sqrt{x} - 2| \geq 0]$$

$$\Rightarrow \quad |\sqrt{x} - 2| = 1$$

$$\Rightarrow \quad \sqrt{x} - 2 = \pm 1$$

$$\Rightarrow \quad \sqrt{x} = 3 \text{ or } 1$$

$$\Rightarrow \quad x = 9 \text{ or } 1$$

$\therefore$ Sum of roots $= 9 + 1 = 10$

70. (d) Given equation is

$$2|\sqrt{x} - 3| + x - 6\sqrt{x} + 6 = 0$$

or $\quad 2|\sqrt{x} - 3| + (\sqrt{x} - 3)^2 - 3 = 0$

Let $\quad |\sqrt{x} - 3| = y$

$$\therefore \quad y^2 + 2y - 3 = 0$$

$$\Rightarrow \quad (y-1)(y+3) = 0$$

$$\Rightarrow \quad y = 1, -3$$

$$\therefore \quad |\sqrt{x} - 3| = 1$$

$$\Rightarrow \quad \sqrt{x} - 3 = \pm 1$$

$$\Rightarrow \quad \sqrt{x} = 2, 4$$

$$\Rightarrow \quad x = 4, 16$$

So, equation has two solutions.

71. (b) $x + 1 - 2\log_2(3 + 2^x) + 2\log_4(10 - 2^{-x}) = 0$

$$\Rightarrow x + 1 - 2\log_2(3 + 2^x) + \log_2\left(\frac{10 \cdot 2^x - 1}{2^x}\right) = 0$$

$$\Rightarrow x + 1 - 2\log_2(3 + 2^x) + \log_2(10 \cdot 2^x - 1) - x = 0$$

$$\Rightarrow \quad 1 + \log_2\left(\frac{10 \cdot 2^x - 1}{(3 + 2^x)^2}\right) = 0$$

$$\Rightarrow \quad \frac{10 \cdot 2^x - 1}{9 + (2^x)^2 + 6 \cdot 2^x} = \frac{1}{2}$$

$$\Rightarrow \quad (2^x)^2 - 14 \cdot 2^x + 11 = 0$$

Let $2^x = y$

$$\Rightarrow \quad y^2 - 14y + 11 = 0$$

$$y = \frac{14 \pm \sqrt{152}}{2} = 7 \pm \frac{\sqrt{152}}{2}$$

$$y_1 = 7 + \frac{\sqrt{152}}{2},$$

$$y_2 = 7 - \frac{\sqrt{152}}{2}$$

$$\Rightarrow \quad 2^{x_1} = 7 + \frac{\sqrt{152}}{2}, \quad 2^{x_2} = 7 - \frac{\sqrt{152}}{2}$$

$$\Rightarrow \quad x_1 = \log_2\left(7 + \frac{\sqrt{152}}{2}\right), \quad x_2 = \log_2\left(7 - \frac{\sqrt{152}}{2}\right)$$

$\therefore$ Sum of roots $= x_1 + x_2 = \log_2\left(49 - \frac{152}{4}\right) = \log_2 11$

72. (b) Let $e^{\sin x} = y$, then equation is

$$y^2 - 4y - 1 = 0$$

$$\Rightarrow \quad y = 2 \pm \sqrt{5}$$

$$\therefore \quad e^{\sin x} = 2 \pm \sqrt{5}$$

Now, $e^{-1} < e^{\sin x} < e$

$\therefore$ $2 - \sqrt{5} < 0$ and $2 + \sqrt{5} > e$

$\therefore$ Equation has no real solution.

73. (a) Let $e^x = t \in (0, \infty)$

Given equation

$$t^4 + t^3 - 4t^2 + t + 1 = 0$$

$$\Rightarrow \quad t^2 + t - 4 + \frac{1}{t} + \frac{1}{t^2} = 0$$

$$\Rightarrow \quad \left(t^2 + \frac{1}{t^2}\right) + \left(t + \frac{1}{t}\right) - 4 = 0$$

Let $\quad t + \frac{1}{t} = y$

$$(y^2 - 2) + y - 4 = 0$$

$$\Rightarrow \quad y^2 + y - 6 = 0$$

$$\Rightarrow \quad y = -3, 2$$

$$\Rightarrow \quad y = 2$$

$$\Rightarrow \quad t + \frac{1}{t} = 2$$

$$\Rightarrow \quad e^x + e^{-x} = 2$$

$x = 0$, is the only solution of the equation.

74. (a) Let $a^{\cos x} = y$, then equation converts to

$$y + \frac{1}{y} = 6$$

$$\Rightarrow \quad y^2 - 6y + 1 = 0$$

$$\therefore \quad y = \frac{6 \pm \sqrt{32}}{2} = 3 \pm 2\sqrt{2}$$

$$\Rightarrow \quad a^{\cos x} = 3 \pm 2\sqrt{2}$$

$$\Rightarrow \quad \cos x = \log_a(3 \pm 2\sqrt{2})$$

Since, $a > 1$ and $-1 \le \cos x \le 1$

$\therefore$ For all real roots, we have must

$$-1 \le \log_a(3 \pm 2\sqrt{2}) \le 1$$

$\therefore \log_a(3 - 2\sqrt{2}) \ge -1$ and $\log_a(3 + 2\sqrt{2}) \le 1$

$$\Rightarrow \quad 3 - 2\sqrt{2} \ge a^{-1} \text{ and } 3 + 2\sqrt{2} \le a$$

$$\Rightarrow \quad \frac{1}{3 + 2\sqrt{2}} \ge \frac{1}{a} \text{ and } 3 + 2\sqrt{2} \le a$$

$$\Rightarrow \quad \frac{1}{3 + 2\sqrt{2}} \ge \frac{1}{a} \text{ and } 3 + 2\sqrt{2} \le a$$

$$\Rightarrow \quad a \ge 3 + 2\sqrt{2} \text{ and } a \ge 3 + 2\sqrt{2}$$

$$\therefore \quad a \in [3 + 2\sqrt{2}, +\infty]$$

75. (b) Let $f(x) = -3 + x - x^2$, then $f(x) < 0 \; \forall \; x$, because coefficient of x^2 is negative and $D < 0$, thus LHS of the given equation is always positive, whereas the RHS is always less than zero.

Hence, there is no solution.

76. (a) The given equation is in $\sin(e^x) = 5^x + 5^{-x}$

We know, 5^x and 5^{-x} both are positive real numbers.

Using AM $\ge$ GM,

$$5^x + \frac{1}{5^x} \ge 2 \Rightarrow 5^x + 5^{-x} \ge 2$$

$\therefore$ RHS of given equation ≥ 2

While $\sin e^x \in [-1, 1]$ i.e. LHS $\in [-1, 1]$

$\therefore$ The equation is not possible for any real value of x.

77. (d) $[e^x]^2 + [e^x + 1] - 3 = 0$

$$\Rightarrow \quad [e^x]^2 + [e^x] + 1 - 3 = 0$$

$$\Rightarrow \quad [e^x]^2 + [e^x] - 2 = 0$$

$$\Rightarrow \quad ([e^x] - 1)([e^x] + 2) = 0$$

$$[e^x] = 1 \text{ or } [e^x] = -2$$

Not possible as $e^x > 0$.

$$\Rightarrow \quad [e^x] = 1$$

$$\Rightarrow \quad 1 \le e^x < 2$$

$$\Rightarrow \quad 0 \le x < \log_e 2$$

78. (b) $2\log_{\frac{1}{25}}(bx + 28) = -\log_5(12 - 4x - x^2)$

$$\Rightarrow \quad -\log_5(bx + 28) = -\log_5(12 - 4x - x^2)$$

$$\Rightarrow \quad bx + 28 = 12 - 4x - x^2$$

$$\Rightarrow x^2 + (b + 4)x + 16 = 0$$

For coincident roots, $D = 0$

$$\Rightarrow \quad (b + 4)^2 - 64 = 0$$

$$\Rightarrow \quad (b + 4) = \pm 8$$

$$\Rightarrow \quad b + 4 = 8 \text{ or } b + 4 = -8$$

$$\Rightarrow \quad b = 4 \text{ or } b = -12$$

For $b = 4$, $x = -4$

For $b = -12$, $x = 4$

Now for $12 - 4x - x^2 > 0$, $x = -4$ and $x = \pm 4$

$\Rightarrow b = 4$

79. (0.5) Let n and $(n + 1)$ be two consecutive roots of $x^2 - bx + c = 0$

Then, $n + (n + 1) = b$

$$\Rightarrow \quad 2n + 1 = b \text{ and } n(n + 1) = c$$

$$\therefore \quad b^2 - 4c = (2n + 1)^2 - 4n(n + 1)$$

$$= 4n^2 + 1 + 4n - 4n^2 - 4n$$

$$\Rightarrow \quad b^2 - 4c = 1$$

$$\therefore \quad \frac{b^2 - 4c}{2} = \frac{1}{2} = 0.5$$

80. (324) Given that, $\alpha + \beta = 1, \alpha\beta = -1$

Let α, β be roots of quadratic equation, then the quadratic equation be

$$x^2 - x - 1 = 0$$

Now, $\quad \alpha^2 - \alpha - 1 = 0$

$$\Rightarrow \quad \alpha^2 = \alpha + 1 \quad \text{...(i)}$$

Similarly, $\quad \beta^2 = \beta + 1 \quad$ ...(ii)

Multiply α^{n-1} in Eq. (i), we get

$$\alpha^{n+1} = \alpha^n + \alpha^{n-1} \quad \text{...(iii)}$$

Multiply β^{n-1} in Eq. (ii), we get

$$\beta^{n+1}=\beta^n+\beta^{n-1} \quad \text{...(iv)}$$

Add Eqs. (iii) and (iv), we get

$$\alpha^{n+1}+\beta^{n+1}=(\alpha^n+\beta^n)+(\alpha^{n-1}+\beta^{n-1})$$

$$p_{n+1}=p_n+p_{n-1}$$

$$29=p_n+11$$

$$\Rightarrow \quad P_n=18$$

$$p_n^2=(18)^2=324$$

81. (6) Given, $x^2-1154x+1=0$

$\therefore \quad \alpha+\beta=1154, \alpha\beta=1$

$$(\sqrt{\alpha}+\sqrt{\beta})^2=\alpha+\beta+2\sqrt{\alpha\beta}$$

$$=1154+2=1156$$

$$\Rightarrow (\sqrt{\alpha}+\sqrt{\beta})^2=(34)^2 \Rightarrow (\sqrt{\alpha}+\sqrt{\beta})=34$$

Again, $\left(\alpha^{\frac{1}{4}}+\beta^{\frac{1}{4}}\right)^2=\sqrt{\alpha}+\sqrt{\beta}+2(\alpha\beta)^{\frac{1}{4}}=34+2=36$

$$\Rightarrow \quad \alpha^{\frac{1}{4}}+\beta^{\frac{1}{4}}=6$$

82. (3) Given equations are

$$ax^2+bx+c=3x-y$$

and $\quad ay^2+by+c=y-3x$

Now, $\quad b^2-4ac\leq 0$

$\therefore\ ax^2+bx+c\leq 0$ and $ay^2+by+c\leq 0$

$\therefore\ 3x-y\leq 0$ and $y-3x\leq 0$

$\Rightarrow \quad 3x=y$, so, $\dfrac{y}{x}=3$

83. (6) Let $f(x)=x^2+2(\lambda+1)x+\lambda^2+\lambda+7$

If both roots of $f(x)=0$ are negative, then

$$D=b^2-4ac\geq 0$$

$$\Rightarrow 4(\lambda+1)^2-4(\lambda^2+\lambda+7)\geq 0$$

$$\Rightarrow \quad \lambda-6\geq 0$$

$$\Rightarrow \quad \lambda\in[6,\infty) \quad \text{...(i)}$$

Sum of roots $=-2(\lambda+1)<0$

$$\Rightarrow \quad \lambda\in(-1,\infty) \quad \text{...(ii)}$$

and product of roots $=\lambda^2+\lambda+7>0, \lambda\in R$ ...(iii)

$\therefore$ From Eqs. (i), (ii) and (iii), we get

$$\lambda\in[6,\infty)$$

Hence, the least value of $\lambda=6$.

84. (1) Let α and β be the roots of the equation

$$x^2-(a-2)x-(a+1)=0$$

Then, $\alpha+\beta=a-2$ and $\alpha\beta=-a-1$

Now, $\alpha^2+\beta^2=(\alpha+\beta)^2-2\alpha\beta=(a-2)^2+2(a+1)$

$$\Rightarrow \quad \alpha^2+\beta^2=a^2+4-4a+2a+2$$

$$\Rightarrow \quad \alpha^2+\beta^2=a^2-2a+6$$

$$\Rightarrow \quad \alpha^2+\beta^2=(a-1)^2+5$$

The value of $\alpha^2+\beta^2$ will be least if $a-1=0$

$$\Rightarrow \quad a=1$$

85. (66) $\dfrac{2}{x-1}-\dfrac{1}{x-2}=\dfrac{2}{k} \Rightarrow x\in R-\{1,2\}$

$$k(2x-4-x+1)=2(x^2-3x+2)$$

$$k(x-3)=2(x^2-3x+2)$$

$$2x^2-(6+k)x+3k+4=0$$

For no real roots $b^2-4ac<0$

$$\therefore \quad (k+6)^2-8\cdot(3k+4)<0$$

$$\Rightarrow \quad k^2-12k-4<0$$

$$\Rightarrow \quad (k-6)^2-32<0$$

$$\Rightarrow \quad (k-6)^2<32$$

$$\Rightarrow \quad -4\sqrt{2}<k-6<4\sqrt{2}$$

$$\Rightarrow \quad 6-4\sqrt{2}<k<6+4\sqrt{2}$$

Integral $k\in\{1,2,3,4,....11\}$

Sum $=66$

86. (9) $\because$ Roots of the equation $x^2+3x+5=0$ are non-rea

Thus, given equations will have two common roots.

$$\Rightarrow \quad \frac{a}{1}=\frac{b}{3}=\frac{c}{5}=\lambda$$

$$\therefore \quad a+b+c=9\lambda$$

Thus, the minimum value of $a+b+c$ is 9. [$\because a, b, c\in N$

87. (0.6) Given that $ax^2+bx+c=0$ and $bx^2+cx+a=0$ have a common root.

$$\therefore \quad (bc-a^2)^2=(ab-c^2)(ac-b^2)$$

$$\Rightarrow b^2c^2+a^4-2a^2bc=a^2bc-ab^3-ac^3+b^2c^2$$

$$\Rightarrow \quad a^4+ab^3+ac^3=3a^2bc$$

$$\Rightarrow \quad \frac{a^3+b^3+c^3}{5abc}=\frac{3}{5}=0.6$$

88. (7.5) We have,

$$x^2+ax+12=0 \quad \text{...(i)}$$

$$x^2+bx+15=0 \quad \text{...(ii)}$$

and $x^2+(a+b)x+36=0$ ...(iii)

On adding Eqs. (i) and (ii), we get

$$2x^2+(a+b)x+27=0 \quad \text{...(iv)}$$

Subtracting Eq. (iii) from Eq. (iv), we get

$$x^2-9=0 \Rightarrow x=3,-3$$

Thus, the common root is 3.

Hence, $9+12+3a=0$ [from Eq. (i)]

$$\Rightarrow \quad a=-7 \text{ and } 9+3b+15=0$$

$$\Rightarrow \quad a=-7 \text{ and } b=-8$$

$$\therefore\ |a+b|=|-7-8|=|-15|=15$$

Thus, $\dfrac{|a+b|}{2}=\dfrac{15}{2}=7.5$

89. (13.334) Let α be the common root.

$\therefore \quad \alpha^2 + a\alpha + 10 = 0 \quad$ ...(i)

and $\quad \alpha^2 + b\alpha - 10 = 0 \quad$ ...(ii)

Subtracting Eq. (ii) from Eq. (i), we get

$$(a-b)\alpha + 20 = 0$$

$$\Rightarrow \quad \alpha = -\frac{20}{a-b}$$

The value of α, substituting in Eq. (i), we have

$$\Rightarrow \quad \left(\frac{-20}{a-b}\right)^2 + a\left(\frac{-20}{a-b}\right) + 10 = 0$$

$$\Rightarrow \quad 400 - 20a(a-b) + 10(a-b)^2 = 0$$

$$\Rightarrow 400 - 20a^2 + 20ab + 10a^2 + 10b^2 - 20ab = 0$$

$$\Rightarrow \quad -10a^2 + 10b^2 = -400$$

$$\Rightarrow \quad a^2 - b^2 = 40$$

$$\therefore \quad \frac{a^2-b^2}{3} = \frac{40}{3} = 13.334$$

90. (18) We have, α is common root of the equations $x^2 - x + 2\lambda = 0$ and $3x^2 - 10x + 27\lambda = 0$.

Now, common root of these equations is

$$(3\alpha^2 - 10\alpha + 27\lambda) - (3\alpha^2 - 3\alpha + 6\lambda) = 0$$

$$\Rightarrow \quad -7\alpha + 21\lambda = 0$$

$$\Rightarrow \quad \alpha = 3\lambda$$

Again, α is root of $x^2 - x + 2\lambda = 0$

$$\therefore \quad \alpha^2 - \alpha + 2\lambda = 0$$

$$\Rightarrow \quad (3\lambda)^2 - 3\lambda + 2\lambda = 0$$

$$\Rightarrow \quad 9\lambda^2 - \lambda = 0$$

$$\Rightarrow \quad \lambda(9\lambda - 1) = 0$$

$$\Rightarrow \quad \lambda = 0, \frac{1}{9}$$

$$\Rightarrow \quad \lambda = \frac{1}{9} \quad [\because \lambda \neq 0]$$

$$\therefore \quad \alpha = 3\lambda = 3 \times \frac{1}{9} = \frac{1}{3}$$

Again, α and β are roots of the equation

$$x^2 - x + 2\lambda = 0$$

$$\therefore \quad \alpha + \beta = \frac{-(-1)}{1} = 1$$

$$\Rightarrow \quad \beta = 1 - \alpha$$

$$= 1 - \frac{1}{3} = \frac{2}{3}$$

And α and γ are the roots of the equation $3x^2 - 10x + 27\lambda = 0$

$$\therefore \quad \alpha + \gamma = \frac{-(-10)}{3} = \frac{10}{3}$$

$$\Rightarrow \quad \gamma = \frac{10}{3} - \alpha = \frac{10}{3} - \frac{1}{3}$$

$$= \frac{9}{3} = 3$$

$$\therefore \quad \frac{\beta\gamma}{\lambda} = \frac{\left(\frac{2}{3}\right) \times (3)}{\left(\frac{1}{9}\right)} = 18$$

91. (3) $\because n^2 \sin^2 x - 2\sin x - (2n+1) = 0$

$$\Rightarrow \quad \sin x = \frac{2 \pm \sqrt{4 + 4n^2(2n+1)}}{2n^2}$$

$$= \frac{1 \pm \sqrt{2n^3 + n^2 + 1}}{n^2}$$

$$\because \quad 0 \le \sin x \le 1$$

$$\Rightarrow \quad 0 \le \frac{1 + \sqrt{2n^3 + n^2 + 1}}{n^2} \le 1$$

$$\Rightarrow \quad 0 \le 1 + \sqrt{2n^3 + n^2 + 1} \le n^2$$

$$\Rightarrow \quad \sqrt{2n^3 + n^2 + 1} \le (n^2 - 1)$$

On squaring both sides,

$$2n^3 + n^2 + 1 \le n^4 - 2n^2 + 1$$

$$\Rightarrow \quad n^4 - 2n^3 - 3n^2 \ge 0$$

$$\Rightarrow \quad n^2 - 2n - 3 \ge 0$$

$$\Rightarrow \quad (n-3)(n+1) \ge 0$$

$$\Rightarrow \quad n \ge 3;\ n = 3, 4, 5, \ldots$$

Hence, the minimum value of n is 3.

92. (3) We have,

$$P(x) = \frac{5}{3} - 6x - 9x^2$$

$$= -(3x+1)^2 + \frac{8}{3}$$

$$\Rightarrow \quad P_{max} = \frac{8}{3} \quad \left[\text{at } x = -\frac{1}{3}\right]$$

Similarly, $Q(y) = -4y^2 + 4y + \frac{13}{2}$

$$= -(2y-1)^2 + \frac{18}{2}$$

$$\Rightarrow \quad Q_{max} = \frac{18}{2} \quad \left[\text{at } y = \frac{1}{2}\right]$$

Now, $P_{max} \times Q_{max} = \frac{8}{3} \times \frac{18}{2} = 20$

So, $(x, y) = \left(-\frac{1}{3}, \frac{1}{2}\right)$

$$\Rightarrow 6x + 10y = 6x - \frac{1}{3} + 10 \times \frac{1}{2}$$

$$= -2 + 5 = 3$$

93. (4) Let $y = \dfrac{3x^2 + mx + n}{x^2 + 1}$

or $x^2(y-3) - mx + y - n = 0$

Since, $x \in R$

$\therefore \quad D \geq 0$

$\Rightarrow \quad m^2 - 4(y-3)(y-n) \geq 0$

$\Rightarrow m^2 - 4(y^2 - ny - 3y + 3n) \geq 0$

$\Rightarrow 4y^2 - 4y(n+3) + 12n - m^2 \leq 0$ ...(i)

Also given, $(y+4)(y-3) \leq 0$

$$y^2 + y - 12 \leq 0 \quad \text{...(ii)}$$

Comparing Eqs. (i) and (ii), we get

$$\frac{4}{1} = -\frac{4(n+3)}{1} = \frac{12n - m^2}{-12}$$

$\Rightarrow \quad m = 0$ and $n = -4$

94. (315) We have,

$$\left(\frac{a^4 + 3a^2 + 1}{a^2}\right)\left(\frac{b^4 + 5b^2 + 1}{b^2}\right)\left(\frac{c^4 + 7c^2 + 1}{c^2}\right)$$

$$= \left(a^2 + \frac{1}{a^2} + 3\right)\left(b^2 + \frac{1}{b^2} + 5\right)\left(c^2 + \frac{1}{c^2} + 7\right)$$

$$= \left\{\left(a - \frac{1}{a}\right)^2 + 5\right\}\left\{\left(b - \frac{1}{b}\right)^2 + 7\right\}\left\{\left(c - \frac{1}{c}\right)^2 + 9\right\}$$

So, minimum value is $5 \times 7 \times 9 = 315$.

95. (3) $x^2 + x(y-a) + y^2 - ay + 1 \geq 0 \; x \in R$

$\Rightarrow \quad (y-a)^2 - 4(y^2 - ay + 1) \leq 0$

$\Rightarrow \quad -3y^2 + 2ay + a^2 - 4 \leq 0 \; \forall \; y \in R$

$\therefore \quad 3y^2 - 2ay + 4 - a^2 \geq 0 \; \forall \; y \in R$

$$D \leq 0$$

$\Rightarrow \quad 4a^2 - 4 \cdot 3(4 - a^2) \leq 0$

$\Rightarrow \quad a^2 - 3(4 - a^2) \leq 0$

$\Rightarrow \quad 4a^2 - 12 \leq 0$

$\therefore$ Range of $a \in [-\sqrt{3}, \sqrt{3}]$

Number of integers $[-1, 0, 1]$

96. (9) Let $f(x) = (a^2 - 14a + 13)x^2 + (a+2)x - 2$

Given that the above equation doesn't have two distinct real.

But $f(0) = -2 < 0$

$\therefore \quad f(0) \leq 0, \forall \; x \in R$

So, $\quad f(-1) \leq 0$

$\Rightarrow \quad (a^2 - 14a + 13) - (a+2) - 2 \leq 0$

$\Rightarrow \quad a^2 - 15a + 9 \leq 0$

$\Rightarrow \quad a^2 - 15a \leq -9$

So, the maximum value of $a^2 - 15a$ is $-9 (= A)$

Then, $|A| = 9$

97. (5) We have,

$$2x^3 - 9x^2 + 12x + k = 0$$

Let the roots are α, α and β

$\therefore \quad 2\alpha + \beta = \dfrac{9}{2}$ ...(i)

$\alpha^2 + 2\alpha\beta = \dfrac{12}{2} = 6$ ...(ii)

and $\quad \alpha^2\beta = -\dfrac{k}{2}$ ...(iii)

Putting $\beta = \left(\dfrac{9}{2} - 2\alpha\right)$ from Eqs. (i) and (ii), we have

$$\alpha^2 + 2\alpha\left(\frac{9}{2} - 2\alpha\right) = 6$$

$\Rightarrow \quad \alpha^2 + 9\alpha - 4\alpha^2 = 6$

$\Rightarrow \quad 3\alpha^2 - 9\alpha + 6 = 0$

$\Rightarrow \quad \alpha^2 - 3\alpha + 2 = 0$

$\Rightarrow \quad (\alpha - 2)(\alpha - 1) = 0$

$\Rightarrow \quad \alpha = 2$ or 1

If $\alpha = 2$, then $\beta = \dfrac{1}{2}$

If $\alpha = 1$, then $\beta = \dfrac{5}{2}$

$\therefore \; k = -2(2^2)\dfrac{1}{2} = -4$

or $\quad k = -2(1^2)\left(\dfrac{5}{2}\right) = -5$

$$|k| = 4 \text{ or } |k| = 5$$

$\therefore$ Maximum value of $|k| = 5$

98. (9) Given, roots of the equation

$10x^3 - cx^2 - 54x - 27 = 0$ are in HP. ...(i)

Now, replacing x by $\dfrac{1}{x}$, we get

$$27x^3 + 54x^2 + cx - 10 = 0 \quad \text{...(ii)}$$

Hence, the roots of Eq. (ii) are in AP.

Let $a-d$, a and $a+d$ are the roots of Eq. (ii),

Then, $a - d + a + a + d = -\dfrac{54}{27}$

$\Rightarrow \quad a = -\dfrac{2}{3}$

$\because$ a is a root of Eq. (ii), then

$$27a^3 + 54a^2 + ca - 10 = 0$$

$\Rightarrow 27\left(\dfrac{-8}{27}\right) + 54\left(\dfrac{4}{9}\right) + c\left(\dfrac{-2}{3}\right) - 10 = 0$

$\Rightarrow \quad 6 - \dfrac{2c}{3} = 0$ or $c = 9$

99. (6) $x^2 + ax + b \equiv (x+1)(x+b) \Rightarrow b + 1 = a$ ...(i)

Also, $x^2 + bx + c \equiv (x+1)(x+c) \Rightarrow c + 1 = b$

or $\quad b + 1 = c + 2$ ...(ii)

Hence, $b+1=a=c+2$

Also, $(x+1)(x+b)(x+c)=x^3-4x^2+x+6$

$\Rightarrow x^3+(1+b+c)x^2+(b+bc+c)x+bc=x^3-4x^2+x+6$

$\Rightarrow \quad 1+b+c=-4$

$\Rightarrow \quad 2c+2=-4$

$\Rightarrow \quad c=-3, b=-2 \text{ and } a=-1$

$\therefore \quad a+b+c=-6$

100. (2) Given equation,

$$e^{4x}-e^{3x}-4e^{2x}-e^x+1=0$$

Let $\quad e^x=t>0$

$$t^4-t^3-4t^2-t+1=0$$

$\Rightarrow \quad t^2-t-4-\frac{1}{t}+\frac{1}{t^2}=0$

$\Rightarrow \quad t^2+\frac{1}{t^2}+2-\left(t+\frac{1}{t}\right)-6=0$

$\Rightarrow \quad \left(t+\frac{1}{t}\right)^2-\left(t+\frac{1}{t}\right)-6=0$

Let $\alpha=t+\frac{1}{t}\geq 2$

$\Rightarrow \quad \alpha^2-\alpha-6=0$

$\Rightarrow \quad \alpha^2-3\alpha+2\alpha-6=0$

$\Rightarrow \quad \alpha(\alpha-3)+2(\alpha-3)=0$

$\Rightarrow \quad (\alpha-3)(\alpha+2)=0$

$\Rightarrow \quad \alpha=3 \text{ or } \alpha=-2$ (not possible)

$\Rightarrow \quad t+\frac{1}{t}=3$

$\Rightarrow \quad t^2-3t+1=0$

$\therefore$ The number of real roots $=2$

101. (9) We have, $x^4-4x^3+6x^2-4x=69$

$\Rightarrow \quad x^4-4x^3+6x^2-4x+1=70$

$\Rightarrow \quad (x-1)^4=70$

$\Rightarrow \quad (x-1)^2=\pm\sqrt{70}$

For the product of non-real roots, consider

$$(x-1)^2=-\sqrt{70}$$

$\Rightarrow x^2-2x+1+\sqrt{70}=0$

Product of roots $=\frac{1+\sqrt{70}}{1}$

$=1+\sqrt{70}$

$\therefore$ Integral part of product of roots $=[1+\sqrt{70}]$

$=[1+8.36]$

$=[9.36]=9$

102. (5) Let $ax^3+bx^2+cx+d=0$ has roots p, q and r.

$\therefore \quad pq+qr+rp=\frac{c}{a}$

But $pq+qr+rp\leq p^2+q^2+r^2$

$=(p+q+r)^2-2\Sigma pq$

$\therefore 3(pq+qr+rp)\leq(p+q+r)^2=16$

$\therefore \quad 3\frac{c}{a}\leq 16$

$\Rightarrow \quad \frac{c}{a}\leq\frac{16}{3}$

$\Rightarrow$ Largest possible integral value of $\frac{c}{a}$ is 5.

103. (0.181) Let $2^{11x}=t$

$\Rightarrow \quad \frac{t^3}{2}+4t=2t^2+1$

$\Rightarrow \quad t^3-8t^2+10t-4=0$

$\Rightarrow \quad t_1t_2t_3=4$

$\Rightarrow \quad 2^{11x_1}\cdot 2^{11x_2}\cdot 2^{11x_3}=4$

$\Rightarrow \quad 2^{11(x_1+x_2+x_3)}=2^2$

$\Rightarrow \quad 11(x_1+x_2+x_3)=2$

$\Rightarrow \quad x_1+x_2+x_3=\frac{2}{11}=0.181$

Level 3

1. (d) $\quad x=1+\cfrac{1}{3+\cfrac{1}{2+\cfrac{1}{3+\cfrac{1}{2\ldots\infty}}}}$

$=1+\cfrac{1}{3+\cfrac{1}{1+x}}$

$=1+\frac{1+x}{4+3x}$ or $3x^2=5$

or $\quad x=\sqrt{\frac{5}{3}}$ (x is positive)

2. (a) Given, $(x^2+x-2)(x^2+x-3)=12$ ...(i)

Putting $x^2+x=y$, so that Eq. (i) becomes

$$(y-2)(y-3)=12$$

$\Rightarrow \quad y^2-5y-6=0$

$\Rightarrow \quad (y-6)(y+1)=0$ or $y=6,-1$

When $y=6$, we get

$$x^2+x-6=0$$

$\Rightarrow \quad (x+3)(x-2)=0$ or $x=-3, 2$

When $y=-1$, we get

$$x^2+x+1=0$$

which has non-real roots and sum of roots is -1.

3. (d) Here, $x = 0$ is not a root, divide both the numerator and denominator by x and put $x + \frac{3}{x} = y$ to obtain

$$\frac{4}{y+1} + \frac{5}{y-5} = -\frac{3}{2}$$

$$\Rightarrow \quad y = -5, 3$$

$x + \frac{3}{x} = -5$ has two irrational roots

and $x + \frac{3}{x} = 3$ has imaginary roots.

4. (c) Clearly, $x = 2$ is a root of the equation and imaginary roots always occur in pairs, therefore, the other roots is also real.

5. (a) We have, $(a-1)x^2 + ax + \sqrt{1-a^2} = 0$

$$\therefore \quad D = a^2 - 4(a-1)\sqrt{1-a^2}$$
$$= a^2 - 4a\sqrt{1-a^2} + 4\sqrt{1-a^2}$$
$$= (a - 2\sqrt{1-a^2})^2 + 4\sqrt{1-a^2}$$
$$(1 - \sqrt{1-a^2}) > 0 \text{ for } a \in (-1, 1)$$

6. (c) Here, a, b and c are positive.

So, $|x| = -b + \sqrt{b^2 + ac}$

Hence, x has two real values, neglecting

$|x| = -b - \sqrt{b^2 - ac}$, as $|x| \geq 0$

7. (b) Given, $(ax^2 + c)y + (a'x^2 + c') = 0$

or $\quad x^2(ay + a') + (cy + c') = 0$

If x is rational, then the discriminant of the above equation must be perfect square.

Hence, $0 - 4(ay + a')(cy + c')$ must be a perfect square.

$\Rightarrow -acy^2 - (ac' + a'c)y - a'c'$ must be a perfect square.

$$\Rightarrow \quad (ac' + a'c)^2 - 4aca'c' = 0 \quad [\because D = 0]$$
$$\Rightarrow \quad (ac' - a'c)^2 = 0$$
$$\Rightarrow \quad ac' = a'c$$
$$\Rightarrow \quad \frac{a}{a'} = \frac{c}{c'}$$

8. (c) Since, α is root of all equations.

$$a\alpha^2 + 2b\alpha + c = 0$$
$$2b\alpha^2 + c\alpha + a = 0 \text{ and } c\alpha^2 + a\alpha + 2b = 0$$

Adding these equations, we get

$$(a + 2b + c)(\alpha^2 + \alpha + 1) = 0$$
$$a + 2b + c \neq 0 \text{ as } a, b, c > 0$$
$$\Rightarrow \quad \alpha^2 + \alpha + 1 = 0 \text{ or } \alpha^2 + \alpha = -1$$

9. (c) $\quad D > 0$

$$\Rightarrow \quad (a-3)^2 + 4(a+2) > 0$$
$$\Rightarrow a^2 - 6a + 9 + 4a + 8 > 0$$
$$\Rightarrow \quad a^2 - 2a + 17 > 0$$
$$\Rightarrow \quad a \in R$$
$$\therefore \quad \frac{a^2+1}{a^2+2} = 1 - \frac{1}{a^2+2} \geq \frac{1}{2}$$

10. (a) $\because \alpha + \beta = 64$ and $\alpha\beta = 256$

$$\frac{\alpha^{\frac{3}{8}}}{\beta^{\frac{5}{8}}} + \frac{\beta^{\frac{3}{8}}}{\alpha^{\frac{5}{8}}} = \frac{\alpha + \beta}{(\alpha\beta)^{\frac{5}{8}}}$$
$$= \frac{64}{(2^8)^{\frac{5}{8}}} = \frac{64}{32} = 2$$

11. (d) Here, $\alpha^4 + \beta^4 = (\alpha^2 + \beta^2)^2 - 2\alpha^2\beta^2$

$$= \{(\alpha+\beta)^2 - 2\alpha\beta\}^2 - 2(\alpha\beta)^2$$
$$= \left(p^2 + \frac{1}{p^2}\right)^2 - \frac{1}{2p^4}$$
$$= p^4 + \frac{1}{2p^4} + 2$$
$$= \left(p^2 - \frac{1}{\sqrt{2}p^2}\right)^2 + 2 + 2\sqrt{2} \geq 2 + \sqrt{2}$$

Thus, the minimum value of $\alpha^4 + \beta^4$ is $2 + \sqrt{2}$.

12. (c) We have,

$$\frac{k+1}{k} + \frac{k+2}{k+1} = -\frac{b}{a} \quad \text{...}$$

and $\quad \frac{k+1}{k}\frac{k+2}{k+1} = \frac{c}{a}$

$$\Rightarrow \quad \frac{k+2}{k} = \frac{c}{a}$$

or $\quad \frac{2}{k} = \frac{c}{a} - 1 = \frac{c-a}{a}$

or $\quad k = \frac{2a}{c-a} \quad \text{...}$

Now eliminating k, putting the value of k in Eq. (i), we g

$$\frac{c+a}{2a} + \frac{2c}{c+a} = -\frac{b}{a}$$
$$\Rightarrow \quad (c+a)^2 + 4ac = -2b(a+c)$$
$$\Rightarrow \quad (a+c)^2 + 2ab(a+c) = -4ac$$

Adding b^2 to both sides, we have

$$(a+b+c)^2 = b^2 - 4ac$$

13. (b) Let equation is $a(x-\alpha)(x-\beta) = 0$, where α and β are prime numbers.

Sum of coefficients $= a - a(\alpha+\beta) + a\alpha\beta$

$= a(\alpha - 1)(\beta - 1)$,

which is prime number.

$\Rightarrow a = 1$ and one of $(\alpha - 1)$ and $(\beta - 1)$ will be 1

Let $\alpha - 1 = 1$ or $\alpha = 2$, then $\beta - 1$ will be 2, so $\beta = 3$

Hence, sum of roots $= 5$

14. (b) Let α and β be the roots of the given quadratic equation,

$$2x^2+2x-1=0 \quad ...(i)$$

Then, $\alpha+\beta=-\frac{1}{2}$

$\Rightarrow -1=2\alpha+2\beta$

and $4\alpha^2+2\alpha-1=0$ [$\because \alpha$ is root of Eq. (i)]

$\Rightarrow 4\alpha^2+2\alpha+2\alpha+2\beta=0$

$\Rightarrow \beta=-2\alpha(\alpha+1)$

15. (c) We have,

$$a\alpha^2+c=-b\alpha \text{ or } a\alpha+b=-\frac{c}{\alpha}$$

Similarly, $a\beta+b=-\frac{c}{\beta}$

Hence, the given expression is

$$\frac{b}{c}(\alpha^2+\beta^2)=\frac{b(b^2-2ac)}{a^2c}$$

16. (d) $\because$ a and b are roots of the equation

$$x^2+20x-2020=0$$

$\therefore$ $a+b=-20;\ ab=-2020$

and c and d are roots of the equation

$$x^2-20x+2020=0$$

$\therefore$ $c+d=20;\ cd=2020$

Now, $ac(a-c)+ad(a-d)+bc(b-c)+bd(b-d)$

$=a^2c-ac^2+a^2d-ad^2+b^2c-bc^2+b^2d-bd^2$

$=a^2(c+d)+b^2(c+d)-c^2(a+b)-d^2(a+b)$

$=(c+d)(a^2+b^2)-(a+b)(c^2+d^2)$

$=(c+d)[(a+b)^2-2ab]-(a+b)[(c+d)^2-2cd]$

$=20[(20)^2+4040]+20[(20)^2-4040]$

$=20\times 800=16000$

17. (c) $x^2-2x\sec\theta+1=0$

$\Rightarrow x=\sec\theta\pm\tan\theta$ and $x^2+2x\tan\theta-1=0$

$\Rightarrow x=-\tan\theta\pm\sec\theta$

$\because -\frac{\pi}{6}<\theta<-\frac{\pi}{12}$

$\Rightarrow \sec\frac{\pi}{6}>\sec\theta>\sec\frac{\pi}{12}$

and $-\tan\frac{\pi}{6}<\tan\theta<-\tan\frac{\pi}{12}$

Also, $\tan\frac{\pi}{12}<-\tan\theta<\tan\frac{\pi}{6}$

$\because \alpha_1$ and β_1 are roots of $x^2-2x\sec\theta+1=0$ and $\alpha_1>\beta_1$

$\therefore \alpha_1=\sec\theta-\tan\theta$ and $\beta_1=\sec\theta+\tan\theta$

$\because \alpha_2$ and β_2 are roots of $x^2+2x\tan\theta-1=0$ and $\alpha_2>\beta_2$

$\therefore \alpha_2=-\tan\theta+\sec\theta$ and $\beta_2=-\tan\theta-\sec\theta$

$\therefore \alpha_1+\beta_2=\sec\theta-\tan\theta-\tan\theta-\sec\theta$

$=-2\tan\theta$

18. (a) Let α and α^2 be the roots of $x^2-x-k=0$, then

$$\alpha+\alpha^2=1 \text{ and } \alpha^3=-k$$

$\Rightarrow (-k)^{\frac{1}{3}}+(-k)^{\frac{2}{3}}=1$

$\Rightarrow -k^{\frac{1}{3}}+k^{\frac{2}{3}}=1$

$\Rightarrow \left(k^{\frac{2}{3}}-k^{\frac{1}{3}}\right)^3=1$

$\Rightarrow k^2-k-3k\left(k^{\frac{2}{3}}-k^{\frac{1}{3}}\right)=1$

$\Rightarrow k^2-k-3k(1)=1$

$\Rightarrow k^2-4k-1=0$

$\Rightarrow k=2\pm\sqrt{5}$

19. (b) We have, $ax^2-bx+c=0$

$$\alpha+\beta=\frac{b}{a} \text{ and } \alpha\beta=\frac{c}{a}$$

Also, $(a+cy)^2=b^2y$

$\Rightarrow c^2y^2-(b^2-2ac)y+a^2=0$

$\Rightarrow \left(\frac{c}{a}\right)^2y^2-\left\{\left(\frac{b}{a}\right)^2-2\left(\frac{c}{a}\right)\right\}y+1=0$

$\Rightarrow (\alpha\beta)^2y^2-(\alpha^2+\beta^2)y+1=0$

$\Rightarrow y^2-(\alpha^{-2}+\beta^{-2})y+\alpha^{-2}\beta^{-2}=0$

$\Rightarrow (y-\alpha^{-2})(y-\beta^{-2})=0$

Hence, the roots are α^{-2} and β^{-2}

20. (b) Given, $a(p+q)^2+2bpq+c=0$

and $a(p+r)^2+2bpr+c=0$

$\Rightarrow$ q and r satisfy the equation $a(p+x)^2+2bpx+c=0$

$\Rightarrow$ q and r are the roots of the equation

$$ax^2+2(ap+bp)x+c+ap^2=0$$

$\Rightarrow qr=$ Product of roots

$$=\frac{c+ap^2}{a}=p^2+\frac{c}{a}$$

21. (a) α and β are roots of $ax^2+bx+c=0$

Hence, $\alpha+\beta=-\frac{b}{a}$ ...(i)

and $\alpha\beta=\frac{c}{a}$ ...(ii)

α^2 and β^2 are roots of $a^2x^2+b^2x+c^2=0$.

Hence, $\alpha^2+\beta^2=-\frac{b^2}{a^2}$ ...(iii)

and $\alpha^2\beta^2=\frac{c^2}{a^2}$ ...(iv)

Now, from Eq. (iii), we have

$$(\alpha+\beta)^2-2\alpha\beta=-\frac{b^2}{a^2}$$

$$\Rightarrow \quad \left(-\frac{b}{a}\right)^2-2\frac{c}{a}=\frac{-b^2}{a^2}$$

$$\Rightarrow \quad 2\frac{b^2}{a^2}=\frac{2c}{a}\Rightarrow b^2=ac$$

$\Rightarrow$ a, b and c are in GP.

22. (a) α and β be the roots of $x^2+px+q=0$

γ and δ be the roots of $x^2+px+r=0$

$$\Rightarrow \quad \alpha+\beta=-p$$

and $\quad \alpha\beta=q$

$$\Rightarrow \quad \gamma+\delta=-p$$

$$\gamma\delta=r$$

Now, $(\alpha-\gamma)(\alpha-\delta)=\alpha^2-\alpha(\gamma+\delta)+\gamma\delta$

$$=\alpha^2-\alpha(\alpha+\beta)+r$$

$$=-\alpha\beta+r=-q+r$$

By symmetry, $(\beta-\gamma)(\beta-\delta)=-q+r$

Hence, $\dfrac{(\alpha-\gamma)(\alpha-\delta)}{(\beta-\gamma)(\beta-\delta)}=1$

23. (b) $a\dfrac{(2x+1)^2}{(x-3)^2}+b\dfrac{(2x+1)}{(x-3)}+c=0$

$$\Rightarrow \quad \frac{2x+1}{x-3}=\alpha \text{ or } \frac{2x+1}{x-3}=\beta$$

or $\quad 2x+1=\alpha x-3\alpha$

$$\Rightarrow \quad x(\alpha-2)=1+3\alpha$$

$$\Rightarrow \quad x=\frac{1+3\alpha}{\alpha-2},\frac{1+3\beta}{\beta-2}$$

24. (c) Given, α and β are roots of equation

$$x^2-2x+3=0$$

$$\Rightarrow \quad \alpha^2-2\alpha+3=0 \quad \text{...(i)}$$

and $\quad \beta^2-2\beta+3=0 \quad$ ...(ii)

$$\Rightarrow \quad \alpha^2=2\alpha-3 \text{ or } \alpha^3=2\alpha^2-3\alpha$$

$$\Rightarrow \quad P=(2\alpha^2-3\alpha)-3\alpha^2+5\alpha-2$$

$$=-\alpha^2+2\alpha-2=3-2=1 \quad \text{[using Eq. (i)]}$$

Similarly, we have

$$Q=2$$

Now, sum of roots is 3 and product of roots is 2.

Hence, the required equation is $x^2-3x+2=0$.

25. (c) Let the roots be α and β.

$\therefore \ \alpha+\beta=-20$ and $\alpha\beta=b$

Given, $\quad |\alpha-\beta|\le 2m$

$$\Rightarrow \quad |\alpha-\beta|^2\le(2m)^2$$

$$\Rightarrow \quad (\alpha+\beta)^2-4ab\le 4m^2$$

$$\Rightarrow \quad 4a^2-4b\le 4m^2$$

$\Rightarrow a^2-m^2\le b$ and discriminant $D>0$ or $4a^2-4b>0$

$\Rightarrow a^2-m^2\le b$ and $b<a^2$

Hence, $b\in[a^2-m^2, a^2)$

26. (b) Let $\quad \dfrac{x}{x^2-5x+9}=y$

$$\Rightarrow \quad yx^2-5yx+9y=x$$

$$\Rightarrow yx^2-(5y+1)x+9y=0$$

Now, x is real, so

$$D\ge 0$$

$$\Rightarrow \{-(5y+1)\}^2-4\cdot y\cdot(9y)\ge 0$$

$$\Rightarrow \quad -11y^2+10y+1\ge 0$$

$$\Rightarrow \quad 11y^2-10y-1\le 0$$

$$\Rightarrow \quad (11y+1)(y-1)\le 0$$

27. (c) According to the question,

$\alpha, \dfrac{1}{\beta}$ are roots of equation $x^2-3ax+2b=0$

and α, β are roots of equation $x^2-3px+2q=0$

Therefore, we have

$$\alpha+\beta=3p, \alpha\beta=2q$$

and $\quad \alpha+\dfrac{1}{\beta}=3a, \dfrac{\alpha}{\beta}=2b$

$$\therefore \quad (2q-2b)^2=\left(\alpha\beta-\frac{\alpha}{\beta}\right)^2=\alpha^2\left(\beta-\frac{1}{\beta}\right)^2$$

$$=\frac{\alpha}{\beta}\cdot\beta\alpha\left[(\alpha+\beta)-\left(\alpha+\frac{1}{\beta}\right)\right]^2$$

$$=2b\cdot 2q[3p-3a]^2$$

$$=36bq(p-a)^2$$

28. (b) Let D_1 and D_2 be discriminants of $x^2+b_1x+c_1=0$ and $x^2+b_2x+c_2=0$, respectively, then

$$D_1+D_2=b_1^2-4c_1+b_2^2-4c_2$$

$$=(b_1^2+b_2^2)-4(c_1+c_2)$$

$$=b_1^2+b_2^2-2b_1b_2 \quad [\because b_1b_2=2(c_1+c_2)]$$

$$=(b_1-b_2)^2\ge 0$$

$\Rightarrow \ D_1\ge 0$ or $D_2\ge 0$ or D_1 and D_2 both are positive.

Hence, atleast one of the equation has real roots.

29. (b) $x^4+x^2+1=(x^2+1)^2-x^2$

$$=(x^2+x+1)(x^2-x+1)$$

$$x^2+x+1=\left(x+\frac{1}{2}\right)^2+\frac{3}{4}\ne 0\ \forall\ x$$

Therefore, we can cancel this factor and we get

$(a-1)(x^2+x+1)=(a+1)(x^2-x+1)$

or $\quad x^2-ax+1=0$

It has real and distinct roots if $D=a^2-4>0$

30. (b) The discriminant of the given equation are

$$D_1 = a^2 + 12b$$

$$D_2 = c^2 - 4b \text{ and } D_3 = d^2 - 8b$$

$\therefore D_1 + D_2 + D_3 = a^2 + c^2 + d^2 \geq 0$

Hence, atleast one of D_1, D_2 and D_3 is non-negative. Therefore, the equation has atleast two real roots.

31. (a) Roots of $ax^2 + bx + c = 0$ and $cx^2 + bx + a = 0$ are reciprocals of each other.

Hence, $\alpha < \frac{1}{2}$ and $\beta > -1$

32. (d) We have, $(x^2 + x + 1)^2 + 1 = (x^2 + x + 1)(x^2 - x - 5)$

Since, $x^2 + x + 1 > 0 \,\forall\, x \in R$

Dividing both sides by $x^2 + x + 1$, we get

$$(x^2 + x + 1) + \frac{1}{(x^2 + x + 1)} = x^2 - x - 5$$

$\therefore$ LHS ≥ 2

But $x^2 - x - 5 < 1$ for $x \in (-2, 3)$

Therefore, the equation has no solution.

33. (d) Quadratic equation with real coefficients are purely imaginary roots, can be considered as

$p(x) = x^2 + a = 0$, where $a > 0$ and $a \in R$

$\because \quad p\{p(x)\} = 0$

$\Rightarrow \quad (x^2 + a)^2 + a = 0 \Rightarrow x^4 + 2ax^2 + (a^2 + a) = 0$

$\Rightarrow \quad x^2 = \dfrac{-2a \pm \sqrt{4a^2 - 4a^2 - 4a}}{2}$

$\Rightarrow \quad x^2 = -a \pm \sqrt{a}i$

$\Rightarrow \quad x = \pm\sqrt{-a \pm \sqrt{a}i}$

$\therefore\ x = \alpha \pm i\beta$, where $\alpha, \beta \neq 0$

$\because\ a > 0$ and $a \in R$

$\therefore\ p\{p(x)\} = 0$, has complex roots, which are neither purely real nor purely imaginary.

34. (b) $\quad x = 2 + \sqrt{3}$

$\Rightarrow \quad (x - 2)^2 = 3$

$\Rightarrow \quad x^2 - 4x + 1 = 0 \quad$ ...(i)

$\Rightarrow \quad (x - 2)^4 = 9$

$\Rightarrow \quad x^4 - 8x^3 + 24x^2 - 32x + 16 = 9$

$\Rightarrow \quad x^4 - 8x^3 + 18x^2 - 8x + 2 + 6(x^2 - 4x + 1) = 0$

Using Eq. (i), we get

$$x^4 - 8x^3 + 18x^2 - 8x + 2 = 1$$

35. (b) $(31 + 8\sqrt{15})^{x^2 - 3} + 1 = (32 + 8\sqrt{15})^{x^2 - 3}$

or $(31 + 8\sqrt{15})^{x^2 - 3} + 1^{x^2 - 3} = (32 + 8\sqrt{15})^{x^2 - 3}$

$\Rightarrow \quad x^2 - 3 = 1$ or $x = \pm 2$

[for $n = 1, a^n + b^n = (a + b)^n$]

36. (a, c) Given, $\quad b^2 = ac$

$\Rightarrow \quad \left(\dfrac{b}{a}\right)^2 = \dfrac{c}{a}$

$\Rightarrow \quad \left(\dfrac{-b}{a}\right)^2 = \dfrac{c}{a}$

$\Rightarrow \quad (\alpha + \beta)^2 = \alpha\beta$

$\Rightarrow \quad \alpha^2 + \beta^2 + \alpha\beta = 0$

$\Rightarrow \left(\dfrac{\alpha}{\beta}\right)^2 + \left(\dfrac{\alpha}{\beta}\right) + 1 = 0$

$\Rightarrow \quad \dfrac{\alpha}{\beta} = \dfrac{-1 \pm i\sqrt{3}}{2} \quad$ [where, $i = \sqrt{-1}$]

37. (a, b, c) $\sec^2\theta + \text{cosec}^2\theta = \sec^2\theta \cdot \text{cosec}^2\theta$

Sum of the roots is equal to their product and the roots are real.

$\therefore \quad -\dfrac{b}{a} = \dfrac{c}{a}$

$\Rightarrow \quad b + c = 0$

Also, $\quad b^2 - 4ac \geq 0$

$\Rightarrow \quad c^2 - 4ac \geq 0$

$\Rightarrow \quad c(c - 4a) \geq 0$

$\Rightarrow \quad c - 4a \geq 0 \quad [\because c > 0]$

Further, $\quad b^2 + 4ab \geq 0$

or $\quad b + 4a \leq 0 \quad [\because b < 0]$

38. (a, d) Given, one root is $\dfrac{\sqrt{a}}{\sqrt{a} + \sqrt{a - b}}$

On rationalizing, we get

$$\frac{\sqrt{a}}{\sqrt{a} + \sqrt{a - b}} \times \frac{\sqrt{a} - \sqrt{a - b}}{\sqrt{a} - \sqrt{a - b}}$$

$$= \frac{\sqrt{a} \cdot (\sqrt{a} - \sqrt{a - b})}{b} = \frac{a - \sqrt{a(a - b)}}{b}$$

$\therefore$ Other root $= \dfrac{a + \sqrt{a(a - b)}}{b}$

On rationalizing, we get

$$\frac{a + \sqrt{a(a - b)}}{b} \times \frac{a - \sqrt{a(a - b)}}{a - \sqrt{a(a - b)}}$$

$$= \frac{a^2 - (a^2 - ab)}{b(a - \sqrt{a(a - b)})} = \frac{\sqrt{a}}{\sqrt{a} - \sqrt{a - b}}$$

39. (a, b) We can write the given equation as

$$\frac{p}{2x} = \frac{(a + b)x + c(b - a)}{x^2 - c^2}$$

$\Rightarrow p(x^2 - c^2) = 2(a + b)x^2 - 2c(a - b)x$

$\Rightarrow (2a + 2b - p)x^2 - 2c(a - b)x + pc^2 = 0$

For this equation to have equal roots,

$$c^2(a-b)^2 - pc^2(2a+2b-p)=0$$

$\Rightarrow \quad (a-b)^2 - 2p(a+b)+p^2=0 \qquad [\because c^2 \neq 0]$

$\Rightarrow [p-(a+b)^2] = (a+b)^2-(a-b)^2 = 4ab$

$\Rightarrow \quad p-(a+b) = \pm\sqrt{ab}$

$\Rightarrow \quad p = a+b \pm 2\sqrt{ab} = (\sqrt{a} \pm \sqrt{b})^2$

40. (a, d) Given, α, β, γ and δ are in AP, then they may be taken as $a-3d$, $a-d$, $a+d$ and $a+3d$, respectively.

$\because x^2-4x+A=0$ having roots α and γ, then we have

$$\alpha\cdot\gamma = A$$

$\Rightarrow \quad (a-3d)(a+d)=A \qquad ...(i)$

and $\quad \alpha+\gamma=4$

$\Rightarrow \quad (a-3d)+(a+d)=4$

$\Rightarrow 2(a-d)=4 \Rightarrow a-d=2 \qquad ...(ii)$

Also, $x^2-6x+B=0$ having roots β and δ, then

$$\beta\cdot\delta = B$$

$\Rightarrow \quad (a-d)(a+3d)=B \qquad ...(iii)$

Also, $\quad \beta+\delta=6$

$\Rightarrow \quad (a-d)+(a+3d)=6$

$\Rightarrow \quad a+d=3 \qquad ...(iv)$

From Eqs. (ii) and (iv), we get

$$a=\frac{5}{2} \text{ and } d=\frac{1}{2}$$

Thus, Eq. (i), $(a-3d)(a+d)=\left(\frac{5}{2}-3\cdot\frac{1}{2}\right)\left(\frac{5}{2}+\frac{1}{2}\right)=3$

and Eq. (iii), $\quad (a-d)(a+3d)=\left(\frac{5}{2}-\frac{1}{2}\right)\left(\frac{5}{2}+\frac{3}{2}\right)=8$

On comparing with Eqs. (i) and (iii), we get

$$A=3 \text{ and } B=8$$

41. (c, d) Product of roots is

$$\frac{a}{bc}<0 \qquad [\because abc<0]$$

Hence, roots are real and of opposite sign.

42. (b, d) Let $f(x)=x^2+ax+b$, then

$$x^2+(2c+a)x+c^2+ac+b=f(x+c)$$

Thus, the roots of $f(x+c)=0$ will be $0, d-c$.

43. (a, d) Given, x_1 and x_2 are roots of $\alpha x^2-x+\alpha=0$

$\therefore \; x_1+x_2=\frac{1}{\alpha}$ and $x_1x_2=1$

Also, $|x_1-x_2|<1$

$\Rightarrow |x_1-x_2|^2<1 \Rightarrow (x_1-x_2)^2<1$

or $\quad (x_1+x_2)^2-4x_1x_2<1$

$\Rightarrow \quad \frac{1}{\alpha^2}-4<1$ or $\frac{1}{\alpha^2}<5$

or $5\alpha^2-1>0$ or $(\sqrt{5}\alpha-1)(\sqrt{5}\alpha+1)>0$

$\therefore \quad \alpha\in\left(-\infty,\frac{-1}{\sqrt{5}}\right)\cup\left(\frac{1}{\sqrt{5}},\infty\right) \qquad ...(i)$

Also, $\quad D>0$

$\Rightarrow \quad 1-4\alpha^2>0$ or $\alpha\in\left(-\frac{1}{2},\frac{1}{2}\right) \qquad ...(ii)$

From Eqs. (i) and (ii), we get

$$\alpha\in\left(-\frac{1}{2},-\frac{1}{\sqrt{5}}\right)\cup\left(\frac{1}{\sqrt{5}},\frac{1}{2}\right)$$

44. (a, b, c)

(a) If $f(1)=0 \Rightarrow a+b+c=0$

So, roots are real

(b) $D=b^2-4ac$

If $ac<0 \Rightarrow D>0$

So, roots are real

(c) $4ac-b^2<0 \Rightarrow b^2-4ac>0$

So, roots are real

(d) This is not the sufficient condition.

45. (a, b) We have, $\cos^2\theta-\sin^2\theta=\cos 2\theta$

$\Rightarrow \quad \cos^4\theta-\sin^4\theta=\cos 2\theta$

$\Rightarrow \quad (-2b)^2-4b=(-4)^2-4\times 2$

[since LHS is difference of roots of first equation and RHS is difference of roots of second equation]

$\Rightarrow \quad 4b^2-4b=16-8=8$

$\Rightarrow \quad 4b^2-4b-8=0$

$\Rightarrow \quad b^2-b-2=0$

$\Rightarrow \quad (b+1)(b-2)=0$

$\Rightarrow \quad b=2,-1$

46. (a, b, c) Since, the roots of $ax^2+bx+c=0$ are non-real.

So, $f(x)=ax^2+bx+c$ will have same sign for every value of x.

Hence, $f(0)=c$, $f(1)=a+b+c$, $f(-1)=a-b+c$ and $\;f(-2)=4a-2b+c$

$\Rightarrow \quad c(a+b+c)>0,\; c(a-b+c)>0,\; c(4a-2b+c)>0$

47. (a, c) We have, $2x^2+6xy+5y^2=1 \qquad ...(i)$

Eq. (i) can be written as

$$2x^2+(6y)x+5y^2-1=0$$

Since, x is real, we have

$$36y^2-8(5y^2-1)\geq 0$$

$\Rightarrow \quad y^2\leq 2 \Rightarrow -\sqrt{2}\leq y\leq\sqrt{2}$

Eq. (i) can also be rewritten as

$$5y^2+(6x)y+2x^2-1=0$$

Since, y is real, we have

$$36x^2-20(2x^2-1)\geq 0$$

$\Rightarrow \quad 36x^2-40x^2+20\geq 0$

$\Rightarrow \quad -4x^2\geq -20$

$\Rightarrow \quad x^2\leq 5$

$\Rightarrow \quad -\sqrt{5}\leq x\leq\sqrt{5}$

8. (a, d) Roots of $4x^2 - x - 1 = 0$ are irrational.

So, one root common implies both roots are common. Therefore,

$$\frac{4}{3} = \frac{-1}{\lambda + \mu} = \frac{-1}{\lambda - \mu}$$

$$\Rightarrow \quad \lambda = -\frac{3}{4} \text{ and } \mu = 0$$

9. (a, d) Equations $x^2 + bx + c = 0$ and $bx^2 + cx + 1 = 0$ have common root.

$$(bc - 1)^2 = (c - b^2)(b - c^2)$$

$$\Rightarrow b^2c^2 + 1 - 2bc = bc - b^3 - c^3 + b^2c^2$$

$$\Rightarrow \quad 1 + c^3 + b^3 = 3bc$$

$$\Rightarrow (1 + c + b)(c^2 + b^2 + 1 - b - c - bc) = 0$$

$$\Rightarrow b + c + 1 = 0 \text{ and } c^2 + b^2 + 1 = c + b + bc$$

0. (a, b, c) We have, $f(x) = Ax^2 + Bx + C$

$$A = a + b - 2c = (a - c) + (b - c) > 0$$

$$\Rightarrow \quad A > 0$$

Hence, the graph is concave upward. Also, $x = 1$ is obvious solution, therefore both roots are rational.

$$b + c - 2a = (b - a) + (c - a) < 0$$

$$\qquad -\text{ve} \qquad -\text{ve}$$

$$\Rightarrow \quad B < 0$$

$$\therefore \quad -\frac{B}{2A} > 0$$

Hence, abscissa of the vertex is positive.

Option (d) need not be correct as with $a = 5, b = 4, c = 2$, $p < 0$ and with $a = 6, b = 3, c = 2, p > 0$.

1. (c, d) We have,

$$D = (b - c)^2 - 4a(a - b - c) > 0$$

$$\Rightarrow b^2 + c^2 - 2bc - 4a^2 + 4ab + 4ac > 0$$

$$\Rightarrow c^2 + (4a - 2b)c - 4a^2 + 4ab + b^2 > 0, \forall \in R$$

Discriminant of the above expression in c must be negative.

Hence, $(4a - 2b)^2 - 4(-4a^2 + 4ab + b^2) < 0$

$$\Rightarrow \quad 4a^2 - 4ab + b^2 + 4a^2 - 4ab - b^2 < 0$$

$$\Rightarrow \quad a(a - b) < 0$$

$$\Rightarrow a < 0 \text{ and } a - b > 0 \text{ or } a > 0 \text{ and } a - b < 0$$

$$\Rightarrow b < a < 0 \text{ and } b > a > 0$$

2. (a, b, c) $y \in [0, 1]$

where, $y = \dfrac{x + 1}{k + x^2}$

$y = 0$, when $x = -1$

If $y \neq 0$, $yx^2 - x + ky - 1 = 0, x \in R$

$$\Rightarrow \quad D \geq 0$$

$$\Rightarrow 1 - 4y(ky - 1) \geq 0$$

$$\Rightarrow 4ky^2 - 4y - 1 \leq 0, \forall\, y \in [0, 1]$$

$f(0) \leq 0$ and $f(1) \leq 0$

$$\Rightarrow -1 \leq 0 \text{ and } 4k - 4 - 1 \leq 0$$

$$\Rightarrow \quad k \leq \frac{5}{4}$$

$$\therefore \quad k \in \left(-\infty, \frac{5}{4}\right]$$

Also, $k > 0$, otherwise graph of $4ky^2 - 4y - 1$ is concave downward and is negative for unbounded interval of values of y.

Hence, $k \in \left[0, \dfrac{5}{4}\right]$

53. (a, d) Let $\dfrac{x^2 + ax + 3}{x^2 + x + a} = y$

$$\Rightarrow \quad x^2(1 - y) - x(y - a) + 3 - ay = 0$$

$\because \; x \in R$

$$\therefore \quad (y - a)^2 - 4(1 - y)(3 - ay) \geq 0$$

$$\Rightarrow (1 - 4a)y^2 + (2a + 12)y + a^2 - 12 \geq 0 \quad \text{...(i)}$$

Now, Eq. (i) is true for all $y \in R$, if $1 - 4a > 0$ and $D \leq 0$.

$$\therefore \quad a < \frac{1}{4} \text{ and } 4(a + b)^2 - 4(a^2 - 12)(1 - 4a) \leq 0$$

$$\Rightarrow \quad a < \frac{1}{4} \text{ and } 4a^3 - 36a + 48 \leq 0$$

$$\Rightarrow \quad a < \frac{1}{4} \text{ and } 4a^3 \leq 36a - 48$$

$$\Rightarrow \quad 4a^3 < 36\left(\frac{1}{4}\right) - 48$$

$$\Rightarrow \quad 4a^3 + 39 < 0 \qquad \left[\because a < \frac{1}{4}\right]$$

54. (b, c) Let $f(x) = x^3 + 3x^2 - 9x + c$ is of the form $(x - \alpha)^2$ and $(x - \beta)$.

Showing that α is a double root so that $f'(x) = 0$ has also one root α

i.e. $3x^2 + 6x - 9 = 0$ has one root α.

Hence, $x^2 + 2x - 3 = 0$ or $(x + 3)(x - 1) = 0$ has the root α which can be either -3 or 1.

If $\alpha = 1$, then $f(x) = 0$ gives $c - 5 = 0$ or $c = 5$.

If $\alpha = -3$, then $f(x) = 0$ gives

$$-27 + 27 + 27 + c = 0$$

$$\therefore \quad c = -27$$

55. (b, c, d) Given equation is

$$x^3 - ax^2 + bx - 1 = 0.$$

If roots of the equation be α, β and γ, then

$$\alpha^2 + \beta^2 + \gamma^2 = (\alpha + \beta + \gamma)^2 - 2(\alpha\beta + \beta\gamma + \gamma\alpha)$$

$$= a^2 - 2b$$

$$\alpha^2\beta^2 + \beta^2\gamma^2 + \gamma^2\alpha^2 = (\alpha\beta + \beta\gamma + \gamma\alpha)^2 - 2\alpha\beta\gamma(\alpha + \beta + \gamma)$$

$$= b^2 - 2a \text{ and } \alpha^2\beta^2\gamma^2 = 1$$

So, the equation whose roots are α^2, β^2 and γ^2 is given by $x^3-(a^2-2b)x^2+(b^2-2a)x-1=0$

It is identical to

$$x^3-ax^2+bx-1=0$$

$\Rightarrow \quad a^2-2b=a$ and $b^2-2a=b$

Eliminating b, we get

$$\frac{(a^2-a)^2}{4}-2a=\frac{a^2-a}{2}$$

$\Rightarrow \quad a\{a(a-1)^2-8-2(a-1)\}=0$

$\Rightarrow \quad a(a^3-2a^2-a-6)=0$

$\Rightarrow \quad a(a-3)(a^2+a+2)=0$

$\Rightarrow \quad a=0$ or $a=3$ or $a^2+a+2=0$

which gives $b=0$ or $b=3$ or $b^2+b+2=0$

So, $a=b=0$

or $a=b=3$ or a and b are roots of $x^2+x+2=0$

56. (a, c, d) Let the roots be $\frac{a}{r}$, a and ar, where $a>0, r>1$.

Now, $$\frac{a}{r}+a+ar=-p \quad \text{...(i)}$$

$$a\left(\frac{a}{r}\right)+a(ar)+(ar)\left(\frac{a}{r}\right)=q \quad \text{...(ii)}$$

and $$\left(\frac{a}{r}\right)(a)(ar)=1 \quad \text{...(iii)}$$

$\Rightarrow \quad a^3=1$

$\Rightarrow \quad a=1$

Hence, Eq. (iii) is correct from Eq. (i), putting $a=1$, we get

$$-p-3>0 \quad \left[\because r+\frac{1}{r}>2\right]$$

$\Rightarrow \quad p<-3$

Hence, Eq. (ii) is not correct.

Also, $$\frac{1}{r}+1+r=-p \quad \text{...(iv)}$$

From Eq. (ii), putting $a=1$, we get

$$\frac{1}{r}+r+1=q \quad \text{...(v)}$$

From Eqs. (iv) and (v), we have

$$-p=q$$

$\Rightarrow \quad p+q=0$

Hence, Eq. (i) is correct.

Now, as $r<1$

$$\frac{a}{r}=\frac{1}{r}<1 \text{ and } ar=r>1$$

Hence, Eq. (iv) is correct.

57. (c) We have,

$$2^{2x}+(a-1)2^{x+1}+a=0$$

Let $2^x=t$, we have

$$t^2+2(a-1)t+a=0$$

Now, $t=1$ should be between the roots of the equation

$$1+2(a-1)+a<0$$

$\Rightarrow \quad 1+2a-2+a<0$

$\Rightarrow \quad 3a<1$

$\Rightarrow \quad a<\frac{1}{3}$

$\therefore \quad a\in\left(-\infty, \frac{1}{3}\right)$

Sol. (Q.Nos. 58-60)

$$(\beta-\alpha)^2=\{(\beta+h)-(\alpha+h)\}^2$$

$$(\beta+\alpha)^2-4\alpha\beta=[(\beta+h)+(\alpha+h)]^2-4(\beta+h)(\alpha+h)$$

$$(-b_1)^2-4c_1=(-b_2)^2-4c_2$$

$$D_1=D_2$$

The least value of $f(x)$ is $\frac{-D_1}{4}=-\frac{1}{4}$

$\Rightarrow \quad D_1=1$ and $D_2=1$

58. (a) Therefore, the least value of $g(x)$ is $-\frac{D_2}{4}=-\frac{1}{4}$

59. (d) The least value of $g(x)$ occurs at $-\frac{b_2}{2}=\frac{7}{2}$ or $b_2=-7$

60. (c) Now, $b_2^2-4c_2=D_2$

or $\quad 49-4c_2=1$ or $\frac{48}{4}=c_2$ or $c_2=12$

$\Rightarrow x^2-7x+12=0$ or $x=3,4$

Sol. (Q.Nos. 61-63)

61. (d) $\because AC=4\sqrt{2}$

$\therefore \quad AB=BC=\frac{4\sqrt{2}}{\sqrt{2}}=4$ units

$$OB=\sqrt{4^2-(2\sqrt{2})^2}=2\sqrt{2}$$

$\therefore \quad A=(-2\sqrt{2},0), C(2\sqrt{2},0)$ and $B(0,-2\sqrt{2})$

Since, $y=ax^2+bx+c$ passes through A, B and C, we get

$$y=\frac{x^2}{2\sqrt{2}}-2\sqrt{2}$$

62. (c) Minimum value of $y=\frac{x^2}{(2\sqrt{2})}-2\sqrt{2}$ is $-2\sqrt{2}$ at $x=0$

63. (c) $f(x)=0$

$\Rightarrow \frac{x^2}{(2\sqrt{2})}-2\sqrt{2}=0$ or $x=\pm 2\sqrt{2}$

Therefore, number of integral values of k for which k lies in $(-2\sqrt{2}, 2\sqrt{2})$ is 5.

Sol. (Q.Nos. 64-66)

Let unknown polynomial be $P(x)$. Let $Q(x)$ and $R(x)$ be the quotient and remainder, respectively.

When it is divided by $(x-3)(x-4)$, then

$$P(x)=(x-3)(x-4)Q(x)+R(x)$$

Then, we have

$$R(x)=ax+b$$

$$\Rightarrow \quad P(x)=(x-3)(x-4)Q(x)+ax+b$$

Given that $P(3)=2$ and $P(4)=1$

Hence, $3a+b=2$ and $4a+b=1$

$$\Rightarrow \quad a=-1 \text{ and } b=5$$

$$\Rightarrow \quad R(x)=5-x$$

64. (d) $5-x=x^2+ax+1$

$$\Rightarrow x^2+(a-1)x-4=0$$

Given that roots are real and distinct, therefore

$$D>0$$

$$\Rightarrow \quad (a+1)^2+16>0$$

which is true for all real a.

65. (c) $-x+5=px^2+(q-1)x+6$

$$\Rightarrow \quad px^2+qx+1=0$$

Now, $p>0$ and equation has no distinct real roots or equation has real and equal or imaginary roots.

Then, $px^2+qx+1\geq 0, \forall x\in R$

$$\Rightarrow \quad f(3)\geq 0$$

$$\Rightarrow \quad 9p+3q+1\geq 0$$

$$\Rightarrow \quad 3p+q\geq -\frac{1}{3}$$

Hence, the least value of $3p+4$ is $-\frac{1}{3}$

66. (c) $f(x)=y=\dfrac{-x+5}{x^2-3x+2}$

$$\Rightarrow \quad yx^2+(1-3y)x+2y-4=0$$

Now, x is real, then $D\geq 0$

$$\Rightarrow \quad (1-3y)^2-4y(2y-5)\geq 0$$

or $\quad y^2+14y+1\geq 0$

or $y\in\left(-\infty,\dfrac{-14-\sqrt{192}}{2}\right)\cup\left(\dfrac{-14+\sqrt{192}}{2},\infty\right)$

or $y\in(-\infty,-7-4\sqrt{3})\cup(-7+4\sqrt{3},\infty)$

67. (i) $\rightarrow$ B, (ii) $\rightarrow$ C, (iii) $\rightarrow$ A, (iv) $\rightarrow$ D

$\because ax^2+2bx+c=0$

$\therefore D=4(b^2-ac)\geq 0$

(i) $f(0)$ and sign of a should be same and $\dfrac{-b}{2a}>0$

$\Rightarrow a, -b$ and c are of same sign.

(ii) $f(0)$ and a should have same sign and $-\dfrac{b}{2a}<0$

$\Rightarrow a, b$ and c are of same sign.

(iii) $f(0)$ and a should have opposite sign.

$\Rightarrow a$ and $-c$ are of opposite sign.

(iv) $D=0$

$\Rightarrow b^2=ac$

$\Rightarrow a, b$ and c are in GP.

68. (i) $\rightarrow$ D, (ii) $\rightarrow$ B, (iii) $\rightarrow$ C, (iv) $\rightarrow$ A

(i) Obviously, when $b\geq 0$, we have no real roots as all the terms becomes positive.

(ii) For $b=-2$, we have,

$$x^2-2|x|+1=0$$

$$(|x|-1)^2=0$$

$$|x|=1$$

$$\Rightarrow \quad x=\pm 1$$

Hence, the given equation has two distinct roots.

(iii) Clearly, the above given equation can never have three distinct real roots for any real value of b.

(iv) Also, when $b<-2$, then

$$|x|=\frac{-b\pm\sqrt{b^2-4}}{2}>0$$

Hence, the given equation has four distinct real roots as $|-b|>\sqrt{b^2-4}$.

69. (i) $\rightarrow$ B, (ii) $\rightarrow$ D, (iii) $\rightarrow$ A, (iv) $\rightarrow$ C

(i) $kx^2-(k+1)x+2k-1=0$

As, $\quad D<0$

$$\Rightarrow (k+1)^2-4k(2k-1)<0$$

$$\Rightarrow \quad -7k^2+6k+1<0$$

$$\Rightarrow \quad (k-1)(7k+1)>0$$

$$\Rightarrow \quad k\in\left(-\infty,-\frac{1}{7}\right)\cup(1,\infty)$$

(ii) $x^2-2(4k-1)x+(15k^2-2k-7)>0$

$$\Rightarrow \quad D<0$$

$$\Rightarrow 4[(4k-1)^2-(15k^2-2k-7)]<0$$

$$\Rightarrow \quad k^2-6k+8<0$$

$$\Rightarrow \quad (k-4)(k-2)<0$$

$$\Rightarrow \quad k\in(2,4)$$

(iii) $2-k-k^2=0$

$$\Rightarrow \quad k=1 \text{ or } -2$$

(iv) $\dfrac{\alpha}{\beta}=\dfrac{1}{2}\Rightarrow 2\alpha=\beta$

$\Rightarrow 3\alpha=1-2k$ and $2\alpha^2=k^2+2$

So, $2\left[\dfrac{1-2k}{3}\right]^2=k^2+2$

So, $2(4k^2+1-4k)=9k^2+18$

$$\Rightarrow \quad k^2+8k+16=0$$

$$\Rightarrow \quad k=-4$$

70. (i) $\rightarrow$ B, (ii) $\rightarrow$ A, (iii) $\rightarrow$ C, (iv) $\rightarrow$ D

(i) $f(x)=(x-a)(x-c)+\pi(x-b)(x-d)=0$

$f(a)=\pi(a-b)(a-d)>0$

$f(b)=(b-a)(b-c)<0$

$f(c)=\pi(c-b)(c-d)<0$

and $f(d)=(d-a)(d-c)>0$

One root in (a,b) and other root in (c,d).

(ii) As, $a>0$, $f(x)=ax^2+bx+c$

$f(1)=a+b+c<0$

$\Rightarrow$ 1 lies in between the roots.

(iii) $x^2+bx+c=0$

As difference of roots is an integer and sum is an integer.

$\Rightarrow$ This is only possible when fractional part of both is $\frac{1}{2}$ or 0.

In $\frac{1}{2}$ case, product cann't be integer.

$\Rightarrow$ Roots are integers.

(iv) $(a^2+b^2+c^2)x^2+2(ab+bc+cd)x+(b^2+c^2+d^2)=0$

$\because$ *a, b, c* and *d* are in GP, let common ratio be *r*.

Then, $(1+r^2+r^4)x^2+2(r+r^3+r^5)x+(r^2+r^4+r^6)=0$

$\Rightarrow D=4r^2(1+r^2+r^4)-4r^2(1+r^2+r^4)^2=0$

71. (i) $\rightarrow$ C, (ii) $\rightarrow$ B, (iii) $\rightarrow$ D, (iv) $\rightarrow$ A

Given, $x^4+x^2(1-2k)+k^2-1=0$

Let $x^2=t$

$t^2+(1-2k)t+(k^2-1)=0$

(I) **Case I** No solution for new equation

$D=(1-2k)^2-4(k^2-1)<0$

$\Rightarrow \quad 4k^2-4k+1-4k^2+4<0$

$\Rightarrow \quad k>\frac{5}{4}$

Case II The new equation have both negative solution.

$\Rightarrow \quad -\frac{b}{2a}<0\Rightarrow\frac{2k-1}{2}<0$

$\Rightarrow \quad k<\frac{1}{2}$

and $\quad f(0)>0\Rightarrow k^2-1>0$

$\Rightarrow \quad k>1$ or $k<-1$

Also, $D\geq 0\Rightarrow k\leq\frac{5}{4}$

Taking intersections of above 3, we get

$k<-1$

$\therefore$ For no solution of original solution,

$k\in(-\infty,-1)\cup\left(\frac{5}{4},\infty\right)$

(ii) For only one solution of original solution, one solution must be zero and other solutions must be negative.

(a) $f(0)=0\Rightarrow k=\pm 2$

(b) $\frac{-b}{2a}<0\Rightarrow k<\frac{1}{2}$

$\therefore k=-1$

(iii) For two solutions (in original equation), the new equation must have one positive and one negative solution or both positive and coincident.

(a) $f(0)<0\Rightarrow k^2-1<0$

$\Rightarrow k\in(-1,1)$

(b) $D=0\Rightarrow k=\frac{5}{4}$

(c) $\frac{-b}{2a}>0\Rightarrow k>\frac{1}{2}$

$\therefore k\in(-1,1)\cup\left\{\frac{5}{4}\right\}$

(iv) For three solutions (in original equation), the new equation must have one positive and one solution as 0.

$\therefore \quad f(0)=k^2-1=0\Rightarrow k\pm 1$

and $\quad \frac{-b}{2a}>0\Rightarrow k>\frac{1}{2}$

$\therefore \quad k=2$

Sol. (Q.Nos. 72-73)

72. (1.25) We have

$x^4-5x^3+ax^2+bx+c=0$

One root is $3+\sqrt{2}$.

$\therefore$ Other roots are $3-\sqrt{2},\alpha,\beta$

Now, $3+\sqrt{2}+3-\sqrt{2}+\alpha+\beta=5$

$\Rightarrow \alpha+\beta=-1$

$(3+\sqrt{2})(3-\sqrt{2})+\alpha\beta+(\alpha+\beta)(3+\sqrt{2}+3-\sqrt{2})=a$

$\Rightarrow 7+\alpha\beta+6(\alpha+\beta)=a$

$\Rightarrow \quad 7+\alpha\beta-6=a \qquad [\because \alpha+\beta=-1]$

$\Rightarrow \quad \alpha\beta=a-1$

$\alpha\beta[(3+\sqrt{2})+3-\sqrt{2}]+(\alpha+\beta)(3+\sqrt{2})(3-\sqrt{2})=-b$

$\Rightarrow \quad 6\alpha\beta+7(\alpha+\beta)=-b$

$\Rightarrow \quad \alpha\beta=\frac{7-b}{6}$

$\Rightarrow \quad \alpha\beta(3+\sqrt{2})(3-2\sqrt{2})=c$

$\Rightarrow \quad \alpha\beta=\frac{c}{7}$

α,β are roots of

$x^2+x+a-1=0$

$D\geq 0\Rightarrow 1-4(a-1)\geq 0$

$a\leq\frac{5}{4}$

$\therefore$ Greatest value of $a=1.25$

3. (5.5) Also α, β are roots of equation

$$x^2 + x - \frac{7-b}{6} = 0$$

$$D \geq 0 \Rightarrow 1 - \frac{2(7-b)}{3} \geq 0 = b \geq \frac{11}{2}$$

$\therefore$ Least value of $b = 5.5$.

4. (1.236) We have α, β are the roots of equation

$$x^2 + (m-4)x + m^2 - 3m + 3 = 0$$

$$(\alpha+\beta)^2 - 2\alpha\beta = \alpha^2 + \beta^2$$

$$(m-4)^2 - 2(m^2 - 3m + 3) = 0 \qquad [\because \alpha^2 + \beta^2 = 6]$$

$$\Rightarrow m^2 + 2m - 4 = 0 \Rightarrow m = -1 \pm \sqrt{5}$$

α, β are real roots.

$$D \geq 0, (m-4)^2 - 4(m^2 - 3m + 3) \geq 0$$

$$\Rightarrow 3m^2 - 4m - 4 \leq 0$$

$$\Rightarrow m \in \left[\frac{-2}{3}, 2\right]$$

$$\therefore m = \sqrt{5} - 1 = 2.236 - 1 = 1.236$$

5. (13.44) $\dfrac{m\alpha^2}{1-\alpha} + \dfrac{m\beta^2}{1-\beta} = \dfrac{m(\alpha^2+\beta^2 - \alpha\beta(\alpha+\beta))}{(1-\alpha)(1-\beta)}$

$$= \frac{m[(m-4)^2 - 2(m^2-3m+3) + (m-4)(m^2-3m+3)]}{1 + (m-4) + m^2 - 3m + 3}$$

$$= \frac{m^3 - 8m^2 + 13m - 2}{m-2} = m^2 - 6m + 1 \qquad m \in \left[\frac{-2}{3}, 2\right]$$

$$\therefore 2^2 - 6(2) + 9 < m^2 - 6m + 9 \leq \left(\frac{-2}{3}\right)^2 - 6\left(\frac{-2}{3}\right) + 9$$

$$1 < m^2 - 6m + 9 \leq \frac{121}{9}$$

$\therefore$ Maximum value of $\dfrac{m\alpha^2}{1-\alpha} + \dfrac{m\beta^2}{1-\beta} = \dfrac{121}{9} = 13.44$

6. (6) $\quad a^2 \geq 8b$ and $4b^2 \geq 4a$

Now, $\quad b^2 \geq a$

$$\Rightarrow b^4 \geq a^2 \geq 8b \qquad [a > 0, b > 0]$$

$$\therefore b^3 \geq 8 \Rightarrow b \geq 2 \qquad \text{...(i)}$$

Again, $a^2 \geq 16$

or $\quad a \geq 4 \qquad$ ...(ii)

From Eqs. (i) and (ii),

$$(a+b)_{\text{least}} = 6$$

7. (6) We have, $\sqrt{2x + \sqrt{2x+4}} = 4$

$$\Rightarrow \sqrt{2x+4} = 16 - 2x$$

$$\Rightarrow 2x + 4 = 4(8-x)^2$$

$$\Rightarrow 2x^2 - 33x + 126 = 0 \Rightarrow (2x-21)(x-6) = 0$$

$$\Rightarrow x = 6, \frac{21}{2}$$

But $x = \dfrac{21}{2}$ does not satisfy the given equation.

78. (12) From the given relations, we can say that a and $\dfrac{1}{b}$ are the roots of the equation $6x^2 + 20x + 15 = 0$

$$\therefore \quad a + \frac{1}{b} = -\frac{10}{3} \text{ and } \frac{a}{b} = \frac{5}{2}$$

$$\frac{b^3}{ab^2 - 9(ab+1)^3} = \frac{1}{a \cdot \frac{1}{b} - 9\left(a + \frac{1}{b}\right)^3}$$

$$= \frac{1}{\frac{5}{2} + 9 \cdot \frac{1000}{27}} = \frac{6}{2015}$$

$$\therefore \quad \frac{4030b^3}{ab^2 - 9(ab+1)^3} = \frac{4030 \times 6}{2015}$$

$$= 12$$

79. (4) Given, $a^2 - 4a + 1 = 4$

$$\Rightarrow \quad a^2 + 1 = 4(1+a)$$

$$y = \frac{(a-1)(1+a^2)}{a^2-1}$$

$$= \frac{a^2+1}{a+1} = \frac{4(a+1)}{a+1} = 4$$

80. (16) Given, $\alpha\beta, \alpha\beta(\alpha+\beta)$ and $\alpha^3 + \beta^3$ are in GP.

$$\alpha + \beta = 4, \alpha\beta = k \text{ and } \alpha\beta^2 + \alpha^2\beta = \alpha\beta(\alpha+\beta) = 4k$$

$$\alpha^3 + \beta^3 = (\alpha+\beta)^3 - 3\alpha\beta(\alpha+\beta)$$

$$= 64 - 3k(4) = 4(16 - 3k)$$

$\therefore$ k, $4k$ and $4(16-3k)$ are in GP.

$$16k^2 = 4k(16 - 3k)$$

$$4k(4k - 16 + 3k) = 0$$

$$k = 0, k = \frac{16}{7}$$

81. (12) We have,

$$\alpha^2 - 6\alpha + 12 = 0$$

$$\therefore \quad 6(\alpha - 2) = \alpha^2$$

and $\quad \beta^2 - 6\beta + 12 = 0$

$$\therefore \quad \beta(\beta - 6) + 12 = 0$$

So, $(\alpha-2)^{24} - \dfrac{(\beta-6)^8}{\alpha^8} + 1$

$$= (\alpha-2)^{24} - \frac{12^8}{\alpha^8\beta^8} + 1 = (\alpha-2)^{24} - \frac{12^8}{12^8} + 1$$

$$= (\alpha-2)^{24}$$

$$= (3 \pm \sqrt{3}i - 2)^{24} \qquad [\because \alpha = 3 \pm \sqrt{3}i]$$

$$= (1 \pm \sqrt{3}i)^{24}$$

$$= \left\{2\left(\cos\frac{\pi}{3} \pm i \sin\frac{\pi}{3}\right)\right\}^{24}$$

$$= 2^{24}(\cos\pi + i\sin 8\pi) = 4^{12}$$

82. (540) Let $\alpha_1 = A, \beta_1 = AR, \alpha_2 = AR^2$ and $\beta_2 = AR^3$

We have, $\alpha_1 + \beta_1 = 6$

$\Rightarrow \quad A(1+R) = 6 \quad$ ...(i)

and $\quad \alpha_1\beta_1 = p$

$\Rightarrow \quad A^2R = p \quad$ ...(ii)

Also, $\quad \alpha_2 + \beta_2 = 54$

$\Rightarrow \quad AR^2(1+R) = 54 \quad$ ...(iii)

and $\quad \alpha_2\beta_2 = q$

$\Rightarrow \quad A^2R^5 = q \quad$ ...(iv)

Now, on dividing Eq. (iii) by Eq. (i), we get

$$\frac{AR^2(1+R)}{A(1+R)} = \frac{54}{6} = 9$$

or $\quad R^2 = 9$

$\therefore \quad R = 3 \quad$ [as it is an increasing GP]

$\therefore$ On putting $R = 3$ in Eq. (i), we get

$$A = \frac{6}{4} = \frac{3}{2}$$

$\therefore \quad p = A^2R = \frac{9}{4} \times 3 = \frac{27}{4}$

and $\quad q = A^2R^5 = \frac{9}{4} \times 243 = \frac{2187}{4}$

Hence, $q - p = \frac{2187 - 27}{4} = \frac{2160}{4} = 540$

83. (8) $x^2 + mx + n = 0$ and $x^2 + px + m = 0$

$2(\alpha + \beta) = -m \quad$ ...(i)

$4\alpha\beta = n \quad$ ...(ii)

and $\quad \alpha + \beta = -p \quad$ ...(iii)

$\alpha\beta = m \quad$ ...(iv)

From Eqs. (i) and (iii), $2p = m$

From Eqs. (ii) and (iv), $4m = n$

$\Rightarrow \quad \frac{n}{p} = \frac{4m}{m/2} = 8$

84. (3) We have, $(\alpha + \beta)^2 = \left(\frac{1}{\alpha} + \frac{1}{\beta}\right)(\alpha^2 + \beta^2)$

$\Rightarrow \quad (\alpha + \beta)^2 = \left(\frac{1}{\alpha} + \frac{1}{\beta}\right)[(\alpha + \beta)^2 - 2\alpha\beta]$

Substituting $\alpha + \beta = -\frac{b}{a}$ and $\alpha\beta = \frac{c}{a}$, we get

$$\frac{b^2}{a^2} = -\frac{b}{c}\left(\frac{b^2}{a^2} - \frac{2c}{a}\right)$$

$\Rightarrow \quad cb^2 + b(b^2 - 2ac) = 0$

$\because \quad b \neq 0$

$\therefore \quad bc + b^2 - 2ac = 0$

$\because$ *a*, *b* and *c* are in AP.

$\therefore \quad b = \frac{a+c}{2}$

Therefore, we have

$$\frac{(a+c)c}{2} + \left(\frac{a+c}{2}\right)^2 - 2ac = 0$$

$\Rightarrow a^2 - 4ac + 3c^2 = 0$ or $(a-c)(a-3c) = 0$

$\therefore \quad a \neq c$

$\therefore \quad a = 3c$

$\therefore \quad \frac{a}{c} = 3$

85. (9) We have, $x^3 + x^2 + ax + b = 0$

Sum of roots $= -1$

$\Rightarrow 1 - \text{cosec}^2\,\alpha = -1$

$\Rightarrow \quad \sin^2\alpha = \frac{1}{2}$

Thus, $\cos^2\alpha = \frac{1}{2}$

So, $P(x) = \left(x - \frac{1}{2}\right)\left(x - \frac{1}{2}\right)(x+2)$

Hence, $P(2) = \frac{3}{2} \cdot \frac{3}{2} \cdot 4 = 9$

86. (1.5) For equation $x^2 + 2x \sin y + 1 = 0$

$$D = 4\sin^2 y - 4 < 0$$

Thus, equation have both roots common.

$\Rightarrow \quad \frac{1}{a} = \frac{2\sin y}{1} = \frac{1}{1}$

$\Rightarrow \quad a = 1$ and $\sin y = \frac{1}{2}$

$\therefore a + \sin y = 1.5$

87. (3) We have,

$$2x^2 + 4x(y-3) + 7y^2 - 2y + t = 0$$

$D = 0$ (for one solution)

$\Rightarrow \quad 16(y-3)^2 - 8(7y^2 - 2y + t) = 0$

$\Rightarrow \quad 2(y-3)^2 - (7y^2 - 2y + t) = 0$

$\Rightarrow 2(y^2 - 6y + 9) - (7y^2 - 2y + t) = 0$

$\Rightarrow \quad -5y^2 - 10y + 18 - t = 0$

$\Rightarrow \quad 5y^2 + 10y + t - 18 = 0$

Again, $D = 0$ (for one solution)

$100 - 20(t - 18) = 0$

$\Rightarrow \quad 5 - t + 18 = 0 \Rightarrow t = 23$

For $t = 23$,

$5y^2 + 10y + 5 = 0$

$\Rightarrow \quad (y+1)^2 = 0$ or $y = -1$

For $y = -1$,

$2x^2 - 16x + 32 = 0$

$\Rightarrow \quad x^2 - 8x + 16 = 0 \Rightarrow x = 4$

$\therefore \quad x + y = 3$

88. (7) Given, $a+b+c=1$ ...(i)

$ab+bc+ca=0$ ...(ii)

and $abc=2$ ...(iii)

Now, $(a+b+c)^2=1$

$a^2+b^2+c^2+2\Sigma ab=1$

$\therefore \quad a^2+b^2+c^2=1$

Now, $a^3+b^3+c^3-3abc=(a+b+c)[\Sigma a^2-\Sigma ab]$

$=1(1-0)=1$

$\Rightarrow \quad a^3+b^3+c^3=1+3abc \Rightarrow a^3+b^3+c^3=7$

89. (36) We have, $x_1+x_2+x_3=8$

$x_1x_2x_3=d$

and $x_1x_2+x_2x_3+x_3x_1=c$

Possible roots 1, 2, 5 or 1, 3, 4

[as, roots are being distinct positive integers]

$\therefore \quad d=10$ or $d=12$

$\Rightarrow \quad c=2+10+5=17$ or $3+12+4=19$

Hence, $d=10$ and $c=17$

or $d=12$ and $c=19$

90. (6621) The given equation

$$x+\frac{1}{x}=3$$

$$\therefore \quad x^2+\frac{1}{x^2}=7$$

$$\Rightarrow \quad x^4+\frac{1}{x^4}=47 \Rightarrow x^8+\frac{1}{x^8}=(47)^2-2$$

$$\therefore \quad x^8+x^{-8}=2207 \quad \text{...(i)}$$

Now, $y=x^9+x^7+x^{-9}+x^{-7}$

$$=x^8\left(x+\frac{1}{x}\right)+x^{-9}+x^{-7}$$

$$=x^8\left(x+\frac{1}{x}\right)+x^{-8}\left(x+\frac{1}{x}\right)$$

$$=\left(x+\frac{1}{x}\right)(x^8+x^{-8}) \quad \text{...(ii)}$$

Substitute, the value of $x^8+x^{-8}=2207$ from Eq. (i)

and $x+\frac{1}{x}=3$

$y=(3)(2207)=6621$

91. (6) Let the roots be $a-3d$, $a-d$, $a+d$ and $a+3d$.

Sum of roots $=4a=0$

$\Rightarrow \quad a=0$

Hence, roots are $-3d$, $-d$, d and $3d$.

Product of roots $=9d^4=m^2$

$$\Rightarrow \quad d^2=\frac{m}{3} \quad \text{...(i)}$$

Again, $\Sigma x_1x_2=3d^2-3d^2-9d^2-d^2-3d^2+3d^2$

or $\frac{10m}{3}=3m+2$ or $10m=9m+6$ or $m=6$

92. (1.667) $\because 4^x-3^{x-\frac{1}{2}}=3^{x+\frac{1}{2}}-2^{2x-1}$

$$\Rightarrow \quad 2^{2x}+2^{2x-1}=3^{x+\frac{1}{2}}+3^{x-\frac{1}{2}}$$

$$\Rightarrow \quad 2^{2x}\left(1+\frac{1}{2}\right)=3^{x-\frac{1}{2}}(1+3)$$

$$\Rightarrow \quad \frac{3}{2}2^{2x}=4\cdot 3^{x-\frac{1}{2}} \Rightarrow 2^{2x-3}=3^{x-\frac{3}{2}}$$

Taking log both sides, we get

$$(2x-3)\log 2=\left(x-\frac{3}{2}\right)\log 3$$

$$\Rightarrow \quad 2x\log 2-3\log 2=x\log 3-\frac{3}{2}\log 3$$

$$\Rightarrow \quad x\log 4-x\log 3=3\log 2-\frac{3}{2}\log 3$$

$$\Rightarrow \quad x\log\left(\frac{4}{3}\right)=\log 8-\log 3\sqrt{3}$$

$$\Rightarrow \quad \log\left(\frac{4}{3}\right)^x=\log\left(\frac{8}{3\sqrt{3}}\right) \Rightarrow \left(\frac{4}{3}\right)^x=\frac{8}{3\sqrt{3}}=\left(\frac{4}{3}\right)^{\frac{3}{2}}$$

$$\therefore \quad x=\frac{3}{2}$$

$\Rightarrow \quad a=3$ and $b=2$

$$\therefore \quad \frac{a+b}{3}=\frac{5}{3}=1.667$$

93. (1.5) $\because 2^{x+2}\cdot 27^{\frac{x}{x-1}}=9 \Rightarrow 2^{x+2}\cdot 3^{\frac{3x}{x-1}}=9$

Taking log both sides, we get

$$(x+2)\log 2+\frac{3x}{x-1}\log 3=\log 9$$

$$\Rightarrow \quad (x+2)\log 2+\frac{3x}{x-1}\log 3=2\log 3$$

$$\Rightarrow (x+2)\log 2+\left(\frac{3x}{x-1}-2\right)\log 3=0$$

$$\Rightarrow \quad (x+2)\log 2+\left(\frac{x+2}{x-1}\right)\log 3=0$$

$$\Rightarrow \quad (x+2)\left[\log 2+\frac{1}{x-1}\log 3\right]=0$$

$$\Rightarrow \quad x=-2 \text{ or } \frac{1}{x-1}\log 3+\log 2=0$$

$$\Rightarrow \quad x=-2 \text{ or } \frac{1}{x-1}=-\frac{\log 2}{\log 3} \Rightarrow x=-2 \text{ or } 1-x=\frac{\log 3}{\log 2}$$

$\Rightarrow \quad x=-2$ or $x=+1-\log_2 3$

$\therefore$ The sum of roots $=-2+1-\log_2 3=-1-\log_2 3$

On comparing, we get

$a=1, b=2$ and $c=3$

Thus, $\frac{ac}{b}=\frac{1.3}{2}=\frac{3}{2}=1.5$

CHAPTER
03

COMPLEX NUMBERS

LEVEL 01 \\ Starter Level

3.1 Algebra of Complex Number

1 Evaluate

(i) i^{135} (ii) $\frac{1}{i^{63}}$

(iii) i^{1998} (iv) i^{-9999}

(v) $\sqrt{-36}+4\sqrt{-9}+2\sqrt{-4}$ (vi) $(-\sqrt{-1})^{4n+3}, n \in N$

2 Find the value of $1+i^2+i^4+i^6+\ldots+i^{2n}$, where $i=\sqrt{-1}, n \in N$.

3 Express each one of the following in the standard form $a+ib$.

(i) $(4+3i)(4-3i)$ (ii) $\frac{5+4i}{4+5i}$

(iii) $\sqrt{-12}(\sqrt{-3}-\sqrt{2})$

4 Find the value of $\frac{i^{592}+i^{590}+i^{588}+i^{586}+i^{584}}{i^{582}+i^{580}+i^{578}+i^{576}+i^{574}}$.

5 If $z=4+i\sqrt{7}$, then find the value of $z^3-4z^2-9z+91$.

6 Find the value of $\sum_{n=0}^{100} i^{n!}$, $[i=\sqrt{-1}]$.

7 Find the value of $(1+i)^6+(1-i)^6$.

3.2 Modulus, Conjugate and Arguments of Complex Number

8 Find the modulus of the following complex number

(i) $3+4i$ (ii) $(1+2i)(-2+3i)$ (iii) $\frac{3-2i}{2i}$

9 Solve $|z| = z + 1 + 2i$.

10 If z is any complex number such that $|z+4| \le 3$, then find the greatest value of $|z+1|$.

11 Find the greatest and the least value of $|z_1 + z_2|$, if $z_1 = 24 + 7i$ and $z_2 = 6$, when $i = \sqrt{-1}$.

12 Find the conjugate of the following complex number.

(i) $3+4i$ (ii) $4i(3-5i)$ (iii) $\frac{1+2i}{1-2i}$

13 Find all non-zero complex numbers z satisfying $\bar{z} = iz^2$

14 Find the square root of $7 - 24i$.

15 Find the arguments of the following

(i) $5+5i$ (ii) $-4+4i$

(iii) $-3-3i$ (iv) $2-2i$

16 If z_1 and z_2 are conjugate to each other, then find the argument of $-z_1 z_2$.

17 If $\arg(z_1) = 170°$ and $\arg(z_2) = 70°$, then find the principal argument of $z_1 z_2$.

18 Find the principal value of the argument of the complex number $z = \frac{2+i}{4i+(1+i)^2}$, where $i = \sqrt{-1}$.

19 If z_1 and z_2 are two non-zero complex numbers such that $|z_1 + z_2| = |z_1| + |z_2|$, then find the value of $\arg(z_1) - \arg(z_2)$.

20 Let $z_1 = 10 + 6i$ and $z_2 = 4 + 6i$. If z is any complex number such that the argument of $\frac{z - z_1}{z - z_2}$ is $\frac{\pi}{4}$, then find the value of $|z - 7 - 9i|$.

21 Find the polar form of the following complex number

(i) $3i$ (ii) $-4+4i$

22 Find the values of θ if $\frac{3+2i\sin\theta}{1-2i\sin\theta}$ is

(i) purely real. (ii) purely imaginary, where $i = \sqrt{-1}$.

23 Let $z = \cos\theta + i\sin\theta$. Then, find the value of $\sum_{m=1}^{15} \text{Im}(z^{2m-1})$ at $\theta = 2°$.

24 Using De-Moivre's theorem, express $(\sqrt{3}+i)^7$ in the form of $a + bi$.

25 Express $\left(\frac{\cos\theta + i\sin\theta}{\sin\theta + i\cos\theta}\right)^4$ in $a + ib$ form.

26 Express $\frac{4(\cos 75° + i\sin 75°)}{0\cdot 4(\cos 30° + i\sin 30°)}$ in $a + ib$ form.

27 Express five fifth-roots of $(\sqrt{3}+i)$ in trigonometric form.

28 If $z = \left(\frac{\sqrt{3}}{2} + \frac{i}{2}\right)^5 + \left(\frac{\sqrt{3}}{2} - \frac{i}{2}\right)^5$, then find the imaginary part of z.

3.3 Cube and *n*th roots of Unity

29 Find the roots of cube root of unity 1.

30 If ω is a cube root of unity, then find the value of $(1-\omega)(1-\omega^2)(1-\omega^4)(1-\omega^8)$.

31 If 1, ω and ω^2 are the cube roots of unity, then find the value of $(3+\omega^2+\omega^4)^6$.

32 If $x = \alpha + \beta$, $y = \alpha\omega + \beta\omega^2$ and $z = \alpha\omega^2 + \beta\omega$, ω is an imaginary cube root of unity, then find the value of xyz in terms of α and β.

33 If ω is a complex cube root of unity, then evaluate

$$\frac{a+b\omega+c\omega^2}{c+a\omega+b\omega^2} + \frac{c+a\omega+b\omega^2}{a+b\omega+c\omega^2} + \frac{b+c\omega+a\omega^2}{b+c\omega^4+a\omega^5}.$$

34 If $a = e^{i\frac{2\pi}{3}}$, then find the equation whose roots are $a + a^{-2}$ and $a^2 + a^{-4}$.

35 If ω is imaginary fifth root of unity. Then, evaluate $\omega^4 + \omega^8 + \omega^{12} + \omega^{16}$.

36 Find the value of $\sum_{\lambda=1}^{12}\left(\sin\frac{2\lambda\pi}{13} - i\cos\frac{2\lambda\pi}{13}\right)$.

37 If ω is the non-real root of $z^7 = -1$, then find the value of $z^{86} + z^{175} + z^{289}$.

3.4 Application of Complex Number in Geometry

38 Find the radius of the circle $\left|\frac{z-i}{z+i}\right| = 5$.

39 Find the locus of the points representing the complex number z for which $|z+5|^2 - |z-5|^2 = 10$.

40 Find the locus of $\left|\frac{z-2}{z-3}\right| = 2$.

41 For all complex numbers z_1 and z_2 satisfying $|z_1| = 12$ and $|z_2 - 3 - 4\lambda| = 5$, then find the minimum value of $|z_1 - z_2|$.

42 If one of the vertices of the square circumscribing the circle $|z-1| = \sqrt{2}$ is $2 + \sqrt{3}i$. Find the other vertices of the square.

43 Let $z = x + iy$ be a complex number, where x and y are integers. Then, find the area of the rectangle whose vertices are the roots of the equation $z\bar{z}^3 + \bar{z}z^3 = 350$.

44 A man walks a distance of 3 units from the origin towards the North-East (N 45° E) directions. From there, he walks a distance of 4 units towards the North-West (N 45° W) direction to reach a point P. Then, find the position P in the argand plane.

45 A particle P starts from the point $z_0 = 1 + 2i$, $i = \sqrt{-1}$. It moves horizontally away form origin by 5 units and then vertically away from origin by 3 units to reach a point z_1. From z_1, the particles moves $\sqrt{2}$ units in the direction of the vector $\hat{i} + \hat{j}$ and then it moves through an angle $\frac{\pi}{2}$ in anti-clockwise direction on a circle with centre at origin, to reach a point z_2. Find the point z_2.

LEVEL 02 \\ JEE Main Level

Single Choice Correct Questions

3.1 Algebra of Complex Number

1 Consider the equation $10z^2 - 3iz - k = 0$, where z is a complex variable and $i^2 = -1$. Which of the following statements is true?
(a) For all complex numbers k, neither root is real
(b) For real positive numbers k, both roots are purely imaginary
(c) For all purely imaginary numbers k, both roots are real and irrational
(d) For real negative numbers k, both roots are purely imaginary

2 If x and y are complex numbers, then the system of equations $(1+i)x + (1-i)y = 1$ and $2ix + 2y = 1 + i$ has
(a) unique solution (b) no solution (c) infinite number of solutions (d) None of these

3 If $a^2 + b^2 = 1$, then $(1 + b + ia)/(1 + b - ia)$ is equal to
(a) 1 (b) 2 (c) 5 (d) $b + ia$

4 If a and b are complex and one of the roots of the equation $x^2 + ax + b = 0$ is purely real whereas the other is purely imaginary, then
(a) $a^2 - (\bar{a})^2 = 4b$ (b) $a^2 - (\bar{a})^2 = 2b$ (c) $b^2 - (\bar{b})^2 = 4a$ (d) $b^2 - (\bar{b})^2 = 2a$

5 Let $A = \left\{\theta \in \left(-\frac{\pi}{2}, \pi\right) : \frac{3 + 2i \sin \theta}{1 - 2i \sin \theta}\right.$ is purely imaginary}. Then, the sum of the elements in A is **[JEE Main 2019]**
(a) $\frac{5\pi}{6}$ (b) π (c) $\frac{3\pi}{4}$ (d) $\frac{2\pi}{3}$

6 Dividing $f(z)$ by $z - i$, we obtain the remainder i and dividing it by $z + i$, we get the remainder $1 + i$, then remainder upon the division of $f(z)$ by $z^2 + 1$ is
(a) $\frac{1}{2}(z+1) + i$ (b) $\frac{1}{2}(iz+1) + i$ (c) $\frac{1}{2}(iz-1) + i$ (d) $\frac{1}{2}(z+1) + i$

7 If the equation $z^4 + a_1 z^3 + a_2 z^2 + a_3 z + a_4 = 0$, where a_1, a_2, a_3 and a_4 are real coefficients different from zero, has a purely imaginary root, then the expression $\frac{a_3}{a_1 a_2} + \frac{a_1 a_4}{a_2 a_3}$ is equal to
(a) 0 (b) 1 (c) -2 (d) 2

8 If $z_1 z_2 \in C$, $z_1^2 + z_2^2 \in R$, $z_1(z_1^2 - 3z_2^2) = 2$ and $z_2(3z_1^2 - z_2^2) = 11$, then the value of $z_1^2 + z_2^2$ is
(a) 10 (b) 12 (c) 5 (d) 8

9 Let n denote the number of solutions of the equation $z^2 + 3\bar{z} = 0$, where z is a complex number. Then, the value of $\sum_{k=0}^{\infty} \frac{1}{n^k}$ is equal to **[JEE Main 2021]**
(a) 1 (b) $\frac{4}{3}$ (c) $\frac{3}{2}$ (d) 2

3.2 Modulus, Conjugate and Arguments of Complex Number

10 If z is a complex number satisfying the equation $z^6 - 6z^3 + 25 = 0$, then the value of $|z|$ is
(a) $5^{\frac{1}{3}}$ (b) $25^{\frac{1}{3}}$ (c) $125^{\frac{1}{3}}$ (d) $625^{\frac{1}{3}}$

11 If $\frac{z-\alpha}{z+\alpha}$ $(\alpha \in R)$ is purely imaginary number and $|z|=2$, then value of α is **[JEE Main 2019]**

(a) 2 (b) 1 (c) $\frac{1}{2}$ (d) $\frac{1}{\sqrt{2}}$

12 If $a>0$ and $z=\frac{(1+i)^2}{a-i}$, has magnitude $\sqrt{\frac{2}{5}}$, then $\bar{z}$ is equal to

(a) $-\frac{1}{5}-\frac{3}{5}i$ (b) $-\frac{3}{5}-\frac{1}{5}i$ (c) $\frac{1}{5}-\frac{3}{5}i$ (d) $-\frac{1}{5}+\frac{3}{5}i$

13 Let z be a complex number such that $|z|+z=3+i$, then $|z|$ is equal to **[JEE Main 2019]**

(a) $\frac{\sqrt{34}}{3}$ (b) $\frac{5}{3}$ (c) $\frac{\sqrt{41}}{4}$ (d) $\frac{5}{4}$

14 If $z=(3+7i)(a+ib)$, where $a, b \in Z-\{0\}$, is purely imaginary, then the minimum value of $|z|^2$ is

(a) 5476 (b) 2025 (c) 3364 (d) 4225

15 If $8iz^3+12z^2-18z+27i=0$, then

(a) $|z|=\frac{3}{2}$ (b) $|z|=\frac{2}{3}$ (c) $|z|=1$ (d) $|z|=\frac{3}{4}$

16 If $|z^2-3|=3|z|$, then the maximum value of $|z|$ is

(a) 1 (b) $\frac{3+\sqrt{21}}{2}$ (c) $\frac{\sqrt{21}-3}{2}$ (d) None of these

17 If $|z|<\sqrt{2}-1$, then $|z^2+2z\cos\alpha|$ is

(a) less than 1 (b) $\sqrt{2}+1$ (c) $\sqrt{2}-1$ (d) None of these

18 Let $|z_r-r|\le r, \forall r=1, 2, 3, \dots n$, then $\left|\sum_{r=1}^{n} z_r\right|$ is less than

(a) n (b) $2n$ (c) $n(n+1)$ (d) $\frac{n(n+1)}{2}$

19 If $\left|z-\frac{4}{z}\right|=2$, then the maximum value of $|z|$ is equal to

(a) $\sqrt{3}+1$ (b) $\sqrt{5}+1$ (c) 2 (d) $2+\sqrt{2}$

20 If $z=x+iy$ and $x^2+y^2=16$, then the range of $||x|-|y||$ is

(a) [0, 4] (b) [0, 2] (c) [2, 4] (d) None of these

21 Number of ordered pairs (c, b) of real numbers such that $(a+ib)^{2008}=a-ib$ holds good, is

(a) 2008 (b) 2009 (c) 2010 (d) 1

22 If $|2z-1|=|z-2|$ and z_1, z_2 and z_3 are complex numbers such that $|z_1-\alpha|<\alpha, |z_2-\beta|<\beta$, then $\left|\frac{z_1+z_2}{\alpha+\beta}\right|$

(a) $<|z|$ (b) $<2|z|$ (c) $>|z|$ (d) $>2|z|$

23 The complex numbers $\sin x+i\cos 2x$ and $\cos x-i\sin 2x$ are conjugate to each other, for

(a) $x=n\pi, n\in Z$ (b) $x=0$

(c) $x=\left(n+\frac{1}{2}\right)\pi, n\in Z$ (d) no value of x

24 The least value of $|z|$, where z is a complex number which satisfies the inequality $\exp\left(\frac{(|z|+3)(|z|-1)}{||z|+1|}\log_e 2\right) \geq \log_{\sqrt{2}}|5\sqrt{7}+9i|.\ i=\sqrt{-1}$, is equal to **[JEE Main 2021]**

(a) 3 (b) $\sqrt{5}$ (c) 2 (d) 8

25 The equation $az^3 + bz^2 + \bar{b}z + \bar{a} = 0$ has a root α, where a, b, z and α belong to the set of complex numbers, the value of $|\alpha|$ is

(a) $\frac{1}{2}$ (b) 1 (c) 2 (d) Cannot be determined

26 The imaginary part of $(3+2\sqrt{-54})^{1/2} - (3-2\sqrt{-54})^{1/2}$ can be **[JEE Main 2020]**

(a) $-\sqrt{6}$ (b) $-2\sqrt{6}$ (c) 6 (d) $\sqrt{6}$

27 If $|z_1| = |z_2|$ and $\arg\left(\frac{z_1}{z_2}\right) = \pi$, then $z_1 + z_2$ is equal to

(a) 0 (b) purely imaginary (c) purely real (d) None of these

28 If for complex numbers z_1 and z_2, $\arg(z_1) - \arg(z_2) = 0$, then $|z_1 - z_2|$ is equal to

(a) $|z_1| + |z_2|$ (b) $|z_1| - |z_2|$ (c) $||z_1| - |z_2||$ (d) 0

29 If $\left|\frac{z_1}{z_2}\right| = 1$ and $\arg(z_1 z_2) = 0$, then

(a) $z_1 = z_2$ (b) $|z_1| = |z_2|$ (c) $|z_1| \neq |z_2|$ (d) None of these

30 If z is a complex number of unit modulus and argument θ, then $\arg\left(\frac{1+z}{1+\bar{z}}\right)$ is equal to

(a) $-\theta$ (b) $\frac{\pi}{2} - \theta$ (c) θ (d) $\pi - \theta$

31 If $\frac{3 + i\sin\theta}{4 - i\cos\theta}$, $\theta \in [0, 2\pi]$ is a real number, then an argument of $\sin\theta + i\cos\theta$ is **[JEE Main 2020]**

(a) $\pi - \tan^{-1}\left(\frac{4}{3}\right)$ (b) $\pi - \tan^{-1}\left(\frac{3}{4}\right)$

(c) $-\tan^{-1}\left(\frac{3}{4}\right)$ (d) $\tan^{-1}\left(\frac{4}{3}\right)$

32 If z_1 and z_2 are complex numbers such that $\text{Re}(z_1) = |z_1 - 1|$, $\text{Re}(z_2) = |z_2 - 1|$ and $\arg(z_1 - z_2) = \frac{\pi}{6}$, then $\text{Im}(z_1 + z_2)$ is equal to **[JEE Main 2020]**

(a) $\frac{2}{\sqrt{3}}$ (b) $2\sqrt{3}$ (c) $\frac{\sqrt{3}}{2}$ (d) $\frac{1}{\sqrt{3}}$

33 Number of solutions of the equation $z^3 + [3(\bar{z})^2/|z|] = 0$, where z is a complex number, is

(a) 2 (b) 3 (c) 6 (d) 5

34 If $z = (i)^{(i)^{(i)}}$, where $i = \sqrt{-1}$, then $|z|$ is equal to

(a) 1 (b) $e^{-\frac{\pi}{2}}$ (c) $e^{-\pi}$ (d) None of these

35 The product of all values of $(\cos\alpha + i\sin\alpha)^{\frac{3}{5}}$ is

(a) 1 (b) $\cos\alpha + i\sin\alpha$

(c) $\cos 3\alpha + i\sin 3\alpha$ (d) $\cos 5\alpha + i\sin 5\alpha$

36 If $z = i \log(2 - \sqrt{3})$, then $\cos z$ is equal to

(a) -1 (b) $-\frac{1}{2}$ (c) 1 (d) 2

37 If α and β are the roots of the equation $u^2 - 2u + 2 = 0$ and if $\cot\theta = x + 1$, then $[(x+\alpha)^n - (x+\beta)^n]/[\alpha - \beta]$ is equal to

(a) $\frac{\sin n\theta}{\sin^n \theta}$ (b) $\frac{\cos n\theta}{\cos^n \theta}$ (c) $\frac{\sin n\theta}{\cos^n \theta}$ (d) $\frac{\cos n\theta}{\sin^n \theta}$

38 If $(\cos\theta + i\sin\theta)(\cos 2\theta + i\sin 2\theta) \ldots (\cos n\theta + i\sin n\theta) = 1$, then the value of θ is

(a) $\frac{2m\pi}{n(n+1)}$ (b) $4m\pi$ (c) $\frac{4m\pi}{n(n+1)}$ (d) $\frac{m\pi}{n(n+1)}$

39 If $\left(\frac{3}{2} + i\frac{\sqrt{3}}{2}\right)^{50} = 3^{25}(x + iy)$, where x and y are real, then the ordered pair (x, y) is

(a) $(-3, 0)$ (b) $(0, 3)$ (c) $(0, -3)$ (d) $\left(\frac{1}{2}, \frac{\sqrt{3}}{2}\right)$

3.3 Cube and *n*th Roots of Unity

40 If z_1 and z_2 are complex roots of the equation $(x-3)^3 + 1 = 0$, then $z_1 + z_2$ is equal to

(a) 1 (b) 3 (c) 5 (d) 7

41 Let z be a complex number such that $|z| < 1$. If $\omega = \frac{5 + 3z}{5(1-z)}$, then **[JEE Main 2019]**

(a) $5\,\text{Re}\,|\omega| > 4$ (b) $4\,\text{Im}\,(\omega) > 5$ (c) $5\,\text{Re}\,|\omega| > 1$ (d) $5\,\text{Im}\,|\omega| < 1$

42 If $\omega (\neq 1)$ is a cube root of unity and $(1+\omega)^7 = A + B\omega$, then (A, B) is equal to

(a) $(-1, 1)$ (b) $(0, 1)$ (c) $(1, 1)$ (d) $(1, 0)$

43 If $x^2 + x + 1 = 0$, then the value of $\left(x + \frac{1}{x}\right)^2 + \left(x^2 + \frac{1}{x^2}\right)^2 + \ldots + \left(x^{27} + \frac{1}{x^{27}}\right)$ is

(a) 27 (b) 72 (c) 45 (d) 54

44 Sum of common roots of the equations $z^3 + 2z^2 + 2z + 1 = 0$ and $z^{1985} + z^{100} + 1 = 0$ is

(a) -1 (b) 1 (c) 0 (d) 1

45 If $|z - 1| \le 2$ and $|\omega z - 1 - \omega^2| = a$ (where ω is a cube root of unity), then complete set of values of a is

(a) $0 \le a \le 2$ (b) $\frac{1}{2} \le a \le \frac{\sqrt{3}}{2}$

(c) $\frac{\sqrt{3}}{2} - \frac{1}{2} \le a \le \frac{1}{2} + \frac{\sqrt{3}}{2}$ (d) $0 \le a \le 4$

46 Suppose A is a complex number and $n \in N$ such that $A^n = (A+1)^n = 1$, then the least value of n is

(a) 3 (b) 6 (c) 9 (d) 12

47 If $5x^3 + Mx + N$, $M, N \in R$ is divisible by $x^2 + x + 1$, then the value of $M + N$ is

(a) 5 (b) 4 (c) -4 (d) -5

48 Which of the following is equal to $\sqrt[3]{-1}$?

(a) $\frac{\sqrt{3} + \sqrt{-1}}{2}$ (b) $\frac{-\sqrt{3} + \sqrt{-1}}{\sqrt{-4}}$ (c) $\frac{\sqrt{3} - \sqrt{-1}}{\sqrt{-4}}$ (d) $-\sqrt{-1}$

49 The value of z satisfying the equation $\log z + \log z^2 + \ldots + \log z^n = 0$ is

(a) $\cos \frac{4m\pi}{n(n+1)} + i \sin \frac{4m\pi}{n(n+1)}$, $m = 0, 1, 2, \ldots$ (b) $\cos \frac{4m\pi}{n(n+1)} - i \sin \frac{4m\pi}{n(n+1)}$, $m = 0, 1, 2, \ldots$

(c) $\sin \frac{4m\pi}{n} + i \cos \frac{4m\pi}{n}$, $m = 0, 1, 2, \ldots$ (d) 0

3.4 Application of Complex Number in Geometry

50 If $|z^2 - 1| = |z^2| + 1$, then z lies on

(a) a circle (b) a parabola (c) an ellipse (d) None of these

51 If $z = (\lambda + 3) - i\sqrt{5 - \lambda^2}$, then the locus of z is

(a) ellipse (b) semicircle (c) parabola (d) straight line

52 Let a be a complex number such that $|a| < 1$ and $z_1, z_2, z_3 \ldots$ be the vertices of a polygon such that $z_k = 1 + a + a^2 + \ldots + a^{k-1}, \forall k = 1, 2, 3, \ldots$, then $z_1, z_2, \ldots$ lies within the circle.

(a) $\left| z - \frac{1}{1-a} \right| = \frac{1}{|a-1|}$ (b) $\left| z + \frac{1}{a+1} \right| = \frac{1}{|a+1|}$

(c) $\left| z - \frac{1}{1-a} \right| = |a - 1|$ (d) $\left| z + \frac{1}{a+1} \right| = |a + 1|$

53 If $|z| = 1$, then the point representing the complex number $-1 + 3z$ will lie on

(a) a circle (b) a straight line (c) a parabola (d) a hyperbola

54 If z is complex number, then the locus of z satisfying the condition $|2z - 1| = |z - 1|$ is

(a) perpendicular bisector of line segment joining $\frac{1}{2}$ and 1

(b) circle

(c) parabola

(d) None of the above

55 The equation $\arg\left(\frac{z-1}{z+1}\right) = \frac{\pi}{4}$ represents a circle with **[JEE Main 2021]**

(a) centre at $(0, -1)$ and radius $\sqrt{2}$ (b) centre at $(0, 1)$ and radius $\sqrt{2}$

(c) centre at $(0, 0)$ and radius $\sqrt{2}$ (d) centre at $(0, 1)$ and radius 2

56 If $|z - 2 - i| = |z| \left| \sin\left(\frac{\pi}{4} - \arg z\right) \right|$, then locus of z is

(a) a pair of straight line (b) circle

(c) parabola (d) ellipse

57 Let C be the set of all complex numbers. Let

$S_1 = \{z \in C \mid |z - 3 - 2i|^2 = 8\}$,

$S_2 = \{z \in C \mid \text{Re}(z) \geq 5\}$ and

$S_3 = \{z \in C \mid |z - \bar{z}| \geq 8\}$.

Then, the number of elements in $S_1 \cap S_2 \cap S_3$ is equal to **[JEE Main 2021]**

(a) 1 (b) 0 (c) 2 (d) Infinite

58 If $|z_1| = |z_2| = |z_3| = 1$ and $z_1 + z_2 + z_3 = 0$, then the area of the triangle whose vertices are z_1, z_2 and z_3, is

(a) $\frac{3\sqrt{3}}{4}$ (b) $\frac{\sqrt{3}}{4}$ (c) 1 (d) 2

59 If $n \in N > 1$, then the sum of real part of roots of $z^n = (z+1)^n$ is equal to

(a) $\frac{n}{2}$ (b) $\frac{(n-1)}{2}$ (c) $-\frac{n}{2}$ (d) $\frac{(1-n)}{2}$

60 If z be a complex number satisfying $|\text{Re}(z)| + |\text{Im}(z)| = 4$, then $|z|$ cannot be **[JEE Main 2020]**

(a) $\sqrt{\frac{17}{2}}$ (b) $\sqrt{10}$ (c) $\sqrt{7}$ (d) $\sqrt{8}$

61 The equation $|z - i| = |z - 1|, i = \sqrt{-1}$ represents **[JEE Main 2019]**

(a) a circle of radius $\frac{1}{2}$

(b) a line through the origin with slope 1

(c) a circle of radius 1

(d) the line through the origin with slope -1

62 z_1 and z_2 are two distinct points in an argand plane. If $a|z_1| = b|z_2|$ (where $a, b \in R$), then the point $\left(\frac{az_1}{bz_2}\right) + \left(\frac{bz_2}{az_1}\right)$ is a point on the

(a) line segment $[-2, 2]$ of the real axis

(b) line segment $[-2, 2]$ of the imaginary axis

(c) unit circle $|z| = 1$

(d) the line with $\arg z = \tan^{-1} 2$

63 If the four complex numbers $z, \bar{z}, \bar{z} - 2\text{Re}(\bar{z})$ and $z - 2\,\text{Re}(z)$ represent the vertices of a square of side 4 units in the argand plane, then $|z|$ is equal to **[JEE Main 2020]**

(a) $4\sqrt{2}$ (b) 4 (c) $2\sqrt{2}$ (d) 2

64 Which of the following represents a point in an argand plane, equidistant from the roots of the equation $(z+1)^4 = 16z^4$?

(a) (0, 0) (b) $\left(-\frac{1}{3}, 0\right)$ (c) $\left(\frac{1}{3}, 0\right)$ (d) $\left(0, \frac{2}{\sqrt{5}}\right)$

65 The roots of the equation $t^3 + 3at^2 + 3bt + c = 0$ are z_1, z_2 and z_3, which represent the vertices of an equilateral triangle, then

(a) $a^2 = 3b$ (b) $b^2 = a$

(c) $a^2 = b$ (d) $b^2 = 3a$

Numerical Value Type Questions

66 If the real part of the complex number $z = \frac{3 + 2i\cos\theta}{1 - 3i\cos\theta}, \theta \in \left(0, \frac{\pi}{2}\right)$ is zero, then the value of $\sin^2 3\theta + \cos^2\theta$ is equal to .... **[JEE Main 2021]**

67 Let $\left(-2 - \frac{i}{3}\right)^3 = \frac{x + iy}{27} (i = \sqrt{-1})$, where x and y are real numbers, then the value of $\frac{y - x}{3}$ is ....

68 If $\left(\frac{1+i}{1-i}\right)^{m/2} = \left(\frac{1+i}{i-1}\right)^{n/3} = 1, (m, n \in N)$, then the greatest common divisor of the least values of m and n is .... **[JEE Main 2020]**

69 Let $z = 9 + bi$, where b is non-zero real and $i^2 = -1$, if the imaginary part of z^2 and z^3 are equal, then b is equal to ....

70 If the expression $(1 + ir)^3$ is of the form of $S(1 + i)$ for some real S, where r is also real and then the sum of all possible values of r is ....

71 Let $|z| = 2$ and $w = \frac{z+1}{z-1}$, where $z, w \in C$ (where C is the set of complex numbers). Then, product of least and greatest value of modulus of w is ....

72 Modulus of non-zero complex number z satisfying $\bar{z} + z = 0$ and $|z|^2 - 4zi = z^2$, is ......

73 The complex number z satisfies $z + |z| = 2 + 8i$, the value of $(|z|)$ is ......

74 If complex number $z (z \neq 2)$ satisfies the equation $z^2 = 4z + |z|^2 + \frac{16}{|z|^3}$, then the value of $|z|^4$ is ....

75 If z is a complex number satisfying $z^4 + z^3 + 2z^2 + z + 1 = 0$, then $|z|$ is equal to ......

76 If $x = \omega - \omega^2 - 2$, then the value of $x^4 + 3x^3 + 2x^2 - 11x - 6$ is (where ω is cube root of unity) ......

77 If a and b are real numbers such that $(2 + \alpha)^4 = a + b\alpha$, where $\alpha = \frac{-1 + i\sqrt{3}}{2}$, then the value of $\frac{a+b}{2}$ is ....

78 The imaginary part of $\left(\frac{1 + \sin\frac{2\pi}{9} + i\cos\frac{2\pi}{9}}{1 + \sin\frac{2\pi}{9} - i\cos\frac{2\pi}{9}} \right)^3$ is ....

79 If z is any complex number satisfying $|z - 3 - 2i| \leq 2$, then the minimum value of $|2z - 6 + 5i|$ is ....

80 Let $i = \sqrt{-1}$. If $\frac{(-1 + i\sqrt{3})^{21}}{(1 - i)^{24}} + \frac{(1 + i\sqrt{3})^{21}}{(1 + i)^{24}} = k$ and $n = [|k|]$ be the greatest integral part of $|k|$. Then, $\sum_{j=0}^{n+5} (j + 5)^2 - \sum_{j=0}^{n+5} (j + 5)$ is equal to .... **[JEE Main 2021]**

LEVEL 03 \\ JEE Advanced Level

Single Choice Correct Questions

1 Let z be a complex number such that the imaginary part of z is non-zero and $a = z^2 + z + 1$ is real, then a cannot take the value

(a) -1 (b) $\frac{1}{3}$ (c) $\frac{1}{2}$ (d) $\frac{3}{4}$

2 A value of θ for which $\frac{2 + 3i\sin\theta}{1 - 2i\sin\theta}$ purely imaginary, is

(a) $\frac{\pi}{6}$ (b) $\sin^{-1}\left(\frac{\sqrt{3}}{4}\right)$ (c) $\sin^{-1}\left(\frac{1}{\sqrt{3}}\right)$ (d) $\frac{\pi}{3}$

3 The number of complex numbers z such that $|z-1|=|z+1|=|z-i|$ is

(a) ∞ (b) 0 (c) 1 (d) 2

4 If z is a complex number such that $|z|\geq 2$, then the minimum value of $\left|z+\frac{1}{2}\right|$

(a) is equal to $\frac{5}{2}$

(b) lies in the interval (1, 2)

(c) is strictly greater than $\frac{5}{2}$

(d) is strictly greater than $\frac{3}{2}$ but less than $\frac{5}{2}$

5 Let α and β be real and z be a complex number. If $z^2+\alpha z+\beta=0$ has two distinct roots on the line $\text{Re}(z)=1$, then it is necessary that

(a) $\beta \in (1, \infty)$ (b) $\beta \in (0, 1)$
(c) $\beta \in (-1, 0)$ (d) $|\beta| = 1$

6 If $\arg(z)<0$, then $\arg(-z)-\arg(z)$ is equal to

(a) π (b) $-\pi$ (c) $-\frac{\pi}{2}$ (d) $\frac{\pi}{2}$

7 If z and ω be complex numbers such that $\bar{z}+i\omega=0$ and $\arg(z\omega)=\pi$, then $\arg[z]$ is equal to

(a) $\frac{\pi}{4}$ (b) $\frac{\pi}{2}$ (c) $\frac{3\pi}{4}$ (d) $\frac{5\pi}{4}$

8 If $z=\sqrt{3}+i$, then the argument of z^2e^{z-i} is equal to

(a) $\frac{\pi}{4}$ (b) $\frac{\pi}{3}$ (c) $\frac{\pi}{6}$ (d) $\frac{\pi}{2}$

9 If $\alpha, \beta \in C$ are the distinct roots of the equation $x^2-x+1=0$, then $\alpha^{101}+\beta^{107}$ is equal to

(a) 2 (b) -1 (c) 0 (d) 1

10 The value of the expression, where ω is cube roots of unity

$2\left(1+\frac{1}{\omega}\right)\left(1+\frac{1}{\omega^2}\right)+3\left(2+\frac{1}{\omega}\right)\left(2+\frac{1}{\omega^2}\right)+\ldots+(n+1)\left(n+\frac{1}{\omega}\right)\left(n+\frac{1}{\omega^2}\right)$ is

(a) $\left[\frac{n(n+1)}{2}\right]^2$ (b) $\left[\frac{n(n+1)}{2}\right]^2+n$

(c) $\left[\frac{n(n-1)}{2}\right]^2$ (d) None of these

11 Let $z=x+iy$ be a complex number, where x and y are integers. Then, the area of the rectangle whose vertices are the roots of the equation $\bar{z}z^3+z\bar{z}^3=350$, is

(a) 48 (b) 32 (c) 40 (d) 80

12 A complex number z is said to be unimodular if $|z|=1$. Suppose, z_1 and z_2 are complex numbers such that $\frac{z_1-2z_2}{2-z_1\bar{z}_2}$ is unimodular and z_2 is non-unimodular. Then, the point z_1 lies on a

(a) straight line parallel to X-axis (b) straight line parallel to Y-axis
(c) circle of radius 2 (d) circle of radius $\sqrt{2}$

13 Let $\theta_1, \theta_2, \ldots, \theta_{10}$ be positive valued angles (in radian) such that $\theta_1 + \theta_2 + \ldots + \theta_{10} = 2\pi$. Define the complex numbers $z_1 = e^{i\theta_1}$, $z_k = z_{k-1}e^{i\theta_k}$ for $k = 2, 3, \ldots, 10$, where $i = \sqrt{-1}$.

Consider the statements P and Q given below :

$$P : |z_2 - z_1| + |z_3 - z_2| + \ldots + |z_{10} - z_9| + |z_1 - z_{10}| \le 2\pi$$

$$Q : |z_2^2 - z_1^2| + |z_3^2 - z_2^2| + \ldots + |z_{10}^2 - z_9^2| + |z_1^2 - z_{10}^2| \le 4\pi$$

Then, **[JEE Advanced 2021]**

(a) P is true and Q is false
(b) Q is true and P is false
(c) Both P and Q are true
(d) Both P and Q are false

14 Let complex numbers α and $\frac{1}{\bar{\alpha}}$ lie on circles $(x - x_0)^2 + (y - y_0)^2 = r^2$ and $(x - x_0)^2 + (y - y_0)^2 = 4r^2$, respectively. If $z_0 = x_0 + iy_0$ satisfies the equation $2|z_0|^2 = r^2 + 2$, then $|\alpha|$ is equal to

(a) $\frac{1}{\sqrt{2}}$ (b) $\frac{1}{2}$ (c) $\frac{1}{\sqrt{7}}$ (d) $\frac{1}{3}$

More than One Correct

15 If α is a complex constant such that $\alpha z^2 + z + \bar{\alpha} = 0$ has real root, then

(a) $\alpha + \bar{\alpha} = 1$
(b) $\alpha + \bar{\alpha} = 0$
(c) $\alpha + \bar{\alpha} = -1$
(d) the absolute value of the real root is 1

16 If $z^3 + (3 + 2i)z + (-1 + ia) = 0$ has one real root, then the value of a lies in the interval ($a \in R$) is

(a) $(-2, 1)$ (b) $(-1, 0)$ (c) $(0, 1)$ (d) $(-2, 3)$

17 If $|z - 3| = \min\{|z - 1|, |z - 5|\}$, then Re$(z)$ is equal to

(a) 2 (b) $\frac{5}{2}$ (c) $\frac{7}{2}$ (d) 4

18 Let a, b, x and y be real numbers such that $a - b = 1$ and $y \ne 0$. If the complex number $z = x + iy$ satisfies. Im$\left(\frac{az + b}{z + 1}\right) = y$, then which of the following is/are possible value(s) of x? **[JEE Advanced 2017]**

(a) $-1 - \sqrt{1 - y^2}$
(b) $1 + \sqrt{1 + y^2}$
(c) $1 - \sqrt{1 + y^2}$
(d) $-1 + \sqrt{1 - y^2}$

19 If $z_1 = a + ib$ and $z_2 = c + id$ are complex numbers such that $|z_1| = |z_2| = 1$ and Re$(z_1\bar{z}_2) = 0$, then the pair of complex numbers $w_1 = a + ic$ and $w_2 = b + id$ satisfies

(a) $|w_1| = 1$
(b) $|w_2| = 1$
(c) Re$(w_1\bar{w}_2) = 0$
(d) Im$(w_1, \bar{w}_2) = 0$

20 For any complex number $w = c + id$, let $\arg(w) \in (-\pi, \pi]$, where $i = \sqrt{-1}$. Let α and β be real numbers such that for all complex numbers $z = x + iy$ satisfying $\arg\left(\frac{z + \alpha}{z + \beta}\right) = \frac{\pi}{4}$, the ordered pair (x, y) lies on the circle $x^2 + y^2 + 5x - 3y + 4 = 0$

Then, which of the following statements is (are) true? **[JEE Advanced 2021]**

(a) $\alpha = -1$
(b) $\alpha\beta = 4$
(c) $\alpha\beta = -4$
(d) $\beta = 4$

21 If z_1 and z_2 are two complex numbers $(z_1 \neq z_2)$ satisfying $|z_1^2 - z_2^2| = |\bar{z}_1^2 + \bar{z}_2^2 - 2\bar{z}_1\bar{z}_2|$, then

(a) $\frac{z_1}{z_2}$ is purely imaginary

(b) $\frac{z_1}{z_2}$ is purely real

(c) $|\arg(z_1) - \arg(z_2)| = \pi$

(d) $|\arg(z_1) - \arg(z_2)| = \frac{\pi}{2}$

22 Let a, b and c be distinct non-zero real numbers satisfying $a^3 + b^3 + 6abc = 8c^3$, then which of the following can be correct? (ω is complex cube root of unity)

(a) a, c and b are in AP
(b) a, c and b are in GP
(c) $a + b\omega - 2c\omega^2 = 0$
(d) $a + b\omega^2 - 2c\omega = 0$

23 If $z = \omega, \omega^2$, where ω is a non-real complex cube root of unity, are two vertices of an equilateral triangle in the argand plane, then the third vertex may be represented by

(a) $z = 1$
(b) $z = 0$
(c) $z = -2$
(d) $z = -1$

24 If $p = a + b\omega + c\omega^2$, $q = b + c\omega + a\omega^2$ and $r = c + a\omega + b\omega^2$, where $a, b, c \neq 0$ and ω is the complex cube root of unity, then

(a) if p, q and r lie on the circle $|z| = 2$, the triangle formed by these points is equilateral

(b) $p^2 + q^2 + r^2 = a^2 + b^2 + c^2$

(c) $p^2 + q^2 + r^2 = 2(pq + qr + rp)$

(d) None of the above

25 Let $P(x)$ and $Q(x)$ be two polynomials suppose that $f(x) = P(x^3) + xQ(x^3)$ is divisible by $x^2 + x + 1$, then

(a) $P(x)$ is divisible by $(x - 1)$ but $Q(x)$ is not divisible by $x - 1$

(b) $Q(x)$ is divisible by $(x - 1)$ but $P(x)$ is not divisible by $x - 1$

(c) Both $P(x)$ and $Q(x)$ are divisible by $x - 1$

(d) $f(x)$ is divisible by $x - 1$

26 Let the equation of a ray be $|z - 2| - |z - 1 - i| = \sqrt{2}$. If it strikes the Y-axis. Then, the equation of reflected ray (including or excluding the point of incidence) is

(a) $\arg(z - 2i) = \frac{\pi}{4}$

(b) $|z - 2i| - |z - 1 - 3i| = \sqrt{2}$

(c) $\arg(z - 2i) = \frac{3\pi}{4}$

(d) $|z - 2i| - |z - 1 - 3i| = 2\sqrt{2}$

27 If $|z_1| = 15$ and $|z_2 - 3 - 4i| = 5$, then

(a) $|z_1 - z_2|_{\min} = 5$

(b) $|z_1 - z_2|_{\min} = 10$

(c) $|z_1 - z_2|_{\max} = 20$

(d) $|z_1 - z_2|_{\max} = 25$

28 $P(z_1), Q(z_2), R(z_3)$ and $S(z_4)$ are four complex numbers representing the vertices of a rhombus taken in order on the complex plane, then which of the following is/are correct?

(a) $\frac{z_1 - z_4}{z_2 - z_3}$ is purely real

(b) $\text{amp}\,\frac{z_1 - z_4}{z_2 - z_4} = \text{amp}\,\frac{z_2 - z_4}{z_3 - z_4}$

(c) $\frac{z_1 - z_3}{z_2 - z_4}$ is purely imaginary

(d) It is not necessary that $|z_1 - z_3| \neq |z_2 - z_4|$

29 A complex number z is rotated in anti-clockwise direction by an angle α and we get z' and if the same complex number z is rotated by an angle α in clockwise direction and we get z'', then

(a) z', z, z'' are in GP
(b) z', z, z'' are in AP
(c) $z' + z'' = 2z \cos \alpha$
(d) $z'^2 + z''^2 = 2z^2 \cos 2\alpha$

30 Let z_1 and z_2 be two distinct complex numbers and let $z = (1-t)z_1 + tz_2$ for some real number t with $0 < t < 1$. If $\arg(w)$ denotes the principal argument of a non-zero complex number w, then

(a) $|z - z_1| + |z - z_2| = |z_1 - z_2|$
(b) $(z - z_1) = (z - z_2)$
(c) $\begin{vmatrix} z - z_1 & \bar{z} - \bar{z}_1 \\ z_2 - z_1 & \bar{z}_2 - \bar{z}_1 \end{vmatrix} = 0$
(d) $\arg(z - z_1) = \arg(z_2 - z_1)$

31 Let $a, b \in R$ and $a^2 + b^2 \neq 0$. Suppose, $S = \left\{ z \in C : z = \frac{1}{a + ibt}, t \in R, t \neq 0 \right\}$, where $i = \sqrt{-1}$.

If $z = x + iy$ and $z \in S$, then (x, y) lies on

(a) the circle with radius $\frac{1}{2a}$ and centre $\left(\frac{1}{2a}, 0\right)$ for $a > 0, b \neq 0$

(b) the circle with radius $-\frac{1}{2a}$ and centre $\left(-\frac{1}{2a}, 0\right)$ for $a < 0, \phi \neq 0$

(c) the X-axis for $a \neq 0, b = 0$

(d) the Y-axis for $a = 0, b \neq 0$

32 If the points $A(z)$, $B(-z)$ and $C(1-z)$ are the vertices of an equilateral ΔABC, then

(a) sum of possible z is $\frac{1}{2}$
(b) sum of possible z is 1
(c) product of possible z is $\frac{1}{4}$
(d) product of possible z is $\frac{1}{2}$

33 Given that the complex numbers which satisfy the equation $|z\bar{z}^3| + |\bar{z}z^3| = 350$ form a rectangle in the argand plane with the length of its diagonal having a integral number of units, then

(a) area of rectangle is 48 sq units
(b) if z_1, z_2, z_3 and z_4 are vertices of rectangle, then $z_1 + z_2 + z_3 + z_4 = 0$
(c) rectangle is symmetrical about the real axis
(d) None of the above

34 System of equations $|z + 3| - |z - 3| = 6$ and $|z - 4| = r$, where $r \in R^+$, has

(a) one solution if $r > 1$
(b) one solution if $r < 1$
(c) two solutions if $r = 1$
(d) atleast one solution

35 If $\dfrac{3}{2+e^{i\theta}} = ax + iby$, then the locus of $P(x,y)$ will represent

(a) ellipse if $a=1, b=2$
(b) circle if $a=b=1$
(c) pair of straight line if $a=1, b=0$
(d) None of these

36 Let $\omega = \dfrac{\sqrt{3}+i}{2}$ and $p = \{\omega^n : n = 1, 2, 3, \ldots\}$. Further, $H_1 = \left\{z \in C : \text{Re}(z) > \dfrac{1}{2}\right\}$ and $H_2 = \left\{z \in C : \text{Re}(z) < \dfrac{-1}{2}\right\}$, where C is the set of all complex numbers. If $z_1 \in P \cap H_1, z_2 \in P \cap H_2$ and O represents the origin, then $< z_1 O \; z_2$ is equal to

(a) $\dfrac{\pi}{2}$
(b) $\dfrac{\pi}{6}$
(c) $\dfrac{\pi}{3}$
(d) $\dfrac{5\pi}{6}$

Linked Comprehension Type

Paragraph I Directions (Q.Nos. 37-40)

Consider the complex numbers $z = \dfrac{(1 - i\sin\theta)}{(1 + i\cos\theta)}$.

37 The value of θ for which z is purely real are

(a) $n\pi - \dfrac{\pi}{4}, n \in I$
(b) $n\pi + \dfrac{\pi}{4}, n \in I$
(c) $n\pi, n \in I$
(d) None of these

38 The value of θ for which z is purely imaginary, are

(a) $n\pi - \dfrac{\pi}{4}, n \in I$
(b) $n\pi + \dfrac{\pi}{4}, n \in I$
(c) $n\pi, n \in I$
(d) no real value of θ

39 The value of θ for which z is unimodular, is given by

(a) $n\pi \pm \dfrac{\pi}{6}, n \in I$
(b) $n\pi \pm \dfrac{\pi}{3}, n \in I$
(c) $n\pi \pm \dfrac{\pi}{4}, n \in I$
(d) no real value of θ

40 If argument of z is $\dfrac{\pi}{4}$, then

(a) $\theta = n\pi, n \in I$
(b) $\theta = (2n+1)\dfrac{\pi}{2}, n \in I$
(c) Both $\theta = n\pi$ and $\theta = (2n+1)\dfrac{\pi}{2}, n \in I$
(d) None of the above

Paragraph II Directions (Q.Nos. 41-44)

Consider the complex numbers z_1 and z_2 satisfying the relation $|z_1 + z_2|^2 = |z_1|^2 + |z_2|^2$.

41 Complex number $z_1 \bar{z}_2$ is

(a) purely real
(b) purely imaginary
(c) zero
(d) None of these

42 Complex number z_1 / z_2 is

(a) purely real
(b) purely imaginary
(c) zero
(d) None of these

43 One of the possible argument of complex number $i\left(\frac{z_1}{z_2}\right)$ is

(a) $\frac{\pi}{2}$ (b) $-\frac{\pi}{2}$ (c) 0 (d) None of these

44 Possible difference between the argument of z_1 and z_2 is

(a) 0 (b) π (c) $-\frac{\pi}{2}$ (d) None of these

Paragraph III Directions (Q.Nos. 45-47)

Consider the equation $az + b\bar{z} + c = 0$, where $a, b, c \in Z$.

45 If $|a| \neq |b|$, then z represents

(a) circle (b) straight line
(c) one point (d) ellipse

46 If $|a| = |b|$ and $\bar{a}c \neq b\bar{c}$, then z has

(a) infinite solutions
(b) no solutions
(c) finite solutions
(d) cannot say anything

47 If $|a| = |b| \neq 0$ and $\bar{a}c = b\bar{c}$, then $az + b\bar{z} + c = 0$ represents

(a) an ellipse (b) a circle
(c) a point (d) a straight line

Matrix Match Type Questions

48 The graph of the quadratic function $y = ax^2 + bx + c$ is as shown in the following figure

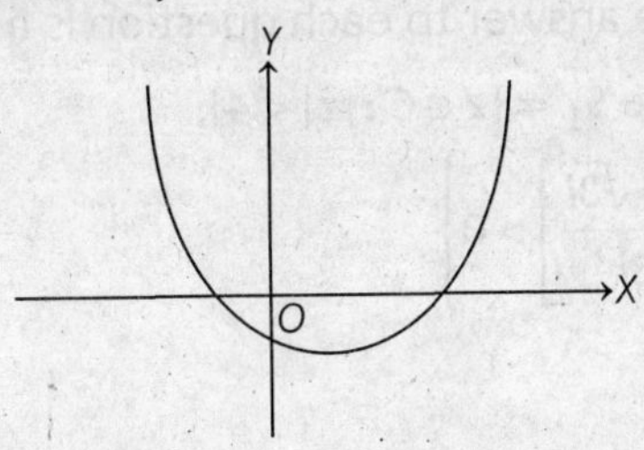

Now, match the complex numbers given in column I and with the corresponding arguments in column II.

	Column-I		Column-II
(i)	$a + ib$	A.	$-\tan^{-1}\frac{b}{a}$
(ii)	$ac - ibc$	B.	$\tan^{-1}\frac{b}{a}$
(iii)	$-ac + ibc$	C.	$\pi + \tan^{-1}\frac{b}{a}$
(iv)	$abc - ib^2c$	D.	$-\pi - \tan^{-1}\frac{b}{a}$

49 Let z_1, z_2 and z_3 be the vertices of a triangle, then match following columns

	Column-I		Column-II
(i)	If $2(z_3 - z_1) + (i\sqrt{3} + 1)(z_1 - z_2) = 0$, then triangle is	A.	right angled but may not be isoceles
(ii)	$\text{Re}\left(\frac{z_3 - z_1}{z_3 - z_2}\right) = 0$, then triangle is	B.	obtuse angle
(iii)	$\text{Re}\left(\frac{z_3 - z_1}{z_3 - z_2}\right) < 0$, then triangle is	C.	isosceles and right angled
(iv)	$\frac{z_3 - z_1}{z_3 - z_2} = 1$, then triangle is	D.	equilateral

50 Complex number z satisfies the equation $||z - 5i| + m|z - 12i|| = n$, then match the values of m and n in column I with the corresponding locus in column II.

	Column-I		Column-II
(i)	$m = 1, n = 13$	A.	ellipse
(ii)	$m = 1, n = 15$	B.	hyperbola
(iii)	$m = -1, n = 13$	C.	line segment
(iv)	$m = -1, n = 12$	D.	two opposite rays

Numerical Value Type Questions

Question Stem (Q.Nos. 51-52) The answer to each question is a **NUMERICAL VALUE.**

Let $S = S_1 \cap S_2 \cap S_3$ where $S_1 = \{z \in C : |z| < 4\}$,

$$S_2 = \left\{z \in C : \text{Im}\left[\frac{(z-1) + \sqrt{3}i}{1 - \sqrt{3}i}\right] > 0\right\}$$

and $S_3 = \{z \in C : \text{Re}(z) > 0\}$

51 Area of S is ....

52 Min $|1 - 3i - z|$ is equal to ....

53 If $|z + 2 - i| = 5$ and maximum value of $|3z + 9 - 7i|$ is M, then the value of M is ....

54 Let $A = \{a \in R\}$, the equation $(1 + 2i)x^3 - 2(3 + i)x^2 + (5 - 4i)x + 2a^2 = 0$ has atleast one real root. Then, the value of Σa^2 is ....

55 The minimum value of the expression $E = |z|^2 + |z - 3|^2 + |z - 6i|^2$ is m, then the value of m is ....

56 Let $z_1 = (8+i)\sin\theta + (7+4i)\cos\theta$ and $z_2 = (1+8i)\sin\theta + (4+7i)\cos\theta$ are two complex numbers. If $z_1 \cdot z_2 = a + ib$, where $a, b \in R$. If M is the greatest value of $(a+b)\ \forall \in R$, then value of M is ....

57 If ω is the imaginary cube root of unity, then the number of pairs of integers (a, b) such that $|a\omega + b| = 1$ is ....

58 Let 1, ω and ω^2 be the cube roots of unity. The least possible degree of a polynomial with real co-efficients having roots 2ω, $(2+3\omega)$, $(2+3\omega^2)$, $(2-\omega-\omega^2)$ is ....

ANSWER KEY

Level 1

1. (i) $-i$ (ii) i (iii) -1 (iv) i (v) $22i$ (vi) i

2. 0(odd), 1, (even)

3. (i) 25 (ii) $\frac{40}{41} - \frac{9}{41}i$ (iii) $-6 - 2i\sqrt{6}$

4. -1 **5.** -1 **6.** $95 + 2i$ **7.** 0

8. (i) 5 (ii) $\sqrt{65}$ (iii) $\frac{\sqrt{13}}{2}$

9. $\frac{3}{2} - 2i$ **10.** 6 **11.** 31 and 19

12. (i) $3 - 4i$ (ii) $20 - 12i$ (iii) $-\frac{3}{5} - \frac{4}{5}i$

13. $i, \frac{\sqrt{3}}{2} - \frac{i}{2}, \frac{-\sqrt{3}}{2} - \frac{i}{2}$ **14.** $\pm(4 - 3i)$

15. (i) $\frac{\pi}{4}$ (ii) $\frac{3\pi}{4}$ (iii) $\frac{5\pi}{4}$ (iv) $-\frac{\pi}{4}$

16. π **17.** $-120°$ **18.** $-\tan^{-1}(2)$

19. 0 **20.** $3\sqrt{2}$

21. (i) $3\left(\cos\frac{\pi}{2} + i\sin\frac{\pi}{2}\right)$ (ii) $4\sqrt{2}\left(\cos\frac{3\pi}{4} + i\sin\frac{3\pi}{4}\right)$

22. (i) $n\pi; n \in I$ (ii) $n\pi \pm \frac{\pi}{3}, n \in I$

23. $\frac{1}{4\sin 2°}$ **24.** $-64\sqrt{3} - 64i$ **25.** $\cos 8\theta + i\sin 8\theta$

26. $\frac{10}{\sqrt{2}} + \frac{i10}{\sqrt{2}}$ **27.** $2^{\frac{1}{5}}(\cos 294° + i\sin 294°)$ **28.** 0

29. $1, -\frac{1}{2} + \frac{i\sqrt{3}}{2}, \frac{-1}{2} - \frac{i\sqrt{3}}{2}$ **30.** 9 **31.** 64 **32.** $\alpha^3 + \beta^3$ **33.** 0

34. $x^2 + 2x + 4 = 0$ **35.** -1 **36.** i **37.** -1 **38.** $\frac{5}{12}$

39. $x = \frac{1}{2}$ **40.** $x^2 + y^2 - \frac{20}{3}x + \frac{32}{3} = 0$ **41.** 2

42. $(1-\sqrt{3}) + i, -i\sqrt{3}$ and $(\sqrt{3}+1) - i$ **43.** 48 unit^2

44. $(3 + 4\hat{i})e^{\frac{i\pi}{4}}$ **45.** $-6 + 7i$

Level 2

1. (d)	**2.** (c)	**3.** (d)	**4.** (a)	**5.** (d)	**6.** (b)	**7.** (b)
8. (c)	**9.** (b)	**10.** (a)	**11.** (a)	**12.** (a)	**13.** (b)	**14.** (c)
15. (a)	**16.** (b)	**17.** (a)	**18.** (c)	**19.** (b)	**20.** (a)	**21.** (c)
22. (b)	**23.** (d)	**24.** (a)	**25.** (b)	**26.** (b)	**27.** (a)	**28.** (c)
29. (b)	**30.** (c)	**31.** (a)	**32.** (b)	**33.** (d)	**34.** (a)	**35.** (c)
36. (d)	**37.** (a)	**38.** (c)	**39.** (d)	**40.** (d)	**41.** (c)	**42.** (c)
43. (d)	**44.** (a)	**45.** (d)	**46.** (b)	**47.** (d)	**48.** (b)	**49.** (a)
50. (d)	**51.** (b)	**52.** (a)	**53.** (a)	**54.** (b)	**55.** (b)	**56.** (c)
57. (a)	**58.** (a)	**59.** (d)	**60.** (c)	**61.** (b)	**62.** (a)	**63.** (c)
64. (c)	**65.** (c)	**66.** (1)	**67.** (30.33)	**68.** (4)	**69.** (15)	**70.** (3)
71. (1)	**72.** (2)	**73.** (17)	**74.** (4)	**75.** (1)	**76.** (1)	**77.** (4.5)
78. (0.5)	**79.** (5)	**80.** (310)				

Level 3

1. (d)	**2.** (c)	**3.** (c)	**4.** (b)	**5.** (a)	**6.** (a)	**7.** (c)
8. (b)	**9.** (d)	**10.** (b)	**11.** (a)	**12.** (c)	**13.** (c)	**14.** (c)
15. (a, c, d)	**16.** (a, b, d)	**17.** (a, d)	**18.** (a, d)	**19.** (a, b, c)	**20.** (b, c)	**21.** (a, d)
22. (a, c, d)	**23.** (a, c)	**24.** (c, a)	**25.** (c, d)	**26.** (a, b)	**27.** (a, d)	**28.** (a, b, c, d)
29. (a, c)	**30.** (a, c, d)	**31.** (a, c, d)	**32.** (a, c)	**33.** (a, b, c)	**34.** (a, c, d)	**35.** (a, b, c)
36. (b, d)	**37.** (a)	**38.** (d)	**39.** (c)	**40.** (d)	**41.** (b)	**42.** (b)
43. (c)	**44.** (c)	**45.** (c)	**46.** (b)	**47.** (d)		

48. (i)–B, (ii)–D, (iii)–A, (iv)–C

49. (i)–D, (ii)–A, (iii)–B, (iv)–C

50. (i)–C, (ii)–A, (iii)–D, (iv)–B

51. (20.95)	**52.** (0.634)	**53.** (20)	**54.** (18)	**55.** (30)	**56.** (125)	**57.** (6)
58. (5)						

SOLUTIONS

Level 1

1. (i) 135 leaves remainder as 3 when it is divided by 4.

Therefore,

$$i^{135} = (i^4)^{33} \cdot i^3$$
$$= (1)^{33} \cdot (-i) \quad [\because i^4 = 1 \text{ and } i^3 = -i]$$
$$\Rightarrow \quad i^{135} = -i$$

(ii) $\dfrac{1}{i^{63}} = \dfrac{1}{(i^4)^{15} \cdot i^3} = \dfrac{1}{i^3} = \dfrac{1}{-i} \times \dfrac{i}{i}$

$$= \frac{-i}{i^2} = \frac{-i}{(-1)} = i \quad [\because i^2 = -1]$$

(iii) $i^{1998} = \dfrac{i^{2000}}{i^2} = \dfrac{1}{(-1)} = -1$

(iv) $i^{-9999} = \dfrac{1}{i^{9999}} = \dfrac{1}{i^3} = i$

(v) $\sqrt{-36} + 4\sqrt{-9} + 2\sqrt{-4}$

$$= 6i + 12i + 2 \times 2i$$
$$= 18i + 4i = 22i$$

(vi) $(-\sqrt{-1})^{4n+3} = (-i)^{4n+3} = (-i)^{4n}(-i)^3$

$$= [(-i)^4]^n(-i^3)$$
$$= -(-i) = i$$

2. Given, $1 + i^2 + i^4 + i^6 + \ldots i^{2n}$

$$= 1 - 1 + 1 - 1 + \ldots + (-1)^n$$

Case I If n is odd, then

$$1 + i^2 + i^4 + i^6 + \ldots + i^{2n}$$
$$= 1 - 1 + 1 - 1 + \ldots + 1 - 1 = 0$$

Case II If n is even, then

$$1 + i^2 + i^4 + i^6 + \ldots + i^{2n}$$
$$= 1 - 1 + 1 - 1 + \ldots + 1 = 1$$

3. (i) Given, $(4 + 3i)(4 - 3i) = 16 - 9i^2$

$$= 16 - 9(-1) = 16 + 9 = 25$$

(ii) Given, $\dfrac{5 + 4i}{4 + 5i} \times \dfrac{4 - 5i}{4 - 5i}$

$$= \frac{(20 + 20) + i(16 - 25)}{16 - 25i^2}$$
$$= \frac{40 - 9i}{41}$$
$$= \frac{40}{41} - \frac{9}{41}i$$

(iii) Given, $\sqrt{-12}(\sqrt{-3} - \sqrt{2})$

$$= i\sqrt{12}(i\sqrt{3} - \sqrt{2})$$
$$= i^2\sqrt{36} - i\sqrt{24}$$
$$= 6i^2 - 2i\sqrt{6}$$
$$= -6 - 2i\sqrt{6}$$

4. $\dfrac{i^{592} + i^{590} + i^{588} + i^{586} + i^{584}}{i^{582} + i^{580} + i^{578} + i^{576} + i^{574}}$

$$= \frac{i^{584}(i^8 + i^6 + i^4 + i^2 + 1)}{i^{574}(i^8 + i^6 + i^4 + i^2 + 1)}$$
$$= \frac{i^{584}}{i^{574}}$$
$$= i^{10}$$
$$= i^4 \cdot i^4 \cdot i^2$$
$$= (1)(1) - 1$$
$$= -1$$

5. We have, $\quad z = 4 + i\sqrt{7}$

$$\Rightarrow \quad z - 4 = i\sqrt{7}$$
$$\Rightarrow \quad z^2 - 8z + 16 = -7$$
$$\Rightarrow \quad z^2 - 8z + 23 = 0$$

On dividing $z^3 - 4z^2 - 9z + 91$ by $z^2 - 8z + 23$, we get

$$z^3 - 4z^2 - 9z + 91 = (z^2 - 8z + 23)(z + 4) - 1$$
$$= 0 - 1 = -1$$

6. $\because$ $n!$ is divisible by 4, $\forall\ n \geq 4$

$$\therefore \sum_{n=4}^{100} i^{n!} = \sum_{n=1}^{97} i^{(n+3)!}$$
$$= i^0 + i^0 + i^0 + \ldots + i^0 \text{ (97 times)}$$
$$= 1 + 1 + 1 + \ldots + 1 \text{ (97 times)}$$
$$= 97 \quad \ldots\text{(i)}$$
$$\therefore \sum_{n=0}^{100} i^{n!} = \sum_{n=0}^{3} i^{n!} + \sum_{n=4}^{100} i^{n!}$$
$$= i^{0!} + i^{1!} + i^{2!} + i^{3!} + 97 \quad \text{[from Eq. (i)]}$$
$$= i^1 + i^1 + i^2 + i^6 + 97$$
$$= i + i - 1 - 1 + 97$$
$$= 95 + 2i$$

7. Given, $(1 + i)^6 + (1 - i)^6 = [(1 + i)^2]^3 + [(1 - i)^2]^3$

$$= [1 + 2i + i^2]^3 + [1 - 2i + i^2]^3$$
$$= (2i)^3 + (-2i)^3$$
$$= (8 - 8)i^3 = 0$$

8. (i) Let $z_1 = 3 + 4i$

$$|z_1| = \sqrt{3^2 + 4^2}$$
$$= \sqrt{9 + 16} = \sqrt{25} = 5$$

(ii) Let $z_2 = (1+2i)(-2+3i)$

$|z_2| = |(1+2i)(-2+3i)|$

$= |1+2i||-2+3i|$ $[\because |z_1 z_2| = |z_1||z_2|]$

$= \sqrt{1+4} \cdot \sqrt{4+9}$

$= \sqrt{5} \cdot \sqrt{13} = \sqrt{65}$

(iii) Let $z_3 = \dfrac{3-2i}{2i}$

$|z_3| = \left|\dfrac{3-2i}{2i}\right|$

$= \dfrac{|3-2i|}{|2i|}$ $\left[\because \left|\dfrac{z_1}{z_2}\right| = \dfrac{|z_1|}{|z_2|}\right]$

$= \dfrac{\sqrt{9+4}}{\sqrt{4}} = \dfrac{\sqrt{13}}{\sqrt{4}}$

$|z_3| = \dfrac{\sqrt{13}}{2}$

9. We have, $|z| = z + 1 + 2i$

$\Rightarrow \sqrt{x^2+y^2} = (x+iy) + (1+2i) = (x+1) + (2+y)i$

$\Rightarrow \sqrt{x^2+y^2} = x+1$ and $0 = 2+y$

$\Rightarrow \sqrt{x^2+y^2} = x+1$ and $y = -2$

$\Rightarrow \sqrt{x^2+(-2)^2} = x+1$ $[\because y = -2]$

$\Rightarrow x^2 + 4 = (x+1)^2 = x^2 + 2x + 1$

$\Rightarrow 2x = 3$

$\Rightarrow x = \dfrac{3}{2}$

$\Rightarrow x + iy = \dfrac{3}{2} - 2i$

10. $\because |z+1| = |(z+4) - 3|$

$= |(z+4) + (-3)| \le |z+4| + |-3|$

$= |z+4| + 3$

$\le 3 + 3 = 6$

$\therefore |z+1| \le 6$

Hence, the greatest value of $|z+1|$ is 6.

11. We have, $z_1 = 24 + 7i$

$\therefore |z_1| = \sqrt{(24)^2 + (7)^2} = 25$

$||z_1| - |z_2|| \le |z_1 + z_2| \le |z_1| + |z_2|$

$\Rightarrow |25 - 6| \le |z_1 + z_2| \le 25 + 6$

$\Rightarrow 19 \le |z_1 + z_2| \le 31$

Hence, the least value of $|z_1 + z_2|$ is 19 and the greatest value of $|z_1 + z_2|$ is 31.

12. (i) Let $z_1 = 3 + 4i$

$\bar{z}_1 = \overline{3+4i} = 3 - 4i$

(ii) Let $z_2 = 4i(3-5i)$

$= 12i - 20i^2$

$= 12i - 20(-1)$

$= 20 + 12i$

$\bar{z}_2 = \overline{20+12i} = 20 - 12i$

(iii) Let $z_3 = \dfrac{1+2i}{1-2i}$

$= \dfrac{1+2i}{1-2i} \times \dfrac{1+2i}{1+2i}$

$= \dfrac{(1+2i)^2}{1+4}$

$= \dfrac{1}{5}(1 + 4i^2 + 4i)$

$= \dfrac{1}{5}(1 - 4 + 4i)$

$= \dfrac{1}{5}(-3+4i) = -\dfrac{3}{5} + \dfrac{4}{5}i$

$\bar{z}_3 = -\dfrac{3}{5} - \dfrac{4}{5}i$

13. Let $z = x + iy$ and $\bar{z} = iz^2$

$\Rightarrow (x - iy) = i(x^2 - y^2 + 2ixy)$

$\Rightarrow x - iy = i(x^2 - y^2) - 2xy$

$\Rightarrow x(1+2y) = 0$ and $x^2 - y^2 + y = 0$

$\Rightarrow x = 0$ or $y = -\dfrac{1}{2} \Rightarrow x = 0, y = 0, 1$

or $y = -\dfrac{1}{2}; x = \pm\dfrac{\sqrt{3}}{2}$

For non-zero complex number z,

$x = 0, y = 1; x = \dfrac{\sqrt{3}}{2}, y = -\dfrac{1}{2}$

$x = -\dfrac{\sqrt{3}}{2}, y = -\dfrac{1}{2}$

$\therefore z = i, \dfrac{\sqrt{3}}{2} - \dfrac{i}{2}, -\dfrac{\sqrt{3}}{2} - \dfrac{i}{2}$

14. Let $\sqrt{7-24i} = x + iy$, then

$7 - 24i = (x+iy)^2$

$\Rightarrow 7 - 24i = (x^2 - y^2) + 2ixy$

$\Rightarrow x^2 - y^2 = 7$ and $2xy = -24$ ...(i)

Now, $(x^2+y^2)^2 = (x^2-y^2)^2 + 4x^2y^2$

$= 49 + 576 = 625$

$\Rightarrow x^2 + y^2 = 25$ $[\because x^2 + y^2 > 0]$

On solving $x^2 - y^2 = 7$ and $x^2 + y^2 = 25$, we get

$x^2 = 16$ and $y^2 = 9$

$\Rightarrow x = \pm 4$ and $y = \pm 3$

As from Eq. (i), $2xy$ is negative, so x and y are of opposite sign.

$\therefore x = 4$ and $y = -3$

or $x = -4$ and $y = 3$

Hence, $\sqrt{7-24i} = \pm(4-3i)$

15. (i) Let $z = 5 + 5i$

Since, z lies in first quadrant.

Therefore, the arguments is given by

$$\arg(z) = \tan^{-1}\left|\frac{5}{5}\right|$$

$$\left[\because \arg z = \tan^{-1}\left|\frac{\text{Im}(z)}{\text{Re}(z)}\right|, z \text{ lies in first quadrant}\right]$$

$$= \tan^{-1}(1) = \frac{\pi}{4}$$

(ii) Let $z = -4 + 4i$, z lies in second quadrant.

$$\therefore \arg(z) = \pi - \tan^{-1}\left|\frac{4}{-4}\right|$$

$$= \pi - \tan^{-1}(1) = \pi - \frac{\pi}{4} = \frac{3\pi}{4}$$

(iii) Let $z = -3 - 3i$, z lies in third quadrant

$$\therefore \arg(z) = \pi + \tan^{-1}\left|\frac{-3}{-3}\right|$$

$$= \pi + \tan^{-1}(1) = \pi + \frac{\pi}{4} = \frac{5\pi}{4}$$

(iv) Let $z = 2 - 2i$, z lies in fourth quadrant.

$$\therefore \quad \arg(z) = -\tan^{-1}\left|\frac{-2}{2}\right| = -\tan^{-1}(1) = -\frac{\pi}{4}$$

16. Given that $z_2 = \bar{z}_1$

$$\therefore \arg(-z_1 z_2) = \arg(-z_1 \bar{z}_1) = \arg(-|z_1|^2)$$

$$= \arg(\text{negative real number})$$

$$= \pi$$

17. $\because \quad \arg(z_1 z_2) = \arg(z_1) + \arg(z_2)$

$$= 170° + 70° = 240°$$

Thus, $z_1 z_2$ lies in third quadrant.

Hence, its principal argument is $-120°$.

18. We have, $z = \dfrac{2+i}{4i + (1+i)^2}$

$$= \frac{2+i}{4i + 1 + i^2 + 2i} = \frac{2+i}{6i}$$

$$\Rightarrow \quad z = \frac{1}{6} - \frac{1}{3}i$$

$\therefore$ z lies in fourth quadrant.

Hence, principal value of $\arg(z) = -\tan^{-1}\left|\dfrac{-\frac{1}{3}}{\frac{1}{6}}\right|$

$$= -\tan^{-1}(2)$$

19. Let $z_1 = r_1 e^{i\theta_1}$ and $z_2 = r_2 e^{i\theta_2}$

$$\therefore \ z_1 + z_2 = r_1 e^{i\theta_1} + r_2 e^{i\theta_2}$$

$$\Rightarrow z_1 + z_2 = (r_1 \cos\theta_1 + r_2 \cos\theta_2) + i(r_1 \sin\theta_1 + r_2 \sin\theta_2)$$

$$[\because e^{i\theta} = \cos\theta + i\sin\theta]$$

$$\Rightarrow |z_1 + z_2|^2 = r_1^2 \cos^2\theta_1 + r_2^2 \cos^2\theta_2 + 2r_1 r_2 \cos\theta_1 \cos\theta_2 + r_1^2 \sin^2\theta_1 + r_2^2 \sin^2\theta_2 + 2r_1 r_2 \sin\theta_1 \sin\theta_2$$

$$\Rightarrow |z_1 + z_2|^2 = r_1^2 + r_2^2 + 2r_1 r_2 \cos(\theta_1 - \theta_2) \quad \text{...(i)}$$

Also, $|z_1 + z_2| = |z_1| + |z_2|$ [given]

$$\Rightarrow |z_1 + z_2|^2 = |z_1|^2 + |z_2|^2 + 2|z_1||z_2|$$

$$\Rightarrow |z_1 + z_2|^2 = r_1^2 + r_2^2 + 2r_1 r_2 \quad \text{...(ii)}$$

From Eqs. (i) and (ii), we get

$$2r_1 r_2 = 2r_1 r_2 \cos(\theta_1 - \theta_2)$$

$$\Rightarrow \cos(\theta_1 - \theta_2) = 1$$

$$\Rightarrow \quad \theta_1 - \theta_2 = 0 \Rightarrow \theta_1 = \theta_2$$

$$\Rightarrow \quad \arg(z_1) = \arg(z_2) \Rightarrow \arg(z_1) - \arg(z_2) = 0$$

20. Given, $z_1 = 10 + 6i$ and $z_2 = 4 + 6i$

Also, $\arg\left(\dfrac{z - z_1}{z - z_2}\right) = \dfrac{\pi}{4}$

$$\Rightarrow \quad \arg(z - z_1) - \arg(z - z_2) = \frac{\pi}{4}$$

$$\Rightarrow \quad \arg[(x+iy) - (10+6i)] - \arg[(x+iy) - (4+6i)] = \frac{\pi}{4}$$

$$\Rightarrow \quad \arg[(x-10) + i(y-6)] - \arg[(x-4) + i(y-6)] = \frac{\pi}{4}$$

$$\Rightarrow \quad \tan^{-1}\left(\frac{y-6}{x-10}\right) - \tan^{-1}\left(\frac{y-6}{x-4}\right) = \frac{\pi}{4}$$

$$\Rightarrow \quad \tan^{-1}\left[\frac{\frac{y-6}{x-10} - \frac{y-6}{x-4}}{1 + \frac{(y-6)^2}{(x-4)(x-10)}}\right] = \frac{\pi}{4}$$

$$\Rightarrow \quad \frac{(x-4)(y-6) - (x-10)(y-6)}{(x-4)(x-10) + (y-6)^2} = \tan\frac{\pi}{4}$$

On simplifying, we have

$$\Rightarrow x^2 + y^2 - 14x - 18y + 112 = 0$$

$$\Rightarrow \quad (x-7)^2 + (y-9)^2 = (3\sqrt{2})^2$$

$$\Rightarrow \quad |z - (7+9i)| = 3\sqrt{2} \quad [\text{where, } z = x + iy]$$

Thus, the value of $|z - (7+9i)|$ is $3\sqrt{2}$.

21. (i) Let $z_1 = 3i$

On comparing with $x + iy$, we get

$$x = 0 \text{ and } y = 3$$

$$\therefore \quad r = \sqrt{x^2 + y^2} = \sqrt{0+9} = \sqrt{9} = 3$$

$$\theta = \tan^{-1}\left|\frac{y}{x}\right| \quad [\because z \text{ lies in first quadrant}]$$

$$= \tan^{-1}\left|\frac{3}{0}\right|$$

$$= \tan^{-1}(\infty)$$

$$= \tan^{-1}\left(\tan\frac{\pi}{2}\right)$$

$$\theta = \frac{\pi}{2}$$

Hence, the polar form $= r(\cos\theta + i\sin\theta)$

$$= 3\left(\cos\frac{\pi}{2} + i\sin\frac{\pi}{2}\right)$$

(ii) Let $z_2 = -4 + 4i$

$$\therefore \quad r = \sqrt{(-4)^2 + (4)^2}$$

$$= \sqrt{16+16} = \sqrt{32} = 4\sqrt{2}$$

$$\theta = \pi - \tan^{-1}\left|\frac{y}{x}\right| \quad [\because z \text{ lies in second quadrant}]$$

$$= \pi - \tan^{-1}\left|\frac{4}{-4}\right|$$

$$= \pi - \tan^{-1}(1) = \pi - \frac{\pi}{4} = \frac{3\pi}{4}$$

$$\therefore \text{Polar form} = 4\sqrt{2}\left(\cos\frac{3\pi}{4} + i\sin\frac{3\pi}{4}\right)$$

22. Let $z = \dfrac{3+2i\sin\theta}{1-2i\sin\theta} \times \dfrac{1+2i\sin\theta}{1+2i\sin\theta}$

$$= \frac{(3-4\sin^2\theta) + (8i\sin\theta)}{1+4\sin^2\theta}$$

$$= \frac{3-4\sin^2\theta}{1+4\sin^2\theta} + \frac{8\sin\theta}{1+4\sin^2\theta}i$$

(i) For purely real, $\text{Im}(z) = 0$

$$\Rightarrow \frac{8\sin\theta}{1+4\sin^2\theta} = 0 \text{ or } \sin\theta = 0$$

$$\therefore \quad \theta = n\pi, n \in I$$

(ii) For purely imaginary, $\text{Re}(z) = 0$

$$\Rightarrow \frac{3-4\sin^2\theta}{1+4\sin^2\theta} = 0$$

or $3 - 4\sin^2\theta = 0$

$$\Rightarrow \quad \sin^2\theta = \frac{3}{4}$$

$$\Rightarrow \quad \sin\theta = \pm\sqrt{\frac{3}{4}} = \pm\frac{\sqrt{3}}{2}$$

$$= \pm\sin\frac{\pi}{3} = \sin\left(\pm\frac{\pi}{3}\right)$$

$$\therefore \quad \theta = n\pi \pm \frac{\pi}{3}, n \in I$$

23. We have, $\quad z = \cos\theta + i\sin\theta = e^{i\theta}$

$$z^{2m-1} = (e^{i\theta})^{2m-1} = e^{i(2m-1)\theta}$$

$$\Rightarrow \quad z^{2m-1} = \cos(2m-1)\theta + i\sin(2m-1)\theta$$

$$\therefore \quad \text{Im}(z^{2m-1}) = \sin(2m-1)\theta$$

$$\therefore \sum_{m=1}^{15}\text{Im}(z^{2m-1}) = \sum_{m=1}^{15}\sin(2m-1)\theta$$

$\Rightarrow \sin\theta + \sin 3\theta + \sin 5\theta + \ldots$ upto 15 terms

$$= \frac{\sin\left[15\times\left(\frac{2\theta}{2}\right)\right]\sin[\theta + 14\times\theta]}{\sin\theta}$$

$[\because \sin\alpha + \sin(\alpha+\beta) + \sin(\alpha+2\beta) + \ldots n$ terms

$$= \frac{\sin\left(\frac{n\beta}{2}\right)\sin[\alpha+(n-1)\beta/2]}{\sin\left(\frac{\beta}{2}\right)}$$

[here, $\alpha = 0, \beta = 20$]

$$= \frac{\sin 15\theta \times \sin 15\theta}{\sin\theta} = \frac{\sin^2 15\theta}{\sin\theta}$$

$$\text{At } \theta = 2^\circ, \frac{\sin^2 30^\circ}{\sin 2^\circ} = \frac{1}{4\sin 2^\circ}$$

24. Let $z = (\sqrt{3}+i)$

$$r = \sqrt{(\sqrt{3})^2 + (1)^2} = \sqrt{3+1} = \sqrt{4} = 2$$

$\because z$ lies in first quadrant.

$$\theta = \tan^{-1}\left|\frac{1}{\sqrt{3}}\right| = \tan^{-1}\left\{\tan\left(\frac{\pi}{6}\right)\right\} = \frac{\pi}{6}$$

$$\therefore \quad (\sqrt{3}+i) = 2\left(\cos\frac{\pi}{6} + i\sin\frac{\pi}{6}\right)$$

$$\therefore (\sqrt{3}+i)^7 = \left[2\left(\cos\frac{\pi}{6} + i\sin\frac{\pi}{6}\right)\right]^7$$

$$= 2^7\left(\cos\frac{\pi}{6} + i\sin\frac{\pi}{6}\right)^7$$

$$= 2^7\left(\cos\frac{7\pi}{6} + i\sin\frac{7\pi}{6}\right)$$

[using De-Moivre's theorem]

$$= 2^7\left(-\frac{\sqrt{3}}{2} + \left(-\frac{1}{2}\right)i\right)$$

$$= -64(\sqrt{3}+i)$$

25. Given, $\dfrac{(\cos\theta + i\sin\theta)^4}{(\sin\theta + i\cos\theta)^4}$

$$= \frac{\cos 4\theta + i\sin 4\theta}{i^4(\cos\theta - i\sin\theta)^4}$$

$$= \frac{\cos 4\theta + i\sin 4\theta}{\cos 4\theta - i\sin 4\theta}$$

$$= \frac{\cos 4\theta + i\sin 4\theta}{\cos 4\theta - i\sin 4\theta} \times \frac{\cos 4\theta + i\sin 4\theta}{\cos 4\theta + i\sin 4\theta}$$

$$= \frac{(\cos 4\theta + i\sin 4\theta)^2}{\cos^2 4\theta - i^2\sin^2 4\theta}$$

$$= \frac{\cos 8\theta + i\sin 8\theta}{\cos^2 4\theta + \sin^2 4\theta}$$

$$= \cos 8\theta + i\sin 8\theta$$

26. Given, $\dfrac{4(\cos 75^\circ + i\sin 75^\circ)}{0.4(\cos 30^\circ + i\sin 30^\circ)}$

$$= \frac{10(\cos 75^\circ + i\sin 75^\circ)}{\cos 30^\circ + i\sin 30^\circ} \times \frac{(\cos 30^\circ - i\sin 30^\circ)}{(\cos 30^\circ - i\sin 30^\circ)}$$

$$= \frac{10(\cos 75^\circ + i\sin 75^\circ)\{\cos(-30^\circ) + i\sin(-30^\circ)\}}{\cos^2 30^\circ + \sin^2 30^\circ}$$

$$= 10[\cos(75^\circ - 30^\circ) + i\sin(75^\circ - 30^\circ)]$$

$$= 10(\cos 45^\circ + i\sin 45^\circ)$$

$$= \frac{10}{\sqrt{2}}(1+i)$$

27. Let $z = \sqrt{3} + i$

$$\therefore \quad r = \sqrt{(\sqrt{3})^2 + (1)^2}$$

$$= \sqrt{3+1} = \sqrt{4} = 2$$

$$\theta = \tan^{-1}\left|\frac{1}{\sqrt{3}}\right| \qquad [\because z \text{ lies in first quadrant}]$$

$$\theta = \tan^{-1}\left(\tan\frac{\pi}{6}\right) = \frac{\pi}{6}$$

$$\therefore \quad z = 2\left(\cos\frac{\pi}{6} + i\sin\frac{\pi}{6}\right) = 2(\cos 30^\circ + i\sin 30^\circ)$$

In general, we can write as

$$z = 2[\cos(30^\circ + 360^\circ k) + i\sin(30^\circ + 360^\circ k)]$$

Applying 5th root, we have

$$z^{\frac{1}{5}} = 2^{1/5}[\cos(30^\circ + 360^\circ k) + i\sin(30^\circ + 360^\circ k)]^{\frac{1}{5}}$$

Using De-Moivre's, we get

$$z^{\frac{1}{5}} = 2^{\frac{1}{5}}\left[\cos\frac{1}{5}(30^\circ + 360^\circ k) + i\sin\frac{1}{5}(30^\circ + 360^\circ k)\right],$$

where $k = 0, 1, 2, 3, 4$

At $k = 0$, $z_1 = 2^{\frac{1}{5}}[\cos 6^\circ + i\sin 6^\circ]$

At $k = 1$, $z_2 = 2^{\frac{1}{5}}[\cos 78^\circ + i\sin 78^\circ]$

At $k = 2$, $z_3 = 2^{\frac{1}{5}}[\cos 150^\circ + i\sin 150^\circ]$

At $k = 3$, $z_4 = 2^{\frac{1}{5}}[\cos 222^\circ + i\sin 222^\circ]$

At $k = 4$, $z_5 = 2^{\frac{1}{5}}[\cos 294^\circ + i\sin 294^\circ]$

28. Given that

$$z = \left(\frac{\sqrt{3}}{2} + \frac{i}{2}\right)^5 + \left(\frac{\sqrt{3}}{2} - i\frac{1}{2}\right)^5$$

$$= \left[\cos\left(\frac{\pi}{6}\right) + i\sin\left(\frac{\pi}{6}\right)\right]^5 + \left[\cos\left(\frac{\pi}{6}\right) - i\sin\left(\frac{\pi}{6}\right)\right]^5$$

$$= \cos\frac{5\pi}{6} + i\sin\frac{5\pi}{6} + \cos\frac{5\pi}{6} - i\sin\frac{5\pi}{6}$$

[using De Moivre's theorem]

$$= 2\cos\frac{5\pi}{6}$$

Hence, imaginary part of z is 0.

29. Suppose the cube root of 1 is a.

$$\Rightarrow \quad a = (1)^{\frac{1}{3}}$$

$$\Rightarrow \quad a^3 = 1$$

$$\Rightarrow \quad a^3 - 1 = 0$$

$$\Rightarrow \quad (a-1)(a^2 + a + 1) = 0$$

$$\therefore \quad a - 1 = 0 \text{ or } a^2 + a + 1 = 0$$

$$a = 1 \text{ or } a = \frac{-1 \pm \sqrt{1-4}}{2 \cdot 1} = \frac{-1 \pm \sqrt{3}i}{2}$$

$$a = -\frac{1}{2} \pm \frac{\sqrt{3}}{2}i$$

$\therefore$ The cube roots of unity are

$$1, -\frac{1}{2} + \frac{i\sqrt{3}}{2}, -\frac{1}{2} - \frac{i\sqrt{3}}{2}$$

30. Given that

$$(1-\omega)(1-\omega^2)(1-\omega^4)(1-\omega^8)$$

$$= (1-\omega)(1-\omega^2)(1-\omega^3 \cdot \omega)[1-(\omega^4)^2]$$

$$= (1-\omega)(1-\omega^2)(1-\omega)(1-\omega^2)$$

$$= (1-\omega)^2(1-\omega^2)^2$$

$$= (1-2\omega+\omega^2)(1-2\omega^2+\omega^4)$$

$$= (1-2\omega+\omega^2)(1-2\omega^2+\omega)$$

$$= (-\omega-2\omega)(-\omega^2-2\omega^2)$$

$$[\because 1+\omega+\omega^2 = 0]$$

$$= (-3\omega)(-3\omega^2)$$

$$= 9\omega^3 = 9$$

31. $\because 1 + \omega + \omega^2 = 0, \omega^3 = 1$

$$\Rightarrow \quad (3+\omega^2+\omega^4)^6 = (3+\omega^2+\omega)^6$$

$$= (3-1)^6$$

$$= 2^6 = 64$$

32. $xyz = (\alpha+\beta)(\alpha\omega+\beta\omega^2)(\alpha\omega^2+\beta\omega)$

$$= (\alpha+\beta)[\alpha^2 + \alpha\beta(\omega^2+\omega) + \beta^2] \qquad [\because \omega^3 = 1]$$

$$= (\alpha+\beta)(\alpha^2 - \alpha\beta + \beta)^2 \qquad [\because 1+\omega+\omega^2 = 0]$$

$$= \alpha^3 + \beta^3$$

33. Given, $\dfrac{a+b\omega+c\omega^2}{c+a\omega+b\omega^2} + \dfrac{c+a\omega+b\omega^2}{a+b\omega+c\omega^2} + \dfrac{b+c\omega+a\omega^2}{b+c\omega^4+a\omega^5}$

$$= \frac{\omega^2(a+b\omega+c\omega^2)}{(a+b\omega+c\omega^2)} + \frac{\omega(a\omega+b\omega^2+c)}{(a\omega+b\omega^2+c)} + \frac{b+c\omega+a\omega^2}{(b+c\omega+a\omega^2)}$$

$$= \omega^2 + \omega + 1$$

$$= 0$$

34. Here, $a = e^{i\frac{2\pi}{3}} = \omega$

$\therefore\ a + \frac{1}{a^2} = \omega + \frac{1}{\omega^2} = \omega + \omega = 2\omega$

Similarly, $a^2 + \frac{1}{a^4} = \omega^2 + \frac{1}{\omega^4} = 2\omega^2$

Now, $a + \frac{1}{a^2} + a^2 + \frac{1}{a^4} = 2\omega + 2\omega^2 = -2$

and $\left(a + \frac{1}{a^2}\right)\left(a^2 + \frac{1}{a^4}\right) = 2\omega \cdot 2\omega^2 = 4$

So, the required equation is $x^2 + 2x + 4 = 0$

35. Given, ω is imaginary 5th root of unity

$\Rightarrow \quad \omega^5 = 1$

So, the roots are $1, \omega, \omega^2, \omega^3, \omega^4$

$\because\ 1 + \omega + \omega^2 + \omega^3 + \omega^4 = 0$

$\Rightarrow \omega^4 + \omega^8 + \omega^{12} + \omega^{16}$

$= \omega^4 + \omega^5 \cdot \omega^3 + \omega^{10} \cdot \omega^2 + \omega^{15} \cdot \omega$

$= \omega^4 + \omega^3 + \omega^2 + \omega$

$= -1$

36. Given, $\sum_{\lambda=1}^{12}\left(\sin\frac{2\lambda\pi}{13} - i\cos\frac{2\lambda\pi}{13}\right)$

$= \sum_{\lambda=1}^{12}\left(\sin\frac{2\lambda\pi}{13} - i\cos\frac{2\lambda\pi}{13}\right)$

$= \sum_{\lambda=1}^{12}(-i)\left(\cos\frac{2\lambda\pi}{13} + i\sin\frac{2\lambda\pi}{13}\right)$

$= (-i)\sum_{\lambda=1}^{12} e^{\frac{i2\lambda\pi}{13}} \qquad [\because e^{i\theta} = \cos\theta + i\sin\theta]$

By nth roots of unity,

$z^n = 1$

$\Rightarrow \quad z = e^{\frac{i2k\pi}{n}}, k \in N$

$\Rightarrow \quad z^{13} = 1, z = e^{\frac{i2\lambda\pi}{13}}$

$\Rightarrow \quad \sum_{\lambda=1}^{12} e^{\frac{2i\lambda\pi}{13}} = -1$

$\Rightarrow \quad (-i)\sum_{\lambda=1}^{12} e^{\frac{2i\lambda\pi}{13}} = (-i)(-1) = i$

37. Given, $z^7 = -1$

We write, $z^{86} + z^{175} + z^{289}$ in terms of 7th root of unity

$\Rightarrow z^{86} + z^{175} + 2^{289} = (z^7)^{12} \cdot z^2 + (z^7)^{25} + (z^7)^{41} \cdot z^2$

$= z^2 + (-1) + (-z^2) = -1$

38. Given circle, $\left|\frac{z - i}{z + i}\right| = 5$

On comparing the given circle $\left|\frac{z - \alpha}{z - \beta}\right| = k$, we get

Radius $= \left|\frac{k(\alpha - \beta)}{1 - k^2}\right| = \left|\frac{5(i + 1)}{1 - 25}\right| = \frac{5}{12}$

39. We have, $\quad |z + 5|^2 - |z - 5|^2 = 10$

$\Rightarrow (z + 5)\overline{(z + 5)} - (z - 5)\overline{(z - 5)} = 10$

$\Rightarrow (z + 5)(\bar{z} + 5) - (z - 5)(\bar{z} - 5) = 10$

$\Rightarrow 5(z + \bar{z}) + 25 + 5(z + \bar{z}) - 25 = 10$

$\Rightarrow \quad 10(z + \bar{z}) = 10$

$\Rightarrow \quad 2x = 1$

$\Rightarrow \quad x = \frac{1}{2}$

which is the equation of a straight line.

40. We have, $\left|\frac{z - 2}{z - 3}\right| = 2$

$|z - \bar{z}|^2 = 4|z - 3|^2$

$\Rightarrow \quad |x - 2 + iy|^2 = 4|x - 3 + iy|^2$

$\Rightarrow \quad (x - 2)^2 + y^2 = 4[(x - 3)^2 + y^2]$

$\Rightarrow 3x^2 + 3y^2 - 24x + 4x + 36 - 4 = 0$

$\Rightarrow \quad x^2 + y^2 - \frac{20}{3}x + \frac{32}{3} = 0$

This represents a circle with centre $\left(\frac{10}{3}, 0\right)$

and radius $= \sqrt{\frac{100}{9} - \frac{32}{3}} = \sqrt{\frac{4}{9}} = \frac{2}{3}$

41. Two circles whose centre and radius are $C_1(0, 0)$, $r_1 = 12$, $C_2(3, 4)$ and $r_2 = 5$ and it passes through origin.

$C_1C_2 = \sqrt{3^2 + 4^2} = 5$

and $\ r_1 - r_2 = 12 - 5 = 7$

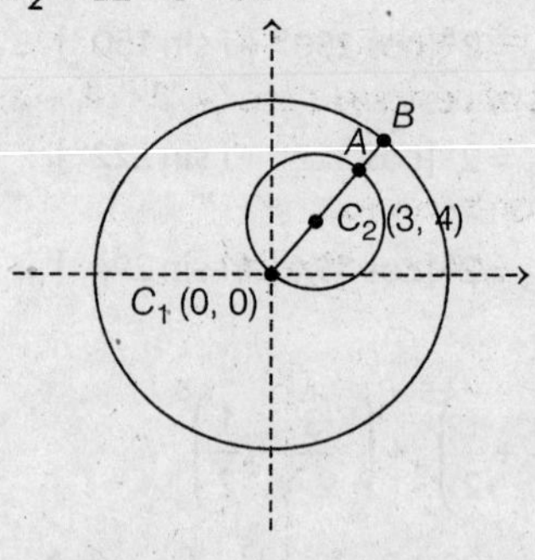

$\therefore \quad C_1C_2 < r_1 - r_2$

Hence, circle C_2 lies inside the circle C_1.

From figure, the minimum distance between them is

$AB = C_1B - C_1A = r_1 - (C_1C_2 + C_2A)$

$= 12 - 10 = 2$

42. The given circle is $|z-1|=\sqrt{2}$,

where $z_0=1$ is the centre and $\sqrt{2}$ is the radius.

Let $z_1=2+\sqrt{3}i$ is one of the vertex of square inscribed in the given circle.

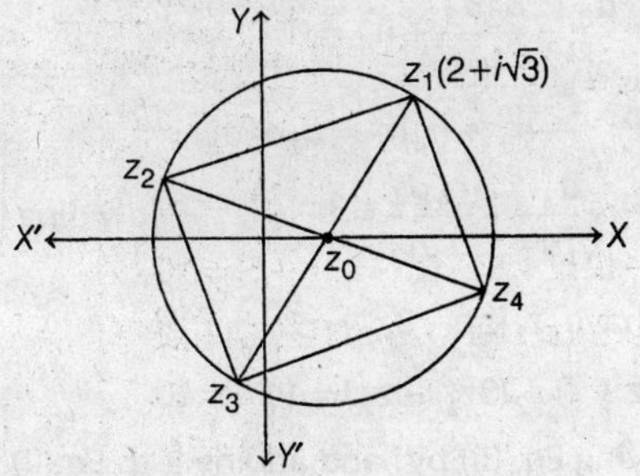

Clearly, z_2 can be obtained by rotating z_1 by an angle 90° in anti-clockwise direction about centre z_0.

Thus, $z_2-z_0=(z_1-z_0)e^{i\frac{\pi}{2}}$

or $\quad z_2-1=(2+i\sqrt{3}-1)i \qquad [\because e^{i\theta}=\cos\theta+i\sin\theta]$

$\Rightarrow \quad z_2=i-\sqrt{3}+1$

$\Rightarrow \quad z_2=(1-\sqrt{3})+i$

Again, rotating z_2 by 90° about z_0, we get

$z_3-z_0=(z_2-z_0)i$

$\Rightarrow \quad z_3-1=[(1-\sqrt{3})+i-1]i$

$\Rightarrow \quad z_3=-i\sqrt{3}$

Similarly, we find $z_4=(\sqrt{3}+1)-i$

43. Given that $z=x+iy$; x and y are integers.

Also, $\quad z\bar{z}^3+\bar{z}z^3=350$

$\Rightarrow \quad |z|^2(\bar{z}^2+z^2)=350$

$\Rightarrow (x^2+y^2)(x^2-y^2)=175$

$\Rightarrow (x^2+y^2)(x^2-y^2)=25\times 7$

or $x^2+y^2=25$ and $x^2-y^2=7$

$\Rightarrow \quad x^2=16$ and $y^2=9$

$\Rightarrow \quad x=\pm 4$ and $y=\pm 3$

$\therefore$ Vertices of rectangle are (4, 3), (4, – 3), (– 4, 3) and (– 4, –3).

Using Heron's formula,

Area = 48 unit^2

44.

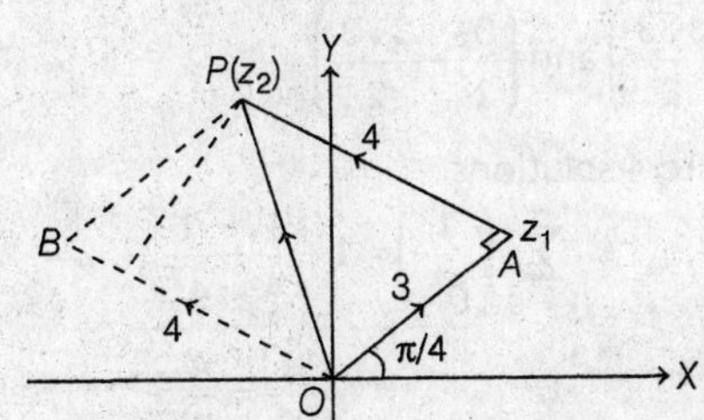

From the figure

$\overline{OP}=\overline{OA}+\overline{AP}$

$\Rightarrow \quad \overline{OP}=\overline{OA}+\overline{OB}$

$\Rightarrow \quad \overline{OP}=3e^{i\frac{\pi}{4}}+4e^{i\left(\frac{\pi}{2}+\frac{\pi}{4}\right)}$

$\Rightarrow \quad \overline{OP}=3e^{i\frac{\pi}{4}}+4e^{i\frac{\pi}{2}}\times e^{i\frac{\pi}{4}}$

$=3e^{i\frac{\pi}{4}}+4ie^{i\frac{\pi}{4}} \qquad [\because e^{i\frac{\pi}{2}}=i]$

$\overline{OP}=(3+4i)e^{i\frac{\pi}{4}}$

Thus, the position of P in the argand plane is $(3+4\hat{i})e^{i\frac{\pi}{4}}$.

45. The initial position of point is

$z_0=1+2i$

$z_1=(1+5)+(2+3)i=6+5i$

Now, z_1 is moved through a distance of $\sqrt{2}$ units in the direction, $\hat{i}+\hat{j}$.

$\therefore$ It becomes $z_1{}'=z_1+(1+i)=7+6i$

Now, $z_1{}'$ is rotated through an angle $\frac{\pi}{2}$ in anti-clockwise direction.

$\therefore \quad z_2=iz_1{}'=i(7+6i) \Rightarrow z_2=-6+7i$

Thus, the point is $z_2=-6+7i$.

Level 2

1. (d) We have, $10z^2-3iz-k=0$

$\Rightarrow \quad z=\dfrac{3i\pm\sqrt{-9+40k}}{20}$

Now, $D=-9+40k$

If $k=1$, then $D=31$

So, (a) is false.

If k is negative real number, then D is a negative real number.

So, (d) is true.

If $k=i$, then $D=-9+40i=16+40i-25=(4+5i)^2$

and the roots are $\left(\frac{1}{5}\right)+\left(\frac{2}{5}\right)i$ and $-\frac{1}{5}-\frac{1}{10}i$.

So, (c) is false.

If $k=0$ (which is a complex numbers), then the roots are 0 and $\left(\frac{3}{10}\right)i$.

So, (b) is false.

2. (c) Observing carefully the system of equations, we find

$\dfrac{1+i}{2i}=\dfrac{1-i}{2}=\dfrac{1}{1+i}$

Hence, there are infinite number of solutions.

3. (d) Given that $a^2+b^2=1$, therefore

$$\frac{1+b+ia}{1+b-ia}=\frac{(1+b+ia)(1+b+ia)}{(1+b-ia)(1+b+ia)}$$

$$=\frac{(1+b)^2-a^2+2ia(1+b)}{1+b^2+2b+a^2}$$

$$=\frac{(1-a^2)+2b+b^2+2ia(1+b)}{2(1+b)}$$

$$=\frac{2b^2+2b+2ia(1+b)}{2(1+b)}=b+ia$$

4. (a) Let α be the real root and $i\beta$ be the imaginary root of the given equation, then

$$\alpha+i\beta=-a\Rightarrow\alpha-i\beta=-\bar{a}$$

So, $2\alpha=-(a+\bar{a})$ and $2i\beta=-(a-\bar{a})$

Multiplying these, we get

$$4i\alpha\beta=a^2-(\bar{a})^2$$

$$\therefore \quad 4b=a^2-(\bar{a})^2$$

5. (d) Let $z=\dfrac{3+2i\sin\theta}{1-2i\sin\theta}$

$\because z$ is purely imaginary, then $z+\bar{z}=0$

$$\Rightarrow \frac{3+2i\sin\theta}{1-2i\sin\theta}+\frac{3-2i\sin\theta}{1+2i\sin\theta}=0$$

$$\Rightarrow \frac{(3+2i\sin\theta)(1+2i\sin\theta)+(3-2i\sin\theta)(1+2i\sin\theta)}{1+4\sin^2\theta}=0$$

$$\Rightarrow \sin^2\theta=\frac{3}{4} \quad \text{[on simplification]}$$

$$\Rightarrow \sin\theta=\pm\frac{\sqrt{3}}{2}$$

$$\Rightarrow \theta=-\frac{\pi}{3},\frac{\pi}{3},\frac{2\pi}{3}(\in A)$$

Now, the sum of elements in A

$$=\frac{-\pi}{3}+\frac{\pi}{3}+\frac{2\pi}{3}=\frac{2\pi}{3}$$

6. (b) $f(z)=g(z)(z-i)(z+i)+az+b,\ a,b\in C$

Given, $f(i)=i$

$$\Rightarrow ai+b=i \quad \text{...(i)}$$

and $f(-i)=1+i$

$$\Rightarrow a(-i)+b=1+i \quad \text{...(ii)}$$

From Eqs. (i) and (ii), we have

$$a=\frac{1}{2}i \text{ and } b=\frac{1}{2}+i$$

Hence, the required remainder is

$$az+b=\left(\frac{1}{2}\right)iz+\left(\frac{1}{2}\right)+i$$

7. (b) Let xi be the root, where $x\neq0$ and $x\in R$

$$x^4-a_1x^3i-a_2x^2+a_3xi+a_4=0$$

$$\Rightarrow x^4-a_2x^2+a_4=0 \quad \text{...(i)}$$

and $a_1x^3-a_3x=0 \quad$ ...(ii)

From Eq (ii), $a_1x^2-a_3=0$

$$\Rightarrow x^2=\frac{a_3}{a_1} \quad [\because x\neq0]$$

Putting the value of x^2 in Eq. (i), we get

$$\frac{a_3^2}{a_1^2}-\frac{a_2a_3}{a_1}+a_4=0$$

$$\Rightarrow a_3^2+a_4a_1^2=a_1a_2a_3$$

$$\Rightarrow \frac{a_3}{a_1a_2}+\frac{a_1a_4}{a_2a_3}=1 \quad \text{[dividing by } a_1a_2a_3]$$

8. (c) We have, $z_1(z_1^2-3z_2^2)=2 \quad$ ...(i)

$$z_2(3z_1^2-z_2^2)=11 \quad \text{...(ii)}$$

Multiplying Eq. (ii) by i and adding it to Eq. (i), we get

$$z_1^3-3z_2^2z_1+i(3z_1^2z_2-z_2^3)=2+11i$$

$$\Rightarrow (z_1+iz_2)^3=2+11i \quad \text{...(iii)}$$

Multiplying Eq. (ii) by i and subtracting it from Eq. (i), we get

$$z_1^3-3z_2^2z_1-i(3z_1^2z_2-z_2^3)=2-11i$$

$$(z_1-iz_2)^3=2-11i \quad \text{...(iv)}$$

Multiplying Eqs. (iii) and (iv), we get

$$(z_1^2+z_2^2)^3=2^2-121i^2=4+121=125$$

$$\Rightarrow z_1^2+z_2^2=5$$

9. (b) $z^2+3z=0$

$$z=x+iy$$

$$\Rightarrow (x^2-y^2)+i(2xy)+3(x-iy)=0$$

$$\Rightarrow (x^2-y^2+3x)+i(2xy-3y)=0$$

$$\begin{cases}x^2-y^2+3x=0\\ y(2x-3)=0\end{cases}$$

$$y=0 \text{ or } x=\frac{3}{2}$$

If $y=0$,

$$x^2+3x=0$$

$$\Rightarrow x(x+3)=0\Rightarrow x=0,-3$$

So, (0, 0) and (−3, 0) are solutions, when $y=0$.

When $x=\frac{3}{2},\ \frac{9}{4}-y^2+\frac{9}{2}=0\Rightarrow y^2=\frac{27}{4}$

$$\Rightarrow y=\pm\frac{3\sqrt{3}}{2}$$

$$\therefore \left(\frac{3}{2},\frac{3\sqrt{3}}{2}\right) \text{ and } \left(\frac{3}{2},-\frac{3\sqrt{3}}{2}\right)$$

There are 4 solutions.

$$\sum_{k=0}^{\infty}\left(\frac{1}{n^k}\right)=1+\frac{1}{4}+\frac{1}{4^2}+\dots\infty$$

$$=\frac{1}{1-\frac{1}{4}}=\frac{4}{3}$$

10. (a) Let $z^3 = t$

$\therefore \quad t^2 - 6t + 25 = 0$

$\Rightarrow \quad t = \dfrac{6 \pm \sqrt{-64}}{2} = 3 \pm 4i = z^3$

So, $|z^3| = 5 \Rightarrow |z| = 5^{\frac{1}{3}}$

11. (a) Let $t = \dfrac{z-\alpha}{z+\alpha}$

$\because$ t is purely imaginary.

$\therefore \quad t + \bar{t} = 0$

$\Rightarrow \quad \dfrac{z-\alpha}{z+\alpha} + \dfrac{\bar{z}-\alpha}{\bar{z}+\alpha} = 0$

$\Rightarrow (z-\alpha)(\bar{z}+\alpha) + (\bar{z}-\alpha)(z+\alpha) = 0$

$\Rightarrow \quad z\bar{z} - \alpha^2 + z\bar{z} - \alpha^2 = 0$

$\Rightarrow \quad z\bar{z} = \alpha^2$

$\Rightarrow \quad \alpha^2 = 4 \quad [\because |z| = 2;\ z\bar{z} = |z|^2]$

$\Rightarrow \quad \alpha = \pm 2$

12. (a) $z = \dfrac{(1+i)^2}{a-i} \times \dfrac{a+i}{a+i}$

On simplifying, we get

$z = \dfrac{2ai - 2}{a^2+1}$

$\therefore \ |z| = \sqrt{\left(\dfrac{-2}{a^2+1}\right)^2 + \left(\dfrac{2a}{a^2+1}\right)^2} = \sqrt{\dfrac{4+4a^2}{(a^2+1)^2}}$

$\Rightarrow |z| = \dfrac{2}{\sqrt{1+a^2}}$ ...(i)

$\because$ It is given that

$|Z| = \sqrt{\dfrac{2}{5}}$ ...(ii)

From Eqs. (i) and (ii), we get

$\dfrac{2}{5} = \dfrac{4}{(1+a^2)} \Rightarrow 1 + a^2 = 10 \Rightarrow a = \pm 3$

$\because a > 0, a = 3$

Then, $\quad z = \dfrac{(1+i)^2}{a-i} = \dfrac{2i}{3-i} \times \dfrac{3+i}{3+i} = \dfrac{-1+3i}{5}$

$\therefore \quad \bar{z} = \dfrac{-1}{5} - \dfrac{3}{5}i$

13. (b) Let $z = a + ib$

$\Rightarrow \quad |a+ib| + (a+ib) = 3 + i$

$\Rightarrow \sqrt{a^2+b^2} + (a+ib) = 3 + i$

On comparing, we have

$b = 1,\ \sqrt{a^2+b^2} + a = 3$

$\Rightarrow \sqrt{a^2+1} + a = 3$

$\Rightarrow \quad a^2 + 1 = a^2 + 9 - 6a \Rightarrow a = \dfrac{4}{3}$

Then, $|z| = \sqrt{\left(\dfrac{4}{3}\right)^2 + (1)^2} = \sqrt{\dfrac{16}{9} + 1} = \dfrac{5}{3}$

14. (c) We have,

$z = (3+7i)(a+ib)$

$z = (3a - 7b) + i(3b + 7a)$

For z to be purely imaginary, we have

$3a = 7b$

$\Rightarrow \quad a = 7 \text{ or } b = 3$ [for least value of $|z|$]

$\Rightarrow \quad |z| = |i(9+49)|$

$= \sqrt{(58)^2}$

$\Rightarrow \quad |z|^2 = (58)^2 = 3364$

15. (a) We have,

$8iz^3 + 12z^2 - 18z + 27i = 0$

$\Rightarrow 4iz^2(2z - 3i) - 9(2z - 3i) = 0$

$\Rightarrow \quad (2z - 3i)(4iz^2 - 9) = 0$

$\Rightarrow \quad z = \dfrac{3i}{2}$ and $z^2 = \dfrac{9}{4i}$

$\Rightarrow \quad |z| = \dfrac{3}{2}$ and $|z^2| = \dfrac{9}{4}$

$\Rightarrow \quad |z| = \dfrac{3}{2}$

16. (b) $|z^2 - 3| \geq |z|^2 - 3$

$\Rightarrow \quad 3|z| \geq |z|^2 - 3$

$\Rightarrow \quad |z|^2 - 3|z| - 3 \leq 0$

$\Rightarrow \quad 0 < |z| \leq \dfrac{3+\sqrt{21}}{2}$

17. (a) $|z^2 + 2z\cos\alpha| \leq |z^2| + |2z\cos\alpha|$

$= |z|^2 + 2|z||\cos\alpha| \leq |z|^2 + 2|z| < (\sqrt{2}-1)^2 + 2(\sqrt{2}-1)$

$\Rightarrow \quad |z^2 + 2z\cos\alpha| < 1 \quad [\because |\cos\alpha| \leq 1] = 1$

18. (c) $\left|\displaystyle\sum_{r=1}^{n} z_r\right| \leq \displaystyle\sum_{r=1}^{n} |z_r| \leq \displaystyle\sum_{r=1}^{n} |z_r - r| + \displaystyle\sum_{r=1}^{n} r \leq 2\displaystyle\sum_{r=1}^{n} r$

19. (b) $\because |z| = \left|z - \dfrac{4}{z} + \dfrac{4}{z}\right|$

$\leq \left|z - \dfrac{4}{z}\right| + \dfrac{4}{|z|}$

$\leq 2 + \dfrac{4}{|z|}$

$\Rightarrow \quad |z|^2 - 2|z| - 4 \leq 0$

$\Rightarrow [|z| - (\sqrt{5}+1)][|z| - (1-\sqrt{5})] \leq 0$

$\Rightarrow \quad 0 \leq |z| \leq \sqrt{5} + 1$

20. (a) Here, $x = 4\cos\theta$ and $y = 4\sin\theta$

$\therefore ||x|-|y|| = |4|\cos\theta| - 4|\sin\theta||$

$= 4||\cos\theta| - |\sin\theta||$

$= 4\sqrt{1-2|\cos\theta||\sin\theta|}$

$= 4\sqrt{1-|\sin 2\theta|}$

Hence, the range is [0, 4].

21. (c) Let $z = a + ib$

$\Rightarrow \quad \bar{z} = a - ib$

So, we have, $z^{2008} = \bar{z}$

$\therefore \quad |z|^{2008} = |\bar{z}| = |z|$

$\Rightarrow |z|[|z|^{2007} - 1] = 0$

$\Rightarrow \quad |z| = 0$ or $|z| = 1$

If $|z| = 0$, then $z = 0$

If $|z| = 1$, then $z^{2009} = z\bar{z} = |z|^2 = 1$

So, there are 2009 values of z.

Therefore, total number of ordered pairs is 2010.

22. (b) We have,

$|2z-1| = |z-2|$

$\Rightarrow \quad |2z-1|^2 = |z-2|^2$

$\Rightarrow \quad (2z-1)(2\bar{z}-1) = (z-2)(\bar{z}-2)$

$\Rightarrow 4z\bar{z} - 2\bar{z} - 2z + 1 = z\bar{z} - 2\bar{z} - 2z + 4$

$\Rightarrow \quad 3|z|^2 = 3$

$\Rightarrow \quad |z| = 1$

Again, $|z_1 + z_2| = |z_1 - \alpha + z_2 - \beta + \alpha + \beta|$

$\leq |z_1 - \alpha| + |z_2 - \beta| + |\alpha + \beta| \quad [\because \alpha, \beta > 0]$

$< \alpha + \beta + |\alpha + \beta|$

$\therefore \quad \left|\frac{z_1 + z_2}{\alpha + \beta}\right| < 2$

or $\quad \left|\frac{z_1 + z_2}{\alpha + \beta}\right| < 2|z|$

23. (d) Let $z_1 = \sin x + i\cos 2x$ and $z_2 = \cos x - i\sin 2x$

Then, $\bar{z}_1 = z_2$

$\Rightarrow \sin x - i\cos 2x = \cos x - i\sin 2x$

$\Rightarrow \quad \sin x = \cos x$ and $\cos 2x = \sin 2x$

$\Rightarrow \quad \tan x = 1$ and $\tan 2x = 1$

$\Rightarrow \quad x = \frac{\pi}{4}$ and $x = \frac{\pi}{8}$

This is not possible.

Hence, there is no value of x.

24. (a) $\exp\left[\frac{(|z|+3)(|z|-1)}{(|z|+1)} \times \log_e 2\right]$

$\geq \log_{\sqrt{2}} |5\sqrt{7} + 9i|$

$\exp\left[\frac{(|z|+3)(|z|-1)}{(|z|+1)} \times \log_e 2\right] \geq \log_{\sqrt{2}} 16$

$\Rightarrow \quad 2^{\left[\frac{(|z|+3)(|z|-1)}{(|z|+1)}\right]} \geq 2^3$

$\Rightarrow \quad \frac{(|z|+3)(|z|-1)}{(|z|+1)} \geq 3$

$\Rightarrow \quad |z|^2 + 2|z| - 3 \geq 3|z| + 3$

$\Rightarrow \quad |z|^2 - |z| - 6 \geq 0$

$\Rightarrow \quad (|z|-3)(|z|+2) \geq 0$

$\Rightarrow \quad |z| = 3$

25. (b) We have, $az^3 + bz^2 + \bar{b}z + \bar{a} = 0$ ...(i)

Taking conjugate of both sides, we get

$\Rightarrow \quad \bar{a}\bar{z}^3 + \bar{b}\bar{z}^2 + b\bar{z} + a = 0$

Dividing this equation by $\bar{z}^3$ and writing the terms in reverse order, we get

$\frac{a}{\bar{z}^3} + \frac{b}{\bar{z}^2} + \frac{\bar{b}}{\bar{z}} + \bar{a} = 0 \quad$ [for $\bar{z} \neq 0$] ...(ii)

Since, Eqs. (i) and (ii) are identical.

$\therefore \quad z^3 = \frac{1}{\bar{z}^3}$

$\Rightarrow \quad |z|^6 = 1 \Rightarrow |z| = 1$

Therefore, if α is a root of the given equation, then

$|\alpha| = 1$

26. (b) Given, $(3 + 2\sqrt{-54})^{1/2} - (3 - 2\sqrt{-54})^{1/2}$

$= (3 + 2 \times 3\sqrt{6}i)^{1/2} - (3 - 2 \times 3\sqrt{6}i)^{1/2}$

$= (9 - 6 + 2 \times 3\sqrt{6}i)^{1/2} - (9 - 6 - 2 \times 3\sqrt{6}i)^{1/2}$

$= \{3^2 + (\sqrt{6}i)^2 + 2 \times 3 \times \sqrt{6}i\}^{1/2}$

$- \{3^2 + (\sqrt{6}i)^2 - 2 \times 3 \times \sqrt{6}i\}^{1/2}$

$= [(3 + \sqrt{6}i)^2]^{1/2} - [(3 - \sqrt{6}i)^2]^{1/2}$

$= \pm(3 + \sqrt{6}i) \mp (3 - \sqrt{6}i) = \pm 2\sqrt{6}i$

$\therefore$ Imaginary part of $(3 + 2\sqrt{-54})^{1/2} - (3 - 2\sqrt{-54})^{1/2}$

$= I_m(\pm 2\sqrt{6}i) = \pm 2\sqrt{6}$

So, from the option $-2\sqrt{6}$ is correct.

27. (a) We have, $\arg\left(\frac{z_1}{z_2}\right) = \pi$

$\Rightarrow \quad \arg(z_1) - \arg(z_2) = \pi$

$\Rightarrow \quad \arg(z_1) = \arg(z_2) + \pi$

Let $\arg(z_2) = \theta$, then $\arg(z_1) = \pi + \theta$

$\therefore \quad z_1 = |z_1|[\cos(\pi + \theta) + i\sin(\pi + \theta)]$

$= |z_1|(-\cos\theta - i\sin\theta)$

and $\quad z_2 = |z_2|(\cos\theta + i\sin\theta)$

$= |z_1|(\cos\theta + i\sin\theta) \quad [\because |z_1| = |z_2|]$

$= -z_1$

$\Rightarrow z_1 + z_2 = 0$

28. (c) We have, $|z_1 - z_2|^2$

$$= |z_1|^2 + |z_2|^2 - 2|z_1||z_2|\cos(\theta_1 - \theta_2),$$

where $\theta_1 = \arg(z_1)$ and $\theta_2 = \arg(z_2)$.

Given, $\arg(z_1) - \arg(z_2) = 0$

$\Rightarrow |z_1 - z_2|^2 = |z_1|^2 + |z_2|^2 - 2|z_1||z_2| = (|z_1| - |z_2|)^2$

$\Rightarrow \quad |z_1 - z_2| = ||z_1| - |z_2||$

29. (b) Let $z_1 = |z_1|(\cos\theta_1 + i\sin\theta_1)$

Now, $\left|\frac{z_1}{z_2}\right| = 1 \Rightarrow |z_1| = |z_2|$

Also, $\arg(z_1 z_2) = 0$ or $\arg(z_1) + \arg(z_2) = 0$

$$z_2 - 2 = |z_1|(\cos\theta_1 - i\sin\theta_1) = \bar{z}_1$$

$\Rightarrow \quad \bar{z}_2 = (\bar{\bar{z}}_1) = z_1$

$\Rightarrow \quad |z_2|^2 = z_1 z_2$

30. (c) $\because |z| = 1,\ z \cdot \bar{z} = 1$

$$\therefore \quad \arg\left(\frac{1+z}{1+\bar{z}}\right) = \arg\left(\frac{1+z}{1+\frac{1}{z}}\right)$$

$$= \arg\left[z\left(\frac{1+z}{1+z}\right)\right]$$

$$= \arg(z) = \theta$$

31. (a) Let $z = \dfrac{3 + i\sin\theta}{4 - i\cos\theta}$

$$\Rightarrow \quad z = \frac{3 + i\sin\theta}{4 - i\cos\theta} \times \frac{4 + i\cos\theta}{4 + i\cos\theta}$$

$\because$ Given that z is purely real.

$\therefore \quad \text{Im}(z) = 0$

$\Rightarrow \quad 3\cos\theta + 4\sin\theta = 0$

$\Rightarrow \quad \tan\theta = -\dfrac{3}{4}$

$$\therefore \arg(\sin\theta + i\cos\theta) = \pi + \tan^{-1}\left(\frac{\cos\theta}{\sin\theta}\right)$$

$$= \pi + \tan^{-1}\left(-\frac{4}{3}\right)$$

$$= \pi - \tan^{-1}\left(\frac{4}{3}\right)$$

32. (b) Let $z_1 = x_1 + iy_1$

and $\quad z_2 = x_2 + iy_2$

$\because \quad |z_1 - 1| = \text{Re}(z_1)$

$\Rightarrow \quad (x_1 - 1)^2 + y_1^2 = x_1^2$

$\Rightarrow \quad y_1^2 - 2x_1 + 1 = 0 \quad$ ...(i)

Also, $|z_2 - 1| = \text{Re}(z_2)$

$\Rightarrow \quad (x_2 - 1)^2 + y_2^2 = x_2^2$

$\Rightarrow \quad y_2^2 - 2x_2 + 1 = 0 \quad$ ...(ii)

Subtracting Eq. (ii) from Eq. (i), we get

$$y_1^2 - y_2^2 - 2(x_1 - x_2) = 0$$

$\Rightarrow \quad y_1 + y_2 = 2\left(\dfrac{x_1 - x_2}{y_1 - y_2}\right) \quad$ ...(iii)

$\because \quad \arg(z_1 - z_2) = \dfrac{\pi}{6}$

$\Rightarrow \quad \tan^{-1}\left(\dfrac{y_1 - y_2}{x_1 - x_2}\right) = \dfrac{\pi}{6}$

$\Rightarrow \quad \dfrac{y_1 - y_2}{x_1 - x_2} = \tan\dfrac{\pi}{6}$

$\Rightarrow \quad \dfrac{2}{y_1 + y_2} = \dfrac{1}{\sqrt{3}} \quad$ [using Eq. (iii)]

$\therefore \quad y_1 + y_2 = 2\sqrt{3}$

$\therefore \text{Im}(z_1 + z_2) = 2\sqrt{3}$

33. (d) Given, $z^3 + \dfrac{3(\bar{z})^2}{|z|} = 0$

Let $z = re^{i\theta}$

$\Rightarrow r^3 e^{i3\theta} + 3re^{-i2\theta} = 0$

Since, r cannot be zero, so

$$r^2 e^{i\,5\theta} = -3$$

which will hold for $r = \sqrt{3}$ and five distinct values of θ.

Thus, there are five solutions.

34. (a) $\because i = \cos\dfrac{\pi}{2} + i\sin\dfrac{\pi}{2} = e^{\frac{i\pi}{2}}$

$\Rightarrow \quad i^i = \left(e^{\frac{i\pi}{2}}\right)^i = e^{-\frac{\pi}{2}}$

$\Rightarrow \quad z = (i)^{(i)^i} = i^{e^{-\frac{\pi}{2}}}$

$\Rightarrow \quad |z| = 1$

35. (c) $\because (\cos\alpha + i\sin\alpha)^{\frac{3}{5}} = e^{i\frac{3\alpha}{5}} = e^{\frac{i(2n\pi + 3\alpha)}{5}}$

$\therefore$ Required product,

$$= e^{\frac{i3\alpha}{5}} \cdot e^{\frac{i(2\pi + 3\alpha)}{5}} \cdot e^{\frac{i(4\pi + 3\alpha)}{5}} \cdot e^{\frac{i(6\pi + 3\alpha)}{5}} \cdot e^{\frac{i(8\pi + 3\alpha)}{5}}$$

$$= e^{i(4\pi + 3\alpha)}$$

$$= \cos(4\pi + 3\alpha) + i\sin(4\pi + 3\alpha)$$

$$= \cos 3\alpha + i\sin 3\alpha$$

36. (d) We have, $z = i\log(2 - \sqrt{3})$

$\Rightarrow \quad e^{iz} = e^{i^2\log(2 - \sqrt{3})} = e^{-\log(2 - \sqrt{3})}$

$$= e^{\log(2 - \sqrt{3})^{-1}} = e^{\log(2 + \sqrt{3})} = (2 + \sqrt{3})$$

$\Rightarrow \quad \cos z = \dfrac{e^{iz} + e^{-iz}}{2} = \dfrac{(2 + \sqrt{3}) + (2 - \sqrt{3})}{2} = 2$

37. (a) We have, $u^2 - 2u + 2 = 0$

$$\Rightarrow \quad u = 1 \pm i$$

$$\Rightarrow \frac{(x+\alpha)^n - (x+\beta)^n}{\alpha - \beta}$$

$$= \frac{[(\cot\theta - 1) + (1+i)]^n - [(\cot\theta - 1) + (1-i)]^n}{2i}$$

$$[\because \cot\theta - 1 = x]$$

$$= \frac{(\cos\theta + i\sin\theta)^n - (\cos\theta - i\sin\theta)^n}{\sin^n\theta\, 2i}$$

$$= \frac{2i\sin n\theta}{\sin^n\theta 2i} = \frac{\sin n\theta}{\sin^n\theta}$$

38. (c) Given, $(\cos\theta + i\sin\theta)(\cos 2\theta + i\sin 2\theta)\ldots(\cos n\theta + i\sin n\theta) = 1$

$$\therefore \cos(\theta + 2\theta + \ldots + n\theta) + i\sin(\theta + 2\theta + \ldots + n\theta) = 1$$

$$\Rightarrow \quad \cos\left[\frac{n(n+1)}{2}\theta\right] + i\sin\left[\frac{n(n+1)}{2}\theta\right] = 1$$

$$\Rightarrow \quad \cos\left[\frac{n(n+1)}{2}\theta\right] = 1 \text{ and } \sin\left[\frac{n(n+1)}{2}\theta\right] = 0$$

$$\therefore \quad \frac{n(n+1)}{2}\theta = 2m\pi$$

$$\Rightarrow \quad \theta = \frac{4m\pi}{n(n+1)}$$

39. (d) We have, $\left(\frac{3}{2} + i\frac{\sqrt{3}}{2}\right)^{50} = 3^{25}(x+iy)$

$$\Rightarrow \quad (\sqrt{3})^{50}\left(\frac{\sqrt{3}}{2} + i\frac{1}{2}\right)^{50} = 3^{25}(x+iy)$$

$$\Rightarrow 3^{25}\left(\cos\frac{\pi}{6} + i\sin\frac{\pi}{6}\right)^{50} = 3^{25}(x+iy)$$

$$\Rightarrow \quad \cos\frac{50\pi}{6} + i\sin\frac{50\pi}{6} = x + iy$$

$$\Rightarrow \quad x = \cos\frac{25\pi}{3} \text{ and } y = \sin\frac{25\pi}{3}$$

$$\Rightarrow \quad x = \cos\frac{\pi}{3} \text{ and } y = \sin\frac{\pi}{3}$$

$$\Rightarrow \quad x = \frac{1}{2} \text{ and } y = \frac{\sqrt{3}}{2}$$

40. (d) We have, $(x-3)^3 + 1 = 0$

$$\Rightarrow \quad \left(\frac{x-3}{-1}\right)^3 = 1$$

$$\Rightarrow \quad \frac{x-3}{-1} = 1, \omega, \omega^2 \Rightarrow x = 2, 3-\omega, 3-\omega^2$$

Hence, the sum of complex roots is
$6 - (\omega + \omega^2) = 6 + 1 = 7$

41. (c) We have, $\omega = \frac{5+3z}{5(1-z)}$

$$\Rightarrow \quad 5\omega - 5\omega z = 5 + 3z$$

$$\Rightarrow \quad 5\omega - 5 = z(3 + 5\omega)$$

$$\Rightarrow \quad z = \frac{5(\omega - 1)}{3 + 5\omega}$$

$$\because \quad |z| < 1$$

$$\Rightarrow \quad \left|\frac{5\omega - 5}{3 + 5\omega}\right| < 1$$

$$\Rightarrow \quad 5|\omega - 1| < |3 + 5\omega|$$

$$\Rightarrow 25(\omega - 1)(\overline{\omega} - 1) < (3 + 5\omega)(3 + 5\overline{\omega})$$

$$\Rightarrow 25(\omega\overline{\omega} - \omega - \overline{\omega} + 1) < 9 + 25\omega\overline{\omega} + 15\omega + 15\overline{\omega}$$

$$\Rightarrow \quad 16 < 40\omega + 40\overline{\omega}$$

$$\Rightarrow \quad \omega + \overline{\omega} > \frac{2}{5}$$

$$\Rightarrow \quad 2\text{Re}(\omega) > \frac{2}{5}$$

$$\Rightarrow \quad 5\,\text{Re}(\omega) > 1$$

42. (c) We have, $(1+\omega)^7 = A + B\omega$

$$\therefore \quad (-\omega^2)^7 = A + B\omega$$

$$\Rightarrow \quad -\omega^2 = A + B\omega$$

$$\Rightarrow \quad 1 + \omega = A + B\omega \quad [\because 1 + \omega + \omega^2 = 0]$$

$$\therefore \quad A = 1 \text{ and } B = 1$$

$$\Rightarrow \quad (A, B) = (1, 1)$$

43. (d) We have,

$$x^2 + x + 1 = 0$$

$$\Rightarrow \quad x = \omega \text{ or } \omega^2$$

Let $x = \omega$, then

$$x + \frac{1}{x} = \omega + \frac{1}{\omega} = \omega + \omega^2 = -1$$

$$x^2 + \frac{1}{x^2} = \omega^2 + \frac{1}{\omega^2} = \omega^2 + \omega = -1$$

$$x^3 + \frac{1}{x^3} = \omega^3 + \frac{1}{\omega^3} = 2$$

$$x^4 + \frac{1}{x^4} = \omega^4 + \frac{1}{\omega^4} = \omega + \omega^2 = -1, \text{ etc}$$

$$\therefore \left(x + \frac{1}{x}\right)^2 + \left(x^2 + \frac{1}{x^2}\right)^2 + \left(x^3 + \frac{1}{x^3}\right)^2 + \ldots + \left(x^{27} + \frac{1}{x^{27}}\right)^2$$

$$= \left[\left(x + \frac{1}{x}\right)^2 + \left(x^2 + \frac{1}{x^2}\right)^2 + \left(x^4 + \frac{1}{x^4}\right)^2 + \ldots\left(x^{26} + \frac{1}{x^{26}}\right)^2\right]$$

$$+\left[\left(x^3+\frac{1}{x^3}\right)^2+\left(x^6+\frac{1}{x^6}\right)^2+\left(x^9+\frac{1}{x^9}\right)^2 + \ldots \left(x^{27}+\frac{1}{x^{27}}\right)^2\right]$$

$= 18 + 9(2^2) = 54$

44. (a) We have, $z^3 + 2z^2 + 2z + 1 = 0$

$\Rightarrow \quad (z^3 + 1) + 2z(z + 1) = 0$

$\Rightarrow \quad (z + 1)(z^2 + z + 1) = 0 \Rightarrow z = -1, \omega, \omega^2$

Since, $z = -1$ does not satisfy $z^{1985} + z^{100} + 1 = 0$ while $z = \omega, \omega^2$ satisfy it.

Hence, sum is $\omega + \omega^2 = -1$

45. (d) We have, $|\omega z - 1 - \omega^2| = a$

$\Rightarrow \quad |z + 1| = a \Rightarrow |z - 1 + 2| = a$

$\Rightarrow |z - 1| + 2 \geq a \Rightarrow 0 \leq a \leq 4$

46. (b) Let $A = x + iy$.

Given, $|A| = 1 \Rightarrow x^2 + y^2 = 1$

and $|A + 1| = 1 \Rightarrow (x + 1)^2 + y^2 = 1$

$\Rightarrow \quad x = -\frac{1}{2}$ and $y = \pm\frac{\sqrt{3}}{2}$

$\Rightarrow \quad A = \omega$ or ω^2

$\Rightarrow \quad (\omega)^n = (1 + \omega)^n$

$= (-\omega^2)^n$

Therefore, n must be even and divisible by 3.

47. (d) Let $f(x) = 5x^3 + Mx + N$

Also, $x^2 + x + 1 = (x - \omega)(x - \omega^2)$

It is given that $f(x)$ is divisible by $x^2 + x + 1$

$\therefore \quad f(\omega) = 5 + M\omega + N = 0$

and $\quad f(\omega^2) = 5 + M\omega^2 + N = 0$

So, $M = 0$ and $N = -5$

$\therefore \quad M + N = -5$

48. (b) Let $x = \sqrt[3]{-1}$

$\Rightarrow \quad x^3 = -1$

$\Rightarrow (-x^3) = 1$

$\Rightarrow \quad -x = 1, \omega^2, \omega$

$\Rightarrow \quad x = -1, -\omega^2, -\omega$

$= -1, \frac{1+\sqrt{3}i}{2}, \frac{1-\sqrt{3}\cdot i}{2}$

$= -1, \frac{-\sqrt{3}+i}{2i}, \frac{\sqrt{3}+i}{2i}$

$= -1, \frac{-\sqrt{3}+\sqrt{-1}}{\sqrt{-4}}, \frac{\sqrt{3}+\sqrt{-1}}{\sqrt{-4}}$

49. (a) We have,

$\log z + \log z^2 + \log z^3 + \ldots + \log z^n = 0$

$\Rightarrow \quad \log(zz^2z^3 \ldots z^n) = 0$

$\Rightarrow \quad \log\left(z^{\frac{n(n+1)}{2}}\right) = 0$

$\Rightarrow \quad z^{\frac{n(n+1)}{2}} = 1$

$\Rightarrow \quad z = 1^{\frac{2}{n(n+1)}} = (\cos 0° + i \sin 0°)^{\frac{2}{n(n+1)}}$

$= (\cos 2m\pi + i \sin 2m\pi)^{\frac{2}{n(n+1)}}, m = 0, 1, 2, 3, \ldots$

$= \cos\frac{4m\pi}{n(n+1)} + i \sin\frac{4m\pi}{n(n+1)}, m = 0, 1, 2, 3, \ldots$

50. (d) Let $z = x + iy$, then

$|z^2 - 1| = |z|^2 + 1$

$\Rightarrow \quad |(x^2 - y^2 - 1) + 2ixy| = x^2 + y^2 + 1$

$\Rightarrow (x^2 - y^2 - 1)^2 + 4x^2y^2 = (x^2 + y^2 + 1)^2$

$\Rightarrow \quad x = 0$

Hence, z lies on imaginary axis.

51. (b) Let $z = x + iy$, then

$x = \lambda + 3$

and $\quad y = -\sqrt{5 - \lambda^2}$

$\Rightarrow (x - 3)^2 = \lambda^2 \quad$ ...(i)

and $\quad y^2 = 5 - \lambda^2 \quad$ ...(ii)

From Eqs. (i) and (ii),

$(x - 3)^2 = 5 - y^2$ or $(x - 3)^2 + y^2 = 5$

Obviously, it is a semicircle as $y < 0$.

Hence, part of the circle lies below the X-axis.

52. (a) Given, $z_k = 1 + a + a^2 + \ldots + a^{k-1}$

$= \frac{1 - a^k}{1 - a}$

$\Rightarrow \quad z_k - \frac{1}{1 - a} = -\frac{a^k}{1 - a}$

$\Rightarrow \quad \left|z_k - \frac{1}{1 - a}\right| = \frac{|a|^k}{|1 - a|} < \frac{1}{|1 - a|} \quad [\because |a| < 1]$

Hence, z_k lies within the circle.

$\therefore \quad \left|z - \frac{1}{1 - a}\right| = \frac{1}{|1 - a|}$

53. (a) We have, $|z| = 1$

Let $\quad \alpha = -1 + 3z$

$\Rightarrow \quad \alpha + 1 = 3z$

$\Rightarrow \quad |\alpha + 1| = 3|z| = 3$

Hence, α lies on a circle centered at -1 and radius is equal to 3.

54. (b) We have, $2\left|z-\frac{1}{2}\right|=|z-1|$

$$\therefore \quad \frac{|z-1|}{\left|z-\frac{1}{2}\right|}=2$$

So, locus of z is a circle.

55. (b) We have, $\arg\left(\frac{z-1}{z+1}\right)=\frac{\pi}{4}$

$\Rightarrow \arg(z-1)-\arg(z+1)=\frac{\pi}{4}$

Let $z=x+iy$

$$\arg[(x-1)+iy]-\arg[(x+1)+iy]=\frac{\pi}{4}$$

$$\Rightarrow \quad \tan^{-1}\left(\frac{y}{x-1}\right)-\tan^{-1}\left(\frac{y}{x+1}\right)=\frac{\pi}{4}$$

$$\Rightarrow \quad \left(\frac{\frac{y}{x-1}-\frac{y}{x+1}}{1+\frac{y}{x-1}\cdot\frac{y}{x+1}}\right)=\tan\frac{\pi}{4}$$

$$\Rightarrow \quad \frac{y(x+1)-y(x-1)}{(x^2-1)+y^2}=1$$

$$\Rightarrow \quad 2y=x^2+y^2-1$$

$$\Rightarrow \quad x^2+y^2-2y-1=0$$

$$\Rightarrow \quad x^2+(y-1)^2=2$$

$$\Rightarrow \quad x^2+(y-1)^2=(\sqrt{2})^2$$

which is a circle with centre (0, 1) and radius $=\sqrt{2}$ units

56. (c) We have,

$$|(x-2)+i(y-1)|=|z|\left|\frac{1}{\sqrt{2}}\cos\theta-\frac{1}{\sqrt{2}}\sin\theta\right|$$

where $\theta=\arg z$

$$\sqrt{(x-2)^2+(y-1)^2}=\frac{1}{\sqrt{2}}|x-y|$$

which represents a parabola.

57. (a) $S_1:|z-3-2i|^2=8$

$$\Rightarrow \quad |(x+iy)-(3+2i)|^2=8$$

$$\Rightarrow \quad |(x-3)+i(y-2)|^2=8$$

$$\Rightarrow \quad (x-3)^2+(y-2)^2=8$$

$$S_2:\text{Re}(z)\ge 5$$

$$x\ge 5$$

$$S_3:|z-\bar{z}|\ge 8$$

$$|(x+iy)-(x-iy)|\ge 8$$

$$\Rightarrow \quad 2y\ge 8$$

$$\Rightarrow \quad y\ge 4$$

$$S_1:(x-3)^2+(y-2)^2=8$$

$$S_2:x\ge 5$$

$$S_3:y\ge 4$$

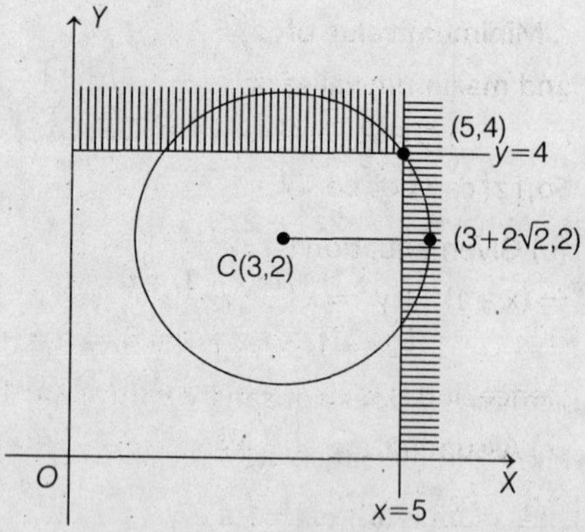

Circle passes through (5,4) as shown in the figure.

$\Rightarrow$ There is exactly one point (5,4) in $S_1\cap S_2\cap S_3$.

58. (a) We have, $|z_1|=|z_2|=|z_3|=1$

Hence, the circumcenter of triangle is origin.

Also, centroid $\frac{(z_1+z_2+z_3)}{3}=0$, which coincides with the circumcenter.

So the triangle is equilateral.

Since, radius is 1 and length of side is $a=\sqrt{3}$

Therefore, the area of the triangle is $\left(\frac{\sqrt{3}}{4}\right)a^2=\left(\frac{3\sqrt{3}}{4}\right)$

59. (d) Given, the equation $z^n=(z+1)^n$ will have exactly $n-1$ roots.

We have, $\left(\frac{z+1}{z}\right)^n=1 \Rightarrow \left|\frac{z+1}{z}\right|=1$

$$\Rightarrow \quad |z+1|=|z|$$

Therefore, z lies on the right bisector of the segment connecting the points (0, 0) and (– 1, 0)

Thus, $\text{Re}(z)=-\frac{1}{2}$. Hence, roots are collinear and will have their real part equal to $-\frac{1}{2}$.

Hence, sum of the real parts of roots is $\left(-\frac{1}{2}\right)(n-1)$.

60. (c) Let $z=x+iy$

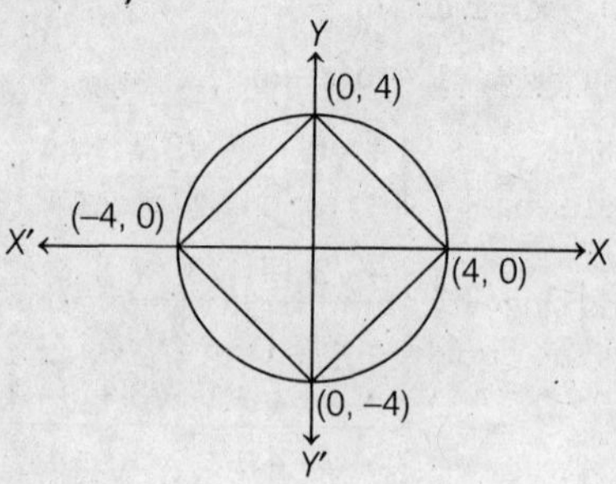

Given that $|x|+|y|=4$

$\because \quad |z|=\sqrt{x^2+y^2}$

$\therefore$ Minimum value of $|z|=2\sqrt{2}$

and maximum value of $|z|=4$

$\therefore \quad |z|\in[2\sqrt{2},4]$ or $|z|\in[\sqrt{8},\sqrt{16}]$

So, $|z|$ can not be $\sqrt{7}$.

61. (b) Given equation is $|z-1|=|z-i|$

$\Rightarrow (x-1)^2+y^2=x^2+(y-1)^2$ [here, $z=x+iy$]

$\Rightarrow \quad 1-2x=1-2y \Rightarrow x-y=0$

Hence, locus is straight line with slope 1.

62. (a) Assuming, $\arg z_1=\theta$ and $\arg z_2=\theta+\alpha$

$$\frac{az_1}{bz_2}+\frac{bz_2}{az_1}=\frac{a|z_1|e^{i\theta}}{b|z_2|e^{i(\theta+\alpha)}}+\frac{b|z_2|e^{i(\theta+\alpha)}}{a|z_1|e^{i\theta}}$$

$$=e^{-i\alpha}+e^{i\alpha}=2\cos\alpha$$

Hence, the point lies on the line segment $[-2,2]$ of the real axis.

63. (c) Let $z=x+iy$

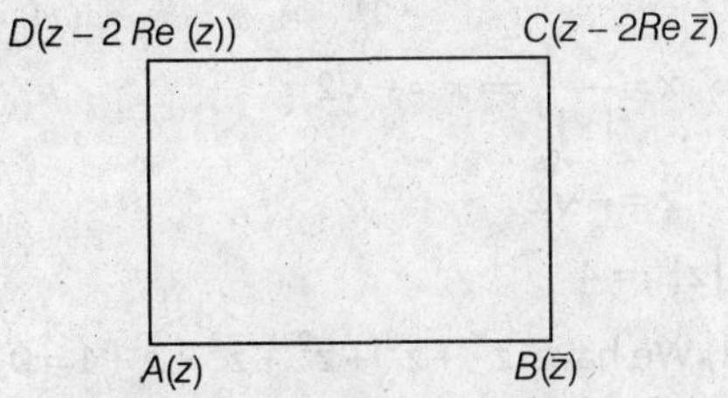

$\because$ Length of side of square = 4 units

$\Rightarrow \quad |z-\bar{z}|=4$

$\Rightarrow \quad |x+iy-x+iy|=4$

$\Rightarrow \quad |2iy|=4$

$\Rightarrow \quad |y|=2$

Also, $|z-\{z-2\text{Re}(z)\}|=4$

$\Rightarrow \quad |2\,\text{Re}(z)|=4$

$\Rightarrow \quad |2x|=4 \Rightarrow |x|=2$

$\therefore \quad |z|=\sqrt{x^2+y^2}=\sqrt{4+4}=\sqrt{8}$

$\therefore \quad |z|=2\sqrt{2}$

64. (c) We have, $\left(\dfrac{z+1}{z}\right)^4=16$

$\Rightarrow \quad \dfrac{z+1}{z}=\pm 2,\pm 2i$

The roots are $1,-\dfrac{1}{3},\left(-\dfrac{1}{5}-\dfrac{2}{5}i\right)$ and $\left(-\dfrac{1}{5}+\dfrac{2}{5}i\right)$.

Note that $\left(-\dfrac{1}{3},0\right)$ and $(1,0)$ are equidistant from $\left(\dfrac{1}{3},0\right)$

and since it lies on the perpendicular bisector of AB, it will be equidistant from A and B also.

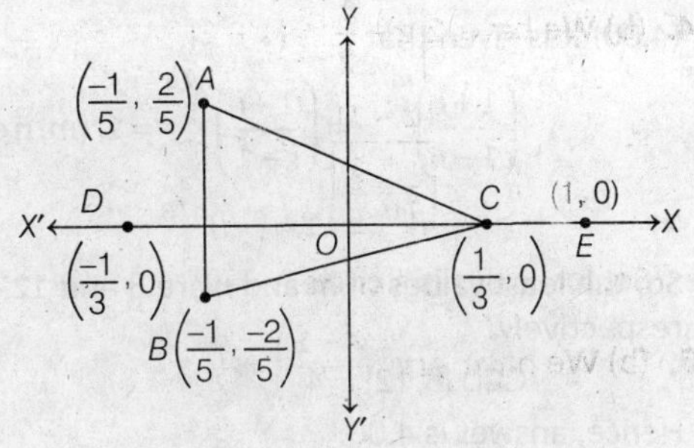

65. (c) $S_1=\Sigma z_1=-3a$ and $S_2=\Sigma z_1z_2=3b$

Since, the triangle is equilateral, we have

$$\Sigma z_1^2=\Sigma z_1z_2$$

$\Rightarrow (\Sigma z_1)^2-2\Sigma z_1z_2=\Sigma z_1z_2$

$\Rightarrow \quad (\Sigma z_1)^2=3\Sigma z_1z_2$

$\Rightarrow \quad (-3a)^2=3(3b)$

$\Rightarrow \quad 9a^2=9b$

$\Rightarrow \quad a^2=b$

66. (1) We have,

$$z=\frac{3+2i\cos\theta}{1-3i\cos\theta}=\frac{3+2i\cos\theta}{1-3i\cos\theta}\times\frac{1+3i\cos\theta}{1+3i\cos\theta}$$

$$=\frac{(3-6\cos^2\theta)+i(9\cos\theta+2\cos\theta)}{1+9\cos^2\theta}$$

$$z=\frac{(3-6\cos^2\theta)+(11\cos\theta)i}{1+9\cos^2\theta}$$

Given, $\quad \text{Re}(z)=0$

$\Rightarrow \quad \dfrac{3-6\cos^2\theta}{1+9\cos^2\theta}=0$

$\Rightarrow \quad 3-6\cos^2\theta=0$

$\Rightarrow \quad \cos\theta=\dfrac{1}{\sqrt{2}}$

$\Rightarrow \quad \theta=\dfrac{\pi}{4},\left\{\theta\in\left(0,\dfrac{\pi}{2}\right)\right\}$

Hence, $\sin^2 3\theta+\cos^2\theta=\sin^2\dfrac{3\pi}{4}+\cos^2\dfrac{\pi}{4}=\dfrac{1}{2}+\dfrac{1}{2}=1$

67. (30.33) We have,

$$\frac{x+iy}{27}=\left(-2-\frac{i}{3}\right)^3=\left[\frac{-1}{3}(6+i)\right]^3$$

$\Rightarrow \quad \dfrac{x+iy}{27}=\dfrac{-1}{27}(216+108i+18i^2+i^3)$

$\Rightarrow \quad x+iy=-(198+107i)$

$\Rightarrow \quad x=-198$ and $y=-107$

$\Rightarrow \quad y-x=-107+198=91$

$\Rightarrow \quad \dfrac{y-x}{3}=\dfrac{91}{3}=30.33$

68. (4.00) It is given that

$$\left(\frac{1+i}{1-i}\right)^{m/2}=\left(\frac{1+i}{i-1}\right)^{n/3}=1, (m, n\in N)$$

$$\Rightarrow \quad i^{m/2}=1=(-i)^{n/3}$$

So, the least values of m and n are 8 and 12, respectively.

$\therefore \quad$ GCD $\{8,12\}=4$

Hence, answer is 4.00.

69. (15) We have,

$$z=9+bi$$

$$z^2=81-b^2+18bi$$

$$z^3=729+243bi-27b^2-b^3i$$

$$\Rightarrow 243b-b^3=18b$$

$$\Rightarrow \quad b=15$$

70. (3) $(1+ri)^3=S(1+i)$

$$\Rightarrow 1+3ri+3r^2i^2+i^3r^3=S(1+i)$$

$$\Rightarrow 1-3r^2+i(3r-r^3)=S+Si$$

$$\Rightarrow \quad 1-3r^2=S=3r-r^3$$

Hence, $1-3r^2=3r-r^3$

$$\Rightarrow \quad r^3-3r^2-3r+1=0$$

Thus, sum of three roots is 3.

71. (1) Let $z=a+ib$

Given, $|z|=2$

$$\Rightarrow \quad w=\frac{(a+1)+ib}{(a-1)+ib}$$

$$\Rightarrow \quad |w|=\sqrt{\frac{(a+1)^2+b^2}{(a-1)^2+b^2}}$$

$$=\sqrt{\frac{a^2+b^2+2a+1}{a^2+b^2-2a+1}}$$

$$=\sqrt{\frac{5+2a}{5-2a}}$$

$$|w|_{max}=\sqrt{\frac{5+4}{1}}=3 \text{ (when } a=2)$$

$$|w|_{min}=\sqrt{\frac{5-4}{9}}=\frac{1}{3} \text{ (when } a=-2)$$

Hence, required product is 1.

72. (2) We have, $\bar{z}+z=0$

$$\Rightarrow \quad \bar{z}=-z \quad \ldots\text{(i)}$$

Now, $|z|^2-4zi=z^2$

$$\Rightarrow \quad -z^2-4zi=z^2 \quad \text{[from Eq. (i)]}$$

$$\Rightarrow \quad 2z=-4i$$

$$\Rightarrow \quad z=-2i$$

$$\Rightarrow \quad |z|=2$$

73. (17) Let $z=a+bi$

$$\Rightarrow \quad |z|^2=a^2+b^2$$

Now, $z+|z|=2+8i$

$$\Rightarrow \quad a+bi+\sqrt{a^2+b^2}=2+8i$$

$$\Rightarrow a+\sqrt{a^2+b^2}=2 \text{ and } b=8$$

$$\Rightarrow a+\sqrt{a^2+64}=2$$

$$\Rightarrow \quad a^2+64=(2-a)^2=a^2-4a+4$$

$$\Rightarrow \quad 4a=-60\Rightarrow a=-15$$

Thus, $a^2+b^2=225+64=289$

$$\therefore \quad |z|=\sqrt{a^2+b^2}=\sqrt{289}=17$$

74. (4) We have, $|z|^2+\frac{16}{|z|^3}=z^2-4z=\bar{z}^2-4\bar{z}$

$$\Rightarrow \quad (z-\bar{z})(z+\bar{z}-4)=0$$

$$\Rightarrow \quad z=\bar{z}=x(x\neq 2)$$

So, $x^2=4x+x^2+\frac{16}{|x|^3}$

$$\Rightarrow x=\frac{-4}{|x|^3}\Rightarrow x=-\sqrt{2}$$

$$\therefore \quad z=-\sqrt{2}$$

$$\therefore |z|^4=4$$

75. (1) We have, $z^4+z^3+z^2+z^2+z+1=0$

$$\Rightarrow \quad z^2(z^2+z+1)+(z^2+z+1)=0$$

$$\Rightarrow \quad (z^2+z+1)(z^2+1)=0$$

$\therefore z=i, -i, \omega, \omega^2$ for each $|z|=1$

76. (1) We have, $x=\omega-\omega^2-2$

$$\Rightarrow \quad x+2=\omega-\omega^2$$

Squaring, $x^2+4x+4=\omega^2+\omega^4-2\omega^3$

$$=\omega^2+\omega-2=-3$$

$$\Rightarrow \quad x^2+4x+7=0$$

Dividing $x^4+3x^3+2x^2-11x-6$ by x^2+4x+7, we get

$$x^4+3x^3+2x^2-11x-6=(x^2+4x+7)(x^2-x-1)+1$$

$$=(0)(x^2-x-1)+1$$

$$=0+1=1$$

77. (4.5) Given that

$$\alpha=\frac{-1+i\sqrt{3}}{2}=\omega$$

$$\therefore \quad (2+\omega)^4=a+b\omega$$

$$\Rightarrow \quad (4+\omega^2+4\omega)^2=a+b\omega$$

$$\Rightarrow \{\omega^2+4(1+\omega)\}^2=a+b\omega$$

$$\Rightarrow \quad (\omega^2-4\omega^2)^2=a+b\omega$$

$$\Rightarrow \quad (-3\omega^2)^2=a+b\omega$$

$$\Rightarrow \quad 9\omega^4 = a + b\omega$$

$$\Rightarrow \quad 9\omega = a + b\omega$$

On comparing, $a = 0$ and $b = 9$

$$\therefore \quad \frac{a+b}{2} = \frac{0+9}{2} = \frac{9}{2} = 4.5$$

78. (0.5) $\left[\dfrac{1+\cos\left(\frac{\pi}{2}-\frac{2\pi}{9}\right)+i\sin\left(\frac{\pi}{2}-\frac{2\pi}{9}\right)}{1+\cos\left(\frac{\pi}{2}-\frac{2\pi}{9}\right)-i\sin\left(\frac{\pi}{2}-\frac{2\pi}{9}\right)}\right]^3$

$$= \left[\frac{1+\cos\frac{5\pi}{18}+i\sin\frac{5\pi}{18}}{1+\cos\frac{5\pi}{18}-i\sin\frac{5\pi}{18}}\right]^3$$

$$= \left[\frac{2\cos^2\frac{5\pi}{36}+i\times 2\sin\frac{5\pi}{36}\cos\frac{5\pi}{36}}{2\cos^2\frac{5\pi}{36}-i\times 2\sin\frac{5\pi}{36}\cos\frac{5\pi}{36}}\right]^3$$

$$= \left[\frac{\cos\frac{5\pi}{36}+i\sin\frac{5\pi}{36}}{\cos\frac{5\pi}{36}-i\sin\frac{5\pi}{36}}\right]^3$$

$$= \left(\cos\frac{5\pi}{36}+i\sin\frac{5\pi}{36}\right)^6$$

$$= \cos\left(6\times\frac{5\pi}{36}\right)+i\sin\left(6\times\frac{5\pi}{36}\right)$$

$$= \cos\frac{5\pi}{6}+i\sin\frac{5\pi}{6}$$

$$= -\frac{\sqrt{3}}{2}+i\times\frac{1}{2}$$

$\therefore$ Imaginary part $= \frac{1}{2} = 0.5$

79. (5) Given that $|z-3-2i| \le 2$, represents a circular region with centre (3, 2) and radius 2.

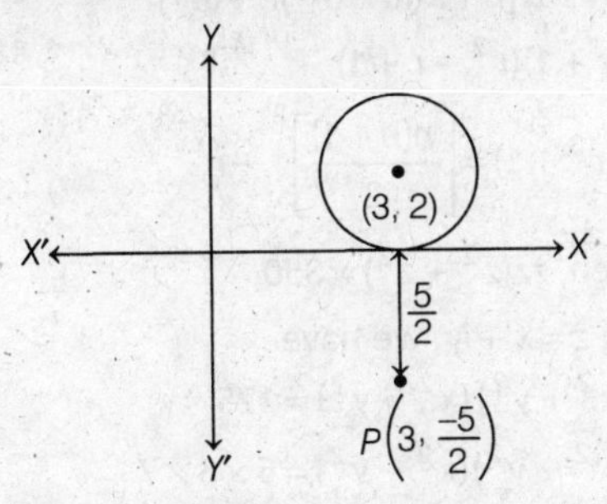

Now, $|2z-6+5i| = 2\left|z-\left(3-\frac{5}{2}\right)i\right|$

$= 2 \times$ distance of z from P

[where z lies in or on the circle]

Also, minimum distance of z from $P = \frac{5}{2}$.

$\therefore$ Minimum value of $|2z-6+5i| = 5$

80. (310) Given, $\dfrac{(-1+i\sqrt{3})^{21}}{(1-i)^{24}} + \dfrac{(1+i\sqrt{3})^{21}}{(1+i)^{24}} = k$

$$\because \quad -1+i\sqrt{3} = 2e^{i2\pi/3}$$

$$1+i\sqrt{3} = 2e^{i\pi/3}$$

$$1-i = \sqrt{2}e^{-i\pi/4}$$

$$1+i = \sqrt{2}e^{i\pi/4}$$

Now, $\dfrac{\left(2e^{i\frac{2\pi}{3}}\right)^{21}}{(\sqrt{2}e^{-i\pi/4})^{24}} + \dfrac{(2e^{i\pi/3})^{21}}{(\sqrt{2}e^{i\pi/4})^{24}}$

$$= \frac{2^{21}\cdot e^{i14\pi}}{2^{12}\cdot e^{-i6\pi}} + \frac{2^{21}\cdot e^{i7\pi}}{2^{12}\cdot e^{i6\pi}}$$

$$= 2^9\cdot e^{i20\pi} + 2^9\cdot e^{i\pi}$$

$$= 2^9(1) + 2^9(-1)$$

$\Rightarrow \quad 2^9 - 2^9 = 0 = k$ **[given]**

$\therefore \quad n = [|k|] = [101] = 0$

Now, $\sum_{j=0}^{5}(j+5)^2 - \sum_{j=0}^{5}(j+5)$ **[$\because n = 0$]**

$$= [5^2+6^2+7^2+8^2+9^2+10^2] - [5+6+7+8+9+10]$$

$$= [(1^2+2^2+3^2+\ldots+10^2) - (1^2+2^2+\ldots+4^2)] - [(1+2+3+\ldots+10) - (1+2+3+4)]$$

$$= \left[\frac{10\times 11\times 21}{6} - \frac{4\times 5\times 9}{6}\right] - \left[\frac{10\times 11}{2} - \frac{4\times 5}{2}\right]$$

$$= (385-30) - (55-10)$$

$$= 385 - 45$$

$$= 310$$

Level 3

1. (d) Given equation is

$$z^2 + z + 1 - a = 0.$$

Clearly, this equation do not have real roots if $D < 0$

$$\Rightarrow \quad 1 - 4(1-a) < 0$$

$$\Rightarrow \quad 4a < 3$$

$$a < \frac{3}{4}$$

2. (c) Given, $\frac{2+3i\sin\theta}{1-2i\sin\theta}$

$$=\frac{2+3i\sin\theta}{1-2i\sin\theta}\times\frac{1+2i\sin\theta}{1+2i\sin\theta}$$

$$=\frac{(2-6\sin^2\theta)+i(7\sin\theta)}{1+4\sin^2\theta}$$

For purely imaginary number,

$$\sin^2\theta=\frac{1}{3} \text{ or } \sin\theta=\frac{1}{\sqrt{3}}$$

$$\therefore \quad \theta=\sin^{-1}\frac{1}{\sqrt{3}}$$

3. (c) Let $z=x+iy$.

So, $|z-1|=|z+1| \Rightarrow \text{Re}(z)=0 \Rightarrow x=0$

$|z-1|=|z+i| \Rightarrow x=y$

$|z+1|=|z-i| \Rightarrow y=-x$

Only (0, 0) will satisfy all conditions.

Therefore, the number of complex numbers (z) is 1.

4. (b) Given, $|z|\geq 2$

$$\therefore \quad \left|z+\frac{1}{2}\right|\geq\left||z|-\left|\frac{1}{2}\right|\right|\geq\left|2-\frac{1}{2}\right|\geq\frac{3}{2}$$

Hence, minimum distance between z and $\left(-\frac{1}{2},0\right)$ is $\frac{3}{2}$.

5. (a) Let the roots of the given equation be $1+ip$ and $1-ip$, where $p\in R$.

$\Rightarrow \beta=$ Product of roots

$=(1+ip)(1-ip)=1+p^2>1, \forall\, p\in R$

$\Rightarrow \beta\in(1,\infty)$

6. (a) Let $z=r(\cos\theta+i\sin\theta)$

$\arg(z)=\theta<0$ (as given) ...(i)

$\therefore \quad -z=-r(\cos\theta+i\sin\theta)$

$-z=r[\cos(\pi+\theta)+i\sin(\pi+\theta)]$

$\Rightarrow \quad \arg(-z)=\pi+\theta$ ...(ii)

From Eqs. (i) and (ii), we get

$$\arg(-z)-\arg(z)=\pi+\theta-\theta=\pi$$

7. (c) Given, $\bar{z}+i\omega=0$

$\Rightarrow \quad \bar{z}=-i\omega$

$\Rightarrow \quad z=i\omega$

$\Rightarrow \quad \omega=-iz$

Also, $\arg(z\omega)=\pi$

$\Rightarrow \arg(-iz^2)=\pi$

$\Rightarrow \arg(-i)+2\arg(z)=\pi$

$\Rightarrow \quad \frac{-\pi}{2}+2\arg(z)=\pi$

$\Rightarrow \quad 2\arg(z)=\frac{3\pi}{2}$

$\Rightarrow \quad \arg(z)=\frac{3\pi}{4}$

8. (b) Given, $z=\sqrt{3}+i$

$\therefore \arg(z^2e^{z-i})=\arg[(3-i+2\sqrt{3}i)e^{\sqrt{3}}]$

$=\arg[2+2\sqrt{3}i)e^{\sqrt{3}}]$

$=\tan^{-1}\left[\frac{2\sqrt{3}}{2}\right]=\frac{\pi}{3}$

9. (d) Given, $x^2-x+1=0$

$\Rightarrow \quad x=\frac{1\pm\sqrt{-3}}{2}$

$=\frac{1\pm i\sqrt{3}}{2}$

$=-\left[\frac{-1\mp i\sqrt{3}}{2}\right]$

$\therefore \quad x=-\omega,-\omega^2,$

where ω is imaginary cube root of unity.

Let $\alpha=-\omega$ and $\beta=-\omega^2$

$\therefore \alpha^{101}+\beta^{107}=(-\omega)^{101}+(-\omega^2)^{107}$

$=-[\omega^{101}+\omega^{214}]$

$=-[\omega^{99}\omega^2+\omega^{213}\omega]$

$=-[\omega^2+\omega]$

$=-(-1)=1 \quad [\because 1+\omega+\omega^2=0$

If we consider $\alpha=-\omega^2$ and $\beta=-\omega$, we get the same results.

10. (b) $2\left(1+\frac{1}{\omega}\right)\left(1+\frac{1}{\omega^2}\right)+3\left(2+\frac{1}{\omega}\right)\left(2+\frac{1}{\omega^2}\right)$

$+\ldots+(n+1)\left(n+\frac{1}{\omega}\right)\left(n+\frac{1}{\omega^2}\right)$

$=2(1+\omega)(1+\omega^2)+3(2+\omega)(2+\omega^2)+\ldots+(n+1)$

$(n+\omega)(n+\omega^2$

$=\Sigma(r+1)(r+\omega)(r+\omega^2)$

$=\Sigma(r+1)[r^2+(\omega+\omega^2)r+\omega^3]$

$=\Sigma(r+1)(r^2-r+1)$

$=\Sigma(r^3+1)=\left[\frac{n(n+1)}{2}\right]^2+n$

11. (a) Given, $z\bar{z}(\bar{z}^2+z^2)=350$

Putting $z=x+iy$, we have

$(x^2+y^2)(x^2-y^2)=175$

$\Rightarrow \quad (x^2+y^2)(x^2-y^2)=5\times5\times7$

$\Rightarrow \quad x^2+y^2=25$ and $x^2-y^2=7$

[as other combinations give non-integra values of x and y

$\therefore \quad x=\pm4$ and $y=\pm3\ (x,y\in I)$

Hence, area is $8\times6=48$ sq units.

12. (c) $|z_1|=1; |z_2| \neq 1$

Given, $\left|\dfrac{z_1-2z_2}{2-z_1\bar{z}_2}\right|=1$

$\therefore \quad |z_1-2z_2|^2=|2-z_1\bar{z}_2|^2$

$\Rightarrow |z_1|^2+4|z_2|^2-2\bar{z}_1z_2-2z_1\bar{z}_2$

$= 4+|z_1|^2|z_2|^2-2\bar{z}_1z_2-2z_1\bar{z}_2$

$\Rightarrow (|z_2|^2-1)(|z_1|^2-4)=0$

$\because \quad |z_2| \neq 1$

$\therefore \quad |z_1|=2$

13. (c) Both P and Q are true.

$\because$ Length of direct distance $\leq$ length of arc

i.e. $|z_2-z_1|$ = length of line $AB \leq$ length of arc AB.

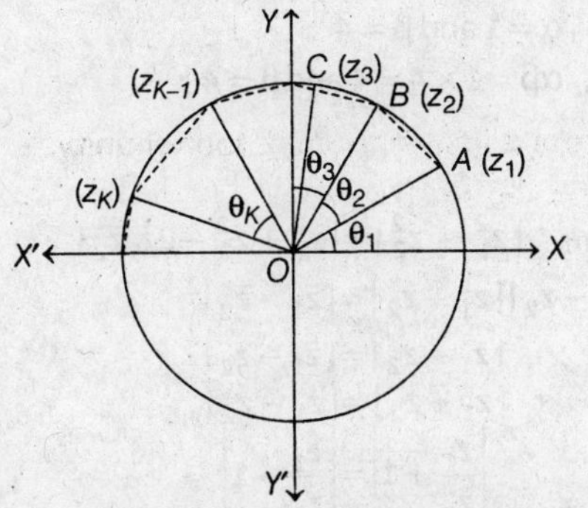

$|z_3-z_2|$ = length of line $BC \leq$ length of arc BC.

$\therefore$ Sum of length of these 10 lines $\leq$ sum of length of arcs (i.e. 2π)

(because $\theta_1+\theta_2+\theta_3+\ldots+\theta_{10}=2\pi$ given)

$\therefore |z_2-z_1|+|z_3-z_2|+\ldots+|z_1-z_{10}| \leq 2\pi \rightarrow P$ is true.

and $|z_k^2-z_{k-1}^2|=|z_k-z_{k-1}||z_k+z_{k-1}|$

As we know that

$|z_k+z_{k-1}| \leq |z_k|+|z_{k-1}| \leq 2$

$\therefore |z_2^2-z_1^2|+|z_3^2-z_2^2|+\ldots+|z_1^2-z_{10}^2| \leq 2$

$(|z_2-z_1|+|z_3-z_2|+\ldots+|z_1-z_{10}|)$

$\leq 2(2\pi)$

$\leq 4\pi \rightarrow Q$ is true.

14. (c) Given circles are

$$(x-x_0)^2+(y-y_0)^2=r^2$$

and $(x-x_0)^2+(y-y_0)^2=4r^2$

or $\quad |z-z_0|=r \quad \ldots$(i)

and $\quad |z-z_0|=2r \quad \ldots$(ii)

where $z_0=x_0+iy_0$

Now, α and $\dfrac{1}{\alpha}$ lies on circle (i) and (ii) respectively, then

$$|\alpha-z_0|=r \text{ and } \left|\frac{1}{\alpha}-z_0\right|=2r$$

$\Rightarrow \quad |\alpha-z_0|=r$ and $|1-\bar{\alpha}z_0|=2r|\bar{\alpha}|$

$\Rightarrow \quad |\alpha-z_0|^2=r^2$ and $|1-\bar{\alpha}z_0|^2=4r^2|\alpha|^2$

Subtracting, we get

$$|1-\bar{\alpha}z_0|^2-|\alpha-z_0|^2=4r^2|\alpha|^2-r^2$$

$\Rightarrow 1+|\alpha z_0|^2-\bar{\alpha}z_0-\alpha\bar{z}_0-(|\alpha|^2+|z_0|^2-\bar{\alpha}z_0-\alpha\bar{z}_0)$

$=4r^2|\alpha|^2-r^2$

$\Rightarrow 1+|\alpha|^2|z_0|^2-|\alpha|^2-|z_0|^2=4r^2|\alpha|^2-r^2$

$\Rightarrow \quad (1-|\alpha|^2)(1-|z_0|^2)=4r^2|\alpha|^2-r^2$

Given, $2|z_0|^2=r^2+2$

$\Rightarrow (1-|\alpha|^2)\left(1-\dfrac{r^2+2}{2}\right)=4r^2|\alpha|^2-r^2$

$\Rightarrow \quad (1-|\alpha|^2)\left(\dfrac{-r^2}{2}\right)=4r^2|\alpha|^2-r^2$

$\Rightarrow \quad |\alpha|^2-1=8|\alpha|^2-2$

$\Rightarrow \quad |\alpha|^2=\dfrac{1}{7} \Rightarrow |\alpha|=\dfrac{1}{\sqrt{7}}$

15. (a, c, d)

Let $z=c$ be a real root. Then,

$$\alpha c^2+c+\bar{\alpha}=0 \quad \ldots\text{(i)}$$

Putting $\alpha=p+iq$, we have

$$(p+iq)c^2+c+p-iq=0$$

$\Rightarrow pc^2+c+p=0$ and $qc^2-q=0$

$\Rightarrow \quad c=\pm 1 \quad [\because q \neq 0]$

From Eq. (i), $\alpha \pm 1+\bar{\alpha}=0$

Also, $|c|=1$

16. (a, b, d)

Let $z=\alpha$ be a real root, then

$$\alpha^3+(3+2i)\alpha+(-1+ia)=0$$

$\Rightarrow \quad (\alpha^3+3\alpha-1)+i(a+2\alpha)=0$

$\Rightarrow \alpha^3+3\alpha-1=0$ and $\alpha=-\dfrac{a}{2}$

$\Rightarrow \quad -\dfrac{a^3}{8}-\dfrac{3a}{2}-1=0$

$\Rightarrow \quad a^3+12a+8=0$

Let $f(a)$ cuts X-axis only once.

Now, $f(-1)<0$, $f(0)>0$, $f(-2)<0$, $f(1)>0$ and $f(3)>0$

Hence, $a \in (-1, 0)$ or $a \in (-2, 1)$ or $a \in (-2, 3)$

17. (a, d)

If $|z-3|=|z-1|$, then z lies on perpendicular bisector of line segment joining '1' and '3'.

So, $\text{Re}(z)=2$

If $|z-3|=|z-5|$, then z lies on perpendicular bisector of line segment joining '3' and '5'.

So, $\text{Re}(z)=4$

18. (a, d)

We have, $\text{Im}\left(\dfrac{az+b}{z+1}\right) = y$ and $z = x + iy$

$$\therefore \quad \text{Im}\left(\frac{a(x+iy)+b}{x+iy+1}\right) = y$$

$$\Rightarrow \quad \text{Im}\left(\frac{(ax+b+iay)(x+1-by)}{(x+1)^2+y^2}\right) = y$$

$$\Rightarrow \quad y(ax+b)+ay(x+1) = y[(x+1)^2+y^2]$$

$$\Rightarrow \quad (a-b)y = y[(x+1)^2+y^2]$$

$$\because \quad y \neq 0 \text{ and } a-b=1$$

$$\therefore \quad (x+1)^2+y^2=1$$

$$\Rightarrow \quad x = -1 \pm \sqrt{1-y^2}$$

19. (a, b, c)

$$|z_1| = |z_2| = 1$$

$$\Rightarrow \quad a^2+b^2 = c^2+d^2 = 1 \qquad \text{...(i)}$$

and $\quad \text{Re}(z_1\bar{z}_2) = 0$

$$\Rightarrow \quad \text{Re}\{(a+ib)(c-id)\} = 0$$

$$\Rightarrow \quad ac+bd = 0 \qquad \text{...(ii)}$$

Now, from Eqs. (i) and (ii), we get

$$a^2+b^2=1$$

$$\Rightarrow \quad a^2 + \frac{a^2c^2}{d^2} = 1$$

$$\Rightarrow \quad a^2 = d^2 \qquad \text{...(iii)}$$

Also, $c^2+d^2=1$

$$\Rightarrow c^2+d^2=1 \Rightarrow c^2+\frac{a^2c^2}{b^2}=1$$

$$\Rightarrow \quad b^2 = c^2 \qquad \text{...(iv)}$$

$|w| = \sqrt{a^2+c^2} = \sqrt{a^2+b^2} = 1$ [from Eqs. (i) and (iv)]

and $|w_2| = \sqrt{b^2+d^2} = \sqrt{c^2+d^2} = 1$

[from Eqs. (i) and (iv)]

Further $\text{Re}(w_1\bar{w}_2) = \text{Re}\{(a+ic)(b-id)\}$

$= ab+cd = ab - \dfrac{ac^2}{b}$ [from Eq. (ii)]

$= \dfrac{ab^2-ac^2}{b} = 0$ [from Eq. (iv)]

Also, $\text{Im}(w_1\bar{w}_2) = bc-ad = bc - a\left(\dfrac{-ac}{b}\right)$

$$= \frac{(a^2+b^2)c}{b} = \frac{c}{b} = \pm 1$$

$\therefore \ |w_1| = 1, |w_2| = 1$ and $\text{Re}(w_1, \bar{w}_2) = 0$

20. (b, c) Circle $x^2+y^2+5x-3y+4=0$ cuts the real axis (X-axis) at $(-4, 0)$, $(-1, 0)$.

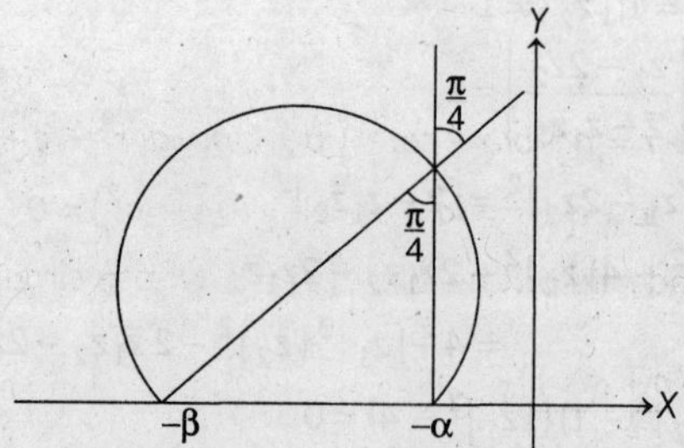

$\arg\left(\dfrac{z+\alpha}{z+\beta}\right) = \dfrac{\pi}{4}$ implies z is on arc and $(-\alpha, 0)$ and $(-\beta, 0)$ subtend $\dfrac{\pi}{4}$ on z.

So, $\quad \alpha = 1$ and $\beta = 4$

Hence, $\alpha\beta = 1 \times 4 = 4$ and $\beta = 4$

21. (a, d)

$$|z_1^2 - z_2^2| = |\bar{z}_1^2 + \bar{z}_2^2 - 2\bar{z}_1\bar{z}_2|$$

$$\Rightarrow |z_1-z_2||z_1+z_2| = |\bar{z}_1-\bar{z}_2|^2$$

$$\Rightarrow \quad |z_1+z_2| = |\bar{z}_1-\bar{z}_2|$$

$$\Rightarrow \quad |z_1+z_2| = |z_1-z_2|$$

$$\Rightarrow \quad \left|\frac{z_1}{z_2}+1\right| = \left|\frac{z_1}{z_2}-1\right|$$

$\Rightarrow \dfrac{z_1}{z_2}$ lies on perpendicular bisector of 1 and – 1

$\Rightarrow \dfrac{z_1}{z_2}$ lies on imaginary axis

$\Rightarrow \dfrac{z_1}{z_2}$ is purely imaginary

$$\Rightarrow \quad \arg\left(\frac{z_1}{z_1}\right) = \pm\frac{\pi}{2}$$

$$|\arg(z_1) - \arg(z_2)| = \frac{\pi}{2}$$

22. (a, c, d)

We have, $a^3+b^3-8c^3+6abc=0$

$$\Rightarrow a^3+b^3+(-2c)^3-3ab(-2c)=0$$

$$\Rightarrow (a+b-2c)(a+b\omega-2c\omega^2)(a+b\omega^2-2c\omega)=0$$

23. (a, c)

Clearly, we have to find it for real z.

Let $z = x$, then,

$$|x-\omega| = |x-\omega^2| = |\omega-\omega^2|$$

$$\Rightarrow \left(x+\frac{1}{2}\right)^2 + \frac{3}{4} = \left|\frac{-1+\sqrt{3}i}{2} - \frac{-1-\sqrt{3}i}{2}\right|^2 = 3$$

$$\Rightarrow \quad x+\frac{1}{2} = \pm\frac{3}{2} \Rightarrow x = 1, -2$$

24. (c, a)

$$p+q+r=a+b\omega+c\omega^2+b+c\omega+a\omega^2+c+a\omega+b\omega^2$$

$$\therefore \quad p+q+r=(a+b+c)(1+\omega+\omega^2)=0 \quad \text{...(i)}$$

p, q and r lie on the circle $|z|=2$, whose circumcentre is origin.

Also, $\dfrac{(p+q+r)}{3}=0$

Hence, the centroid coincides with circumcentre.

So, the triangle is equilateral.

Now, $(p+q+r)^2=0$

$$\Rightarrow p^2+q^2+r^2=-2pqr\left[\frac{1}{p}+\frac{1}{q}+\frac{1}{r}\right]$$

$$=-2pqr\left[\frac{1}{a+b\omega+c\omega^2}+\frac{1}{b+c\omega+a\omega^2}+\frac{1}{c+a\omega+b\omega^2}\right]$$

$$=-2pqr\left[\frac{1}{\omega^2(a\omega+b\omega^2+c)}+\frac{1}{\omega(b\omega^2+c+a\omega)}+\frac{1}{c+a\omega+b\omega^2}\right]$$

$$=-\frac{2pqr}{a\omega+b\omega^2+c}\left[\frac{1}{\omega^2}+\frac{1}{\omega}+\frac{1}{1}\right]=0 \quad \text{...(ii)}$$

Hence, $p^2+q^2+r^2=2(pq+qr+rp)$

25. (c, d)

We have, $x^2+x+1=(x-\omega)(x-\omega^2)$

Since, $f(x)$ is divisible by x^2+x+1, $f(\omega)=0$,

$$f(\omega^2)=0$$

So, $\quad P(\omega^3)+\omega Q(\omega^3)=0$

$$\Rightarrow \quad P(1)+\omega Q(1)=0 \quad \text{...(i)}$$

$$P(\omega^6)+\omega^2 Q(\omega^6)=0$$

$$\Rightarrow \quad P(1)+\omega^2 Q(1)=0 \quad \text{...(ii)}$$

Solving Eqs. (i) and (ii), we obtain

$$P(1)=0 \text{ and } Q(1)=0$$

Therefore, both $P(x)$ and $Q(x)$ are divisible by $x-1$.

Hence, $P(x^3)$ and $Q(x^3)$ are divisible by x^3-1 and so by $x-1$.

Since, $f(x)=P(x^3)+xQ(x^3)$, we get

$f(x)$ is divisible by $x-1$.

26. (a, b)

Incident ray is $|z-2|-|z-1-i|=\sqrt{2}$

$$=|2-(1+i)|$$

This ray is emanating from the point '2' and moving to the left of '2'.

It strikes Y-axis at point $2i$.

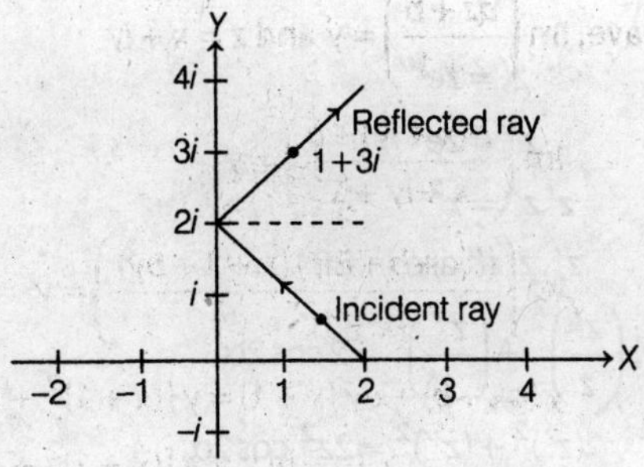

Then, equation of reflected ray is $\arg(z-2i)=\dfrac{\pi}{4}$

or $\quad |z-2i|-|z-1-3i|=\sqrt{2}$

27. (a, d)

We have, $|z_1|=15$ and $|z_2-3-4i|=5$

Minimum value of $|z_1-z_2|$ is

$$AB=OB-OA=15-10=5$$

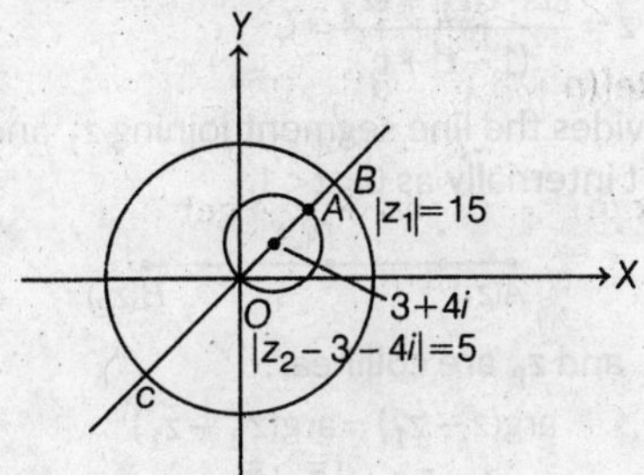

Maximum value of $|z_1-z_2|$ is

$$CA=OA+OC=10+15=25$$

28. (a, b, c, d)

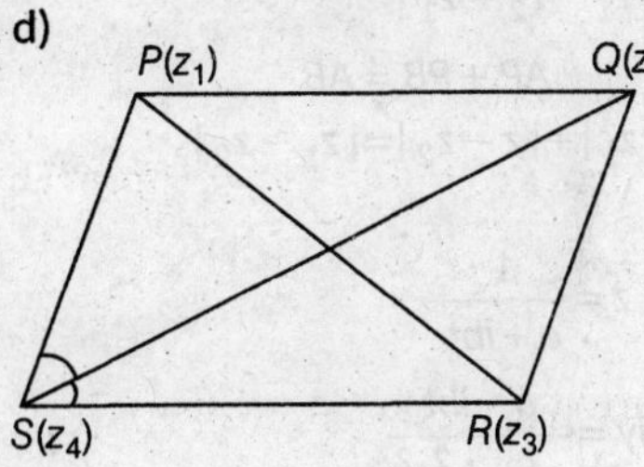

(a) $PS \parallel QR \Rightarrow \arg\left(\dfrac{z_1-z_4}{z_2-z_3}\right)=0$

$$\Rightarrow \frac{z_1-z_4}{z_3-z_4} \text{ is purely real}$$

(b) Since, diagonals bisect the angle

$$\Rightarrow \quad \text{amp}\left(\frac{z_1-z_4}{z_2-z_4}\right)=\text{amp}\left(\frac{z_2-z_4}{z_3-z_4}\right)$$

(c) Diagonals of rhombus are perpendicular.

Hence, $(z_1-z_3/(z_2-z_4)$ is purely imaginary.

(d) Diagonals of rhombus are not necessarily equal.

Hence, $|z_1-z_3|\neq|z_2-z_4|$ not necessarily.

29. (a, c)

$$z' = ze^{i\alpha} \quad ...(i)$$
$$z'' = ze^{-i\alpha} \quad ...(ii)$$
$$\therefore \quad z'z'' = z^2$$
$\Rightarrow$ z', z, z'' are in GP

Also, $\left(\frac{z'}{z}\right)^2 + \left(\frac{z''}{z}\right)^2 = 2\cos 2\alpha$

$$\Rightarrow \quad (z')^2 + (z'')^2 = 2z^2 \cos 2\alpha$$
$$= 2z^2 (2\cos^2 \alpha - 1)$$
$$\text{or} \quad (z')^2 + (z'')^2 + 2z^2 = 4z^2 \cos^2 \alpha$$
$$\text{or} \quad (z')^2 + (z'')^2 + 2z'z'' = 4z^2 \cos^2 \alpha$$
$$\text{or} \quad (z' + z'')^2 = 4z^2 \cos^2 \alpha$$

30. (a, c, d)

Given, $z = (1-t)z_1 + tz_2$

$$\Rightarrow z = \frac{(1-t)z_1 + tz_2}{(1-t)+t}$$

$\Rightarrow z$ divides the line segment joining z_1 and z_2 in ratio $(1-t):t$ internally as $0 < t < 1$.

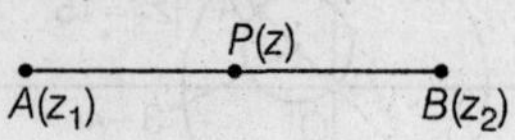

$\Rightarrow z, z_1$ and z_2 are collinear.

$$\Rightarrow \quad \arg(z - z_1) = \arg(z_2 - z_1)$$
$$\Rightarrow \quad \frac{z - z_1}{z_2 - z_1} = \frac{\bar{z} - \bar{z}_1}{\bar{z}_2 - \bar{z}_1}$$
$$\Rightarrow \begin{vmatrix} z - z_1 & \bar{z} - \bar{z}_1 \\ z_2 - z_1 & \bar{z}_2 - \bar{z}_1 \end{vmatrix} = 0$$
$$AP + PB = AB$$
$$\Rightarrow |z - z_1| + |z - z_2| = |z_1 - z_2|$$

31. (a, c, d)

$$z = \frac{1}{a + ibt}$$
$$\Rightarrow \quad x + iy = \frac{a - ibt}{a^2 + b^2t^2}$$
$$\Rightarrow \quad x = \frac{a}{a^2 + b^2t^2}$$
$$\text{and} \quad y = \frac{-bt}{a^2 + b^2t^2}$$

Eliminating t, we get

$$x^2 + y^2 = \frac{x}{a}$$
$$\Rightarrow \left(x - \frac{1}{2a}\right)^2 + y^2 = \left(\frac{1}{2a}\right)^2$$

$\therefore$ Option (a) is correct.

(c), (d) can be verified by putting $b = 0$ and $a = 0$, respectively.

32. (a, c)

ΔABC is equilateral.

$$\therefore \quad z^2 + (-z)^2 + (1-z)^2 = z(-z) + z(1-z) + (-z)(1-z)$$
$$\text{or} \quad 3z^2 - 2z + 1 = -z^2$$
$$\text{or} \quad 4z^2 - 2z + 1 = 0$$

Sum of roots $= \frac{1}{2}$ and product of roots $= \frac{1}{4}$

33. (a, b, c)

Let $z = x + iy$, where x and y satisfy the given equation.

$\therefore (x^2 + y^2)(x^2 - y^2) = 175$

$\Rightarrow x^2 + y^2 = 25$ and $x^2 - y^2 = 7$

[as all other possibilities will give non-integral solutions]

Hence, possible values of z will be $4 + 3i$, $4 - 3i$, $-4 + 3i$ and $-4 - 3i$.

Clearly, it will form a rectangle having length of the diagonal 10.

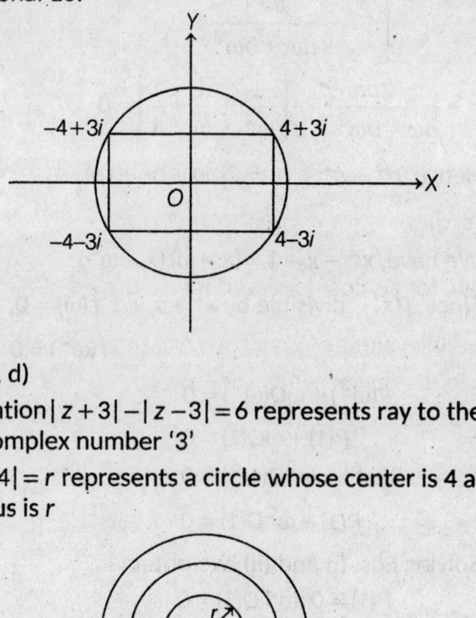

34. (a, c, d)

Equation $|z + 3| - |z - 3| = 6$ represents ray to the right of complex number '3'

$|z - 4| = r$ represents a circle whose center is 4 and radius is r

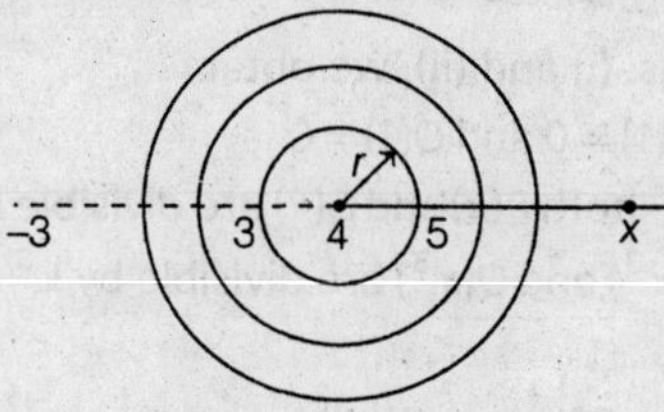

When radius of circle $r = 1$, it cuts ray at two points '3' and '5' for radius $r < 1$, circle cuts ray at two points.

When radius of circle $r > 1$, it cuts ray at only one point.

35. (a, b, c)

We have, $\frac{3}{2 + e^{i\theta}} = ax + iby$

$$\Rightarrow \quad \frac{3}{ax + iby} - 2 = e^{i\theta}$$
$$\Rightarrow \quad \frac{3 - 2ax - 2iby}{ax + iby} = e^{i\theta}$$

$$\Rightarrow \left|\frac{3-2ax-2iby}{ax+iby}\right| = |e^{i\theta}| = 1$$

$$\Rightarrow (3-2ax)^2 + 4b^2y^2 = a^2x^2 + b^2y^2$$

When $a=1$ and $b=2$, we have

$$(3-2x)^2 + 16y^2 = x^2 + 4y^2$$

$$\Rightarrow 3x^2 + 12y^2 - 12x + 9 = 0$$

$\Rightarrow \quad x^2 - 4x + 4y^2 + 3 = 0$, which is ellipse

When $a=b=1$, we have

$$(3-2x)^2 + 4y^2 = x^2 + y^2$$

$$\Rightarrow \quad 3x^2 + 3y^2 - 12x + 9 = 0$$

or $x^2 + y^2 - 4x + 3 = 0$, which is circle

When $a=1$ and $b=0$, we have

$(3-2x)^2 = x^2$, which is pair of straight lines.

36. (b, d)

$$\omega = \frac{\sqrt{3}+i}{2} = e^{i\frac{\pi}{6}}, \text{so } \omega^n = e^{\frac{in\pi}{6}}, n = 0, 1, 2, 3 \ldots$$

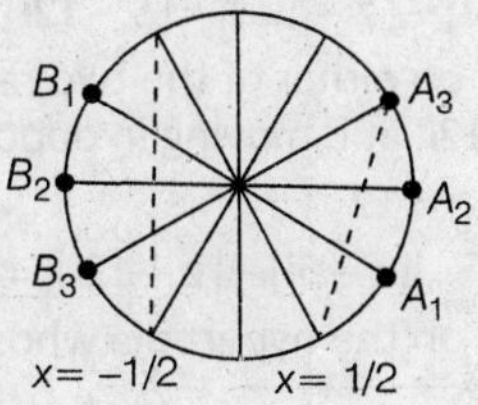

Now, for z_1, $\cos\frac{\pi}{6} > \frac{1}{2}$ and for z_2, $\cos\frac{n\pi}{6} < \frac{-1}{2}$

Possible values of z_1 are A_1, A_2, A_3 wheras of z_2 are B_1, B_2, B_3 (as shown in the figure).

So, possible values of $< z_1 0 z_2$ according to the given option is $\frac{2\pi}{3}$ or $\frac{5\pi}{6}$.

Sol. (Q.Nos. 37-40)

37. (a), **38.** (d), **39.** (c), **40.** (d)

We have, $z = \frac{1-i\sin\theta}{1+i\cos\theta}$

$$= \frac{(1-i\sin\theta)(1-i\cos\theta)}{(1+i\cos\theta)(1-i\cos\theta)}$$

$$= \frac{(1-\sin\theta\cos\theta) - i(\cos\theta+\sin\theta)}{(1+\cos^2\theta)}$$

If z is purely real, then

$$\cos\theta + \sin\theta = 0 \text{ or } \tan\theta = -1$$

$$\Rightarrow \quad \theta = n\pi - \frac{\pi}{4}, n \in I$$

If z is purely imaginary, then $1 - \sin\theta\cos\theta = 0$ or $\sin\theta\cos\theta = 1$, which is not possible.

$$|z| = \left|\frac{1-i\sin\theta}{1+i\cos\theta}\right| = \frac{\sqrt{1+\sin^2\theta}}{\sqrt{1+\cos^2\theta}}$$

If $|z| = 1$, then $\cos^2\theta = \sin^2\theta$

$$\Rightarrow \quad \tan^2\theta = 1$$

$$\Rightarrow \quad \theta = n\pi \pm \frac{\pi}{4}, n \in I$$

We have, $\arg(z) = \tan^{-1}\left(\frac{-(\cos\theta+\sin\theta)}{(1-\sin\theta\cos\theta)}\right)$

Now, $\arg(z) = \frac{\pi}{4}$

$$\Rightarrow \quad \frac{-(\cos\theta+\sin\theta)}{(1-\sin\theta\cos\theta)} = 1$$

$$\Rightarrow \cos^2\theta + \sin^2\theta + 2\sin\theta\cos\theta = 1 + \sin^2\theta\cos^2\theta - 2\sin\theta\cos\theta$$

$$\Rightarrow 1 + 4\sin\theta\cos\theta = 1 + \sin^2\theta\cos^2\theta$$

$$\Rightarrow \sin^2\theta\cos^2\theta - 4\sin\theta\cos\theta = 0$$

$$\Rightarrow \quad \sin\theta\cos\theta(\sin\theta\cos\theta - 4) = 0$$

$\Rightarrow \quad \sin\theta\cos\theta = 0 \quad$ [$\because \sin\theta\cos\theta = 4$ is not possible]

$\Rightarrow \theta = (2n+1)\pi$ or $\theta = (4n-1)\frac{\pi}{2}, n \in I$ [$\because \cos\theta - \sin\theta > 0$]

Sol. (Q.Nos. 41-44)

41. (b), **42.** (b), **43.** (c), **44.** (c)

Given that

$$|z_1 + z_2|^2 = |z_1|^2 + |z_2|^2$$

$$\Rightarrow |z_1|^2 + |z_2|^2 + z_1\bar{z}_2 + \bar{z}_1 z_2 = |z_1|^2 + |z_2|^2$$

$\Rightarrow \quad z_1\bar{z}_2 + \bar{z}_1 z_2 = 0 \quad$ ...(i)

$\Rightarrow \quad \frac{z_1}{z_2} + \frac{\bar{z}_1}{\bar{z}_2} = 0 \quad$ [dividing by $z_2\bar{z}_2$]

$\Rightarrow \quad \frac{z_1}{z_2} + \overline{\left(\frac{z_1}{z_2}\right)} = 0 \quad$ ...(ii)

From Eq. (i), $z_1\bar{z}_2$ is purely imaginary.

From Eq. (i), $\frac{z_1}{z_2}$ is purely imaginary.

Hence, $\arg\left(\frac{z_1}{z_2}\right) = \pm\frac{\pi}{2}$ or $\arg(z_1) - \arg(z_2) = \pm\frac{\pi}{2}$

Also, $i\left(\frac{z_1}{z_2}\right)$ is purely real.

Hence, its possible arguments are 0 and π.

Sol. (Q.Nos. 45-47)

45. (c), **46.** (b), **47.** (d)

We have, $\quad az + b\bar{z} + c = 0 \quad$ ...(i)

or $\quad \bar{a}\bar{z} + \bar{b}z + \bar{c} = 0 \quad$ ...(ii)

Eliminating $\bar{z}$ from Eqs. (i) and (ii), we get

$$z = \frac{c\bar{a} - b\bar{c}}{|b|^2 - |a|^2}$$

If $|a| \neq |b|$, then z represents one point on the argand plane.

If $|a|=|b|$ and $\bar{a}c \neq b\bar{c}$, then no such z exists.

Adding Eqs. (i) and (ii),

$$(\bar{a}+b)\bar{z}+(a+\bar{b})z+(c+\bar{c})=0$$

This is of the from $A\bar{z}+\bar{A}z+B=0$, where $c+\bar{c}$ is real.

Hence, locus of z is a straight line.

48. (i) $\to$ B, (ii) $\to$ D, (iii) $\to$ A, (iv) $\to$ C

Since, graph is concave upward, $a>0$

Graph intersects Y-axis at negative Y-axis.

So, $c<0$.

Also, vertex lies to the right of Y-axis.

So, $-\dfrac{b}{2a}>0$

$\therefore \quad b<0$

(i) Complex number $a+ib$ lies in fourth quadrant

$\therefore \arg(a+ib)=\tan^{-1}\dfrac{b}{a}$

(ii) Complex number $ac-ibc$ lies in third quadrant

$\therefore \arg(ac-ibc)=-\pi-\tan^{-1}\dfrac{b}{a}$

(iii) Complex number $-ac+ibc$ lies in first quadrant

$\therefore \arg(-ac+ibc)=-\tan^{-1}\dfrac{b}{a}$

(iv) Complex number $abc-ib^2c$ lies in second quadrant

$\therefore \arg(abc-ib^2c)=\pi+\tan^{-1}\dfrac{b}{a}$

49. (i) $\to$ D, (ii) $\to$ A, (iii) $\to$ B, (iv) $\to$ C

(i) $\dfrac{z_3-z_1}{z_2-z_1}=\dfrac{1+i\sqrt{3}}{2}$

$\Rightarrow \dfrac{z_3-z_1}{z_2-z_1}=e^{i\frac{\pi}{3}}$

$\Rightarrow \left|\dfrac{z_3-z_1}{z_2-z_1}\right|=1$ and $\arg\left(\dfrac{z_3-z_1}{z_2-z_1}\right)=\dfrac{\pi}{3}$

So, triangle is equilateral.

(ii) $\text{Re}\left(\dfrac{z_3-z_1}{z_3-z_2}\right)=0$

$\Rightarrow \dfrac{z_3-z_1}{z_3-z_2}$ is purely imaginary

$\Rightarrow \arg\left(\dfrac{z_3-z_1}{z_3-z_2}\right)=\pm\dfrac{\pi}{2}$

So, triangle is right angled.

(iii) Let $\dfrac{z_3-z_1}{z_3-z_2}=r(\cos\theta+i\sin\theta)$

Since, $\cos\theta<0$, $\arg\left(\dfrac{z_3-z_1}{z_3-z_2}\right)>\dfrac{\pi}{2}$

So, triangle is obtuse angled.

(iv) $\dfrac{z_3-z_1}{z_3-z_2}=1$

$\Rightarrow |z_3-z_1|=|z_3-z_2|$ and $\arg\left(\dfrac{z_3-z_1}{z_3-z_2}\right)=\dfrac{\pi}{2}$

So, triangle is isosceles right angled.

50. (i) $\to$ C, (ii) $\to$ A, (iii) $\to$ D, (iv) $\to$ B

We have equation,

$$||z-5i|+m|z-12i||=n$$

(i) $m=1$ and $n=13$

$\therefore |z-5i|+|z-12i|=13$

or $|z-5i|+|z-12i|=|5-12i|$

So, z lies on the line segment joining complex numbers '5' and '12i'.

(ii) $m=1, n=15$

$\therefore |z-5i|+|z-12i|=15>13$

So, z lies on the ellipse whose foci are '5' and '12i'

(iii) $m=-1$ and $n=13$

$\therefore ||z-5i|-|z-12i||=13$

or $||z-5i|-|z-12i||=|5-12i|$

So, z lies on either of the two rays emanating from '5' and '12i' and moving in opposite directions.

(iv) $m=-1, n=12$

$\therefore \quad ||z-5i|-|z-12i||=12<13$

So, z lies on the hyperbola whose foci are '5' and '12i'.

Sol. (Q.Nos. 51-52)

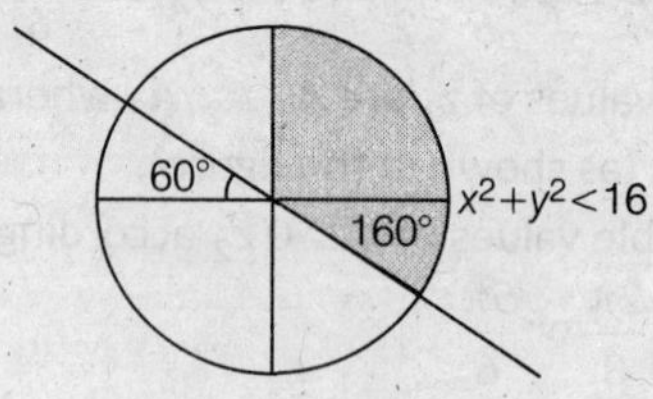

$S_1=|z|<4$, z lies inside the circle or radius 4.

$S_2:\sqrt{3}x+y>0$, z lies above the line $\sqrt{3}x+y=0$

$S_3:\text{Re}(z)>0$, z lies right of the imaginary axis.

51. (20.95) Area of region $S_1\cap S_2\cap S_3=$ Shaded area

$$=\frac{\pi\times 4^2}{4}+\frac{4^2\times\pi}{6}=\pi 4^2\left[\frac{1}{4}+\frac{1}{6}\right]$$

$$=\frac{20\pi}{3}=\frac{20\times 22}{3\times 7}=\frac{440}{21}=20.95$$

52. (0.634) Distance of $(1,-3)$ from $y+\sqrt{3}x=0$ is

$$\left|\frac{\sqrt{3}\times 1-3}{2}\right|=\frac{3-\sqrt{3}}{2}=\frac{3-1.732}{2}=0.634$$

53. (20) $|3z+9-7i|=|(3z+6-3i)+(3-4i)|$

$\leq|3z+6-3i|+|3-4i|$

$=3|z+2-i|+5=20$

54. (18) $A = (1+2i)x^3 - 2(3+i)x^2 + (5-4i)x + 2a^2 = 0$

Let the real root of equation be α, then

$(1+2i)\alpha^3 - 2(3+i)\alpha^2 + (5-4i)\alpha + 2a^2 = 0$

Equating imaginary part zero, we get

$$2\alpha^3 - 2\alpha^2 - 4\alpha = 0$$

$$\Rightarrow \quad \alpha(\alpha^2 - \alpha - 2) = 0 \Rightarrow \alpha = 0 \text{ or } \alpha = -1, 2$$

Now, equating real part zero, we have

$$\alpha^3 - 6\alpha^2 + 5\alpha + 2a^2 = 0$$

$$\alpha = 0 \Rightarrow a = 0$$

$$\alpha = -1 \Rightarrow a = \pm\sqrt{6}$$

$$\alpha = 2$$

$$\Rightarrow \quad a = \pm\sqrt{3}$$

$$\Rightarrow \quad \Sigma a^2 = (0)^2 + (+\sqrt{6})^2 + (-\sqrt{6})^2 + (+\sqrt{3})^2 + (-\sqrt{3})^2$$

$$= 18$$

55. (30) Let $z = x + iy$

$$\therefore \quad E = z\bar{z} + (z-3)(\bar{z}-3) + (z-6i)(\bar{z}+6i)$$

$$= 3z\bar{z} - 3(z+\bar{z}) + 9 + 6(z-\bar{z})i + 36$$

$$= 3(x^2+y^2) - 6x - 12y + 45$$

$$= 3[x^2+y^2-2x-4y+15]$$

$$= 3[(x-1)^2 + (y-2)^2 + 10]$$

$\therefore \quad E_{min} = 30$, when $x = 1$ and $y = 2$

56. (125) $z_1 = (8\sin\theta + 7\cos\theta) + i(\sin\theta + 4\cos\theta)$

$$z_2 = (\sin\theta + 4\cos\theta) + i(8\sin\theta + 7\cos\theta)$$

Hence, $z_1 = x + iy$ and $z_2 = y + ix$

where $x = (8\sin\theta + 7\cos\theta)$ and $y = (\sin\theta + 4\cos\theta)$

$$z_1 \cdot z_2 = (xy - xy) + i(x^2+y^2) = i(x^2+y^2) = a + ib$$

$$\Rightarrow \quad a = 0 \text{ and } b = x^2 + y^2$$

Now, $x^2 + y^2 = (8\sin\theta + 7\cos\theta)^2 + (\sin\theta + 4\cos\theta)^2$

$$= 65\sin^2\theta + 65\cos^2\theta + 120\sin\theta\cdot\cos\theta$$

$$= 65 + 60\sin 2\theta$$

$$\Rightarrow \quad z_1 \cdot z_2\big|_{max} = 125$$

57. (6) We have, $|a\omega + b| = 1$

$$\Rightarrow \quad |a\omega + b|^2 = 1$$

$$\Rightarrow \quad (a\omega + b)(a\bar{\omega} + b) = 1$$

$$\Rightarrow a^2 + ab(\omega + \bar{\omega}) + b^2 = 1$$

$$\Rightarrow \quad a^2 - ab + b^2 = 1$$

$$\Rightarrow \quad (a-b)^2 + ab = 1 \quad \text{...(i)}$$

When $(a-b)^2 = 0$ and $ab = 1$, then $(1, 1), (-1, -1)$

When $(a-b)^2 = 1$ and $ab = 0$, then $(0, 1), (1, 0), (0, -1), (-1, 0)$

Hence, there are 6 ordered pairs.

58. (5) Roots are $2\omega, (2+3\omega), (2+3\omega^2)$ and $(2-\omega-\omega^2)$.

$2+3\omega$ and $2+\omega^2$ are conjugate to each other.

2ω is complex root, then other root must be $2\omega^2$ (as complex roots occur in conjugate pair)

$2-\omega-\omega^2 = 2-(-1) = 3$ which is real.

Hence, least degree of the polynomial is 5.

CHAPTER

04

SEQUENCE AND SERIES

LEVEL 01 \ Starter Level

4.1 Introduction to Sequence and Series and Arithmetic Progression

1 Find the first five terms of each sequence

(i) $a_n = n^2 - 1$

(ii) $a_n = (-1)^n + 1$

(iii) $a_n = \sin(\pi n)$

(iv) $a_n = (-1)^n \frac{2^n}{n}$

2 Find the sequence of numbers defined as

(i) $a_n = \begin{cases} n; & n \text{ is odd} \\ -n; & n \text{ is even} \end{cases}$

(ii) $a_n = \begin{cases} \dfrac{n+2}{n+3}; & n \text{ is odd} \\ 5^n; & n \text{ is even} \end{cases}$

(iii) $a_n = \begin{cases} 1/n; & n \text{ is odd} \\ -\dfrac{1}{n^2}; & n \text{ is even} \end{cases}$

3 Write the first four terms defined as

(i) $a_1 = 1$ and $a_n = a_{n-1} + n$

(ii) $a_1 = 2$ and $a_{n+1} = \dfrac{a_n}{a_n + 2}$

(iii) $a_1 = 1, a_2 = 1$ and $a_{n+2} = a_n + a_{n+1}$

4 Find a formula for the general term a_n of each of the following sequences.

(i) 1, 0, – 1, 0, 1, 0, – 1, 0 .....   (ii) $1, \frac{-1}{3}, \frac{1}{5}, \frac{-1}{7}, \frac{1}{9}, ...$

(iii) 5, 5, 5, 5, 5, ...

5 Find the sum of series

(i) $\sum_{i=1}^{5} 2^i$ (ii) $\sum_{i=1}^{10} (-1)^i \cdot i$

(iii) $\sum_{i=1}^{4} (i+2)^2$ (iv) $\sum_{i=1}^{4} 3i^3 + 2i - 4$

6 Let $a_1, a_2, \ldots$ be integers such that $a_1 - a_2 + a_3 - a_4 + \ldots + (-1)^{n-1} a_n = n$, for all $n \geq 3$. Then, find the sum of $a_{51} + a_{52} + \ldots + a_{1023}$.

7 Find the 75th term of the sequence 7, 10, 13, 16, ... ,

8 If the mth term of an AP is $\frac{1}{n}$ and the nth term is $\frac{1}{m}$, then find the mnth term of an AP.

9 Find the first negative term of the sequence 121, 117, 113, ..., also find the value of the term.

10 The sum of three numbers in AP is 45. If the sum of their squares is 683, Find the numbers.

11 The sum of the four integers in AP is 24 and their product is 945. Find these integers.

12 Let α and β be the roots of the equation $x^2 - 2x + P = 0$ and γ and δ be the roots of the equation $x^2 - 18x + Q = 0$. If $\alpha < \beta < \gamma < \delta$ are in AP, then find the values of P and Q.

13 Find the number of terms common to both the arithmetic progressions 2, 5, 8, 11, ..., 179 and 3, 5, 7, 9, ..., 101

14 Find the sum of first twenty five terms of the series

$20 + 19\frac{1}{3} + 18\frac{2}{3} + ...$

15 If the sum of n terms of a series is given by $S_n = 5n^2 + 4n$, find its 20th term.

16 An AP has first term ln 2 and common difference ln 4. Then, find the sum of first n terms.

17 Find the sum of all natural numbers lying between 100 and 1000, which are multiples of 5.

18 Find the sum of first 24 terms of an AP $a_1, a_2, a_3, \ldots$, if.

$a_1 + a_5 + a_{10} + a_{15} + a_{20} + a_{24} = 225$

19 The sum of n terms of two AP's are in the ratio $\frac{3n - 13}{5n + 21}$. Find the ratio of their 24th terms.

20 The difference between any two consecutive interior angles of a polygon is 5°. If the smallest angle is 120°, find the number of the sides of the polygon.

21 Find the degree of the expression $(1 + x)(1 + x^6)(1 + x^{11}) ..... (1 + x^{101})$.

22 150 workers were engaged to finish a piece of work in a certain number of days. 4 workers dropped the second day. 4 more workers dropped the third day and so on it takes eight more days to finish the work now. Find the number of days in which of the work was completed.

4.2 Geometric Progression

23 Find the formulae for the *n*th term and hence find the seventh term of the following geometric sequence.

(i) 3, 6, 12, 24, ... (ii) 36, 18, 9, ...

(iii) 6, – 6, 6, – 6, ... (iv) $\frac{1}{4}, \frac{1}{12}, \frac{1}{36}, \frac{1}{108}, \ldots$

24 If the *p*th and *q*th terms of a GP are *q* and *p* respectively, then find its $(p+q)$th term.

25 If the third term of a GP is 4, then find the product of first five terms of the GP.

26 The sum of first three terms of a GP is $\frac{13}{12}$ and their product is – 1. Find the terms.

27 The sum of three numbers in GP is 56. If we subtract 1, 7, 21 from these numbers in that order the resulting numbers are in AP. Find the numbers.

28 Evaluate $\sum_{k=1}^{9} (7+5^k)$

29 How many terms of GP $3, \frac{3}{2}, \frac{3}{4}$ are needed to give the sum $\frac{3069}{512}$?

30 Find the sum of *n* terms of the series :

(i) 2 + 22 + 222 + ... (*n* terms) (ii) 0.6 + 0.66 + 0.666 + ... (*n* terms)

31 Find the sum of *n* terms of the series

$\frac{5}{2}+\frac{9}{4}+\frac{17}{8}+\frac{33}{16}+\ldots$

32 If *f* is a function satisfying $f(x+y)=f(x)\,f(y)$ for all $x, y \in N$, such that $f(1)=3$ and $\sum_{k=1}^{n} f(k)=1092$, find the value of *n*.

33 Find the following sums

(i) $\frac{1}{3}+\frac{1}{9}+\frac{1}{27}+\ldots\infty$ (ii) $1-\frac{1}{2}+\frac{1}{4}-\frac{1}{8}+\ldots\infty$

34 The sum of infinite number of terms in GP is 20 and the sum of their squares is 100. Then, find the common ratio of GP.

35 Find the sum of the following series

(i) $\sum_{n=1}^{\infty} 5\left(\frac{1}{2}\right)^{n-1}$ (ii) $(\sqrt{2}+1)+1+(\sqrt{2}-1)+\ldots\infty$

(iii) $\frac{1}{2}+\frac{1}{3^2}+\frac{1}{2^3}+\frac{1}{3^4}+\frac{1}{2^5}+\frac{1}{3^6}+\ldots\infty$

36 Using geometrical progression, express in $\frac{p}{q}$ form

(i) 0.3333333 .... (ii) 1.233333 ....

37 A ball is dropped from the height of 200 m. After striking a floor, the ball rebounds $\frac{3}{5}$th of the height from which it was dropped. Then, find the total distance that it travels before coming to rest.

38 If S_n and *S* denote the sum of *n* terms and to infinity of the series $2+1\frac{1}{2}+1\frac{1}{8}+\ldots$, then find the value of *n* such that $S-S_n<\frac{1}{1000}$.

4.3 Arithmetic and Geometric Mean

39 Insert six numbers between 3 and 24 such that resulting sequence is an AP.

40 Insert three geometric means between the numbers 1 and 256.

41 The sum of four numbers in GP is 60 and the arithmetic mean of the first and last numbers is 18. Find the numbers.

4.4 Arithmetic Geometric Progression

42 Find the sum of the series $1+\frac{4}{5}+\frac{7}{5^2}+\frac{10}{5^3}+\ldots$

(i) to n terms (ii) to infinity.

43 Find the sum of the series

$1+3x+6x^2+10x^3+\ldots\infty$

44 Find the sum of infinite series,

$1+\left(1+\frac{1}{5}\right)\left(\frac{1}{2}\right)+\left(1+\frac{1}{5}+\frac{1}{5^2}\right)\left(\frac{1}{2^2}\right)+\ldots$

45 Find the sum of

$1^2+2^2x+3^2x^2+4^2x^3+\ldots\infty$ terms.

4.5 Summation of Series

46 Find $\sum_{r=1}^{n}(6r^2+4r-1)$.

47 Find the sum

$2\times5+5\times9+8\times13+11\times17+\ldots n$ terms

48 Find the sum of n terms of the series

$0.7+7.7+0.77+77.7+0.777+777.7+\ldots.. n$ terms (where n is even)

49 Find the value of the expression

$\sum_{i=1}^{n}\sum_{j=1}^{i}\sum_{k=1}^{j}1$

4.6 Inequalities

50 If a, b, c are positive, then verify that $(a^2b+b^2c+c^2a)(ab^2+bc^2+ca^2)\geq 9a^2b^2c^2$.

51 If $2a+b+3c=1$ and $a>0, b>0, c>0$. Find the greatest value of $a^4b^2c^2$ and obtain the corresponding values of a, b, c.

52 Is $\frac{\sec^4\alpha}{\tan^2\beta}+\frac{\sec^4\beta}{\tan^2\alpha}\geq 8$, if each term in the expression is well defined? If yes, then give the proper justification.

53 Find the greatest value of xyz for positive values of x, y, z subject to the condition $xy+yz+xz=12$.

54 Find the minimum value of $4^{\sin^2 x}+4^{\cos^2 x}$ for $x\in R$.

LEVEL 02 \\ JEE Main Level

Single Choice Correct Questions

4.1 Introduction to Sequence and Series and Arithmetic Progression

1 If $\frac{1}{a}, \frac{1}{b}, \frac{1}{c}$ are in AP, then $\left(\frac{1}{a}+\frac{1}{b}-\frac{1}{c}\right)\left(\frac{1}{b}+\frac{1}{c}-\frac{1}{a}\right)$ is equal to

(a) $\frac{4}{ac}-\frac{3}{b^2}$ (b) $\frac{b^2-ac}{a^2b^2c^2}$ (c) $\frac{4}{ac}-\frac{1}{b^2}$ (d) None of these

2 The fourth term of an AP is three times of the first term and the seventh term exceeds the twice of the third term by one, then the common difference of the progression is

(a) 2 (b) 3 (c) $\frac{3}{2}$ (d) -1

3 If 8, -4 and 13 be three (not neccessarily consecutive term) of an AP. How many such AP are possible?

(a) 1 (b) 2 (c) Infinitely many (d) No such AP possible

4 Let a_n be the nth term of an AP If $\sum_{r=1}^{100} a_{2r}=\alpha$ and $\sum_{r=1}^{100} a_{2r-1}=\beta$, then common difference of the AP is

(a) $\alpha\cdot\beta$ (b) $\beta-\alpha$ (c) $\frac{\alpha-\beta}{2}$ (d) None of these

5 If the coefficients of rth, $(r+1)$th, and $(r+2)$th terms in the binomial expansion of $(1+y)^m$ are in AP, then m and r satisfy the equation.

(a) $m^2-m(4r-1)+4r^2-2=0$ (b) $m^2-m(4r+1)+4r^2+2=0$

(c) $m^2-m(4r+1)+4r^2-2=0$ (d) $m^2-m(4r-1)+4r^2+2=0$

6 If $S_n=n^2a+\frac{n}{4}(n-1)d$ is the sum of first n terms of an AP, then common difference is equal to

(a) $a+2d$ (b) $2a+d$ (c) $\frac{a+d}{2}$ (d) $2a+\frac{d}{2}$

7 If $\frac{b+c-a}{a}, \frac{c+a-b}{b}, \frac{a+b-c}{c}$ are in AP, then $\frac{1}{a}, \frac{1}{b}, \frac{1}{c}$ are in

(a) GP (b) AP (c) AGP (d) None of these

8 If a^2, b^2, c^2 are in AP consider two statements

(i) $\frac{1}{b+c}, \frac{1}{c+a}, \frac{1}{a+b}$ are in AP (ii) $\frac{a}{b+c}, \frac{b}{c+a}, \frac{c}{a+b}$ are in AP

(a) (i) and (ii) both correct (b) (i) and (ii) both incorrect

(c) (i) correct (ii) incorrect (d) (i) incorrect (ii) correct

9 For any three positive real numbers a, b and c, $9(25a^2+b^2)+25(c^2-3ac)=15b(3a+c)$. Then **[JEE Main 2017]**

(a) a, b, c are in GP (b) a, b, c are in AP

(c) b, c, a are in GP (d) b, c, a are in AP

10 If 1, $\log_9(3^{1-x}+2)$, $\log_3(4\cdot 3^x-1)$ are in AP, then x equals

(a) $\log_3 4$ (b) $1-\log_3 4$ (c) $1-\log_4 3$ (d) $\log_4 3$

11 If $a_1, a_2, \ldots, a_n$ are in arithmetic progression, where $a_i > 0$ for all i, then the value of $\frac{1}{\sqrt{a_1}+\sqrt{a_2}}+\frac{1}{\sqrt{a_2}+\sqrt{a_3}}+\ldots+\frac{1}{\sqrt{a_{n-1}}+\sqrt{a_n}}$ is

(a) $\frac{n-1}{\sqrt{a_1}+\sqrt{a_{n-1}}}$
(b) $\frac{n-1}{\sqrt{a_1}+\sqrt{a_n}}$
(c) $\frac{n+1}{\sqrt{a_1}+\sqrt{a_{n-1}}}$
(d) $\frac{n+1}{\sqrt{a_1}+\sqrt{a_{n-1}}}$

12 The sum of each of two sets of three terms in AP is 15. The common difference of the first set is greater than that of the second by 1 and the ratio of the products of the terms in the first set and that of the second set is 7 : 8. The ratio of the smallest terms in two sets of term is

(a) $\frac{3}{4}$ or $\frac{11}{12}$
(b) $\frac{2}{3}$ or $\frac{1}{2}$
(c) $\frac{2}{3}$ or $\frac{3}{4}$
(d) None of these

13 The perimeter of a triangle whose sides are in AP is 21 cm and the product of length of the shortest side and the longest side exceeds the length of the other side by 6 cm. The longest side the triangle is

(a) 1 cm (b) 7 cm (c) 13 cm (d) None of these

14 Let $a_1, a_2, \ldots\ldots, a_{21}$ be an AP such that $\sum_{n=1}^{20} \frac{1}{a_n a_{n+1}} = \frac{4}{9}$. If the sum of this AP is 189, then $a_6 a_{16}$ is equal to **[JEE Main 2021]**

(a) 57 (b) 72 (c) 48 (d) 36

15 In the sum of the first $2n$ terms of the AP 2, 5, 8, ... is equal to the sum of the first n terms of the AP 57, 59, 61, ... then n equals

(a) 10 (b) 12 (c) 11 (d) 13

16 For an AP, $a_1, a_2, a_3, \ldots. a_n, a_1 = \frac{5}{2}, a_{10} = 16$. If $a_1 + a_2 + \ldots. + a_n = 110$. Then, n is equal to

(a) 9 (b) 10 (c) 11 (d) 12

17 If sum of the first 21 terms of the series $\log_{9^{1/2}} x + \log_{9^{1/3}} x + \log_{9^{1/4}} x + \ldots.$, where $x > 0$ is 504, then x is equal to **[JEE Main 2021]**

(a) 243 (b) 9 (c) 7 (d) 81

18 If S_n denotes the sum of n terms of an AP, then $S_{n+3} - 3S_{n+2} + 3S_{n+1} - S_n$ is equal to

(a) 0 (b) 1 (c) $\frac{1}{2}$ (d) 2

19 A person is to count 4500 currency notes. Let a_n denote the number of notes he counts in the nth minutes. If $a_1 = a_2 = \ldots. a_{10} = 150$ and $a_{10}\, a_{11}$ ...., are in AP with common difference -2, then the time taken by him to count all notes is

(a) 135 min
(b) 24 min
(c) 34 min
(d) 125 min

20 A man saves ₹ 200 in each of the first three months of his services. In each of the subsequent months his saving increases by ₹ 40 more than the saving of immediately previous month. His total saving from the start of service will be ₹ 11040 after

(a) 21 months
(b) 18 months
(c) 19 months
(d) 20 months

21 The interior angles of a convex non-equiangular polygon of 9 sides are in AP The least positive integer that limits the upper value of the common difference between the measures of the angles in degree is

(a) 9 (b) 10 (c) 11 (d) 12

22 The internal angles of a polygon are in AP. The smallest angle is 120° and the common difference is 5°. The number of sides of the polygon is

(a) 8 (b) 9 (c) 10 (d) 16

4.2 Geometric Progression

23 The first two terms of a geometric progression add upto 12. The sum of the third and the fourth term is 48. If the terms of the geometric progression are alternately positive and negative, then the first term is

(a) -4 (b) -12 (c) 12 (d) 4

24 a, b, c are three distinct real number and they are in a GP if $a+b+c=xb$, then

(a) $x \le -1$ or $x \ge 3$ (b) $x < -1$ or $x > 3$

(c) $x \le 1$ or $x > 3$ (d) $x < -3$ or $x > 2$

25 Suppose a, b, c are in AP and a^2, b^2, c^2 are in GP, if $a > b > c$ and $a+b+c=\frac{3}{2}$, then the value of a is

(a) $\frac{1}{2}+\frac{1}{\sqrt{2}}$ (b) $\frac{1}{2}-\frac{1}{\sqrt{2}}$

(c) $\frac{1}{\sqrt{2}}-\frac{1}{2}$ (d) $1+\frac{1}{\sqrt{2}}$

26 If $(a^2+b^2+c^2)p^2-2p(ab+bc+ca)+(a^2+b^2+c^2)\le 0$, where a, b, c, p are non-zero numbers, then a, b, c are in

(a) AP (b) GP (c) AGP (d) None of these

27 In the quadratic equation $ax^2+bx+c=0$, $\Delta=b^2-4ac$ and $\alpha+\beta, \alpha^2+\beta^2, \alpha^3+\beta^3$ are in GP, where α, β are the root of the $ax^2+bx+c=0$, then

(a) $\Delta \ne 0$ (b) $b\Delta=0$ (c) $c\Delta=0$ (d) $\Delta=0$

28 If $\log a, \log b, \log c$ are in AP and also $\log a-\log 2b, \log 2b-\log 3c, \log 3c-\log a$ are in AP, then

(a) a, b, c are in HP (b) $a, 2b, 3c$ are in AP

(c) a, b, c are the sides of a triangle (d) None of these

29 Let α, β be the roots of $x^2-x+p=0$ and γ, δ be the roots of $x^2-4x+q=0$ if $\alpha, \beta, \gamma, \delta$ are in GP, then the integral values of p and q respectively, are

(a) $-2, -32$ (b) $-2, 3$ (c) $-6, 3$ (d) $-6, -32$

30 If a, b, c, d are in GP, then

(a) $a+b, b+c, c+d$ are in GP

(b) $(b-c)^2+(c-a)^2+(d-b)^2=(a-d)^2$

(c) $(a^2+b^2+c^2)(b^2+c^2+d^2)=(ab+bc+cd)^2$

(d) All are correct

31 If the pth, qth and rth terms of both AP and GP be respectively a, b, c, then $a^{b-c}\cdot b^{c-a}\cdot c^{a-b}$ is equal to

(a) 1 (b) abc (c) pqr (d) $a^p b^q c^r$

32 If a, b, c are in AP, $\frac{1}{p}, \frac{1}{q}, \frac{1}{r}$ be in AP and ap, bq, cr be in GP, then $\frac{p}{r}+\frac{r}{p}$ is equal to

(a) $\frac{a^2+c^2}{ac}$ (b) $\frac{2ac}{a+c}$ (c) $\frac{1}{a}+\frac{1}{c}$ (d) $\frac{a+c}{2ac}$

33 If x, y and z are pth, qth are rth terms respectively of an AP and also of a GP, then $x^{y-z}y^{z-x}z^{x-y}$ is equal to

(a) xyz (b) 0 (c) 1 (d) None of these

34 If $0 < x < 1$, then $\frac{3}{2}x^2+\frac{5}{3}x^3+\frac{7}{4}x^4+......$, is equal to **[JEE Main 2021]**

(a) $x\left(\frac{1+x}{1-x}\right)+\log_e(1-x)$ (b) $x\left(\frac{1-x}{1+x}\right)+\log_e(1-x)$

(c) $\frac{1-x}{1+x}+\log_e(1-x)$ (d) $\frac{1+x}{1-x}+\log_e(1-x)$

35 The pollution in a normal atmosphere is less, than 0.01%. Due to leakage of a gas from a factory, the pollution is increased to 20%. If everyday 80% of the pollution is neutralised, in how many days the atmosphere will be normal?

(a) 3 (b) 4 (c) 5 (d) 6

36 A series is such that its every even term is 'a' times the term before it and every odd term is 'c' times the term before it. The sum of 2n term of the series is (the first term is unity)

(a) $\frac{(1-c^n)(1-a^n)}{1-ac}$ (b) $\frac{(1+a)(1-c^na^n)}{1-ac}$ (c) $\frac{(1+C^n)(1+a^n)}{1-ac}$ (d) $\frac{(1+a)(1+c^na^n)}{1+ac}$

37 Bijay writes letters to four of his friends. He asked each of them to copy the letter and mail to four different persons with the request that they continue the chain similarly. Assuming that the chain is not broken and that it costs 25 paise to mail one letter, the total money spent on postage till the 8th set of letter is mailed is

(a) ₹ 18245 (b) ₹ 21845 (c) ₹ 18522 (d) ₹ 21854

38 The sum of the 20 terms of the sequence 0.7, 0.77, 0.777, ... is **[JEE Main 2013]**

(a) $\frac{7}{81}(179-10^{-20})$ (b) $\frac{7}{9}(99-10^{-20})$ (c) $\frac{7}{81}(179+10^{-20})$ (d) $\frac{7}{9}(99+10^{-20})$

39 Select the correct statement

(a) $0.5+0.55+0.555+....$ to n terms $=\frac{5n}{9}-\frac{5}{81}(1-10^{-n})$

(b) $8+88+888+....$ to n terms $=\frac{80}{81}(10^n-1)-\frac{8n}{9}$

(c) $1^2+(1^2+2^2)+(1^2+2^2+3^2)+....$ to n terms $\frac{n(n+1)^2(n+2)}{12}$

(d) All are correct

40 Sum of infinite number of terms of GP is 20 and sum of their square is 100. The common ratio of GP is

(a) 5 (b) $\frac{3}{5}$ (c) $\frac{8}{5}$ (d) $\frac{1}{5}$

41 If $S_1, S_2, S_3, ..., S_p$ be the sum of infinite geometric series whose first terms are 1, 2, 3, ..., p and whose common ratio are $\frac{1}{2}, \frac{1}{3}, \frac{1}{4}, ..., \frac{1}{p+1}$ respectively, then

$S_1+S_2+S_3+...+S_p=........$ is

(a) $\frac{p(p+1)}{2}$ (b) $\frac{p(p+2)}{2}$ (c) $\frac{p(p+3)}{2}$ (d) $\frac{p(p-1)}{2}$

42 An infinite GP has first term 'x' and sum is '5' then x belongs to

(a) $x < -10$ (b) $-10 < x < 0$ (c) $0 < x < 10$ (d) $x > 10$

43 Consider an infinite geometric series with first term a and common ratio r, if its sum is 4 and the second term is 3/4, then

(a) $a = \frac{4}{7}, r = \frac{3}{7}$ (b) $a = 2, r = \frac{3}{8}$ (c) $a = \frac{3}{2}, r = \frac{1}{2}$ (d) $a = 3, r = \frac{1}{4}$

44 The rational number, which equals the number $2.\overline{357}$ with recurring decimal is

(a) $\frac{2355}{1001}$ (b) $\frac{2379}{997}$ (c) $\frac{2355}{999}$ (d) None of these

45 Let $x = 1 + a + a^2 + \dots$ to ∞

$y = 1 + b + b^2 + \dots$ to ∞

If $1 + ab + a^2b^2 + \dots$ to $\infty = S$, then S is equal to

(a) $x + y$ (b) xy (c) $\frac{x+y}{xy-1}$ (d) $\frac{xy}{x+y-1}$

$[0 < a < 1, 0 < b < 1]$

46 The first term of an infinite GP is 1 and each term is twice the sum of the succeeding terms then the sum of the series is

(a) 2 (b) 3 (c) $\frac{3}{2}$ (d) $\frac{5}{2}$

47 If $a = 1 + (\sqrt{3}-1) + (\sqrt{3}-1)^2 + (\sqrt{3}-1)^3 + \dots$ and $ab = 1$, then a and b are roots at the equation

(a) $x^2 + 4x - 1 = 0$ (b) $x^2 - 4x - 1 = 0$

(c) $x^2 + 4x + 1 = 0$ (d) $x^2 - 4x + 1 = 0$

48 After striking the floor a certain ball rebounds $\frac{4}{5}$th of its height from which it has fallen. The total distance that the ball travels before coming to rest if it is gently released from a height of 120 m is

(a) 960 m (b) 1000 m (c) 1080 m (d) infinite

49 If $A = 1 + r^a + r^{2a} + r^{3a} + \dots \infty$ and $B = 1 + r^b + r^{2b} + r^{3b} + \dots \infty$, then $\frac{a}{b}$ is equal to

(a) $\log_n(A)$ (b) $\log_{1-n}(1-A)$

(c) $\log_{\frac{B-1}{B}}\left(\frac{A-1}{A}\right)$ (d) None of these

4.3 Arithmetic and Geometric Mean

50 The arithmetic mean of 7 consecutive integers starting with a is m. Then, the arithmetic mean of 11 consecutive integers starting with $a+2$ is

(a) $2a$ (b) $2m$ (c) $a+4$ (d) $m+4$

51 If AM and GM of x and y are in the ratio $p:q$, then $x:y$ is

(a) $p:q$ (b) $p+\sqrt{p^2+q^2} : p-\sqrt{p^2-q^2}$

(c) $q+\sqrt{p^2-q^2} : q=\sqrt{p^2-q^2}$ (d) $p+\sqrt{p^2+q^2} : q-\sqrt{p^2-q^2}$

4.4 Arithmetic Geometric Progression

52 If the sum of infinity of the series $1+4x+7x^2+10x^3$ .... is $\frac{35}{16}$, then the value of x is

(a) $\frac{1}{3}$ (b) $\frac{1}{4}$ (c) $\frac{1}{5}$ (d) $\frac{1}{6}$

53 The value of $2^{1/4}\cdot 4^{1/8}\cdot 8^{1/16}\ldots\infty$ is

(a) 1 (b) 2 (c) $\frac{3}{2}$ (d) $\frac{1}{3}$

54 The sum of infinite series of $1+\frac{4}{5}+\frac{7}{5^2}+\frac{10}{5^3}+\ldots$ is

(a) $\frac{16}{35}$ (b) $\frac{35}{16}$ (c) $\frac{16}{39}$ (d) $\frac{39}{16}$

4.5 Summations of Series

55 The sum of the series $\frac{1}{1\cdot 2}-\frac{1}{2\cdot 3}+\frac{1}{3\cdot 4}\ldots$ up to ∞ is equal to

(a) $\log_e\left(\frac{4}{e}\right)$ (b) $2\log_e 2$ (c) $\log_e 2-1$ (d) $\log_e 2$

56 The sum of the first n terms of the series $1^2+2\cdot 2^2+3^2+2\cdot 4^2+5^2+2\cdot 6^2+\ldots$ is $\frac{n(n+1)^2}{2}$ when n is even. When n is odd the sum is

(a) $\left[\frac{n(n+1)}{2}\right]^2$ (b) $\frac{n^2(n+1)}{2}$ (c) $\frac{n(n+1)^2}{4}$ (d) $\frac{3n(n+1)}{2}$

57 $1^3-2^3+3^3-4^3+\ldots.+9^3$ is equal to

(a) 425 (b) – 425 (c) 475 (d)– 475

58 If $S_n=\sum_{r=0}^{n}\frac{1}{{}^nC_r}$ and $t_n=\sum_{r=0}^{n}\frac{r}{{}^nC_r}$, then $\frac{t_n}{S_n}$ is equal to

(a) $\frac{2n-1}{2}$ (b) $\frac{1}{2}n-1$ (c) $n-1$ (d) $\frac{1}{2}n$

59 The value of $\sum_{n=1}^{\infty}\frac{7n+32}{n(n+2)}\frac{3^n}{4^n}$ is

(a) $\frac{33}{2}$ (b) $\frac{144}{49}$ (c) $\frac{13}{7}$ (d) $\frac{25}{49}$

60 The limiting sum of infinite series $\frac{1}{10}+\frac{2}{10^2}+\frac{3}{10^3}+\ldots.$ whose nth term is $\frac{n}{10^n}$ is equal to

(a) $\frac{1}{9}$ (b) $\frac{10}{81}$ (c) $\frac{1}{8}$ (d) $\frac{17}{72}$

61 Sum of the first n terms of the series $\frac{1}{2}+\frac{3}{4}+\frac{7}{8}+\frac{15}{16}+\ldots.$ is equal to

(a) 2^n-n-1 (b) $1-2^{-n}$ (c) $n+2^{-n}-1$ (d) 2^n+1

62 If in a series $S_n=an^2+bn+c$, where S_n denotes the sum of n terms, then

(a) the series is always arithmetic
(b) the series is arithmetic from the second term onwards
(c) the series may or may not be arithmetic
(d) the series cannot be arithmetic

63 The 20th terms of the series $2+3+5+9+16+\dots$ is

(a) 950 (b) 975 (c) 990 (d) 1010

64 The maximum sum of the series $20+19\frac{1}{3}+18\frac{2}{3}+18+\dots$ is

(a) 300 (b) 310 (c) $311\frac{2}{3}$ (d) $333\frac{1}{3}$

65 $\sum_{K=1}^{n} K\left(1+\frac{1}{n}\right)^{(K-1)}$ is equal to

(a) $n(n-1)$ (b) $n(n+1)$ (c) n^2 (d) $(n+1)^2$

66 Let $S_n (1 \le n \le 9)$ denotes the sum of n terms of series $1+22+333+\dots+9999999999$, then for $2 \le n \le 9$

(a) $S_n - S_{n-1} = \frac{1}{9}(10^n - n^2 + n)$ (b) $S_n = \frac{1}{9}(10^n - n^2 + 2n - 2)$

(c) $9(S_n - S_{n-1}) = n(10^n - 1)$ (d) None of these

67 The sum of an infinite geometric series is 2 and the sum of the geometric series made from the cubes of this infinite series is 24, then the series is

(a) $3+\frac{3}{2}-\frac{3}{4}+\frac{3}{8}.....$ (b) $3+\frac{3}{2}+\frac{3}{4}+\frac{3}{8}+....$

(c) $3-\frac{3}{2}+\frac{3}{4}-\frac{3}{8}+....$ (d) None of these

68 Ths sum of $\frac{\frac{1}{2}\cdot\frac{2}{2}}{1^3}+\frac{\frac{2}{2}\cdot\frac{3}{2}}{1^3+2^3}+\frac{\frac{3}{2}\cdot\frac{4}{2}}{1^3+2^3+3^3}+\dots$ up to n terms is equal to

(a) $\frac{n-1}{n}$ (b) $\frac{n}{n+1}$ (c) $\frac{n+1}{n+2}$ (d) $\frac{n+1}{n}$

69 The sum of the series $1+\frac{1}{4\cdot 2!}+\frac{1}{16\cdot 4!}+\frac{1}{64\cdot 6!}+\dots\infty$. is

(a) $\frac{e-1}{\sqrt{e}}$ (b) $\frac{e+1}{\sqrt{e}}$ (c) $\frac{e-1}{2\sqrt{e}}$ (d) $\frac{e+1}{2\sqrt{e}}$

70 There are 25 trees at equal distances of 5 metres in a line with a well, the distance of the well from the nearest tree being 10 m. A gardener waters all the traces separately starting from the well and he returns to the well after waking each tree to get water for the next. The total distance the gardener will cover in order to water all the tree is

(a) 3550 m (b) 3434 m (c) 3370 m (d) 3200 m

71 In the sum of the first n terms of an AP is cn^2, then the sum of squares of these n terms is

(a) $\frac{n(4n^2-1)c^2}{6}$ (b) $\frac{n(4n^2+1)c^2}{3}$

(c) $\frac{n(4n^2-1)c^2}{3}$ (d) $\frac{n(4n^2+1)c^2}{6}$

72 Let $a_1, a_2, a_3, \dots, a_{49}$ be in AP such that $\sum_{k=0}^{12} a_{4k+1} = 416$ and $a_9 + a_{43} = 66$. If $a_1^2 + a_2^2 + \dots + a_{17}^2 = 140\,m$, then m is equal to **[JEE Main 2018]**

(a) 33 (b) 66 (c) 68 (d) 34

4.6 Inequalities

73 Two consecutive numbers from 1, 2, 3, ..., n are removed. The arithmetic mean of the remaining number is $\frac{105}{4}$. Then, the value of n lies in

(a) [45, 50] (b) [52, 60] (c) [41, 49] (d) None of these

74 If p and q are positive real numbers such that $p^2 + q^2 = 1$, then the maximum value of $(p + q)$ is

(a) $\frac{1}{2}$ (b) $\frac{1}{\sqrt{2}}$ (c) $\sqrt{2}$ (d) 2

75 x and y are positive number. Let g and a be GM and AM of these numbers. Also let G be GM of $x + 1$ and $y + 1$. If G and g are roots of equation $x^2 - 5x + 6 = 0$, then

(a) $x = 2, y = \frac{3}{4}$ (b) $x = \frac{3}{4}, y = 12$ (c) $x = \frac{5}{2}, y = \frac{8}{5}$ (d) $x = y = 2$

76 If m is the AM of two distinct real numbers and $n(l, n > 1)$ and G_1, G_2 and G_3 are geometric means between l and n, then $(G_1)^4 + 2(G_2)^4 + (G_3)^4$ equals

(a) $4l^2mn$ (b) $4lm^2n$ (c) $4lmn^2$ (d) $4l^2m^2n^2$

Numerical Value Type Questions

77 Let $\{a_n\}$ consists of positive numbers and for any positive integer n, $\frac{a_n + 2}{2} = \sqrt{2S_n}$, where $S_n = \sum_{i=1}^{n} a_i$, then the value of $\frac{a_{14} - a_{13}}{5}$ is ....

78 Suppose that all terms of an AP are natural numbers. If the ratio of the sum of first seven terms to the sum of first eleven term is 6 : 11 and the seventh term lies in between 130 and 140, then if the common difference is d, then $\frac{d}{4}$ is equal to ....

79 Let $a_1, a_2, a_3, \ldots, a_{100}$ be an arithmetic progression, with $a_1 = 3$ and $S_p = \sum_{i=1}^{p} a_i, 1 \le p \le 100$.

For any integer n with $1 \le n < 20$, let $m = 5n$. If $\frac{S_m}{S_n}$ does not depend on n, then a_2 is equal to ....

[JEE Main 2011]

80 The 5th and 8th terms of a geometric sequence of real numbers are 7! and 8! respectively. If the sum of first n terms of the GP is 2205, then n is equal to ....

81 For $k \in N$, let $\frac{1}{\alpha(\alpha + 1)(\alpha + 2) \ldots (\alpha + 20)} = \sum_{k=0}^{20} \frac{A_k}{\alpha + k}$, where $\alpha > 0$. Then, the value of $100\left(\frac{A_{14} + A_{15}}{A_{13}}\right)^2$ is equal to ....

[JEE Main 2021]

82 Let $a_n = 16, 4, 1, \ldots$ be a geometric sequence. Define P_n as the product of first n terms. The value of $\sum_{n=1}^{\infty} \sqrt[n]{P_n}$ is ....

83 Let $a_1, a_2, a_3, \ldots, a_{101}$ are in GP, with $a_{101} = 25$ and $\sum_{i=1}^{201} a_i = 625$. Then, the value of $\sum_{i=1}^{201} \frac{1}{a_i}$ is ....

84 The sum of infinite arithmetic geometric progression 3, 4, 4, ... is ....

85 The sum $\frac{7}{2^2 \times 5^2} + \frac{13}{5^2 \times 8^2} + \frac{19}{8^2 \times 11^2} + ...$ 10 terms is S, then the value of 1024 S is ....

86 The value of the sum $\sum_{i=1}^{20} i\left(\frac{1}{i} + \frac{1}{i+1} + ... + \frac{1}{20}\right)$ is ....

87 The value of $\sum_{n=0}^{\infty} \frac{2n+3}{3^n}$ is equal to ....

88 Let $S_k, k = 1, 2, ..., 100$, denote the sum of the infinite geometric series whose first term is $\frac{k-1}{k!}$ and the common ratio is $\frac{1}{k}$, then the value of $\frac{100^2}{100!} + \sum_{k=2}^{100} |(k^2 - 3k + 1)S_k|$ is ....

89 For any odd integer $n \geq 1$
$n^3 - (n-1)^3 + ... (-1)^{n-1} 1^3 = A(n+1)^2 (2n-1)$, then the value of A is ....

90 For $x \geq 0$, the smallest value of the function $f(x) = \frac{4x^2 + 8x + 13}{6(1+x)}$, is ....

91 If x, y and z are positive real numbers and $x = \frac{12 - yz}{y + z}$. The maximum value of (xyz) is ....

92 Given that x, y, z are positive reals such that $xyz = 32$. The minimum value of $x^2 + 4xy + 4y^2 + 2z^2$ is ....

93 If $x, y \in IR^+$ satisfying $x + y = 3$, then the maximum value of x^2y is ....

94 The minimum value of the sum of real numbers $a^{-5}, a^{-4}, 3a^{-3}, 1, a^8, a^{10}$ with $a > 0$ is ....

95 Given, a, b, c, d, e are such that $a + b + c + d + e = 8$ and $a^2 + b^2 + c^2 + d^2 + e^2 = 16$, then the greatest value of e is ....

LEVEL 03\\ JEE Advanced Level

Single Choice Correct Questions

1 If $a_1, a_2, ... a_n$ be in AP with common difference d and $\sec a_1 \sec a_2 + \sec a_2 \sec a_3 + \sec a_3 \sec a_4 .... + \sec a_{n-1} \sec a_n = A(\tan a_n - \tan a_1)$, then A equal to

(a) $\sin d$ (b) $\cos d$ (c) $\operatorname{cosec} d$ (d) $\sin d \cos d$

2 Suppose α, β are roots of $ax^2 + bx + c = 0$ and γ and δ are roots of $Ax^2 + Bx + C = 0$. If $\alpha, \beta, \gamma, \delta$ are in AP, then common difference of AP is

(a) $\frac{1}{4}\left(\frac{b}{a} - \frac{B}{A}\right)$ (b) $\frac{1}{3}\left(\frac{b}{a} - \frac{B}{A}\right)$ (c) $\frac{1}{2}\left(\frac{c}{a} - \frac{B}{A}\right)$ (d) $\frac{1}{3}\left(\frac{c}{a} - \frac{C}{A}\right)$

3 Let $b_i > 1$ for $i = 1, 2, ..., 101$. Suppose $\log_e b_1, \log_e b_2, ... \log_e b_{101}$, are in AP with the common difference $\log_e 2$. Suppose $a_1, a_2, ..., a_{101}$ are in A.P such that $a_1 = b_1$ and $a_{51} = b_{51}$. If $t = b_1 + b_2 + ... + b_{51}$ and $s = a_1 + a_2 + ... + a_{51}$, then **[JEE Adv. 2016]**

(a) $s > t$ and $a_{101} > b_{101}$ (b) $s > t$ and $a_{101} < b_{101}$
(c) $s < t$ and $a_{101} > b_{101}$ (d) $s < t$ and $a_{101} < b_{101}$

4 Let $a_1, a_2, \ldots$ be in AP and $q_1, q_2, \ldots$ be in GP. If $a_1 = q_1 = 2$ and $a_{10} = q_{10} = 3$, then

(a) $a_7 \cdot a_{19}$ is not an integer (b) $a_{19} q_7$ is an integer
(c) $a_7 q_{19} = a_{19} q_{10}$ (d) None of these

5 In a GP of positive terms, if any term is equal to the sum of the next two terms, then the common ratio of the GP is

(a) $\sin 18°$ (b) $2\cos 18°$ (c) $\cos 18°$ (d) $2\sin 18°$

6 Let S be the sum, P be the product and R be the sum of the reciprocals of 3 terms of a GP, then $P^2R^3 : S^3$ is equal to

(a) 1 : 1 (b) common ratio : 1
(c) (first term)^2 : (common ratio)^2 (d) (common ration)^n : 1

7 If a, b, c, d are in GP, then $(a^3 + b^3)^{-1}, (b^3 + c^3)^{-1}, (c^3 + d^3)^{-1}$ are in

(a) AP (b) GP (c) AGP (d) None of these

8 If $S_1, S_2, S_3, \ldots. S_n$ are the sum of infinite geometric series whose first terms are 1, 2, 3, ... n and whose common ratio are $\frac{1}{2}, \frac{1}{3}, \frac{1}{4}, \ldots, \frac{1}{(n+1)}$ respectively, then the value of $S_1^2 + S_2^2 + S_3^2 + \ldots. + S_{2n-1}^2$ is equal to

(a) $\frac{1}{3}[n(2n+1)(4n+1) - 3]$ (b) $\frac{1}{3}[n(2n+1)(4n+1) + 3]$
(c) $\frac{1}{3}[n(2n-1)(4n+1) - 3]$ (d) None of these

9 If the sum of the series $\sum_{n=1}^{\infty} r^n, |r| < 1$, is S, then $\sum_{n=1}^{\infty} r^{2n}$ is

(a) S^2 (b) $\frac{2S}{S^2 - 1}$ (c) $\frac{S^2}{2S + 1}$ (d) None of these

10 If $|r| > 1$ and $x = a + \frac{a}{r} + \frac{a}{r^2} + \ldots$ to ∞, $y = b - \frac{b}{r} + \frac{b}{r^2} \ldots$ to ∞ and $z = c + \frac{c}{r^2} + \frac{c}{r^4} + \ldots.$ to ∞, then $\frac{xy}{z}$ is equal to

(a) $\frac{ab}{c}$ (b) $\frac{ac}{b}$ (c) $\frac{bc}{a}$ (d) None of these

11 If $\frac{b+c}{a+d} = \frac{bc}{ad} = 3\left(\frac{b-c}{a-d}\right)$, then $\pi^{\frac{1}{a}-1}, \pi^{\frac{1}{b}-1}, \pi^{\frac{1}{c}-1}, \pi^{\frac{1}{d}-1}$ are in

(a) AP (b) GP (c) AGP (d) None of these

12 If a, b, x, y are positive rational numbers such that $\frac{1}{x} + \frac{1}{y} = 1$, then $\left(\frac{a^x}{x} + \frac{b^y}{y} - ab\right)$ is

(a) < 0 (b) > 1 (c) ≥ 0 (d) None of these

13 Sum of 20 terms of the series $1 \cdot 3^2 + 2 \cdot 5^2 + 3 \cdot 7^2 + \ldots.$ is

(a) 178090 (b) 168090 (c) 188090 (d) 190090

14 If in a series $t_n = \frac{n}{(n+1)!}$, then $\sum_{n=1}^{20} t_n$ is equal to

(a) $\frac{20! - 1}{20!}$ (b) $\frac{21! - 1}{21!}$ (c) $\frac{1}{2(20-1)!}$ (d) None of these

15 The value of $\frac{1}{1\cdot3\cdot5}+\frac{1}{3\cdot5\cdot7}+\frac{1}{5\cdot7\cdot9}+\frac{1}{9\cdot11\cdot13}$ equals

(a) $\frac{70}{249}$ (b) $\frac{53}{249}$ (c) $\frac{35}{429}$ (d) $\frac{31}{429}$

16 The value of $\lim_{n\to\infty}\left[\frac{3}{1^2}+\frac{5}{1^2+2^2}+\frac{7}{1^2+2^2+3^2}+\frac{9}{1^2+2^2+3^2+4^2}+\dots n \text{ terms}\right]$

(a) 6 (b) 7 (c) 8 (d) 9

17 Let S_n be the sum of the series $\frac{8}{5}+\frac{16}{65}+\frac{24}{325}+\dots$, then the value of $\lim_{n\to\infty} S_n$ is

(a) 0 (b) $\frac{1}{2}$ (c) 2 (d) 4

18 Sum of the series $\frac{1}{1+1^2+1^4}+\frac{2}{1+2^2+2^4}+\frac{3}{1+3^2+3^4}+\dots$ upto n terms is equal to

(a) $\frac{n^2+n-1}{2(n^2+n+1)}$ (b) $\frac{n^2+n}{2(n^2+n+1)}$ (c) $\frac{n^2-n+1}{n^2+n+1}$ (d) $\frac{n^2-n}{2(n^2+n+1)}$

19 Consider the pattern shown below

Row 1 1 2

Row 2 2 3 5

Row 3 3 7 9 11

Row 4 4 13 15 17 19

The number at the end of row 60 is

(a) 3659 (b) 3519 (c) 3681 (d) 3731

20 $\sum_{r=1}^{n} r^2 - \sum_{m=1}^{n}\sum_{r=1}^{m} r$ is equal to

(a) 0 (b) $\frac{1}{2}\left(\sum_{r=1}^{n} r^2+\sum_{r=1}^{n} r\right)$ (c) $\frac{1}{2}\left(\sum_{r=1}^{n} r^2-\sum_{r=1}^{n} r\right)$ (d) None of these

21 If $a+b+c=3$ and $a>0, b>0, c>0$, then the greatest value of $a^2b^3c^2$ is

(a) $\frac{3^{10}\cdot2^4}{7^7}$ (b) $\frac{3^9\cdot2^4}{7^7}$ (c) $\frac{3^8\cdot2^4}{7^7}$ (d) None of these

More Than One Correct

22 Which of the following can be terms (not necessarily consecutive) of any AP

(a) 1, 6, 19 (b) $\sqrt{2}, \sqrt{50}, \sqrt{98}$

(c) log2, log16, log128 (d) $\sqrt{2}, \sqrt{3}, \sqrt{7}$

23 For an increasing AP $a_1, a_2, \dots a_n$ if $a_1+a_3+a_5=-12$ and $a_1a_3a_5=80$, then which of the following is/are true?

(a) $a_1=-10$ (b) $a_2=-1$ (c) $a_3=-4$ (d) $a_5=2$

24 If the sum of n terms of an AP is given by $S_n=a+bn+cn^2$, where a, b, c are independent of n, then

(a) $a=0$ (b) common difference of AP must be $2b$

(c) common difference of AP must be $2c$ (d) first term of AP is $b+c$

25 If a, b, c and d are four unequal positive numbers which are in AP, then

(a) $\frac{1}{a}+\frac{1}{d}>\frac{1}{b}+\frac{1}{c}$ (b) $\frac{1}{a}+\frac{1}{d}<\frac{1}{b}+\frac{1}{c}$ (c) $\frac{1}{b}+\frac{1}{c}>\frac{4}{a+d}$ (d) $\frac{1}{a}+\frac{1}{d}=\frac{1}{b}+\frac{1}{c}$

26 Let $a_1, a_2, a_3, \ldots, a_n$ be in GP such that $3a_1+7a_2+3a_3-4a_5=0$, then common ratio of GP can be

(a) 2 (b) $\frac{3}{2}$ (c) $\frac{5}{2}$ (d) $\frac{-1}{2}$

27 The next term of the GP x, x^2+2 and x^3+10 is

(a) $\frac{729}{16}$ (b) 6 (c) 0 (d) 54

28 If $P(x)=\frac{1+x^2+x^4+\ldots+x^{2n-2}}{1+x+x^2+\ldots+x^{n-1}}$ is a polynomial in x, then n can be

(a) 5 (b) 10 (c) 20 (d) 17

29 If sum of an infinite GP, $p, 1, \frac{1}{p}, \frac{1}{p^2}, \ldots$ is $\frac{9}{2}$, then value of p is

(a) 2 (b) $\frac{3}{2}$ (c) 3 (d) $\frac{9}{2}$

30 The terms of an infinitely decreasing GP having common ratio r in which all the terms are positive, the first term is 4, and the difference between the third and fifth term is 32/81, then

(a) $r=\frac{1}{3}$ (b) $r=\frac{2\sqrt{2}}{3}$

(c) sum of infinite term is 6 (d) None of these

31 If $1+2x+3x^2+4x^3+\ldots\infty\geq 4$ then,

(a) least value of x is 1/2 (b) greatest value of x is 4/3

(c) least value of x is $\frac{2}{3}$ (d) greatest value of x does not exist

32 In an arithmetic progression whose first term is α and common difference is β, $\alpha, \beta \neq 0$, the ratio r of the sum of the first n terms to the sum of n terms succeeding them, does not depend on n. Then, which of the following is/are correct.

(a) $\alpha:\beta=2:1$

(b) if α and β are roots of the equation $ax^2+bx+c=0$, then $2b^2=9ac$

(c) the sum of infinite GP, $1+r+r^2+\ldots$ is $\frac{3}{2}$

(d) if $\alpha=1$, then sum of 10 terms of AP is 100

33 If $\frac{1}{a}, \frac{1}{b}, \frac{1}{c}$ are in AP and $a, b, -2c$ are in GP. where a, b, c are non zero then,

(a) $a^3+b^3+c^3=3abc$ (b) $-2a, b, -2c$ are in AP

(c) $a^2, b^2, 4c^2$ are in GP (d) roots of the equation $ax^2+bx+c=0$ are real

34 The numbers 1, 4, 16 can be three terms (not necessarily consecutive) of

(a) no AP (b) only one GP

(c) infinite number of AP's (d) infinite number of GP's

35 For $0<\phi<\frac{\pi}{2}$, if $x=\sum_{n=0}^{\infty}\cos^{2n}\phi$, $y=\sum_{n=0}^{\infty}\sin^{2n}\phi$ and $z=\sum_{n=0}^{\infty}\cos^{2n}\phi\sin^{2n}\phi$, then

(a) $xyz=xz+y$ (b) $xyz=xy+z$ (c) $xyz=x+y+z$ (d) $xyz=yz+x$

36 Sum of infinite GP is 2 and sum of its first two term is 1. If its second term is negative, then which of the following is/are true?

(a) One of the positive values of the first term is $(2-\sqrt{2})$

(b) One of the possible values of the first term is $(2+\sqrt{2})$

(c) One of the possible values of the common ratio is $\sqrt{2}-1$

(d) One of the possible values of the common ratio is $\frac{1}{\sqrt{2}}$

37 Let S_1, S_2 .... be squares such that for each $n \geq 1$, the length of a side of S_n equals the length of a diagonal of S_{n+1}, if the length of a side of S_1 is 1 cm, then for which of the following values of n is the area of S_n less than 1 sq cm.

(a) 7 (b) 8 (c) 9 (d) 10

38 If a, b, c are in GP and x, y respectively, be arithmetic means between a, b and b, c, then

(a) $\frac{a}{x}+\frac{c}{y}=2$ (b) $\frac{a}{x}+\frac{c}{y}=\frac{c}{a}$ (c) $\frac{1}{x}+\frac{1}{y}=\frac{2}{b}$ (d) $\frac{1}{x}+\frac{1}{y}=\frac{2}{ac}$

39 For the series,

$$S=1+\frac{1}{1+3}(1+2)^2+\frac{1}{(1+3+5)}(1+2+3)^2+\frac{1}{(1+3+5+7)}(1+2+3+4)^2+....$$

(a) 7th term is 6 (b) 7th term is 16

(c) Sum of first 10 terms is $\frac{505}{4}$ (d) Sum of first 10 term is $\frac{405}{4}$

40 If $\sum_{r=1}^{n} r(r+1)(2r+3)=an^4+bn^3+cn^2+dn+e$, then

(a) $a-b=d-c$ (b) $e=0$ (c) $a, b-\frac{2}{3}, c-1$ are in AP (d) $\frac{(b+d)}{a}$ is an integer

41 Let $S_n=\sum_{k=1}^{4n}(-1)^{\frac{k(k+1)}{2}}k^2$. Then, S_n can take value(s) **[JEE Adv. 2013]**

(a) 1056 (b) 1088 (c) 1120 (d) 1332

42 A straight line through the vertex P of a ΔPQR intersects the side QR at the point S and the circumcircle of the ΔPQR at the point T. If S is not the centre of the circumcircle, then

(a) $\frac{1}{PS}+\frac{1}{ST}<\frac{2}{\sqrt{QS \times QR}}$ (b) $\frac{1}{PS}+\frac{1}{ST}>\frac{2}{\sqrt{QS \times SR}}$

(c) $\frac{1}{PS}+\frac{1}{ST}<\frac{4}{QR}$ (d) $\frac{1}{PS}+\frac{1}{ST}>\frac{4}{QR}$

Linked Comprehension Type

Paragraph I Directions (Q.Nos. 43-44)

Let $M=\{(x, y) \in R \times R : x^2+y^2 \leq r^2\}$, where $r>0$. Consider the geometric progression $a_n=\frac{1}{2^{n-1}}, n=1, 2, 3, ...$ Let $S_0=0$ and for $n \geq 1$, let S_n denote the sum of the first n terms of this progression. For $n \geq 1$, let C_n denote the circle with center $(S_{n-1}, 0)$ and radius a_n and D_n denote the circle with centre (S_{n-1}, S_{n-1}) and radius a_n. **[JEE Adv. 2021]**

43 Consider M with $r = \frac{1025}{513}$. Let k be the number of all those circles C_n that are inside M. Let l be the maximum possible number of circles among these k circles such that no two circles intersect. Then,

(a) $k + 2l = 22$ (b) $2k + l = 26$
(c) $2k + 3l = 34$ (d) $3k + 2l = 40$

44 Consider M with $r = \frac{(2^{199} - 1)\sqrt{2}}{2^{198}}$. The number of all those circles D_n that are inside M is

(a) 198 (b) 199
(c) 200 (d) 201

Paragraph II Directions (Q.Nos. 45-47)

The real roots of the equation $2x^3 - 19x^2 + 57x + k = 0$ are the first three terms of a geometric progression.

45 The value of k equals

(a) 216 (b) 108
(c) – 54 (d) – 108

46 If the geometric progression is increasing, then the sum of its first n terms equals

(a) $\left(\frac{3}{2}\right)^n - 1$ (b) $4\left(\left(\frac{3}{2}\right)^n - 1\right)$
(c) $6\left(\left(\frac{3}{2}\right)^n - 1\right)$ (d) $4(2^n - 1)$

47 If the geometric progression is decreasing, then the sum of its infinite number of term is

(a) $\frac{27}{2}$ (b) 9
(c) $\frac{9}{2}$ (d) 12

Matrix Match Type

48 If α and β are roots of the equation $x^2 - 8x + 4 = 0$, then match the following list

	List-I		List-II
(i)	If $\alpha, a_1, a_2, a_3, a_4, a_5, \beta$ are in AP then $\sum_{i=1}^{5} a_i$ is divisible by	A.	3
(ii)	If $\alpha, a_1, a_2, a_3, a_4, a_5, \beta$ in GP, then $\sum_{i=1}^{5} \log_2 a_i$ is divisible by	B.	4
(iii)	If $(\alpha)^2, (a_1)^2, (a_2)^2, (a_3)^2, (a_4)^2, (a_5)^2, \beta^2$ are in AP, then $\sum_{i=1}^{5} a_i^2$ is divisible by	C.	5
(iv)	If $1/\alpha, 1/a_1, 1/a_2, 1/a_3, 1/a_4, 1/a_5, 1/\beta$ are in AP, then $\left(\frac{10}{\sum_{i=1}^{5} \frac{1}{a_i}} + 1\right)$ is divisible by	D.	7

49 Match the columns

	Column-I		Column-II
(i)	The sequence $a, b, 10, c, d$ is an arithmetic progression, then the value of $a+b+c+d$ is	A.	10
(ii)	The sides of right triangle form a three term geometric sequence. The shortest side has length 2. The length of the hypotenuse is of the form $a+\sqrt{b}$ where $a \in N$ and $\sqrt{b}$ is surd, then a^2+b^2 equals	B.	20
(iii)	The sum of first three consecutive numbers of an infinite GP is 70. If the two extremes be multiplied each by 4 and the mean by 5. The products are in AP. The second term of the GP is	C.	26
(iv)	The diagonals of a parallelogram have a measure of 4m and 6 m. They cut off forming an angle of 60°. If the perimeter of the parallelogram is $2(\sqrt{a}+\sqrt{b})$ where $a, b \in N$, then$(a+b)$ equals	D.	40

50 Match the columns

	Column-I		Column-II
(i)	The equation $x^{x\sqrt{x}} = (x\sqrt{x})^x$ has two solutions in positive real number x. One obvious solution is $x=1$ the other one is	A.	8/3
(ii)	Suppose a triangle ABC is inscribed in a circle of radius 10 cm. If the perimeter of the triangle is 32 cm, then the value of $\sin A + \sin B + \sin C$	B.	9/4
(iii)	Sum of infinite series $1 + \frac{3}{4} + \frac{7}{16} + \frac{15}{64} + \frac{31}{256} + \ldots$ equals	C.	5/4
(iv)	The sum of $\Sigma\left(\frac{r+3}{r(r+1)(r+2)}\right)$ equals	D.	8/5

Numerical Value Type Questions

Question Stem Directions (Q.Nos. 51-53) The answer to each question is a **NUMERICAL VALUE.**

If the set of natural numbers is partitioned into subsets $S_1 = \{1\}$, $S_2 = \{2, 3\}$, $S_3 = \{4, 5, 6\}$, ...

51 Last term of S_{50} is ....

52 Sum of the elements in the subset S_{50} is ....

53 First term of subset S_7 is ....

Question Stem Directions (Q.Nos. 54-55) The answer to each question is a **NUMERICAL VALUE.**

Let $f(n)$ denote the nth term of the sequence 3, 6, 11, 18, 27, ... and $g(n)$ denote the nth term of the sequence 3, 7, 13, 21, ...

Let $F(N)$ and $G(N)$ denote the sum of n terms of the above sequences, respectively. Now answer the following.

54 $\lim_{n\to\infty} \frac{f(n)}{g(n)}$ is equal to ....

55 $\lim_{n\to\infty} \frac{F(n)}{G(n)}$ is equal to ....

56 The first four terms of a sequence are given by $T_1 = 0, T_2 = 1, T_3 = 1, T_4 = 2$. The general term of the sequence is given by $T_n = A\alpha^{n-1} + B\beta^{n-1}$, where A, B, α, β are independent of n and A is positive, then the value of $A^2 + B^2$ is ....

57 Let $a_1, a_2, a_3, \ldots, a_{11}$ be real numbers satisfying $a_1 = 15, 27/2 - a_2 > 0$ and $a_k = 2a_{k-1} - a_{k-2}$ for $k = 3, 4, \ldots, 11$. If $\frac{a_1^2 + a_2^2 + \ldots + a_{11}^2}{11} = 90$, then the value of $\frac{a_1 + a_2 + \ldots + a_{11}}{11}$ is equal to ....

58 The sides of a right angled triangle are in AP. If the triangle has area 24, then what is the length of the smallest side? **[JEE Adv. 2017]**

59 An arithmetic progression has the following property for any even number of terms, the ratio of the sum of first half of the terms to the sum of second half is always equal to a constant k. Let the first term of AP is 1, then the sum of all possible values of k is ....

60 Suppose that all the terms of an AP are natural numbers. If the ratio of the sum of the first seven terms to the sum of the first eleven terms is 6 : 11 and the seventh term lies in between 130 and 140, then the common difference of this AP is .... **[JEE Adv. 2015]**

61 If the coefficients of x in the expansion of $\sum_{r=1}^{110} (1 + rx)$ is $\lambda(1 + 10)(1 + 10 + 10^2)$, then the value of λ is ....

62 Let X be the set consisting of the first 2018 terms of the AP 1, 6, 11, ..., and Y be the set consisting of the first 2018 terms of the AP 9, 16, 23, ... Then, the sum of elements in the set $X \cup Y$ is .... **[JEE Adv. 2018]**

63 The sequence $a_1, a_2, \ldots$ is geometric sequence with common ratio r. The sequence $b_1, b_2, \ldots$ is also geometric sequence. If $b_1 = 1, b_2 = \sqrt[4]{7} - \sqrt[4]{28} + 1, a_1 = \sqrt[4]{28}$ and $\sum_{n=1}^{\infty} \frac{1}{a_n} = \sum_{n=1}^{\infty} b_n$, then the value of $(1 + r^2 + r^4)$ is ....

64 Let $F_0 = 0, F_1 = 1$ and $F_n = F_{n-1} + F_{n-2}, (n \geq 2)$, then the value of the sum $\sum_{n=1}^{\infty} \frac{F_n}{3^n}$ is ....

65 Let a, b, c be positive integers such that $\frac{b}{a}$ is an integer. If a, b, c are in GP and the arithmetic mean of a, b, c is $b + 2$, then the value of $\frac{a^2 + a - 14}{a + 1}$ is .... **[JEE Adv. 2014]**

66 Let the harmonic mean of two positive real numbers a and b be 4. If q is positive real number such that $a, 5, q, b$ is an AP, then the smallest value of $|q - a|$ is ....

67 If n is a positive integer satisfying the equation $2 + (6 \cdot 2^2 - 4 \cdot 2) + (6 \cdot 3^2 - 4 \cdot 3) + \ldots + (6n^2 - 4n) = 140$, then the value of n is ....

68 A pack contains n cards numbered from 1 to n. Two consecutive numbered cards are removed from the pack and the sum of the numbers on the remaining cards is 1224. If the smaller of the numbers on the removed cards is k, then $k - 20$ is equal to **[JEE Adv. 2013]**

69 Let x, y, z are positive real numbers satisfy $2x - 2y + \frac{1}{z} = \frac{1}{2018}$; $2y - 2z + \frac{1}{x} = \frac{1}{2018}$; $2z - 2x + \frac{1}{y} = \frac{1}{2018}$, then the value of $\frac{x + y + z}{2x}$ is .... ?

ANSWER KEY

Level 1

1. (i) $a_1 = 0, a_2 = 3, a_3 = 8, a_4 = 15, a_5 = 24$ (ii) $a_1 = 0, a_2 = 2, a_3 = 0, a_4 = 2, a_5 = 0$
(iii) $a_1 = 0, a_2 = 0, a_3 = 0, a_4 = 0, a_5 = 0$ (iv) $a_1 = -2, a_2 = 2, a_3 = -\frac{8}{3}, a_4 = 4, a_5 = \frac{-32}{5}$

2. (i) $1, -2, 3 - 4, 5 - 6, \ldots$ (ii) $\frac{3}{4}, 25, \frac{5}{6}, 625, \ldots$ (iii) $1, -\frac{1}{4}, \frac{1}{3}, -\frac{1}{16}, \frac{1}{5}, \ldots$

3. (i) $a_1 = 1, a_2 = 2, a_3 = 5, a_4 = 9$ (ii) $a_1 = 2, a_2 = 1/2, a_3 = 1/5, a_4 = 1/11$
(iii) $a_1 = 1, a_2 = 1, a_3 = 2, a_4 = 3$

4. (i) $\sin\left(\frac{n\pi}{2}\right)$ (ii) $a_n = \frac{(-1)^{n+1}}{2n-1}$ (iii) 5

5. (i) 62 (ii) 5 (iii) 86 (iv) 304

6. 1 7. 229 8. (1) 9. -3 10. 17, 15, 13

11. 3, 5, 7, 9 12. $P = -3$ and $Q = 77$ 13. $n = 17$ 14. 300 15. 199

16. $n^2 \ln 2$ 17. 9850 18. 900 19. 1 : 12 20. 9 21. 1011 22. 25 days

23. (i) $3(2)^6$ (ii) $\frac{9}{16}$ (iii) 6 (iv) $\frac{1}{2916}$

24. $\left(\frac{q^p}{p^q}\right)^{1/p-q}$ 25. 4^5 26. $\frac{3}{4}, -1, \frac{4}{3}$ if $r = -\frac{4}{3}$ or $\frac{4}{3}, -1, \frac{3}{4}$ if $r = -\frac{3}{4}$

27. 8, 16, 32 for $r = 2$ or 32, 16, 8 for $r = \frac{1}{2}$ 28. 2441468 29. $n = 10$

30. (i) $\frac{2}{81}[10(10^n - 1) - 9n]$ (ii) $\frac{2}{3}n - \frac{2}{27}(1 - 10^{-n})$

31. $2n + 1 - \frac{1}{2^n}$ 32. $n = 6$

33. (i) $\frac{1}{2}$ (ii) $\frac{2}{3}$

34. $\frac{3}{5}$

35. (i) 10 (ii) $\frac{4 + 3\sqrt{2}}{2}$ (iii) $\frac{19}{24}$

36. (i) $x = \frac{1}{3}$ (ii) $x = \frac{37}{30}$

37. 800 m 38. $n = 32$ 39. 3, 6, 9, 12, 15, 18, 21, 24 40. 4, 16, 64

41. 4, 8, 16, 32 if $r = 2$ and $a = 4$ or 32, 16, 8, 4 if $r = \frac{1}{2}$ and $a = 32$

42. (i) $1 + \frac{3}{4}\left[1 - \left(\frac{1}{5}\right)^{n-1}\right] - (3n-2)\left(\frac{1}{5}\right)^n$ (ii) $\frac{35}{16}$

43. $\frac{1}{(1-x)^3}$ 44. $\frac{20}{9}$ 45. $\frac{1+x}{(1-x)^3}$ 46. $n(n+2)(2n+1)$ 47. $\frac{n}{2}[8n^2 + 11n + 1]$

48. $\frac{7}{9}\left[\frac{n}{2} - \frac{1}{9}\left(1 - \frac{1}{10^{n/2}}\right)\right] + \frac{7}{90}\left[\frac{100}{9}(10^{n/2} - 1) - \frac{n}{2}\right]$

49. $\frac{n(n+1)(n+2)}{6}$

50. $(a^2b + b^2c + c^2a)(ab^2 + bc^2 + ca^2) \geq 9a^2b^2c^2$

51. $a = \frac{1}{4}, b = \frac{1}{4}$ and $c = \frac{1}{12}$

52. Yes 53. 8 54. 4

Level 2

1. (a)	**2.** (a)	**3.** (c)	**4.** (d)	**5.** (c)	**6.** (d)	**7.** (b)
8. (a)	**9.** (d)	**10.** (b)	**11.** (b)	**12.** (a)	**13.** (c)	**14.** (b)
15. (c)	**16.** (c)	**17.** (d)	**18.** (a)	**19.** (c)	**20.** (a)	**21.** (a)
22. (b)	**23.** (b)	**24.** (b)	**25.** (a)	**26.** (b)	**27.** (c)	**28.** (c)
29. (a)	**30.** (d)	**31.** (a)	**32.** (a)	**33.** (c)	**34.** (a)	**35.** (d)
36. (b)	**37.** (b)	**38.** (c)	**39.** (d)	**40.** (b)	**41.** (c)	**42.** (c)
43. (d)	**44.** (c)	**45.** (d)	**46.** (c)	**47.** (d)	**48.** (c)	**49.** (c)
50. (d)	**51.** (b)	**52.** (c)	**53.** (b)	**54.** (b)	**55.** (a)	**56.** (b)
57. (a)	**58.** (d)	**59.** (a)	**60.** (b)	**61.** (c)	**62.** (b)	**63.** (c)
64. (b)	**65.** (c)	**66.** (c)	**67.** (c)	**68.** (b)	**69.** (d)	**70.** (c)
71. (c)	**72.** (d)	**73.** (a)	**74.** (c)	**75.** (d)	**76.** (b)	**77.** (0.8)
78. (2.25)	**79.** (9)	**80.** (3)	**81.** (9)	**82.** (32)	**83.** (1)	**84.** (27)
85. (85)	**86.** (115)	**87.** (6)	**88.** (3)	**89.** (0.25)	**90.** (2)	**91.** (8)
92. (96)	**93.** (4)	**94.** (8)	**95.** (3.2)			

Level 3

1. (c)	**2.** (a)	**3.** (b)	**4.** (c)	**5.** (d)	**6.** (a)	**7.** (b)
8. (a)	**9.** (c)	**10.** (a)	**11.** (b)	**12.** (c)	**13.** (c)	**14.** (b)
15. (c)	**16.** (a)	**17.** (c)	**18.** (b)	**19.** (a)	**20.** (c)	**21.** (a)
22. (a, b, c)	**23.** (a, c, d)	**24.** (a, c, d)	**25.** (a, [illegible]	**26.** (b, d)	**27.** (a, d)	**28.** (a, d)
29. (b, c)	**30.** (a, b, c)	**31.** (a, d)	**32.** (b, d)	**33.** (a, b, c, d)	**34.** (c, d)	**35.** (b, c)
36. (a, b, d)	**37.** (b, c, d)	**38.** (a, c)	**39.** (b, [illegible]	**40.** (a, b, c, d)	**41.** (a, d)	**42.** (b, d)
43. (d)	**44.** (b)	**45.** (c)	**46.** (b)	**47.** (a)		

48. (i)–B (ii)–C, (iii)–B, C, D (iv)–A

49. (i)–D, (ii)–C, (iii)–B, (iv)–C

50. (i)–B, (ii)–D, (iii)–A, (iv)–C

51. (1275)	**52.** (62525)	**53.** (22)	**54.** (1)	**55.** (1)	**56.** (0.4)	**57.** (0)
58. (6)	**59.** (1.333)	**60.** (9)	**61.** (5)	**62.** (3748)	**63.** (7)	**64.** (0.6)
65. (4)	**66.** (2)	**67.** (4)	**68.** (5)	**69.** (1.5)		

SOLUTIONS

Level 1

1. (i) $a_n = n^2 - 1$

Replacing n with 1, 2, 3, 4 and 5 gives

$a_1 = 0, a_2 = 3, a_3 = 8, a_4 = 15, a_5 = 24$

(ii) $a_n = (-1)^n + 1$

$a_1 = (-1) + 1 = 0$

$a_2 = (-1)^2 + 1 = 1 + 1 = 2$

$a_3 = (-1)^3 + 1 = -1 + 1 = 0$

$a_4 = (-1)^4 + 1 = 1 + 1 = 2$

$a_5 = (-1)^5 + 1 = -1 + 1 = 0$

(iii) $a_n = \sin(n\pi)$

$a_1 = \sin\pi = 0$

$a_2 = \sin 2\pi = 0$

$a_3 = \sin 3\pi = 0$

$a_4 = \sin 4\pi = 0$

$a_5 = \sin 5\pi = 0$

(iv) $a_n = (-1)^n \dfrac{2^n}{n}$

$a_1 = (-1)^1 \dfrac{2}{1} = -2$

$a_2 = (-1)^2 \dfrac{2^2}{2} = \dfrac{4}{2} = 2$

$a_3 = (-1)^3 \dfrac{2^3}{3} = -\dfrac{8}{3}$

$a_4 = (-1)^4 \dfrac{2^4}{4} = 4$

$a_5 = (-1)^5 \dfrac{2^5}{5} = -\dfrac{32}{5}$

2. Putting $n = 1, 2, 3, 4, ...$ successively, we get

(i) $a_1 = 1 ; a_2 = -2$

$a_3 = 3 ; a_4 = -4$

$a_5 = 5 ; a_6 = -6$

....

Hence, we obtain a sequence

$1, -2, 3, -4, 5, -6, ...$

(ii) $a_1 = \dfrac{1+2}{1+3} = \dfrac{3}{4}$

$a_2 = 5^2 = 25$

$a_3 = \dfrac{3+2}{3+3} = \dfrac{5}{6}$

$a_4 = 5^4 = 625$

Hence, the sequence is

$\dfrac{3}{4}, 25, \dfrac{5}{6}, 625, ...$

(iii) $a_1 = \dfrac{1}{1} = 1$

$a_2 = -\dfrac{1}{2^2} = -\dfrac{1}{4}$

$a_3 = \dfrac{1}{3}$

$a_4 = -\dfrac{1}{4^2} = -\dfrac{1}{16}$

$a_5 = \dfrac{1}{5}$

Hence, we obtain a sequence

$1, -\dfrac{1}{4}, \dfrac{1}{3}, \dfrac{-1}{16}, \dfrac{1}{5}, ...$

3. (i) $a_1 = 1$ and $a_n = a_{n-1} + n$

Putting $n = 2$, we get

$a_2 = a_1 + 1 = 1 + 1 = 2$ $[\because a_1 = 1]$

Similarly,

$a_3 = a_2 + 3 = 2 + 3 = 5$

$a_4 = a_3 + 4 = 5 + 4 = 9$

(ii) $a_1 = 2$ and $a_{n+1} = \dfrac{a_n}{a_n + 2}$

Putting $n = 1$ in recursive formulation, we get

$a_2 = \dfrac{a_1}{a_1 + 2} = \dfrac{2}{2+2} = \dfrac{2}{4} = \dfrac{1}{2}$ $[\because a_1 = 2]$

Similarly,

$a_3 = \dfrac{a_2}{a_2 + 2} = \dfrac{\frac{1}{2}}{\frac{1}{2} + 2} = \dfrac{\frac{1}{2}}{\frac{5}{2}} = \dfrac{1}{5}$

$a_4 = \dfrac{a_3}{a_3 + 2} = \dfrac{\frac{1}{5}}{\frac{1}{5} + 2} = \dfrac{\frac{1}{5}}{\frac{11}{5}} = \dfrac{1}{11}$

(iii) $a_1 = 1, a_2 = 1, a_{n+2} = a_n + a_{n+1}$

Putting $n = 1$, we get

$a_{1+2} = a_1 + a_2$

$a_3 = a_1 + a_2 = 1 + 1 = 2 [\because a_1 = 1, a_2 = 1]$

Similarly, $a_4 = a_2 + a_3 = 1 + 2 = 3$

4. (i) A possible formula for the general term is $\sin\left(\dfrac{n\pi}{2}\right)$.

(ii) A possible formula for the general term is

$a_n = \dfrac{(-1)^{n+1}}{2n-1}$

(iii) The general term $a_n = 5$

5. (i) $\sum_{i=1}^{5} 2^i = 2^1 + 2^2 + 2^3 + 2^4 + 2^5$

$= 2 + 4 + 8 + 16 + 32 = 62$

(ii) $\sum_{i=1}^{10} (-1)^i i = (-1)^1 \cdot 1 + (-1)^2 \cdot 2 + (-1)^3 \cdot 3 + \ldots$

$+ (-1)^{10} \cdot 10$

$= -1 + 2 - 3 + 4 - 5 + 6 - 7 + 8 - 9 + 10$

$= 5$

(iii) $\sum_{i=1}^{4} (i+2)^2 = (1+2)^2 + (2+2)^2 + (3+2)^2 + (4+2)^2$

$= 3^2 + 4^2 + 5^2 + 6^2 = 86$

Alternate Method

$\sum_{i=1}^{4} (i+2)^2 = \sum_{i=1}^{4} (i^2 + 4i + 4)$

$= \sum_{i=1}^{4} i^2 + 4 \sum_{i=1}^{4} i + \sum_{i=1}^{4} 4$

$= \frac{4 \times (4+1)(2 \times 4 + 1)}{6} + 4 \times \frac{4 \times (4+1)}{2} + 4 \times 4$

$= \frac{180}{6} + 40 + 16 = 30 + 56 = 86$

(iv) $\sum_{i=1}^{4} (3i^3 + 2i - 4) = 3 \sum_{i=1}^{4} i^3 + 2 \sum_{i=1}^{4} i - \sum_{i=1}^{4} 4$

$= 3 \times \left[\frac{4 \times (4+1)}{2}\right]^2 + 2 \frac{4 \times (4+1)}{2} - 4 \times 4$

$= 300 + 20 - 16 = 304$

6. Given,

$a_1 - a_2 + a_3 - a_4 + \ldots + (-1)^{n-1} a_n = n$

Lets take $n = 1$, so $a_1 = 1$

Similarly,

$a_1 - a_2 = 2$

$\Rightarrow \quad a_2 = 1 - 2 = -1$

$\Rightarrow \quad a_1 - a_2 + a_3 = 3$

$\Rightarrow \quad 2 + a_3 = 3$

$\Rightarrow \quad a_3 = 1$

So, the terms proceeds as $1, -1, 1, -1, \ldots$.

Then, $a_{51} + a_{52} + \ldots + a_{1023}$

$= 1 + (-1) + \ldots + 1$

$= 1$

7. $a_1 = 7, a_2 = 10, a_3 = 13$

$d = a_2 - a_1 = a_3 - a_2 = 3$

$a_n = a + (n-1)d$

$= 7 + (n-1)3$

$= 7 + 3n - 3$

$a_n = 4 + 3n$

$\therefore a_{75} = 4 + 3 \times 75$

$= 4 + 225$

$= 229$

8. Let a be the first term and d be the common **difference** of the AP.

$m\text{th term} = \frac{1}{n}$

$a + (m-1)d = \frac{1}{n}$ **...(i)**

and $\quad n\text{th term} = \frac{1}{m}$

$a + (n-1)d = \frac{1}{m}$ **...(ii)**

Subtract Eq. (ii) from (i), we get

$(m-1)d - (n-1)d = \frac{1}{n} - \frac{1}{m} = \frac{m-n}{mn}$

$d[m - 1 - n + 1] = \frac{m-n}{mn}$

$\Rightarrow \quad d = \frac{1}{mn}$

$\therefore \quad a + (m-1) \times \frac{1}{mn} = \frac{1}{n} \quad \left[\because d = \frac{1}{mn}, \text{using Eq. (i)}\right]$

$a = \frac{1}{n} - \frac{m-1}{mn} = \frac{m - m + 1}{mn}$

$a = \frac{1}{mn}$

$\therefore \quad a_{mn} = a + (mn-1)d$

$= \frac{1}{mn} + (mn-1) \times \frac{1}{mn}$

$= \frac{1 + mn - 1}{mn}$

$\Rightarrow \quad a_{mn} = 1$

9. $a = 121, d = 117 - 121 = -4 (= 113 - 117)$

Let the nth term of the AP be its first **negative term.**

$a_n < 0$

$a + (n-1)d < 0$

$121 + (n-1)(-4) < 0$

$121 - 4n + 4 < 0$

$125 - 4n < 0$

$4n > 125$

$n > \frac{125}{4} = 31.25$

$\because$ The first negative term is greater than 31.25.

$\therefore$ It is 32nd term.

$\therefore a_{32} = 121 + (32-1)(-4)$

$= 121 - 128 + 4 = -3$

10. Let the three numbers in AP be

$a - d, a, a + d$

Given, $a - d + a + a + d = 45$

or $\quad 3a = 45$

$\therefore \quad a = 15$

and $(a-d)^2 + a^2 + (a+d)^2 = 683$

or $a^2+d^2-2ad+a^2+a^2+d^2+2ad=683$

or $3a^2+2d^2=683$

or $3\cdot(15)^2+2d^2=683$

or $2d^2=683-675=8$

or $d^2=4$

$d=\pm 2$

If $d=2$, the three numbers are 13, 15, 17

and if $d=-2$, the three numbers are 17, 15, 13.

11. Let the integers that are in AP are $a-3d, a-d, a+d, a+3d$.

Given,

$$a-3d+a-d+a+d+a+3d=24$$

$\Rightarrow$ $4a=24$

$\Rightarrow$ $a=6$

Also, it is given that

$$(a-3d)(a-d)(a+d)(a+3d)=945$$

or $(a^2-9d^2)(a^2-d^2)=945$

or $(36-9d^2)(36-d^2)=945$

or $(4-d^2)(36-d^2)=105$

On simplifying,

$$d^4-40d^2+39=0$$

or $d^4-d^2-39d^2+39=0$

or $d^2(d^2-1)-39(d^2-1)=0$

$\Rightarrow$ $(d^2-1)(d^2-39)=0$

$d^2-1=0$ or $d^2=39$

$d=\pm 1$ or $d=\pm\sqrt{39}$

$\because$ d must be an integer.

So, the sequence of number is 9, 7, 5, 3 if $d=-1$ and 3, 5, 7, 9 if $d=1$.

12. $\because$ $\alpha, \beta, \gamma, \delta$ are in AP

Let $\beta=\alpha+d, \gamma=\alpha+2d, \delta=\alpha+3d, d>0$

Given, $\alpha+\beta=2, \alpha\beta=P$

$\Rightarrow$ $2\alpha+d=2, \alpha\beta=P$ ...(i)

and $\gamma+\delta=18, \gamma\delta=Q$

$\Rightarrow$ $2\alpha+5d=18, \gamma\delta=Q$ ...(ii)

From Eqs. (i) and (ii), we get

$d=4, \alpha=-1$

$\beta=3, \gamma=7, \delta=11$

$\Rightarrow$ $P=\alpha\beta=(-1)(3)=-3$

and $Q=\gamma\delta=(7)(11)=77$

13. Let r terms be identical.

Now, the sequence of identical term is

5, 11, 17, ...; $a=5, d=11-5=6$

Let 101 be the last term

$\therefore$ $101=a+(n-1)d$

or $101=5+(n-1)\times d$

or $101=5+(n-1)\times 6$

or $101=5+6n-6$

or $101=6n-1$

or $6n=102$

$\therefore$ $n=17$

14. Here, $a=20$, $d=19\frac{1}{3}-20=\frac{58}{3}-20=\frac{-2}{3}$

$\because$ $S_n=\frac{n}{2}[2a+(n-1)d]$

$$S_{25}=\frac{25}{2}\left[2\times 20+(25-1)\times\left(-\frac{2}{3}\right)\right]$$

$$=\frac{25}{2}\left[40+24\left(-\frac{2}{3}\right)\right]$$

$$=\frac{25}{2}[40-16]$$

$$=\frac{25}{2}\times 24$$

$$=300$$

15. Let a_n be the nth term of the series,

So, $a_n=S_n-S_{n-1}$

$=5n^2+4n-[5(n-1)^2+4(n-1)]$

$=5n^2+4n-5n^2+10n-5-4n+4$

$\therefore$ $a_n=10n-1$

So, the 20th term is $a_{20}=10\times 20-1$

$a_{20}=199$

16. Given, $a=\ln 2$, $d=\ln 4$

$\because$ $S_n=\frac{n}{2}[2a+(n-1)d]$

$$=\frac{n}{2}[2\ln 2+(n-1)\ln 4]$$

$$=\frac{n}{2}[2\ln 2+(n-1)2\ln 2]$$

$$=\frac{n}{2}\times 2\ln 2[1+n-1]$$

$$=n^2\ln 2$$

17. The natural number lying between 100 and 1000, which are multiples of 5 are 105, 110, ..., 995.

Here, the first term, $a=105$

and common difference, $d=5$

last term, $l=995$

$\therefore$ $S_n=\frac{n}{2}[a+l]$

$\because$ $l=a+(n-1)d$

or $995=105+(n-1)\times 5$

or $n-1=178$

$\Rightarrow$ $n=179$

$\therefore$ $S_{179}=\frac{179}{2}[105+995]=98450$

18. We know that in an AP the sum of terms equidistant from the beginning and end is always same and is equal to the sum of first and last term.

Then, $a_1 + a_{24} = a_5 + a_{20} = a_{10} + a_{15}$

Given,

$$a_1 + a_5 + a_{10} + a_{15} + a_{20} + a_{24} = 225$$

$$\Rightarrow (a_1 + a_{24}) + (a_5 + a_{20}) + (a_{10} + a_{15}) = 225$$

$$\Rightarrow \quad 3(a_1 + a_{24}) = 225$$

$$\Rightarrow \quad a_1 + a_{24} = 75$$

$$\therefore \quad S_{24} = \frac{24}{2}(a_1 + a_{24})$$

$$= \frac{24}{2} \times 75 = 900$$

19. Let the two AP's be $a_1, a_1 + d_1, a_1 + 2d_1, \ldots$ and $a_2, a_2 + d_2, a_2 + 2d_2, \ldots$

Let the sum be S'_n and S''_n respectively.

$$\because \quad S_n = \frac{n}{2}[2a + (n-1)d]$$

$$\Rightarrow \quad \frac{S'_n}{S''_n} = \frac{3n-13}{5n+21}$$

$$\Rightarrow \quad \frac{\frac{n}{2}[2a_1 + (n-1)d_1]}{\frac{n}{2}[2a_2 + (n-1)d_2]} = \frac{3n-13}{5n+21}$$

$$\Rightarrow \quad \frac{a_1 + \left(\frac{n-1}{2}\right)d_1}{a_2 + \left(\frac{n-1}{2}\right)d_2} = \frac{3n-13}{5n+21} \quad \ldots\text{(i)}$$

The 24th term of two AP's are respectively.

$$T'_{24} = a_1 + 23d_1 \text{ and } T''_{24} = a_2 + 23d_2$$

$$\therefore \quad \frac{T'_{24}}{T''_{24}} = \frac{a_1 + 23d_1}{a_2 + 23d_2} \quad \ldots\text{(ii)}$$

On comparing Eqs. (i) and (ii), we get

$$\frac{n-1}{2} = 23$$

$$\Rightarrow \quad n = 47$$

From Eq. (i),

$$\frac{a_1 + 23d_1}{a_2 + 23d_2} = \frac{3 \times 47 - 13}{5 \times 47 + 21} = \frac{128}{256} = \frac{1}{2}$$

$\therefore$ The ratio of their 24th terms is 1 : 2.

20. Given, difference between two consecutive angles $= 5°$ and smallest angle $= 120°$

$\therefore$ Second smallest angle $= 120° + 5° = 125°$, then the angles are 120°, 125°, 130°, ...

The angles form an AP whose common difference is 5°.

$$\because \quad S_n = \frac{n}{2}[2a + (n-1)d]$$

$$\Rightarrow \quad S_n = \frac{n}{2}[2 \times 120° + (n-1)5°]$$

$$\Rightarrow \quad S_n = \frac{n}{2}[240° + 5n - 5°]$$

$$\Rightarrow \quad S_n = \frac{n}{2}[235° + 5n] \quad \ldots\text{(i)}$$

$\because$ Sum of all angles of a polygon with n sides

$$= 180°(n-2) \quad \ldots\text{(ii)}$$

$\therefore$ From Eqs. (i) and (ii), we get

$$180°(n-2) = \frac{n}{2}(235 + 5n)$$

$$\text{or} \quad 360(n-2) = 235n + 5n^2$$

$$\Rightarrow 5n^2 - 125n + 720 = 0$$

$$\Rightarrow \quad n^2 - 25n + 144 = 0$$

$$\Rightarrow \quad (n-9)(n-16) = 0$$

$$\Rightarrow \quad n = 9 \text{ or } n = 16$$

The number of sides of the polygon is 9 as if $n = 16$, then $a_{16} = 195° > 180°$.

21. The degree of the expression is $1 + 6 + 11 + \ldots + 101$, which is an AP

Now, $101 = 1 + 5(n-1)$

$$\Rightarrow \quad n = 21$$

$$\Rightarrow 1 + 6 + 11 + \ldots + 101 = \frac{21}{2}[1 + 101]$$

$$= 21 \times 51$$

$$= 1071$$

Therefore, the degree of the expression is 1071.

22. We can write

$150 + 146 + 142 + \ldots$ to n terms $= (n-8) \times 150$

$$\Rightarrow \frac{n}{2}[300 + (n-1)(-4)] = (n-8) \times 150$$

$$\Rightarrow \quad n^2 - n - 600 = 0 \Rightarrow n = 25 \text{ days}$$

23. (i) $a = 3, r = \frac{6}{3} = \frac{12}{6} = 2$

$$\because \quad a_n = ar^{n-1}$$

$$a_n = 3(2)^{n-1}$$

$$\therefore \quad a_7 = 3(2)^{7-1} = 3(2)^6$$

(ii) $a = 36, r = \frac{18}{36} = \frac{9}{18} = \frac{1}{2}$

$$a_n = ar^{n-1}$$

$$a_n = 36\left(\frac{1}{2}\right)^{n-1}$$

$$\therefore \quad a_7 = 36\left(\frac{1}{2}\right)^{7-1} = 36\left(\frac{1}{2}\right)^6 = 36 \times \frac{1}{64} = \frac{9}{16}$$

(iii) $a = 6, r = \frac{-6}{6} = \frac{6}{-6} = -1$

$$\Rightarrow \quad a_n = ar^{n-1}$$

$$\Rightarrow \quad a_n = 6(-1)^{n-1}$$

∴ $a_7 = (-1)^{7-1}6 = (-1)^6 6 = 6$

(iv) $a = \frac{1}{4},\ r = \frac{1/12}{1/4} = \frac{1/36}{1/12} = \frac{1}{3}$

$$\Rightarrow \quad a_n = ar^{n-1}$$

$$\Rightarrow \quad a_n = \frac{1}{4}\left(\frac{1}{3}\right)^{n-1}$$

$$\therefore \quad a_7 = \frac{1}{4}\left(\frac{1}{3}\right)^{7-1} = \frac{1}{4}\left(\frac{1}{3}\right)^{6} = \frac{1}{2916}$$

24. Given, $T_p = q$ and $T_q = p$

Let the first term and common ratio of GP be a and r respectively.

$$\because \quad T_p = q$$

$$ar^{p-1} = q \quad \text{...(i)}$$

and

$$T_q = p$$

$$ar^{q-1} = p \quad \text{...(ii)}$$

$$\frac{ar^{p-1}}{ar^{q-1}} = \frac{q}{p}$$

$$r^{p-1-q+1} = \frac{q}{p}$$

$$r^{p-q} = \frac{q}{p}$$

$$\therefore \quad r = \left(\frac{q}{p}\right)^{1/p-q} \quad \text{...(iii)}$$

From Eq. (i), $a = q\left(\frac{p}{q}\right)^{\frac{p-1}{p-q}}$ [using Eq. (ii)]

∴ $(p+q)$th term $= ar^{p+q-1}$

$$= q\left(\frac{p}{q}\right)^{\frac{p-1}{p-q}}\left(\frac{q}{p}\right)^{\frac{p+q-1}{p-q}}$$

$$= q^{\left(1-\frac{p-1}{p-q}+\frac{p+q-1}{p-q}\right)} p^{\left(\frac{p-1}{p-q}-\frac{p+q-1}{p-q}\right)}$$

$$= q^{\frac{p}{p-q}} p^{-\frac{q}{p-q}} = \left(\frac{q^p}{p^q}\right)^{1/p-q}$$

25. Let a be the first term and r be the common ratio of the GP.

Given that the third term is 4.

$$\therefore \quad ar^2 = 4$$

Product of first five terms $= a \times ar \times ar^2 \times ar^3 \times ar^4$

$$= a^5 r^{10} = (ar^2)^5 = 4^5$$

26. Let $\frac{a}{r}, a, ar$ be three terms of the GP.

Given, $\frac{a}{r} + a + ar = \frac{13}{12}$ and

$$a\left(\frac{1}{r} + 1 + r\right) = \frac{13}{12}$$

$$a\left(\frac{r^2 + r + 1}{r}\right) = \frac{13}{12}$$

$$a(r^2 + r + 1) = \frac{13}{12}r \quad \text{...(i)}$$

Also, $\frac{a}{r} \times a \times ar = -1$

$$a^3 = -1$$

$$a = -1 \quad \text{...(ii)}$$

Putting the value of a in Eq. (i), we get

$$r^2 + r + 1 = -\frac{13}{12}r$$

$$\Rightarrow \quad 12r^2 + 25r + 12 = 0$$

or $12r^2 + 16r + 9r + 12 = 0$

$$\Rightarrow \quad (3r+4)(4r+3) = 0$$

$$\Rightarrow \quad 3r + 4 = 0 \text{ or } 4r + 3 = 0$$

$$r = -\frac{4}{3} \text{ or } r = -\frac{3}{4}$$

Hence, the numbers are $\frac{3}{4}, -1, \frac{4}{3}$, if $r = -\frac{4}{3}$

or $\frac{4}{3}, -1, \frac{3}{4}$ if $r = \frac{-3}{4}$

27. Let the numbers in GP be a, ar and ar^2.

$$a + ar + ar^2 = 56 \quad \text{[given that]}$$

$$\Rightarrow \quad ar^2 = 56 - a - ar \quad \text{...(i)}$$

Also, when 1, 7, 21 subtracting from these numbers respectively, the new numbers are in AP.

So, $(a-1)(ar-7)(ar^2-21)$ are in AP.

$$\Rightarrow \quad (ar-7) - (a-1) = (ar^2 - 21) - (ar - 7)$$

$$\Rightarrow \quad ar^2 - 2ar + a - 8 = 0$$

or $(56 - a - ar) - 2ar + a - 8 = 0$ [using Eq. (i)]

or $\quad ar = \frac{48}{3} = 16$

$$\Rightarrow \quad a = \frac{16}{r}$$

$$\because \quad a + ar + ar^2 = 56$$

$$\Rightarrow \quad a(1 + r + r^2) = 56$$

$$\Rightarrow \quad \frac{16}{r}(1 + r + r^2) = 56 \quad \left[\because a = \frac{16}{r}\right]$$

On simplifying, we have

$$2r^2 - 5r + 2 = 0$$

$$\Rightarrow \quad (2r-1)(r-2) = 0$$

$$2r - 1 = 0 \text{ or } r - 2 = 0$$

$$r = \frac{1}{2} \text{ or } r = 2$$

If $r=\frac{1}{2}$, then $a=\frac{16}{r}=32$

If $r=2$, then $a=8$

Thus, the numbers are 8, 16, 32 for $r=2$

and 32, 16, 8 for $r=\frac{1}{2}$

28. $\sum_{k=1}^{9}(7+5^k)=\sum_{k=1}^{9}7+\sum_{k=1}^{9}5^k$

$$=7\times9+(5+5^2+5^3+...+5^9)$$

$$=63+\frac{5(5^9-1)}{(5-1)}$$

$$=63+\frac{5}{4}(5^9-1)$$

$$=63+2441405=2441468$$

29. First term $a=3$,

Common ratio $=\frac{3/2}{3}=\frac{3/4}{3/2}=\frac{1}{2}$

$\because \quad S_n=\frac{a(1-r^n)}{1-r}$

$$\Rightarrow \quad \frac{3069}{512}=\frac{3\left[1-\left(\frac{1}{2}\right)^n\right]}{1-\frac{1}{2}}$$

$$\Rightarrow \quad \frac{3069}{512}=6\left[1-\left(\frac{1}{2}\right)^n\right]$$

or $\quad 1-\left(\frac{1}{2}\right)^n=\frac{3069}{3072}$

or $\quad \left(\frac{1}{2}\right)^n=1-\frac{3069}{3072}=\frac{3}{3072}$

or $\quad \left(\frac{1}{2}\right)^n=\frac{1}{1024}$

or $\quad \left(\frac{1}{2}\right)^n=\left(\frac{1}{2}\right)^{10}$

$\Rightarrow \quad n=10$

30. (i) $2+22+222+....$ to n terms

$=2[1+11+111+....$ to n terms$]$

$=\frac{2}{9}[9+99+999+...$ to n terms$]$

$$=\frac{2}{9}[(10-1)+(10^2-1)+(10^3-1)+....+(10^n-1)]$$

$$=\frac{2}{9}[(10+10^2+10^3+....+10^n)-(1+1+1+.....+n\text{ times})]$$

$$=\frac{2}{9}\left[10\times\frac{10^n-1}{10-1}-n\right]$$

$$=\frac{2}{9}\left[\frac{10}{9}(10^n-1)-n\right]$$

$$=\frac{2}{81}[10(10^n-1)-9n]$$

(ii) $0.6+0.66+0.666+....n$ terms

$=6[0.1+0.11+0.111+...$ to n terms$]$

$=\frac{6}{9}[0.9+0.99+0.999+...$ to n terms$]$

$$=\frac{6}{9}\left[\left(1-\frac{1}{10}\right)+\left(1-\frac{1}{10^2}\right)+...\text{ to } n \text{ terms}\right]$$

$$=\frac{6}{9}\left[(1+1+1+....n\text{ times})-\frac{1}{10}\left(1+\frac{1}{10}+\frac{1}{10^2}+....n\text{ terms}\right)\right]$$

$$=\frac{2}{3}\left[n-\frac{1}{10}\frac{\left[1-\left(\frac{1}{10}\right)^n\right]}{\left(1-\frac{1}{10}\right)}\right]$$

$$=\frac{2}{3}\left[n-\frac{1}{9}\left(1-\left(\frac{1}{10}\right)^n\right)\right]$$

$$=\frac{2}{3}n-\frac{2}{27}(1-10^{-n})$$

31. We have,

$$S_n=\left(2+\frac{1}{2}\right)+\left(2+\frac{1}{4}\right)+\left(2+\frac{1}{8}\right)+....+n\text{ terms}$$

$$S_n=(2+2+....n\text{ times})+\left(\frac{1}{2}+\frac{1}{4}+\frac{1}{8}+....\right)$$

$$S_n=2n+\frac{1}{2}\left(\frac{1-\frac{1}{2^n}}{1-\frac{1}{2}}\right)$$

$$S_n=2n+1-\frac{1}{2^n}$$

32. Given,

$f(x+y)=f(x)\cdot f(y)$ and $f(1)=3$

$\Rightarrow \quad f(1+1)=f(1)\cdot f(1)$

$\Rightarrow \quad f(2)=3\cdot3=3^2$

Also, $f(2+1)=f(3)=f(2)\cdot f(1)$

$f(3)=3^2\cdot3=3^3$

Similarly, $f(4)=3^4$

In general $f(n)=3^n$

Given, $\sum_{k=1}^{n}f(k)=1092$

$\Rightarrow f(1)+f(2)+f(3)+...+f(n)=1092$

or $\quad 3+3^2+3^3+...+3^n=1092$

$$\frac{3(3^n-1)}{(3-1)}=1092 \quad \text{[using sum of } n \text{ terms of GP]}$$

$\Rightarrow \quad 3^n-1=\frac{1092\times 2}{3}=728$

or $\quad 3^n=729$

$\Rightarrow \quad 3^n=3^6$

$\Rightarrow \quad n=6$

33. (i) We have, $a=\frac{1}{3}, r=\frac{1}{3}$

Thus, $S_\infty=\frac{a}{1-r}=\frac{1/3}{1-\frac{1}{3}}=\frac{1/3}{2/3}=\frac{1}{2}$

(ii) $a=1, r=-\frac{1}{2}$

Thus, $S_\infty=\frac{a}{1-r}=\frac{1}{1-\left(-\frac{1}{2}\right)}=\frac{1}{1+\frac{1}{2}}=\frac{2}{3}$

34. $a+ar+ar^2+...\infty=20$

$\Rightarrow \quad \frac{a}{1-r}=20 \quad$ ...(i)

and $a^2+a^2r^2+...+...\infty=100$

$$\frac{a^2}{1-r^2}=100 \quad \text{...(ii)}$$

On squaring Eq. (i), we have

$$\frac{a^2}{(1-r)^2}=400 \quad \text{...(iii)}$$

On dividing Eq. (iii) by Eq. (ii), we get

$$\frac{a^2/(1-r)^2}{a^2/(1-r^2)}=\frac{400}{100}$$

or $\quad \frac{1-r^2}{(1-r)^2}=4$

or $\quad \frac{1+r}{1-r}=4$

or $\quad 1+r=4-4r$

$\Rightarrow \quad 5r=3$

$\Rightarrow \quad r=\frac{3}{5}$

35. (i) $\sum_{n=1}^{\infty} 5\left(\frac{1}{2}\right)^{n-1}=5+\frac{5}{2}+\frac{5}{2^2}+\frac{5}{2^3}+....+\infty$ terms

$$\left[\text{first term }=5, r=\frac{1}{2}\right]$$

$$=\frac{5}{1-\frac{1}{2}}=\frac{5}{1/2}=10$$

(ii) $(\sqrt{2}+1)+1+(\sqrt{2}-1)+...\infty$

The given series is geometric series with first term $a=\sqrt{2}+1$

and the common ratio $=\frac{1}{\sqrt{2}+1}$

$$=\frac{\sqrt{2}-1}{(\sqrt{2}+1)(\sqrt{2}-1)}=\frac{\sqrt{2}-1}{2-1}$$

$r=\sqrt{2}-1$

$$S_\infty=\frac{a}{1-r}=\frac{\sqrt{2}+1}{1-(\sqrt{2}-1)}=\frac{\sqrt{2}+1}{2-\sqrt{2}}\times\frac{2+\sqrt{2}}{2+\sqrt{2}}$$

$$=\frac{2\sqrt{2}+2+2+\sqrt{2}}{4-2}=\frac{4+3\sqrt{2}}{2}$$

(iii) $\frac{1}{2}+\frac{1}{3^2}+\frac{1}{2^3}+\frac{1}{3^4}+\frac{1}{2^5}+\frac{1}{3^6}+...\infty$

$$=\left(\frac{1}{2}+\frac{1}{2^3}+\frac{1}{2^5}+...\infty\right)+\left(\frac{1}{3^2}+\frac{1}{3^4}+\frac{1}{3^6}+....\infty\right)$$

$$=\frac{1/2}{\left(1-\frac{1}{2^2}\right)}+\frac{1/3^2}{\left(1-\frac{1}{3^2}\right)}=\frac{2}{3}+\frac{1}{8}=\frac{19}{24}$$

36. (i) Let $x=0.333333...=0.\overline{3}$

$\therefore \quad x=0.3+0.03+0.003+0.0003+...$

$$=\frac{3}{10}+\frac{3}{10^2}+\frac{3}{10^3}+\frac{3}{10^4}+....$$

$$=\frac{3/10}{1-\frac{1}{10}}=\frac{3}{9}$$

$\therefore \quad x=\frac{1}{3}$

(ii) Let $x=1.233333...=1.2\overline{3}$

$x=1.2+0.03+0.003+0.0003+...$

$$=1.2+\frac{3}{10^2}+\frac{3}{10^3}+\frac{3}{10^4}+....$$

$$=1.2+\frac{3/10^2}{\left(1-\frac{1}{10}\right)}$$

$$=1.2+\frac{1}{30}=\frac{12}{10}+\frac{1}{30}$$

$\therefore \quad x=\frac{37}{30}$

37. Initially the ball falls from the height of 200 m. After striking the floor, it rebounds and rises to a height of $\frac{3}{5}\times 200$ m.

Now, it falls from a height of $\frac{3}{5}\times 200$ m and after rebounding, rises to a height of $\frac{3}{5}\times\frac{3}{5}\times 200$ m.

The process is continued till the ball come rest.

$\therefore$ The total distance travelled

$$= 200 + 2\left[\frac{3}{5}\times 200 + \left(\frac{3}{5}\right)^2 \times 200 + \ldots \infty\right]$$

$$= 200 + 2\left[\frac{\frac{3}{5}\times 200}{1-\frac{3}{5}}\right] = 200 + \frac{2\times 120}{2/5}$$

$$= 200 + 600 = 800 \text{ m}$$

38. It is given that $a = 2, r = \frac{3}{4}$

$$S = \frac{a}{1-r} = \frac{2}{1-\frac{3}{4}} = 8$$

$$S_n = \frac{a(1-r^n)}{1-r} = \frac{2\left[1-\left(\frac{3}{4}\right)^n\right]}{1-\frac{3}{4}} = 8 - 8\left(\frac{3}{4}\right)^n$$

$\therefore \quad S - S_n = 8\left(\frac{3}{4}\right)^n < \frac{1}{1000}$ (given that)

If $(0.75)^n < 0.000125$

If $n\log_{10} 0.75 < \log_{10} 0.000125$

If $n(\bar{1}.8751) < \bar{4}.0969$

$$n > \frac{\bar{4}.0969}{\bar{1}.8751} \approx 31.25$$

$\therefore \quad n = 32$

39. We know that to insert n numbers between a and b.

The common difference $= d = \frac{b-a}{n+1}$

Here, $a = 3, b = 24, n = 6$

$\therefore \quad d = \frac{24-3}{6+1} = \frac{21}{7} = 3$

First inserted number $= a + d$

$= 3 + 3 = 6$

Second number $= a + 2d = 3 + 2\cdot 3$

$= 3 + 6 = 9$

Similarly, the number between 3 and 24 is

3, 6, 9, 12, 15, 18, 21, 24.

40. Let three geometric means between 1 and 256 be g_1, g_2 and g_3.

So, 1, g_1, g_2, g_3, 256 are in GP.

$\therefore \quad 256 = 1(r)^{5-1} = r^4$

$\Rightarrow \quad r^4 = 256 = 4^4$

$\therefore \quad r = 4$

$\Rightarrow \quad g_1 = ar = 1(4) = 4 \quad$ [$\because a = 1$ (first term)]

$g_2 = 1(4)^2 = 16$

$g_3 = 1(4)^3 = 64$

41. Let the first term and common ratio of GP is a and r, respectively.

Let the four numbers be a, ar, ar^2, ar^3.

Given that,

$$a + ar + ar^2 + ar^3 = 60$$

$$\Rightarrow \quad \frac{a(r^4-1)}{r-1} = 60$$

$$\Rightarrow \quad \frac{a(r^2+1)(r+1)(r-1)}{r-1} = 60$$

$$a(r^2+1)(r+1) = 60 \quad \ldots\text{(i)}$$

and $\quad \frac{a + ar^3}{2} = 18 \quad$ [given]

$$a(1+r^3) = 36$$

$$\therefore \quad a = \frac{36}{(1+r^3)} \quad \ldots\text{(ii)}$$

Putting the value of a in Eq. (i),

$$\frac{36}{r^3+1}(r^2+1)(r+1) = 60$$

$$36(r+1)(r^2+1) = 60(r^3+1)$$

On rearranging the terms on both sides, we get

$$24r^3 - 36r^2 - 36r + 24 = 0$$

On solving, we get $r = 2, -1, \frac{1}{2}$

$r = -1$, is neglected in view of Eq. (ii).

For $r = 2$, we get $a = 4$ and the four numbers are 4, 8, 16, 32

and for $r = \frac{1}{2}$, we get $a = 32$ and the four numbers are 32, 16, 8, 4.

42. (i) $T_n = $ [nth terms of 1, 4, 7, 10, ...]

[nth terms of $1, \frac{1}{5}, \frac{1}{5^2}, \ldots$]

$$T_n = (3n-2)\left(\frac{1}{5}\right)^{n-1}$$

$$\text{Let } S_n = 1 + 4\left(\frac{1}{5}\right) + 7\left(\frac{1}{5}\right)^2 + 10\left(\frac{1}{5}\right)^3 \ldots$$

$$+ (3n-2)\left(\frac{1}{5}\right)^{n-1} \quad \ldots\text{(i)}$$

$$\frac{S_n}{5} = \frac{1}{5} + 4\left(\frac{1}{5}\right)^2 + 7\left(\frac{1}{5}\right)^3 + \ldots + (3n-2)\left(\frac{1}{5}\right)^n \quad \ldots\text{(ii)}$$

On substracting Eq. (ii) from Eq. (i), we get

$$S_n - \frac{S_n}{5} = 1 + 3$$

$$\left\{\frac{1}{5} + \left(\frac{1}{5}\right)^2 + \left(\frac{1}{5}\right)^3 + \ldots\right\} - (3n-2)\left(\frac{1}{5}\right)^n$$

$$= 1 + 3\left[\frac{\frac{1}{5}\left[1-\left(\frac{1}{5}\right)^{n-1}\right]}{1-\frac{1}{5}}\right] - (3n-2)\left(\frac{1}{5}\right)^n$$

$$= 1 + \frac{3}{4}\left[1-\left(\frac{1}{5}\right)^{n-1}\right] - (3n-2)\left(\frac{1}{5}\right)^n$$

(ii) $S_\infty = 1 + 4\left(\frac{1}{5}\right) + 7\left(\frac{1}{5}\right)^2 + \ldots$ upto ∞ ...(A)

$\frac{1}{5}S_\infty = \left(\frac{1}{5}\right) + 4\left(\frac{1}{5}\right)^2 + 7\left(\frac{1}{5}\right)^3 + \ldots$ ...(B)

On substracting Eq. (B) from Eq. (A), we get

$$\frac{4}{5}S_\infty = 1 + 3\left[\left(\frac{1}{5}\right) + \left(\frac{1}{5}\right)^2 + \ldots\infty\right]$$

$$\Rightarrow \quad \frac{4}{5}S_\infty = 1 + 3\left(\frac{1/5}{1-\frac{1}{5}}\right)$$

$$\Rightarrow \quad \frac{4}{5}S_\infty = 1 + \frac{3}{4}$$

$$\Rightarrow \quad \frac{4}{5}S_\infty = \frac{7}{4}$$

$$\Rightarrow \quad S_\infty = \frac{35}{16}$$

43. Let $S = 1 + 3x + 6x^2 + 10x^3 + \ldots$ ...(i)

$xS = x + 3x^2 + 6x^3 + \ldots$ ...(ii)

On subtracting Eq. (ii) from Eq. (i), we get

$S(1-x) = 1 + 2x + 3x^2 + 4x^3 + \ldots\infty$ ...(iii)

$x(1-x)S = x + 2x^2 + 3x^3 + \ldots\infty$ ... iv)

Again, subtracting Eq. (iv) from Eq. (iii), we get

$S[(1-x) - x(1-x)] = 1 + x + x^2 + \ldots\infty$

$$= \frac{1}{1-x}$$

$$\Rightarrow S[(1-x)(1-x)] = \frac{1}{1-x}$$

$$\Rightarrow \quad S = \frac{1}{(1-x)^3}$$

44. $S = 1 + \left(1+\frac{1}{5}\right)\left(\frac{1}{2}\right) + \left(1+\frac{1}{5}+\frac{1}{5^2}\right)\left(\frac{1}{2^2}\right) + \ldots$ ...(i)

$\frac{S}{2} = \frac{1}{2} + \left(1+\frac{1}{5}\right)\left(\frac{1}{2^2}\right) + \ldots$ ...(ii)

Subtracting Eq. (ii) from Eq. (i), we get

$$\frac{S}{2} = 1 + \frac{1}{5}\times\frac{1}{2} + \frac{1}{5^2}\times\frac{1}{2^2} + \ldots$$

$$\frac{S}{2} = \frac{1}{1-\frac{1}{10}} = \frac{10}{9}$$

$$\Rightarrow \quad S = \frac{20}{9}$$

45. Let $S = 1^2 + 2^2x + 3^2x^2 + 4^2x^3 + \ldots$

$(1-x)S = 1 + 3x + 5x^2 + 7x^3 + \ldots$

$(1-x)^2 S = 1 + 2x + 2x^2 + 2x^3 + \ldots$

$(1-x)^2 S = 1 + 2x \times \frac{1}{1-x}$, when $|x| < 1$

$\therefore (1-x)^2 S = \frac{1-x+2x}{1-x} = \frac{1+x}{1-x} \Rightarrow S = \frac{1+x}{(1-x)^3}$

46. $\sum_{r=1}^{n}(6r^2 + 4r - 1) = \sum_{r=1}^{n} 6r^2 + \sum_{r=1}^{n} 4r - \sum_{r=1}^{n} 1$

$$= \frac{6\times n\times(n+1)(2n+1)}{6} + \frac{4n(n+1)}{2} - n$$

$= n(n+1)(2n+1) + 2n(n+1) - n$

$= n[(n+1)(2n+1) + 2(n+1) - 1]$

$= n(2n^2 + 3n + 1 + 2n + 2 - 1)$

$= n(2n^2 + 5n + 2) = n(n+2)(2n+1)$

47. The first factors of terms are 2, 5, 8, 11, ...

This is an AP whose nth terms is $(3n-1)$

The second factor of terms are 5, 9, 13, 17, ...

This is also an AP whose nth terms is $(4n+1)$.

So, rth term of the given series.

$T_r = (3r-1)(4r+1) = 12r^2 - r - 1$

$\therefore \quad S = \sum_{r=1}^{n} T_r = \sum_{r=1}^{n} 12r^2 - r - 1$

$\Rightarrow \quad S = 12\sum_{r=1}^{n} r^2 - \sum_{r=1}^{n} r - \sum_{r=1}^{n} 1$

$\Rightarrow \quad S = 12\frac{n(n+1)(2n+1)}{6} - \frac{n(n+1)}{2} - n$

On rearranging, we get

$$S = \frac{n}{2}[8n^2 + 11n + 1]$$

48. $\because n$ is even.

Let $n = 2$ m

$\therefore \quad S = (0.7 + 0.77 + 0.777 + \ldots m \text{ terms})$
$+ (7.7 + 77.7 + 777.7 + \ldots m \text{ terms})$

$S = \frac{7}{9}(0.9 + 0.99 + 0.999 + \ldots m \text{ terms})$

$+ \frac{7}{90}[(10^2 - 1) + (10^3 - 1) + (10^4 - 1) + \ldots m \text{ terms}]$

$$S = \frac{7}{9}\left[\frac{n}{2} - \frac{1}{9}\left(1 - \frac{1}{10^{n/2}}\right)\right]$$
$$+ \frac{7}{90}\left[\frac{100}{9}(10^{n/2} - 1) - \frac{n}{2}\right]\left[\because m = \frac{n}{2}\right]$$

49. $\sum_{i=1}^{n}\sum_{j=1}^{i}\sum_{k=1}^{j} 1 = \sum_{i=1}^{n}\sum_{j=1}^{i} j$

$$= \sum_{i=1}^{n} \frac{i(i+1)}{2}$$

$$= \frac{1}{2}\left[\sum_{i=1}^{n} i^2 + \sum_{i=1}^{n} i\right]$$

$$= \frac{1}{2}\left[\frac{n(n+1)(2n+1)}{6} + \frac{n(n+1)}{2}\right]$$

On rearranging, we get

$$= \frac{n(n+1)(n+2)}{6}$$

50. Using AM $\geq$ GM

$$\frac{a^2b + b^2c + c^2a}{3} \geq (a^2b \times b^2c \times c^2a)^{1/3}$$

Also,

$$\frac{ab^2 + bc^2 + ca^2}{3} \geq (ab^2 \times bc^2 \times ca^2)^{1/3}$$

On multiplying,

$$\frac{(a^2b + b^2c + c^2a)}{3}\frac{(ab^2 + bc^2 + ca^2)}{3} \geq (a^3b^3c^3)^{1/3}(a^3b^3c^3)^{1/3}$$

Thus, $(a^2b + b^2c + c^2a)(ab^2 + bc^2 + ca^2) \geq 9a^2b^2c^2$.

51. Consider $\frac{a}{2}, \frac{a}{2}, \frac{a}{2}, \frac{a}{2}, \frac{b}{2}, \frac{b}{2}, \frac{3c}{2}, \frac{3c}{2}$

Using AM $\geq$ GM

$$\frac{\frac{a}{2} + \frac{a}{2} + \frac{a}{2} + \frac{a}{2} + \frac{b}{2} + \frac{b}{2} + \frac{3c}{2} + \frac{3c}{2}}{8} \geq \left(\frac{a^4}{16} \times \frac{b^2}{4} \times \frac{9c^2}{4}\right)^{1/8}$$

or $$\frac{2a + b + 3c}{8} \geq \left(\frac{a^4}{16} \times \frac{b^2}{4} \times \frac{9c^2}{4}\right)^{1/8}$$

or $$\frac{1}{8} \geq \left(\frac{9a^4b^2c^2}{256}\right)^{1/8}$$

or $$\frac{9a^4b^2c^2}{256} \leq \left(\frac{1}{8}\right)^8$$

$$a^4b^2c^2 \leq \left(\frac{1}{8}\right)^8 \times \frac{256}{9} = \frac{1}{4^8 \times 2^8} \times \frac{4^4}{9}$$

$$\Rightarrow \quad a^4b^2c^2 \leq \frac{1}{9 \cdot 4^8}$$

Thus, the greatest value of $a^4b^2c^2 = \frac{1}{9 \cdot 4^8}$

When the equality holds, the greatest value is attained, when all the numbers are equal.

i.e. $\frac{a}{2} = \frac{b}{2} = \frac{3c}{2} \Rightarrow a = b = 3c$

$$\frac{a}{3} = \frac{b}{3} = \frac{c}{1} = k$$

$$\Rightarrow \quad a = 3k, b = 3k, c = k$$

$$\therefore \quad 2a + b + 3c = 1$$

$$\Rightarrow \quad 6k + 3k + 3k = 1$$

$$12k = 1$$

$$\Rightarrow \quad k = \frac{1}{12}$$

$$\therefore \quad a = \frac{3}{12}, b = \frac{3}{12}, c = \frac{1}{12}$$

i.e., $$a = \frac{1}{4}, b = \frac{1}{4}, c = \frac{1}{12}$$

52. Let $\tan^2\alpha = a$ and $\tan^2\beta = b$

$$\frac{\sec^4\alpha}{\tan^2\beta} + \frac{\sec^4\beta}{\tan^4\alpha} = \frac{(a+1)^2}{b} + \frac{(b+1)^2}{a}$$

$$= \frac{a^2}{b} + \frac{b^2}{a} + \frac{1}{a} + \frac{1}{b} + 2\frac{a}{b} + 2\frac{b}{a}$$

Using AM $\geq$ GM

$$\frac{\frac{a^2}{b} + \frac{b^2}{a} + \frac{1}{a} + \frac{1}{b} + \frac{a}{b} + \frac{a}{b} + \frac{b}{a} + \frac{b}{a}}{8}$$

$$\geq \left(\frac{a^2}{b} \times \frac{b^2}{a} \times \frac{1}{a} \times \frac{1}{b} \times \frac{a}{b} \times \frac{a}{b} \times \frac{b}{a} \times \frac{b}{a}\right)^{1/8}$$

$$\Rightarrow \frac{a^2}{b} + \frac{b^2}{a} + \frac{1}{a} + \frac{1}{b} + \frac{a}{b} + \frac{a}{b} + \frac{b}{a} + \frac{b}{a} \geq 8$$

$$\therefore \quad \frac{\sec^4\alpha}{\tan^2\beta} + \frac{\sec^4\beta}{\tan^2\alpha} \geq 8$$

53. Let the numbers be xy, yz and zx.

Using AM $\geq$ GM

$$\frac{xy + yz + zx}{3} \geq (xy \times yz \times zx)^{1/3}$$

or $$\frac{12}{3} \geq (x^2y^2z^2)^{1/3}$$

or $$(xyz)^{2/3} \leq 4$$

or $$(xyz)^2 \leq (4)^3 = 64$$

$$\Rightarrow \quad (xyz) \leq 8$$

$\therefore$ The maximum value of xyz is 8.

54. Using AM $\geq$ GM

$$\frac{4^{\sin^2 x} + 4^{\cos^2 x}}{2} \geq \sqrt{4^{\sin^2 x} \times 4^{\cos^2 x}}$$

$$4^{\sin^2 x} + 4^{\cos^2 x} \geq 2\sqrt{4^{\sin^2 x + \cos^2 x}}$$

$$\Rightarrow \quad 4^{\sin^2 x} + 4^{\cos^2 x} \geq 2\sqrt{4^1} = 4$$

Hence, the minimum value of $4^{\sin^2 x} + 4^{\cos^2 x}$ is 4.

Level 2

1. (a) $\frac{1}{a}-\frac{1}{b}=\frac{1}{b}-\frac{1}{c}$

$\therefore \left(\frac{1}{a}+\frac{1}{b}-\frac{1}{c}\right)\left(\frac{1}{b}+\frac{1}{c}-\frac{1}{a}\right)$

$=\left(\frac{2}{a}-\frac{1}{b}\right)\left(\frac{2}{c}-\frac{1}{b}\right)=\frac{4}{ac}-\frac{1}{b}\left(\frac{2}{a}+\frac{2}{c}\right)+\frac{1}{b^2}$

$=\frac{4}{ac}-\frac{2}{b}\left(\frac{2}{b}\right)+\frac{1}{b^2}=\frac{4}{ac}-\frac{3}{b^2}$

2. (a) Let the progression be $a, a+d, a+2d$.

Then, $x_4=3x_1 \Rightarrow a+3d=3a \Rightarrow 3d=2a$ ...(i)

Again $x_7=2x_3+1$

$\Rightarrow a+6d=2(a+2d)+1 \Rightarrow 2d=a+1$ ...(ii)

Solving Eqs. (i) and (ii), we get

$a=3, d=2$

3. (c) Since an AP is either increasing or decreasing, if possible let -4 be the first term of an AP, whose mth and nth terms are respectively 8 and 13, then

$8=-4+(m-1)d$ and $13=-4+(n-1)d$

$\Rightarrow \quad \frac{12}{m-1}=\frac{17}{n-1}=d$

Let $\frac{m-1}{12}=\frac{n-1}{17}=K$, then $m=12K+1$, and $n=17K+1$

$\therefore$ For $K=1, 2, 3, \ldots$, we get different pairs of values of m and n, which shows that infinite number of AP's can be obtained.

4. (d) Let d be the common difference of the AP, then

$a_{2r}=a_{2r-1}+d$

$\therefore \sum_{r=1}^{100} a_{2r}=\sum_{r=1}^{100}(a_{2r-1}+d)=\sum_{r=1}^{100} a_{2r-1}+100d$

$\Rightarrow \alpha=\beta+100d \Rightarrow d=\frac{\alpha-\beta}{100}$

5. (c) Given ${}^mC_{r-1}, {}^mC_r, {}^mC_{r+1}$ are in AP.

$2\,{}^mC_r={}^mC_{r-1}+{}^mC_{r+1}$

$\Rightarrow 2=\frac{{}^mC_{r-1}}{{}^mC_r}+\frac{{}^mC_{r+1}}{{}^mC_r}=\frac{r}{m-r+1}+\frac{m-r}{r+1}$

$\Rightarrow m^2-m(4r+1)+4r^2-2=0$

6. (d) Let d_1 be the common difference.

So, $\quad S_1=a=T_1$

$S_2=4a+\frac{d}{2}$

$\Rightarrow \quad T_1+T_2=4a+\frac{d}{2}$

$\Rightarrow \quad T_2=3a+\frac{d}{2}$

$\Rightarrow \quad d=T_2-T_1=2a+\frac{d}{2}$

7. (b) $\frac{b+c-a}{a}, \frac{c+a-b}{b}, \frac{a+b-c}{c}$ are in AP

$\frac{b+c-a}{a}+2, \frac{c+a-b}{b}+2, \frac{a+b-c}{c}+2$ are in AP

$\frac{a+b+c}{a}, \frac{a+b+c}{b}, \frac{a+b+c}{c}$ are in AP

$\Rightarrow \frac{1}{a}, \frac{1}{b}, \frac{1}{c}$ are in AP.

8. (a) Given a^2, b^2, c^2 are in AP

$\Rightarrow a^2+(ab+bc+ca), b^2+(ab+bc+ca)$ and $c^2+(ab+bc+ca)$ are in AP

$\Rightarrow (a+b)(a+c), (b+c)(b+a), (c+a)(c+b)$ are in AP

$\Rightarrow \frac{1}{(b+c)}, \frac{1}{(c+a)}, \frac{1}{(a+b)}$ are in AP

[divided by $(a+b)(b+c)(c+a)$]

Again a^2, b^2, c^2 are in AP.

$\frac{1}{b+c}, \frac{1}{c+a}, \frac{1}{a+b}$ are in AP

$\Rightarrow \frac{a+b+c}{b+c}, \frac{a+b+c}{c+a}, \frac{a+b+c}{a+b}$ are in AP

$\Rightarrow \frac{a}{b+c}+1, \frac{b}{c+a}+1$ and $\frac{c}{a+b}+1$ are in AP.

$\Rightarrow \frac{a}{b+c}, \frac{b}{c+a}, \frac{c}{a+b}$ are in AP

9. (d) $9(25a^2+b^2)+25(c^2-3ac)=15b(3a+c)$

$\Rightarrow (15a)^2+(3b)^2+(5c)^2-45ab-15bc-75ac=0$

$\Rightarrow \frac{1}{2}[(15a-3b)^2+(3b-5c)^2+(5c-15a)^2]=0$

$\Rightarrow (15a-3b)^2+(3b-5c)^2+(5c-15a)^2=0$

or $15a-3b=0, 3b-5c=0; 5c-15a=0$

$\because b=5a, c=3a$

$\Rightarrow 5a, 3a, a$ are in AP

$\Rightarrow b, c, a$ are in AP

10. (b) $1, \log_9(3^{1-x}+2), \log_3(4\cdot 3^x-1)$ are in AP

$\Rightarrow 2\log_9(3^{1-x}+2)=1+\log_3(4\cdot 3^x-1)$

$\Rightarrow \log_3(3^{1-x}+2)=\log_3 3+\log_3(4\cdot 3^x-1)$

$\Rightarrow \log_3(3^{1-x}+2)=\log_3[3\cdot(4\cdot 3^x-1)]$

$\Rightarrow \quad 3^{1-x}+2=3(4\cdot 3^x-1)$

$\Rightarrow \quad 3\cdot 3^{-x}+2=12\cdot 3^x-3$

Let $3^x=t$

$\Rightarrow \quad \frac{3}{t}+2=12t-3$

$\Rightarrow \quad 12t^2-5t-3=0$

$\Rightarrow \quad t = \frac{-1}{3}, \frac{3}{4} \Rightarrow 3^x = \frac{3}{4}$ [as $3^x \neq -$ve]

$\Rightarrow \quad x = \log_3\left(\frac{3}{4}\right) = \log_3 3 - \log_3 4 \Rightarrow x = 1 - \log_3 4$

11. (b) It is given that

$a_1, a_2, \ldots, a_n$ are in AP, $\forall\, a_i > 0$.

$\therefore\ a_2 - a_1 = a_3 - a_2 = \ldots = a_n - a_{n-1} = d$

Now, $\frac{1}{\sqrt{a_1}+\sqrt{a_2}} + \frac{1}{\sqrt{a_2}+\sqrt{a_3}} + \ldots + \frac{1}{\sqrt{a_{n-1}}+\sqrt{a_n}}$

$= \frac{\sqrt{a_1}-\sqrt{a_2}}{a_1 - a_2} + \frac{\sqrt{a_2}-\sqrt{a_3}}{a_2 - a_3} + \ldots + \frac{\sqrt{a_{n-1}}-\sqrt{a_n}}{a_{n-1} - a_n}$

$= \frac{\sqrt{a_1}-\sqrt{a_2}}{-d} + \frac{\sqrt{a_2}-\sqrt{a_3}}{-d} + \ldots + \frac{\sqrt{a_{n-1}}-\sqrt{a_n}}{-d}$

$= \frac{\sqrt{a_1}-\sqrt{a_2}+\sqrt{a_2}-\sqrt{a_3}+\ldots+\sqrt{a_{n-1}}-\sqrt{a_n}}{-d}$

$= \frac{\sqrt{a_1}-\sqrt{a_n}}{-d} = \frac{\sqrt{a_n}-\sqrt{a_1}}{d}$

$= \frac{\sqrt{a_n}-\sqrt{a_1}}{d} \times \frac{\sqrt{a_n}+\sqrt{a_1}}{\sqrt{a_n}+\sqrt{a_1}} = \frac{a_n - a_1}{d(\sqrt{a_n}+\sqrt{a_1})}$

$= \frac{(n-1)d}{d(\sqrt{a_n}+\sqrt{a_1})}$ [$\because a_n = a_1 + (n-1)\,d$]

$= \frac{n-1}{\sqrt{a_1}+\sqrt{a_n}}$

12. (a) Let the first set of number be $a-d, a, a+d$.

Then, $a - d + a + a + d = 15 \Rightarrow a = 5$

The second set of numbers will be

$b-(d-1), b, b+(d-1)$,

Again, $b-(d-1)+b+b+(d-1) = 15 \Rightarrow b = 5$

Hence, the sets of numbers are

$5-d, 5, 5+d$ and $6-d, 5, 4+d$

Further, from the given condition

$\frac{(5-d)5(5+d)}{(6-d)5(4+d)} = \frac{7}{8} \Rightarrow \frac{25-d^2}{24+2d-d^2} = \frac{7}{8}$

$\Rightarrow d^2 + 14d - 32 = 0 \Rightarrow d = 2, -16$

$\therefore$ The two sets are 3, 5, 7 and 4, 5, 6 or 21, 5, -11 and 22, 5, -12.

$\therefore$ Ratio of their smallest term is $\frac{3}{4}$ or $\frac{-11}{-12}$ i.e. $\frac{3}{4}$ or $\frac{11}{12}$.

13. (c) Let the sides of the triangle be $a-d, a, a+d$, then perimeter $= (a+d)+a+(a-d) = 21$

$\therefore \quad a = 7$.

Again, $(a-d)(a+d) = a+6$

$\Rightarrow a^2 - d^2 = a + 6 \Rightarrow 49 - d^2 = 13$

$\therefore \quad d = \pm 6$.

Hence, the side of the triangle are 1 cm, 7 cm, 13 cm.

14. (b) Let d be the common difference of an AP $a_1, a_2, \ldots, a_{21}$ and $\sum_{n=1}^{20} \frac{1}{a_n a_{n+1}} = \frac{4}{9}$

$\Rightarrow \quad \sum_{n=1}^{20} \frac{1}{a_n(a_n+d)} = \frac{4}{9}$

$\Rightarrow \quad \frac{1}{d}\sum_{n=1}^{20}\left(\frac{1}{a_n} - \frac{1}{a_n+d}\right) = \frac{4}{9}$

$\Rightarrow \frac{1}{d}\left[\frac{1}{a_1} - \frac{1}{a_2} + \frac{1}{a_3} - \frac{1}{a_4} + \ldots + \frac{1}{a_{20}} - \frac{1}{a_{21}}\right] = \frac{4}{9}$

$\Rightarrow \quad \frac{1}{d}\left(\frac{1}{a_1} - \frac{1}{a_{21}}\right) = \frac{4}{9} \Rightarrow \frac{1}{d}\left(\frac{a_{21}-a_1}{a_1 a_{21}}\right) = \frac{4}{9}$

$\Rightarrow \quad a_1 a_{21} = 45 \Rightarrow a_1(a_1 + 20d) = 45$ ...(i)

Also sum of first 21 terms = 189

$\Rightarrow \quad \frac{21}{2}(2a_1 + 20d) = 189$

$\Rightarrow \quad a_1 + 10d = 9$ ...(ii)

From Eqs. (i) and (ii), we get $a_1 = 3, d = 3/5$

or $\quad a_1 = 15, d = -\frac{3}{5}$

So, $a_6 a_{16} = (a_1 + 5d)(a_1 + 15d) = 72$

15. (c) According to the question,

$2+5+8+\ldots+2n$ terms $= 57+59+61+\ldots+n$ terms

$\Rightarrow \quad \frac{2n}{2}[4+(2n-1)3] = \frac{n}{2}[114+(n-1)2]$

$\Rightarrow \quad 6n+1 = n+56$

$\Rightarrow \quad 5n = 55 \Rightarrow n = 11$

16. (c) $a_1, a_2, \ldots, a_n$, are in AP

and $\quad a_1 = \frac{5}{2}, a_{10} = 16$

$\therefore \quad a_1 + a_2 + \ldots + a_n = 110$

$\Rightarrow \quad \frac{n}{2}(a_1 + a_n) = 110$

$\therefore \quad \frac{n}{2}\left[\frac{5}{2} + \frac{5}{2} + (n-1)d\right] = 110$

Now, $d = \frac{a_{10}-a_1}{10-1} = \frac{16-\frac{5}{2}}{9} = \frac{27}{9\times 2} = \frac{3}{2}$

$\Rightarrow \frac{n}{2}\left[5 + (n-1)\frac{3}{2}\right] = 110$ $\left[\because d = \frac{3}{2}\right]$

$\Rightarrow 3n^2 + 7n - 440 = 0$

$\Rightarrow 3n^2 + 40n - 33n - 440 = 0$

$\Rightarrow n(3n+40) - 11(3n+40) = 0$

$\Rightarrow \quad (3n+40)(n-11) = 0$

$n = \frac{-40}{3}$ or $n = 11$

$\therefore \quad n = 11$ (as $n \in N$)

Hence, option (c) is correct.

17. (d) Let $S=\log_{9^{1/2}} x+\log_{9^{1/3}} x+\ldots$

Using property, $\log_{a^b} x=\frac{1}{b}\log_a x$

$$S=2\log_9 x+3\log_9 x+\ldots+22\log_9 x$$
$$=\log_9 x(2+3+4+\ldots+22)$$
$$=\log_9 x\left[\frac{21}{2}(4+20)\right]=\log_9 x(21\times 12)$$

$\therefore S=252\log_9 x$

Given, $S=504$, then

$$252\log_9 x=504$$
$$\Rightarrow \log_9 x=2$$
$$\Rightarrow x=(9)^2=81$$

18. (a) $S_{n+3}-3S_{n+2}+3S_{n+1}-S_n$
$$=(S_{n+3}-S_{n+2})-2(S_{n+2}-S_{n+1})+S_{n+1}-S_n$$
$$=t_{n+3}-2t_{n+2}+t_{n+1}=0$$
$[\text{in AP } t_{n+3}-t_{n+2}=t_{n+2}-t_{n+1}]$

19. (c) Till 10th min, the number of counted notes is 1500.

$$\therefore \quad 3000=\frac{n}{2}[2\times 148+(n-1)(-2)]$$
$$=n(148-n+1)$$
$$\Rightarrow n^2-149n+3000=0$$
$$\therefore \quad n=125, 24$$
$\because$ $n=125$ is not possible.

Total time required is $24+10=34$ min

20. (a) Total savings $=200+200+200+240+280+\ldots+n$ months

Given that, total savings $=11040$

$$\Rightarrow 400+\frac{n-2}{2}[400+(n-3)\cdot 40]=11040$$
$$\Rightarrow (n-2)(140+20n)=10640$$
$$\Rightarrow 20n^2+100n-280=10640$$
$$\Rightarrow n^2+5n-546=0$$
$$(n-21)(n+26)=0$$
$$n=21;$$
$\because$ $n=-26$ (not possible)

21. (a) Let the first angle $=a$

Common difference $=d$

Number of sides, $n=9$

$\therefore$ Sum of interior angles $=(n-2)\times 180°$

$$\frac{n}{2}[2a+(n-1)d]=(n-2)\times 180°$$
$$\Rightarrow \frac{9}{2}[2a+8d)=7\times 180°$$
$$\Rightarrow a+4d=140°$$

and largest angle $T_9=a+8d$

$$a+8d<180°$$
$$\Rightarrow 4d<40°$$
$$d<10°$$
$$\Rightarrow d=9°$$

22. (b) From geometry, the sum of all internal angles $=(n-2)\times 180°$

where, n is the number of sides of the polygon.

$$\therefore \frac{n}{2}[2\times 120°+(n-1)\times 5°]=(n-2)\times 180°$$
$$\Rightarrow n^2-25n+144=0 \Rightarrow (n-16)(n-9)=0$$

If $n=16$, then 16th internal angle

$$=120°+(16-1)\times 5°=195°>180°$$

$\therefore$ $n\neq 16$. Hence, $n=9$

23. (b) As per question,

$$a+ar=12 \quad \ldots(i)$$
$$ar^2+ar^3=48 \quad \ldots(ii)$$
$$\Rightarrow \frac{ar^2(1+r)}{a(1+r)}=\frac{48}{12}$$
$$\Rightarrow r^2=4, \Rightarrow r=-2$$
[$\because$ terms are + ve and – ve alternately]
$$\Rightarrow a=-12$$

24. (b) Let $b=ar$ and $c=ar^2$, so that a,b,c are in GP

$$\therefore \quad a+b+c=xb$$
$$\Rightarrow a+ar+ar^2=x\cdot ar \Rightarrow r^2+(1-x)r+1=0 \quad \ldots(i)$$

If r is real, then

Discriminant of (i) ≥ 0

$$\Rightarrow (1-x)^2-4\cdot 1\cdot 1\geq 0 \Rightarrow x^2-2x-3\geq 0$$
$$\Rightarrow (x+1)(x-3)\geq 0 \Rightarrow x\leq -1 \text{ or } x\geq 3$$

Now, for $x=3$, we get $r=1$, which will make $a=b=c$

Also, for $x=-1$, we get $r=-1$, for which $a=c$, thus $x<-1$ or $x>3$.

25. (a) Since a,b,c are in AP and sum of a,b,c is given.

Let $a=b-D$, $c=b+D$ $\quad [D<0, a>b>c]$

and $a+b+c=\frac{3}{2}$

$$\Rightarrow b-D+b+b+D=\frac{3}{2}$$
$$\therefore \quad b=\frac{1}{2}$$

Then, $a=\frac{1}{2}-D$ and $C=\frac{1}{2}+D$

Also, a^2,b^2,c^2 are in GP, then

$$(b^2)^2=a^2c^2$$
$$\Rightarrow \pm b^2=ac \Rightarrow \pm\frac{1}{4}=\frac{1}{4}-D^2$$
$$D^2=\frac{1}{4}\pm\frac{1}{4}=\frac{1}{2} \quad [\because D\neq 0]$$
$$\therefore \quad D=\pm\frac{1}{\sqrt{2}} \Rightarrow D=-\frac{1}{\sqrt{2}} \quad [\because D<0]$$

Hence, $a=\frac{1}{2}+\frac{1}{\sqrt{2}}$

26. (b) $(a^2+b^2+c^2)p^2-2p(ab+bc+ca)+(a^2+b^2+c^2)\le 0$

$\Rightarrow (ap-b)^2+(bp-c)^2+(cp-a)^2\le 0$

$\Rightarrow ap-b=0, bp-c=0$ and $cp-a=0$

$$p=\frac{b}{a}, p=\frac{c}{b}, p=\frac{a}{c}$$

$$\Rightarrow \quad \frac{b}{a}=\frac{c}{b}=\frac{a}{c}$$

$\Rightarrow a, b, c$ are in GP.

27. (c) In the quadratic equation $ax^2+bx+c=0$

$\Delta=b^2-4ac$ and $\alpha+\beta=-\frac{b}{a}, \alpha\beta=\frac{c}{a}$

$$\alpha^2+\beta^2=(\alpha+\beta)^2-2\alpha\beta=\frac{b^2}{a^2}-\frac{2c}{a}=\frac{b^2-2ac}{a^2}$$

and $\alpha^3+\beta^3=-\frac{b^3}{a^3}-\frac{3c}{a}\left(-\frac{b}{a}\right)=-\left(\frac{b^3-3abc}{a^3}\right)$

According to question, $\alpha+\beta, \alpha^2+\beta^2, \alpha^3+\beta^3$ are in GP.

$\Rightarrow -\frac{b}{a}, \frac{b^2-2ac}{a^2}, -\frac{b^3-3abc}{a^3}$ are in GP

$$\Rightarrow \left(\frac{b^2-2ac}{a^2}\right)^2=\frac{b}{a}\left(\frac{b^3-3abc}{a^3}\right)$$

$\Rightarrow b^4+4a^2c^2-4ab^2c=b^4-3ab^2c$

$\Rightarrow 4a^2c^2-ab^2c=0 \Rightarrow ac\,\Delta=0$

$\Rightarrow \quad c\Delta=0 \quad$ [$\because$ in quadratic $a\ne 0$]

28. (c) $\log a, \log b, \log c$ are in AP

$\Rightarrow 2\log b=\log a+\log c \Rightarrow \log b^2=\log(ac)$

$\Rightarrow b^2=ac \Rightarrow a, b, c$ are in GP

$\log a-\log 2b, \log 2b-\log 3c, \log 3c-\log a$ are in AP

$\Rightarrow 2(\log 2b-\log 3c)=(\log a-\log 2b)+(\log 3c-\log a)$

$\Rightarrow 3\log 2b=3\log 3c \Rightarrow 2b=3c$

Now, $b^2=ac \Rightarrow b^2=a.\frac{2b}{3} \Rightarrow b=\frac{2a}{3}, c=\frac{4a}{9}$

i.e. $a=a, b=\frac{2a}{3}, c=\frac{4a}{9} \Rightarrow a:b:c=1:\frac{2}{3}:\frac{4}{9}=9:6:4$

Thus, a, b, c form a triangle.

29. (a) α, β are the roots of $x^2-x+p=0$

$\therefore \quad \alpha+\beta=1 \quad$ ...(i)

$\alpha\beta=p \quad$ ...(ii)

γ, δ are the roots of $x^2-4x+q=0$

$\therefore \quad \gamma+\delta=4 \quad$ ...(iii)

$\gamma\delta=q \quad$ ...(iv)

$\alpha, \beta, \gamma, \delta$ are in GP.

$\therefore$ Let $\alpha=a, \beta=ar, \gamma=ar^2, \delta=ar^3$

Substituting these values in Eqs. (i), (ii), (iii) and (iv), we get

$a+ar=1 \quad$ ...(v)

$a^2r=p \quad$ ...(vi)

$ar^2+ar^3=4 \quad$ ...(vii)

$a^2r^5=q \quad$ ...(viii)

Dividing Eq. (vii) by Eq. (v), we get

$$\frac{ar^2(1+r)}{a(1+r)}=\frac{4}{1} \Rightarrow r^2=4 \Rightarrow r=2, -2$$

$$\Rightarrow \quad a=\frac{1}{1+r}=\frac{1}{1+2} \text{ or } \frac{1}{1-2}=\frac{1}{3} \text{ or } -1$$

As p is an integer (given), r is also an integer (2 or -2)

$\Rightarrow \quad a\ne\frac{1}{3}.$

Hence, $a=-1$ and $r=-2$

$\therefore \quad p=(-1)^2\times(-2)=-2$

and $\quad q=(-1)^2\times(-2)^5=-32$

30. (d) We have, $\frac{b}{a}=\frac{c}{b}=\frac{d}{c}=r$

$b=ar, c=br=ar^2, d=cr=br^2=ar^3$

Now, $\frac{b+c}{a+b}=\frac{ar+ar^2}{a+ar}=r$ and $\frac{c+d}{b+c}=\frac{br+cr}{b+c}=r$

$\therefore a+b, b+c, c+d$ are in GP.

Again, $(b-c)^2+(c-a)^2+(d-b)^2$

$$=(ar-ar^2)^2+(ar^2-a)^2+(ar^3-ar)^2$$
$$=a^2r^2(1-r)^2+a^2(1-r^2)^2+a^2r^2(1-r^2)^2$$
$$=a^2(1-r)^2[r^2+(1+r)^2+r^2(1+r)^2]$$
$$=a^2(1-r)^2[r^2+(1+r)^2(1+r^2)]$$
$$=a^2(1-r)^2[r^2+(1+r^2)^2+2r(1+r^2)]$$
$$=a^2(1-r)^2(r+1+r^2)^2$$
$$=a^2(1-r^3)^2=(a-ar^3)^2=(a-d)^2$$

Further, $(a^2+b^2+c^2)(b^2+c^2+d^2)$

$$=(a^2+a^2r^2+a^2r^4)(a^2r^2+a^2r^4+a^2r^6)$$
$$=a^4r^2(1+r^2+r^4)(1+r^2+r^4)=[a^2r(1+r^2+r^4)]^2$$
$$=(a\cdot ar+ar\cdot ar^2+ar^2\cdot ar^3)^2=(ab+bc+cd)^2$$

31. (a) If the AP be $\alpha, \alpha+\beta, \alpha+2\beta, \ldots$ then

$a=\alpha+(p-1)\beta, b=\alpha+(q-1)\beta$ and $c=\alpha+(r-1)\beta$

From above three, we get

$b-c=(q-r)\beta \quad$ ...(i)

$c-a=(r-p)\beta \quad$ ...(ii)

$a-b=(p-q)\beta \quad$ ...(iii)

Let the GP be $A, AR, AR^2, \ldots$, then

$a=AR^{p-1} \quad$ ...(iv)

$b = AR^{q-1}$ ...(v)

$c = AR^{r-1}$ ...(vi)

Now, $\log[a^{b-c} \cdot b^{c-a} \cdot c^{a-b}]$

$= (b-c)\log a + (c-a)\log b + (a-b)\log c]$

$= (q-r)\beta\log(AR^{p-1}) + (r-p)\beta\log(AR^{q-1})$

$+ (p-q)\beta(AR^{r-1})]$

$= \beta[(q-r)\{\log A + (p-1)\log R\} + (r-p)$

$\{\log A + (q-1)\log R\} + (p-q)\{\log A + (r-1)\log R\}]$

$= \beta[\log A(q-r+r+p-p-q) + \log R$

$\{(q-r)(p-1) + (r-p)(q-1) + (p-q)(r-1)\}]$

$= \beta[\log A \times 0 + \log R \times 0] = 0$

$\therefore a^{b-c} \cdot b^{c-a} \cdot c^{a-b} = 1$

32. (a) $\because a, b, c$ are in AP

$\Rightarrow \quad 2b = a + c \Rightarrow b = \frac{a+c}{2}$ ...(i)

Also, $\frac{1}{p}, \frac{1}{q}, \frac{1}{r}$ are in AP

$\Rightarrow \quad \frac{2}{q} = \frac{1}{p} + \frac{1}{r}$

$\Rightarrow \quad q = \frac{2pr}{p+r}$

and ap, bq, cr are in GP.

$\Rightarrow \quad (bq)^2 = (ap)(cr)$

$\Rightarrow \quad \left[\left(\frac{a+c}{2}\right)\left(\frac{2pr}{p+r}\right)\right]^2 = (ac)(pr)$

$\Rightarrow \quad \frac{(a+c)^2}{ac} = \frac{(p+r)^2}{pr}$

$\Rightarrow \quad \frac{a}{c} + \frac{c}{a} + 2 = \frac{p}{r} + \frac{r}{p} + 2$

$\Rightarrow \quad \frac{p}{r} + \frac{r}{p} = \frac{a^2+c^2}{ac}$

33. (c) $\because x, y, z$ are the pth, qth and rth terms of an AP

$\therefore x = A + (p-1)D, y = A + (q-1)D,$

$z = A + (r-1)D$

where, A is the first term and D is the common difference.

$\Rightarrow x - y = (p-q)D, y - z = (q-r)D,$

$z - x = (r-p)D$

Also, x, y, z are the pth, qth and rth terms of a GP.

$\therefore x = AR^{p-1}, y = AR^{q-1}, z = AR^{r-1}$

$\therefore x^{y-z}y^{z-x}z^{x-y} = (AR^{p-1})^{y-z}(AR^{q-1})^{z-x}(AR^{r-1})^{x-y}$

$= A^{y-z+z-x+x-y}R^{(p-1)(y-z)+(q-1)(z-x)+(r-1)(x-y)}$

$= A^0R^{(p-1)(q-r)D+(q-1)(r-p)D+(r-1)(p-q)D}$

$= A^0R^0 = 1$

34. (a) We have,

$\frac{3}{2}x^2 + \frac{5}{3}x^3 + \frac{7}{4}x^4 + \ldots = \left(2-\frac{1}{2}\right)x^2 + \left(2-\frac{1}{3}\right)x^3 + \left(2-\frac{1}{4}\right)x^4 + \ldots$

$= 2(x^2 + x^3 + x^4 + \ldots) - \left(\frac{x^2}{2} + \frac{x^3}{3} + \frac{x^4}{4} + \ldots\right)$

$= 2 \cdot \frac{x^2}{1-x} - [-\log_e(1-x) - x]$

[using sum of infinite GP $= \frac{a}{1-r}$ and logarithmic series]

$= \frac{2x^2}{1-x} + x + \log_e(1-x)$

$= \frac{2x^2 + x - x^2}{1-x} + \log_e(1-x)$

$= \frac{x^2+x}{1-x} + \log_e(1-x)$

$= x\left(\frac{1+x}{1-x}\right) + \log_e(1-x)$

35. (d) Let the pollution on 1st day = 20

The pollution of 2nd day $= 20 \times 20\% = 20(0.20)$

The pollution of 3rd day $= 20 \times (0.20)^2$

Let in n days the atmosphere will be normal.

$\Rightarrow \quad 20(0.20)^{n-1} < 0.01$

$\left(\frac{2}{10}\right)^{n-1} < \frac{1}{2000}$

$(n-1)[\log 2 - \log 10] < [\log 1 - \log 2000]$

$\Rightarrow \quad n - 1 > \frac{3.3010}{0.6990} \Rightarrow n > 5.722$

Hence, the atmosphere will be normal in 6 days.

36. (b) Clearly, the required series is

$1 + a + ca + a(ca) + c(aca) + a(caca) + \ldots$ to $2n$ terms

$= 1 + a + ca + ca^2 + c^2a^2 + c^2a^3 + \ldots$ to $2n$ terms

$= (1 + ca + c^2a^2 + \ldots$ to n terms$) + (a + ca^2 + c^2a^3 + \ldots$ to n terms$)$

$= \frac{1\{1-(ca)^n\}}{1-ca} + \frac{a\{1-(ca)^n\}}{1-ca} = \frac{(1+a)(1-c^na^n)}{1-ca}$

37. (b) 1st set = 4 (number of letters, sent by Bijay)

2nd set = 4 + 4 + 4 + 4 = 16 (number of letters)

Similarly in the third set

$= 4 + 4 + 4 + \ldots + (16 \text{ terms}) = 64$

$\therefore$ The numbers of letters sent in the 1st, 2nd, 3rd set, ... are respectively 4, 16, 64, ..., which is a GP, with $a = 4$ and common ratio $= \frac{16}{4} = 4$

$\therefore$ Total number of letters in all the first 8 set

$$=\frac{4(4^8-1)}{4-1}=87380$$

$\therefore$ Total money spent on letters $=87380\times\frac{25}{100}$

$$=₹\,21845$$

38. (c) Given, sequence can be written as

$$\frac{7}{10}+\frac{77}{100}+\frac{777}{10^3}+\ldots+\text{ up to 20 terms}$$

$$=7\left[\frac{1}{10}+\frac{11}{100}+\frac{111}{10^3}+\ldots\text{up to 20 terms}\right]$$

Multiply and divided by 9,

$$\frac{7}{9}\left[\frac{9}{10}+\frac{99}{100}+\frac{999}{1000}+\ldots\text{up to 20 terms}\right]$$

$$=\frac{7}{9}\left[\left(1-\frac{1}{10}\right)+\left(1-\frac{1}{10^2}\right)+\left(1-\frac{1}{10^3}\right)+\ldots\text{up to 20 terms}\right]$$

$$=\frac{7}{9}\left[20-\frac{\frac{1}{10}\left(1-\left(\frac{1}{10}\right)^{20}\right)}{1-\frac{1}{10}}\right]$$

$$=\frac{7}{9}\left[\frac{179}{9}+\frac{1}{9}\left(\frac{1}{10}\right)^{20}\right]=\frac{7}{81}[179+(10)^{-20}]$$

39. (d) We have,

(i) $0.5+0.55+0.555+=\frac{5}{9}[0.9+0.99+0.999+\ldots]$

$$=\frac{5}{9}[(1-0.1)+(1-0.01)+(1-0.001)+\ldots\text{ to } n \text{ terms}]$$

$$=\frac{5}{9}\left[(1+1+\ldots+\text{ to } n\text{ terms})-\left(\frac{1}{10}+\frac{1}{10^2}+\frac{1}{10^3}+\ldots+\text{to } n \text{ terms}\right)\right]$$

$$=\frac{5}{9}\left[n-\frac{\frac{1}{10}\left\{1-\frac{1}{10^n}\right\}}{1-\frac{1}{10}}\right]=\frac{5n}{9}-\frac{5}{81}(1-10^{-n})$$

(ii) $8+88+888+\ldots$ to n terms.

$$=\frac{8}{9}[9+99+999+\ldots\text{ to } n\text{ terms}]$$

$$=\frac{8}{9}[(10-1)+(10^2-1)+(10^3-1)+\ldots\text{ to } n\text{ terms}]$$

$$=\frac{8}{9}[(10+10^2+10^3+\ldots\text{ to } n\text{ terms})-(1+1+1\ldots\text{ to } n\text{ terms})]$$

$$=\frac{8}{9}\left[\frac{10(10^n-1)}{10-1}-n\right]=\frac{80}{81}(10^n-1)-\frac{8n}{9}$$

(iii) The nth terms in the sequence is

$$x_n=1^2+2^2+3^2+\ldots+n^2$$

$$=\frac{n(n+1)(2n+1)}{6}=\frac{1}{3}n^3+\frac{1}{2}n^2+\frac{1}{6}n$$

$\therefore$ The required sum $=\Sigma x_n$

$$=\frac{1}{3}\Sigma n^3+\frac{1}{2}\Sigma n^2+\frac{1}{6}\Sigma n$$

$$=\frac{1}{3}\left[\frac{n(n+1)}{2}\right]^2+\frac{1}{2}\cdot\frac{n(n+1)(2n+1)}{6}+\frac{1}{6}\cdot\frac{n(n+1)}{2}$$

$$=\frac{n(n+1)}{12}[n(n+1)+(2n+1)+1]$$

$$=\frac{n(n+1)}{12}[n^2+3n+2]$$

$$=\frac{n(n+1)}{12}(n+1)(n+2)=\frac{n(n+1)^2(n+2)}{12}$$

40. (b) Let $a=$ first term of GP and $r=$ common ratio of GP, then GP is a, ar, ar^2.

Given, $S_\infty=20\Rightarrow\frac{a}{1-r}=20\Rightarrow a=20(1-r)$ ...(i)

Also, $a^2+a^2r^2+a^2r^4+\ldots$ to $\infty=100$

$$\Rightarrow\frac{a^2}{1-r^2}=100\Rightarrow a^2=100(1-r)(1+r)\quad\ldots\text{(ii)}$$

From Eq. (i), $a^2=400(1-r)^2$,

From Eq. (ii), we get $100(1-r)(1+r)=400(1-r)^2$

$$\Rightarrow\quad 1+r=4-4r\Rightarrow 5r=3\Rightarrow r=\frac{3}{5}$$

41. (c) $\because S_p=\frac{p}{1-\frac{1}{p+1}}=(p+1)$

$\therefore\quad S_1=2, S_2=3, S_3=4,\ldots$

$$\Rightarrow S_1+S_2+S_3+\ldots+S_p$$

$$=2+3+4+\ldots+(p+1)$$

$$=\frac{p}{2}(2+p+1)=\frac{p(p+3)}{2}$$

42. (c) $\frac{x}{1-r}=5\Rightarrow r=1-\frac{x}{5}$

Since GP contains infinite terms.

$\therefore\quad -1<r<1$

$$\Rightarrow\quad -1<1-\frac{x}{5}\Rightarrow -2<-\frac{x}{5}<0$$

$$\Rightarrow\quad 0<x<10$$

43. (d) Sum $=4$ and second term $=3/4$, it is given that first term is a and common ratio r.

$$\Rightarrow\quad\frac{a}{1-r}=4\text{ and } ar=3/4\Rightarrow r=\frac{3}{4a}$$

Therefore, $\frac{a}{1-\frac{3}{4a}}=4\Rightarrow\frac{4a^2}{4a-3}=4$

or $a^2 - 4a + 3 = 0 \Rightarrow (a-1)(a-3) = 0$

$\Rightarrow \quad a = 1$ or 3

When $a = 1$, $r = \frac{3}{4}$ and when $a = 3$, $r = \frac{1}{4}$

44. (c) $2.\overline{357} = 2 + .357 + 0.000357 + \dots \infty$

$$= 2 + \frac{357}{10^3} + \frac{357}{10^6} + \dots \infty$$

$$= 2 + \frac{\frac{357}{10^3}}{1 - \frac{1}{10^3}} = 2 + \frac{357}{999} = \frac{2355}{999}$$

45. (d) We have, $x = \frac{1}{1-a} \Rightarrow a = \frac{x-1}{x}$

and $y = \frac{1}{1-b} \Rightarrow b = \frac{y-1}{y}$

Now $\quad S = 1 + ab + a^2b^2 + \dots$ to ∞

$$= \frac{1}{1-ab} = \frac{1}{1 - \left(\frac{x-1}{x}\right)\left(\frac{y-1}{y}\right)} = \frac{xy}{x+y-1}$$

46. (c) Let GP be $1, r, r^2, \dots \infty$

Given $x_n = 2(x_{n+1} + x_{n+2} + \dots \infty)$

$\therefore \quad x_n = 2\frac{x_{n+1}}{1-r}$ [common ratio is r]

$\therefore \quad \frac{x_{n+1}}{x_n} = \frac{1-r}{2} \Rightarrow r = \frac{1-r}{2}, \therefore r = \frac{1}{3}$

The required series is $1 + \frac{1}{3} + \left(\frac{1}{3}\right)^2 + \dots \infty = \frac{1}{1 - \frac{1}{3}} = \frac{3}{2}$

47. (d) $a = \frac{1}{1 - (\sqrt{3} - 1)} = \frac{1}{2 - \sqrt{3}} = 2 + \sqrt{3}$

and $ab = 1 \Rightarrow b = 2 - \sqrt{3}$

So, a and b are roots of $x^2 - 4x + 1 = 0$.

48. (c) Clearly total distance described

$$= 120 + 2\left[120 \times \frac{4}{5} + 120 \times \frac{4}{5} \times \frac{4}{5} + \dots \text{ to } \infty\right]$$

Except in the first fall the same ball will travel twice in each step the same distance one upward and second downward travel.

$\therefore$ Distance travelled

$$= 120 + 2 \times 120\left[\frac{4}{5} + \left(\frac{4}{5}\right)^2 + \dots \text{ to } \infty\right]$$

$$= 120 + 240\left[\frac{\frac{4}{5}}{1 - \frac{4}{5}}\right] = 120 + 240 \times 4 = 1080 \text{ m}$$

49. (c) $A = \frac{1}{1-r^a} \Rightarrow 1 - r^a = \frac{1}{A} \Rightarrow r^a = 1 - \frac{1}{A} = \frac{A-1}{A}$

$B = \frac{1}{1-r^b} \Rightarrow 1 - r^b = \frac{1}{B} \Rightarrow r^b = 1 - \frac{1}{B} = \frac{B-1}{B}$

$\therefore \; a \log r = \log\left(\frac{A-1}{A}\right)$ and $b \log r = \log\left(\frac{B-1}{B}\right)$

$$\therefore \quad \frac{a}{b} = \frac{\log\left(\frac{A-1}{A}\right)}{\log\left(\frac{B-1}{B}\right)} = \log_{\frac{B-1}{B}}\left(\frac{A-1}{A}\right)$$

50. (d) $\frac{a + (a+1) + (a+2) + \dots + (a+6)}{7} = m$

$\Rightarrow 7a + 21 = 7m \Rightarrow a + 3 = m$

$\therefore \quad \frac{(a+2) + (a+3) + \dots + (a+12)}{11} = \frac{11a + 77}{11}$

$\Rightarrow \quad a + 7 = m - 3 + 7 = m + 4$

51. (b) $\because \quad \frac{x+y}{2\sqrt{xy}} = \frac{p}{q} \Rightarrow \frac{(x+y)^2}{4xy} = \frac{p^2}{q^2}$ ...(i)

$\Rightarrow \quad \frac{(x+y)^2}{4xy} - 1 = \frac{p^2}{q^2} - 1$

$\Rightarrow \quad \frac{(x-y)^2}{4xy} = \frac{p^2 - q^2}{q^2}$ ...(ii)

From Eqs. (i) and (ii), we get

$$\left(\frac{x+y}{(x-y)}\right)^2 = \frac{p^2}{p^2 - q^2}$$

$$\frac{x+y}{x-y} = \frac{p}{\sqrt{p^2 - q^2}}$$

$$\frac{2x}{2y} = \frac{p + \sqrt{p^2 + q^2}}{p - \sqrt{p^2 - q^2}}$$

[using componendo-dividendo]

$\therefore \; x : y = p + \sqrt{p^2 + q^2} : p - \sqrt{p^2 - q^2}$

52. (c) Let $S_\infty = 1 + 4x + 7x^2 + 10x^3 + \dots + \infty$ ...(i)

$\therefore \; x S_\infty = x + 4x^2 + 7x^3 + 10x^4 + \dots + \infty$ ...(ii)

Subtracting Eqs. (ii) from Eq. (i), we get

$(1-x)S_\infty = 1 + 3x + 3x^2 + 3x^3 + \dots \infty$

$\Rightarrow (1-x)S_\infty = 1 + 3(x + x^2 + x^3 + \dots \infty)$

$\Rightarrow (1-x)S_\infty = 1 + 3\left(\frac{x}{1-x}\right)$

$\therefore \quad S_\infty = \frac{(1+2x)}{(1-x)^2} = \frac{35}{16}$ [given that]

$\Rightarrow \quad 16 + 32x = 35 - 70x + 35x^2$

$\Rightarrow 35x^2 - 102x + 19 = 0$

$$\Rightarrow (7x-19)(5x-1)=0$$

$$x \neq \frac{19}{7} \quad [\because |x|<1 \text{ for infinite series}]$$

Hence, $x=\frac{1}{5}$

3. (b) $2^{\frac{1}{4}}\cdot 4^{\frac{1}{8}}\cdot 8^{\frac{1}{16}}\ldots\infty = 2^{\frac{1}{4}+\frac{2}{8}+\frac{3}{16}+\ldots\infty}$

$$=2^{\frac{1}{4}\left\{1+\frac{2}{2}+\frac{3}{2^2}+\frac{4}{2^3}+\ldots\infty\right\}}$$

$$=2^{\left(\frac{1}{4}\right)\left\{\frac{1}{1-\frac{1}{2}}+\frac{1\cdot\frac{1}{2}}{\left(1-\frac{1}{2}\right)^2}\right\}}$$

[using sum of AGP Formula]

$$=2^{\frac{1}{4}(2+2)}=2^{\frac{4}{4}}$$

$$=2$$

4. (b) $S_\infty = 1+4\left(\frac{1}{5}\right)+7\left(\frac{1}{5}\right)^2+\ldots\infty$

$$\therefore \frac{1}{5}S_\infty=\left(\frac{1}{5}\right)+4\left(\frac{1}{5}\right)^2+7\left(\frac{1}{5}\right)^3+\ldots\infty$$

$$\Rightarrow \left(1-\frac{1}{5}\right)S_\infty = 1+3\left[\frac{1}{5}+\left(\frac{1}{5}\right)^2+\left(\frac{1}{5}\right)^3+\ldots\infty\right]$$

$$=1+3\left(\frac{\frac{1}{5}}{1-\frac{1}{5}}\right)=1+\frac{3}{4}$$

$$\Rightarrow \quad \frac{4}{5}S_\infty=\frac{7}{4}$$

$$\therefore \quad S_\infty=\frac{35}{16}$$

5. (a) $\frac{1}{1\cdot 2}-\frac{1}{2\cdot 3}+\frac{1}{3\cdot 4}\ldots\ldots\infty$

$$|T_n|=\frac{1}{n(n+1)}=\left(\frac{1}{n}-\frac{1}{n+1}\right)$$

$$S=T_1-T_2+T_3-T_4+T_5\ldots\infty$$

$$=\left(\frac{1}{1}-\frac{1}{2}\right)-\left(\frac{1}{2}-\frac{1}{3}\right)+\left(\frac{1}{3}-\frac{1}{4}\right)-\left(\frac{1}{4}-\frac{1}{5}\right)$$

$$=1-2\left[\frac{1}{2}-\frac{1}{3}+\frac{1}{4}-\frac{1}{5}\ldots\infty\right]$$

$$=1-2[-\log(1+1)+1]=2\log 2-1=\log\left(\frac{4}{e}\right)$$

6. (b) If n is odd, the required sum is

$$1^2+2\cdot 2^2+3^2+2\cdot 4^2+\ldots 2(n-1)^2+n^2$$

$$=\frac{(n-1)(n-1+1)^2}{2}+n^2 \quad [\because (n-1) \text{ is even}$$

$\because$ Using given formula for the sum of $(n-1)$ terms]

$$=\left(\frac{n-1}{2}+1\right)n^2=\frac{n^2(n+1)}{2}$$

57. (a) $1^3-2^3+3^3-4^3+\ldots+9^3$

$$=(1^3+2^3+3^3+4^3+\ldots+9^3)-2(2^3+4^3+6^3+8^3)$$

$$=\left[\frac{9\times 10}{2}\right]^2-2\times 2^3(1^3+2^3+3^3+4^3)$$

$$=(45)^2-16\left(\frac{4\times 5}{2}\right)^2=425$$

58. (d) $S_n=\frac{1}{{}^nC_0}+\frac{1}{{}^nC_1}+\frac{1}{{}^nC_2}+\ldots+\frac{1}{{}^nC_n}$ ...(i)

$$t_n=\frac{0}{{}^nC_0}+\frac{1}{{}^nC_1}+\frac{2}{{}^nC_2}+\ldots+\frac{n}{{}^nC_n} \quad \ldots\text{(ii)}$$

$$t_n=\frac{n}{{}^nC_n}+\frac{n-1}{{}^nC_{n-1}}+\frac{n-2}{{}^nC_{n-2}}+\ldots+\frac{0}{{}^nC_0} \quad \ldots\text{(iii)}$$

Add Eqs (ii) and (iii), we get

$$2t_n=(n)\left[\frac{1}{{}^nC_0}+\frac{1}{{}^nC_1}+\ldots+\frac{1}{{}^nC_n}\right]=nS_n \text{ [from Eq. (i)]}$$

$$\therefore \frac{t_n}{S_n}=\frac{n}{2}$$

59. (a) $\because \sum_{n=1}^{\infty}\frac{7n+32}{n(n+2)}\cdot\frac{3^n}{4^n}=\sum_{n=1}^{\infty}\frac{16(n+2)-9n}{n(n+2)}\cdot\frac{3^n}{4^n}$

$$=\sum_{n=1}^{\infty}\left[\left(\frac{3^n}{n\cdot 4^{n-2}}\right)-\left(\frac{3^{n+2}}{(n+2)4^n}\right)\right]$$

$$=\left(\frac{3}{1\cdot 4^{-1}}-\frac{3^3}{3\cdot 4^1}\right)+\left(\frac{3^2}{2\cdot 4^0}-\frac{3^4}{4\cdot 4^2}\right)$$

$$+\left(\frac{3^3}{3\cdot 4^1}-\frac{3^3}{5\cdot 4^3}\right)+\ldots$$

$$=\frac{3}{1\cdot 4^{-1}}+\frac{3^2}{2\cdot 4^0}+0=12+\frac{9}{2}$$

$$=\frac{33}{2} \quad \left[\because \lim_{x\to\infty}\frac{3^{n+2}}{(n+2)4^n}=0\right]$$

60. (b) Let $S=\frac{1}{10}+\frac{2}{10^2}+\frac{3}{10^3}+\ldots$ ...(i)

$$\frac{S}{10}=\frac{1}{10^2}+\frac{2}{10^3}+\ldots\ldots \quad \ldots\text{(ii)}$$

Subtracting Eq. (ii) from Eq. (i), we have

$$\frac{9S}{10}=\frac{1}{10}+\frac{1}{10^2}+\frac{1}{10^3}+\ldots$$

$$\Rightarrow \quad \frac{9S}{10}=\frac{1/10}{1-\frac{1}{10}}=\frac{1}{9}$$

$$\Rightarrow \quad S=\frac{1}{9}\left(\frac{10}{9}\right)=\frac{10}{81}$$

61. (c) Let $S = \frac{1}{2} + \frac{3}{4} + \frac{7}{8} + \frac{15}{16} + ... n$ terms

$$= \left(1-\frac{1}{2}\right) + \left(1-\frac{1}{4}\right) + \left(1-\frac{1}{8}\right) + \left(1-\frac{1}{16}\right) ... n \text{ terms}$$

$$= (1+1+1+...n \text{ terms}) - \left(\frac{1}{2} + \frac{1}{2^2} + \frac{1}{2^3} + \frac{1}{2^4} + ... + \frac{1}{2^n}\right)$$

$$= n - \left[\frac{\frac{1}{2}\left(1-\frac{1}{2^n}\right)}{1-\frac{1}{2}}\right] = n - 1 + 2^{-n}$$

62. (b) $S_n = an^2 + bn + c$

$\therefore S_{n-1} = a(n-1)^2 + b(n-1) + c$ for $n \geq 2$

$\therefore \quad t_n = S_n - S_{n-1}$

$= a\{n^2 - (n-1)^2\} + b\{n - (n-1)\} = a(2n-1) + b$

$\therefore \quad t_n = 2an + b - a; n \geq 2$

$\therefore t_{n-1} = 2a(n-1) + b - a$ for $n \geq 3$

$\therefore t_n - t_{n-1} = 2a(n-n+1) = 2a$ for $n \geq 3$

$\therefore \quad t_3 - t_2 = t_4 - t_3 = ... = 2a$

Now $t_2 - t_1 = (S_2 - S_1) - S_1 = S_2 - 2S_1$

$= (a \cdot 2^2 + b \cdot 2 + c) - 2\{a \cdot 1^2 + b \cdot 1 + c\} = 2a - c \neq 2a$

$\therefore$ Series is arithmetic from second term onwards.

63. (c) Let $S_1 = 2 + 3 + 5 + 9 + 16 + ... + x_n$

$$\begin{array}{l} S_1 = 2 + 3 + 5 + 9 + ... x_{n-1} + x_n \\ \hline 0 = 2 + [1 + 2 + 4 + 7 + ... + \text{to } (n-1) \text{ term}] - x_n \end{array}$$

$\therefore x_n = 2 + [1 + 2 + 4 + 7 + ... \text{to } (n-1) \text{ terms}]$

again let $S_2 = 1 + 2 + 4 + 7 + ... + t_{n-1}$

$$\begin{array}{l} S_2 = 1 + 2 + 4 + 7 + ... + t_{n-2} + t_{n-1} \\ \hline 0 = 1 + [1 + 2 + 3 + ... + (n-2) \text{term}] - t_{n-1} \end{array}$$

$$t_{n-1} = 1 + \frac{(n-2)(n-1)}{2} = \frac{n^2 - 3n + 4}{2}$$

$$\therefore S_n = \sum_{n=1}^{n-1} t_{n-1} = \frac{1}{2}\Sigma n^2 - \frac{3}{2}\Sigma n + 2\Sigma 1$$

$$= \frac{1}{2}\frac{(n-1)\,n\,(2n-1)}{6} - \frac{3}{2}\frac{n(n-1)}{2} + 2(n-1)$$

$$= (n-1)\left[\frac{2n^2 - n}{12} - \frac{3n}{4} + 2\right]$$

$$= \frac{n-1}{12}[2n^2 - n - 9n + 24]$$

$$= \frac{n-1}{6}[n^2 - 5n + 12] = \frac{n^3 - 6n^2 + 17n - 12}{6}$$

$$\therefore \quad x_n = 2 + S_2 = 2 + \frac{n^3 - 6n^2 + 17n - 12}{6}$$

$$= \frac{n^3 - 6n^2 + 17n}{6}$$

So, $x_{20} = 990$

64. (b) The given series is arithmetic whose first term = 20 and common difference $= -\frac{2}{3}$

As the common difference is negative the terms will become negative after some stage. So the sum is maximum when all positive terms are added now, for the positive terms $x_n \geq 0$

$$\Rightarrow \quad 20 + (n-1)\left(-\frac{2}{3}\right) \geq 0$$

$$\Rightarrow \quad 60 - 2(n-1) \geq 0 \Rightarrow n \leq 31$$

$\therefore$ The first 31 terms are non-negative

$\therefore$ Maximum sum

$$= S_{31} = \frac{31}{2}\left[2 \times 20 + (31-1) \times \left(-\frac{2}{3}\right)\right] = 310$$

65. (c) $S_n = 1 + 2a + 3a^2 + ... na^{n-1}$

$a \cdot S_n = a + 2a^2 + ... + (n-1)a^{n-1} + na^n$ where, $a = 1 + \frac{1}{n}$

$\therefore (1-a)S_n = 1 + a + a^2 + ... a^{n-1} - na^n$

$$(1-a)S_n = \frac{a^n - 1}{a - 1} - na^n$$

$$\Rightarrow \quad -\frac{1}{n}S_n = \frac{\left(1+\frac{1}{n}\right)^n - 1}{\frac{1}{n}} - n\left(1+\frac{1}{n}\right)^n = -n \Rightarrow S_n = n^2$$

66. (c) $S_n = \frac{1}{9}(9) + \frac{2}{9}(99) + \frac{3}{9}(999) + ...$

$$= \frac{1}{9}[10 + 2\cdot 10^2 + 3\cdot 10^3 + ...] - \frac{1}{9}[1 + 2 + 3 ....]$$

$$= \frac{1}{9}S - \frac{1}{9}\frac{n(n+1)}{2}$$

$$S = 10 + 2\cdot 10^2 + 3\cdot 10^3 + ... + n\cdot 10^n$$

$$\begin{array}{l} 10S = 10^2 + 2\cdot 10^3 + ... + (n-1)10^n + n\cdot 10^{n+1} \\ \hline -9S = (10 + 10^2 + 10^3 + ... + 10^n) - n10^{n+1} \end{array}$$

$$S = \frac{n}{9}10^{n+1} - \frac{10^{n+1} - 10}{81}$$

$$\therefore \quad S_n = \frac{n}{81}10^{n+1} - \frac{10^{n+1} - 10}{9\cdot 81} - \frac{1}{9}\frac{n(n+1)}{2}$$

$$\therefore 9S_n = \frac{(9n-1)\,10^{n+1}}{81} + \frac{10}{81} - \frac{n(n+1)}{2}$$

$$\therefore 9(S_n - S_{n-1}) = \frac{10^n}{81}$$

$$\{10(9n-1) - (9n-10)\} - n = n(10^n - 1)$$

67. (c) Let first term = a, common ratio = r, where $-1 < r < 1$

Then, $\frac{a}{1-r} = 2$ and $\frac{a^3}{1-r^3} = 24$

$$\therefore \quad \frac{1-r^3}{(1-r)^3} = \frac{1}{3}$$

i.e. $1-2r+r^2=3(1+r+r^2)$ or $2r^2+5r+2=0$

$\therefore \quad r=-2$ or $-\dfrac{1}{2}$

As $\quad -1<r<1$

$\therefore$ We have $r=-\dfrac{1}{2}$

$\therefore$ The series is $3-\dfrac{3}{2}+\dfrac{3}{4}-\dfrac{3}{8}+\ldots$

68. (b) The general term is

$$T_n=\frac{\frac{n}{2}\cdot\frac{n+1}{2}}{1^3+2^3+3^3+\ldots+n^3}=\frac{1}{n(n+1)}=\frac{1}{n}-\frac{1}{n+1}$$

$$\therefore S_n=1-\frac{1}{n+1}=\frac{n}{n+1}$$

69. (d) $\dfrac{e^x+e^{-x}}{2}=1+\dfrac{x^2}{2!}+\dfrac{x^4}{4!}+\dfrac{x^6}{6!}+\ldots$

Putting $x=\dfrac{1}{2}$ we get

$$1+\frac{1}{4\cdot 2!}+\frac{1}{16\cdot 4!}+\frac{1}{64\cdot 6!}+\ldots$$

$$=\frac{e^{1/2}+e^{-1/2}}{2}=\frac{\sqrt{e}+\frac{1}{\sqrt{e}}}{2}=\frac{e+1}{2\sqrt{e}}$$

70. (c) Obviously, the well (W) must be on one side of the trees $T_1, T_2, \ldots\ldots, T_{25}$.

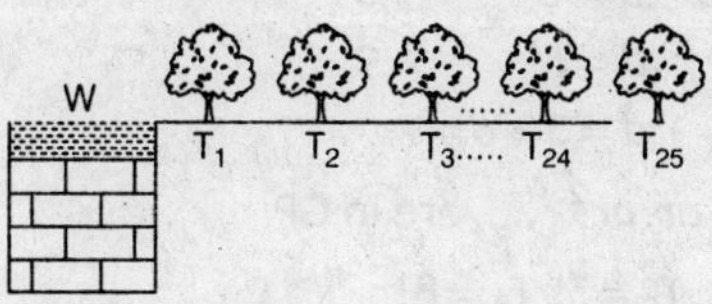

The total distance covered by the gardener

$=WT_1+(2WT_1+T_1T_2)+(2WT_2+T_2T_3)+\ldots\ldots$
$\qquad\qquad +(2WT_{24}+T_{24}T_{25})$

$=10+(2\times 10+5)+(2\times 15+5)+\ldots.+$ to 25 terms.

$=10+(25+35+45+\ldots.$ to 24 terms)

$=10+\dfrac{24}{2}[2\times 25+(24-1)\times 10]=10+12[50+230]$

$=3370$ m

71. (c) Given that for an AP, $S_n=cn^2$

Then $T_n=S_n-S_{n-1}=cn^2-c(n-1)^2=(2n-1)c$

$\therefore$ Sum of squares of n terms of this AP

$=\Sigma T_n^2=\Sigma(2n-1)^2\cdot c^2$

$=c^2[4\Sigma n^2-4\Sigma n+n]$

$=c^2\left[\dfrac{4n(n+1)(2n+1)}{6}-\dfrac{4n(n+1)}{2}+n\right]$

$=c^2 n\left[\dfrac{4n^2-1}{3}\right]=\dfrac{n(4n^2-1)c^2}{3}$

72. (d) Let the first term of AP be 'a' and the common difference be 'd'.

$$\sum_{k=0}^{12} a_{4k+1}=416$$

$\Rightarrow a_1+a_5+a_9+\ldots+a_{49}=416$

$\Rightarrow \quad \dfrac{13}{2}\times(a_1+a_{49})=416$

$a_1+a_1+48d=64$

$\therefore \quad a_1+24d=32 \qquad \ldots$(i)

Also, $a_9+a_{43}=66$

$\Rightarrow a+8d+a+42d=66$

$\Rightarrow \quad a+25d=33 \qquad \ldots$(ii)

On solving Eqs. (i) and (ii), we get

$d=1, a=8$

Now, $140m=a_1^2+a_2^2+\ldots+a_{17}^2$

$=8^2+9^2+\ldots+24^2$

$=(1^2+2^2+\ldots 24^2)-(1^2+2^2+\ldots.7^2)$

$=\dfrac{24\times 25\times 49}{6}-\dfrac{7\times 8\times 15}{6}$

$=4900-140$

$=140(35-1)=140\times 34$

$\therefore m=34$

73. (a) Let m and $(m+1)$ be the removed numbers from 1, 2, 3, ..., n.

Then, the sum of the remaining numbers is

$$\frac{n(n+1)}{2}-(2m+1)$$

As given, $\dfrac{105}{4}=\dfrac{\frac{n(n+1)}{2}-(2m+1)}{(n-2)}$

or $\quad 2n^2-103n-8m+206=0$

$\because m$ and n are integers, so n must be even.

Let $n=2k$, then

$$m=\frac{4k^2+103(1-k)}{4}$$

$\because m$ is an integer, then $(1-k)$ must be divisible by 4.

Let $k=1+4t$.

Then we get $n=8t+2$

and $m=16t^2-95t+1$

Now, $1\le m<n$

$1\le 16t^2-95t+1<8t+2$

On solving, we get $t=6$

Hence $n=50$ and $m=7$

74. (c) Since, p and q are positive real numbers

$p^2+q^2=1$ (given)

Using AM $\geq$ GM

$\therefore \left(\frac{p+q}{2}\right)^2 \geq (\sqrt{pq})^2 = \frac{p^2+q^2+2pq}{4} \geq pq$

$\frac{1+2pq}{4} \geq pq$ or $1+2pq \geq 4pq$

$1 \geq 2pq$ or $pq \leq \frac{1}{2}$

Now, $(p+q)^2 = p^2+q^2+2pq$

$\Rightarrow \quad (p+q)^2 \leq 1+2\times\frac{1}{2}$

$\Rightarrow \quad p+q \leq \sqrt{2}$

75. (d) The roots of equation are 2 and 9

$\therefore \quad g = \sqrt{xy} = 2$

$\Rightarrow \quad xy = 4 \quad \quad$ ...(i)

$G = \sqrt{(x+1)(y+1)} = 3$

$\Rightarrow \quad (x+1)(y+1) = 9 \quad \quad$ ...(ii)

On solving Eqs. (i) and (ii), we get

$\therefore\ x = y = 2$

76. (b) $m = \frac{l+n}{2}$ and l, G_1, G_2, G_3, n are in GP

$\therefore \quad r = \left(\frac{n}{l}\right)^{1/4}$

$\therefore\ G_1 = l\left(\frac{n}{l}\right)^{1/4}, G_2 = l\left(\frac{n}{l}\right)^{1/2}; G_3 = l\left(\frac{n}{l}\right)^{3/4}$

$(G_1)^4 + 2(G_2)^4 + (G_3)^4$

$= l^4\frac{n}{l} + 2l^4\left(\frac{n}{l}\right)^2 + l^4\left(\frac{n}{l}\right)^3$

$= l^3n + 2l^2n^2 + ln^3$

$= nl(l^2+2nl+n^2)$

$= nl(n+l)^2$

$= 4m^2nl$

77. (0.8) $\because \frac{a_1+2}{2} = \sqrt{2a_1}$

$\Rightarrow (a_1-2)^2 \Rightarrow 0 \Rightarrow a_1 = 2$

and $(a_n+2)^2 = 8S_n$

$(a_{n-1}+2)^2 = 8S_{n-1}$

$\Rightarrow \quad (a_n+2)^2 - (a_{n-1}+2)^2 = 8a_n$

$\Rightarrow (a_n - a_{n-1})(a_n + a_{n-1} + 4) = 8a_n$

$\Rightarrow a_n^2 - a_{n-1}^2 = 4(a_n + a_{n-1})$

$\Rightarrow a_n - a_{n-1} = 4$

$\therefore \quad a_{14} - a_{13} = 4$

$\Rightarrow \quad \frac{a_{14}-a_{13}}{5} = \frac{4}{5} = 0.8$

78. (2.25) Given that $\frac{\frac{7}{2}[2a+6d]}{\frac{11}{2}[2a+10d]} = \frac{6}{11}$

$\Rightarrow \quad \frac{7(2a+6d)}{(2a+10d)} = 6$

$\Rightarrow \quad 2a = 18d$

$\Rightarrow \quad a = 9d$

Also, $130 < a+6d < 140$

$130 < 9d+6d < 140 \quad\quad (\because a = 9d)$

$130 < 15d < 140$

$\frac{26}{3} < d < \frac{28}{3}$

$\Rightarrow d = 9$ (as all the terms of the AP are natural number)

$\Rightarrow \frac{d}{4} = \frac{9}{4} = 2.25$

79. (9) $a_1, a_2, a_3, \ldots, a_{100}$ is an AP, $a_1 = 3$, $S_p = \sum_{i=1}^{p} a_i$,

$1 \leq p \leq 100$

$\frac{S_m}{S_n} = \frac{S_{5n}}{S_n} = \frac{\frac{5n}{2}[6+(5n-1)d]}{\frac{n}{2}[6+(n-1)d]}$

$\frac{S_m}{S_n}$ is independent of n

$\Rightarrow \quad 6-d = 0$

$d = 6$

$\therefore a_2 = a_1 + d = 3+6 = 9$

80. (3) Let $a, ar, ar^2, \ldots$ are in GP

$T_5 = 7!, T_8 = 8!$

$\Rightarrow \quad ar^4 = 7!; ar^7 = 8!$

or $\quad \frac{ar^7}{ar^4} = \frac{8!}{7!}$ or $r^3 = 8$ or $r = 2$

Hence, $a(2^4) = 7! = 5040$

$\therefore \quad a = \frac{5040}{16} = 315$

So, 315, 630, 1260, ... are in GP.

$\therefore \quad S_3 = 2205 \Rightarrow n = 3$

81. (9) Given,

$\frac{1}{\alpha(\alpha+1)\ldots+(\alpha+20)} = \sum_{k=0}^{20} \frac{A_k}{\alpha+k}, \alpha > 0$

$\Rightarrow \quad \frac{1}{\alpha(\alpha+1)\ldots(\alpha+20)} = \frac{A_0}{\alpha} + \frac{A_1}{\alpha+1} + \ldots + \frac{A_{20}}{\alpha+20}$

$A_{14} = \frac{1}{(-14)(-13)\ldots(-1)(1)\ldots(6)} = \frac{1}{14!6!}$

$A_{15} = \frac{1}{(-15)(-14)\ldots(-1)(1)\ldots(5)} = \frac{-1}{15!5!}$

$$A_{13}=\frac{1}{(-13)...(-1)(1)...(7)}=\frac{-1}{13!7!}$$

$$\frac{A_{14}}{A_{13}}=\frac{-13!7!}{14!6!}=\frac{-7}{14}=\frac{-1}{2}$$

$$\Rightarrow \frac{A_{15}}{A_{13}}=\frac{13!7!}{15!5!}=\frac{42}{15\times 14}=\frac{1}{5}$$

$$\therefore 100\left(\frac{A_{14}}{A_{13}}+\frac{A_{15}}{A_{13}}\right)^2=100\left(-\frac{1}{2}+\frac{1}{5}\right)^2$$

$$=100\left(\frac{-3}{10}\right)^2=9$$

82. (32) Let the GP be $a, ar, ar^2, ...$

$$P_n=a(ar)(ar^2)....(ar^{n-1})$$

$$=a^n r^{1+2+....+(n-1)}=a^n r^{\frac{(n-1)n}{2}}$$

$$\therefore \quad S=\sum_{n=1}^{\infty}\sqrt[n]{P_n}=\sum_{n=1}^{\infty} ar^{\frac{n-1}{2}}$$

Now, $\sum_{n=1}^{\infty} ar^{\frac{n-1}{2}}=a[1+\sqrt{r}+r+r\sqrt{2}+....\infty]=\frac{a}{1-\sqrt{r}}$

Given, $a=16$, $r=\frac{1}{4}$;

$$\therefore \quad S=\frac{16}{1-\left(\frac{1}{2}\right)}=32$$

83. (1) Let a be the first term and r be the common ratio of GP. Then,

$$625=\frac{a(1-r^{201})}{1-r}$$

Now, $\sum_{i=1}^{201}\frac{1}{a_i}=\frac{1}{a_1}+\frac{1}{a_2}+...+\frac{1}{a_{201}}$

$$=\frac{1}{a}+\frac{1}{ar}+...+\frac{1}{ar^{200}}$$

$$=\frac{\frac{1}{a}\left[\left(\frac{1}{r}\right)^{201}-1\right]}{\left(\frac{1}{r}-1\right)}$$

$$=\frac{1}{a}\left(\frac{1-r^{201}}{1-r}\right)\frac{1}{r^{200}}$$

$$=\frac{1}{a}\times\frac{625}{a}\times\frac{1}{r^{200}}$$

$$=\frac{625}{(ar^{100})^2}=\frac{625}{(a_{101})^2}$$

$$=\frac{625}{25\times 25}=1$$

84. (27) Let AGP be $a, (a+d)r, (a+2d)r^2, ...$

On comparing, with given expression, we get

$$a=3, (3+d)r=4, (3+2d)r^2=4$$

$$\therefore \quad (3+d)=\frac{4}{r};\ (3+2d)=\frac{4}{r^2}$$

$$\Rightarrow \quad \frac{4}{r}-3=d;\ d=\left(\frac{4}{r^2}-3\right)\frac{1}{2}$$

$$\Rightarrow \quad \frac{4}{r}-3=\left(\frac{4}{r^2}-3\right)\frac{1}{2}$$

$$\Rightarrow 3r^2-8r+4=0$$

$$\Rightarrow (3r-2)(r-2)=0$$

$\because$ We have infinite AGP, $r=\frac{2}{3}$

$$\therefore \quad (3+d)\frac{2}{3}=4 \Rightarrow d=3$$

$\therefore$ Sum of infinite AGP is $S_\infty=\frac{a}{1-r}+\frac{dr}{(1-r)^2}$

$$S_\infty=\frac{3}{\left(1-\frac{2}{3}\right)}+\frac{3\times\frac{2}{3}}{\left(1-\frac{2}{3}\right)^2}$$

$$\therefore \quad S_\infty=27$$

85. (85) $S=\frac{7}{2^2\times 5^2}+\frac{13}{5^2\times 8^2}+\frac{19}{8^2\times 11^2}+....10$ terms

$$=\sum_{r=1}^{10}\frac{6r+1}{(3r-1)^2(3r+2)^2}$$

$$=\frac{1}{3}\sum_{r=1}^{10}\frac{(3r+2)^2-(3r-1)^2}{(3r-1)^2(3r+2)^2}$$

$$=\frac{1}{3}\sum_{r=1}^{10}\left[\frac{1}{(3r-1)^2}-\frac{1}{(3r+2)^2}\right]$$

$$=\frac{1}{3}\left[\frac{1}{2^2}-\frac{1}{2^{10}}\right]=\frac{1}{3}\left(\frac{2^8-1}{2^{10}}\right)$$

$$\therefore \quad S=\frac{85}{1024}$$

$$\Rightarrow 1024S=85$$

86. (115) Given sum is

$$1\left(\frac{1}{1}+\frac{1}{2}+\frac{1}{3}+...+\frac{1}{20}\right)+2\left(\frac{1}{2}+\frac{1}{3}+...+\frac{1}{20}\right)$$

$$+3\left(\frac{1}{3}+\frac{1}{4}+....+\frac{1}{20}\right)+20\left(\frac{1}{20}\right)$$

$$=\frac{1}{1}(1)+\frac{1}{2}(1+2)+\frac{1}{3}(1+2+3)+.....$$

$$+\frac{1}{20}(1+2+3+....+20)$$

$$= \sum_{r=1}^{20} \frac{1}{r}\left(\frac{r(r+1)}{2}\right) = \sum_{r=1}^{20} \frac{r+1}{2}$$

$$= \frac{1+1}{2} + \frac{2+1}{2} + \frac{3+1}{2} + \ldots + \frac{20+1}{2}$$

$$= \frac{20 \times 21}{4} + 10 = 115$$

87. (6) We have $S = \sum_{n=0}^{\infty} \frac{2n+3}{3^n}$

$$S = 3 + \sum_{n=1}^{\infty} \frac{2n+3}{3^n}$$

$$= 3 + \sum_{n=1}^{\infty} \frac{1}{3^{n-1}} - \sum_{n=1}^{\infty} \frac{2n}{3^n}$$

$$= 3 + \frac{1}{\left(1-\frac{1}{3}\right)} + \sum_{n=1}^{\infty} \frac{2n}{3^n} \quad \ldots\text{(i)}$$

$$S_1 = \sum_{n=1}^{\infty} \frac{2n}{3^n} = \frac{2}{3} + \frac{4}{3^2} + \frac{6}{3^3} + \ldots$$

$$\frac{S_1}{3} = \frac{2}{3^2} + \frac{4}{3^3} + \frac{6}{3^4} + \ldots$$

On subtracting,

$$\frac{2S_1}{3} = \frac{2}{3}\left[1 + \frac{1}{3} + \frac{1}{3^2} + \ldots \infty\right]$$

$$\therefore \quad S_1 = \frac{1}{\left(1-\frac{1}{3}\right)} = \frac{3}{2}$$

From Eq. (i)

$$S = 3 + \frac{3}{2} + \frac{3}{2} \Rightarrow S = 6$$

88. (3) $S_k = \dfrac{\frac{k-1}{k!}}{1-\frac{1}{k}} = \dfrac{1}{(k-1)!}$

$$= \sum_{k=2}^{100} \left|(k^2 - 3k + 1)\frac{1}{(k-1)!}\right|$$

$$= \sum_{k=2}^{\infty} \left|\frac{(k-1)^2 - k}{(k-1)!}\right|$$

$$= \sum_{k=2}^{\infty} \left|\frac{k-1}{(k-2)!} - \frac{k}{(k-1)!}\right|$$

$$= \left|\frac{1}{0!} - \frac{2}{1!}\right| + \left|\frac{2}{1!} - \frac{3}{2!}\right| + \left|\frac{3}{2!} - \frac{4}{3!}\right| + \ldots$$

$$= \frac{2}{1!} - \frac{1}{0!} + \frac{2}{1!} - \frac{3}{2!} + \frac{3}{2!} - \frac{4}{3!} + \ldots + \frac{99}{98!} - \frac{100}{99!}$$

$$= 3 - \frac{100}{99!}$$

$$= 3 - \frac{100 \times 100}{100 \times 99!}$$

$$= 3 - \frac{(100)^2}{100!}$$

$$\therefore \quad \frac{(100)^2}{100!} + \Sigma\left|k^2 - 3k + 1\frac{1}{(k-1)!}\right| = 3$$

89. (0.25) $\because n$ is an odd integer, $(-1)^{n-1} = 1$ and $n-1, n-3$, $n-5$ are even.

$$= n^3 - (n-1)^3 + (n-2)^3 + \ldots + (-1)^{n-1} \cdot 1$$

$$= n^3 + (n-1)^3 + (n-2)^3 + \ldots + 1^3 - 2\left[(n-1)^3 + (n-3)^3 + \ldots 2^3\right]$$

$$= n^3 + (n-1)^3 + (n-2)^3 + \ldots + 1^3 - 2 \times 2^3 \left[\left(\frac{n-1}{2}\right)^3 + \left(\frac{n-3}{2}\right)^3 + \ldots + 1^3\right]$$

[$\because n-1, n-3, n-5$ are even integers]

$$= \frac{1}{4}n^2(n+1)^2 - 16\frac{(n-1)^2(n+1)^2}{16 \times 4}$$

$$= \frac{1}{4}(n+1)^2[n^2 - (n-1)^2]$$

$$= \frac{1}{4}(n+1)^2 (2n-1)$$

Thus, $A = \frac{1}{4} = 0.25$

90. (2) $\because f(x) = \dfrac{4x^2 + 8x + 13}{6(1+x)} = \dfrac{4(x+1)^2 + 9}{6(1+x)}$

$$f(x) = \frac{2}{3}(x+1) + \frac{3}{2(x+1)}$$

Using AM $\geq$ GM

$$\frac{\frac{2}{3}(x+1) + \frac{3}{2(x+1)}}{2} \geq \sqrt{\frac{2}{3}(x+1)\frac{3}{2(x+1)}}$$

$$\frac{2}{3}(x+1) + \frac{3}{2(x+1)} \geq 2$$

$\therefore$ The minimum value of $f(x)$ is 2.

91. (8) Consider the values are xy, yz and zx using AM $\geq$ GM

$$\frac{xy + yz + zx}{3} \geq (x^2y^2z^2)^{1/3}$$

$$\because \quad x = \frac{12 - yz}{y+z}$$

$$\Rightarrow \quad xy + xz = 12 - yz$$

$$xy + yz + xz = 12$$

$$\therefore \quad \frac{12}{3} \geq (x^2y^2z^2)^{1/3}$$

$$4 \geq (xyz)^{2/3}$$

$$\therefore (xyz) \leq (4)^{3/2} = 8$$

Thus, the maximum value of xyz is 8.

2. (96) Using AM $\geq$ GM for $x^2, 2xy, 2xy, 4y^2, z^2, z^2$

$$\therefore \frac{x^2+2xy+2xy+4y^2+z^2+z^2}{6} \geq [16(xyz)^4]^{1/6}$$

$$x^2+4xy+4y^2+2z^2 \geq 6[16(32)^4]^{1/6}$$

$$= 6[16(2)^{20}]^{1/6}$$

$$= 6[(2)^{24}]^{1/6}$$

$$= 6 \times 16 = 96$$

Thus, the minimum value of $x^2+4xy+4y^2+2z^2$ is 96.

3. (4) Using AM $\geq$ GM for $\frac{x}{2}, \frac{x}{2}$ and y.

$$\frac{\frac{x}{2}+\frac{x}{2}+y}{3} \geq \left[\frac{x^2}{4} \cdot y\right]^{1/3}$$

$$\frac{x+y}{3} \geq \left(\frac{x^2 y}{4}\right)^{1/3}$$

$$\frac{3}{3} \geq \left(\frac{x^2 y}{4}\right)^{1/3}$$

$$\Rightarrow \quad x^2 y \leq 4$$

$\therefore$ Maximum value of x^2y is 4.

4. (8) Using AM $\geq$ GM for $a^{-5}, a^{-4}, a^{-3}, a^{-3}, a^{-3}$, 1, a^8 and a^{10}

$$\frac{a^{-5}+a^{-4}+a^{-3}+a^{-3}+a^{-3}+1+a^8+a^{10}}{8}$$

$$\geq [a^{-5} \times a^{-4} \times a^{-3} \times a^{-3} \times a^{-3} \times 1 \times a^8 \times a^{10}]^{1/8}$$

$$\frac{1}{8}[a^{-5}+a^{-4}+3a^{-3}+1+a^8+a^{10}] \geq [1]^{1/8}$$

$$\Rightarrow a^{-5}+a^{-4}+3a^{-3}+1+a^8+a^{10} \geq 8$$

Thus, the minimum value of the sum is 8.

5. (3.2) Applying root mean square $\geq$ AM on a, b, c, d, we get

$$\sqrt{\frac{a^2+b^2+c^2+d^2}{4}} \geq \frac{a+b+c+d}{4}$$

$$\Rightarrow \quad \sqrt{\frac{16-e^2}{4}} \geq \frac{a+b+c+d}{4} = \frac{8-e}{4}$$

$$\Rightarrow \quad \frac{16-e^2}{4} \geq \frac{(8-e)^2}{16}$$

$$\Rightarrow \quad 64-4e^2 \geq 64+e^2-16e$$

$$\Rightarrow \quad 5e^2-16e \leq 0$$

$$\Rightarrow \quad e \in \left[0, \frac{16}{5}\right]$$

$$\therefore \quad e_{max} = \frac{16}{5} = 3.2$$

Level 3

1. (c) The general term is

$$x_k = \sec a_k \sec a_{k+1} = \frac{1}{\sin d} \cdot \frac{\sin d}{\cos a_k \cos a_{k+1}}$$

$$= \frac{1}{\sin d}\left[\frac{\sin(a_{k+1}-a_k)}{\cos a_k \cos a_{k+1}}\right]$$

$$= \operatorname{cosec} d\,[\tan a_{k+1} - \tan a_k] = \operatorname{cosec} d\,[f(k+1) - f(k)]$$

$$\therefore \sum_{k=1}^{n-1} = \operatorname{cosec} d\,[f(n) - f(1)] = \operatorname{cosec} d\,[\tan a_n - \tan a_1]$$

2. (a) $\because$ α, β, γ and δ are in AP.

Let $\beta = \alpha + D$, $\gamma = \alpha + 2D$ and $\delta = \alpha + 3D$

$$\because \quad \alpha + \beta = \frac{-b}{a} \Rightarrow \alpha + \alpha + D = -\frac{b}{a} \quad \text{...(i)}$$

$$\text{or} \quad 2\alpha + D = \frac{-b}{a}$$

$$\text{and } \gamma + \delta = \frac{-B}{A} \Rightarrow 2\alpha + 5D = \frac{-B}{A} \quad \text{...(ii)}$$

From Eqs. (i) and (ii), we get

$$4D = \left(\frac{b}{a} - \frac{B}{A}\right) \text{ or, } D = \frac{1}{4}\left(\frac{b}{a} - \frac{B}{A}\right)$$

3. (b) We have, $\log b_2 - \log b_1 = \log(2)$

$$\Rightarrow \quad \frac{b_2}{b_1} = 2$$

$\Rightarrow b_1, b_2, \ldots\ldots$, are in GP with common ratio 2.

$$\therefore \quad t = b_1 + 2b_1 + \ldots + 2^{50}b_1 = b_1(2^{51}-1)$$

$$s = a_1 + a_2 + \ldots + a_{51}$$

$$= \frac{51}{2}(a_1 + a_{51})$$

$$= \frac{51}{2}(b_1 + b_{51})$$

$$= \frac{51}{2}b_1(1+2^{50})$$

$$\Rightarrow \quad s - t = b_1\left(\frac{51}{2} + 51 \times 2^{49} - 2^{51} + 1\right)$$

$$= b_1\left(\frac{53}{2} + 2^{49} \times 47\right) > 0$$

$$\Rightarrow \quad s > t$$

$$b_{101} = 2^{100} b_1$$

$$\therefore \quad a_{101} = a_1 + 100d$$

$$= 2(a_1 + 50d) - a_1$$

$$= 2a_{51} - a_1$$

$$= 2b_{51} - b_1 \quad [\because a_{51} = b_{51}]$$

$$= (2 \times 2^{51} - 1)\, b_1$$

$$= (2^{52} - 1)\, b_1$$

$$\therefore \quad b_{101} > a_{101}$$

4. (c) $\because a_1, a_2, \ldots$, are in A.P and $q_1, q_2, \ldots$ are in GP
$a_1 = q_1 = 2$ and $a_{10} = q_{10} = 3$.
Let d be the common difference of AP

i.e., $d = \dfrac{3-2}{9} = \dfrac{1}{9}$

Then, $a_7 = a_1 + 6d = 2 + 6d = 2 + 6 \times \dfrac{1}{9} = \dfrac{8}{3}$

$a_{19} = a_1 + 18d = 2 + 18d = 2 + 18 \times \dfrac{1}{9} = 4$

Let r be the common ratio of GP, i.e. $r = \left(\dfrac{3}{2}\right)^{1/9}$

Then, $q_7 = q_1 r^6 = 2r^6 = 2\left(\dfrac{3}{2}\right)^{6 \times \frac{1}{9}} = 2\left(\dfrac{3}{2}\right)^{2/3}$

$q_{10} = q_1 r^9 = 2r^9 = 2 \times \left(\dfrac{3}{2}\right)^{9 \times \frac{1}{9}} = 3$

$q_{19} = q_1 r^{18} = 2\left(\dfrac{3}{2}\right)^{18 \times \frac{1}{9}} = \dfrac{9}{2}$

$a_7 q_{19} = \dfrac{8}{3} \times \dfrac{9}{2} = 12$, which is an integer

and $a_{19}\, q_7 = 4 \times 2 \times \left(\dfrac{3}{2}\right)^{2/3}$, which is not an integer.

Hence, $a_{19} q_{10} = 4 \times 3 = 12$
Thus, $a_7\, q_{19} = a_{19} q_{10}$

5. (d) Given, $t_n = t_{n+1} + t_{n+2}$

$\Rightarrow \quad ar^{n-1} = ar^n + ar^{n+1}$

$\Rightarrow \quad 1 = r + r^2 \Rightarrow r^2 + r - 1 = 0$

$\Rightarrow \quad r = \dfrac{\pm\sqrt{1+4}}{2}$

$\therefore \quad r = \dfrac{-1 \pm \sqrt{5}}{2}$,

since $r > 0$

Therefore, $r = 2\sin 18°$ $\quad \left[\because \sin 18° = \dfrac{\sqrt{5}-1}{4}\right]$

6. (a) If the three terms of the GP be $\dfrac{a}{r}$, a and ar, then

$$S = \frac{a}{r} + a + ar = \frac{a}{r}(1 + r + r^2)$$

$$P = a^3 \text{ and } R = \frac{r}{a} + \frac{1}{a} + \frac{1}{ar} = \frac{1}{ar}(r^2 + r + 1)$$

$$\text{Now, } \frac{P^2R^3}{S^3} = \frac{a^6 \dfrac{1}{a^3 r^3}(r^2 + r + 1)^3}{\dfrac{a^3}{r^3}(r^2 + r + 1)^3} = 1$$

So, the required ratio is 1 : 1.

7. (b) Let A, Ar, Ar^2, Ar^3 are respectively represented as a, b, c and d.

$$(a^3 + b^3)^{-1} = \frac{1}{A^3(1 + r^3)} \quad \ldots$$

$$(b^3 + c^3)^{-1} = \frac{1}{A^3 r^3 (1 + r^3)} \quad \ldots$$

and $$(c^3 + d^3)^{-1} = \frac{1}{A^3 r^6 (1 + r^3)} \quad \ldots($$

Thus, from Eqs. (i), (ii) and (iii), they are in GP with common ratio as $\dfrac{1}{r^3}$.

8. (a) We have,

$$S_n = \frac{n}{1 - \dfrac{1}{n+1}}$$

$$= \frac{n(n+1)}{n+1-1} = n + 1$$

$\Rightarrow \quad S_n^{\,2} = (n+1)^2$

So, $S_1^2 + S_2^2 + \ldots S_{2n-1}^2 = 2^2 + 3^3 + \ldots + (2n)^2$

$$= \frac{1}{6}(2n)(2n+1)(4n+1) - 1$$

$$= \frac{1}{3}[n(2n+1)(4n+1) - 3]$$

9. (c) $S = \sum_{n=1}^{\infty} r^n$

$$S = \frac{r}{1-r} \Rightarrow r = \frac{S}{1+S}$$

Now, $$\sum_{n=1}^{\infty} r^{2n} = \frac{r^2}{1-r^2} = \frac{\dfrac{S^2}{(1+S)^2}}{1 - \dfrac{S^2}{(1+S^2)}} = \frac{S^2}{1+2S}$$

10. (a) $x = \dfrac{a}{1 - \dfrac{1}{r}} = \dfrac{ar}{r-1}$

Similarly, $y = \dfrac{b}{1 - \left(-\dfrac{1}{r}\right)} = \dfrac{br}{r+1}$

and $$z = \frac{c}{1 - \dfrac{1}{r^2}} = \frac{cr^2}{r^2 - 1} \quad \ldots$$

$\therefore \quad xy = \dfrac{ar}{r-1} \times \dfrac{br}{r+1} = \dfrac{abr^2}{r^2 - 1} \quad \ldots$

Dividing Eq. (ii) by Eq. (i), we get

$$\frac{xy}{z} = \frac{abr^2}{r^2 - 1} \times \frac{r^2 - 1}{cr^2} = \frac{ab}{c}$$

11. (b) As, $\frac{b+c}{a+d}=\frac{bc}{ad}$

$\Rightarrow \quad \frac{1}{b}+\frac{1}{c}=\frac{1}{a}+\frac{1}{d}$

$\Rightarrow \frac{1}{a},\frac{1}{b},\frac{1}{c},\frac{1}{d}$ are in AP.

$\Rightarrow \frac{1}{a}-1, \frac{1}{b}-1, \frac{1}{c}-1, \frac{1}{d}-1$ are in AP.

$\Rightarrow \pi^{\frac{1}{a}-1}, \pi^{\frac{1}{b}-1}, \pi^{\frac{1}{c}-1}, \pi^{\frac{1}{d}-1}$ are in GP.

12. (c) Using AM $\geq$ GM

$$\frac{a^x+a^x+....(y \text{ times})+(b^y+b^y+...x \text{ times})}{x+y} \geq (a^{xy}\times b^{xy})^{1/x+y}$$

$$\frac{ya^x+xb^y}{x+y} \geq a^{\frac{xy}{x+y}} b^{\frac{xy}{x+y}}$$

$\because \quad \frac{1}{x}+\frac{1}{y}=1 \Rightarrow \frac{x+y}{xy}=1$

$\Rightarrow \quad x+y=xy$

$\Rightarrow \quad \frac{ya^x+xb^y}{xy} \geq ab$

$\Rightarrow \quad \frac{a^x}{x}+\frac{b^y}{y} \geq ab$

$\Rightarrow \quad \frac{a^x}{x}+\frac{b^y}{y}-ab \geq 0$

13. (c) We have,

t_n = [nth term of 1, 2, 3 ...] [nth term of $3^2, 5^2, 7^2$ ...]

$= n(2n+1)^2 = 4n^3+4n^2+n$

$\therefore \quad S_n = \Sigma t_n = 4\Sigma n^3 + 4\Sigma n^2 + \Sigma n$

$= 4\left\{\frac{n(n+1)}{2}\right\}^2 + 4\cdot\frac{n(n+1)(2n+1)}{6}+\frac{n(n+1)}{2}$

$= n^2(n+1)^2 + \frac{2}{3}n(n+1)(2n+1)+\frac{1}{2}n(n+1)$

$\therefore \quad S_{20} = 20^2\cdot 21^2 + \frac{2}{3}\cdot 20\cdot(21)(41)+\frac{1}{2}\cdot 20\cdot 21 = 188090.$

14. (b) $t_n = \frac{n}{(n+1)!} = \frac{1}{n!}-\frac{1}{(n+1)!}$

$\Sigma t_n = \sum_{n=1}^{20}\left[\frac{1}{n!}-\frac{1}{(n+1)!}\right]$

$= \left[\frac{1}{1!}-\frac{1}{2!}+\frac{1}{2!}-\frac{1}{3!}+...+\frac{1}{20!}-\frac{1}{21!}\right]$

$= 1-\frac{1}{21!}$

$= \frac{21!-1}{21!}$

15. (c) $T_r = \frac{1}{(2r-1)(2r+1)(2r+3)}$

$= \frac{1}{4}\left[\frac{1}{(2r-1)(2r+1)}-\frac{1}{(2r+1)(2r+3)}\right]$

Now, $S = \frac{1}{4}\left[\frac{1}{1\cdot 3}-\frac{1}{3\cdot 5}+\frac{1}{3\cdot 5}-\frac{1}{5\cdot 7}+\frac{1}{5\cdot 7}-\frac{1}{7\cdot 9}\right.$

$\left.+\frac{1}{7\cdot 9}-\frac{1}{9\cdot 11}+\frac{1}{9\cdot 11}-\frac{1}{11\cdot 13}\right]$

$= \frac{1}{4}\left[\frac{1}{3}-\frac{1}{11.13}\right] = \frac{35}{429}$

16. (a) $T_r = \frac{(2r+1)}{r(r+1)(2r+1)/6} = 6\left[\frac{1}{r(r+1)}\right] = 6\left[\frac{1}{r}-\frac{1}{r+1}\right]$

$S_n = 6\sum_{r=1}^{\infty}\left(\frac{1}{r}-\frac{1}{r+1}\right) = 6$

17. (c) Let $S_n = \frac{8}{5}+\frac{16}{65}+\frac{24}{325}+...$

$T_r = \frac{8r}{4r^4+1} = \frac{8r}{(2r^2+2r+1)(2r^2-2r+1)}$

$\Rightarrow \quad T_r = 2\left[\frac{(2r^2+2r+1)-(2r^2-2r+1)}{(2r^2+2r+1)(2r^2-2r+1)}\right]$

$\Rightarrow \quad T_r = 2\left[\frac{1}{2r^2-2r+1}-\frac{1}{2r^2+2r+1}\right]$

$\lim_{n\to\infty} S_n = \lim_{n\to\infty}\sum_{r=1}^{n} T_r$

$= 2\lim_{n\to\infty}\sum_{r=1}^{n}\left[\frac{1}{2r^2-2r+1}-\frac{1}{2r^2+2r+1}\right]$

$= 2\lim_{n\to\infty}\left(1-\frac{1}{2n^2+2n+1}\right)$

$= 2(1-0) = 2$

18. (b) The general term is

$x_K = \frac{K}{1+K^2+K^4} = \frac{K}{1+2K^2+K^4-K^2}$

$= \frac{K}{(K^2+1)^2-K^2} = \frac{1}{2}\frac{2K}{(K^2+1+K)(K^2+1-K)}$

$= \frac{1}{2}\left[\frac{1}{K^2-K+1}-\frac{1}{K^2+K+1}\right] = \frac{1}{2}$

$\left\{f(K)-f(K+1), \text{ where } f(K) = \frac{1}{K^2-K+1}\right\}$

$\therefore \quad S_n = \sum_{K=1}^{n} x_K = \frac{1}{2}\sum_{K=1}^{n}[f(K)-f(K+1)]$

$= \frac{1}{2}[f(1)-f(n+1)] = \frac{1}{2}\left[1-\frac{1}{n^2+n+1}\right]$

$= \frac{n^2+n}{2(n^2+n+1)}$

19. (a) If last term of nth row is T_n, then

$$S = 1 + 5 + 11 + 19 + \ldots + T_n$$
$$S = 1 + 5 + 11 + \ldots + T_{n-1} + T_n$$

On subtracting,

$$0 = 1 + 4 + 6 + 8 + \ldots + n \text{ terms} - T_n$$
$$\Rightarrow \quad T_n = 1 + 4 + 6 + 8 + \ldots + n \text{ terms}$$
$$\Rightarrow \quad T_n = 1 + 2[2 + 3 + 4 + \ldots + (n-1) \text{ terms}]$$
$$\Rightarrow \quad T_n = 1 + 2 \times \frac{(n-1)}{2}(n+2)$$
$$\Rightarrow \quad T_n = 1 + n^2 + n - 2$$
$$\therefore \quad T_n = n^2 + n - 1$$
$$\therefore \quad T_{60} = (60)^2 + 60 - 1 = 3600 + 59 = 3659$$

20. (c) $\sum_{r=1}^{n} r^2 - \sum_{m=1}^{n} \sum_{r=1}^{m} r$

$$= \sum_{r=1}^{n} r^2 - \sum_{m=1}^{n} \frac{m(m+1)}{2}$$
$$= \sum_{r=1}^{n} r^2 - \sum_{m=1}^{n} \frac{m^2}{2} - \sum_{m=1}^{n} \frac{m}{2}$$
$$= \frac{1}{2}\left[\sum_{r=1}^{n} r^2 - \sum_{r=1}^{n} m\right] = \frac{1}{2}\left[\sum_{r=1}^{n} r^2 - \sum_{r=1}^{n} r\right]$$

21. (a) Taking AM, GM of 7 numbers

$\frac{a}{2}, \frac{a}{2}, \frac{b}{3}, \frac{b}{3}, \frac{b}{3}, \frac{c}{2}, \frac{c}{2}$, we get

$$\frac{2 \cdot \frac{a}{2} + 3 \cdot \frac{b}{3} + 2 \cdot \frac{c}{2}}{7} \geq \left\{\left(\frac{a}{2}\right)^2 \left(\frac{b}{3}\right)^3 \left(\frac{c}{2}\right)^2\right\}^{\frac{1}{7}}$$
$$\Rightarrow \quad \frac{3^7}{7^7} \geq \left(\frac{a^2b^3c^2}{2^2 3^3 2^2}\right) \Rightarrow \frac{3^7}{7^7} \geq \frac{a^2b^3c^2}{2^2 \cdot 3^3 \cdot 2^2}$$
$$\Rightarrow \quad a^2b^3c^2 \leq \frac{3^{10} \cdot 2^4}{7^7}$$

$\therefore$ Greatest value of $a^2b^3c^2 = \frac{3^{10} \cdot 2^4}{7^7}$

22. (a, b, c)

Let a, b, c be pth, qth and rth terms of AP, then

$a = A + (p-1)D, b = A + (q-1)D, c = A + (r-1)D$

$\Rightarrow \frac{r-q}{q-p} = \frac{c-b}{b-a}$ is rational number.

Now for 1, 6, 19, $\frac{r-q}{q-p} = \frac{19-6}{6-1}$ is a rational number

For $\sqrt{2}, \sqrt{50}, \sqrt{98}$, $\frac{r-q}{q-p} = \frac{\sqrt{98}-\sqrt{50}}{\sqrt{50}-\sqrt{2}} = \frac{7\sqrt{2}-5\sqrt{2}}{5\sqrt{2}-\sqrt{2}}$

$= \frac{1}{2}$ is a rational number

For $\log 2, \log 16, \log 128$,

$\frac{r-q}{q-p} = \frac{\log 128 - \log 16}{\log 16 - \log 2} = \frac{7\log 2 - 4\log 2}{4\log 2 - \log 2} = 1$ is a rational number.

But for $\sqrt{2}, \sqrt{3}, \sqrt{7}$, $\frac{r-q}{q-p}$ is not a rational number.

23. (a, c, d)

$$a_1 + a_3 + a_5 = -12$$
$$a + a + 2d + a + 4d = -12 (d > 0)$$
$$a + 2d = -4$$
$$a_1 a_3 a_5 = 80$$
$$a(a+2d)(a+4d) = 80$$

Or $(-4-2d)(-4+2d) = -20$ or $d = \pm 3$

Since AP is increasing, so $d = +3$, $a = -10$, Hence,

$$a_1 = -10, a_2 = -7$$
$$a_3 = a + 2d = -10 + 6 = -4$$
$$a_5 = a + 4d = -10 + 12 = 2$$

24. (a, c, d)

$$S_n = \frac{n}{2}[2a + (n-1)d] = a + bn + cn^2$$
$$\Rightarrow na + \frac{n(n-1)}{2}d = a + bn + cn^2$$
$$\left(a - \frac{d}{2}\right)n + \frac{n^2 d}{2} = a + bn + cn^2$$

On comparing,

$$a = 0, b = a - \frac{d}{2}, c = \frac{d}{2} \Rightarrow d = 2c$$

25. (a, c)

Let $b = a + p$, $c = a + 2p$, $d = a + 3p$ (where p is common difference) then,

$$\frac{\frac{1}{a} + \frac{1}{d}}{\frac{1}{b} + \frac{1}{c}} = \frac{\frac{1}{a} + \frac{1}{a+3p}}{\frac{1}{a+p} + \frac{1}{a+2p}} = \frac{(a+p)(a+2p)}{a(a+3p)}$$
$$= \frac{a^2 + 3ap + 2p^2}{a^2 + 3ap} > 1$$
$$\therefore \quad \frac{1}{a} + \frac{1}{d} > \frac{1}{b} + \frac{1}{c}$$
$$\left(\frac{1}{b} + \frac{1}{c}\right)(a+d) = \left(\frac{1}{a+p} + \frac{1}{a+2p}\right)(a + a + 3p)$$
$$= \frac{(2a+3p)^2}{a^2 + 3ap + 2p^2} = 4 + \frac{p^2}{a^2 + 3ap + 2p^2} > 4$$

26. (b, d)

Given, $3a_1 + 7a_2 + 3a_3 - 4a_5 = 0$

$$\Rightarrow 7(a_1 + a_2 + a_3) = 4(a_1 + a_3 + a_5)$$
$$\Rightarrow \quad 7(1 + r + r^2) = 4(1 + r^2 + r^4)$$
$$\Rightarrow \quad 7 = 4(r^2 - r + 1)$$

$\Rightarrow \quad 4r^2 - 4r + 1 = 4$

$\Rightarrow \quad (2r-1)^2 = 4$

$\Rightarrow \quad 2r - 1 = \pm 2$

$\Rightarrow \quad r = \frac{3}{2}, -\frac{1}{2}$

27. (a,d)

x, x^2+2, x^3+10 are in GP. Hence,

$$x(x^3+10) = (x^2+2)^2 = x^4 + 4x^2 + 4$$

$\Rightarrow \quad 4x^2 - 10x + 4 = 0$

or $\quad 2x^2 - 5x + 2 = 0$

or $\quad x = 2, \frac{1}{2}$

The 4th term of GP is

$$(x^3+10)r = (x^3+10)\left(\frac{x^2+2}{x}\right)$$

$$= \begin{cases} 54 & \text{when } x = 2 \\ \frac{729}{16} & \text{when } x = \frac{1}{2} \end{cases}$$

28. (a, d)

$$P(x) = \left(\frac{1-x^{2n}}{1-x^2}\right)\left(\frac{1-x}{1-x^n}\right) = \frac{1+x^n}{1+x}$$

As $P(x)$ is a polynomial,

$x = -1$ must be a zero of $1 + x^n$.

Hence, $1 + (-1)^n = 0$. So, n must be odd.

29. (b, c)

We have, $\frac{p}{1-1/p} = \frac{9}{2}$

or $\quad 2p^2 - 9p + 9 = 0$

or $\quad p = \frac{3}{2}, 3$

30. (a, b, c)

Given that

$$a = 4, T_3 - T_5 = \frac{32}{81}.$$

Hence, $\quad a(r^2 - r^4) = \frac{32}{81}$

or $\quad r^4 - r^2 + \frac{8}{81} = 0 \quad$ [given $a = 4$]

or $\quad 81r^4 - 81r^2 + 8 = 0$

or $\quad (9r^2 - 8)(9r^2 - 1) = 0$

$\therefore \quad r^2 = \frac{8}{9}, \frac{1}{9}$

Therefore, the value of r is to be positive since all the terms are positive.

For $r = \frac{1}{3}$,

$$S_\infty = \frac{a}{1-r} = \frac{4}{1-\frac{1}{3}} = \frac{4 \times 3}{2} = 6$$

Similarly, we can find S_∞ when $r = \frac{2\sqrt{2}}{3}$

31. (a, d)

$$S = 1 + 2x + 3x^2 + 4x^3 + \ldots.\infty$$

$\Rightarrow \quad xS = x + 2x^2 + 3x^3 + 4x^4 + \ldots.\infty$

$\Rightarrow (1-x)S = 1 + x + x^2 + \ldots \infty = \frac{1}{1-x}, |x| < 1$

$\Rightarrow \quad S = \frac{1}{(1-x)^2}$

Now, $S \geq 4$

$\Rightarrow \frac{1}{(1-x)^2} \geq 4 \Rightarrow (x-1)^2 \leq \frac{1}{4} \Rightarrow -\frac{1}{2} \leq x - 1 \leq \frac{1}{2}$

$\Rightarrow \quad \frac{1}{2} \leq x \leq \frac{3}{2}$

Also $0 < |x| < 1$

$\Rightarrow \quad \frac{1}{2} \leq x < 1$

32. (b, c, d)

Let S_n denote the sum of first n terms of AP.

According to the question,

$r = \frac{S_n}{S_{2n} - S_n} = \frac{1}{\frac{S_{2n}}{S_n} - 1}$ is independent of n.

Therefore, $\frac{S_{2n}}{S_n}$ is independent of n.

Now, $\frac{S_{2n}}{S_n} = \frac{\frac{2n}{2}(2\alpha + (2n-1)\beta)}{\frac{n}{2}(2\alpha + (n-1)\beta)} = \frac{2(2\alpha - \beta + 2n\beta)}{(2\alpha - \beta + n\beta)}$

This ratio is independent of n, if

$$2\alpha - \beta = 0$$

$\therefore \quad \alpha : \beta = 1 : 2$

$\therefore \quad r = \frac{1}{4-1} = \frac{1}{3}$

α and β are the roots of the equation $ax^2 + bx + c = 0$

$\therefore \quad 2\alpha + \alpha = -\frac{b}{a}$ and $2\alpha \times \alpha = \frac{c}{a}$

Solving those, we get

$$2\left(-\frac{b}{3a}\right)^2 = \frac{c}{a} \Rightarrow 2b^2 = 9ac$$

$$1 + r + r^2 + \ldots. = \frac{1}{1-r} = \frac{1}{1-\frac{1}{3}} = \frac{3}{2}, \alpha = 1, \beta = 2$$

$\therefore$ Sum of 10 terms, AP $= \frac{10}{2}(2 \times 1 + (10-1) \times 2) = 100$

33. (a, b, c, d)

We have, $b=\dfrac{2ac}{a+c}$ and $b^2=-2ac$

$\therefore \quad b=\dfrac{-b^2}{a+c}$

$\Rightarrow \quad a+b+c=0$

$\Rightarrow \quad 2b=-2a-2c$

Therefore, $-2a, b, -2c$ are in AP.

Also, $a^3+b^3+c^3=3abc$

Since $b^2=-2ac$, $b^4=4a^2c^2$

So, $a^2, b^2, 4c^2$ are in GP.

Since, $a+b+c=0$, equation $ax^2+bx+c=0$ has one root $x=1$.

So, equation has real roots.

34. (c, d)

$4=1+(n-1)d, 16=1+(m-1)d$

$\Rightarrow \dfrac{15}{3}=\dfrac{m-1}{n-1}$ or

$\dfrac{n-1}{1}=\dfrac{m-1}{5}=$ positive integer

$\therefore n=p+1, m=5p+1$, So n, m have infinite pairs of values.

Also, $4=1\cdot r^n, 16=1\cdot r^m$

$\Rightarrow r^{2n}=r^m \Rightarrow 2n=m$

$\therefore \dfrac{m}{2}=\dfrac{n}{1}=q=$ positive integer

So, m, n have infinite pairs of values.

35. (b, c)

It is clearly a GP.

Hence, $\quad x=\dfrac{1}{1-\cos^2\phi}=\dfrac{1}{\sin^2\phi}$ ...(i)

Similarly, $y=\dfrac{1}{1-\sin^2\phi}=\dfrac{1}{\cos^2\phi}$ ...(ii)

and $\quad z=\dfrac{1}{1-\sin^2\phi\cos^2\phi}$ ...(iii)

Now, $x+y=\dfrac{1}{\sin^2\phi}+\dfrac{1}{\cos^2\phi}=\dfrac{1}{\sin^2\phi\cos^2\phi}$

Again, $\quad \dfrac{1}{z}=1-\sin^2\phi\cos^2\phi=1-\dfrac{1}{xy}$

$\Rightarrow \quad \dfrac{1}{z}=\dfrac{xy-1}{xy} \Rightarrow xy=xyz-z$

$\Rightarrow \quad xy+z=xyz$

From Eq. (i), $x+y=xy$

$\Rightarrow \quad x+y+z=xyz$

36. (a, b, d)

Let the first term be 'a' and common ratio be 'r'.

According to the question, we have

$$a+ar=1 \text{ and } \frac{a}{1-r}=2$$

Solving these, we get

$$\frac{1}{(1-r)(1+r)}=2$$

$\Rightarrow \quad 1-r^2=\dfrac{1}{2}\Rightarrow r^2=\dfrac{1}{2}\Rightarrow r=\pm\dfrac{1}{\sqrt{2}}$

If $r=\dfrac{1}{\sqrt{2}}, a=2-\sqrt{2}$

If $r=-\dfrac{1}{\sqrt{2}}, a=2+\sqrt{2}$

37. (b, c, d)

We have length of a side of S_n is equal to the length of a diagonal of S_{n+1}.

Hence, Length of a side of $S_n=\sqrt{2}$ [length of a side of S_{n+1}]

or $\dfrac{\text{Length of a side of } S_{n+1}}{\text{Length of a side of } S_n}=\dfrac{1}{\sqrt{2}}$, for all $n\geq 1$

Hence, sides of $S_1, S_2, \ldots. S_n$ form a GP with common ratio $\dfrac{1}{\sqrt{2}}$ and first term 10

$\therefore$ Side of $S_n=10\left(\dfrac{1}{\sqrt{2}}\right)^{n-1}=\dfrac{10}{2^{\frac{n-1}{2}}}$

Area of $S_n=(\text{side})^2=\left(\dfrac{10}{2^{\frac{n-1}{2}}}\right)^2=\dfrac{100}{2^{n-1}}$

Now, area of $S_n<1$

$\Rightarrow \quad 2^{n-1}>100$

$\Rightarrow \quad n>7$

38. (a, c)

a, b, c are in GP.

Hence, $\quad b^2=ac$ ...(i)

x is AM of a and b.

Hence, $\quad 2x=a+b$ ...(ii)

y is AM of b and c.

Hence, $\quad 2y=b+c$ ...(iii)

$\therefore \dfrac{a}{x}+\dfrac{c}{y}=a\times\dfrac{2}{a+b}+c\times\dfrac{2}{b+c}$ [using Eqs. (ii) and (iii)]

$=2\left[\dfrac{ab+ac+ac+bc}{ab+ac+b^2+bc}\right]=2$ [using Eq. (i)]

Again, $\dfrac{1}{x}+\dfrac{1}{y}=\dfrac{2}{a+b}+\dfrac{2}{b+c}$

$=\dfrac{2(a+c+2b)}{ab+ac+b^2+bc}$

$$= \frac{2(a+c+2b)}{ab+2b^2+bc} \quad [\because b^2 = ac]$$

$$= \frac{2(a+c+2b)}{b(a+c+2b)} = \frac{2}{b}$$

39. (b, c)

$$S = 1 + \frac{1}{1+3}(1+2)^2 + \frac{1}{(1+3+5)}$$

$$(1+2+3)^2 + \frac{1}{(1+3+5+7)}(1+2+3+4)^2 + \ldots$$

The rth term is given by

$$T_r = \frac{1}{r^2}(1+2+\ldots+r)^2 = \frac{1}{r^2}\left\{\frac{r(r+1)}{2}\right\}^2 = \frac{r^2+2r+1}{4}$$

$$\therefore T_7 = 16$$

and $S_{10} = \sum_{r=1}^{10} T_r$

$$= \frac{1}{4}\left\{\frac{(10)(10+1)(20+1)}{6} + (10)(10+1) + 10)\right\}$$

$$= \frac{505}{4}$$

40. (a, b, c, d)

$$an^4 + bn^3 + cn^2 + dn + e$$

$$= 2\sum_{r=1}^{n} r(r+1)(r+2) - \sum_{r=1}^{n} r(r+1)$$

$$= \frac{2}{4}n(n+1)(n+2)(n+3) - \frac{1}{3}n(n+1)(n+2)$$

On simplifying, we get

$$= \frac{1}{6}(3n^4 + 16n^3 + 27n^2 + 14n)$$

$$\Rightarrow \quad a = \frac{3}{6} = \frac{1}{2}, b = \frac{16}{6} = \frac{8}{3}, c = \frac{27}{6}, d = \frac{14}{6}$$

$$e = 0$$

41. (a, d)

$$S_n = \sum_{k=1}^{4n} (-1)^{\frac{k(k+1)}{2}} k^2$$

$$= \sum_{r=0}^{n-1} [(4r+4)^2 + (4r+3)^2 - (4r+2)^2 - (4r+1)^2]$$

$$= \sum_{r=0}^{n-1} [2(8r+6) + 2(8r+4)]$$

$$= \sum_{r=0}^{n-1} [32r + 20]$$

$$= 16(n-1)n + 20n$$

$$= 4n(4n+1)$$

$$= \begin{cases} 1056 & \text{for } n=8 \\ 1332 & \text{for } n=9 \end{cases}$$

42. (b, d)

Let a straight line through the vertex P of a given ΔPQR intersects the side QR at the point S, and the circumcircle of ΔPQR at the point T.

Points P, Q, T, R are concyclic.

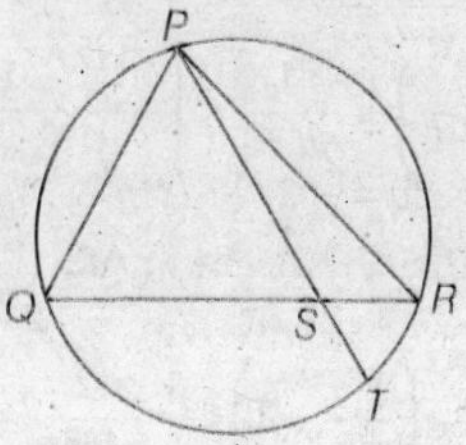

Then,

$$PS \cdot ST = QS \cdot SR$$

$$PT \geq 2\sqrt{PS \cdot ST}$$

Now, $\frac{PS+ST}{2} \geq \sqrt{PS \cdot ST}$ [using AM ≥ GM]

and $\frac{1}{PS} + \frac{1}{ST} \geq \frac{2}{\sqrt{PS \cdot ST}} = \frac{2}{\sqrt{QS \cdot SR}}$

also, $\frac{SQ+SR}{2} \geq \sqrt{SQ \cdot SR}$

$$\Rightarrow \quad \frac{QR}{2} \geq \sqrt{SQ \cdot SR}$$

$$\Rightarrow \quad \frac{1}{\sqrt{SQ . SR}} \geq \frac{2}{QR} \Rightarrow \frac{2}{\sqrt{SQ \cdot SR}} \geq \frac{4}{QR}$$

Hence, $\frac{1}{PS} + \frac{1}{ST} \geq \frac{2}{\sqrt{SQ \cdot SR}} \geq \frac{4}{QR}$

43. (d) $a_n = \frac{1}{2^{n-1}}$ and $S_n = 2\left(1 - \frac{1}{2^n}\right)$

For circle C_n to be inside M.

$$S_{n-1} + a_n < \frac{1025}{513} \Rightarrow S_n < \frac{1025}{513}$$

$$\Rightarrow \quad 2\left(1 - \frac{1}{2^n}\right) < \frac{1025}{513} \Rightarrow 1 - \frac{1}{2^n} < \frac{1025}{1026}$$

$$\Rightarrow \quad 1 - \frac{1}{2^n} < 1 - \frac{1}{1026} \Rightarrow \frac{-1}{2^n} < \frac{-1}{1026}$$

$$\Rightarrow \quad \frac{1}{2^n} > \frac{1}{1026} \Rightarrow 2^n < 1026$$

$$\Rightarrow \quad n \leq 10$$

$\therefore$ Number of circles inside be $10 = k$. Clearly, alternate circle do not intersect each other i.e. C_1, C_3, C_5, C_7, C_9 do not intersect each other as well as C_2, C_4, C_6, C_8 and C_{10} do not intersect each other.

Hence, maximum 5 set of circles do not intersect each other.

$$\therefore \quad l = 5$$

So, $3k + 2l = 40$

44. (b) $\because\ r=\dfrac{(2^{199}-1)\sqrt{2}}{2^{198}}$

Now, $\sqrt{2}\,S_{n-1}+a_n<\left(\dfrac{2^{199}-1}{2^{198}}\right)\sqrt{2}$

$$2\sqrt{2}\left(1-\frac{1}{2^{n-1}}\right)+\frac{1}{2^{n-1}}<\left(\frac{2^{199}-1}{2^{198}}\right)\sqrt{2}$$

$$\therefore\quad 2\sqrt{2}-\frac{\sqrt{2}}{2^{n-2}}+\frac{1}{2^{n-1}}<2\sqrt{2}-\frac{\sqrt{2}}{2^{198}}$$

$$\frac{1}{2^{n-2}}\left(\frac{1}{2}-\sqrt{2}\right)<-\frac{\sqrt{2}}{2^{198}}\Rightarrow\frac{2\sqrt{2}-1}{2\cdot 2^{n-2}}>\frac{\sqrt{2}}{2^{198}}$$

$$\Rightarrow\quad 2^{n-2}<\left(2-\frac{1}{\sqrt{2}}\right)\cdot 2^{197}$$

$$\therefore\quad n\le 199$$

So, number of circles $=199$

Sol. (Q.Nos. 45-47)

Let roots of equation are a, ar, ar^2

$$\Rightarrow\quad a(1+r+r^2)=\frac{19}{2}\qquad\text{...(i)}$$

and $a^2r+a^2r^2+a^2r^3=\dfrac{57}{2}$

$$\Rightarrow\quad a^2r(1+r+r^2)=\frac{57}{2}\qquad\text{...(ii)}$$

From Eqs. (i) and (ii),

$$\frac{1+r+r^2}{r}=\frac{19\times19}{2\times57}=\frac{19}{6}$$

$$\Rightarrow\quad 6r^2-13r+6=0$$

$$\Rightarrow\quad (2r-3)(3r-2)=0$$

$$r=\frac{3}{2}\text{ or }\frac{2}{3}$$

If $r=\dfrac{3}{2}$, then $a=2$

If $r=\dfrac{2}{3}$, then $a=\dfrac{9}{2}$

Roots $\Rightarrow 2,3,\dfrac{9}{2}$

Product $=a^3r^3=\dfrac{-k}{2}\Rightarrow(-2)(3)^3=k\Rightarrow k=-54$

As given that GP is increasing, thus taking $r=\dfrac{3}{2}$

$$S_n=\frac{2\left(\left(\frac{3}{2}\right)^n-1\right)}{\frac{3}{2}-1}=4\left(\left(\frac{3}{2}\right)^n-1\right)$$

As given that GP is decreasing, thus taking

$$r=\frac{2}{3}\Rightarrow S_\infty=\frac{\frac{9}{2}}{1-\frac{2}{3}}=\frac{27}{2}$$

45. (c) **46.** (b) **47.** (a)

48. (i) $\rightarrow$ B, C, (ii) $\rightarrow$ C, (iii) $\rightarrow$ B, C, D, (iv) $\rightarrow$ A

We have, $\alpha+\beta=8$ and $\alpha\beta=4$

(i) If $\alpha, a_1, a_2, a_3, a_4, a_5, \beta$ are in AP, then

$$\sum_{i=1}^{5}a_i=5\left(\frac{\alpha+\beta}{2}\right)=5\left(\frac{8}{2}\right)=20$$

(ii) If $\alpha, a_1, a_2, a_3, a_4, a_5, \beta$ are in GP, then

$$\sum_{i=1}^{5}\log_2 a_i=\log_2 a_1+\log_2 a_2+\log_2 a_3+\log_2 a_4+\log_2 a_5$$

$$=\log_2(a_1a_2a_3a_4a_5)$$

$$=\log_2(\sqrt{\alpha\beta})^5$$

$$=5\log_2\sqrt{4}=5$$

(iii) If $\alpha^2,(a_1)^2,(a_2)^2,(a_3)^2,(a_4)^2,(a_5)^2,\beta^2$ are in AP, then

$$\sum_{i=1}^{5}a_i^2=5\left(\frac{\alpha^2+\beta^2}{2}\right)$$

$$=5\left(\frac{(\alpha+\beta)^2-2\alpha\beta}{2}\right)$$

$$=5\left(\frac{8^2-2\times4}{2}\right)=140$$

(iv) If $1/\alpha, 1/a_1, 1/a_2, 1/a_3, 1/a_4, 1/a_5, 1/\beta$ are in AP, then

$$\sum_{i=1}^{5}\frac{1}{a_i}=5\left(\frac{\frac{1}{\alpha}+\frac{1}{\beta}}{2}\right)=\frac{5}{2}\,\frac{\alpha+\beta}{\alpha\beta}=\frac{8\times5}{2\times4}=\frac{1}{2}$$

$$\therefore\quad \left[\frac{10}{\sum_{i=1}^{5}\frac{1}{a_i}}+1\right]=2+1=3$$

49. (i) $\rightarrow$ D, (ii) $\rightarrow$ C, (iii) $\rightarrow$ B, (iv) $\rightarrow$ C

(i) $b+c=20$ and $a+d=20$

$$\Rightarrow a+b+c+d=40\qquad\left[\because\frac{b+c}{2}=10\text{ and }\frac{a+d}{2}=10,\ \text{as } a,b,10,c,d \text{ are in AP}\right]$$

(ii) Sides are $2, 2r, 2r^2$

Now, $(2r^2)^2=4r^2+4\Rightarrow r^4-r^2-1=0$

$$\Rightarrow\quad r^2=\frac{1+\sqrt{5}}{2}\Rightarrow 2r^2=1+\sqrt{5}$$

$$a^2+b^2=26$$

(iii) $a+ar+ar^2=70$

Now, $4a, 5ar, 4ar^2$ are in AP

$$\Rightarrow\quad 5r-4=4r^2-5r$$

$$\Rightarrow\quad 5r=2(r^2+1)$$

So, $ar + a\left(\frac{5r}{2}\right) = 70$

$\Rightarrow \quad ar = 20$

$\Rightarrow$ GP is 40, 20, 10

(iv) $\cos 60° = \frac{4+9-c_1^2}{12}$

$\Rightarrow \quad c_1^2 = 7$

$\Rightarrow \quad c_1 = \sqrt{7}$

other side is $\cos 120° = \frac{4+9-c_2^2}{12}$

$\Rightarrow \quad c_2^2 = 19$

$\Rightarrow \quad c_2 = \sqrt{19}$

Perimeter $= 2(\sqrt{7} + \sqrt{19})$

$\Rightarrow \quad a + b = 26$

50. (i) $\rightarrow$ B, (ii) $\rightarrow$ D, (iii) $\rightarrow$ A, (iv) $\rightarrow$ C

(i) $x^{x\sqrt{x}} = x^{\frac{3}{2}x}$

$\Rightarrow \quad x\sqrt{x} = \frac{3}{2}x$

$\Rightarrow \quad x = \frac{9}{4}$

(ii) $(\sin A + \sin B + \sin C) = \frac{a+b+c}{2R} = \frac{16}{10} = \frac{8}{5}$

(iii) $S = 1 + \frac{3}{4} + \frac{7}{16} + \frac{15}{64} + \ldots$

$\frac{S}{4} = \frac{1}{4} + \frac{3}{16} + \frac{7}{64} + \ldots$

$\frac{3S}{4} = 1 + \frac{1}{2} + \frac{1}{4} + \frac{1}{8} + \ldots$

$\frac{3S}{4} = 1 + \frac{1/2}{1-\frac{1}{2}} \Rightarrow \frac{3S}{4} = 2 \Rightarrow S = \frac{8}{3}$

(iv) $T_r = \frac{r+3}{r(r+1)(r+2)} = \frac{1}{r(r+2)} - 2\left[\frac{1}{2r} - \frac{1}{r+1} + \frac{1}{2(r+2)}\right]$

$= \frac{1}{2}\frac{1}{r} - \frac{1}{2(r+2)} - \frac{1}{r} + \frac{2}{r+1} - \frac{1}{r+2}$

$T_r = \frac{2}{r+1} - \frac{1}{2r} - \frac{3}{2(r+2)}$

Now, $S = \sum \frac{2}{r+1} - \frac{1}{2r} - \frac{3}{2(r+2)}$

$= 2\sum\left(\frac{1}{r+1} - \frac{1}{r+2}\right) - \frac{1}{2}\sum\left(\frac{1}{r} - \frac{1}{r+2}\right)$

$= \frac{5}{4}$

51. (1275) Last term of $S_{50} = 1 + 2 + 3 + \ldots + 50$

$= \frac{50 \times 51}{2} = 1275$

52. (62525) Last term of $S_{49} = 1 + 2 + \ldots 49$

$= \frac{49 \times 50}{2} = 1225$

$\Rightarrow$ Sum of $S_{50} = 1226 + \ldots 1275$

$= 50 \times 1225 + \frac{50 \times 51}{2}$

$= 62525$

53. (22) Last term of $S_6 = 1 + 2 + \ldots + 6$

$= \frac{6 \times 7}{2} = 21$

First term of $S_7 = 22$

Sol. (Q.Nos. 54-55)

$S = 3 + 6 + 11 + 18 + \ldots t_n$

$S = 3 + 6 + 11 + \ldots + t_{n-1} + t_n$

$\therefore \quad 0 = 3 + (3 + 5 + 7 + 9 + \ldots (n-1) \text{ terms} - t_n$

$\therefore \quad t_n = n^2 + 2;\ f(n) = n^2 + 2$

Similarly, nth term of $g(n) = n^2 + n + 1$,

$\lim_{n\to\infty} \frac{f(n)}{g(n)} = \lim_{n\to\infty} \frac{n^2+2}{n^2+n+1} = 1$

$F(N) = \Sigma(n^2 + 2)$

$= \frac{n(n+1)(2n+1)}{6} + 2n$

$= \frac{n(2n^2 + 3n + 13)}{6}$

$G(N) = \Sigma(n^2 + n + 1)$

$= \frac{n(n^2 + 3n + 5)}{3}$

$\therefore \quad \lim_{n\to\infty} \frac{F(n)}{G(n)} = \frac{\frac{n(2n^2+3n+13)}{6}}{\frac{n(n^2+3n+5)}{3}} = 1$

54. (1) **55.** (1)

56. (0.4) We have, $T_1 = 0 \Rightarrow A + B = 0$ ...(i)

$T_2 = 1 \Rightarrow A\alpha + B\beta = 1$ ...(ii)

$T_3 = 1 \Rightarrow A\alpha^2 + B\beta^2 = 1$ ...(iii)

$T_4 = 2 \Rightarrow A\alpha^3 + B\beta^3 = 2$ ...(iv)

$\Rightarrow \quad A = -B$, [using Eq. (i)]

From Eq. (ii) $\Rightarrow A(\alpha - \beta) = 1$ ...(v)

From Eq. (iii) $\Rightarrow A(\alpha^2 - \beta^2) = 1$ ...(vi)

From Eq. (iv) $\Rightarrow A(\alpha^3 - \beta^3) = 1$ ...(vii)

From Eqs. (v) and (vi) $\Rightarrow \alpha + \beta = 1$ ...(viii)

From Eqs. (iii) and (iv) $\Rightarrow \alpha^2 + \alpha\beta + \beta^2 = 2$ ...(ix)

From Eqs. (viii) and (ix), we get

$$\alpha\beta = -1$$

$\therefore$ α, β are the roots of the equation

$$x^2 - x - 1 = 0$$

$$\Rightarrow \quad \alpha = \frac{\sqrt{5}+1}{2}; \beta = \frac{1-\sqrt{5}}{2}$$

As $(\alpha - \beta) = \sqrt{5}$, so from $A(\alpha - \beta) = 1$, we get

$$A = \frac{1}{\sqrt{5}}; B = \frac{-1}{\sqrt{5}} \Rightarrow A^2 + B^2 = \frac{2}{5} = 0.4$$

57. (0) $a_k = 2a_{k-1} - a_{k-2} \Rightarrow a_1, a_2, \ldots, a_{11}$ are in AP.

$$\therefore \frac{a_1^2 + a_2^2 + \ldots + a_{11}^2}{11} = \frac{11a^2 + 35 \times 11d^2 + 110ad}{11} = 90$$

$$\Rightarrow \quad 225 + 35d^2 + 150d = 90$$

$$35d^2 + 150d + 135 = 0$$

$$\Rightarrow \quad d = -3, -\frac{9}{7}$$

Given, $a_2 < \frac{27}{2}$, we get $d = -3$

$$\Rightarrow \frac{a_1 + a_2 + \ldots + a_{11}}{11} = \frac{11}{2}[30 - 10 \times 3] = 0$$

58. (6) Let the sides be $a - d, a, a + d$, such that $d > 0$.

$\therefore$ By Pythagoras theorem,

$$a^2 + (a-d)^2 = (a+d)^2$$

$$\Rightarrow \quad a = 4d$$

Thus, sides are $3d$, $4d$ and $5d$

Given that area $= \frac{1}{2}(3d)(4d)$

$$\Rightarrow \quad 24 = \frac{1}{2} \times 12d^2 \Rightarrow d = 2$$

Thus, sides are 6, 8 and 10.

Hence, the smallest side is 6.

59. (1.333) Let number of terms $= 2n$

$$\Rightarrow k = \frac{\frac{n}{2}[2a + (n-1)d]}{\frac{n}{2}[2(a+nd) + (n-1)d]} = \frac{2 + (n-1)d}{2 + (3n-1)d}$$

$$= 1 - \frac{2d}{\left(\frac{2-d}{n} + 3d\right)}$$

$\Rightarrow$ for k to be constant, $d = 0$ or $d = 2$

$$\Rightarrow \quad k = 1, \frac{1}{3}$$

Thus, the sum of all possible values $= 1 + \frac{1}{3} = \frac{4}{3} = 1.333$

60. (9) Given that $\frac{S_7}{S_{11}} = \frac{6}{11}$

$$\Rightarrow \quad \frac{\frac{7}{2}(2a+6d)}{\frac{11}{2}(2a+10d)} = \frac{6}{11} \Rightarrow \frac{7(a+3d)}{11(a+5d)} = \frac{6}{11}$$

$$\Rightarrow \quad 7a + 21d = 6a + 30d \Rightarrow a = 9d$$

$$\Rightarrow \quad T_7 = a + 6d = 15d$$

Given that $130 < T_7 < 140$

$$130 < 15d < 140 \Rightarrow d = 9$$

61. (5) $\sum_{r=1}^{110} (1 + rx) = (1+x)(1+2x)(1+3x) \ldots (1+110x)$

$$= 1^{110} + (x + 2x + 3x + \ldots + 110x) \cdot 1^{109} + \ldots$$

So, coefficient of x in

$$\sum_{r=1}^{110} (1+rx) = 1 + 2 + 3 + \ldots + 110 = \frac{110 \times 111}{2} = 55 \times 111$$

$$= 6105$$

Now, $\lambda(1+10)(1+10+10^2) = \lambda(11)(111)$

$$\therefore \quad \lambda(11)(111) = 6105$$

$$\Rightarrow \quad \lambda = 5$$

62. (3748) $X = \{1, 6, 11, 16, \ldots, 2018 \text{ terms}\}$

$Y = \{9, 16, 23, \ldots, 2018 \text{ terms}\}$

$\because n(X \cup Y) = n(X) + n(Y) - n(X \cap Y)$

For $n(X \cap Y)$, we have to find the common terms in two AP's.

Common difference of set X is 5 and that of set Y is 7.

So, common difference of AP of common terms is LCM $(5, 7) = 35$

So, common terms are 16, 51, 86, ...

Now, last term in set X is $1 + (2018 - 1) \times 5 = 10086$ and last term in set Y is $9 + (2018 - 1) \times 7 = 14128$

$\therefore$ last term of the common term series is

$$16 + (n-1) \times 35 \le 10086, n \le 288.7$$

So, there will be 288 terms.

$\therefore$ $n(X \cup Y) = 2018 + 2018 - 288 = 3748$

63. (7) $\because$ $a_1, a_2, a_3, \ldots$ are in GP with common ratio r and $b_1, b_2, b_3, \ldots$ is also a GP.

i.e. $b_1 = 1, b_2 = \sqrt[4]{7} - \sqrt[4]{28} + 1, a_1 = \sqrt[4]{28}$

$$\because \Sigma \frac{1}{a_n} = \Sigma b_n$$

$$\Rightarrow \frac{1}{a_1} + \frac{1}{a_2} + \ldots + \infty = b_1 + b_2 + b_3 + \ldots + \infty$$

$$\Rightarrow \frac{1}{\sqrt[4]{28}} + \frac{1}{\sqrt[4]{28}\,r} + \frac{1}{\sqrt[4]{28}\,r^2} + \ldots + \infty$$

$$= 1 + (\sqrt[4]{7} - \sqrt[4]{28} + 1) + (\sqrt[4]{7} - \sqrt[4]{28} + 1)^2 r + \ldots + \infty$$

$$\Rightarrow \quad \frac{\frac{1}{\sqrt[4]{28}}}{1 - \frac{1}{r}} = \frac{1}{1 - \sqrt[4]{7} + \sqrt[4]{28} - 1}$$

$$\Rightarrow \quad \frac{r}{(r-1)\sqrt[4]{28}} = \frac{1}{\sqrt[4]{28} - \sqrt[4]{7}} = \frac{1}{\sqrt[4]{7}(\sqrt[4]{4} - 1)}$$

$$\Rightarrow \frac{r}{(r-1)} \times \frac{1}{\sqrt[4]{4}} = \frac{1}{(\sqrt[4]{4} - 1)}$$

Let $\sqrt[4]{4}=\alpha$

$$\frac{r}{(r-1)\alpha}=\frac{1}{\alpha-1}$$

$\Rightarrow \quad r\alpha-r=r\alpha-\alpha \Rightarrow r=\alpha \Rightarrow r=\sqrt[4]{4}$

Now, $1+r^2+r^4=1+(\sqrt[4]{4})^2+(\sqrt[4]{4})^4=1+2+4=7$

64. (0.6) If $F_n=F_{n-1}+F_{n-2}$

$\Rightarrow \quad x^2-x-1=0$

Let α, β be the roots of the equation.

$\therefore \quad F_n=k_1\alpha^n+k_2\beta^n$

$\therefore \quad F_0=k_1+k_2=0 \Rightarrow k_2=-k_1$

$\therefore \quad F_n=k_1(\alpha^n-\beta^n)$

$\quad F_1=k_1(\alpha-\beta)=1$

$$\therefore \sum_{n=1}^{\infty}\frac{k_1(\alpha^n-\beta^n)}{3^n}=k_1\left[\sum_{n=1}^{\infty}\left(\frac{\alpha}{3}\right)^n-\sum_{n=1}^{\infty}\left(\frac{\beta}{3}\right)^n\right]$$

$$=k_1\left[\frac{3}{3-\alpha}-\frac{3}{3-\beta}\right]$$

$$=3k_1\left[\frac{\alpha-\beta}{(3-\alpha)(3-\beta)}\right]=\frac{3}{(3-\alpha)(3-\beta)}$$

$[\because k_1(\alpha-\beta)=1]$

$\because x^2-x-1=(x-\alpha)(x-\beta)$

Put $x=3$

$\Rightarrow \quad (3-\alpha)(3-\beta)=5$

Thus, $\displaystyle\sum_{n=1}^{\infty}\frac{F_n}{3^n}=\frac{3}{5}=0.6$

65. (4) $\because \dfrac{b}{a}=\dfrac{c}{b}=$ integer [given that]

$\Rightarrow b^2=ac \Rightarrow c=\dfrac{b^2}{a}$

Also, $\dfrac{a+b+c}{3}=b+2$

$\Rightarrow \quad a+b+c=3b+6 \Rightarrow a-2b+c=6$

$\Rightarrow a-2b+\dfrac{b^2}{a}=6 \Rightarrow 1-\dfrac{2b}{a}+\dfrac{b^2}{a^2}=\dfrac{6}{a}$

$\Rightarrow \quad \left(\dfrac{b}{a}-1\right)^2=\dfrac{6}{a}$ $\quad\left[\because \dfrac{b}{a}\text{ is an integer}\right]$

$\Rightarrow \quad a=6$ only

$\Rightarrow \dfrac{a^2+a-14}{a+1}=4$

66. (2) $\because \dfrac{2ab}{a+b}=4$

$\Rightarrow \quad ab=2a+2b$ ...(i)

Now, $q=10-a$ and $2q=5+b$

$\Rightarrow \quad 20-2a=5+b \Rightarrow 15=2a+b$ ...(ii)

From Eqs. (i) and (ii),

$\quad a(15-2a)=2a+2(15-2a)$

$\Rightarrow \quad 15a-2a^2=-2a+30 \Rightarrow 2a^2-17a+30=0$

$\Rightarrow a=6, \dfrac{5}{2}$, according $q=4, \dfrac{15}{2}$

$\Rightarrow \quad |q-a|=2, 5$

$\therefore$ The smallest value of $|q-a|$ is 2.

67. (4) $\because 2+(6\cdot 2^2-4\cdot 2)+(6\cdot 3^2-4\cdot 3)+\ldots.$

$+(6\cdot n^2-4\cdot n)=140$

$\Rightarrow 2+6(2^2+3^2+\ldots+n^2)-4(2+3+\ldots.+n)=140$

$$\Rightarrow 2+6\left[\frac{n(n+1)(2n+1)}{6}-1\right]-4\left[\frac{n(n+1)}{2}-1\right]=140$$

$\Rightarrow \quad 2+n(n+1)(2n+1)-6-2n(n+1)+4=140$

$\Rightarrow \quad 2n^3+3n^2+n-2n^2-2n-140=0$

$\Rightarrow \quad 2n^3+n^2-n-140=0$

$\Rightarrow \quad (n-4)(2n^2+9n+35)=0$

$n=4$ or $n=\dfrac{9\pm\sqrt{-199}}{4}$ [neglected]

Thus, only positive integer value of n is 4.

68. (5) Clearly, $1+2+3+\ldots.+(n-2)\le 1224\le 3+4+\ldots.+n$

$\Rightarrow \dfrac{(n-2)(n-1)}{2}\le 1224\le\dfrac{(n-2)(3+n)}{2}$

$\Rightarrow \quad n^2-3n-2446\le 0$ and $n^2+n-2454\ge 0$

$\Rightarrow \quad 49<n<51 \Rightarrow n=50$

$\therefore \quad \dfrac{n(n+1)}{2}-(2k+1)=1224$

or $\quad k=25$

$\therefore \quad k-20=5$

69. (1.5) On adding, we have

$$\frac{1}{x}+\frac{1}{y}+\frac{1}{z}=\frac{3}{2018}$$

Also $2xz-2yz+1=\dfrac{z}{2018}$

$2xy-2zx+1=\dfrac{x}{2018}$

$2yz-2xy+1=\dfrac{y}{2018}$

On adding, $x+y+z=3\times 2018$

$$\because \frac{3}{\frac{1}{x}+\frac{1}{y}+\frac{1}{z}}\le\frac{x+y+z}{3}$$

$\Rightarrow \quad x=y=z$ $\quad[\because x, y, z$ are positive$]$

$\Rightarrow \quad x=y=z=2018$

Then, $\dfrac{x+y+z}{2x}=\dfrac{3}{2}=1.5$

CHAPTER

05

PERMUTATIONS & COMBINATIONS

LEVEL 01 \ Starter Level

5.1 General Arrangements and Factorial Notations

1 A cinema hall has 20 gates. In how many ways can a person enter the hall through one gate and come out through a different gate?

2 There are 100 History books and 75 Geography books in a library. In how many ways can we choose one History and one Geography books?

3 A student has 7 different books that he wishes to arrange in a row. How many different arrangements are possible?

4 There are 50 students in a group in which 30 boys and 20 girls. The head of the group select either a boy or a girl for captain of the group. In how many ways the head of the group can make this selection?

5 Evaluate

(i) $7!$ (ii) $6! - 4!$ (iii) $\frac{12!}{9!\,2!}$

6 If $\frac{1}{8!}+\frac{1}{9!}=\frac{x}{10!}$, find x.

7 Find n if $(n+2)! = 60 \times (n-1)!$.

8 Find the remainder when $\sum_{r=1}^{100} r!$ is divided by 15.

9 Simplify the expression $\frac{3\cdot 6\cdot 9\cdot 12 .....(3n)}{3^n}$.

10 Find the exponent of 3 in 100!.

11 Find the number of zeroes at the end of 100!.

5.2 Permutations

12 Find the value of n such that

(i) ${}^{n}P_{4} = 5040$

(ii) ${}^{n}P_{5} = 42 \cdot {}^{n}P_{3}$

13 If ${}^{56}P_{r+6} : {}^{54}P_{r+3} = 30800 : 1$, find ${}^{r}P_{2}$.

14 Suppose 10 people enter an event in a quiz competitions. In how many ways could the first, second and third prizes be awarded?

15 Determine the number of permutations of the letters of word 'COVID' taken all at a time.

16 A child has four pockets and three marbles. In how many ways can the child put the marbles in its pockets?

17 How many 3-letter code words can be made from the word 'PRIME' if

(i) we can't repeat letters.

(ii) we can repeat letter.

18 How many role numbers can be made, using one digit followed by six letters (e.g. 3 IDIOTS)?

19 Find the total number of 9-digit numbers which have all different digits.

20 How many different signals can be given using any number of flags from 3 flags of different colours?

21 How many words (with or without meaning) can be formed with the letters of the word 'SHAKUNTALA' by rearranging them.

22 How many words can be formed with the letters of the word 'PATALIPUTRA' without changing the relative positions of vowels and consonants?

23 If the letters of the word are arranged as in dictionary (with or without meaning). Find the rank of word 'INDIA'.

24 Find the number of ways in which 5 boys and 5 girls can be arranged in a row.

(i) If no two boys are together.

(ii) If boys and girls are alternate.

(iii) All the girls sit together and all the boys sit together.

(iv) All the girls are never together.

25 18 guests have to be seated half on each side of a long table. Four particular guests desire to sit on one particular side and three other on the other side. Determine the number of ways in which the sitting arrangements can be made.

26 In how many ways can five boys and five girls form a circle such that the boys and girls alternate?

27 Six persons A, B, C, D, E and F are to be seated at a circular table. Find the number of ways, this can be done if A must have either B or C on his right and B must have either C or D on his right side.

28 Find the number of ways in which 12 different beads can be arranged to form a necklace.

5.3 Combinations

29 If ${}^{n}C_{9} = {}^{n}C_{7}$, then find the value of n.

30 If ${}^{2n}C_{3} : {}^{n}C_{3} = 11 : 1$, find the value of n.

31 If ${}^{n}C_{r-1} = 36$, ${}^{n}C_{r} = 84$ and ${}^{n}C_{r+1} = 126$, then find the values of n and r.

32 A sales representative has 10 accounts in a certain city.

(i) In how many ways can 3 accounts to be selected to call on?

(ii) In how many ways can calls be scheduled for 3 of the 10 accounts?

33 How many different selections of 6 books can be made from 11 different books, if

(i) two particular books are always selected?

(ii) two particular books are never selected?

34 7 relatives of a man comprises 4 ladies and 3 gentleman, his wife has also 7 relatives; 3 of them are ladies and 4 gentlemen. In how many ways can they invite a dinner party of 3 ladies and 3 gentlemen so that there are 3 of man's relatives and 3 of the wife's relatives?

35 A box contains two white balls, three black balls and four red balls. In how many ways can three balls be drawn from the box, if atleast one black ball is to be included in the draw?

36 A committee of 12 is to formed from 9 women and 8 men. In how many ways this can be done if at least 5 women have to be included in a committee? In how many of these committee the men are in majority?

37 A student is allowed to select atmost n books from collection of $(2n+1)$ books. If the total number of ways in which he can select atleast one book is 63, find the value of n.

38 In a high school, a committee has to be formed from a group of 6 boys and 5 girls.

(i) Find the number of ways in which the committee can be formed such that it has 5 members, having exactly 3 boys and 2 girls.

(ii) Find the number of ways in which the committee can be formed such that committee has atleast 2 members and having an equal number of boys and girls.

(iii) Find the number of ways in which the committee can be formed such that it has 5 members, atleast 2 of them being girls.

(iv) Find the number of ways in which the committee can be formed such that it has 4 members, having atleast 2 girls such that one particular boy and a particular girl are not in the committee together.

39 There are three books of Physics, four of Chemistry and five of Mathematics. How many different collections can be made such that each collection consists of

(i) one book of each subject?

(ii) atleast one book of each subject?

(iii) atleast one book of Mathematics?

40 In how many ways can the letters 'AAABCD' be arrange so that

(i) the two B's are together but no two A's are together?

(ii) no two B's and no two A's are together?

41 There are ten points in a plane. Of these ten points four points are in a straight line and with the exception of these four points no other three points are in same straight line. Find

(i) the number of triangles formed.

(ii) the number of straight lines formed.

42 If m parallel lines are intersected by n other parallel lines, then find the number of parallelograms thus formed.

43 Find the number of rectangles excluding squares from a rectangle of size 9×6.

44 How many ways are there to place 2 identical kings on an 8×8 chessboard so that the kings are not in adjacent squares?

45 How many selections of atleast one red ball from a bag containing 4 red balls and 5 black balls, balls of the same colour being identical?

46 For Number $N = 35700$, find

(i) number of divisors. (ii) number of proper divisors.

(iii) number of even divisors. (iv) sum of all divisors.

5.4 Division, Distribution, De-arrangement and Multinomial Theorem

47 In how many ways can a pack of 52 cards be equally divided among 4 players?

48 In how many ways can 12 balls be divided between 2 boys, on receiving 5 balls and other 7 balls?

49 In how many ways 5 different balls can be arranged into 3 different boxes so that no box remains empty?

50 (i) In how many ways can 5 people be divided into three groups?

(ii) In how many ways, 5 people be distributed in three different rooms if no room must be empty?

(iii) In how many ways, can 5 people be arranged in three different rooms if no room must be empty?

51 In how many ways can 5 different books be tied up in three bundles?

52 If $n(A) = 5$ and $n(B) = 3$, find the number of onto mappings from A to B.

53 Four boys picked up by 30 mangoes. In how many ways can they divide them, if all mangoes be identical?

54 Find the positive number of solutions of $x + y + z + w = 20$. Under the following conditions.

(i) Zero values of x, y, z and w are included.

(ii) Zero values are excluded.

55 How many integral solutions are there to $x + y + z + t = 29$, when $x \geq 1, y \geq 2, z \geq 3$ and $t \geq 0$.

56 How many integral solutions are there to the system of equations $x_1 + x_2 + x_3 + x_4 + x_5 = 20$ and $x_1 + x_2 = 15$, when $x_i \geq 0, i = 1, 2, \ldots, 5$.

57 Find the number of non-negative integral solutions of $3x + y + z = 24$.

58 In how many ways we can put 5 differently coloured balls into 5 boxes of colours same as that of the balls so that no ball goes to the box of the same colour?

59 Find the number of ways to put 4 letters in 4 addressed envelope so that all are in wrong envelopes.

60 In how many ways 16 identical things can be distributed among 4 persons if each person gets atleast 3 things.

61 Find the number of integers which lies between the number 786 and 999786, if the sum of their digit is 25.

LEVEL 02\\ JEE Main Level

Single Choice Correct Questions

5.1 General Arrangement and Factorial Notations

1 A college planning committee consists of 3 freshmen, 4 sopohomore, 5 juniors and 2 seniors. A subcommittee of 4 consists 1 person from each class. How many choices are possible?

(a) 20 (b) 14 (c) 120 (d) 60

2 Anil goes to a restaurant and choose to create his own burger. He has 2 different kind of cheese, 3 different breads and 3 different sauces, he can choose from, but he can only choose one of each category. How many different ways can he create the burger?
(a) 8 (b) 18 (c) 12 (d) 10

3 How many 3-digit numbers can be formed from the digits 2, 3, 4, 5, 7 and 9.
(a) 196 (b) 216 (c) 184 (d) 120

4 The number less than 1000 that can be formed using the digits 0, 1, 2, 3, 4 and 5 when repetition is not allowed, is equal to
(a) 130 (b) 131 (c) 156 (d) 155

5 If the number of five digit numbers with distinct digits and 2 at the 10th place is 336 k, then k is equal to **[JEE Main 2020]**
(a) 4 (b) 6 (c) 7 (d) 8

6 The number of four digit numbers strictly greater than 4321, then can be formed using the digits 0, 1, 2, 3, 4 and 5 (repetition of digit is allowed), is **[JEE Main 2019]**
(a) 288 (b) 360 (c) 306 (d) 310

7 If $n!, 3\times(n!)$ and $(n+1)!$ are in GP, then the value of n will be
(a) 3 (b) 4 (c) 8 (d) 10

8 The sum of n terms of the series whose nth term is $n\cdot n!$, is
(a) $(n!)^2$ (b) $(n+1)!-1!$ (c) $(n+1)!$ (d) $(n+2)!$

9 The greatest of d so that 12 divides in 100! is equal to $(a\in N)$
(a) 47 (b) 48 (c) 49 (d) 97

5.2 Permutations

10 The range of the function $f(x) = {}^{7-x}P_{x-3}$ is
(a) $\{1, 2, 3\}$ (b) $\{1, 2, 3, 4, 5, 6\}$ (c) $\{1, 2, 3, 4\}$ (d) $\{1, 2, 3, 4, 5\}$

11 The value of $(2\cdot{}^1P_0 - 3\cdot{}^2P_1 + 4\cdot{}^3P_2 - \dots$ up to 51th terms) $+ (1! - 2! + 3! \dots\dots$ upto 51th term) is equal to **[JEE Main 2020]**
(a) $1-51(51)!$ (b) $1+(51)!$ (c) $1+(52)!$ (d) 1

12 Let $X = \{1, 2, 3, 4, 5\}$, the number of different ordered pairs (Y, Z) that can be formed such that $Y\subseteq X$, $Z\subseteq X$ and $Y\subseteq Z$ is empty, is
(a) 5^2 (b) 3^5 (c) 2^5 (d) 5^3

13 The total number not more than 20 digits that are formed by using the digits 0, 1, 2, 3 and 4 is
(a) 5^{20} (b) $5^{20}-1$ (c) $5^{20}+1$ (d) None of these

14 Two teams are to play a series of five matches between them. A match ends in a win, loss, or draw for a team. A number of people forecast the result of each match and no two people make the same forecast for the series of matches. The smallest group of people in which one person forecast correctly for all the matches will contain n people, where n is
(a) 81 (b) 243 (c) 486 (d) None of these

15 The number of integers greater than 6000 that can be formed, using the digits 3, 5, 6, 7 and 8, without repitition is
(a) 216 (b) 192 (c) 120 (d) 72

16 How many numbers can be made with the digits 3, 4, 5, 6, 7 and 8 lying between 3000 and 4000, which are divisible by 5 while repetition of any digits is not allowed any number?
(a) 60 (b) 12 (c) 120 (d) 24

17 Total number of 6-digit numbers in which only and all five digits 1, 3, 5, 7 and 9 appear, is **[JEE Main 2020]**

(a) $\frac{1}{2}6!$ (b) $6!$ (c) 5^6 (d) $\frac{5}{2}(6!)$

18 The number of 6-digit numbers that can be formed using the digits 0, 1, 2, 5, 7 and 9 which are divisible by 11 and no digit is repeated, is **[JEE Main 2019]**

(a) 72 (b) 60 (c) 48 (d) 36

19 How many ways are there to arrange the letters in the word 'GARDEN' with the vowels in alphabetical order?

(a) 120 (b) 240 (c) 360 (d) 480

20 In a city no two persons have identical set of teeth and there is no person without a tooth. Also, no person has more than 32 teeth. If we disregard the shape and size of tooth and consider only the positioning of the teeth, then the maximum population of the city is

(a) 2^{32} (b) $(32)^2 - 1$ (c) $2^{32} - 1$ (d) 2^{32-1}

21 In a room, there are 12 bulbs of the same watt, each having a separate switch. The number of ways to light the room with different amount of illumination is

(a) $12^2 - 1$ (b) 2^{12} (c) $2^{12} - 1$ (d) 12^2

22 If all the words formed from the letters of the word 'HORROR' are arranged in the opposite order as they are in a dictionary, then the rank of the word 'HORROR' is

(a) 56 (b) 57 (c) 58 (d) 59

23 If all the words (with or without meaning) having five letters, formed using the letters of the word 'SMALL' and arranged as in a dictionary, then the position of the word 'SMALL' is

(a) 59 (b) 52 (c) 58 (d) 46

24 Three boys of class X, four boys of class XI and five boys of class XII sit in a row. The total number of ways in which these boys can sit so that all the boys of same class sit together, is equal to

(a) $(3!)^2 (4!) (5!)$ (b) $(3!) (4!)^2 (5!)$

(c) $(3!)(4!)(5!)$ (d) $(3!) (4!) (5!)^2$

25 The number of possible outcomes in a throw of n ordinary dice in which atleast one of the dice shows an odd number is

(a) $6^n - 1$ (b) $3^n - 1$ (c) $6^n - 3^n$ (d) None of these

26 The number of 5-digit numbers that contain 7 exactly once is

(a) $(41)(9^3)$ (b) $(37)(9^3)$ (c) $(7)(9^4)$ (d) $(41)(9^4)$

27 There are $(n+1)$ white and $(n+1)$ black balls, each set numbered 1 to $n+1$. The number of ways in which the balls can be arranged in a row so that the adjacent balls are of different colours, is

(a) $(2n+2)!$ (b) $(2n+2)! \times 2$ (c) $(n+1)! \times 2$ (d) $2\{(n+1)!\}^2$

28 A train timetable must be compiled for various days of week so that two trains twice a day depart for three days. One train daily for two days and three trains once a day for two days. How many different timetables can be compiled?

(a) 140 (b) 210 (c) 133 (d) 72

29 A seven digit number without repetition and divisible by 9 is to be formed by using seven digits out of 1, 2, 3, 4, 5, 6, 7, 8 and 9, the number of ways in which this can be done, is

(a) $9!$ (b) $2(7!)$ (c) $4(7!)$ (d) None of these

30 A tea party is arranged for 16 people along two sides of a large table with 8 chairs on each side. Four men want to sit on one particular side and two on the other side. The number of ways in which they can be seated, is

(a) $\frac{6!\,8!\,10!}{4!\,6!}$ (b) $\frac{8!\,8!\,10!}{4!\,6!}$

(c) $\frac{8!\,8!\,6!}{6!\,4!}$ (d) None of these

31 Two families with three members each and one family with four members are to be seated in row. In how many ways can they be seated so that the same family members are not separated? [JEE main 2020]

(a) $2!\,3!\,4!$ (b) $(3!)^3\,(4!)$

(c) $(3!)^2\,4!$ (d) $3!\,(4!)^3$

32 In a three-storey building, there are four rooms on the ground floor, two on the first and two on the second floor. If the rooms are to be alloted to six persons, one person occupying one room only. The number of ways in which this can be done so that no floor remains empty is

(a) ${}^8P_6 - 2\,(6!)$ (b) ${}^8P_6$ (c) ${}^8P_5\,(6!)$ (d) None of these

33 Six teachers and six students have to sit round a circular table such that there is a teacher between any two students. The number of ways in which they can sit, is

(a) $6! \times 6!$ (b) $5! \times 6!$ (c) $5! \times 5!$ (d) None of these

5.3 Combinations

34 If $\alpha = {}^mC_2$, then ${}^{\alpha}C_2$ is equal to

(a) ${}^{m+1}C_4$ (b) ${}^{m-1}C_4$ (c) $8 \cdot {}^{m+2}C_4$ (d) $3 \cdot {}^{m+1}C_4$

35 If ${}^nP_r = {}^nP_{r+1}$ and ${}^nC_r = {}^nC_{r-1}$, then the value of r is equal to [JEE Main 2021]

(a) 1 (b) 4 (c) 2 (d) 3

36 If ${}^nC_3 + {}^nC_4 > {}^{n+1}C_3$, then

(a) $n > 6$ (b) $n > 7$ (c) $n < 6$ (d) None of these

37 The number of ordered pairs (r, k) for which $6 \cdot {}^{35}C_r = (k^2 - 3) \cdot {}^{36}C_{r+1}$, where k is an integer, is [JEE Main 2020]

(a) 3 (b) 2 (c) 6 (d) 4

38 If the difference of the numbers of arrangements of three things from a certain number of dissimilar things and the number of selections of the same number of things from them exceeds 100, then the least number of dissimilar things is

(a) 8 (b) 6 (c) 5 (d) 7

39 In how many ways can a team of 11 players be formed out of 25 players, if 6 out of them are always to be included and 5 always to be excluded?

(a) 2020 (b) 2002 (c) 2008 (d) 8002

40 Among 10 persons A, B, C are to speak at a function. The number of ways in which it can be done if A wants to speak before B and B wants to speak before C, is

(a) 10!/24 (b) 9!/6 (c) 10!/6 (d) None of these

41 Consider a class of 5 girls and 7 boys. The number of different teams consisting of 2 girls and 3 boys that can be formed from this class, if there are two specific boys A and B who refuse to be members of the same team, is [JEE Main 2019]

(a) 500 (b) 200 (c) 300 (d) 350

42 A committee of 11 members is to be formed from 8 males and 5 females. If m is the number of ways the committee is formed with atleast 6 males and n is the number of ways the committee is formed with atleast 3 females, then **[JEE Main 2019]**

(a) $m+n=68$ (b) $m=n=78$
(c) $n=m-8$ (d) $m=n=68$

43 There are 3 sections in a question paper and each section contains 5 questions. A candidate has to answer a total of 5 questions, choosing atleast one question from each section. Then, the number of ways, in which the candidate can choose the question, is **[JEE Main 2020]**

(a) 3000 (b) 1500
(c) 2255 (d) 2250

44 The number of ways of choosing 10 objects out of 31 objects of which 10 are identical and the remaining 21 are distinct, is **[JEE Main 2019]**

(a) $2^{20}-1$ (b) 2^{21} (c) 2^{20} (d) $2^{20}+1$

45 The total number of ways of selecting two numbers from the set $\{1, 2, 3, 4 \dots 3n\}$ so that their sum is divisible by 3, is equal to

(a) $\frac{2n^2-n}{2}$ (b) $\frac{3n^2-n}{2}$ (c) $2n^2-n$ (d) $3n^2-n$

46 Two packs of 52 cards are shuffled by together. The number of ways in which a man can be dealt 26 cards so that he does not get two cards of the same suit and same denomination, is

(a) ${}^{52}C_{26}\times 2^{26}$ (b) ${}^{104}C_{26}$ (c) $2\times {}^{52}C_{26}$ (d) None of these

47 The total number of ways in which three distinct numbers in AP can be selected from the set $\{1, 2, 3 \dots\dots 24\}$, is equal to

(a) 66 (b) 132 (c) 198 (d) None of these

48 Total number of six digit numbers that can be formed having the property that every succeeding digit is greater than the preceding digit, is equal to

(a) ${}^{9}C_3$ (b) ${}^{10}C_3$ (c) ${}^{9}P_3$ (d) ${}^{10}P_3$

49 The sum of all the 4-digit distinct numbers that can be formed with the digits 1, 2, 2 and 3 is **[JEE Main 2021]**

(a) 26664 (b) 122664 (c) 122234 (d) 22264

50 The number of ways in which we can select four numbers from 1 to 30 so as to exclude every selection of four consecutive numbers, is

(a) 27378 (b) 27405 (c) 27399 (d) None of these

51 There were two women participating in a chess tournament. Every participant played two games with the other participates, the number of games that the men played among themselves, proved to exceed by 66 number of games that the men played with the women. The number of participants is

(a) 6 (b) 11 (c) 13 (d) None of these

52 A is a set containing n different elements. A subset P of A is chosen. The set A is reconstructed by replacing the elements of P. A subset Q of A is again chosen the number of ways of choosing P and Q so that $P\cap Q$ contains exactly two elements, is

(a) ${}^{n}C_3\cdot 2^n$ (b) ${}^{n}C_2\cdot 3^{n-2}$ (c) 3^{n-2} (d) None of these

53 The number of words of four letters containing equal number of vowels and consonants, where repetition is allowed, is

(a) 105^2 (b) 210×243 (c) 105×243 (d) 150×21^2

54 The number of different seven digit numbers that can be written using only three digits 1, 2 and 3 with the condition that the digit 2 occurs twice in each number, is
(a) ${}^7P_2 \cdot 2^5$ (b) ${}^7C_2 \cdot 2^5$ (c) ${}^7C_2 \cdot 5^2$ (d) None of these

55 The total number of three letter words that can be formed from the letter of the word 'SAHARANPUR', is equal to
(a) 210 (b) 237 (c) 247 (d) 227

56 From six different novels and three different dictionaries. Four novels and one dictionary are to be selected and arranged in a row on the shelf so that the dictionary is always in the middle. Then, the number of such arrangements is **[JEE Main 2018]**
(a) less than 500 (b) atleast 500 but less than 750
(c) atleast 750 but less than 1000 (d) atleast 1000

57 There are two urns, urn *A* has three distinct red balls and urn *B* has nine distinct blue balls. From each urn two balls are taken out at random and then transferred to the other. The number of ways in which this can be done, is **[JEE Main 2018]**
(a) 108 (b) 3 (c) 36 (d) 66

58 A person predicts the outcome of 20 cricket matches of his home team. Each match can result in either a win, loss or tie for the home team, total number of ways in which he can make the predictions so that exactly 10 predictions are correct, is equal to
(a) ${}^{20}C_{10} \times 2^{10}$ (b) ${}^{20}C_{10} \times 3^{20}$ (c) ${}^{20}C_{10} \times 3^{10}$ (d) ${}^{20}C_{10} \times 2^{20}$

59 The number of ways of choosing a committee of two women and three men from five women and six men, if Mr. *A* refuses to serve on the committee if Mr. *B* is a member and Mr. *B* can only serve, if Ms. *C* is the member of the committee, is
(a) 60 (b) 84 (c) 124 (d) None of these

60 To fill 12 vacancies, there are 25 candidates of which 5 are from scheduled caste. If three of the vacancies are reserved for scheduled caste candidates while the rest are open to all, the number of ways in which the selection can be made, is
(a) ${}^5C_3 \times {}^{22}C_9$ (b) ${}^{22}C_9 - {}^5C_3$ (c) ${}^{22}C_3 \times {}^5C_3$ (d) None of these

61 In how many ways can a team of 6 horses be selected out of a stud of 16 so that there shall always be three out of $ABC A'B'C'$, but never AA', BB' or CC' together?
(a) 840 (b) 1260 (c) 960 (d) 720

62 $2m$ white counters and $2n$ red counters are arranged in a straight line with $(m+n)$ counters on each side of a central mark. The number of ways of arranging the counters so that the arrangements are symmetrical with respect to the central mark is
(a) ${}^{m+n}C_m$ (b) ${}^{2m+2n}C_{2m}$ (c) $\frac{1}{2}\frac{(m+n)!}{m!n!}$ (d) None of these

63 There are 10 points in a plane, out of these 6 are collinear. If N is the number of triangles formed by joining these points, then
(a) $N > 190$ (b) $N \leq 100$ (c) $100 < N \leq 140$ (d) $140 < N < 190$

64 Let T_n be the number of all possible triangles formed by joining vertices of an n-sided regular polygon. If $T_{n+1} - T_n = 10$, then the value of n is
(a) 7 (b) 5 (c) 10 (d) 8

65 Consider a rectangle $ABCD$ having 5, 7, 6, 9 points in the interior of the line segments AB, CD, BC, DA, respectively. Let α be the number of triangles having these points from different sides as vertices and β be the number of quadrilaterals having these points from different sides as vertices. Then, $(\beta - \alpha)$ is equal to **[JEE Main 2021]**
(a) 795 (b) 1173 (c) 1890 (d) 717

66 If m parallel lines in a plane are intersected by a family of n parallel lines. The number of parallelograms that can be formed, is

(a) $\frac{1}{4}mn(m-1)(n-1)$ (b) $\frac{1}{2}mn(m-1)(n-1)$

(c) $\frac{1}{4}m^2n^2$ (d) None of these

67 A natural number has prime factorisation given by $n=2^x3^y5^z$, where y and z are such that $y+z=5$ and $y^{-1}+z^{-1}=\frac{5}{6}, y>z$. Then, the number of odd divisors of n, including 1, is **[JEE Main 2021]**

(a) 11 (b) 6 (c) $6x$ (d) 12

68 Assuming the balls to be identical except for difference in colours, the number of ways in which one or more balls can be selected from 10 white, 9 green and 7 black balls, is **[JEE Main 2018]**

(a) 880 (b) 629 (c) 630 (d) 879

69 The number of divisors of the numbers of 38808 (excluding 1 and the number itself) is

(a) 70 (b) 71 (c) 72 (d) 73

70 If a, b, c, d and e are prime integers, then the number of divisors of ab^2c^2de excluding 1 as a factor is

(a) 94 (b) 72 (c) 36 (d) 71

5.4 Division, Distribution, De-arrangement and Multinomial

71 A man has three friends. The number of ways he can invite one friend everyday for dinner on six successive nights so that no friend is invited more than three times, is

(a) 640 (b) 320 (c) 420 (d) 510

72 The total number of ways in which n^2 number of identical balls can be put in n numbered boxes (1, 2, 3 ...., n) such that ith box contains atleast i number of balls, is

(a) ${}^{n^2}C_{n-1}$ (b) ${}^{n^2-1}C_{n-1}$ (c) ${}^{\frac{n^2+n-2}{2}}C_{n-1}$ (d) None of these

73 The total number of ways in which 15 identical blankets can be distributed among four persons so that each of them gets atleast two blankets, is equal to

(a) ${}^{10}C_3$ (b) ${}^{9}C_3$ (c) ${}^{11}C_3$ (d) None of these

74 Number of ways in which 25 identical things be distributed among five persons if each gets odd number of things is

(a) ${}^{25}C_4$ (b) ${}^{12}C_8$ (c) ${}^{14}C_{10}$ (d) ${}^{13}C_3$

75 Fifteen identical balls have to be put in five different boxes. Each box can contain any number of balls, the total number of ways of putting the balls into the boxes so that each box contains atleast two balls, is equal to

(a) ${}^{9}C_5$ (b) ${}^{10}C_5$ (c) ${}^{6}C_5$ (d) ${}^{10}C_5$

76 The number of integral solutions of $x+y+z=0$ with $x\geq -5, y\geq 5$ and $z\geq -5$ is

(a) 134 (b) 136 (c) 138 (d) 140

77 Number of ways in which ₹ 18 can be distributed amongst four persons such that nobody receives less than ₹ 3, is

(a) 4^2 (b) 2^4 (c) $4!$ (d) 84

78 The total numbers of positive integral solution of $15<x_1+x_2+x_3\leq 20$ is equal to

(a) 685 (b) 785 (c) 1125 (d) None of these

79 Six cards and six envelopes are numbered 1, 2, 3, 4, 5, 6 and cards are to be placed in envelopes so that each envelope contains exactly one card and no card is placed in the envelope bearing the same number and more over the card numbered 1 is always placed in envelope numbered 2. Then, the total number of ways it can be done, is

(a) 264 (b) 265 (c) 53 (d) 67

80 If 4 dice are rolled, then the number of ways of getting the sum 10 is

(a) 56 (b) 64

(c) 72 (d) 80

Numerical Value Type Questions

81 The number of six letter words (with or without meaning), formed using all the letters of the word 'VOWELS', so that all the consonants never come together, is **[JEE Main 2021]**

82 Number of ways in which $5A$'s and $6B$'s can be arranged in a row which reads the same backwards and forwards is N, then the value of $N/2$ is ....

83 There are n distinct white and n distinct black balls. If the number of ways of arranging them in a row so that neighbouring balls are of different colour, is 1152. Then, value of n is ....

84 Number of four digit numbers of the form $N = abcd$ which satisfy following three conditions

(i) $4000 \leq N < 6000$

(ii) N is a multiple of 5

(iii) $3 \leq b < C \leq 6$

is equal to ....

85 The number of three digits number having only two consecutive digits identical, is ....

86 A number is called a palindrome if it reads the same backward as well as forward e.g., 285582 is a six digit palindrome. The number of six digit palindromes, which are divisible by 55, is **[JEE Main 2021]**

87 A class has three teachers, Mr. P, Ms. Q, Mrs. R and six students A, B, C, D, E and F number of ways in which they can be seated in a line of 9 chairs, if between any two teachers, they are exactly two students, is $K \times 6!$. Then, the value of K is ....

88 Consider the five points comprising the vertices of a square and the intersection point of its diagonals. The number of triangles that can be formed using these point, is ....

89 There are 15 players in a cricket team, out of which 6 are bowlers, 7 are batsmen and 2 are wicketkeepers. The number of ways, a team of 11 players be selected from them so as to include atleast 4 bowlers, 5 batsman and 1 wicketkeeper, is .... **[JEE Main 2021]**

90 Total number of ways in which six '+' and four '–' signs can be arranged in a line such that no two '–' signs occur together is ....

91 An urn contains 5 red marbles, 4 black marbles and 3 white marbles, then the number of ways in which 4 marbles can be drawn so that at the most three of them are red, is .... **[JEE Main 2020]**

92 The number of ways in which a person can walk up a stairway which has 7 steps, if he can take 1 or 2 steps up the stairs at a time, is ....

93 The sum of all three-digit numbers less than or equal to 500, that are formed without using the digit 1 and they all are multiple of 11, is **[JEE Main 2021]**

94 Number of ways in which 7 people can occupy 6 seats, 3 seats on each side in a first class railway compartment if two specified persons are to be always included and occupy adjacent seats on the same side, is ....

95 There are 20 books on Algebra and Calculus in one library. For the greatest number of selections each of which consists of 5 books on each topic. If the possible number of Algebra books are N, then the value of $N/2$ is ....

96 Number of permutations of 1, 2, 3, 4, 5, 6, 7, 8 and 9 taken all at a time are such that the digit. 1 appearing some where to the left of 2, 3 appearing to the left of 4 and 5 some where to the left of 6, is $K \times 7!$, then the value of K is ..... .

97 The total number of 3-digit numbers, whose sum of digit is 10, is .... . **[JEE Main 2020]**

98 There are four balls of different colour and four boxes of colours, same as those of the balls. The number of ways in which the balls, one each in a box, could be placed such that a ball does not go to a box of its own colour, is ....

LEVEL 03\\ JEE Advanced Level

Single Choice Correct Questions

1 If $\frac{1}{1!\times 11!}+\frac{1}{3!\times 9!}+\frac{1}{5!\times 7!}=\frac{2^m}{n!}$ and ${}^kP_3=9240$, then

(a) $m+n-k=0$ (b) $m+n+k=12$ (c) $m+k-n=0$ (d) $n+k-m=0$

2 Total number of four digit odd numbers that can be formed using 0, 1, 2, 3, 5 and 7 (using repetition allowed) are

(a) 216 (b) 375 (c) 400 (d) 720

3 Total number of four digit even numbers in which repetition is not allowed.

(a) 504 (b) 1792 (c) 2296 (d) 2284

4 The number of seven digit numbers divisible by 9 formed with digits 1, 2,3,4,5,6,7,8 and 9 without repetition, is

(a) 7! (b) ${}^9P_7$ (c) 3(7!) (d) 4(7!)

5 Ten different letters of an alphabet are given, words with five letters are formed from these given letters. Then, the number of words which have atleast one letter repeated, is

(a) 69760 (b) 30240 (c) 99784 (d) None of these

6 The number of four digit numbers that can be made with the digits 1, 2, 3, 4 and 5 in which atleast two digits are identical, is

(a) $4^4-5!$ (b) 505 (c) 600 (d) None of these

7. The total number of 5-digit numbers of different digits in which the digit in the middle is the largest, is

(a) $\sum_{n=4}^{9} {}^nP_4$ (b) $\sum_{n=4}^{9} {}^nP_4 - \sum_{n=3}^{8} {}^nP_3$ (c) 30(3!) (d) 20(3!)

8 If the letters of the word 'MOTHER' are written in all possible orders and these words are written out as in a dictionary, then the rank of the word 'MOTHER' is

(a) 240 (b) 261 (c) 308 (d) 309

9 A debate club consists of six girls and four boys. A team of four members is to be selected from this club including the selection of a captain (from among these 4 members) for the team, if the team has to include atmost one boy, then the number of ways of selecting the team is

(a) 380 (b) 320 (c) 260 (d) 95

10 Let $S = \{1,2,3,\ldots, 9\}$ for $k = 1,2,\ldots, 5$, let N_k be the number of subsets of S, each containing five elements out of which exactly k are odd, then $N_1 + N_2 + N_3 + N_4 + N_5$ is equal to

[JEE Advanced 2017]

(a) 125 (b) 252 (c) 210 (d) 126

11 Number of ways in which a lawn-tennis mixed double be made from seven married couples if no husband and wife play in the same set, is

(a) 240 (b) 420 (c) 720 (d) None of these

12 Ten IIT and 2 DCE students sit in a row. The number of ways in which exactly 3 IIT students sit between 2 DCE students, is

(a) ${}^{10}C_3 \times 2! \times 3! \times 8!$ (b) $10! \times 2! \times 3! \times 8!$

(c) $5! \times 2! \times 9! \times 8!$ (d) None of these

13 In an examination of nine papers, a candidate has to pass in more papers than the number of papers in which he fails in order to be successful. The number of ways in which he can be unsuccessful, is

(a) 255 (b) 256 (c) 193 (d) 319

14 Two players P_1 and P_2 play a series of $2n$ games. Each game can result in either a win or a loss. For P_1 the total number of ways in which P_1 can win the series of these games, is equal to

(a) $\frac{1}{2}(2^{2n} - {}^{2n}C_n)$ (b) $\frac{1}{2}(2^{2n} - 2 \times {}^{2n}C_n)$ (c) $\frac{1}{2}(2^n + {}^{2n}C_n)$ (d) $\frac{1}{2}(2^n - 2 \times {}^{2n}C_n)$

15 Let $A = \{x_1, x_2, x_3, \ldots\ldots\ldots, x_7\}$ and $B = \{y_1, y_2, y_3\}$. The total number of functions $f: A \to B$ that are onto and there are exactly three elements x in A such that $f(x) = y_2$ is equal to

(a) 490 (b) 510 (c) 630 (d) None of these

16 n is selected from the set $\{1,2,3,\ldots\ldots\ldots 10\}$ and the number $2^n + 3^n + 5^n$ is formed. Total number of ways of selecting n so that the formed number is divisible by 4, is equal to

(a) 50 (b) 49 (c) 48 (d) None of these

17 The number of even divisors of the number $N = 12600$ is

(a) 72 (b) 54 (c) 18 (d) None of these

18 The total number of divisors of 480 that are of the from $4n + 2, n \geq 0$, is equal to

(a) 2 (b) 3 (c) 4 (d) None of these

19 The number of ways to give 16 different things to three persons A, B *and* C so that B gets one more than A and C gets two more than B is

(a) $\frac{16!}{4!5!7!}$ (b) $4!5!7!$ (c) $\frac{16!}{3!5!8!}$ (d) None of these

20 There are three copies each of four different books. The number of ways in which they can be arranged in a shelf, is

(a) $\frac{12!}{(3!)^4}$ (b) $\frac{12!}{(4!)^3}$ (c) $\frac{21!}{(3!)^4 4!}$ (d) $\frac{12!}{(4!)^3 3!}$

21 Sixteen players $S_1, S_2, S_3, \ldots\ldots\ldots\ldots S_{16}$ play in a tournament. Number of ways in which they can be grouped into eight pairs so that S_1 and S_2 are in different pairs, is

(a) $\frac{(14)!}{2^6 \cdot 6!}$ (b) $\frac{(15)!}{2^7 \cdot 7!}$ (c) $\frac{(14)!}{2^7 \cdot 6!}$ (d) $\frac{(14)!}{2^6 \cdot 7!}$

22 The number of ways in which an examiner can assign 30 marks to 8 questions, giving not less than 2 marks to any question, is

(a) ${}^{30}C_7$ (b) ${}^{21}C_8$ (c) ${}^{21}C_7$ (d) ${}^{30}C_8$

23 Five cards and five envelopes are numbered 1, 2, 3, 4, 5 and cards to be placed in envelope so that each envelope contains exactly one card and no card is placed in envelope bearing the same score and moreover the card numbered 1 is never placed in envelope number 2. If n is the total number of ways it can be done, then the value of $\frac{n}{11}$ is

(a) 1 (b) 2 (c) 3 (d) 4

24 In a cricket match between two team X and Y. The team X requires 10 runs to win in the last 3 balls. If the possible runs that can be made from a ball be 0, 1, 2, 3, 4, 5 and 6. The number of sequence of runs made by the batsman, is

(a) 12 (b) 18 (c) 21 (d) 36

More Than One Correct

25 If $p = 21(21^2 - 1^2)(21^2 - 2^2)(21^2 - 3^2)......(21^2 - 10^2)$, then p is divisible by

(a) 22! (b) 21! (c) 19! (d) 20!

26 If $102! = 2^{\alpha}3^{\beta}5^{\gamma}7^{\delta}$ ....., then

(a) $\alpha = 98$ (b) $\beta = 2\gamma + 1$ (c) $\alpha = 2\beta$ (d) $2\gamma = 3\delta$

27 A decimal code is declared legal if it has an even number of zeroes for example 1900200 is a legal code, but 10002 is not. Let a_n be the number of legal decimal codes of length 'n'. Then,

(a) $a_n = 8a_{n-1} + 10^{n-1}$ (b) $a_n = \frac{8^n + 10^n}{2}$

(c) $a_n = 9a_{n-1} + 10^n$ (d) $a_n = a^n + 10^n$

28 Number of points of intersection of n straight lines if n satisfies ${}^{n+5}P_{n+1} = \frac{11(n-1)}{2} \times {}^{n+3}P_n$ is

(a) 15 (b) 28 (c) 21 (d) 10

29 If n is the number of necklaces which can be formed using 17 identical pearls and two identical diamonds and similarly m is number of necklaces which can be formed using 17 identical pearls and different diamonds, then

(a) $n = 9$ (b) $m = 18$ (c) $n = 18$ (d) $m = 9$

30 A forecast is to be made of the results of five cricket matches, each of which can be a win or a draw or a loss for Indian team.

Let $p =$ Number of forecasts with exactly 1 error

$q =$ Number of forecasts with exactly 3 errors

$r =$ Number of forecasts with all five errors

Then, the correct statement(s) is/are

(a) $2q = 5r$ (b) $8p = q$ (c) $8p = 5r$ (d) $2(p + r) > q$

31 Kanchan has 10 friends among whom two are married to each other. She wishes to invite five of them for a party. If the married couple refuse to attend separately, then the number of different ways in which she can invite five friends, is

(a) ${}^8C_5$ (b) $2 \times {}^8C_3$ (c) ${}^{10}C_5 - 2 \times {}^8C_4$ (d) None of these

32 Ten persons numbered 1, 2, ..., 10 play a chess tournament. Each player playing against every other player exactly one game. It is known that no game ends in a draw, if $w_1, w_2, ... w_{10}$ are the number of games won by the players 1, 2, 3, ..., 10, respectively and $l_1, l_2 ... l_{10}$ are the number of games lost by the players 1, 2, ..., 10, respectively, then

(a) $\sum w_i = \sum l_i = 45$ (b) $w_i + l_i = 9$

(c) $\sum w_i^2 = 81 + \sum l_i^2$ (d) $\sum w_i^2 = \sum l_i^2$

33 Number of ways of selecting three integers from $\{1,2,3,\ldots n\}$ if their sum is divisible by 2, is

(a) $3({}^{n/3}C_2)+(n/3)^3$ if $n=3k, k\in N$

(b) $2({}^{(n-1)/3}C_3)+({}^{(n+2)/3}C_3)+\{(n-1)/3\}^2\{(n+2)/3\}$ if $n=3k+1, k\in N$

(c) $2({}^{(n-1)/3}C_3)+({}^{(n+2)/3}C_3)+\{(n-1)/3\}^2/\{(n+2)/3\}$

(d) Independence of n

34 Let $p=2520$, $x=$ number of divisors of p which are multiple of 6 and $y=$ number of divisors of p which are multiple of 9, then

(a) $x=12$ (b) $x=24$ (c) $y=12$ (d) $y=16$

35 Number of ways in which 200 people can be divided in 100 couples, is

(a) $\dfrac{(200)!}{2^{100}(100)!}$ (b) $1\times3\times5\times\ldots\times199$

(c) $\left(\dfrac{101}{2}\right)\left(\dfrac{102}{2}\right)\ldots\left(\dfrac{200}{2}\right)$ (d) $\dfrac{(200)!}{(100)!}$

36 Number of ways in which 30 identical things are distributed among six persons, is

(a) ${}^{17}C_5$ if each gets odd number of things

(b) ${}^{16}C_{11}$ if each gets odd number of things

(c) ${}^{14}C_5$ if each gets even number of things (excluding 0)

(d) ${}^{15}C_{10}$ if each gets even number of things (excluding 0)

37 The number of ways of choosing triplet (x,y,z) such that $z\geq\max\{x,y\}$ and $x, y, z \in\{1,2,\ldots, n, n+1\}$ is

(a) ${}^{n+1}C_3+{}^{n+2}C_3$ (b) $n(n+1)(2n+1)/6$

(c) $1^2+2^2+\ldots+n^2$ (d) $2({}^{n+2}C_3)-{}^{n+1}C_2$

38 For non-negative integers s and r, let

$$\binom{s}{r}=\begin{cases}\dfrac{s!}{r!(s-r)!} & \text{if } r\leq s\\ 0 & \text{if } r>s\end{cases}$$

For positive integers m and n, let

$$g(m,n)=\sum_{p=0}^{m+n}\frac{f(m,n,p)}{\binom{n+p}{p}}$$

[JEE Adv. 2020]

where for any non-negative integer p.

$$f(m,n,p)=\sum_{i=0}^{p}\binom{m}{i}\binom{n+i}{p}\binom{p+n}{p-i}$$

Then, which of the following statements is/are true?

(a) $g(m, n)=g(n,m)$ for all positive integers m, n

(b) $g(m, n+1)=g(m+1,n)$ for all positive integers m, n

(c) $g(2m, 2n)=2g(m,n)$ for all positive integers m, n

(d) $g(2m, 2n)=[g(m,n)]^2$ for all positive integers m, n

39 If N denotes the number of ways of selecting r objects out of n distinct $(r\geq n)$ with unlimited repetition but with each object included atleast once in selection, the N is equal to

(a) ${}^{r-1}C_{r-n}$ (b) ${}^{r-1}C_n$ (c) ${}^{r-1}C_{n-1}$ (d) None of these

40 A is a set containing n elements. A subset p_1 of A is chosen. The set A is reconstructed by replacing the element of p_1, next a subset p_2 of A is chosen and again the set is reconstructed by replacing the element of p_2. In this way, $(m > 1)$ subsets $p_1 . p_2, ..., p_m$ of A are chosen. The number of ways of choosing $p_1, p_2, ... p_m$ is

(a) $(2^m - 1)^n$ if $p_1 \cap p_2 \cap ... p_m = \phi$
(b) 2^{mn} if $p_1 \cup p_2 \cup .... \cup p_m = A$
(c) 2^{mn} if $p_1 \cap p_2 \cap .... \cap p_m = \phi$
(d) $(2^m - 1)^n$ if $p_1 \cup p_2 \cup .... \cup p_m = A$

41 Let N denotes the number of ways in which $3n$ letters can be selected from $2nA$'s, $2nB$'s and $3nC$'s. Then,

(a) $3 | (N-1)$
(b) $n | (N-1)$
(c) $(n+1) | (N-1)$
(d) $3n(n+1) | (N-1)$

Linked Comprehension Type

Paragraph I Directions (Q.Nos. 42-43)

Different words are formed by arranging the letters of the word 'SUCCESS'.

42 The number of words in which C's are together but S's are separated, is

(a) 120 (b) 96 (c) 24 (d) 420

43 The number of words in which no two C's and no two S's are together is

(a) 120 (b) 96 (c) 24 (d) 180

Paragraph II Directions (Q.Nos. 44-46)

Consider the letters of the word 'MATHEMATICS'.

44 Possible number of words taking all letters at a time such that atleast one repeating letter is at odd position in each word, is

(a) $\frac{11!}{2!2!2!} - \frac{9!}{2!2!}$ (b) $\frac{9!}{2!2!2!}$ (c) $\frac{9!}{2!2!}$ (d) $\frac{11!}{2!2!2!}$

45 Possible number of words taking all letters at a time such that in each word both M's are together and both T's are together but both A's are not together, is

(a) $\frac{11!}{2!2!2!} - \frac{10!}{2!2!}$ (b) $7! \cdot {}^8C_2$ (c) $\frac{6!4!}{2!2!} - \frac{10!}{2!2!}$ (d) $\frac{9!}{2!2!2!}$

46 Possible number of words in which no two vowels are together, is

(a) $7! \cdot {}^8C_4 \cdot \frac{4!}{2!}$ (b) $\frac{7!}{2!} \cdot {}^8C_4 \cdot \frac{4!}{2!}$ (c) $\frac{7!}{2!2!} \cdot {}^8C_4 \cdot \frac{4!}{2!}$ (d) $\frac{7!}{2!2!2!} \cdot {}^8C_4 \cdot \frac{4!}{2!}$

Paragraph III Directions (Q.Nos. 47-49)

Twelve persons to be arranged around two round tables such that one table can accomodate seven persons and another five persons only. Answer the following questions.

47 The number of ways in which these 12 persons can be arranged, is

(a) ${}^{12}P_5 6!4!$ (b) $6!4!$ (c) ${}^{12}C_5 6!4!$ (d) None of these

48 The number of ways of arrangements if two particular persons A and B do not want to be on the same table, is

(a) ${}^{10}C_4 6!4!$ (b) $2 \cdot {}^{10}C_6 6!4!$ (c) ${}^{11}C_6 6!4!$ (d) None of these

49 The number of ways of arrangements if two particular persons A and B want to be together and consecutive, is

(a) ${}^{10}C_7 6!3!2! + {}^{10}C_5 4!5!2!$
(b) ${}^{10}C_5 6!3! + {}^{10}C_7 4!5!$
(c) ${}^{10}C_7 6!2! + {}^{10}C_5 5!2!$
(d) None of these

Matrix Match Type

50 Consider all possible permutations of the letters of the word 'ENDEANOEL'. Match the statements/expressions in Column I with the statements/expressions in Column II.

	Column I		Column II
(i)	The number of permutations containing the world 'ENDEA', is	A.	5!
(ii)	The number of permutations in which the letter E occurs in the first and last positions, is	B.	$2\times5!$
(iii)	The number of permutations in which none of the letters D, L, N occur in the last five positions, is	C.	$7\times5!$
(iv)	The number of permutations in which the letters A, E, O occur only in odd positions is	D.	$21\times5!$

51 Match the following columns.

	Column I		Column II
(i)	The number of five digit numbers having the product of digits 20, is	A.	> 70
(ii)	A closest has five pairs of shoes. The number of ways in which four shoes can be drawn from it such that there will be no complete pair, is	B.	< 60
(iii)	Three ladies have each brought their one child for admission to a school. The principal wants to interview the six persons one by one subject to the condition that no mother is interviewed before her child. The number of ways in which interview can be arranged, is	C.	$\in(50,110)$
(iv)	The figures 4, 5, 6, 7 and 8 are written in every possible order. The number of numbers greater than 56000 is	D.	$\in(40,70)$

52 In a high school, a committee has to be formed from a group of 6 boys $M_1, M_2, M_3, M_4, M_5, M_6$ and 5 girls G_1, G_2, G_3, G_4, G_5.

(i) Let α_1 be the total number of ways in which the committee can be formed such that committee has 5 members having exactly 3 boys and girls.

(ii) Let α_2 be the total number of ways in which the committee can be formed such that the committee has atleast 2 members and having an equal numbers of boys and girls.

(iii) Let α_3 be the total number of ways in which the committee can be formed such that the committee has 5 members atleast 2 of them being girls.

(iv) Let α_4 be the total number of ways in which the committee can be formed such that the committee has 4 members, having atleast 2 girls and such that both M_1 and G_1 are not in the committee together.

[JEE Advanced 2018]

	Column I		Column II
(i)	The value of α_1, is	A.	136
(ii)	The value of α_2, is	B.	189
(iii)	The value of α_3, is	C.	192
(iv)	The value of α_4, is	D.	200
		E.	381
		F.	461

53 Match the following columns.

	Column I		Column II
(i)	Number of straight lines joining any two of 10 points if which four points are collinear, is	A.	30
(ii)	Maximum number of points of intersection of 10 straight lines in the plane, is	B.	60
(iii)	Maximum number of points of intersection of 6 circles in the plane, is	C.	40
(iv)	Maximum number of points of intersection of 6 parabolas, is	D.	45

54 Match the following columns.

	Column I		Column II
(i)	Total number of functions $f: \{1, 2, 3, 4, 5\} \rightarrow \{1, 2, 3, 4, 5\}$ that are onto and $f(i) \neq i$ is equal to	A.	divisible by 11
(ii)	If $x_1 x_2 x_3 = 2 \times 5 \times 7^2$, then the number of solution sets for (x_1, x_2, x_3), where $x_i \in N, x_i > 1$, is	B.	divisible by 7
(iii)	Number of factors of 3780 are divisible by either 3 or 2 or both, is	C.	divisible by 3
(iv)	Total number of divisors of $n = 2^5 \times 3^4 \times 5^{10}$ that are of the form $4\lambda + 2, \lambda \geq 1$, is	D.	divisible by 4

55 Consider as 6×6 chessboard. Then, match the following columns.

	Column I		Column II
(i)	Number of rectangles is	A.	$^{10}C_5$
(ii)	Number of squares is	B.	441
(iii)	Number of ways three squares can be selected if they are not in same row or column, is	C.	91
(iv)	In how many ways eleven '+' sign can be arranged in the squares, if no row remains empty	D.	2400

Numerical Value Type Questions

Question Stem Directions (Q.Nos. 56-58) The answer to each question is a **NUMERICAL VALUE.**

Five balls are to be placed in three boxes. Each box can hold all the five and no box remains empty.

56 The number of ways if balls are different but hoxes are identical, is ....

57 The number of ways if balls and boxes are identical, is ....

58 The number of ways if balls as well as boxes are identical but boxes are kept in a row, is ....

Question Stem Directions (Q.Nos. 59-61) The answer to each question is a **NUMERICAL VALUE.**

Consider the word W = TERRORIST

59 Number of four letters word that can be made using only the letters from the word W, if each word must contains atleast one vowel, is ....

60 Number of arrangements of the word W, if no two R's are together, is ....

61 Number of arrangements of the word W, if R's as well as T's are separated, is ....

62 The number of 5-digit numbers which are divisible by 4, with digits from the set {1, 2, 3, 4, 5} and the repetition of digits is allowed, is .... **[JEE Advanced 2018]**

63 Let $n_1 < n_2 < n_3 < n_4 < n_5$ be positive integers such that $n_1 + n_2 + n_3 + n_4 + n_5 = 20$, then the number of such distinct arrangements, is...........

64 Number of ways six faces of cube can be painted with 6 different colours, is............

65 Five persons A, B, C, D and E are seated in a circular arrangement. If each of them is given a hat of one it three colours red, blue and green, then the number of ways of distributing the hats such that the persons seated in a adjacent seats get different colured hat, is **[JEE Advanced 2019]**

66 Let P_n denotes the number of ways in which three people can be selected out of n people sitting in a row, if no two of them are consecutive, if $P_{n+1} - P_n = 15$, then the value of n is........

67 Let $n \geq 2$ be an integer, take n distinct points on a circle and join each pair of points by a line segment colour the line segment joining every pair of adjacent points by blue and the rest by red. If the number of red and blue line segments are equal, then the value of n, is.......

68 There are 3 men and 7 women taking a dance class. The number of different ways in which each man be paired with a woman partner and the four remaining women be paired into two pairs each of two, is.........

69 Words of length 10 are formed using the letters A, B, C, D, E, F, G, H, I and J. Let x be the number of such words where no letter is repeated and let y be the number of such words where exactly one letter is repeated twice and no other letter is repeated. Then, $\frac{y}{9x}$ is .... **[JEE Advanced 2017]**

70 In a hotel, four rooms are available, six persons are to be accomodated in these four rooms in such a way that each of these rooms contains atleast one person and atmost two persons. Then, the number of all possible ways in which this can be done is.......... **[JEE Advanced 2020]**

71 The number of divisors of 5400 which are divisible by 6 but not by 9, is ....

72 An engineer is required to visit a factory for exactly four days during the first 15 days of every month and it is mandatory that no two visits take place on consecutive days. Then, the number of all possible ways in which such visits to the factory can be made by the engineer during 1-15 June 2021, is .... **[JEE Advanced 2020]**

ANSWER KEY

Level 1

1. 380 days **2.** 7500 ways **3.** 5040 **4.** 50 ways
5. (i) 5040 (ii) 696 (iii) 660
6. 100 **7.** $n = 3$ **8.** 3 **9.** $n!$ **10.** 48 **11.** 24
12. (i) $n = 10$ (ii) $n = 10$
13. 1640 **14.** 720 **15.** 120 **16.** 64 ways
17. (i) 60 (ii) 125
18. $26^6 \times 10$ **19.** $9 \times 9!$ **20.** 15 **21.** $\frac{10!}{3!}$ **22.** 3600 says **23.** 46
24. (i) $5! \times {}^6P_5$ (ii) $2 \times 5! \times 5!$ (iii) $2! \times 5! \times 5!$ (iv) $10! - (5! \times 6!)$
25. ${}^9P_4 \times {}^9P_3 \times 11!$ **26.** 2880 **27.** 18 **28.** $\frac{11!}{2}$ **29.** 16 **30.** $n = 6$
31. $n = 9$ and $r = 3$
32. (i) 120 ways (ii) 720 ways
33. (i) 126 (ii) 84
34. 485 **35.** 64 **36.** 1008 **37.** $n = 3$
38. (i) 200 (ii) 461 (iii) 381 (iv) 189
39. (i) 60 (ii) 3255 (iii) 3968
40. (i) 24 ways (ii) 96 ways
41. (i) 116 (ii) 40
42. $\frac{1}{4}mn(m-1)(n-1)$ **43.** $792 \times 48 \times ({}^8C_2 - 7)$ **45.** 24
46. (i) 72 (ii) 70 (iii) 48 (iv) 124992
47. $\frac{52!}{(13!)^4}$ **48.** 1584 **49.** 720
50. (i) 25 (ii) 150 (iii) 720
51. 25 **52.** 150 **53.** 5456
54. (i) 1771 (ii) 969
55. 2600 **56.** 336 **57.** 17 **58.** 44 **59.** 9 **60.** 35
61. 50262

Level 2

1. (c)	**2.** (b)	**3.** (b)	**4.** (b)	**5.** (d)	**6.** (d)	**7.** (c)
8. (b)	**9.** (b)	**10.** (a)	**11.** (c)	**12.** (b)	**13.** (a)	**14.** (b)
15. (b)	**16.** (b)	**17.** (d)	**18.** (b)	**19.** (c)	**20.** (c)	**21.** (c)
22. (c)	**23.** (c)	**24.** (a)	**25.** (c)	**26.** (a)	**27.** (d)	**28.** (b)
29. (c)	**30.** (b)	**31.** (b)	**32.** (a)	**33.** (b)	**34.** (d)	**35.** (c)
36. (a)	**37.** (d)	**38.** (d)	**39.** (b)	**40.** (c)	**41.** (c)	**42.** (b)
43. (d)	**44.** (c)	**45.** (b)	**46.** (a)	**47.** (b)	**48.** (a)	**49.** (a)
50. (a)	**51.** (c)	**52.** (b)	**53.** (d)	**54.** (b)	**55.** (c)	**56.** (d)
57. (a)	**58.** (a)	**59.** (c)	**60.** (a)	**61.** (c)	**62.** (a)	**63.** (b)
64. (b)	**65.** (d)	**66.** (a)	**67.** (d)	**68.** (a)	**69.** (a)	**70.** (d)
71. (d)	**72.** (c)	**73.** (a)	**74.** (c)	**75.** (a)	**76.** (b)	**77.** (d)
78. (a)	**79.** (c)	**80.** (d)	**81.** (576)	**82.** (5)	**83.** (4)	**84.** (24)
85. (162)	**86.** (100)	**87.** (18)	**88.** (8)	**89.** (777)	**90.** (35)	**91.** (490)
92. (21)	**93.** (7744)	**94.** (960)	**95.** (5)	**96.** (9)	**97.** (55)	**98.** (9)

Level 3

1. (a) 2. (d) 3. (c) 4. (d) 5. (a) 6. (b) 7. (b)
8. (d) 9. (a) 10. (d) 11. (b) 12. (a) 13. (b) 14. (a)
15. (a) 16. (b) 17. (b) 18. (c) 19. (a) 20. (a) 21. (a)
22. (c) 23. (a) 24. (d) 25. (b, c, d) 26. (a, b, c, d) 27. (a, b) 28. (a, c)
29. (a,d) 30. (a, b, d) 31. (b, c) 32. (a, b, d) 33. (a, b, c) 34. (b, d) 35. (a, b, c)
36. (a, c) 37. (b, c, d) 38. (a, b, d) 39. (a, c) 40. (a, d) 41. (a, b, c, d)
42. (c) 43. (b) 44. (d) 45. (b) 46. (c) 47. (c) 48. (b)
49. (a)
50. (i)–A, (ii)–D, (iii)–B, (iv)–B
51. (i)–B, D, (ii)–A, C, (iii)–A, C, (iv)–A, C
52. (i)–D, (ii)–F, (iii)–E, (iv)–B
53. (i)–C, (ii)–D, (iii)–A, (iv)–B
54. (i)–A, D, (ii)–B, C (iii)–A, D, (iv)–C
55. (i)–B, (ii)–C, (iii)–D, (iv)–A
56. (25) 57. (2) 58. (6) 59. (588) 60. (12600) 61. (10200) 62. (625)
63. (7) 64. (30) 65. (30) 66. (8) 67. (5) 68. (630) 69. (5)
70. (1080) 71. (9) 72. (495)

SOLUTIONS

Level 1

1. Since, there are 20 ways of entering into the hall. After entering into the hall, the person come out through a different gate in 19 ways.

By fundamental principle of multiplication, total number of ways $= 20 \times 19 = 380$ ways.

2. Choosing a History book is event A and choosing a Geography book is event B.

Since, both events are independent for choosing History and Geography books.

Thus, the multiplication rule applies here and two books can together be choosen in $100 \times 75 = 7500$ ways.

3. For the first spot, teacher has 7 choices. After a choice has been made, the teacher has 6 choices for the second spot.

Similarly, choices are occured accordingly.

↓	↓	↓	↓	↓	↓	↓
7	6	5	4	3	2	1

(Choices)

∴ By the multiplication rule, there are

$7 \times 6 \times 5 \times 4 \times 3 \times 2 \times 1 = 5040$

4. Since, there are 30 ways to select a boy, so there are 20 ways to select a girl.

By fundamental principle of addition, either a boy or a girl can be selected in $30 + 20 = 50$ ways.

5. (i) $7! = 7 \times 6 \times 5 \times 4 \times 3 \times 2 \times 1 = 5040$

(ii) $6! - 4! = 6 \times 5 \times 4 \times 3 \times 2 \times 1 - 4 \times 3 \times 2 \times 1$

$= 4 \times 3 \times 2 \times 1 (30 - 1) = 24 \times 29 = 696$

(iii) $\dfrac{12!}{9!2!} = \dfrac{12 \times 11 \times 10 \times 9!}{9! \times 2 \times 1}$

$= 6 \times 11 \times 10 = 660$

6. We have, $\dfrac{1}{8!} + \dfrac{1}{9!} = \dfrac{x}{10!}$

$\Rightarrow \quad \dfrac{1}{8!} + \dfrac{1}{9 \times 8!} = \dfrac{x}{10 \times 9 \times 8!}$

$\Rightarrow \quad 1 + \dfrac{1}{9} = \dfrac{x}{90}$

$\Rightarrow \quad \dfrac{10}{9} = \dfrac{x}{90}$

$\Rightarrow \quad x = 100$

7. $\because (n+2)! = (n+2)(n+1)n(n-1)!$

$\Rightarrow (n+2)(n+1)n(n-1)! = 60 \times (n-1)!$

$\Rightarrow \quad 60 = (n+2)(n+1)n$

$\Rightarrow \quad 5 \times 4 \times 3 = (n+2)(n+1)n$

$\therefore \quad n = 3$

8. Let $N = \sum_{r=1}^{100} r! = 1! + 2! + 3! + \ldots + 100!$

$\Rightarrow N = (1! + 2! + 3! + 4!) + (5! + 6! + 7! + \ldots + 100!)$

$\Rightarrow \dfrac{N}{15} = \left(\dfrac{1! + 2! + 3! + 4!}{15}\right) + \left(\dfrac{5! + 6! + 7! + \ldots + 100!}{15}\right)$

$= \dfrac{1 + 2 + 6 + 24}{15} + \text{Integer}$

[as 5!, 6!, ..., 100! are divisible by 15]

$= \dfrac{33}{15} + \text{Integer}$

$= 2 + \dfrac{3}{15} + \text{Integer}$

$= \dfrac{3}{15} + \text{Integer}$ $\quad [\because 2 + \text{integer} = \text{integer}]$

Hence, remainder is 3.

9. $\dfrac{3 \cdot 6 \cdot 9 \ldots (3n)}{3^n} = \dfrac{(3 \times 1) \times (3 \times 2) \times (3 \times 3) \times \ldots (3 \times n)}{3^n}$

$= \dfrac{3^n \times (1 \times 2 \times 3 \times 4 \times \ldots \times n)}{3^n} = n!$

10. $E_3(100!) = \left[\dfrac{100}{3}\right] + \left[\dfrac{100}{3^2}\right] + \left[\dfrac{100}{3^3}\right] + \left[\dfrac{100}{3^4}\right] + \ldots$

$= 33 + 11 + 3 + 1 + 0 + 0 + \ldots$

$= 48$

Hence, exponent of 3 is 48.

11. 100! can be written as $2^a \cdot 3^b \cdot 5^c \cdot 7^d \ldots$

Now, $E_2(100!) = \left[\dfrac{100}{2}\right] + \left[\dfrac{100}{2^2}\right] + \left[\dfrac{100}{2^3}\right] + \left[\dfrac{100}{2^4}\right] + \left[\dfrac{100}{2^5}\right] + \left[\dfrac{100}{2^6}\right]$

$= 50 + 25 + 12 + 6 + 3 + 1 = 97$

and $E_5(100!) = \left[\dfrac{100}{5}\right] + \left[\dfrac{100}{5^2}\right] + \ldots$

$= 20 + 4 = 24$

$\therefore \quad 100! = 2^{97} \cdot 3^b \cdot 5^{24} \cdot 7^d \ldots$

$= (2 \times 5)^{24} \times 2^{73} \times 7^d \times \ldots$

Hence, number of zeroes at the end of 100! is 24.

or exponent of 10 in 100! = Min (97, 24) = 24.

12. (i) $\because {}^nP_4 = 5040$

or ${}^nP_4 = 10 \times 504$

or ${}^nP_4 = 10 \times 9 \times 8 \times 7$

or ${}^nP_4 = {}^{10}P_4$

$\Rightarrow \quad n = 10$ $\quad$ [on comparing]

(ii) $\because {}^nP_5 = 42 \cdot {}^nP_3$

or $n(n-1)(n-2)(n-3)(n-4) = 42n(n-1)(n-2)$

$\because \quad n > 4$, so $n(n-1)(n-2) \neq 0$

$\Rightarrow \quad (n-3)(n-4) = 42$

$\Rightarrow \quad n^2 - 7n - 30 = 0$

$\Rightarrow \quad n^2 - 10n + 3n - 30 = 0$

$n(n-10) + 3(n-10) = 0$

$\Rightarrow \quad (n-10)(n+3) = 0$

$n = 10$, or $n = -3$ (neglected)

$\therefore \quad n = 10$

13. We have, $\dfrac{{}^{56}P_{r+6}}{{}^{54}P_{r+3}} = \dfrac{30800}{1}$

$\Rightarrow (56)(55)\dfrac{{}^{54}P_{r+4}}{{}^{54}P_{r+3}} = \dfrac{30800}{1} \quad [\because {}^nP_r = n({}^{n-1}P_{r-1})]$

$\Rightarrow \quad \dfrac{{}^{54}P_{r+4}}{{}^{54}P_{r+3}} = 10$

$\Rightarrow 54 - (r+4) + 1 = 10 \quad \left[\because \dfrac{{}^nP_r}{{}^nP_{r-1}} = n - r + 1\right]$

$\Rightarrow \quad r = 41$

$\therefore \quad {}^rP_2 = {}^{41}P_2 = 41 \times 40 = 1640$

14. The required number of ways is arrangement of 3 people out of 10 i.e.

$${}^{10}P_3 = \frac{10!}{7!} = \frac{10 \times 9 \times 8 \times 7!}{7!}$$

$$= 10 \times 9 \times 8 = 720$$

15. There are 5 letters in the word 'COVID' and all the 5 letters are different.

Hence, the number of permutations taking all the letters at a time $= {}^5P_5 = 5! = 120$

16. The first marble can be put into the pocket in 4 ways, so can the second. Thus, the number of ways in which child can put the marbles $= 4 \times 4 \times 4 = 4^3 = 64$ ways.

17. (i) | 5 | 4 | 3 |

$5 \times 4 \times 3 = 60$ code words without repeats.

(ii) | 5 | 5 | 5 | (total number of letters = 5)

$5^3 = 125$ code words with repeats.

18.

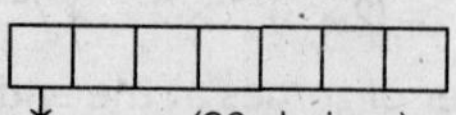

We have a choice of 10 digits for the first place.

From 2nd spot to 7th spot,

we have 26 choices each time.

$\therefore$ The total number of permutations is

$(10 \times 26 \times 26 \times 26 \times 26 \times 26 \times 26) = 26^6 \times 10.$

19. Number of digits are 10.

Total number of 9- digit numbers $= {}^{10}P_9$

Out of these, the number of numbers having zero at the first place $= {}^9P_8$

Hence, required number of numbers $= {}^{10}P_9 - {}^9P_8$

$= 10 \times {}^9P_8 - {}^9P_8 = 9 \times {}^9P_8 = 9 \times 9!$

20. The signals can be made by using one or more flags at a time.

$\therefore$ By fundamental principle of addition, the total number of signals

$= {}^3P_1 + {}^3P_2 + {}^3P_3$

$= 3 + 3 \times 2 + 3 \times 2 \times 1$

$= 3 + 6 + 6 = 15$

21. Given the word 'SHAKUNTALA'

Here, there are 3 A's and rest letters are once.

Total number of letters = 10

$\therefore$ The number of words formed with the letters

$= \dfrac{10!}{3!}$

22. The word 'PATALIPUTRA' has 11 letters, in which 2P's, 3A's, 2T's, 1L, 1U, 1 R and 1 I.

Vowels are AAIUA.

These vowels can be arranged themselves in $\dfrac{5!}{3!}$

$= 20$ ways

The consonants are PTLPTR

These consonants can be arranged in $\dfrac{6!}{2!2!} = 180$ ways

$\therefore$ Required number of words $= 180 \times 20 = 3600$ ways.

23. The letters in alphabetical order are A, D, I, I N.

$\therefore$ The first word is ADIIN.

Number of words begining with A $= \dfrac{4!}{2!} = 12$

Number of words begining with D $= \dfrac{4!}{2!} = 12$

Number of words begining with IA $= 3! = 6$

Number of words begining with ID $= 6$

Number of words begining with II $= 6$

Number of words begining with INA $= 2! = 2$

Number of words begining with INDA $= 1! = 1$

Number of words begining with INDIA $= 1$

$\therefore$ Rank of word 'INDIA'

$= 12 + 12 + 6 + 6 + 6 + 2 + 1 + 1 = 46$

24. (i) | G | G | G | G | G |

Here, there is no condition for arranging the girls.

Now, 5 girls can be arranged in 5! ways.

When girls are arranged, 6 gapes are created.

$\therefore$ 5 boys can be arranged in six gapes $= {}^6P_5$ ways.

Hence, total number of arrangements $= 5! \times {}^6P_5$

(ii) Let us first arrange 5 girls amongs themselves.

$\therefore$ Number of ways $= 5!$

$\because$ Boys and girls are alternate, the cases are

| G | G | G | G | G or G | G | G | G |G |

In each the above cases, boys can arrange among themselves in 5! ways.

Hence, total number of arrangements
$= 5! \times 5! + 5! \times 5! = 2 \times 5! \times 5!$

(iii) B_1, B_2, B_3, B_4, B_5 G_1, G_2, G_3, G_4, G_5

The required number of arrangements
$= 2! \times 5! \times 5!$

(iv) Number of arrangements that all the girls are not together = Total arrangements without restriction – arrangement when all the girls are together.

$= 10! - (5! \times 6!)$

25. Let the two sides be *A* and *B*. Assume that 4 particular guests wish to sit on side *A*. 4 guests who wish to sit on side A, can be accommodated on 9 chairs in ${}^9P_4$ ways and three guests who wish to sit on side *B*, can be accomodated in ${}^9P_3$ ways. Now, the remaining guests are left who can sit on 11 chairs on both the sides of table in 11! ways.

Hence, the total number of ways in which is 18 persons, can be seated $= {}^9P_4 \times {}^9P_3 \times 11!$

26.

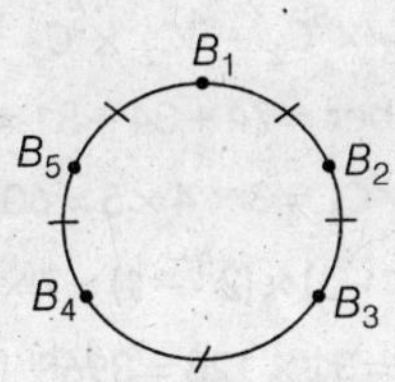

After fixing up one boy on the circle, the remaining can be arranged in $(5-1)! = 4!$ ways

But boys and girls are to alternate.

There will be 5 places, one place each between 2 boys, these five places can be filled by 5 girls in 5! ways.

$\therefore$ By principle of multiplication the required number of ways $= 4! \times 5! = 2880$

27. When *A* has *B* or *C* to his right, we have the order *AB* or *AC*.

When *B* has *C* or *D* to his right, we have the order *BC* or *BD*.

Taking these two possiblities together, we must have *ABC* or *ABD* or *AC* and *BD*.

For *ABC*, *D*, *E* and *F* to arrange along a circle, the number of ways $= (4-1)! = 3! = 6$

For *ABD*, *C*, *E* and *F*, the number of ways $= 6$

For *AC*, *BD*, *E* and *F*, the number of ways $= 6$

Hence, total number of ways $= 6 + 6 + 6 = 18$

28. 12 different beads can be arranged among themselves in a circular order in $(12-1) = 11!$ ways

Now, in the case of necklace, there is no distinction between clockwise and anti-clockwise arrangements.

So, the required number of arrangements $= \dfrac{11!}{2}$

29. $\because {}^nC_9 = {}^nC_7 \Rightarrow n = 9 + 7 = 16$ $\quad [\because 9 \neq 7]$

30. We have,

$$\frac{{}^{2n}C_3}{{}^nC_3} = \frac{11}{1}$$

$$\Rightarrow \frac{\dfrac{2n(2n-1)(2n-2)}{3\cdot 2\cdot 1}}{\dfrac{n(n-1)(n-2)}{3\cdot 2\cdot 1}} = \frac{11}{1}$$

$$\Rightarrow \frac{4(2n-1)}{n-2} = 11$$

$$\Rightarrow 8n - 4 = 11n - 22$$

$$\Rightarrow 3n = 18$$

$$\Rightarrow n = 6$$

31. We know that $\dfrac{{}^nC_r}{{}^nC_{r-1}} = \dfrac{n-r+1}{r}$

$$\Rightarrow \frac{84}{36} = \frac{7}{3} = \frac{n-r+1}{r} \quad \text{[given]}$$

$$\Rightarrow 3n - 10r + 3 = 0 \quad \text{...(i)}$$

Also given, $\dfrac{{}^nC_r}{{}^nC_{r+1}} = \dfrac{84}{126}$

$$\Rightarrow \frac{r+1}{n-r} = \frac{2}{3}$$

$$\Rightarrow 2n - 5r - 3 = 0 \quad \text{...(ii)}$$

On solving Eqs. (i) and (ii), we get

$n = 9$ and $r = 3$

32. (i) Within a selection of 3 accounts, the arrangement of the visits is not important.

So, there are ${}^{10}C_3 = \dfrac{10 \times 9 \times 8}{3 \times 2 \times 1} = 120$ ways to select 3 accounts.

(ii) To schedule calls, the sales representative must order each selection of 3 accounts. Use permutation here, since order is important.

$${}^{10}P_3 = \frac{10!}{(10-3)!} = \frac{10!}{7!}$$
$$= 10 \times 9 \times 8 = 720$$

33. (i) Since, two particular books are always selected, it means that $6 - 2 = 4$ books are selected out of $11 - 2 = 9$ books

$\therefore$ The required number of ways

$$= {}^9C_4 = \frac{9 \times 8 \times 7 \times 6}{4 \times 3 \times 2 \times 1} = 126$$

(ii) Since, two particular books are never selected, it means that 6 books are selected out of the remaining $11-2=9$ books

$\therefore$ The required number of ways $= {}^9C_6 = {}^9C_3$

$$= \frac{9\times8\times7}{3\times2\times1} = 84$$

34. The possible cases are as follows

Case I A man invites 3 ladies and woman invites 3 gentlemen.

$\therefore$ Number of ways $= {}^4C_3 \times {}^4C_3 = 16$

Case II A man invites (2 ladies, 1 gentleman) and woman invites (2 gentlemen, 1 lady).

$\therefore$ The number of ways

$$= ({}^4C_2 \times {}^4C_1) \times ({}^3C_1 \times {}^4C_2) = 324$$

Case III A man invites (1 lady, 2 gentlemen) and woman invites (2 ladies, 1 gentleman).

$\therefore$ The number of ways

$$= ({}^4C_1 \times {}^3C_2) \times ({}^3C_2 \times {}^4C_1) = 144$$

Case IV A man invites (3 gentlemen) and woman invites (3 ladies)

The number of ways $= {}^3C_3 \times {}^3C_3 = 1$

$\therefore$ Total number of ways

$$= 16 + 324 + 144 + 1 = 485$$

35. The possible cases are as follows

Case I If one black and two other balls are drawn.

Number of ways $= {}^3C_1 \times {}^6C_2 = 45$

Case II If two black and one other balls are drawn.

Number of ways $= {}^3C_2 \times {}^6C_1 = 18$

Case III If all three black balls are drawn.

Number of ways $= {}^3C_3 = 1$

$\therefore$ Total number of ways $= 45 + 18 + 1 = 64$

36. Given that there are 9 women and 8 men, a committee of 12 is to be formed including atleast 5 women.

This can be done in as

$$= (5\text{ W and }7\text{ M}) + (6\text{ W and }6\text{ M}) + (7\text{ W and }5\text{ M}) + (8\text{ W and }4\text{ M}) + (9\text{ W and }3\text{ M})$$

$$= ({}^9C_5 \times {}^8C_7) + ({}^9C_6 \times {}^8C_6) + ({}^9C_7 \times {}^8C_5) + ({}^9C_8 \times {}^8C_4) + ({}^9C_9 \times {}^8C_3)$$

$$= 1008 + 2352 + 2016 + 630 + 56 = 6062$$

For the case, where men are in majority

$$= (5\text{ W and }7\text{ M}) = {}^9C_5 \times {}^8C_7 = 1008$$

37. $\because$ Student is allowed to select atmost n books out of $(2n+1)$ books.

$\therefore \quad {}^{2n+1}C_1 + {}^{2n+1}C_2 + \dots + {}^{2n+1}C_n = 63$ [given] ...(i)

We know that, ${}^nC_0 + {}^nC_1 + {}^nC_2 + \dots + {}^nC_n = 2^n$

$\therefore {}^{2n+1}C_0 + {}^{2n+1}C_1 + {}^{2n+1}C_2 + \dots + {}^{2n+1}C_{2n+1} = 2^{2n+1}$

$\Rightarrow \quad {}^{2n+1}C_0 = {}^{2n+1}C_{2n+1} = 1$

$\Rightarrow \quad {}^{2n+1}C_1 = {}^{2n+1}C_{2n}$

$\Rightarrow 1 + 2({}^{2n+1}C_1 + {}^{2n+1}C_2 + \dots + {}^{2n+1}C_n) + 1 = 2^{n+1}$

$\Rightarrow \quad {}^{2n+1}C_1 + {}^{2n+1}C_2 + \dots + {}^{2n+1}C_n = 2^{2n} - 1$ ...(ii)

From Eqs. (i) and (ii), we get

$\Rightarrow \quad 2^{2n} - 1 = 63$

$\Rightarrow \quad 2^{2n} = 64 \Rightarrow 2^{2n} = 2^6$

$\Rightarrow \quad 2n = 6 \Rightarrow n = 3$

38. (i) The number of ways selecting 3 boys and 2 girls from 6 boys and 5 girls.

$\therefore$ The number of ways $= {}^6C_3 \times {}^5C_2 = 20 \times 10 = 200$

(ii) The number of ways selecting atleast 2 members and having an equal number of boys and girls i.e.

$${}^6C_1{}^5C_1 + {}^6C_2{}^5C_2 + {}^6C_3{}^5C_3 + {}^6C_4{}^5C_4 + {}^6C_5{}^5C_5$$
$$= 30 + 150 + 200 + 75 + 6 = 461$$

(iii) (2 G, 3 B) or (3 G , 2 B) or (4G, 1 B) or (5G, 0B)

$$= {}^5C_2 \times {}^6C_3 + {}^5C_3 \times {}^6C_2 + {}^5C_4 \times {}^6C_1 + {}^5C_5 \times {}^6C_0$$
$$= 200 + 150 + 30 + 1 = 381$$

(iv) Let particular boy is B_1 and a particular girl is G_1

Case I G_1 is included

$\Rightarrow {}^4C_1 \times {}^5C_2 + {}^4C_2 \times {}^5C_1 + {}^4C_3 = 40 + 30 + 4 = 74$

Case II B_1 is included

$\Rightarrow {}^4C_2 \times {}^5C_1 + {}^4C_3 = 30 + 4 = 34$

Case III G_1 and B_1 both are not included

$\Rightarrow {}^4C_4 + {}^4C_3 \times {}^5C_1 + {}^4C_2 \times {}^5C_2 = 1 + 20 + 60 = 81$

$\therefore$ Total number $= 74 + 34 + 81 = 189$

39. (i) ${}^3C_1 \times {}^4C_1 \times {}^5C_1 = 3 \times 4 \times 5 = 60$

(ii) $(2^3 - 1) \times (2^4 - 1) \times (2^5 - 1) = 7 \times 15 \times 31 = 3255$

(iii) $(2^5 - 1) \times 2^7 = 31 \times 128 = 3968$

40. (i) Consider two B's as one single object.

The letters BB, C, D can be arranged in 3! ways

|BB|C|D|

While the three A's can be placed in any one of the four places in ${}^4C_3$ ways.

Hence, required number of arrangement

$$= 3! \times {}^4C_3 = 24 \text{ ways}$$

(ii) The letters 2B's 1C and 1D can be arranged

permutation $= \frac{4!}{2!}$ ways

While the other three A's can be placed in any of the five places in ${}^5C_3$ ways.

But this includes those cases in which the 2B's are together.

Hence, the required number of arrangements

$$= \frac{4!}{2!} \times {}^5C_3 - 24$$
$$= 120 - 24 = 96$$

41. (i) The following cases are occured for number of points selected out 4 collinear points and from points selected out of remaining 6 points as follows

(0, 3) or (1, 2) or (2, 1) or (3, 0)

$\therefore$ Number of triangles formed

$$= {}^4C_0 \times {}^6C_3 + {}^4C_1 \times {}^6C_2 + {}^4C_2 \times {}^6C_1 + 0$$

$$= 20 + 60 + 36 + 0 = 116$$

(ii) The following cases are occured as follows

(0, 2) or (1, 1) or (2, 0)

$\therefore$ Number of straight lines formed

$$= {}^4C_0 \times {}^6C_2 + {}^4C_1 \times {}^6C_1 + 1$$

$$= 15 + 24 + 1 = 40$$

42. Two parallel lines can be selected from the first set of m parallel lines in ${}^mC_2$ ways.

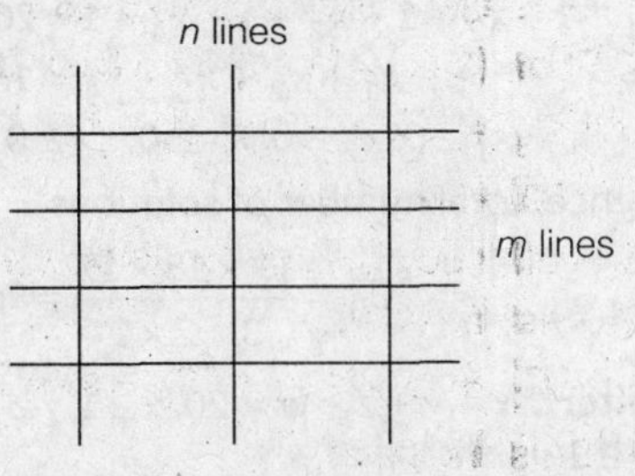

Again, two parallel lines can be selected from other set of n parallel lines in ${}^nC_2$ ways

$\therefore$ The number of selection of 4 lines, 2 from first set and 2 from second set $= {}^mC_2 \times {}^nC_2$

and each such selection of a lines gives a parallelogram.

$\therefore$ The number of parallelogram then formed

$$= {}^mC_2 \times {}^nC_2 = \frac{m!}{2!(m-2)!} \times \frac{n!}{2!(n-2)!}$$

$$= \frac{1}{4} mn(m-1)(n-1)$$

43. Number of rectangles excluding square

$$= \frac{6 \times 9}{4}(6+1)(9+1) - \sum_{r=1}^{6} (7-r)(10-r)$$

[$\because$ in a rectangle of $n \times P$ $(n < P)$ number of rectangles of any size is $\frac{nP}{4}(n+1)(P+1)$ and number of squares of any size is $\sum_{r=1}^{n} [(n+1-r)(P+1-r)]$]

$$= 945 - \sum_{r=1}^{6} (70 - 17r + r^2) = 945 - 154 = 791$$

44. A row can be chosen in 8 ways.

In a row, there are all in ${}^8C_2$ pairs of squares in which 7 are pairs of adjacent squares.

So, there are $({}^8C_2 - 7)$ pairs of places in any of which the two identical kings can be placed.

Hence, the kings can be placed in a row in $8 \times ({}^8C_2 - 7)$ ways and in the same manner in a column.

Thus, the total number of ways is $2 \times 8 \times ({}^8C_2 - 7)$.

45. Number of selection of atleast one red ball from 4 identical red balls = 4

Number of selection of any number of black balls from 5 identical black balls $= 5 + 1 = 6$

$\therefore$ Required number of selection of balls $= 4 \times 6 = 24$

46. $N = 35700 = 2^2 \times 3^1 \times 7^1 \times 5^2 \times 17^1$

(i) Number of divisors

$$= (2+1) \times (1+1) \times (1+1) \times (2+1) \times (1+1)$$

$$= 3 \times 2 \times 2 \times 3 \times 2 = 72$$

(ii) Number of proper divisors $= 72 - 2 = 70$

(iii) Number of even divisors $= 3 \times 2 \times 2 \times 2 \times 2 = 48$ (as 2 must occur atleast once)

(iv) Sum of divisors $= (2^0 + 2^1 + 2^2)(3^0 + 3^1)(7^0 + 7^1)$
$(5^0 + 5^1 + 5^2)(17^0 + 17^1)$

$$= 7 \times 4 \times 8 \times 31 \times 18 = 124992$$

47. Here, order of group is important, then the number of ways in which 52 different cards can be divided equally into 4 players is

$$\frac{52!}{4!(13!)^4} \times 4! = \frac{52!}{(13!)^4}$$

48. $\because$ The order is important, thus the number of ways

$$= \frac{12!}{5! \times 7!} \times 2!$$

$$= \frac{12 \times 11 \times 10 \times 9 \times 8 \times 7!}{(5 \cdot 4 \cdot 3 \cdot 2 \cdot 1) \times 7!} \times 2 = 1584$$

49.

Box	I	II	III
Number of balls	1	1	3

or

Box	I	II	III
Number of balls	1	2	2

All 5 balls can be arranged by 5! ways and boxes arranged in each system by $\frac{3!}{2!}$.

$$\therefore \text{Required number of ways} = 5! \times \frac{3!}{2!} + 5! \times \frac{3!}{2!}$$

$$= 120 \times 3 + 120 \times 3 = 720$$

Aliter

The required number of ways

$$= 5! \times {}^{5-1}C_{3-1}$$

$$= 5! \times {}^4C_2$$

$$= 120 \times \frac{4 \times 3}{1 \times 2} = 720$$

50. (i) 5 people can be divided into three groups in the following ways:

1	1	3	or	1	2	2

Hence, total number of ways

$$= \frac{5!}{3!} \times \frac{1}{2!} + \frac{5!}{(2!)^2} \times \frac{1}{2!}$$

$$= 10 + 15 = 25$$

(ii)

Room	I	II	III
People	1	1	3

or

Room	I	II	III
People	1	2	2

Number of ways = Ways in which such groups can be formed × number of ways in which the groups be arranged in three different rooms

$$= \left(\frac{5!}{3!} \times \frac{1}{2!}\right) \times 3! + \left(\frac{5!}{(2!)^2} \times \frac{1}{2!}\right) \times 3!$$

$$= 60 + 90 = 150$$

(iii) In this case, the positioning of the people amongst themselves is also be taken into account.

∴ The required number of ways

$$= 60 \times 3! + 90 \times 2! \times 2!$$

$$= 360 + 360$$

$$= 720$$

51. The required number of ways

$$= \frac{1}{3!}(3^5 - {}^3C_1 \times 2^5 + {}^3C_2 \times (1)^5)$$

$$= \frac{150}{6} = 25$$

52. We know that in onto mapping, each image must be assigned atleast one pre-image.

Thus, the total number of onto mappings from from A to B

$$= 3^5 - {}^3C_1 (3-1)^5 + {}^3C_2 (3-2)^5$$

$$= 243 - 96 + 3 = 150$$

53. Clearly, 30 mangoes can be distributed among 4 boys such that each boy can receive any number of mangoes.

Hence, total number of ways

$$= {}^{30+4-1}C_{4-1}$$

$$= {}^{33}C_3$$

$$= \frac{33 \times 32 \times 31}{1 \times 2 \times 3} = 5456$$

54. (i) $\because x + y + z + w = 20$

Here, $x \geq 0, y \geq 0, z \geq 0$ and $w \geq 0$

The number of solutions of the given equations in this case is same as the number of ways of distributing 20 things among 4 different groups.

$\therefore$ Total number of solutions $= {}^{20+4-1}C_{4-1}$

$$= {}^{23}C_3 = \frac{23 \cdot 22 \cdot 21}{1 \cdot 2 \cdot 3} = 1771$$

(ii) $\because x + y + z + w = 20$ ...(i)

Here, $x \geq 1, y \geq 1, z \geq 1$ and $w \geq 1$

or $x - 1 \geq 0, y - 1 \geq 0, z - 1 \geq 0, w - 1 \geq 0$

$$x_1 \geq 0, x_2 \geq 0, x_3 \geq 0, x_4 \geq 0$$

where $x_1 = x - 1, x_2 = y - 1, x_3 = z - 1$ and $x_4 = w - 1$

Then, the Eq. (i) becomes

$x_1 + 1 + x_2 + 1 + x_3 + 1 + x_4 + 1 = 20$

$\Rightarrow \quad x_1 + x_2 + x_3 + x_4 = 16$

$\because \quad x_1 \geq 0, x_2 \geq 0, x_3 \geq 0, x_4 \geq 0$

Hence, total number of solutions $= {}^{16+4-1}C_{4-1}$

$$= {}^{19}C_3 = \frac{19 \times 18 \times 17}{3 \times 2 \times 1} = 969$$

Aliter $\because x + y + z + w = 20, x \geq 1, y \geq 1, z \geq 1$ and $w \geq 1$

Hence, total number of solutions

$$= {}^{20-1}C_{4-1} = {}^{19}C_3 = 969$$

55. $\because x + y + z + t = 29$

and x, y, z, t are integers

$\therefore \quad x \geq 1, y \geq 2, z \geq 3, t \geq 0$

$\Rightarrow x - 1 \geq 0, y - 2 \geq 0, z - 3 \geq 0, t \geq 0$

Let $x_1 = x - 1, x_2 = y - 2, x_3 = z - 3$

$\Rightarrow \quad x_1 + 1 + x_2 + 2 + x_3 + 3 + t = 29$

$\Rightarrow \quad x_1 + x_2 + x_3 + t = 23$

where $x_1 \geq 0, x_2 \geq 0, x_3 \geq 0, t \geq 0$

Hence, total number of solutions $= {}^{23+4-1}C_{4-1}$

$$= {}^{26}C_3$$

$$= \frac{26 \times 25 \times 24}{1 \times 2 \times 3} = 2600$$

56. We have, $x_1 + x_2 + x_3 + x_4 + x_5 = 20$

and $\quad x_1 + x_2 = 15$

$$x_3 + x_4 + x_5 = 5 \quad \text{...(i)}$$

and $\quad x_1 + x_2 = 15 \quad$ ...(ii)

$$\forall x_i \geq 0, (i) = 1, 2, \ldots 5.$$

$\therefore$ The number of solutions of Eq. (i), we have

$$= {}^{5+3-1}C_{3-1} = {}^7C_2 = 21$$

Also, the number of solutions of Eq. (ii), we have

$$= {}^{15+2-1}C_{2-1} = {}^{16}C_1 = 16$$

Hence, total number of solutions $= 21 \times 16 = 336$

57. We have, $3x + y + z = 24, x \geq 0, y \geq 0, z \geq 0$

Let $x = k$

$\therefore \quad y + z = 24 - 3k \quad$ ...(i)

Here, $\quad 24 \geq 24 - 3k \geq 0 \quad [\because x \geq 0]$

Hence, $\quad 0 \leq k \leq 8$

The total number of integral solutions of Eq. (i) is

$^{(24-3k)+2-1}C_{2-1} = {}^{25-3k}C_1 = 25 - 3k$

Hence, the total number of solutions of the original

equation $= \sum_{k=0}^{8} (25 - 3k)$

$= 25 \sum_{k=0}^{8} 1 - 3 \sum_{k=0}^{8} k$

$= 25 \times 9 - 3 \times \frac{8 \times 9}{2} = 225 - 108 = 17$

58. The problem is based on the number of dearrangements of 5 objects.

$\therefore$ Number of ways

$= 5!\left(1 - \frac{1}{1!} + \frac{1}{2!} - \frac{1}{3!} + \frac{1}{4!} - \frac{1}{5!}\right) = 44$

59. The number of ways = (number of ways without restrictions) – (number of ways in which all are in incorrect envelopes)– (number of ways in which 1 letter is in incorrect envelope) – (number of ways in which 2 letters are in incorrect envelope) – (number of ways in which 3 letters are in incorrect envelopes)

$= 4! - 1 - {}^4C_1 \times 2 - {}^4C_2 \times 1 - 0$

$= 24 - 1 - 8 - 6 - 0 = 9$

Aliter

The problem is based on the number of dearrangements of 4 objects

$\therefore$ Number of ways $= D_4$

$= 4! = 4!\left(1 - \frac{1}{1!} + \frac{1}{2!} - \frac{1}{3!} + \frac{1}{4!}\right) = 9$

60. The required number

$=$ Coefficient of x^{16} in $(x^3 + x^4 + \ldots + x^7)^4$

$=$ Coefficient of x^{16} in $x^{12}(1 + x + \ldots\ldots x^4)^4$

$=$ Coefficient of x^4 in $(1 + x + \ldots + x^4)^4$

$=$ Coefficient of x^4 in $\left(\frac{1 - x^5}{1 - x}\right)^4$

$=$ Coefficient of x^4 in $[(1 - x^5)^4 (1 - x)^{-4}]$

$=$ Coefficient of x^4 in

$[(1 - 4x^5 + 6x^{10} \ldots)(1 - x)^{-4}]$

$=$ Coefficient of x^4 in $(1 - x)^{-4}$

[leaving terms containing powers of x greater than 4]

$= {}^7C_3 = \frac{4 \cdot 5 \cdot 6 \cdot 7}{4!} = 35$

61. The required number of integers (note that the number will be of 3, 4, 5 and 6 digits)

$=$ Coefficient of x^{25} in $(1 + x + x^2 + \ldots + x^9)^6$

$=$ Coefficient of x^{25} in $\left[\frac{1 - x^{10}}{1 - x}\right]^6$.

$=$ Coefficient of x^{25} in $[(1 - x^{10})^6 (1 - x)^{-6}]$

$=$ Coefficient of x^{25} in

$[(1 - 6x^{10} + 15x^{20} + \ldots)(1 - x)^{-6}]$

$= {}^{30}C_5 - 6 \times {}^{15}C_5 + 15 \times {}^{10}C_5$

$= 142506 - 96024 + 3780 = 50262$

Level 2

1. (c) Using fundamental principle of counting, the number of required choices

3 4 5 2

$= 3 \times 4 \times 5 \times 2 = 120$

2. (b) Using fundamental principle of counting, the number of required choices

2 3 3

$= 2 \times 3 \times 3 = 18$

3. (b) Since, there are six digits available.

$\therefore$ Number of 3-digit numbers $= 6 \times 6 \times 6 = 216$

4. (b) The number of one digit number is 6

The number of 2-digit numbers is $5 \times 5 = 25$

The number 3-digit numbers is $5 \times 5 \times 4 = 100$

Hence, the total number

$= 6 + 25 + 100 = 131$

5. (d) Number of 5-digit numbers with 2 at 10th place

$= 8 \times 8 \times 7 \times 1 \times 6 = 2688$

· · · 2 ·

8 8 7 1 6

(Choices)

Given that $2688 = 336\,k$

$\Rightarrow \quad k = 8$

6. (d) Number of 4-digit numbers starting with 5 is

$= 6 \times 6 \times 6 = 216$

5 – – –

6 6 6

Similarly,

number of 4-digit numbers starting with 45 is

$= 6 \times 6 = 36$

Number of 4-digit numbers starting with 44 is

$$= 6 \times 6 = 36$$

Number of 4-digit numbers starting with 43 and greater than $4321 = 3 \times 6 = 18$

4	3		

↓ 3 ↓ 6 (5/4/3)

Number of 4-digit numbers starting with 432 and greater than $4321 = 4$

4	3	2	–

↓ 4 (2/3/4/5)

Hence, required numbers

$$= 216 + 36 + 36 + 18 + 4 = 310$$

7. (c) Given, $n!, 3(n)!$ and $(n+1)$ are in GP

$$\Rightarrow \quad (3 \times n!)^2 = n! \times (n+1)!$$
$$9 \times (n!)^2 = n! \times (n+1)!$$
$$9 \times n! = (n+1)!$$
$$9 \times n! = (n+1) \cdot n!$$
$$\Rightarrow \quad n+1 = 9 \Rightarrow n = 8$$

8. (b) The required sum $= 1! + 2(2!) + 3(3!) + \ldots + n \cdot n!$

$$= (2-1)1! + (3-1)2! + (4-1)3! + \ldots + [(n+1)-n]n!$$
$$= (2!-1!) + (3!-2!) + (4!-3!) + \ldots + [(n+1)! - n!]$$
$$= (n+1)! - 1!$$

9. (b) Exponent of 2 in $100! = \left[\frac{100}{2}\right] + \left[\frac{100}{2^2}\right] + \left[\frac{100}{2^3}\right] + \left[\frac{100}{2^4}\right] + \left[\frac{100}{2^5}\right] + \left[\frac{100}{2^6}\right]$

$$= 97$$

$$E_3(100!) = \left[\frac{100}{3}\right] + \left[\frac{100}{3^2}\right] + \left[\frac{100}{3^3}\right] + \left[\frac{100}{3^4}\right]$$
$$= 48$$

$\therefore$ Exponent of 12 in $100! = 48$

10. (a) Given that, $f(x) = {}^{7-x}P_{x-3}$

The above function is defined,

if $7 - x \geq 0$ and $x - 3 \geq 0 \Rightarrow 7 - x \geq x - 3$

$\Rightarrow \quad x \leq 7, x \geq 3$ and $x \leq 5$

$\therefore \quad D_f = \{3, 4, 5\}$ (domain of f)

Now, $f(3) = {}^4P_0 = 1$

$f(4) = {}^3P_1 = 3$

$f(5) = {}^2P_2 = 2$

$\therefore$ Range of $f = \{1, 2, 3\}$

11. (c) We know that

$$(r+1)\,{}^rP_{r-1} = (r+1) \cdot \frac{r!}{1!} = (r+1)!$$

Thus, $(2 \cdot {}^1P_0 - 3 \cdot {}^2P_1 + 4 \cdot {}^3P_2 \ldots$ upto 51 terms)

$+ (1! - 2! + 3! \ldots +$ upto 51 terms)

$$= (2! - 3! + 4! \ldots + 52!) + (1! - 2! + 3! - 4! + \ldots + 51!)$$
$$= 52! + 1!$$

12. (b) Every element has 3 options either set Y or set Z or none.

So, number of ordered pairs $= 3^5$

13. (a) Each position can be filled in 5 ways.

Hence, the total number of numbers is 5^{20}.

14. (b) The smallest number of people

= Total number of possible forecasts

= Total number of possible results

$= 3 \times 3 \times 3 \times 3 \times 3 = 243$

15. (b) Number of integers greater than 6000 may be 4-digit or 5-digit.

Number of 4-digit numbers

$= 3 \times 4 \times 3 \times 2 \times 1 = 72$

($\because$ first digit can be 6, 7 or 8)

Number of 5-digit numbers $= 5! = 120$

$\therefore$ Total numbers $= 72 + 120 = 192$

16. (b) 3 must be at thousand's place and since the number should be divisible by 5 or 5 must be at unit's place. Now, we have to fill two places (tens and hundreds) i.e. ${}^4P_2 = 12$

17. (d) Given, 5-digits are 1, 3, 5, 7, 9 for selection of one digit, we have = 5 choice (for digit to repeat) and six digits can be arrange in $\frac{6!}{2!}$ ways.

+	+	+	+	+	–

→ (Repeat one digit)

Hence, total such numbers $= 5 \times \frac{6!}{2!} = \frac{5}{2}(6!)$

18. (b) Given, digits 0, 1, 2, 5, 7, and 9

a	b	c	d	e	f

For divisibility of 11,

$$(a + c + e) - (b + d + f) = 11k$$

$\therefore$ The choices would be $(1, 2, 9)$ and $(0, 5, 7)$

Number of ways to arranging them

$$= 3! \times 3! + 3! \times 2 \times 2 = 36 + 24 = 60$$

19. (c) Total number of ways in which all letters can be arranged in alphabetical order $= 6!$

$\because$ There are two vowels (E, A) in the word 'GARDEN'. Total number of ways in which these two vowels can be arranged $= 2!$

$\therefore$ Total number of required ways $= \frac{6!}{2!} = 360$

20. (c) We have 32 places for teeth for each place, we have two choices, either there is a tooth or there is no tooth. Therefore, the number of ways to fill up these places is 2^{32}.

As there is no person without a tooth.

The maximum population is $2^{32}-1$.

21. (c) For each bulb, there are two possibilities. It will be switched either on or off.

Hence, total number of ways in which the room can be illuminated is $2^{12}-1$.

22. (c) R, R, R, O, O, H

Words starting with $R=\dfrac{5!}{2!\times 2!}=30$

Words starting with $O=\dfrac{5!}{3!}=20$

Words starting with $HR=\dfrac{4!}{2!\times 2!}=6$

The word following the above words will be 'HORRO', the next word will be 'HORROR'.

∴ Rank of the word 'HORROR'

$=30+20+6+1+1=58$

23. (c) 'SMALL'

Words starting with $A=\dfrac{4!}{2!}=12$

Words starting with $L=4!=24$

Words starting with $M=\dfrac{4!}{2!}=12$

Words starting with $SA=\dfrac{3!}{2!}=3$

Words starting with $SL=3!=6$

Next word is 'SMALL'.

∴ Its position is 58th.

24. (a) We can think of three packets. One consisting of 3 boys of class X, other consisting of 4 boys of class XI and last one consisting of 5 boys of class XII. These packets can be arranged in 3! ways and contents of these packets can be further arranged in 3!, 4! and 5! ways, respectively.

Hence, the total number of ways is $3!\times 3!\times 4!\times 5!$.

25. (c) The total number of ways is $6\times 6\times ....$ to n times $=6^n$

The total number of ways to show only even numbers is $3\times 3\times ....$ to n times $=3^n$

Therefore, the required number of ways is 6^n-3^n.

26. (a) If 7 is used at first place, then the number of numbers is 9^4 and for any other four places it is $8\times 4\times 9^3$

So, total number of numbers

$=9^4+4\times 8\times 9^3=41\times 9^3$

27. (d) Since, the balls are to be arranged in a row so that the adjacent balls are of different colours. We can therefore, begin with a white ball or a black ball. If we begin with a white ball, we find that $n+1$ white balls numbered 1 to $n+1$ can be arranged in a row in $(n+1)!$ ways.

× W × W × W × W ...... × W

Since, white and black balls are alternate, we have exactly $(n+1)$ place for black balls marked with 'X'.

So, the total number of arrangements in which adjacent balls are of different colours and first ball is a white ball is $(n+1)!\times(n+1)!=[(n+1)!]^2$

But we can begin with a black ball also.

Hence, the required number of arrangements is $2[(n+1)!]^2$.

28. (b) The number of trains a day (the digits 1, 2, 3) are three groups of like elements from which a sample must be formed.

In the time-table for a week, the number 1 is repeated twice, the number 2 is repeated 3 times and the number 3 is repeated twice.

The number of different time-table is given by

$$P(2,3,2)=\frac{7!}{2!3!2!}=210$$

29. (c) Sum of seven digits is a multiple of 9.

Now, sum of numbers 1, 2, 3, 4, 5, 6, 7, 8 and 9 is 45.

So, we have to remove to two numbers whose sum is 9.

So, pairs of removed numbers are (1, 8), (2, 7) (3, 6) and (4, 5) with each pair left number of 7-digit number is 7!.

So with all 4 pairs, total number are $4\times 7!$.

30. (b) There are 8 chairs on each side of the table. Let the sides be separated by A and B. Let 4 persons sit on side A, the number of ways of arranging 4 persons on 8 chairs on side $A={}^8P_4$

and then two persons sit on side B. The number of ways of arranging 2 persons on 8 chairs on side $B={}^8P_2$

and the remaining 10 persons can be arranged in remaining 10 chairs in 10! ways.

Hence, the total number of ways in which the person can be arranged

$$={}^8P_4\times{}^8P_2\times 10!$$

$$=\frac{8!\times 8!\times 10!}{4!\times 6!}$$

31. (b)

F – 1	F – 2	F – 3
(3)	(3)	(4)

Total number of ways = Arrangement of 3 families × internal arrangement of family members.

$$=(3!\times 3!\times 4!)\times 3!$$

$$=(3!)^3\times 4!$$

32. (a) The number of ways of allotment without any restriction is $^8P_6$.

Now, it is possible that none of the rooms of 2nd floor or 3rd floor are occupied.

Thus, there are two ways in which one floor remains unoccupied. Hence, the number of ways of allotement in which a floor is unoccupied is $2 \times 6!$.

Hence number of ways in which none of the floor remains unoccupied is $^8P_6 - 2(6!)$.

33. (b) Six students $S_1, S_2, \ldots S_6$ can be arranged in a round circular table in 5! ways. Among these 6 students there are six vacant place, in which six teachers can sit in 6! ways.

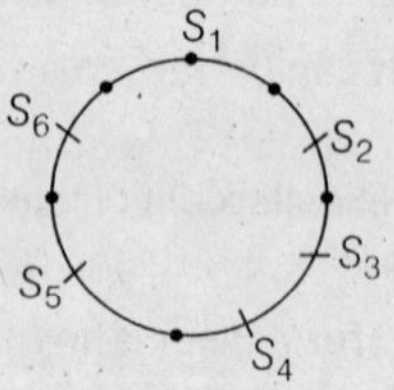

Hence, number of arrangement $= 5! \times 6!$

34. (d) $\alpha = {}^mC_2 \Rightarrow \alpha = \dfrac{m(m-1)}{2}$

$$\therefore\ {}^{\alpha}C_2 = \frac{\alpha(\alpha-1)}{2} = \frac{1}{2}\frac{m(m-1)}{2}\left\{\frac{m(m-1)}{2} - 1\right\}$$

$$= \frac{1}{8}m(m-1)(m-2)(m+1)$$

$$= \frac{1}{8}(m+1)m(m-1)(m-2)$$

$$= \frac{(m+1)m(m-1)(m-2)}{4\cdot 3\cdot 2\cdot 1} \times 3$$

$$= \frac{(m+1)! \times 3}{4!(m-3)!} = 3\cdot {}^{m+1}C_4$$

35. (c) Given, $^nP_r = {}^nP_{r+1}$

$$\Rightarrow \quad \frac{n!}{(n-r)!} = \frac{n!}{(n-r-1)!}$$

$$\Rightarrow \quad \frac{n!}{(n-r)(n-r-1)!} = \frac{n!}{(n-r-1)!}$$

$$\Rightarrow \quad n - r = 1 \quad \ldots(i)$$

and $\quad {}^nC_r = {}^nC_{r-1}$

$$\Rightarrow \quad \frac{n!}{r!(n-r)!} = \frac{n!}{(r-1)!(n-r+1)!}$$

$$\Rightarrow \quad \frac{1}{r(n-r)!} = \frac{1}{(n-r+1)(n-r)!}$$

$$\Rightarrow \quad n - r + 1 = r$$

From Eq. (i),

$$1 + 1 = r \Rightarrow r = 2$$

36. (a) $^nC_3 + {}^nC_4 > {}^{n+1}C_3$

$$\Rightarrow \quad {}^{n+1}C_4 > {}^{n+1}C_3 \qquad [\because\ {}^nC_r + {}^nC_{r+1} = {}^{n+1}C_{r+1}]$$

or $\quad \dfrac{^{n+1}C_4}{^{n+1}C_3} > 1$

or $\quad \dfrac{n-2}{4} > 1$

or $\quad n > 6$

37. (d) $6 \times {}^{35}C_r = (k^2 - 3) \times \dfrac{36}{r+1}\,{}^{35}C_r \left[\because\ {}^nC_{r+1} = \dfrac{n}{r+1}\,{}^{n-1}C_r\right]$

$$\Rightarrow \quad 6 = \frac{36}{(r+1)}(k^2 - 3)$$

$$\Rightarrow \quad (k^2 - 3) = \frac{r+1}{6} \Rightarrow k^2 = 3 + \frac{r+1}{6}$$

r can be 5, 35 for $k \in I$ (integers)

$\therefore \quad r = 5, k = \pm 2$ and if $r = 35, k = \pm 3$.

Hence, number of ordered pairs $= 4$

38. (d) Here, $^nP_3 - {}^nC_3 > 100$

$$\Rightarrow \quad \frac{n!}{(n-3)!} - \frac{n!}{3!(n-3)!} > 100$$

$$\Rightarrow \quad \frac{5}{6}n(n-1)(n-2) > 100$$

$$n(n-1)(n-2) > 120$$

$$\Rightarrow \quad n(n-1)(n-2) > 6 \times 5 \times 4$$

$$\Rightarrow \quad n = 7, 8, \ldots\ldots$$

$\therefore$ Least smaller things $= 7$

39. (b) Since, 5 players are always to be excluded and 6 players are always to be included, therefore 5 players are to be chosen form 14.

Hence, required number of ways $= {}^{14}C_5 = 2002$

40. (c) Places for A, B and C can be chosen in $^{10}C_3$ ways. Remaining 7 persons can speak in 7! ways.

Hence, the number of ways in which they can speak, is

$$^{10}C_3 \times 7! = 10!/6$$

41. (c) Since, the number of ways to select 2 girls is $^5C_2$.

Now, 3 boys can be selected in 3 ways.

(i) Selection of A and any two other boys (except B) in $^5C_2$ ways.

(ii) Selection of B and any two other boys (except A) in $^5C_2$ ways.

(iii) Selection of 3 boys (except A and B) in $^5C_3$ ways.

Hence, required number of different teams

$$= {}^5C_2({}^5C_2 + {}^5C_2 + {}^5C_3)$$

$$= 300$$

42. (b) $\because m =$ Number of ways the committee is formed with atleast 6 males

$$\Rightarrow \quad m = (6M, 5F) \text{ or } (7M, 4F) \text{ or } (8M, 3F)$$

$$\Rightarrow \quad m = ({}^8C_6 \times {}^5C_5) + ({}^8C_7 \times {}^5C_4) + ({}^8C_8 \times {}^5C_3)$$

$\Rightarrow \quad m=78$

and $n=$ Number of ways the committee is formed with atleast 3 females

$\Rightarrow \quad n=$ (3F, 8M) or (4F, 7M) or (5F, 6M)

$=({}^5C_3\times{}^8C_8)+({}^5C_4\times{}^8C_7)+({}^5C_5\times{}^8C_6)$

$\Rightarrow \quad n=78$

Hence, $m=n=78$

43. (d) $\because$ Each section has 5 questions.

5 – 1	5 – 2	5 – 3
(5)	(5)	(5)

$\therefore$ Total number of selection of 5 questions

$$=3\times({}^5C_1\times{}^5C_1\times{}^5C_3)+3\times({}^5C_1\times{}^5C_2\times{}^5C_2)$$
$$=3\times5\times5\times10+3\times5\times10\times10$$
$$=750+1500=2250$$

44. (c) Number of ways of selecting 10 objects

= (10*I*, 0*D*) or (9*I*, 1*D*) or (8*I*, 2*D*) or ...... (0*I*, 10*D*)

where $I=$ identical object and $D=$ distinct object

0 identical +10 distinct, number of ways $=1\times{}^{21}C_{10}$

1 identical +9 distinct, number of ways $=1\times{}^{21}C_9$

2 identical +8 distinct, number of ways $=1\times{}^{21}C_8$

$\therefore$ Total ways $={}^{21}C_{10}+{}^{21}C_9+{}^{21}C_8+\ldots+{}^{21}C_0$

Let $S={}^{21}C_0+{}^{21}C_2+\ldots+{}^{21}C_{10}$ ...(i)

$\because {}^nC_r={}^nC_{n-r}$

$\Rightarrow S={}^{21}C_{11}+{}^{21}C_{12}+{}^{21}C_{13}+\ldots{}^{21}C_{21}$ ...(ii)

Adding Eqs. (i) and (ii), we get

$$2S={}^{21}C_0+{}^{21}C_2+\ldots+{}^{21}C_{21}$$

$$\Rightarrow \quad 2S=2^{21}$$

$$\therefore \quad S=\frac{2^{21}}{2}=2^{20}$$

45. (b) Given numbers can be grouped as $(\lambda\in N)$

1, 4, 7 ......, $3n-2\rightarrow 3\lambda-2$

2, 5, 8 ......, $3n-1\rightarrow 3\lambda-1$

3, 6, 9, ..., $3n\rightarrow 3\lambda$

That means, we must take two numbers from last row or one number each from first and second rows.

Therefore, the total number of ways is

$${}^nC_2+{}^nC_1\times{}^nC_1=\frac{n(n-1)}{2}+n^2$$
$$=\frac{3n^2-n}{2}$$

46. (a) 26 cards can be chosen out of 52 cards in ${}^{52}C_{26}$ ways, there are two ways in which each card can be dealt, because a card can be either from the first pack or from the second.

Hence, the total number of ways is ${}^{52}C_{26}\times2^{26}$.

47. (b) Let the numbers selected be x_1, x_2 and x_3, we must have, $2x_2=x_1+x_3$

$\Rightarrow \quad x_1+x_3=$ Even

Therefore, x_1 and x_3 both are odd or both are even.

If x_1 and x_3 both are even, then we can selected them in ${}^{12}C_2$ ways.

Similarly, if x_1 and x_3 both are odd, then we can again select them in ${}^{12}C_2$ ways.

Thus, the total number of ways is $2\times{}^{12}C_2=132$

48. (a) $x_1<x_2<x_3<x_4<x_5<x_6$, when the number is $x_1x_2x_3x_4x_5x_6$.

Clearly, no digit can be zero. Also all the digits are distinct.

So, let us first select six digits from the list of digits 1, 2, 3, 4, 5, 6, 7, 8 and 9, which can be done in ${}^9C_6$ ways.

After selecting these digits, they can be put only in one order.

Thus, total number of such numbers is

$${}^9C_6\times1={}^9C_3 \quad [\because {}^nC_r={}^nC_{n-r}]$$

49. (a) Given, digits are $=1,2,2,3$

$\therefore$ Total distinct numbers $=4!/2!=12$

1 at unit place $\Rightarrow$ Number of such numbers $=\frac{3!}{2!}=3$

2 at unit place $\Rightarrow$ Number of such numbers $=3!=6$

3 at unit place $\Rightarrow$ Number of such numbers $=\frac{3!}{2!}=3$

$\therefore$ Sum of digits at unit place is

$$3\times1+6\times2+3\times3=24$$

Hence, sum of all 4 digit such numbers

$$=(3+12+9)(10^3+10^2+10+1)$$
$$=1111\times24=26664$$

50. (a) The number of ways of selecting four numbers from 1 to 30 without any restriction is ${}^{30}C_4$.

The number of ways of selecting four consecutive [i.e (1, 2, 3, 4), (2, 3, 4, 5), ..., (27, 28, 29, 30)] numbers is 27. Hence, the number of ways of selecting four integers which excludes consecutive four selections, is

$${}^{30}C_4-27=\frac{30\times29\times28\times27}{24}-27=27378$$

51. (c) Let there be n men participants. Then, the number of games that the men play between themselves, is $2\times{}^nC_2$ and the number of games that the men played with the women is $2\times(2n)$.

$\therefore \quad 2\times{}^nC_2-2\times2n=66$ [by hypothesis]

or $\quad n^2-5n-66=0$

or $\quad n=11$

Hence, the number of participants is

11 men +2 women = 13

52. (b) Two elements for set $P \cap Q$ can be selected in ${}^nC_2$ ways. Each of the remaining $(n-2)$ elements can be put in any of the three sets $(P \cap Q')$, $(P' \cap Q)$ or $(P \cup Q)'$

So, total number of subsets $= {}^nC_2 \times 3^{n-2}$

53. (d) Let us first select two places for vowels, which can be selected from 4 places in ${}^4C_2$ ways.

Now, this places can be filled by vowels in $5 \times 5 = 25$ ways as repetition is allowed. The remaining two places can be filled by consonants in 21×21 ways.

Then, the total number of words is

$${}^4C_2 \times 25 \times 21^2 = 150 \times 21^2$$

54. (b) Other than two, remaining five places can be filled by 1 and 3 for each place.

The number of ways for five places is $2 \times 2 \times 2 \times 2 \times 2 = 2^5$.

For 2, selecting two places out of 7 is ${}^7C_2$.

Hence, the required number of ways is ${}^7C_2 \times 2^5$.

55. (c) 1S, 3A, 1H, 2R, 1 N, 1P, 1U

when all letters are different, corresponding ways is ${}^7P_3 = 210$.

When two letters are of one kind and other is different, corresponding number of ways is ${}^2C_1 \times {}^6C_1 \times (3!/2!) = 36$

When all letters are alike, corresponding number of ways is 1.

Thus, total number of words that can be formed is $210 + 36 + 1 = 247$

56. (d) Four novels can be selected from six novels in ${}^6C_4$ ways, one dictionary can be selected from three dictionaries in ${}^3C_1$ ways.

As the dictionary selected is fixed in the middle, the remaining four novels can be arranged in 4! ways.

Hence, the required number of ways of arrangement is

$${}^6C_4 \times {}^3C_1 \times 4! = 1080$$

57. (a) Number of ways,

= Number of ways in which we select two balls from each box

$$= {}^3C_2 \times {}^9C_2 = 3 \times 36 = 108$$

58. (a) Matches whose predictions are correct, can be selected in ${}^{20}C_{10}$ ways.

Now, each wrong prediction can be made in 2 ways.

Thus, the total number of ways is ${}^{20}C_{10} \times 2^{10}$.

59. (c) (i) Ms. C is taken

(A) B included $\Rightarrow A$ excluded

$\Rightarrow \quad {}^4C_1 \times {}^4C_2 = 24$

(B) B excluded $\Rightarrow {}^4C_1 \times {}^5C_3 = 40$

(ii) Ms. C is not taken

$\Rightarrow B$ does not come $= {}^4C_2 \times {}^5C_3 = 60$

$\therefore$ Total combinations $= 24 + 40 + 60 = 124$

60. (a) The selection can be made in ${}^5C_3 \times {}^{22}C_9$ ways (since 3 vacancies are filled from 5 candidates in ${}^5C_3$ ways and now remaining candidates are 22 and remaining seats are 9)

61. (c)

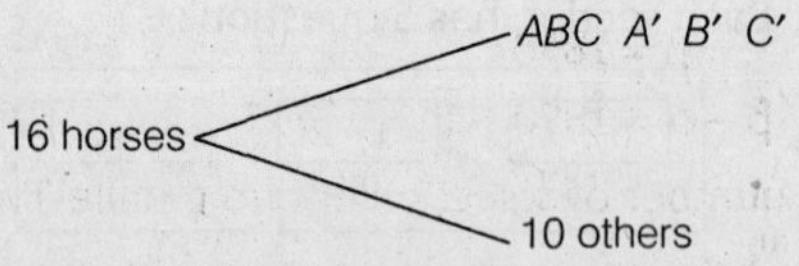

The number of ways is ${}^{10}C_3 \times$ number of ways of choosing out of $ABC\,A'B'C'$, so that $AA'\,BB'$ or CC' are not together,

$$= {}^{10}C_3 \times 8 = \frac{10 \times 9 \times 8}{1 \times 2 \times 3} \times 8 = 960$$

62. (a) $(m+n)$ counters on one side can be arranged in $\frac{(m+n)!}{m!\,n!}$ ways.

For each arrangement on one side, corresponding arrangement on the other side is just one as arrangements are symmetrical.

Hence, the total number of arrangements is

$$\frac{(m+n)!}{m!\,n!} = {}^{m+n}C_m$$

63. (b) If out of n points m are collinear, then number of triangles $= {}^nC_3 - {}^mC_3$

$\therefore$ Required number of triangles $= {}^{10}C_3 - {}^6C_3 = 120 - 20$

$$N = 100$$

64. (b) $\because T_n = {}^nC_3$, hence $T_{n+1} = {}^{n+1}C_3$

Now, $\quad T_{n+1} - T_n = 10$

$\Rightarrow \quad {}^{n+1}C_3 - {}^nC_3 = 10 \quad$ [given]

$\Rightarrow \quad \frac{(n+1)n(n-1)}{3!} - \frac{n(n-1)(n-2)}{3!} = 10$

$\Rightarrow \quad \frac{n(n-1)}{3!}[n+1-n+2] = 10$

$\Rightarrow \quad \frac{n(n-1)}{3 \times 2} \times 3 = 10$

$\Rightarrow \quad n^2 - n - 20 = 0 \Rightarrow n = 5$

65. (d)

Number of triangles that can be formed from the points on 3 of the sides.

$$^5C_1\,^7C_1\,^6C_1 + {^5C_1}\,^7C_1\,^9C_1 + {^5C_1}\,^6C_1\,^9C_1 + {^6C_1}\,^7C_1\,^9C_1$$

$$= 210 + 315 + 270 + 378$$

$$\Rightarrow \alpha = 1173$$

Number of quadrilaterals that can be formed by taking one point from each of the four vertex

$$^5C_1\,^7C_1\,^6C_1\,^9C_1 = 5 \times 6 \times 7 \times 9 = 1890$$

$$\Rightarrow \quad \beta = 1890$$

$$\therefore \quad \beta - \alpha = 1890 - 1173 = 717$$

66. (a) The number of selection of two parallel lines from m lines is $^mC_2$.

The number of selection of two parallel lines from n lines is $^nC_2$.

Hence, the number of parallelograms is

$$^mC_2 \times {^nC_2} = \frac{1}{4} mn(m-1)(n-1)$$

67. (d) Given, $n = 2^x 3^y 5^z$

and $y + z = 5$

$$\frac{1}{y} + \frac{1}{z} = \frac{5}{6} \text{ or } \frac{y+z}{yz} = \frac{5}{6}$$

This implies,

$$y + z = 5 \text{ and } yz = 6$$

Put $y = \frac{6}{z}$ in $y + z = 5$

$$\Rightarrow \quad \frac{6}{z} + z = 5 \text{ or } z^2 - 5z + 6 = 0$$

$$\Rightarrow \quad z^2 - 3z - 2z + 6 = 0$$

$$(z-3)(z-2) = 0$$

$$\Rightarrow \quad z = 3 \text{ or } 2$$

Using $y = \frac{6}{z}$, we get $y = 2$ or 3

For calculating the odd divisor x must be 0 i.e. $x = 0$

$$\therefore \quad n = 2^0\, 3^3 5^2 \text{ or } n = 2^0 3^2 5^3$$

Total number of odd divisor of n is equal to

$$= (3+1)(2+1)$$

$$= (4)(3) = 12$$

68. (a) Number of ways of selecting zero or more balls from 10 identical white balls = 1 + 1 + 1 ... (11 times) = 11

Similarly, number of ways of selecting zero or more balls from 9 identical green balls and 7 identical black balls will be $(9+1)$ and $(7+1)$ ways, respectively.

So, total number of ways of selecting one or more balls from all the balls $= 11 \times 10 \times 8 = 880$

69. (a) $\because\ 38808 = 2^3 \times 3^2 \times 7^2 \times 11^1$

$\therefore$ The number of divisors (excluding 1 and number itself)

$$= 4 \times 3 \times 3 \times 2 - 2 = 72 - 2 = 70$$

70. (d) The number of divisors of ab^2c^2de (excluding 1)

$$= (1+1)(2+1)(2+1)(1+1)(1+1) - 1$$

$$= 2 \times 3 \times 3 \times 2 \times 2 - 1 = 72 - 1 = 71$$

71. (d) Let x, y, z be the friends and a, b, c denote the case, when x is invited a times, y is invited b times and z is invited c times.

Now, we have the following possibilities,

$(a, b, c) = (1, 2, 3)$ or (330) or (222)

[grouping of 6 days of week]

Hence, the total number of ways is

$$\frac{6!}{1!2!3!}3! + \frac{6!}{3!3!2!}3! + \frac{6!}{(2!2!2!)3!}3!$$

$$= 360 + 60 + 90 = 510$$

72. (c) Given problem can be written as

$$x_1 + x_2 + \ldots + x_n = n^2, x_i \ge i, i \in 1 \text{ to } n.$$

Let $x_i - i = y_i$

$$\because \quad x_i \ge i \Rightarrow y_i \ge 0$$

$$\Rightarrow (y_1 + y_2 + \ldots + y_n) + (1 + 2 + \ldots + n) = n^2$$

$$\Rightarrow \quad y_1 + y_2 + \ldots + y_n + \frac{n(n+1)}{2} = n^2$$

$$\Rightarrow \quad y_1 + y_2 + \ldots + y_n = n^2 - \frac{n(n+1)}{2} = \frac{2n^2 - n^2 - n}{2}$$

$$\Rightarrow \quad y_1 + y_2 + \ldots + y_n = \frac{n^2 - n}{2} = \frac{n(n-1)}{2}$$

$\therefore$ Total number of possibilities is

$$^{n + \frac{n(n-1)}{2} - 1}C_{n-1}$$

$$= {^{\frac{n^2+n-2}{2}}C_{n-1}}$$

73. (a) Let the blankets received by the persons are x_1, x_2, x_3 and x_4, we have,

$$x_1 + x_2 + x_3 + x_4 = 15 \text{ and } x_i \ge 2$$

$$\Rightarrow (x_1 - 2) + (x_2 - 2) + (x_3 - 2) + (x_4 - 2) = 7$$

$$\Rightarrow \quad y_1 + y_2 + y_3 + y_4 = 7$$

(where $y_i = x_i - 2 \ge 0$)

The required number is equal to the number of non-negative integral solutions of this equation, which is equal to

$$^{4+7-1}C_7 \text{ i.e. } {^{10}C_7} = {^{10}C_3}$$

74. (c) Let person P_i gets x_i number of things such that

$$x_1 + x_2 + x_3 + x_4 + x_5 = 25$$

Let $x_i = 2\lambda_i + 1$, where $\lambda_i \ge 0$, then

$$2(\lambda_1 + \lambda_2 + \lambda_3 + \lambda_4 + \lambda_5) + 5 = 25$$

or $\quad \lambda_1 + \lambda_2 + \lambda_3 + \lambda_4 + \lambda_5 = 10$

We have to simply obtain the number of non-negative integral solutions of this equation which is equal to $^{14}C_{10}$.

75. (a) Let the balls put in x_1, x_2, x_3, x_4 and x_5, we have

$$x_1 + x_2 + x_3 + x_4 + x_5 = 15, x_i \geq 2$$

or $(x_1 - 2) + (x_2 - 2) + (x_3 - 2) + (x_4 - 2) + (x_5 - 2) = 5$

$\Rightarrow \quad y_1 + y_2 + y_3 + y_4 + y_5 = 5, y_i = x_i - 2 \geq 0$

The total number of ways is equal to the number of non-negative integral of the last equation, which is equal to

$$^{5+5-1}C_5 = {}^9C_5$$

76. (b) Let $x = p - 5, y = q - 5$ and $z = r - 5$, where $p, q, r \geq 0$

Then, the given equation reduces to

$$p + q + r = 15 \quad \text{...(i)}$$

Now, we have to find non-negative integral solution of Eq. (i), the total number of such solutions is

$$^{15+3-1}C_{3-1} = {}^{17}C_2 = 136$$

77. (d) Suppose, ith person receives ₹ $x_i, i = 1, 2, 3, 4$

then, $x_1 + x_2 + x_3 + x_4 = 18$, where $x_i \geq 3$

Let $y_i = x_i - 3, i = 1, 2, 3, 4$, then

$$y_1 + y_2 + y_3 + y_4 = 6$$

The total number of ways is equal to number of solutions of the above equation, which is given by

$$^{6+4-1}C_{4-1} = {}^9C_3 = 84$$

78. (a) $15 < x_1 + x_2 + x_3 \leq 20$

$$x_1 + x_2 + x_3 = 16 + r, r = 0, 1, 2, 3, 4, x_i \geq 1$$

$$y_1 + y_2 + y_3 = 13 + r \quad [y_i = x_i - 1]$$

$$y_i \geq 0$$

Now, the number of positive integral solution of

$y_1 + y_2 + y_3 = 13 + r$ is $^{13+r+3-1}C_{13+r}$

i.e. $\quad ^{15+r}C_{13+r} = {}^{15+r}C_2$

Thus, the total number of solution is

$$\sum_{r=0}^{4} {}^{15+r}C_2 = {}^{15}C_2 + {}^{16}C_2 + {}^{17}C_2 + {}^{18}C_2 + {}^{19}C_2$$

$$= \frac{1}{2}(15 \times 14 + 16 \times 15 + 17 \times 16 + 18 \times 17 + 19 \times 18) = 685$$

79. (c) $\because$ Card number 1 is always placed in envelope numbered 2, we can consider two cases.

Case I Card number 2 is placed in envelope number 1.

Then, it is dearrangement of 4 objects, which can be done in

$$4!\left(1 - \frac{1}{1!} + \frac{1}{2!} - \frac{1}{3!} + \frac{1}{4!}\right) = 9 \text{ ways.}$$

Case II Card number 2 is not placed in envelope number 1.

Then, it is dearrangement of 5 objects, which can be done in

$$5!\left(1 - \frac{1}{1!} + \frac{1}{2!} - \frac{1}{3!} + \frac{1}{4!} - \frac{1}{5!}\right) = 44 \text{ ways}$$

$\therefore$ Total ways $= 44 + 9 = 53$ ways.

80. (d) Coefficient of x^{10} in $(x + x^2 + \ldots + x^6)^4$

$= $ Coefficient of x^6 in $(1 + x + \ldots . + x^5)^4$

$= $ Coefficient of x^6 in $(1 - x^6)^4 (1 - x)^{-4}$

$= $ Coefficient of x^6 in $(1 - 4x^6 + \ldots)(1 + {}^4C_1 x + \ldots .)$

Hence, coefficient of x^6 is $^9C_6 - 4 = 80$.

81. (576) VOWELS (2 Vowel + 4 consonant)

All consonants must not be together

Total possibility of formation of 6 letter word $= 6!$

The number of arrangement when all the consonant comes together $= 3! \times 4!$

Number of arrangement when all the consonants never come together

$= $ Total $-$ All consonant together

$= 6! - 3!\,4!$

$= 576$

82. (5) *AAAAA* | *BBBBBB*

Since, word reads the same backwards and forwards, the middle digit must be *A*.

$$\begin{array}{c} M \\ \text{XXXXX}\downarrow\ \text{XXXXX} \end{array}$$

So, that even number of *A*'s and *B*'s are available for arrangement about middle position *M* in the above figure, Take *AABBB* on one side of *M* (6th place) and then their image about *M* in a unique way.

$\therefore$ Number of ways, $N = \dfrac{5!}{2!3!} = 10$

$\therefore \quad \dfrac{N}{2} = 5$

83. (4) Number of arrangements are $2n!\,n!$

Given that $2n!\,n! = 1152$

$\Rightarrow \quad (n!)^2 = 576$

$\Rightarrow \quad n! = 24$

$\Rightarrow \quad n = 4$

84. (24) We have, $N = \boxed{a}\boxed{b}\boxed{c}\boxed{d}$

First place a can be filled in 2 ways i.e. 4, 5 $(4000 \leq N < 6000)$

For b and c, total possibilities are $6 (3 \leq b < c \leq 6)$ i.e. 34, 35, 36, 45, 46, 56

Last place d can be filled in 2 ways i.e. 0, 5 (*N* is a multiple of 5)

Hence, total numbers $= 2 \times 6 \times 2 = 24$

85. (162)

$\boxed{x}\boxed{x}\boxed{\ }$ When two consecutive digits are 11, 22, etc

$= 9 \times 9 = 81$

$\boxed{\ }\boxed{0}\boxed{0}$ When two consecutive digits are $0, 0 = 9$

$\boxed{\ }\boxed{x}\boxed{x}$ When two consecutive digits are 11, 22, 33, ...

$= 9 \cdot 8 = 72$

$\therefore$ Total $= 81 + 9 + 72 = 162$

86. (100) Form of six digit palindrome number

xyzzyx

This will be divisible by 55

Hence, $x=5$

and $5yzzy5$ will be divisible by 11.

$\Rightarrow (5+z+y)-(y+z+5)$ is divisible by 11

which is true for all values of y and z.

$\Rightarrow y$ and z can be chosen in 10×10 ways

Number of such number $=100$

87. (18) Let T and S denote teacher and student, respectively.

Then, we have following possible patterns according to question,

(i) T SS T SS T SS

(ii) S T SS T SST S

(iii) SS T SS T SS T

Hence, total number of arrangements are

$$3 \times (3!)6! = 18 \times 6!$$

88. (8) To form a triangle, 3 points out of 5 can be chosen in ${}^5C_3 = 10$ ways

But of these, the three points lying on the 2 diagonals will be collinear.

So, $10-2=8$ triangles can be formed.

89. (777) Total number of players $=15$

Bowlers $=6$, Batsman $=7$, Wicket keepers $=2$

Bowlers	Batsman	Wicket Keepers	Total
4 + 1	5	1	${}^6C_5 \times {}^7C_5 \times {}^2C_1 = 252$
4	5 + 1	1	${}^6C_4 \times {}^7C_6 \times {}^2C_1 = 210$
4	5	1 + 1	${}^6C_4 \times {}^7C_5 + {}^2C_2 = 315$

Total $=252+210+315=777$

90. (35) '+' signs can be put in a row in 1 way, creating 7 blank places to keep '–' sign so that no two '–' signs occur together.

□ + □ + □ + □ + □ + □ + □

Out of these 7 places 4 can be chosen in ${}^7C_4$ ways.

$\therefore$ Required number of arrangements $= {}^7C_4 = 35$

91. (490) 0 red, 1 red, 2 red, 3 red

5R 4B 3W

$\therefore$ Number of ways

$$= {}^7C_4 + {}^5C_1 \times {}^7C_3 + {}^5C_2 \times {}^7C_2 + {}^5C_3 \times {}^7C_1$$

$$= 35 + 175 + 210 + 70 = 490$$

92. (21) If x denotes the number of times, he can take unit step and y denotes the number of times, he can take 2 steps, then $x+2y=7$

We must have $x=1,3,5$

If $x=1$, the steps will be 1 2 2 2

$\Rightarrow$ Number of ways $\dfrac{4!}{3!}=4$

If $x=3$, the steps will be 1 1 1 2 2

$\Rightarrow$ Number of ways $\dfrac{5!}{2!3!}=10$

If $x=5$, the steps will be 1 1 1 1 1 2

$\Rightarrow$ Number of ways $= {}^6C_1 = 6$

If $x=7$, the steps will be 1 1 1 1 1 11

$\Rightarrow \quad {}^7C_0 = 1$

Hence, total number of ways is 21.

93. (7744) Multiples of 11 such that they are of 3-digit and less than 500.

121, 132, ..., 495

$$n = \frac{495-121}{11} + 1 = 35$$

$$S = \frac{35}{2}(121+495) = 10780$$

Again, multiplies of 11 which are 3-digits, less than 500 and having 1 at hundred's place are 121, 132, ..., 198

$$n_1 = \left(\frac{198-121}{11}\right) + 1 = 8$$

$$S_1 = \frac{8}{2}(121+198) = 1276$$

The multiply of 11 which are of 3-digits, less than 500 and having 1 at ten's place are 319, 418

$\therefore \quad S_2 = 319 + 418 = 737$

The multiple of 11 which are 3-digits, less than 500 and having 1 at unit place are 231, 341, 451

$\therefore \quad S_3 = 231 + 341 + 451 = 1023$

$\therefore$ Required sum $= S - S_1 - S_2 - S_3 = 7744$

94. (960) Including the two specified people, 4 others can be selected in ${}^5C_4$ ways.

The two adjacent seats can be taken in 4 ways and the two specified people can be arranged in 2! ways, remaining 4 people can be arranged in 4! ways.

Therefore, ${}^5C_4 \cdot 4 \cdot 2!\,4! = 960$

95. (5) Let r number of books of Algebra and $20-r$ of calculus.

Number of selections $= {}^rC_5 \times {}^{20-r}C_5$

which has maximum value, when

$$r = 10$$

96. (9) Number of digits are 9, select 2 places for the digits 1 and 2 in ${}^9C_2$ ways.

From the remaining 7 places select any two places for 3 and 4 in ${}^7C_2$ ways and from the remaining 5 places, select any two for 5 and 6 in ${}^5C_2$ ways.

Now, the remaining 3 digits can be filled in 3! ways

$\therefore$ Total ways $= {}^9C_2 \cdot {}^7C_2 \cdot {}^5C_2 \cdot 3!$

$$= \frac{9!}{2!7!} \cdot \frac{7!}{2!5!} \cdot \frac{5!}{2!\cdot 3!} \cdot 3!$$

$$= \frac{9!}{8} = \frac{9\cdot 8\cdot 7!}{8} = 9\cdot 7!$$

97. (55) Let xyz be 3-digit number

$x+y+z=10, x\geq 1, y\geq 0, z\geq 0$

Let $x-1=t \Rightarrow x=1+t, x-1\geq 0, t\geq 0$

$\Rightarrow t+y+z=10-1=9, t\geq 0, y\geq 0, z\geq 0$

$\therefore$ Total number of non-negative integral solutions

$$= {}^{9+3-1}C_{3-1} = {}^{11}C_2 = \frac{11\times 10}{2} = 55$$

98. (9) Number of ways putting all the 4 balls into boxes of different colour

$$= 4!\left(1-\frac{1}{1!}+\frac{1}{2!}-\frac{1}{3!}+\frac{1}{4!}\right) = 9$$

Level 3

1. (a) We have, $\frac{1}{1!\times 11!}+\frac{1}{3!\times 9!}+\frac{1}{5!\times 7!}=\frac{2^m}{n!}$ ...(i)

$$= \frac{1}{1!\times 11!}+\frac{1}{3!\times 9!}+\frac{1}{5!\times 7!}+\frac{1}{7!\times 5!}+\frac{1}{9!\times 3!}+\frac{1}{11!\times 1!}$$

$$= \frac{1}{12!}\left[\frac{12!}{1!\,11!}+\frac{12!}{3!\,9!}+\frac{12!}{5!\,7!}+\frac{12!}{7!\,5!}+\frac{12!}{9!\,3!}+\frac{12!}{11!\,1!}\right]$$

$$= \frac{1}{12!}[{}^{12}C_1+{}^{12}C_3+{}^{12}C_5+{}^{12}C_7+{}^{12}C_9+{}^{12}C_{11}]$$

$$= \frac{1}{12!}[2^{12-1}] = \frac{2^{11}}{12!}$$

$$\therefore\ 2\left[\frac{1}{1!\times 11!}+\frac{1}{3!\times 9!}+\frac{1}{5!\times 7!}\right] = \frac{2^{11}}{12!}$$

$$\Rightarrow\ \frac{1}{1!\times 11!}+\frac{1}{3!\times 9!}+\frac{1}{5!\times 7!} = \frac{2^{10}}{12!}$$

On comparing with Eq (i), we have

$m=10$ and $n=12$

Given, ${}^kP_3 = 9240$

$\Rightarrow k(k-1)(k-2) = 9240$

$= 22\times 21\times 20$

$\Rightarrow k = 22$

$\therefore m+n-k = 10+12-22 = 0$

2. (d) Required number of numbers

$= 5\times 6\times 6\times 4 = 36\times 20 = 720$

□ □ □ □ ↑ {1,3,5,7}

3. (c) $\because$ Number is even, unit place can be filled with 0, 2, 4, 6 or 8

Th H T U

Also, thousand's place can not be filled with '0'

Case I Zero at unit place

Th H T U — (0 at U); ↑9 ↑8 ↑7

So, number of numbers $= 9\times 8\times 7\times 1$

$= 504$

Case II Zero is not at unit place

Th H T U — ↑8 ↑8 ↑7 ↑2,4,6,8

So, number of numbers

$= 8\times 8\times 7\times 4 = 1792$

Total numbers $= 504+1792 = 2296$

4. (d) The number is divisible by 9 if sum is divisible by 9.

Now, $1+2+3+4+5+\ldots\ldots+9 = 45$, is divisible by 9.

So, seven digit number divisible by 9 should not contain (1,8) or (2,7) or (3,6) or (4,5) and digits can be arranged in 7! ways.

$\therefore$ Number of such numbers = 4(7!)

5. (a) Total number of words that can be formed $= 10^5$

Number of words which no letter is repeated $= {}^{10}P_5$

So, number of words in which atleast one letter is repeated $= 10^5 - {}^{10}P_5 = 69760$

6. (b) The number of numbers when repetition is allowed, is 5^4.

The number of numbers when digit can not be repeated, is ${}^5P_5 = 5!$

Therefore, the required number of numbers is $5^4 - 5!$.

$= 625-120 = 505$

7. (b) Since, the largest digit is in the middle, the middle digit is greater than or equal to 4, the number of numbers with 4 in the middle $= {}^4P_4 - {}^3P_3$

($\because$ the other ten's place are to be filled by 0, 1, 2 and 3, and a number can not begin with 0)

Similarly, the number of numbers with 5 in the middle

$= {}^5P_4 - {}^4P_3$

$\therefore$ The required number of ways

$$= ({}^4P_4 - {}^3P_3) + ({}^5P_4 - {}^4P_3) + \ldots + ({}^9P_4 - {}^8P_3)$$

$$= \sum_{n=4}^{9} {}^nP_4 - \sum_{n=3}^{8} {}^nP_3$$

8. (d) The number of words starting from E = 5! = 120
The number of words starting from H = 5! = 120
The number of words starting from ME = 4! = 24
The number of words starting from MH = 4! = 24
The number of words starting from MOE = 3! = 6
The number of words starting from MOH = 3! = 6
The number of words starting from MOR = 3! = 6
The number of words starting from MOTE = 2! = 2
The number of words starting from MOTHER = 1! = 1
Hence, the rank of the word 'MOTHER'
$= 2(120) + 2(24) + 3(6) + 2 + 1 = 309$

9. (a) Number of ways = (selecting 4 girls and a captain) or (selecting 3 girls, 1 boy and a captain)
$= {}^6C_4 \times 4 + {}^6C_3 \times {}^4C_1 \times 4$
$= 380$

10. (d) $N_1 + N_2 + N_3 + N_4 + N_5$
= Total number of ways – N number of ways when no odd number is selected
Since, only available even numbers are 2, 4, 6 and 8, the number of ways when no odd number is selected, is zero.
$\therefore N_1 + N_2 + N_3 + N_4 + N_5 = {}^9C_5 - 0 = 126$

11. (b) We first select 2 men out of 7 in ${}^7C_2$ ways.
Now, we exclude the wives of these two selected men and so select 2 ladies from remaining 5 ladies in ${}^5C_2$ ways.
Let A, B be two men and X, Y be the ladies playing in one set. Then, we can have.
(i) A and X playing against B and Y
(ii) A and Y playing against B and X
Then, the total number of ways is
$${}^7C_2 \times {}^5C_2 \times 2 = 21 \times 10 \times 2 = 420$$

12. (a) 3 IIT students who will be between the DCE students can be selected in ${}^{10}C_3$ ways.
Now, 2 DCE students having 3 IIT students between them can be arranged in $2! \times 3!$ ways.
Finally, a group of above 5 students and the remaining 7 students together can be arranged in 8! ways.
Hence, total number of ways is ${}^{10}C_3 \times 2! \times 3! \times 8!$

13. (b) The candidate is unsuccessful if he fails in 9 or 8 or 7 or 6 or 5 papers. Therefore, the number of ways to be unsuccessful is ${}^9C_9 + {}^9C_8 + {}^9C_7 + {}^9C_6 + {}^9C_5$
$= {}^9C_0 + {}^9C_1 + {}^9C_2 + {}^9C_3 + {}^9C_4$
$= \frac{1}{2}({}^9C_0 + {}^9C_1 + \ldots + {}^9C_9)$
$= \frac{1}{2} \times 2^9 = 2^8 = 256$

14. (a) P_1 must win atleast $n+1$ games.
Let P_1 win $(n+r)$ games $(r = 1, 2, \ldots, n)$. Therefore, corresponding number of ways is ${}^{2n}C_{n+r}$.
The total number of ways is
$$\sum_{r=1}^{n} {}^{2n}C_{n+r} = {}^{2n}C_{n+1} + {}^{2n}C_{n+2} + \ldots + {}^{2n}C_{2n} = \frac{2^{2n} - {}^{2n}C_n}{2}$$

15. (a) Three elements from set A can be selected in ${}^7C_3$ ways. Their image has to be y_2. Remaining r images can be assigned to remaining 4 pre-images in 2^4 ways.
But the function is onto, hence the number of ways is $2^4 - 2 = 14$.
Then, the total number of functions is ${}^7C_3 \times 14 = \mathbf{490}$.

16. (b) If n is odd, $3^n = 4\lambda_1 - 1$ and $5^n = 4\lambda_2 + 1$
$\Rightarrow 2^n + 3^n + 5^n$ is divisible by 4 if $n \geq 2$
Thus, $n = 3, 5, 7, 9, \ldots, 99$ i.e. n can take 49 different values.
If n is even, $3^n = 4\lambda_1 + 1$ and $5^n = 4\lambda_2 + 1$
$\Rightarrow 2^n + 3^n + 5^n$ is not divisible by 4.
As, $2^n + 3^n + 5^n$ will be in the form of $4\lambda + 2$
Thus, the total number of ways of selecting 'n' is equal to 49.

17. (b) $N = 12600 = 2^3 \times 3^2 \times 5^2 \times 7^1$
The number of even divisors is equal to number of ways in which one or more '2', zero or more '3', zero or more '5' and zero or more '7' can be selected and is given by $(3)(2+1)(2+1)(1+1) = 54$

18. (c) $480 = 2^5 \times 3 \times 5$
Now, $4n + 2 = 2(2n+1)$ = Odd multiple of 2.
Thus, the total number of such divisors is $1 \times 2 \times 2 = \mathbf{4}$

19. (a) Obviously, A, B and C get 4, 5 and 7 objects, respectively. Then, number of distribution ways is equal to number of ways of division, which is given by $\frac{16!}{4!5!7!}$.

20. (a) Here, we have to divide 12 books into 4 sets of 3 books each
Therefore, the required number of ways is
$$\frac{12!}{(3!)^4 4!} \times 4! = \frac{12}{(3!)^4}$$

21. (a) Required number of ways
= (number of ways in which 16 players can be divided into 8 pairs) – (number of pairs in which S_1 and S_2 are together)
$$= \frac{16!}{2^8 \cdot 8!} - \frac{(14)!}{2^7 \cdot 7!} = \frac{16 \cdot 15 \cdot 14!}{2 \cdot 2^7 \cdot 8 \cdot 7!} - \frac{(14)!}{2^7 \cdot 7!}$$
$$= \frac{(14)!}{2^7 \cdot 7!}\left[\frac{16 \cdot 15}{16} - 1\right] = \frac{(14) \cdot (14)!}{14 \cdot (6!) \cdot 2^6} = \frac{(14)!}{2^6 \cdot 6!}$$

22. (c) Let $x_1, x_2, ..., x_8$ denote the marks assign to 8 questions.

$\therefore x_1 + x_2 + ... + x_8 = 30, x_i > 2, i = 1$ to 8

Let $u_i = x_i - 2,\ u_i \geq 0$

$\Rightarrow u_1 + u_2 + ... + u_8 = 30 - 16 = 14$

$\therefore$ Total number of solutions $= {}^{14+8-1}C_{8-1} = {}^{21}C_7$

23. (a) **Case I** Card number 2 is not placed in envelope number 1, then it is dearrangements of 4 objects.

$\Rightarrow \quad 4!\left(1 - \frac{1}{1!} + \frac{1}{2!} - \frac{1}{3!} + \frac{1}{4!}\right) = 9$ ways

Case II Card number 2 is placed in envelope number 1, then it is dearrangement of 3 objects.

$\Rightarrow \quad 3!\left(1 - \frac{1}{1!} + \frac{1}{2!} - \frac{1}{3!}\right) = 2$ ways

$\therefore$ Total ways $= n = 9 + 2 = 11$

Hence, $\frac{n}{11} = \frac{11}{11} = 1.$

24. (d) Required number is the coefficient of x^{10} in

$(1 + x + x^2 + ... + x^6)^3$

$= (1 - x^7)^3 (1 - x)^{-3} = (1 - 3x^7 + ...)$

$(1 + {}^3C_1 x + {}^4C_2 x^2 ...)$

Hence, coefficient of x^{10} is

${}^{12}C_{10} - 3({}^5C_3) = 36$

25. (b,c,d)

$p = 21(21+1)(21-1)(21+2)(21-2)$
$...(21+10)(21-10)$

$= (21-20)(21-9)...(21-1)21(21+1)$
$...(21+10)$

$= 11 \times 12 \times ... 30 \times 31$

which is divisible by 21! and hence by 20! and 19!

26. (a,b,c,d)

$E_2(102!) = \left[\frac{102}{2}\right] + \left[\frac{102}{2^2}\right] + \left[\frac{102}{2^3}\right] + \left[\frac{102}{2^4}\right] + \left[\frac{102}{2^5}\right] + \left[\frac{102}{2^6}\right]$

$= 51 + 25 + 12 + 6 + 3 + 1$

$E_2(102!) = 98$

$E_3(103!) = \left[\frac{102}{3}\right] + \left[\frac{102}{3^2}\right] + \left[\frac{102}{3^3}\right] + \left[\frac{102}{3^4}\right]$

$= 34 + 11 + 3 + 1 = 49$

Similarly, we get $E_5(102!) = 24$ and $E_7(102)! = 16$

$\therefore 102! = 2^{98} \times 3^{49} \times 5^{24} \times 7^{16} ...$

$\alpha = 98, \beta = 49, \gamma = 24, \delta = 16$

27. (a,b) There are two possibilities for a_n, either append a non-zero digit to a legal $(n-1)$ string or append a zero to an illegal $(n-1)$ string.

$\Rightarrow a_n = 9a_{n-1} + (10^{n-1} - a_{n-1})$

$\Rightarrow a_n = 8a_{n-1} + 10^{n-1}$

$\Rightarrow a_n = 8(8a_{n-1} + 10^{n-2}) + 10^{n-1}$

$\Rightarrow a_n = 8^2 a_{n-2} + 8 \times 10^{n-2} + 10^{n-1}$

$\Rightarrow a_n = 8^2(8a_{n-3} + 10^{n-3}) + 8 \times 10^{n-2} + 10^{n-1}$

$\Rightarrow a_n = 8^3 a_{n-3} + 8^2 \times 10^{n-3} + 8 \times 10^{n-2} + 10^{n-1}$

$\Rightarrow a_n = 8^{n-1} a_{n-(n-1)} + [10^{n-1} + 8 \times 10^{n-2} + ...(n-1)$ terms]

$\Rightarrow a_n = 8^{n-1} a_1 + \frac{10^{n-1}\left(\left(\frac{8}{10}\right)^{n-1} - 1\right)}{\frac{8}{10} - 1}$

$\Rightarrow a_n = 8^{n-1} \times 9 + \frac{10(10^{n-1} - 8^{n-1})}{2} \quad [\because a_1 = 9]$

$\Rightarrow a_n = 9 \times 8^{n-1} + \frac{10^n - 10 \times 8^{n-1}}{2}$

$\Rightarrow a_n = \frac{10^n + 8^n}{2}$

28. (a,c)

${}^{n+5}P_{n+1} = \frac{11(n-1)}{2} \times {}^{n+3}P_n$

or ${}^{n+5}P_{n+1} = \frac{(n+5)!}{4!} = \frac{11(n-1)(n+3)!}{2 \cdot 3!}$

or $(n+5)(n+4) = 22(n-1)$

After solving, we get $n = 6$ or $n = 7$

The number of points of intersection of lines is ${}^6C_2$ or ${}^7C_2 = 15$ or 21.

29. (a,d)

Problem is same as dividing 17 identical things in two groups.

$\therefore \quad n = \frac{17+1}{2} = 9$

There is not effect if two diamonds are different as necklace can be flipped over.

Hence, $n = m = 9$

30. (a,b,d)

$p = {}^5C_4 \times {}^2C_1 = 10$

$q = {}^5C_2 \times ({}^2C_1)^3 = 80$

$r = {}^5C_0 \times ({}^2C_1)^5 = 32$

$\Rightarrow \quad 2q = 5r, 8p = q$ and $2(p + r) > q$

31. (b,c)

The number of ways of inviting with the couple not included, is ${}^8C_5$. The number of ways of inviting with the couple included, is ${}^8C_3$.

Therefore, the required number of ways is

$${}^8C_5 + {}^8C_3 = {}^8C_3 + {}^8C_3 \quad [\because {}^8C_5 = {}^8C_3]$$

Also, $${}^{10}C_5 - 2\times {}^8C_4 = \frac{10}{5}\cdot {}^9C_4 - 2\times {}^8C_4$$
$$= 2({}^9C_4 - {}^8C_4)$$
$$= 2({}^8C_3 + {}^8C_4 - {}^8C_4)$$
$$= 2\times {}^8C_3$$

32. (a,b,d)

Clearly, each player will play 9 games and total number of games is ${}^{10}C_2 = 45$

Clearly,

$$w_i + l_i = 9 \text{ and } \Sigma w_i = \Sigma l_i = 45$$
$$\Rightarrow \quad w_i = 9 - l_i \Rightarrow w_i^2 = 81 + l_i^2 - 18 l_i$$
$$\Rightarrow \quad \Sigma {w_i}^2 = 81\times 10 + \Sigma {l_i}^2 - 18\times 45 = \Sigma {l_i}^2$$

33. (a,b,c)

When $n = 3k$, there are exactly n/3 integers of each type $3p, 3p+1$ and $3p+2$.

Now, sum of three selected integers are of the same type $3p, 3p+1$ or $3p+2$ or one- one integer from each type. Then, number of selection ways is

$${}^{n/3}C_3 + {}^{n/3}C_3 + {}^{n/3}C_3 + ({}^{n/3}C_1)({}^{n/3}C_1)({}^{n/3}C_1)$$
$$= 3({}^{n/3}C_3) + (n/3)^3$$

If $n = 3k+1$, then there are $(n-1)/3$ integers of the type $3p$, $3p+2$ and $(n+2)/3$ integers of the type $3p+1$. Then, number of selection ways is

$$2({}^{(n-1)/3}C_3) + ({}^{(n+2)/3}C_3) + ((n-1)/3)^2\{(n+2)/3\}$$

When $n = 3K+3$, the number of selection ways are same as in the case of $n = 3k+1$

34. (b,d)

$$\because \quad p = 2520 = 2^3\times 3^2\times 5^1\times 7^1$$
$$= 6\times 2^2\times 3^1\times 5^1\times 7^1 = 9\times 2^3\times 5^1\times 7^1$$
$$\therefore \quad x = (2+1)(1+1)(1+1)(1+1) = 24$$
$$y = (3+1)(1+1)(1+1) = 16$$

35. (a,b,c)

The number of ways $= \dfrac{(200)!}{2^{100}\,(100)!}$

Also, $$\frac{(200)!}{\underbrace{2!2!\ldots 2!}_{100 \text{ times}}\,(100)!} = \frac{(200)!}{100!\,2^{100}} = 1\times 3\times 5\ldots 199$$

Also, $$\frac{(200)!}{100!2^{100}} = \frac{(100)!(101)(102)(103)\ldots(200)}{2^{100}\cdot(100)!}$$
$$= \left(\frac{101}{2}\right)\left(\frac{102}{2}\right)\cdots\left(\frac{200}{2}\right)$$

36. (a,c)

Let person P_i get x_i number of things such that

$$x_1 + x_2 + x_3 + x_4 + x_5 + x_6 = 30$$

If x_i is odd or $x_i = 2\lambda_i + 1$, where $\lambda_i \ge 0$, then

$$2(\lambda_1 + \lambda_2 + \lambda_3 + \lambda_4 + \lambda_5 + \lambda_6) + 6 = 30$$
$$\Rightarrow \quad \lambda_1 + \lambda_2 + \lambda_3 + \lambda_4 + \lambda_5 + \lambda_6 = 12$$

Then, number of solutions is ${}^{12+6-1}C_{6-1} = {}^{17}C_5$

If x_i is even or $x_i = 2\lambda_i$, where $\lambda_i \ge 1$, then

$$2(\lambda_1 + \lambda_2 + \lambda_3 + \lambda_4 + \lambda_5 + \lambda_6) = 30$$
$$\text{or } \lambda_1 + \lambda_2 + \lambda_3 + \lambda_4 + \lambda_5 + \lambda_6 = 15$$

Therefore, required number of ways is ${}^{15-1}C_{6-1} = {}^{14}C_5$

37. (b,c,d)

When $z = n+1$, we can choose x and y from $\{1,2,\ldots\ldots\ldots\ldots,n\}$

When $z = n+1$, x and y can be chosen in n^2 ways and when $z = n$, x and y can be chosen in $(n-1)^2$ ways and so on.

Therefore, the number of ways of choosing triplets is

$$n^2 + (n-1)^2 + \ldots\ldots\ldots + 1^2 = \frac{1}{6}n(n+1)(2n+1)$$

Alternatively triplets with $x = y < z$, $x < y < z$ and $y < x < z$ can be chosen in ${}^{n+1}C_2$, ${}^{n+1}C_3$, ${}^{n+1}C_3$ ways

Therefore, $${}^{n+1}C_2 + 2({}^{n+1}C_3) = {}^{n+2}C_2 + {}^{n+1}C_3$$
$$= 2({}^{n+2}C_3) - {}^{n+1}C_2$$

38. (a,b, d)

Given that

$$f(m,n,p) = \sum_{i=0}^{p}\binom{m}{i}\frac{(n+i)!}{(p)!(n+i-p)!}\times\frac{(p+n)!}{(n+i)!(p-i)!}$$
$$= \frac{(n+p)!}{(p)!}\sum_{i=0}^{p}\binom{m}{i}\frac{1}{(p-i)!\{n-(p-i)\}!}$$
$$= \frac{(n+p)!}{n!p!}\sum_{i=0}^{p}{}^mC_i\times{}^nC_{p-i}$$
$$= \frac{(n+p)!}{n!p!}\times{}^{m+n}C_p \quad \left[\because \sum_{i=0}^{p}{}^mC_i\cdot{}^nC_{p-i} = {}^{m+n}C_p\right]$$
$$= {}^{n+p}C_p\times{}^{m+n}C_p$$
$$\because g(m,n) = \sum_{p=0}^{m+n}\frac{f(m,n,p)}{{}^{n+p}C_p} = \sum_{p=0}^{m+n}{}^{m+n}C_p = 2^{m+n}$$

39. (a,c)

Let $x_i (1 \le i \le n)$ be the number of objects, selected of the ith type

Since each object is to be selected atleast once, we must have $x_i \ge 1$ and $x_1 + x_2 + \dots x_n = r$

We have to find the number of positive integral solutions of the above equation.

$\therefore$ Total number of such solutions is ${}^{r-1}C_{n-1} = {}^{r-1}C_{r-n}$

40. (a,d)

Let $A = \{a_1, a_2, \dots\dots a_n\}$ for each $a_i (1 \le i \le n)$, we have either $a_i \in p_j$ of $a_i \notin p_j (1 \le j \le m)$. i.e. there are 2^m choices in which $a_i (1 \le i \le n)$ may belong to the p_j's.

One of these, there is only one choice in which $a_i \in p_j$ $\forall j = 1, 2, \dots$ which is not favourable for $p_1 \cap p_2 \cap \dots \cap p_m$ to be ϕ.

Thus, $a_1 \in p_1, \cap p_2 \cap \dots\dots \cap p_m$ in $2^m - 1$ ways.

Since, there are n elements in set A, the total number of choices is $(2^m - 1)^n$

Also, there is exactly one choice in which $a_i \notin p_j$, $\forall_j = 1, 2, \dots, m$ which is not favourable for $p_1 \cup p_2 \cup \dots \cup p_m$ to be equal to A.

Thus, a_i can belong to $p_1 \cup p_2 \cup \dots \cup p_m$ in $(2^m - 1)$ ways.

Since there are n elements in set A, the number of ways in which $p_1 \cup p_2 \cup \dots \cup p_m$ can be equal to A, is $(2^m - 1)^n$.

41. (a,b,c,d)

$\because x + y + z = 3n$

$\Rightarrow N =$ Coefficient of α^{3n} in $(1 + \alpha + \alpha^2 \dots\dots + \alpha^{2n})^3$

$\Rightarrow N =$ Coefficient of α^{3n} in $(1 - \alpha^{2n+1})^3 (1 - \alpha)^{-3}$

$\Rightarrow N =$ Coefficient of α^{3n} in $(1 - 3\alpha^{2n+1})(1 + {}^3C_1\alpha + \dots..)$

$= {}^{3n+2}C_2 - 3 \cdot {}^{n+1}C_2$

$= 3n^2 + 3n + 1$

$\therefore N - 1 = 3n(n + 1)$

Sol. (Q.Nos. 42-43)

42. (c) |U| CC |E|

There are four gap to place SSS.

Hence, required no of ways $= {}^4C_3 \times 3! = 24$

43. (b) |U| C | C |E|

There are five available places for SSS.

$\therefore$ Total number of ways no two S's are together

$$= {}^5C_3 \times \frac{4!}{2!} = 120$$

Hence, number of words having CC separated and SSS separated $= 120 - 24 = 96$

Sol. (Q.Nos. 44-46)

44. (d) Since, there are 5 even places and 3 pairs of repeated letters, therefore atleast one of these must be at an odd place.

Therefore, the number of ways is 11!/ (2!2!2!)

45. (b) Make a group of both M's and another group of T's, then except A's, we have 5 letter remaining.

So, M's, T's and the letters except A's can be arranged in 7! ways. Therefore, total number of arrangements is $7! \times {}^8C_2$.

46. (c) Consonants can be placed in 7!/(2!2!) ways, then there are 8 places and 4 vowels.

Therefore, number of ways is $\frac{7!}{2!2!} \cdot {}^8C_4 \cdot \frac{4!}{2!}$.

Sol. (Q.Nos. 47-49)

47. (c) Seven persons can be selected for the first table in ${}^{12}C_7$ ways.

Now, these seven persons can be arranged in 6! ways, the remaining five persons can be arranged on the second table in 4! ways.

Hence, total number of ways is ${}^{12}C_5 6!4!$.

48. (b) Here, A can sit on first table and B on the second or A on the second table and B on the first table. If A is on the first table, then remaining six for first table can be selected in ${}^{10}C_6$ ways.

Now, these seven persons can be arranged in 6! ways, remaining five can be arranged on the other table in 4! ways.

Hence, total number of ways is $2\,{}^{10}C_6 6!4!$.

49. (a) If A and B are on the first table, then remaining five can be selected in ${}^{10}C_5$ ways.

Now, seven persons including A and B can be arranged on the first table in which A and B are together in 2!5! ways, remaining five can be arranged in the second table in 4! ways.

Total number of ways is ${}^{10}C_5 4!5!2!$

If A and B are on the second table, then remaining three can be selected in ${}^{10}C_3$ ways.

Now, five persons including A and B can be arranged on the second table in which A and B are together in 2!3! ways, remaining seven can be arranged on the first table in 6! ways.

Hence, total number of ways is ${}^{10}C_7 6!3!2!$.

50. (i) $\rightarrow$ A, (ii) $\rightarrow$ D, (iii) $\rightarrow$ B , (iv) $\rightarrow$ B

(i) For the permutations, containing the word 'ENDEA' We consider 'ENDEA' as single letter. Then, we have total 'ENDEA' : N, O, E, L

i.e. 5 letter which can be arranged in 5! ways

(ii) If E occupies the first and last positions, then the middle 7 positions can be filled by N, D, E, A, N, O, I in $\frac{7!}{2!} = 7\times6\times5\times4\times3 = 21\times120 = 21\times5!$ ways

(iii) If none of the letters D, L, N occur in the last 5 positions then we should arrange D, L, N, N at first four positions and rest five i.e. E, E, E, A, O at last five positions. This can be done in

$\frac{4!}{2!}\times\frac{5!}{3!} = 4\times3\times\frac{5!}{3\times2} = 2\times5!$ ways

(iv) As per question, A, E, E, E, O can be arranged at 1st, 3rd, 5th, 7th and 9th positions and rest D, L, N, N at rest 4 positions. This can be done in

$\frac{5!}{3!}\times\frac{4!}{2!} = 2\times5!$ ways

51. (i) → B, D, (ii) → A, C, (iii) → A, C, (iv) → A, C

(i) There are two cases

(i) 5, 4, 1, 1, 1

Number of ways = 5!/3! = 20

5, 2, 2, 1, 1

Number of ways = 5!/2!2! = 30

Hence, total number is 20 + 30 = 50

(ii) Select 4 pairs in $^5C_4 = 5$ ways

Now, select exactly one shoe from each of the pairs selected in $(^2C_1)^4$ ways. This will fulfill the condition.

Hence, required answer is $5\times16 = 80$

(iii) The first child C_1 can be chosen in 3 ways, his/her mother can be interviewed in 5 ways. The second child C_2 can be chosen in 2 ways and his/her mother can be interviewed in 3 ways.

Hence, total number of ways is $3\times5\times2\times3 = 90$

(iv) Required number of ways is

$5! - 4! - 3! = 120 - 24 - 6 = 90$

(number will be less than 56000 only if either 4 occurs on the first place or 5, 4 occurs on the first two places)

52. (i) → D, (ii) → F, (iii) → E, (iv) → B

Given 6 boys $M_1, M_2, M_3, M_4, M_5, M_6$ and 5 girls G_1, G_2, G_3, G_4, G_5

(i) α_1 = Total number of ways of selecting 3 boys and 2 girls from 6 boys and 5 girls

i.e. $^6C_3\times{}^5C_2 = 20\times10 = 200$

$\therefore \quad \alpha_1 = 200$

(ii) α_2 = Total number of ways selecting atleast 2 members having equal number of boys and girls

i.e. $^6C_1{}^5C_1 + {}^6C_2{}^5C_2 + {}^6C_3{}^5C_3 + {}^6C_4{}^5C_4 + {}^6C_5{}^5C_5$

$= 30 + 150 + 200 + 75 + 6 = 461$

$\therefore \quad \alpha_2 = 461$

(iii) α_3 = Total number of ways of selecting 5 members in which atleast 2 of them are girls

i.e. $^5C_2{}^6C_3 + {}^5C_3{}^6C_2 + {}^5C_4{}^6C_1 + {}^5C_5{}^6C_0$

$= 200 + 150 + 30 + 1 = 381$

$\Rightarrow \quad \alpha_3 = 381$

(iv) α_4 = Total number of ways for selecting 4 members having atleast 2 girls such that M_1 and G_1 are not included together

G_1 is included → $^4C_1.{}^5C_2 + {}^4C_2.{}^5C_1 + {}^4C_3$

$= 40 + 30 + 4 = 74$

M_1 is included → $^4C_2.{}^5C_1 + {}^4C_3 = 30 + 4 = 34$

G_1 and M_1 both are not included

$^4C_4 + {}^4C_3\cdot{}^5C_1 + {}^4C_2\cdot{}^5C_2$

$1 + 20 + 60 = 81$

$\therefore$ Total number = 74 + 34 + 81 = 189

$\alpha_4 = 189$

53. (i) → C, (ii) → D, (iii) → A, (iv) → B

(i) $^{10}C_2 - {}^4C_2 + 1 = 45 - 6 + 1 = 40$

(ii) $1\times{}^{10}C_2 = 45$

(iii) $2\times{}^6C_2 = 30$

(iv) $^6C_2\times4 = 60$

54. (i) → A, D, (ii) → B, C, (iii) → A, D, (iv) → C

(i) Total number of required functions is equal to number of dearrangement of 5 objects, which is given by

$$5!\left(\frac{1}{2!} - \frac{1}{3!} + \frac{1}{4!} - \frac{1}{5!}\right) = 44$$

(ii) $x_1x_2x_3 = 2\times35\times7$

$= 2\times49\times5 = 10\times7\times7$

$= 14\times7\times5$

So, total number of solution set is

$3\times3! + 3!/2! = 21$

(iii) $3780 = 2^2\times3^2\times5\times7$

Number of divisors which are divisible by 2 but not by 3 is $2\times2\times2 = 8$

Number of divisors which are divisible by 3 but not 2 is $3\times2\times2 = 12$

Number of divisors which are divisible by 2 as well as 3 is $2\times3\times2\times2 = 24$

Hence, total number of divisors is 44.

(iv) $4\lambda + 2 = 2(2\lambda + 1)$ = odd multiple of 2.

Thus, total number of divisors is $1\times5\times11 - 1 = 54$

(1 is subtracted and powers of three and five are zero each and this will make $\lambda = 0$)

55. (i) → B, (ii) → C, (iii) → D, (iv) → A

(i) Number of rectangles = $^7C_2\times{}^7C_2 = 441$

(select two vertical and two horizontal lines)

(ii) Number of squares

$= 1^2 + 2^2 + 3^2 + 4^2 + 5^2 + 6^2 = 91$

(iii) First square can be selected in 36 ways, second square can be selected in $(36-6-5)=25$ ways and third square can be selected in $(25-5-4)=16$ ways

Number of ways of selection $=36\times25\times16$

But in this one-by-one type of selection order of selection is also considered.

Hence, actual number of ways is

$$\frac{(36\times25\times16)}{3!}=2400$$

(iv) $a_1+a_2+a_3+a_4+a_5+a_6=11$

where $a_1,a_2,a_3,a_4,a_5,a_6\geq1$

$\therefore$ Required ways $={}^{11-1}C_{6-1}={}^{10}C_5$

Sol. (Q.Nos. 56-58)

56. (25) When balls are different and boxes are identical, number of distribution is equal to number of divisors in (1,1,3) or (1,2,2) ways.

Hence, total number of ways is

$$\frac{5!}{(1!)^2 3!2!}+\frac{5!}{1!(2!)^2 2!}=25$$

57. (2) When balls as well as boxes are identical, we have only two ways (1,1,3) and (1,2,2).

Hence, number of ways is 2.

58. (6) When boxes are kept in a row, they will be treated as different.

In this case, the number of ways will be

$${}^{5-1}C_{3-1}={}^4C_2=6$$

Sol. (Q.Nos. 59-61)

59. (588) There are 9 letters T, T, R, R, R, E, O, I, S

$\lambda=$ Number of four lettered words (no restrictions)

$=$ Coefficient of x^4 in

$$4!\left(1+\frac{x}{1!}+\frac{x^2}{2!}\right)\left(1+\frac{x}{1!}+\frac{x^2}{2!}+\frac{x^3}{3!}\right)(1+x)^4$$

$=626$

$\mu=$ Number of four lettered words (no vowels)

$=$ Coefficient of x^4 in

$$4!\left(1+\frac{x}{1!}+\frac{x^2}{2!}\right)\left(1+\frac{x}{1!}+\frac{x^2}{2!}+\frac{x^3}{3!}\right)(1+x)$$

$=38$

$\therefore$ Required ways (at least one vowel)

$=626-38=588$

60. (12600) | T | E | O | I | S | T |

There are 7 available places of RRR

$$\therefore \text{Required ways}={}^7C_3\times\frac{6!}{2!}$$

$=12600$

61. (10200) |E|TT|O|I|S

Number of ways having TT together and RRR separated

$$={}^6C_3\times5!=2400$$

Hence, number of arrangements of the word W, if R's as well as T's are separated $=12600-2400=10200$

62. (625)

The last two digits, in 5-digit number divisible by 4, can be 12, 24, 32, 44 or 52.

| – | – | – | • | • |

Also, each of the first 3-digits can be any of {1, 2, 3, 4, 5}

$\therefore$ 5 options for each of the first 3 place and total 5 options for last 2 place.

$\therefore$ Required number of 5 digits $=5\times5\times5\times5$

$=625$ ways

63. (7) Possible solutions are

1, 2, 3, 4, 10

1, 2, 3, 5, 9

1, 2, 3, 6, 8

1, 2, 4, 5, 8

1, 2, 4, 6, 7

1, 3, 4, 5, 7

2, 3, 4, 5, 6

Hence, 7 solutions are there.

64. (30)

First, we will consider a face to be fixed with a specific colours and find out the relative arrangements of the rest. When we have fixed a colour, the opposite face can be painted with the rest 5 colours.

Thus, there are 5 choices.

Now the rest 4 faces we can calculate using the circular permutations.

i.e. 4 colours can be painted on the remaining 4 faces in $(4-1)!=3!$

Thus, total number of ways $=5\times3!=30$

65. (30)

5 persons *A*, *B*, *C*, *D* and *E* are seated in circular arrangement. Let A be given red hat, then the following two cases will be there

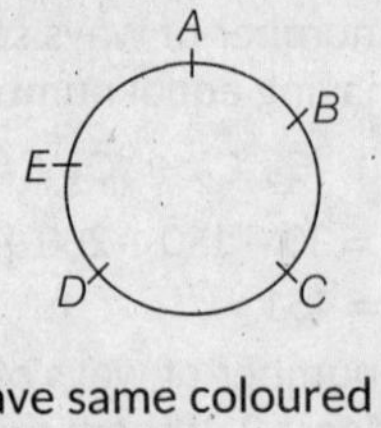

Case I *B* and *E* have same coloured hat blue/green.

Let *B* and *E* have blue hat.

Then, C and D have either red and green or green and red i.e two ways.

Similarly, if B and E have different green hat, there will be 2 ways for C to B.

$\therefore$ Total ways $= 2 + 2 = 4$ ways

Case II Let B has blue and E has green.

If C has green, then D can have red or blue

If C has red, then D can have only blue.

$\therefore$ 3 ways

Similarly, 3 ways will be there, when B has green and E has blue.

$\therefore$ Total ways $= 3 + 3 = 6$ ways

On combining two cases $= 4 + 6 = 10$ ways

Similarly, if A as blue and green hat , we get 10 ways for each

$\therefore$ In all there will be $= 10 + 10 + 10 = 30$ ways

66. (8) Here, $P_n = {}^{n-2}C_3$ and $P_{n+1} = {}^{n-1}C_3$

Hence, ${}^{n-1}C_3 - {}^{n-2}C_3 = 15$

or $\frac{(n-1)(n-2)(n-3)}{3!} - \frac{(n-1)(n-3)(n-4)}{3!} = 15$

On solving, we get

$$n^2 - 5n + 6 = 30$$

$$\Rightarrow \quad (n-8)(n+3) = 0$$

$$\Rightarrow \quad n = 8, n = -3 \text{ (negative)}$$

67. (5)

Number of red lines $= {}^nC_2 - n$

Number of blue lines $= n$

Hence, ${}^nC_2 - n = n \Rightarrow {}^nC_2 = 2n$

$$\Rightarrow \frac{n(n-1)}{2} = 2n \Rightarrow n - 1 = 4 \Rightarrow n = 5$$

68. (630)

3 women can be selected in ${}^7C_3$ ways, can be paired with 3 men in 3! ways.

Remaining 4 women can be grouped into two couples in $\frac{4!}{2!2!2!} = 3$

$\therefore$ Total $= {}^7C_3 \cdot 3! \cdot 3 = 630$

69. (5)

According to question, $x = 10!$

Now for y, we can select one letter which will be repeated twice in ${}^{10}C_1$ ways.

Remaining eight letters can be selected in ${}^9C_8$ ways.

So, $y = {}^{10}C_1 \times {}^9C_8 \times \frac{10!}{2!}$

$$\Rightarrow \frac{y}{9x} = \frac{{}^{10}C_1 \times {}^9C_8}{9 \times 2!}$$

$$= \frac{10 \times 9}{9 \times 2} = 5$$

70. (1080)

Groups can be possible in only 2, 2,1,1 ways

Number of ways of dividing persons in group

$$= \frac{6!}{(2!)^2 (2!)^2 (1!)(1!)} = \frac{6!}{(2!)^4}$$

++	++	+	+

$\therefore$ Number of ways after arranging rooms $= \frac{6!}{(2!)^4} \times 4!$

$$= 1080$$

71. (9)

$5400 = 2^3 \times 3^3 \times 5^2$

Since, divisor is divided by 6 but not by 9, divisor must be even

i.e. 2 must occur atleast once, Also, 3 must occur exactly once.

So, number of required divisor $= 3 \times 1 \times 3 = 9$

72. (495)

We know that total number of ways of selection of r days out of n days such that no two of them are consecutive $= {}^{n-r+1}C_r$

$\therefore$ Selection of 4 days out of 15 days such that no two of them are consecutive $= {}^{15-4+1}C_4 = {}^{12}C_4$

$$= \frac{12 \times 11 \times 10 \times 9}{4 \times 3 \times 2}$$

$$= 11 \times 5 \times 9$$

$$= 495$$

CHAPTER

06

BINOMIAL THEOREM & MATHEMATICAL INDUCTION

LEVEL 01 \ Starter Level

6.1 Binomial Expansion for Positive Index

1 Expand and simplify the expansion of the following

(i) $(x+a)^5$ (ii) $\left(2x-\frac{3}{y}\right)^5$

2 Write the expansion of $(1+x)^7$.

3 Find the 7th term of the following

(i) $\left(\frac{4x}{5}-\frac{5}{2x}\right)^9$ (ii) $\left(4x-\frac{1}{2\sqrt{x}}\right)^{13}$

4 Find the 25th term of $(10x+5y)^{30}$.

5 Find the coefficient of the following

(i) x^7 in the expansion of $\left(ax^2+\frac{1}{bx}\right)^{11}$.

(ii) x^{-7} in the expansion of $\left(ax-\frac{1}{bx^2}\right)^{11}$.

6 Simplify $\left(x+\sqrt{x^2-1}\right)^6+\left(x-\sqrt{x^2-1}\right)^6$.

7 Find the 4th term from the end in the expansion of $\left(\frac{x^3}{2}-\frac{2}{x^2}\right)^7$.

8 Find the term independent of x in the expansion of $\left(x^2+\frac{1}{x}\right)^9$.

9 Find the sum of all rational terms in the expansion of $(3^{1/5}+2^{1/3})^{15}$.

10 Find the number of irrational terms in the expansion of $(\sqrt[8]{3}+\sqrt[6]{2})^{100}$.

11 Find the middle term in the expansion of

(i) $\left(\frac{a}{x}+bx\right)^{12}$ (ii) $\left(3x-\frac{x^3}{6}\right)^9$.

12 Find numerically the greatest term in the expansion of $(2+3x)^9$, when $x=\frac{3}{2}$.

13 The coefficients of three consecutive terms in the expansion of $(1+a)^n$ are in the ratio 1:7:42. Find n.

6.2 Applications of Binomial Theorem

14 The integral part of $(5+2\sqrt{6})^n$ is odd? Justify your answer.

15 If $x+t=(7+4\sqrt{3})^n$, where n and x are positive integers and t is a proper fraction, then find the value of $(1-t)(x+t)$.

16 Find the approximation of $(0.99)^5$, using the first three terms of its expansion.

17 Which is the large number $99^{50}+100^{50}$ or 101^{50}? Justify your answer.

18 If 7^{103} is divided by 25, then find the remainder.

19 Find the last two digits of 3^{400}.

6.3 Properties of Binomial Coefficients

20 If $a_n=\sum_{r=0}^{n}\frac{1}{{}^nC_r}$, find the value of $\sum_{r=0}^{n}\frac{r}{{}^nC_r}$.

21 Find the value of $\sum_{r=1}^{n}\frac{r\,.\,{}^nC_r}{{}^nC_{r-1}}$.

22 Evaluate $\sum_{r=0}^{n}{}^{n+r}C_n$.

23 Find the value of $\frac{1}{(n-1)!}+\frac{1}{(n-3)!\,3!}+\frac{1}{(n-5)!\,5!}+\ldots$

24 If $(1+x)^n=C_0+C_1x+C_2x^2+\ldots$, then find the values of

(i) $C_0-C_2+C_4-C_6+\ldots$ (ii) $C_1-C_3+C_5-C_7+\ldots$

(iii) $C_0+C_3+C_6+\ldots$

25 If $(1+x)^n = C_0 + C_1x + C_2x^2 + \ldots + C_n x^n$, then find the values of

(i) $C_1 + 2C_2 + 3C_3 + \ldots + nC_n$ (ii) $C_0 + 2C_1 + 3C_2 + \ldots + (n+1)C_n$

26 If $(1+x)^n = C_0 + C_1x + C_2x^2 + \ldots + C_nx^n$, then find the values of

(i) $C_0 + \frac{C_1}{2} + \frac{C_2}{3} + \ldots \frac{C_n}{n+1}$ (ii) $C_0 - \frac{C_1}{2} + \frac{C_2}{3} + \ldots + (-1)^n \frac{C_n}{n+1}$

27 If $(1+x)^n = C_0 + C_1x + C_2x^2 + \ldots + C_nx^n$, then find the value of

$C_0 \cdot {}^{2n}C_n - C_1 \cdot {}^{2n-2}C_n + C_2 \cdot {}^{2n-4}C_n \ldots$

28 If $(1+x)^n = C_0 + C_1x + C_2x^2 + \ldots C_nx^n$, then find the value of $\sum_{i=0}^{n} \sum_{j=0}^{n} (C_i + C_j)$.

6.4 Binomial Theorem for Negative/Rational Index

29 Find the values of x for which $\frac{1}{\sqrt{5+4x}}$ can be expanded as an infinite series.

30 If $|x| < 1$, then find the expansion of $(1-x)^{1/2}$.

31 Find the 4th term in the expansion of $(1-2x)^{3/2}$.

32 Expand $\frac{1}{(3+2x)^2}$ to three terms

(i) in ascending powers of x (ii) in descending powers of x

33 Find the coefficient of x^5 in the expansion of $\frac{1+x}{1-x}$, if $|x| < 1$.

34 Find the sum of series $1 + \frac{1}{3} + \frac{1 \cdot 3}{3 \cdot 6} + \frac{1 \cdot 3 \cdot 5}{3 \cdot 6 \cdot 9} + \ldots \infty$.

6.5 Multinomial Theorem and Sum of Coefficients

35 Find the coefficient of x^7 in the expansion of $(1+3x-2x^3)^{10}$.

36 Find the coefficient of x^8 in the expansion of $(1+x^2-x^3)^9$.

37 Find the coefficient of $a^3b^4c^7$ in the expansion of $(bc+ca+ab)^8$.

38 Find the sum of rational terms in $(\sqrt{2} + \sqrt[3]{3} + \sqrt[6]{5})^{10}$.

39 Find the total number of distinct or dissimilar terms in the expansion of $(x+y+z+w)^n$, $n \in N$.

40 Find the sum of coefficients in the expansion of $(1-2x+3x^2)^{50}$.

41 If the sum of the coefficients in the expansion of $(\alpha^2x^2 - 3\alpha x + 2)^{51}$ is zero, then find the value of α.

42 If the sum of the coefficient in the expansion of $(\alpha x^2 - 2x + 1)^{35}$ is equal to the sum of the coefficients in the expansion of $(x - \alpha y)^{35}$, then find the value of α.

6.6 Mathematical Induction

43 Use mathematical induction to verify that $\sum_{r=0}^{n} r \cdot {}^nC_r = n \cdot 2^{n-1}, \forall n \in N$.

44 Use the principle of mathematical induction to verify that $5^{2n+1} + 3^{n+2} \cdot 2^{n-1}$ is divisible by 19 for all natural numbers n.

45 Use mathematical induction to verify that

$$\tan^{-1}\left(\frac{1}{3}\right) + \tan^{-1}\left(\frac{1}{7}\right) + \ldots + \tan^{-1}\left(\frac{1}{n^2+n+1}\right) = \tan^{-1}\left(\frac{n}{n+2}\right)$$

46 Let $I_m = \int_0^{\pi} \left(\frac{1-\cos mx}{1-\cos x}\right) dx$. Use mathematical induction to verify that $I_m = m\pi, m = 1, 2, \ldots$

LEVEL 02 \ JEE Main Level

Single Choice Correct Questions

6.1 Binomial Expansion for Positive Index

1 If $(1+ax)^n = 1 + 8x + 24x^2 + \ldots$, then the values of a and n are

(a) 2 and 4 (b) 2 and 3 (c) 3 and 6 (d) 1 and 2

2 If the sum of odd numbered terms and the sum of even numbered terms in the expansion of $(x+a)^n$ are A and B respectively, then the value of $(x^2 - a^2)^n$ is

(a) $A^2 - B^2$ (b) $A^2 + B^2$ (c) $4AB$ (d) None of these

3 Let $(x+10)^{50} + (x-10)^{50} = a_0 + a_1 x + a_2 x^2 + \ldots + a_{50} x^{50}$, for all $x \in R$, then $\frac{a_2}{a_0}$ is equal to **[JEE Main 2019]**

(a) 12.50 (b) 12.00 (c) 12.25 (d) 12.75

4 If α and β be the coefficients of x^4 and x^2, respectively in the expansion of $\left(x+\sqrt{x^2-1}\right)^6 + \left(x-\sqrt{x^2-1}\right)^6$, then **[JEE Main 2020]**

(a) $\alpha + \beta = 60$ (b) $\alpha - \beta = 60$ (c) $\alpha + \beta = -30$ (d) $\alpha - \beta = -132$

5 The expression $\left(\sqrt{2x^2+1}+\sqrt{2x^2-1}\right)^6 + \left(\frac{2}{\sqrt{2x^2+1}+\sqrt{2x^2-1}}\right)^6$ is a polynomial of degree

(a) 6 (b) 8 (c) 10 (d) 12

6 If $(1+x)^5 = \alpha_0 + \alpha_1 x + \alpha_2 x^2 + \alpha_3 x^3 + \alpha_4 x^4 + \alpha_5 x^5$, then the value of $(\alpha_0 - \alpha_2 + \alpha_4)^2 + (\alpha_1 - \alpha_3 + \alpha_5)^2$ is equal to

(a) 243 (b) 32 (c) 1 (d) 2^{10}

7 If the 7th term in the binomial expansion of $\left(\frac{3}{\sqrt[3]{84}} + \sqrt{3}\ln x\right)^9$, $x > 0$ is equal to 729, then x can be

(a) e^2 (b) e (c) $\frac{e}{2}$ (d) $2e$

8 If the 3rd term in the binomial expansion of $(1+x^{\log_2 x})^5$ equals 2560, then the possible value of x is [JEE Main 2019]

(a) $\frac{1}{4}$ (b) $4\sqrt{2}$ (c) $\frac{1}{8}$ (d) $2\sqrt{2}$

9 If the coefficient of 5th, 6th and 7th terms in the expansion of $(1+x)^n$ are in AP, then n is equal to

(a) 7 only (b) 14 only (c) 7 or 14 (d) None of these

10 If in the expansion of $(a-2b)^n$, the sum of 5th and 6th terms is 0,then the value of $\frac{a}{b}$ is equal to

(a) $\frac{n-4}{5}$ (b) $\frac{2(n-4)}{5}$ (c) $\frac{5}{n-4}$ (d) $\frac{5}{2(n-4)}$

11 If the coefficient of rth and $(r+1)$th terms in the expansion of $(3+7x)^{29}$ are equal, then r is equal to

(a) 15 (b) 21 (c) 14 (d) None of these

12 If the 6th term in the expansion of $\left(\frac{1}{x^{8/3}}+x^2\log_{10}x\right)^8$ is 5600, then x is equal to

(a) 1 (b) $\log_e 10$ (c) 10 (d) x does not exist

13 If the 4th term in the binomial expansion of $\left(\frac{2}{x}+x^{\log_8 x}\right)^6$ $(x>0)$ is 20×8^7, then the value of x is [JEE Main 2019]

(a) 8^3 (b) 8^2 (c) 8 (d) 8^{-2}

14 The 4th term from the end in $\left(\frac{x^3}{2}-\frac{2}{x^2}\right)^9$ is

(a) $-\frac{672}{x^3}$ (b) $-\frac{64}{x^2}$ (c) $\frac{672}{x^3}$ (d) $-\frac{64}{x^3}$

15 If the ratio of the 7th term from the beginning to the 7th term from the end in $\left(\sqrt[3]{2}+\frac{1}{\sqrt[3]{3}}\right)^n$ is $\frac{1}{6}$, then n is equal to

(a) 10 (b) 9 (c) 8 (d) 12

16 A ratio of the 5th term from the beginning to the 5th term from the end in the binomial expansion of $\left[2^{1/3}+\frac{1}{2(3)^{1/3}}\right]^{10}$ is [JEE Main 2019]

(a) $1:2(6)^{1/3}$ (b) $1:4(16)^{1/3}$ (c) $4(36)^{1/3}:1$ (d) $2(36)^{1/3}$

17 If the middle term in the expansion of $\left(\frac{1}{x}+x\sin x\right)^{10}$ equals $7\frac{7}{8}$, then x is equal to

(a) $2n\pi\pm\frac{\pi}{6}$ (b) $n\pi+\frac{\pi}{6}$ (c) $n\pi+(-1)^n\frac{\pi}{6}$ (d) $n\pi+(-1)^n\frac{5\pi}{6}$

18 The coefficient of the middle term in the binomial expansion in powers of x of $(1+\alpha x)^4$ and of $(1-\alpha x)^6$ is the same, if α is equal to

(a) $-\frac{5}{3}$ (b) $\frac{10}{3}$ (c) $-\frac{3}{10}$ (d) $\frac{3}{5}$

19 If the constant term in the binomial expansion of $\left(\sqrt{x}-\frac{k}{x^2}\right)^{10}$ is 405, then $|k|$ is equal to **[JEE Main 2020]**

(a) 9 (b) 1 (c) 3 (d) 2

20 If the term independent of x in the expansion of $\left(\frac{3}{2}x^2-\frac{1}{3x}\right)^9$ is k, then $18k$ is equal to

(a) 5 (b) 9 (c) 7 (d) 11

21 The term independent of x in expansion of $\left(\frac{x+1}{x^{2/3}-x^{1/3}+1}-\frac{x-1}{x-x^{1/2}}\right)^{10}$ is

(a) 4 (b) 120 (c) 210 (d) 310

22 The term independent of x in the expansion of $\{(t^{-1}-1)x+(t^{-1}+1)^{-1}x^{-1}\}^8$ is

(a) $56\left(\frac{1-t}{1+t}\right)^3$ (b) $56\left(\frac{1+t}{1-t}\right)^3$ (c) $70\left(\frac{1-t}{1+t}\right)^4$ (d) $70\left(\frac{1+t}{1-t}\right)^4$

23 The coefficient of x^n in the expansion of $(1+x)^{2n}$ and $(1+x)^{2n-1}$ are in the ratio

(a) 1 : 2 (b) 1 :3 (c) 3 : 1 (d) 2 : 1

24 If the coefficient of x^7 in $\left[ax^2+\left(\frac{1}{bx}\right)\right]^{11}$ equal the coefficient of x^{-7} in $\left[ax-\left(\frac{1}{bx^2}\right)\right]^{11}$, then a and b satisfy the relation

(a) $a+b=1$ (b) $a-b=1$ (c) $ab=1$ (d) $\frac{a}{b}=1$

25 In the expansion of $\left(\frac{x}{\cos\theta}+\frac{1}{x\sin\theta}\right)^{16}$, if l_1 is the least value of the term independent of x, when $\frac{\pi}{8}\le\theta\le\frac{\pi}{4}$ and l_2 is the least value of the term independent of x, when $\frac{\pi}{16}\le\theta\le\frac{\pi}{8}$, then the ratio of $l_2 : l_1$ is equal to **[JEE Main 2020]**

(a) 1 : 8 (b) 16 : 1 (c) 8 : 1 (d) 1 : 16

26 The total number of irrational terms in the expansion of $(7^{1/5}-3^{1/10})^{60}$ is **[JEE Main 2019]**

(a) 55 (b) 49 (c) 48 (d) 54

27 If the number of integral terms in the expansion of $(3^{1/2}+5^{1/8})^n$ is exactly 33, then the least value of n is **[JEE Main 2020]**

(a) 264 (b) 128 (c) 256 (d) 248

28 The number of real negative terms in the binomial expansion of $(1+ix)^{4n-2}$, $n\in N$, $x>0$ is

(a) n (b) $n+1$ (c) $n-1$ (d) $2n$

29 If $x+y=1$, then $\sum_{r=0}^{n} r\cdot {}^nC_r\cdot x^r\cdot y^{n-r}$ is equal to

(a) 1 (b) n (c) nx (d) ny

30 The coefficient of x^{53} in the expansion of $\sum_{m=0}^{100} {}^{100}C_m (x-3)^{100-m} \cdot 2^m$ is

(a) $(2) \cdot {}^{100}C_{47}$ (b) ${}^{100}C_{53}$ (c) $(-1) \cdot {}^{100}C_{53}$ (d) ${}^{100}C_{100}$

31 The coefficient of x^7 in the expansion of $(1-x-x^3+x^4)^8$ is equal to

(a) -648 (b) 792 (c) -792 (d) 648

32 The coefficient of x^5 in the expansion of $(x^2-x-2)^5$ is

(a) -83 (b) -82 (c) -86 (d) -81

33 If the coefficient of x^n in $(1+x)^{101}(1-x+x^2)^{100}$ is non-zero, then n can not be of the form

(a) $3r+1$ (b) $3r$ (c) $3r+2$ (d) None of these

34 The coefficient of x^{28} in the expansion of $(1+x^3-x^6)^{30}$ is

(a) 1 (b) 0 (c) ${}^{30}C_6$ (d) ${}^{30}C_3$

35 In the expansion of $\left(1+x+\frac{7}{x}\right)^{11}$, the term not containing x is

(a) $\sum_{r=0}^{5} {}^{11}C_r \cdot {}^{11-r}C_{2r} \cdot 7^r$ (b) $\sum_{r=1}^{5} {}^{11}C_r \cdot {}^{11-r}C_{11-2r} \cdot 7^r$

(c) $\sum_{r=0}^{5} {}^{11}C_r \cdot {}^{11-r}C_{11-2r} \cdot 7^r$ (d) None of these

36 The coefficient of x^7 in the expansion of $(1-x-x^2+x^3)^6$ is

(a) 132 (b) 144 (c) -132 (d) -144

37 The coefficient of $x^2\,[0 \le r \le (n-1)]$ in the expansion of $(x+3)^{n-1}+(x+3)^{n-2}(x+2)$ $+(x+3)^{n-3}(x+2)^2+\ldots+(x+2)^{n-1}$ is

(a) ${}^nC_r(3^r-2^n)$ (b) ${}^nC_r(3^{n-r}-2^{n-r})$

(c) ${}^nC_r(3^r+2^{n-r})$ (d) None of these

38 If the coefficients of x^3 and x^4 in the expansion of $(1+ax+bx^2)(1-2x)^{18}$ in power of x are both zero, then (a, b) is equal to

(a) $\left(16, \frac{251}{3}\right)$ (b) $\left(14, \frac{251}{3}\right)$ (c) $\left(14, \frac{272}{3}\right)$ (d) $\left(16, \frac{272}{3}\right)$

39 If for some positive integer n, the coefficients of three consecutive terms in the binomial expansion of $(1+x)^{n+5}$ are in the ratio 5 : 10 : 14, then the largest coefficients in the expansion is **[JEE Main 2020]**

(a) 462 (b) 330 (c) 792 (d) 252

40 The interval in which x must lies so that numerically greatest term in the expansion of $(1-x)^{21}$ has the greatest coefficient is $a > 0$

(c) $\left[\frac{5}{6}, \frac{6}{5}\right]$ (b) $\left(\frac{5}{6}, \frac{6}{5}\right)$ (c) $\left(\frac{4}{5}, \frac{5}{4}\right]$ (d) $\left[\frac{4}{5}, \frac{5}{4}\right]$

41 The greatest term in the expansion of $(1+3x)^{54}$, where $x=\frac{1}{3}$ is

(a) 28th term (b) 25th term (c) 26th term (d) 24th term

6.2 Applications of Binomial Theorem

42 If $n > (8+3\sqrt{7})^{10}$, $n \in N$, then the least value of n is

(a) $(8+3\sqrt{7})^{10} - (8-3\sqrt{7})^{10}$ (b) $(8+3\sqrt{7})^{10} + (8-3\sqrt{7})^{10}$

(c) $(8+3\sqrt{7})^{10} - (8-3\sqrt{7})^{10} + 1$ (d) $(8+3\sqrt{7})^{10} - (8-3\sqrt{7})^{10} - 1$

43 If $R = (6\sqrt{6}+14)^{2n+1}$ and $f = R - [R]$, where $[R]$ is integer and $0 \le f < 1$, the value of $Rf = k^{2n+1}$, then the value of k is

(a) 10 (b) 20 (c) 15 (d) 30

44 If the fractional part of the number $\frac{2^{403}}{15}$ is $\frac{k}{15}$, then k is equal to

(a) 6 (b) 8 (c) 4 (d) 14

45 If $\{P\}$ denotes the fractional part of the number p, then $\left\{\frac{3^{200}}{8}\right\}$ is equal to **[JEE Main 2020]**

(a) $\frac{5}{8}$ (b) $\frac{7}{8}$ (c) $\frac{3}{8}$ (d) $\frac{1}{8}$

46 If 25^{15} is divided by 13, then the remainder is

(a) 1 (b) 4 (c) 9 (d) 12

47 If $7^9 + 9^7$ is divided by 64, then the remainder is

(a) 0 (b) 1 (c) 2 (d) 63

48 If $A = (1000)^{1000}$ and $B = (1001)^{999}$, then

(a) $A > B$ (b) $A = B$ (c) $A < B$ (d) None of these

6.3 Properties of Binomial Coefficients

49 If ${}^{20}C_1 + (2^2)\,{}^{20}C_2 + (3)^2\,{}^{20}C_3 + \ldots + (20)^2\,{}^{20}C_{20} = a\,(2^b)$, then the ordered pair (a, b) is equal to **[JEE Main 2019]**

(a) (420, 19) (b) (420, 18) (c) (380, 18) (d) (380, 19)

50 The sum of the series $2\cdot{}^{20}C_0 + 5\cdot{}^{20}C_1 + 8\cdot{}^{20}C_2 + 11\cdot{}^{20}C_3 + \ldots + 62\cdot{}^{20}C_{20}$ is equal to **[JEE Main 2019]**

(a) 2^{26} (b) 2^{25} (c) 2^{23} (d) 2^{24}

51 The value of ${}^{20}C_0 + {}^{20}C_1 + {}^{20}C_2 + {}^{20}C_3 + {}^{20}C_4 + {}^{20}C_{12} + {}^{20}C_{13} + {}^{20}C_{14} + {}^{20}C_{15}$ is

(a) $2^{19} - \frac{({}^{20}C_{10} + {}^{20}C_9)}{2}$ (b) $2^{19} - \frac{({}^{20}C_{10} + 2\times{}^{20}C_9)}{2}$

(c) $2^{19} - \frac{{}^{20}C_{10}}{2}$ (d) None of these

52 The value of $\sum_{r=0}^{11} (r)\cdot{}^{20}C_r$ is equal to

(a) $20\,(2^{18} + {}^{19}C_{10})$ (b) $10\,(2^{18} + {}^{19}C_{10})$

(c) $20\,(2^{18} + {}^{19}C_{11})$ (d) $10\,(2^{18} + {}^{19}C_{11})$

53 The value of $({}^{21}C_1 - {}^{10}C_1) + ({}^{21}C_2 - {}^{10}C_2) + ({}^{21}C_3 - {}^{10}C_3) + ({}^{21}C_4 - {}^{10}C_4) + \ldots + ({}^{21}C_{10} - {}^{10}C_{10})$ is **[JEE Main 2017]**

(a) $2^{20} - 2^{10}$ (b) $2^{21} - 2^{11}$ (c) $2^{21} - 2^{10}$ (d) $2^{20} - 2^9$

54 For $r = 0, 1, \ldots, 10$, let A_r, B_r and C_r denote respectively, the coefficients of x^r in the expansions of $(1+x)^{10}$, $(1+x)^{20}$ and $(1+x)^{30}$, then $\sum_{r=1}^{10} A_r(B_{10}B_r - C_{10}A_r)$ is equal to **[JEE Main 2010]**

(a) $B_{10} - C_{10}$ (b) $A_{10}(B_{10}^2 - C_{10}A_{10})$ (c) 0 (d) $C_{10} - B_{10}$

55 If $(1+x)^{15} = C_0 + C_1x + C_2x^2 + \ldots + C_{15}x^{15}$, then $C_2 + 2C_3 + 3C_4 + \ldots + 14C_{15}$ is equal to

(a) $14 \cdot 2^{14}$ (b) $13 \cdot 2^{14} + 1$ (c) $13 \cdot 2^{14} - 1$ (d) None of these

56 The value of ${}^{4n}C_0 + {}^{4n}C_4 + {}^{4n}C_8 + \ldots + {}^{4n}C_{4n}$, $n \in N$ is equal to

(a) $2^{4n-2} + (-1)^n 2^{2n-1}$ (b) $2^{4n-2} + 2^{2n-1}$

(c) $2^{2n-1} + (-1)^n 2^{4n-2}$ (d) $2^{4n-2} - 2^{2n-1}$

57 $[({}^nC_0 + {}^nC_3 + \ldots) - \left(\frac{1}{2}\right)({}^nC_1 + {}^nC_2 + {}^nC_4 + {}^nC_5 + \ldots)]^2 + \left(\frac{3}{4}\right)({}^nC_1 - {}^nC_2 + {}^nC_4 - {}^nC_5 + \ldots)^2$ is equal to

(a) 3 (b) 4 (c) 2 (d) 1

58 The value of $\sum_{r=1}^{n+1}\left(\sum_{k=1}^{n} {}^kC_{r-1}\right)$ (where $r, k, n \in N$) is

(a) $2^{n+1} - 2$ (b) $2^{n+1} - 1$ (c) 2^{n+1} (d) None of these

6.4 Binomial Theorem for Negative/Rational Index

59 The value of the expression $1 - \frac{n}{1} \cdot \frac{1+x}{1+nx} + \frac{n(n-1)}{1 \cdot 2} \frac{1+2x}{(1+nx)^2} - \frac{n(n-1)(n-2)}{1 \cdot 2 \cdot 3} \cdot \frac{1+3x}{(1+nx)^3} + \ldots$ is

(a) 2 (b) 1 (c) 3 (d) 0

60 If b is very small as compared to the value of a, so that the cube and other higher powers of $\frac{b}{a}$ can be neglected in the identity $\frac{1}{a-b} + \frac{1}{a-2b} + \frac{1}{a-3b} + \ldots + \frac{1}{a-nb} = \alpha n + \beta n^2 + \gamma n^3$, then the value of γ is **[JEE Main 2021]**

(a) $\frac{a^2+b}{3a^3}$ (b) $\frac{a+b}{3a^2}$ (c) $\frac{b^2}{3a^3}$ (d) $\frac{a+b^2}{3a^3}$

61 If x is positi.e, the first negative term in the expansion of $(1+x)^{27/5}$ is

(a) 7th term (b) 5th term (c) 8th term (d) 6th term

62 The term independent of a in the expansion of $\left(1 + \sqrt{a} + \frac{1}{\sqrt{a}-1}\right)^{-30}$ is

(a) ${}^{30}C_{20}$ (b) 0 (c) ${}^{30}C_{10}$ (d) None of these

2.5 Multinomial Theorem and Sum of Coefficients

63 The coefficient of x^2y^3 in the expansion of $(1-x+y)^{20}$ is

(a) $\frac{20!}{2!\,3!}$ (b) $-\frac{20!}{2!\,3!}$ (c) $\frac{20!}{5!\,2!\,3!}$ (d) $\frac{20!}{15!\,2!\,3!}$

64 If the coefficient of $a^2b^3c^4$ in $(a+b+c)^m$ (where $m \in N$) is L ($L \neq 0$), then in same expansion coefficient of $a^4b^4c^1$ will be

(a) L (b) $\frac{L}{3}$ (c) $\frac{mL}{4}$ (d) $\frac{L}{2}$

65 The number of dissimilar terms in the expansion of $(a+b)^n$ is $n+1$, therefore number of dissimilar terms in the expansion of $(a+b+c)^{12}$ is

(a) 13 (b) 39 (c) 78 (d) 91

66 If the number of terms in the expansion of $(x-2y+3z)^n$ is 45, then n is equal to

(a) 7 (b) 8 (c) 9 (d) 6^{10}

67 Sum of coefficients in the expansion of $(x+2y+3z)^{10}$ is

(a) 2^{10} (b) 3^{10} (c) 1 (d) 6^{10}

68 Sum of the coefficients of terms of degree 13 in the expansion of $(1+x)^{11}(1+y^2-z)^{10}$ is

(a) ${}^{10}C_3$ (b) ${}^{10}C_4$ (c) ${}^{11}C_3$ (d) ${}^{11}C_4$

2.6 Mathematical Induction

69 Consider the statement '$P(n): n^2-n+41$ is prime'. Then, which one of the following is true?

(a) Both $P(3)$ and $P(5)$ are true (b) $P(3)$ is false but $P(5)$ is true
(c) Both $P(3)$ and $P(5)$ are false (d) $P(5)$ is false but $P(3)$ is true

70 Let $S(k)=1+3+5+\ldots+(2k-1)=3+k^2$. Then, which of the following is true?

(a) $S(1)$ is correct
(b) $S(k) \Rightarrow S(k+1)$
(c) $S(k) \not\Rightarrow S(k+1)$
(d) Principle of mathematical induction can be used to prove the formula

71 The inequality $n! > 2^{n-1}$ is true for

(a) $n>2$ (b) $n \in N$ (c) $n>3$ (d) $n>1$

Numerical Value Type Questions

72 The natural number m, for which the coefficients of x in the binomial expansion of $\left(x^m+\frac{1}{x^2}\right)^{22}$ is 1540, is ....

73 If the 2nd term of the expansion $\left[a^{1/13}+\frac{a}{\sqrt{a^{-1}}}\right]^n$ is $14a^{5/2}$, then the value of $\frac{{}^nC_3}{{}^nC_2}$ is ....

74 $3\times 7^{22}+2\times 10^{22}-44$ when divided by 18, leaves the remainder .... **[JEE Main 2021]**

75 The largest value of x for which the 4th term in the expansion of $\left(5^{2/5 \log_5 \sqrt{4^x+44}} + \frac{1}{5^{\log_5 \sqrt[3]{2^{x-1}+7}}}\right)^8$ is 336, is ....

76 If the coefficient of x^8 in $\left(ax^2 + \frac{1}{bx}\right)^{13}$ is equal to the coefficient of x^{-8} in $\left(ax - \frac{1}{bx^2}\right)^{13}$, then the value of $ab+3$ is ....

77 If in the expansion of $\left(3x - \frac{2}{x^2}\right)^{15}$, rth term is independent of x, then the value of r is ....

78 If the sum of the coefficient in the expansion of $(x+y)^n$ is 4096, then the greatest coefficient in the expansion is ....

79 The coefficient of x^4 in the expansion of $(1+x+x^2+x^3)^6$ in power of x, is ....

80 Let a and b be the coefficient of x^3 in $(1+x+2x^2+3x^3)^4$ and $(1+x+2x^2+3x^3+4x^4)^4$ respectively, then the value of $\frac{4a}{b}$ is ....

81 For a positive n, $\left(1+\frac{1}{x}\right)^n$ is expanded in increasing powers of x. If three consecutive coefficients in this expansion are in the ratio $2:5:12$, then n is equal to ....

82 If the three consecutive coefficients in the expansion of $(1+x)^n$ are 28, 56 and 70, then the value of n is ....

83 Remainder when $6^{83}+8^{83}$ is divided by 49, is ....

84 Least positive integer just greater than $(1+0.00002)^{50000}$ is ....

85 If $S_n = ({}^nC_0)^2 + ({}^nC_1)^2 + ({}^nC_3)^2 + \ldots + ({}^nC_n)^2$, then maximum value of $\left[\frac{S_{n+1}}{S_n}\right]$ is ....

86 The value of $\sum_{0 \le i < j \le 5} \sum ({}^5C_j)({}^jC_i)$ is equal to ....

87 The value of $\lim_{n \to \infty} \sum_{r=1}^{n} \left(\sum_{t=0}^{n} 5^{\frac{1}{n}} \cdot {}^nC_r \cdot {}^rC_t \cdot 3^t \right)$ is ....

88 If $\sum_{r=0}^{n} \left(\frac{r+2}{r+1}\right) {}^nC_r = \frac{2^8 - 1}{6}$, then n is equal to.........

89 The value of $50 \sum_{r=1}^{49} \frac{2r^2 - 48r + 1}{(50-r) \cdot {}^{50}C_r}$ is.........

90 If $(1-x-x^2)^{20} = \sum_{r=0}^{40} a_r \cdot x^r$, then the value of $a_1 + 3a_3 + 5a_5 + \ldots + 39a_{39}$ is ....

91 If the sum of the coefficients of all even powers of x in the product $(1+x+x^2+\ldots+x^{2n})(1-x+x^2-x^3+\ldots+x^{2n})$ is 61, then n is equal to ....

92 In the expansion of $(a-b+c+d)^6$, the coefficient of a^3b^2c is ....

93 The coefficient of x^3 in the expansion of $(1-x)^{-3}$ is ....

LEVEL 03 \\ JEE Advanced Level

Single Choice Correct Questions

1 If $(1+2x+x^2)^n = \sum_{r=0}^{2n} a_r \cdot x^r$, then a_r is equal to

(a) $({}^nC_r)^2$ (b) ${}^nC_r \cdot {}^nC_{r+1}$ (c) ${}^{2n}C_r$ (d) ${}^{2n}C_{r+1}$

2 If $(4x^2+1)^n = \sum_{r=0}^{n} a_r(1+x^2)^{n-r} x^{2r}$, then the value of $\sum_{r=0}^{n} a_r$ is

(a) 3^n (b) 4^n (c) 5^n (d) 6^n

3 If the term independent of x in $\left(\sqrt{x}-\frac{k}{x^2}\right)^{10}$ is 405, then k is equal to

(a) 2, – 2 (b) 3, – 3 (c) 4, – 4 (d) 1, – 1

4 The greatest term in the expansion of $\sqrt{3}\left(1+\frac{1}{\sqrt{3}}\right)^{20}$ is

(a) $\frac{26840}{9}$ (b) $\frac{24840}{9}$ (c) $\frac{25840}{9}$ (d) None of these

5 In the expansion of $\left(\frac{1}{x^2}-x^3\right)^n$, $n \in N$, if the sum of the coefficients of x^5 and x^{10} is 0, then n is equal to

(a) 25 (b) 20 (c) 15 (d) None of these

6 Coefficient of x^{11} in the expansion of $(1+x^2)^4(1+x^3)^7(1+x^4)^{12}$ is **[JEE Advanced 2014]**

(a) 1051 (b) 1106 (c) 1113 (d) 1120

7 The coefficient of x^9 in the expansion of $(1+x)(1+x^2)(1+x^3)+\ldots(1+x^{100})$ is .... **[JEE Advanced 2015]**

(a) 7 (b) 8 (c) 9 (d) 10

8 The coefficient of x^5 in the expansion of $(1+x)^{21}+(1+x)^{22}+\ldots+(1+x)^{30}$ is

(a) ${}^{51}C_5$ (b) ${}^{9}C_5$ (c) ${}^{31}C_6 - {}^{21}C_6$ (d) ${}^{30}C_5 + {}^{20}C_5$

9 The smallest integer, larger than $(\sqrt{3}+\sqrt{2})^6$ is

(a) 969 (b) 970 (c) 971 (d) 972

10 The fractional part of $\frac{2^{4n}}{15}$ is $(n \in N)$

(a) $\frac{1}{15}$ (b) $\frac{2}{15}$ (c) $\frac{4}{15}$ (d) None of these

11 The remainder when the number $3^{256}-3^{12}$ is divided by 8, is

(a) 0 (b) 3 (c) 4 (d) 7

12 The remainder left out when $8^{2n}-(62)^{2n+1}$ is divided by 9, is

(a) 0 (b) 2 (c) 7 (d) 8

13 The value of $2 \times {}^nC_1 + 2^3 \times {}^nC_3 + 2^5 \times {}^nC_5 + \ldots$ is

(a) $\frac{3^n + (-1)^n}{2}$ (b) $\frac{3^n - (-1)^n}{2}$ (c) $\frac{3^n + 1}{2}$ (d) $\frac{3^n - 1}{2}$

14 If $(1+x)^n = \sum_{r=0}^{n} C_r \cdot x^r$, then the value of

$C_0 + (C_0 + C_1) + (C_0 + C_1 + C_2) + \ldots + (C_0 + C_1 + C_2 + \ldots + C_n)$ is

(a) $n \cdot 2^{n-1}$ (b) $(n+2) \cdot 2^n$ (c) 2^n (d) $(n+2) \cdot 2^{n-1}$

15 The sum of rational terms in the expansion of $(\sqrt{2} + \sqrt[5]{3})^{10}$ is

(a) 46 (b) 42 (c) 41 (d) 40

16 Let $P(n): n^2 + n + 1 (n \in N)$ is an even integer. Therefore, $P(n)$ is true

(a) for $n > 1$ (b) for all n (c) for $n > 2$ (d) None of these

17 If $a_n = \sqrt{7 + \sqrt{7 + \sqrt{7 + \ldots}}}$ having n radical signs. Then, by mathematical induction which one is true?

(a) $a_n > 7, \forall n \geq 1$ (b) $a_n > 3, \forall n \geq 1$

(c) $a_n < 4, \forall n \geq 1$ (d) $a_n < 3, \forall n \geq 1$

More than One Correct

18 In the expansion of $(x+a)^n$, if the sum of odd terms is p and the sum of even terms is q, then

(a) $p^2 - q^2 = (x^2 - a^2)n$ (b) $4pq = (x+a)^{2n} - (x-a)^{2n}$

(c) $2(p^2 + q^2) = (x+a)^{2n} + (x-a)^{2n}$ (d) None of these

19 If the 4th term in the expansion of $\left(ax + \frac{1}{x}\right)^n$ is $\frac{5}{2}$, then

(a) $a = \frac{1}{2}$ (b) $n = 8$ (c) $a = \frac{2}{3}$ (d) $n = 6$

20 The value/values of x in the expression $(x + x^{\log_{10} x})^5$, if the 3rd term in the expansion is 1000000 is/are

(a) 10 (b) 100 (c) $10^{-5/2}$ (d) $10^{-3/2}$

21 If the coefficients of x^{39} and x^{40} are equal in the expansion of $(p + qx)^{49}$, when the possible values of p and q respectively, are

(a) 4 and 16 (b) 1 and 4 (c) 5 and 20 (d) 3 and 9

22 The value of x, for which the 6th term in the expansion of

$$\left[2^{\log_2(\sqrt{9^{x-1}+7})} + \frac{1}{\frac{1}{2^5}\log_2(3^{x-1}+1)}\right]^7$$ is 84, is equal to

(a) 4 (b) 3 (c) 2 (d) 1

23 The middle term in the expansion of $\left(\frac{x}{2} + 2\right)^8$ is 1120, then $x \in R$ is equal to

(a) -2 (b) 3 (c) -3 (d) 2

24 In the expansion of $(a+b)^n$, if two consecutive terms are equal, then which of the following is/are always integer?

(a) $\frac{(n+1)b}{a+b}$ (b) $\frac{(n+1)a}{a+b}$ (c) $\frac{na}{a-b}$ (d) $\frac{na}{a+b}$

25 For natural numbers m and n, if $(1-y)^m(1+y)^n = 1+a_1y+a_2y^2+\ldots$, and $a_1 = a_2 = 10$, then

(a) $m<n$ (b) $m>n$ (c) $m+n=80$ (d) $m-n=20$

26 For which of the following values of x, 5th term is the numerically greatest term in the expansion of $\left(1+\frac{x}{3}\right)^{10}$?

(a) -2 (b) 1.8 (c) 2 (d) -1.9

27 If $\left(x+\frac{1}{x}+1\right)^6 = a_0+\left(a_1x+\frac{b_1}{x}\right)+\left(a_2x^2+\frac{b_2}{x^2}\right)+\ldots+\left(a_6x^6+\frac{b_6}{x^6}\right)$, then

(a) $a_0 = 141$ (b) $a_5 = 6$ (c) $\sum_{i=1}^{6}(a_i+b_i) = 588$ (d) $\sum_{i=1}^{6}(a_i+b_i) = 3^6$

28 The sum of the coefficient in the expansion of $(1+ax-2x^2)^n$ is

(a) positive, when $a<1$ and $n=2k, k \in N$ (b) negative, when $a<1$ and $n=2k+1, k \in N$
(c) positive, when $a>1$ and $n \in N$ (d) zero, when $a=1$

29 If $\sum_{r=0}^{n}\frac{r}{{}^nC_r} = \sum_{r=0}^{n}\frac{n^2-3n+3}{2\cdot{}^nC_r}$, then

(a) $n=1$ (b) $n=2$ (c) $n=3$ (d) None of these

30 In the expansion of $\left(x^2+1+\frac{1}{x^2}\right)^n, n \in N$,

(a) number of terms is $2n+1$ (b) constant term is 2^{n-1}
(c) coefficient of x^{2n-2} is n (d) coefficient of x^2 is n

31 If $\sum_{r=0}^{n}(pr+2)\cdot{}^nC_r = (25)(64)$, where $n, p \in N$, then

(a) $p=3$ (b) $p=4$ (c) $n=7$ (d) $n=6$

32 If $f(m) = \sum_{i=0}^{m}\binom{30}{30-i}\binom{20}{m-i}$, where $\binom{p}{q} = {}^pC_q$, then

(a) maximum value of $f(m)$ is ${}^{50}C_{25}$ (b) $f(m)$ is always divisible by 50 $(1 \le m \le 49)$
(c) $f(0)+f(1)+\ldots+f(50) = 2^{50}$ (d) The value of $\sum_{m=0}^{50}\{f(m)\}^2 = {}^{100}C_{50}$

33 If $(1+x)^n = C_0+C_1x+C_2x^2+\ldots C_nx^n, n \in N$, then $C_0-C_1+C_2+\ldots+(-1)^{m-1}C_{m-1}$ is equal to $(m<n)$

(a) $\frac{(n-1)(n-2)\ldots(n-m+1)}{(m-1)!}$ (b) ${}^{n-1}C_{m-1}(-1)^{m-1}$
(c) $\frac{(n-1)(n-2)\ldots(n-m)}{(m-1)!}(-1)^{m-1}$ (d) ${}^{n-1}C_{n-m}(-1)^{m-1}$

34 If for z as real or complex, $(1+z^2+z^4)^8 = C_0 + C_1 z^2 + C_2 z^4 + \ldots + C_{16} z^{32}$, then

(a) $C_0 - C_1 + C_2 - C_3 + \ldots + C_{16} = 1$

(b) $C_0 + C_3 + C_6 + C_9 + C_{12} + C_{15} = 3^7$

(c) $C_2 + C_5 + C_8 + C_{11} + C_{14} = 3^6$

(d) $C_1 + C_4 + C_7 + C_{10} + C_{13} + C_{16} = 3^7$

35 If $(1+x)^n = C_0 + C_1 x + C_2 x^2 + \ldots C_n x^n$, then $C_0 - (C_0 + C_1) + (C_0 + C_1 + C_2) - (C_0 + C_1 + C_2 + C_3) + \ldots (-1)^{n-1} (C_0 + C_1 + \ldots + C_{n-1})$ is (where n is even integer and $C_r = {}^nC_r$)

(a) a positive value (b) a negative value (c) divisible by 2^{n-1} (d) divisible by 2^n

36 The value of ${}^nC_0 \times {}^{2n}C_r - {}^nC_1 \times {}^{2n-2}C_r + {}^nC_2 \times {}^{2n-4}C_r + \ldots$ is equal to

(a) ${}^nC_{r-n} \times 2^{2n-r}$, if $r \geq n$

(b) 0, if $r < n$

(c) ${}^nC_{r-n} \times 2^{n-r}$, if $r \geq n$

(d) $-{}^nC_{r-n} \times 2^{2n-r}$, if $r < n$

37 Consider, $(1+x+x^2)^{2n} = \sum_{r=0}^{2n} a_r x^r$, where $a_0, a_1, a_2, \ldots, a_{2n}$ are real numbers and n is a positive correct. Then, the value of a_2 is

(a) ${}^{2n}C_1$ (b) ${}^{2n}C_2$ (c) ${}^{2n}C_2 + {}^{2n}C_1$ (d) ${}^{2n+1}C_2$

Linked Comprehension Type

Paragraph I Directions (Q.Nos. 38-40)

An equation $a_0 + a_1 x + a_2 x^2 + \ldots + a_{99} x^{99} + x^{100} = 0$ has roots ${}^{99}C_0, {}^{99}C_1, {}^{99}C_2, \ldots {}^{99}C_{99}$.

38 The value of a_{99} is equal to

(a) 2^{98}

(b) 2^{99}

(c) -2^{99}

(d) None of these

39 The value of a_{98} is

(a) $\dfrac{2^{198} - {}^{198}C_{99}}{2}$ (b) $\dfrac{2^{198} + {}^{198}C_{99}}{2}$ (c) $2^{99} - {}^{99}C_{49}$ (d) None of these

40 The value of $({}^{99}C_0)^2 + ({}^{99}C_1)^2 + \ldots + ({}^{99}C_{99})^2$ is equal to

(a) $2a_{98} - a_{99}^2$ (b) $a_{99}^2 - a_{98}$ (c) $a_{99}^2 - 2a_{98}$ (d) None of these

Paragraph II Directions (Q.Nos. 41-43)

Let $P = \sum_{r=1}^{50} \dfrac{{}^{50+r}C_r (2r-1)}{{}^{50}C_r (50+r)}$, $Q = \sum_{r=0}^{50} ({}^{50}C_r)^2$ and $R = \sum_{r=0}^{100} (-1)^r ({}^{100}C_r)^2$

41 The value of $P - Q$ is equal to

(a) 1 (b) -1 (c) 2^{50} (d) 2^{100}

42 The value of $P - R$ is equal to

(a) 1 (b) -1 (c) 2^{50} (d) 2^{100}

43 The value of $Q+R$ is equal to

(a) $2P+1$ (b) $2P-1$ (c) $2P+2$ (d) $2P-2$

Matrix Match Type Questions

44 Consider, set $A=\{T(r+1)={}^nC_r(3)^{n-r}(5x)^r,\ r=0,1,2,3,\dots,n\}$. Match the following lists.

	List I		List II
(i)	For $x=1$, if $T(12)$ is the greatest, then the value of n can be	A.	10
(ii)	For $n=25$, if $T(21)$ is the greatest, then the value of $5x$ can be	B.	18
(iii)	For $n=20$, if $T(10)=T(11)$, then the value of $33x$ can be	C.	17
(iv)	For $x=2$, if $T(8)$ is the second largest, then the value of n is	D.	12

45 List I contains the different sum of the series and List II contains the maximum value of these sums. Match the following lists.

	List I		List II
(i)	$\sum_{r=0}^{14}\frac{(15-r)}{r+1}\,{}^{15}C_r\,{}^{20}C_{m-r}$	A.	${}^{34}C_{17}$
(ii)	$\lvert{}^{35}C_0-{}^{35}C_1+{}^{35}C_2-{}^{35}C_3+\dots+(-1)^r\,{}^{35}C_r+\dots\rvert$	B.	${}^{35}C_{18}$
(iii)	${}^{31}C_{r-5}+5\times{}^{31}C_{r-4}+10\times{}^{31}C_{r-3}+10\times{}^{31}C_{r-2}+5\times{}^{31}C_{r-1}+{}^{31}C_r$	C.	${}^{37}C_{18}$
(iv)	$\sum_{r=0}^{k}2^{k-r}\cdot(-1)^r\cdot{}^{37}C_r\,{}^{37-r}C_{37-k}$	D.	${}^{36}C_{18}$

46. Match the given lists.

	List I		List II
(i)	If $m,n>r$ and $m+n<r$, then	A.	${}^mC_r+{}^mC_{r-1}\times{}^nC_1+{}^mC_{r-2}\times{}^nC_2+\dots+{}^nC_r={}^{m+n}C_r$
(ii)	If $m,n>r$, then	B.	${}^mC_m\times{}^nC_{r-m}+{}^mC_{m-1}\times{}^nC_{r-m+1}+{}^mC_{m-2}\times{}^nC_{r-m+2}+\dots+{}^mC_{r-m}\times{}^nC_m={}^{m+n}C_r$
(iii)	If $m,n<r<m+n$, then	C.	${}^mC_m\times{}^nC_{r-m}+{}^mC_{m-1}\times{}^nC_{r-m+1}+{}^mC_{m-2}\times{}^nC_{r-m+2}+\dots+{}^nC_r={}^{m+n}C_r$
(iv)	If $m<r<n$, then	D.	${}^mC_r+{}^mC_{r-1}\times{}^nC_1+{}^mC_{r-2}\times{}^nC_2+\dots+{}^nC_r=0$

Numerical Value Type Questions

Question Stem Directions (Q.Nos. 47-49) The answer to each question is a **NUMERICAL VALUE.**

The 6th term in the expansion of $\left[\sqrt{\{2^{\log(10-3^x)}\}}+\sqrt[5]{\{2^{(x-2)\log 3}\}}\right]^m$ is equal to 21, if it is known that the binomial coefficient of the 2nd, 3rd and 4th terms in the expansion represent respectively, the 1st, 3rd and 5th terms of an AP (the symbol log stands for logarithm to the base 10).

47 The value of m is ....

48 The sum of possible values of x is ....

49 The minimum value of expression is ....

Question Stem Directions (Q.Nos. 50-52) The answer to each question is a **NUMERICAL VALUE.**

The 2nd, 3rd and 4th terms in the expansion of $(x+a)^n$ are 240, 720 and 1080, respectively.

50 The value of $(x-a)^n$ can be ....

51 The value of least term in the expansion is ....

52 The sum of odd numbered terms is ....

53 The coefficient of three consecutive terms of $(1+x)^{n+5}$ are in the ratio 5 : 10 : 14, then n is equal to .... **[JEE Advanced 2013]**

54 Given, $(1-2x+5x^2-10x^3)(1+x)^n=1+a_1x+a_2x^2+\ldots$ and that $a_1^2=2a_2$, then the value of n is ....

55 Let $X=({}^{10}C_1)^2+2({}^{10}C_2)^2+3({}^{10}C_3)^2+\ldots+10({}^{10}C_{10})^2$, where ${}^{10}C_r, r\in\{1,2,\ldots,10\}$ denotes binomial coefficient. Then, the value of $\frac{1}{1430}X$ is .... **[JEE Advanced 2018]**

56 Let m be the smallest positive integer such that the coefficient of x^2 in the expansion of $(1+x)^2+(1+x)^3+\ldots+(1+x)^{49}+(1+mx)^{50}$ is $(3n+1)\,{}^{51}C_3$ for some positive integer n, then the value of n is .... **[JEE Advanced 2016]**

57 The coefficient of x^{103} in $(1+x+x^2+x^3+x^4)^{199}(x-1)^{201}$ is ....

58 The constant term in the expansion of $[\log(x^{\log x})-\log_{x^2}100]^{12}$ is (base of log is 10) ....

59 The total number of different terms in the product $({}^{101}C_0-{}^{101}C_1x+{}^{101}C_2x^2+\ldots-{}^{101}C_{101}x^{101})(1+x+x^2+\ldots+x^{100})^{101}$ is ....

60 The value of $\sum_{r=0}^{3}{}^8C_r({}^5C_{r+1}-{}^4C_r)$ is ....

61 The sum of the series $\frac{{}^{101}C_1}{{}^{101}C_0}+\frac{2.{}^{101}C_2}{{}^{101}C_1}+\frac{3.{}^{101}C_3}{{}^{101}C_2}+\ldots+\frac{101.{}^{101}C_{101}}{{}^{101}C_{100}}$ is ... .

62 Let $a=3^{\frac{1}{223}}+1$ and for all $n\geq 3$, let $f(n)={}^nC_0\cdot a^{n-1}-{}^nC_1\cdot a^{n-2}+{}^nC_2\cdot a^{n-3}-\ldots+(-1)^{n-1}\cdot{}^nC_{n-1}\cdot a^0$. If the value of $f(2007)+f(2008)=3^k$, where $k\in N$, then the value of k is ....

63 The remainder, if $1+2+2^2+2^3+\ldots+2^{1999}$ is divided by 5 is ....

64 In the expansion of $(1+3x+2x^2)^6$, the coefficient of x^{11} is ....

ANSWER KEY

Level 1

1. (i) $x^5 + 5x^4a + 10x^3a^2 + 10x^2a^3 + 5xa^4 + a^5$

(ii) $32x^5 - \frac{240}{y}x^4 + \frac{720x^3}{y^2} - \frac{1080x^2}{y^3} + \frac{810x}{y^4}x - \frac{243}{y^5}$

2. $x^7 + 7x^6 + 21x^5 + 35x^4 + 35x^3 + 21x^2 + 7x + 1$

3. (i) $\frac{10500}{x^3}$ (ii) $2^8 \frac{13!}{6! \times 7!} x^4$ 4. $\frac{30!}{24!\,6!}(10)^6(5)^{24}(x)^6(y)^{24}$

5. (i) ${}^{11}C_5 \cdot \frac{a^6}{b^5}$ (ii) ${}^{11}C_5 \cdot \frac{a^5}{b^6}$

6. $2(32x^6 - 48x^4 + 18x^2 - 1)$ 7. $70x$ 8. 84 9. 59 10. 97

11. (i) $924a^6b^6$ (ii) $\frac{189}{8}x^{17}$ and $\frac{-21}{16}x^{19}$

12. $\frac{7 \cdot 3^{13}}{2}$ 13. 55 15. 1 16. 0.951 17. $(101)^{50} > (100)^{50} + (99)^{50}$

18. 18 19. 01 20. $\frac{n}{2}a_n$ 21. $\frac{n(n+1)}{2}$ 22. ${}^{2n+1}C_{n+1}$ 23. $\frac{1}{n!}(2)^{n-1}$

24. (i) $2^{n/2} \cdot \cos\frac{n\pi}{4}$ (ii) $2^{n/2} \cdot \sin\frac{n\pi}{4}$ (iii) $\frac{1}{3}\left(2^n + 2\cos\frac{n\pi}{3}\right)$

25. (i) $n(2)^{n-1}$ (ii) $(n+2)(2)^{n-1}$

26. (i) $\frac{(2)^{n+1} - 1}{n+1}$ (ii) $\frac{1}{n+1}$

27. 2^n 28. $(n+1)2^{n+1}$ 29. $|x| < \frac{5}{4}$ 30. $1 - \frac{x}{2} - \frac{x^2}{8} - \frac{x^3}{16} - \ldots\infty$ 31. $\frac{x^3}{2}$

32. (i) $\frac{1}{9} - \frac{4}{27}x + \frac{4}{27}x^2$ (ii) $\frac{1}{4x^2} - \frac{3}{4x^3} + \frac{27}{16x^4}$

33. 2 34. $\sqrt{3}$ 35. 62640 36. 378 37. 280 38. 12632

39. $\frac{(n+1)(n+2)(n+3)}{6}$ 40. 2^{50} 41. 1 or 2 42. 1

Level 2

1. (a)	2. (a)	3. (c)	4. (d)	5. (a)	6. (b)	7. (b)
8. (a)	9. (c)	10. (b)	11. (b)	12. (c)	13. (b)	14. (c)
15. (b)	16. (c)	17. (c)	18. (c)	19. (c)	20. (c)	21. (c)
22. (c)	23. (d)	24. (c)	25. (b)	26. (d)	27. (c)	28. (a)
29. (c)	30. (c)	31. (c)	32. (d)	33. (c)	34. (b)	35. (c)
36. (d)	37. (b)	38. (d)	39. (a)	40. (b)	41. (a)	42. (b)
43. (b)	44. (b)	45. (d)	46. (d)	47. (a)	48. (a)	49. (b)
50. (b)	51. (b)	52. (a)	53. (a)	54. (d)	55. (b)	56. (a)
57. (d)	58. (a)	59. (d)	60. (c)	61. (c)	62. (b)	63. (d)
64. (d)	65. (d)	66. (b)	67. (d)	68. (b)	69. (a)	70. (b)
71. (c)	72. (13)	73. (4)	74. (15)	75. (1)	76. (2)	77. (6)
78. (924)	79. (120)	80. (4)	81. (118)	82. (8)	83. (35)	84. (3)
85. (3)	86. (211)	87. (1)	88. (5)	89. (2499)	90. (40)	91. (30)
92. (60)	93. (10)					

Level 3

1. (c)	**2.** (b)	**3.** (b)	**4.** (c)	**5.** (c)	**6.** (c)	**7.** (b)
8. (c)	**9.** (b)	**10.** (a)	**11.** (a)	**12.** (b)	**13.** (b)	**14.** (d)
15. (c)	**16.** (d)	**17.** (c)	**18.** (a, b, c)	**19.** (a, d)	**20.** (a, c)	**21.** (a, b, c)
22. (c, d)	**23.** (a, d)	**24.** (a, b)	**25.** (a, c)	**26.** (a b, c,d)	**27.** (a, b, c)	**28.** (a, b, c, d)
29. (a, c)	**30.** (a, c)	**31.** (a, c)	**32.** (a, c, d)	**33.** (b, c, d)	**34.** (a, b, d)	**35.** (b, c)
36. (a, b)	**37.** (c, d)	**38.** (c)	**39.** (a)	**40.** (c)	**41.** (b)	**42.** (b)
43. (c)						

44. (i)–C, (ii)–D, (iii)–B, (iv)–A

45. (i)–B, (ii)–A, (iii)–D, (iv)–C

46. (i)–D, (ii)–A, (iii)–B, (iv)–C

47. (7)	**48.** (2)	**49.** (128)	**50.** (−1)	**51.** (32)	**52.** (1562)	**53.** (6)
54. (6)	**55.** (646)	**56.** (5)	**57.** (0)	**58.** (495)	**59.** (102)	**60.** (220)
61. (5151)	**62.** (9)	**63.** (0)	**64.** (576)			

SOLUTIONS

Level 1

1. (i) Using binomial theorem,

$$(x+a)^n = \sum_{r=0}^{n} {}^nC_r \cdot x^{n-r} \cdot a^r, \text{ we get}$$

$$(x+a)^5 = {}^5C_0 \cdot x^5 + {}^5C_1 \cdot x^4(a)^1 + {}^5C_2 \cdot x^3(a)^2 + {}^5C_3 \cdot x^2(a)^3 + {}^5C_4 \cdot x(a)^4 + {}^5C_5 \cdot a^5$$

$$\Rightarrow (x+a)^5 = x^5 + 5x^4a + 10x^3a^2 + 10x^2a^3 + 5xa^4 + a^5$$

(ii) $$\left(2x - \frac{3}{y}\right)^5 = {}^5C_0(2x)^5\left(-\frac{3}{y}\right)^0 + {}^5C_1(2x)^4\left(-\frac{3}{y}\right)^1 + {}^5C_2(2x)^3\left(-\frac{3}{y}\right)^2 + {}^5C_3(2x)^2\left(-\frac{3}{y}\right)^3 + {}^5C_4(2x)^1\left(-\frac{3}{y}\right)^4 + {}^5C_5(2x)^0\left(-\frac{3}{y}\right)^5$$

$$\Rightarrow \left(2x - \frac{3}{y}\right)^5 = 32x^5 - \frac{240}{y}x^4 + \frac{720x^3}{y^2} - \frac{1080x^2}{y^3} + \frac{810}{y^4}x - \frac{243}{y^5}$$

2. $$(x+1)^7 = \sum_{r=0}^{7} {}^7C_r \cdot x^{7-r}$$

$$= {}^7C_0 \cdot x^7 + {}^7C_1 \cdot x^6 + {}^7C_2 \cdot x^5 + {}^7C_3 \cdot x^4 + {}^7C_4 \cdot x^3 + {}^7C_5 \cdot x^2 + {}^7C_6 \cdot x + {}^7C_7$$

$$= x^7 + 7x^6 + \frac{7\cdot 6}{2}x^5 + \frac{7\cdot 6\cdot 5}{2\cdot 3}x^4 + \frac{7\cdot 6\cdot 5}{2\cdot 3}x^3 + \frac{7\cdot 6}{2}x^2 + 7x + 1$$

$$= x^7 + 7x^6 + 21x^5 + 35x^4 + 35x^3 + 21x^2 + 7x + 1$$

3. (i) $$T_7 = T_{6+1} = {}^9C_6\left(\frac{4x}{5}\right)^{9-6}\left(\frac{-5}{2x}\right)^6$$

$$= \frac{9!}{3!6!}\left(\frac{4x}{5}\right)^3(-1)^6\frac{5^6}{2^6x^6} = \frac{10500}{x^3}$$

(ii) $$T_7 = T_{6+1} = {}^{13}C_6(4x)^{13-6}\left(-\frac{1}{2\sqrt{x}}\right)^6$$

$$= \frac{13!}{6!7!} \times 4^7 \times x^7 \times (-1)^6\frac{1}{2^6x^3}$$

$$= 2^8 \times \frac{13!}{6!\times 7!}x^4$$

4. $$T_{25} = T_{24+1} = {}^{30}C_{24}(10x)^{30-24}(5y)^{24}$$

$$= {}^{30}C_{24}(10x)^6(5y)^{24}$$

$$= \frac{30!}{24!\times 6!}(10)^6(5)^{24}(x)^6(y)^{24}$$

5. (i) $$T_{r+1} = {}^{11}C_r(ax^2)^{11-r}\left(\frac{1}{bx}\right)^r$$

$$= {}^{11}C_r\frac{a^{11-r}}{b^r}x^{22-3r} \quad \text{...(i)}$$

For the coefficient of x^7, $22-3r$ must be 7

i.e. $22-3r=7 \Rightarrow r=5$

Hence, putting $r=5$, in Eq. (i), we get

$$\therefore \text{Required coefficient} = {}^{11}C_5\frac{a^6}{b^5}$$

(ii) $$T_{r+1} = {}^{11}C_r(ax)^{11-r}\left(-\frac{1}{bx^2}\right)^r$$

$$= {}^{11}C_r(a)^{11-r}(x)^{11-r}(-1)^r\frac{1}{b^r}x^{-2r}$$

$$= {}^{11}C_r(a)^{11-r}\frac{(-1)^r}{b^r}(x)^{11-3r} \quad \text{...(i)}$$

For the coefficient of x^{-7}, $11-3r$ must be -7.

i.e. $11-3r=-7 \Rightarrow r=6$

Hence, putting $r=6$, in Eq. (i), we get

$$\therefore \text{Required coefficient} = (-1)^6\,{}^{11}C_6\frac{a^5}{b^6} = {}^{11}C_5\frac{a^5}{b^6}$$

6. Since, $(x+y)^n + (x-y)^n = 2[x^n + {}^nC_2(x)^{n-2}y^2 + {}^nC_4(x)^{n-4}y^4 + \dots]$

Putting $y = \sqrt{x^2-1}$ and $n=6$, we get

$$(x+\sqrt{x^2-1})^6 + (x-\sqrt{x^2-1})^6$$

$$= 2[x^6 + {}^6C_2(x)^4(x^2-1) + {}^6C_4(x)^2(x^2-1)^2 + {}^6C_6(x^2-1)^3]$$

$$= 2[x^6 + 15x^4(x^2-1) + 15x^2(x^2-1)^2 + (x^2-1)^3]$$

$$= 2(32x^6 - 48x^4 + 18x^2 - 1)$$

7. 4th term from the end = (7 – 4 + 2) th term from the beginning

⇒ 4th term from the end = 5th term from the beginning

$$\therefore T_5 = T_{4+1} = {}^7C_4\left(\frac{x^3}{2}\right)^{7-4}\left(-\frac{2}{x^2}\right)^4$$

$$= \frac{7\cdot 6\cdot 5}{1\cdot 2\cdot 3}\left(\frac{x^3}{2}\right)^3(-1)^4\frac{(2)^4}{x^8} = 70x$$

8. $\because T_{r+1} = {}^9C_r(x^2)^{9-r}\left(\frac{1}{x}\right)^r$

$= {}^9C_r(x)^{18-2r}(x)^{-r}$

$= {}^9C_r(x)^{18-3r}$ ...(i)

For independent term of x, we must have

$$18 - 3r = 0 \Rightarrow r = 6$$

Putting $r = 6$ in Eq. (i), we get

$$T_{6+1} = {}^9C_6 = \frac{9!}{6! \times 3!}$$

$$\Rightarrow \quad T_7 = \frac{9 \times 8 \times 7 \times 6!}{6! \times 3 \times 2} = 84$$

$\therefore$ 7th term is independent of x and the value of independent term of x is 84.

9. The general term in the expansion of $(3^{1/5} + 2^{1/3})^{15}$ is

$$T_{r+1} = {}^{15}C_r(3^{1/5})^{15-r}(2^{1/3})^r$$

$$= {}^{15}C_{(r)} 3^{3-\frac{r}{5}} (2)^{\frac{r}{3}}$$

Now, $0 \le r \le 15$

For $r = 0, 15$, rational terms are

$$T_{0+1} \text{ and } T_{15+1}.$$

Then, $T_1 = {}^{15}C_0(3)^3(2^0) = 27$

and $T_{16} = {}^{15}C_{15}(3)^0(2)^5 = 32$

$\therefore$ Sum of all rational terms $= 27 + 32 = 59$

10. General term,

$$T_{r+1} = {}^{100}C_r(3^{1/8})^{100-r}(2^{1/6})^r$$

$$= {}^{100}C_r(3)^{\frac{100-r}{8}}(2)^{\frac{r}{6}}$$

As, 2 and 3 are co-prime.

$\therefore T_{r+1}$ will be rational if $100 - r$ is a multiple of 8 and r is a multiple of 6.

Also, $0 \le r \le 100$

$\therefore r = 0, 6, 12, 18, \ldots, 96$

Now, $100 - r = 4, 10, 16, \ldots, 100$ ...(i)

and $100 - r = 0, 8, 16, 24, \ldots, 96$ ...(ii)

The common terms in Eqs. (i) and (ii) are 16, 40, 64 and 88.

$\therefore r = 84, 60, 36, 12$ gives rational terms.

$\therefore$ The number of irrational terms $= 101 - 4 = 97$

11. (i) The number of terms $= 13$ (odd)

Its middle term is $\left(\frac{12}{2} + 1\right)$th i.e. 7th term

$$\therefore \text{Required term, } T_7 = T_{6+1} = {}^{12}C_6\left(\frac{a}{x}\right)^6(bx)^6$$

$$= {}^{12}C_6 \cdot a^6 \cdot b^6$$

$$= 924a^6 \cdot b^6$$

(ii) The number of terms $= 10$ (even)

So, there are two middle terms

i.e. $\left(\frac{9+1}{2}\right)$th and $\left(\frac{9+3}{2}\right)$th terms.

$\Rightarrow T_5$ and T_6 are middle terms.

$$\therefore \quad T_5 = {}^9C_4(3x)^5\left(-\frac{x^3}{6}\right)^4$$

$$= \frac{9 \cdot 8 \cdot 7 \cdot 6}{1 \cdot 2 \cdot 3 \cdot 4} 3^5(x^5)\left(\frac{x^{12}}{6^4}\right)$$

$$= \frac{189}{8} x^{17}$$

and $T_6 = {}^9C_5(3x)^4\left(-\frac{x^3}{6}\right)^5$

$$= -{}^9C_5 \cdot 3^4 \cdot x^4 \cdot \frac{x^{15}}{6^5}$$

$$= -\frac{9 \cdot 8 \cdot 7 \cdot 6}{1 \cdot 2 \cdot 3 \cdot 4} \cdot 3^4 \cdot \frac{x^{19}}{6^5}$$

$$= -\frac{21}{16} x^{19}$$

12. Let T_{r+1} be the greatest term in the expansion of $(2 + 3x)^9$, we have

$$\frac{T_{r+1}}{T_r} = \left(\frac{9 - r + 1}{r}\right)\left|\frac{3x}{2}\right|$$

$$= \left(\frac{10 - r}{r}\right)\left|\frac{3}{2} \times \frac{3}{2}\right| \quad \left[\because x = \frac{3}{2}\right]$$

$$= \frac{90 - 9r}{4r}$$

$$\therefore \quad \frac{T_{r+1}}{T_r} \ge 1$$

$$\Rightarrow \quad \frac{90 - 9r}{4r} \ge 1 \Rightarrow 90 \ge 13r$$

$$\therefore \quad r \le \frac{90}{13} = 6\frac{12}{13} \text{ or } r \le 6\frac{12}{13}$$

$\therefore$ Maximum value of r is 6.

So, the greatest term $= T_{6+1} = {}^9C_6(2)^{9-6}(3x)^6$

$$= {}^9C_3 \cdot 2^3\left(3 \times \frac{3}{2}\right)^6$$

$$= \frac{9 \cdot 8 \cdot 7}{3 \cdot 2 \cdot 1} \frac{2^3 \cdot 3^{12}}{2^6} = \frac{7 \cdot 3^{13}}{2}$$

13. Suppose the three consecutive terms in the expansion of $(1 + a)^n$ are $(r - 1)$th, rth and $(r + 1)$th terms.

Since, the coefficients are in the ratio $1 : 7 : 42$, so we have

$$\frac{{}^nC_r}{{}^nC_{r-1}} = \frac{42}{7}$$

$$\Rightarrow \quad \frac{n-r+1}{r}=6 \qquad \left[\because \frac{{}^nC_r}{{}^nC_{r-1}}=\frac{n-r+1}{r}\right]$$

$$\Rightarrow \quad n-r+1=6r$$

$$\Rightarrow \quad n-7r+1=0 \qquad \text{...(i)}$$

Also, $\dfrac{{}^nC_{r-1}}{{}^nC_{r-2}}=\dfrac{7}{1}$

$$\Rightarrow \quad \frac{n-r+2}{r-1}=7 \qquad \left[\because \frac{{}^nC_r}{{}^nC_{r-1}}=\frac{n-r+1}{r}\right]$$

$$\Rightarrow \quad 2+n-r=7r-7$$

$$\Rightarrow \quad n-8r+9=0 \qquad \text{...(ii)}$$

On solving Eqs. (i) and (ii), we get

$$n=55$$

14. Let $(5+2\sqrt{6})^n=I+f$, where I is an integer and $0\le f<1$

Now, let $g=(5-2\sqrt{6})^n, 0<g<1$

$$I+f+g=(5+2\sqrt{6})^n+(5-2\sqrt{6})^n$$

$$=2[5^n+{}^nC_2(5)^{n-2}(2\sqrt{6})^2+{}^nC_4 5^{n-4}(2\sqrt{6})^4+\dots]$$

$=$ even integer

$=2k(k\in N)$

Hence, $f+g=2k-I$ is an integer, but $0<f+g<2$, therefore $f+g=1$

$$1=2k-I$$

$$\Rightarrow \quad I=2k-1=\text{odd integer}$$

Hence, the integral part of $(5+2\sqrt{6})^n$ is an odd integer.

15. $\because$ t is a proper fraction $0<t<1$.

Given, $(7+4\sqrt{3})^n=x+t$

Now, let $t'=(7-4\sqrt{3})^n; 0<t'<1$

Also, $x+t+t'=(7+4\sqrt{3})^n+(7-4\sqrt{3})^n$

$=2[(7)^n+{}^nC_2(7)^{n-2}(4\sqrt{3})^2+{}^nC_4(7)^{n-4}(4\sqrt{3})^4+\dots]$

$=2$ (integer) $=2k, k\in N$

$=$ even integer

$\therefore t+t'=$ even integer $-x$, but $0<t+t'<2$

$\therefore t+t'=1 \Rightarrow t'=(1-t)$

Hence, $(1-t)(x+t)=t'(x+t)=(7-4\sqrt{3})^n(7+4\sqrt{3})^n$

$$=(49-48)^n=1$$

16. $(0.99)^5=(1-0.01)^5$

$={}^5C_0(1)^5-{}^5C_1(1)^4(0.01)+{}^5C_2(1)^3(0.01)^2$ (approximately)

$=1-5(0.01)+10(0.01)^2$

$=1-0.05+0.001$

$=1.001-0.05=0.951$

Thus, the value of $(0.99)^5$ is approximately 0.951.

17. We have, $(101)^{50}=(100+1)^{50}$

$={}^{50}C_0(100)^{50}+{}^{50}C_1(100)^{49}+{}^{50}C_2(100)^{48}+\dots$

$=(100)^{50}+50(100)^{49}+\dfrac{50\times 49}{1\times 2}(100)^{48}+\dots \qquad \text{...(i)}$

and $(99)^{50}=(100-1)^{50}$

$={}^{50}C_0(100)^{50}-{}^{50}C_1(100)^{49}+{}^{50}C_2(100)^{48}+\dots$

$=(100)^{50}-50\times(100)^{49}+\dfrac{50\times 49}{2\times 1}(100)^{48}+\dots \qquad \text{...(ii)}$

Subtracting Eq. (ii) from Eq. (i), we get

$(101)^{50}-(99)^{50}=(100)^{50}+2\times\dfrac{50\times 49\times 48}{3\times 2\times 1}(100)^{47}+\dots$

$>(100)^{50}$.

Hence, $(101)^{50}>(100)^{50}+(99)^{50}$

18. We have, $7^{103}=7\cdot 7^{102}=7\cdot(7^2)^{51}=7(49)^{51}$

$$\Rightarrow \quad 7^{103}=7(50-1)^{51}$$

$=7[(50)^{51}-{}^{51}C_1(50)^{50}+{}^{51}C_2(50)^{49}\dots-1]$

$=7[(50)^{51}-{}^{51}C_1(50)^{50}+{}^{51}C_2(50)^{49}\dots+{}^{51}C_{50}(50)]-7+18-18$

$=7[50k]-25+18$ [where, k is an integer]

$=25[14k-1]+18$

$=25p+18$ [where, p is an integer]

Now, $\dfrac{7^{103}}{25}=p+\dfrac{18}{25}$, hence remainder is 18.

19. $3^{400}=(3^2)^{200}=(9)^{200}=(10-1)^{200}$

$=(10)^{200}-{}^{200}C_1(10)^{199}+{}^{200}C_2(10)^{198}\dots-{}^{200}C_{199}(10)+1$

$=100k-{}^{200}C_{199}(10)+1; k\in I$

$=100k-{}^{200}C_1(10)+1$

$=100k-2000+1$

$=100(k-20)+1=100p+1; p\in I$

Hence, last two digits of 3^{400} is $00+1=01$.

20. Let $x=\displaystyle\sum_{r=0}^{n}\frac{r}{{}^nC_r} \qquad \text{...(i)}$

Replacing r by $n-r$ in Eq. (i), we get

$$x=\sum_{r=0}^{n}\frac{(n-r)}{{}^nC_{n-r}}=\sum_{r=0}^{n}\frac{(n-r)}{{}^nC_r} \quad [\because {}^nC_r={}^nC_{n-r}] \text{...(ii)}$$

On adding Eqs. (i) and (ii), we get

$$2x=\sum_{r=0}^{n}\frac{n}{{}^nC_r}=n\sum\frac{1}{{}^nC_r}=na_n$$

$$\therefore \quad x=\frac{n}{2}a_n$$

21. $\because \dfrac{^nC_r}{^nC_{r-1}} = \dfrac{n-r+1}{r}$

$\therefore \dfrac{r \cdot {^nC_r}}{^nC_{r-1}} = n - r + 1$

$\therefore \sum_{r=1}^{n} \dfrac{r \cdot {^nC_r}}{^nC_{r-1}} = \sum_{r=1}^{n} (n-r+1) = \sum_{r=1}^{n} n - \sum_{r=1}^{n} r + \sum_{r=1}^{n} 1$

$= n\sum 1 - \dfrac{n(n+1)}{2} + n$

$= n \cdot n - \dfrac{n(n+1)}{2} + n = \dfrac{n(n+1)}{2}$

22. $\because \sum_{r=0}^{n} {^{n+r}C_n} = {^nC_n} + {^{n+1}C_n} + {^{n+2}C_n} + \ldots + {^{2n}C_n}$

$=$ Coefficient of x^n in $[(1+x)^n + (1+x)^{n+1} + \ldots (1+x)^{2n}]$

$=$ Coefficient of x^n in $\left[\dfrac{(1+x)^n [(1+x)^{n+1} - 1]}{(1+x) - 1}\right]$

$=$ Coefficient of x^{n+1} in $[(1+x)^{2n+1} - (1+x)^n]$

$= {^{2n+1}C_{n+1}} - 0 = {^{2n+1}C_{n+1}}$

23. $\dfrac{1}{(n-1)!\,1!} + \dfrac{1}{(n-3)!\,3!} + \dfrac{1}{(n-5)!\,5!} + \ldots$

$\because$ Sum of values of each terms in factorial are equal

i.e. $(n-1)+1 = (n-3)+3 = (n-5)+5 = n$

$= \dfrac{1}{n!}\left[\dfrac{n!}{(n-1)!} + \dfrac{n!}{(n-3)!\,3!} + \dfrac{n!}{(n-5)!\,5!} + \ldots\right]$

$= \dfrac{1}{n!}[{^nC_1} + {^nC_3} + {^nC_5} + \ldots] = \dfrac{1}{n!} \times 2^{n-1}$

24. $\because (1+x)^n = C_0 + C_1x + C_2x^2 + C_3x^3 + \ldots$

Putting $x = i$, where $i = \sqrt{-1}$, then

$(1+i)^n = C_0 + C_1 i + C_2 i^2 + C_3 i^3 + C_4 i^4 + \ldots$

$(1+i)^n = (C_0 - C_2 + C_4 \ldots) + i(C_1 - C_3 + C_5 - \ldots)$ ...(i)

Also, $(1+i)^n = \left[\sqrt{2}\left(\dfrac{1}{\sqrt{2}} + \dfrac{i}{\sqrt{2}}\right)\right]^n$

$\Rightarrow \quad (1+i)^n = 2^{n/2}\left[\cos\dfrac{n\pi}{4} + i\sin\dfrac{n\pi}{4}\right]$ ...(ii)

From Eqs. (i) and (ii), we get

$(C_0 - C_2 + C_4 - \ldots) + i(C_1 - C_3 + C_5 \ldots)$

$= 2^{n/2}\cos\dfrac{n\pi}{4} + i \cdot 2^{n/2}\sin\dfrac{n\pi}{4}$

On comparing real and imaginary parts, we get

(i) $C_0 - C_2 + C_4 - \ldots = 2^{n/2}\cos\dfrac{n\pi}{4}$

(ii) $C_1 - C_3 + C_5 \ldots = 2^{n/2}\sin\dfrac{n\pi}{4}$

(iii) $\because (1+x)^n = C_0 + C_1x + C_2x^2 + C_3x^3 + \ldots$

Putting $x = 1, \omega, \omega^2$ (cube roots of unity) and adding, we get

$3(C_0 + C_3 + C_6 + \ldots) = 2^n + (1+\omega)^n + (1+\omega^2)^n$

$= 2^n + (-\omega^2)^n + (-\omega)^n$

$= 2^n + (-1)^n [\omega^{2n} + \omega^n]$

$= 2^n + (-1)^n \left[e^{\frac{4\pi in}{3}} + e^{\frac{2\pi in}{3}}\right]$

$= 2^n + (-1)^n \left[e^{n\pi i} \cdot 2\cos\dfrac{n\pi}{3}\right]$

$= 2^n + (-1)^n (-1)^n \cdot 2\cos\dfrac{n\pi}{3}$

$= 2^n + (-1)^{2n}\, 2\cos\dfrac{n\pi}{3} = 2^n + 2\cos\dfrac{n\pi}{3}$

$\Rightarrow C_0 + C_3 + C_6 + \ldots = \dfrac{1}{3}\left[2^n + 2\cos\dfrac{n\pi}{3}\right]$

25. (i) $\because (1+x)^n = C_0 + C_1x + C_2x^2 + \ldots + C_nx^n$

Differentiating both sides w.r.t. x, we get

$n(1+x)^{n-1} = 0 + C_1 + 2C_2x + 3C_3x^2 + \ldots + nC_nx^{n-1}$

Putting $x = 1$, we get

$n \cdot 2^{n-1} = C_1 + 2C_2 + 3C_3 + \ldots + nC_n$

(ii) $\because (1+x)^n = C_0 + C_1x + C_2x^2 + \ldots + C_nx^n$

On multiplying by x on both sides, we get

$x(1+x)^n = xC_0 + C_1x^2 + C_2x^3 + \ldots + C_nx^{n+1}$

On differentiating w.r.t. x, we get

$x \cdot n(1+x)^{n-1} + (1+x)^n = C_0 + 2C_1x + 3C_2x^2 + \ldots + (n+1)C_nx^n$

Putting $x = 1$, we get

$n(2)^{n-1} + 2^n = C_0 + 2C_1 + 3C_2 + \ldots + (n+1)C_n$

$\Rightarrow C_0 + 2C_1 + 3C_2 + \ldots + (n+1)C_n = (n+2)2^{n-1}$

26. (i) $\because (1+x)^n = C_0 + C_1x + C_2x^2 + .. + C_nx^n$ ...(i)

Integrating both sides within limits 0 to 1, we get

$\int_0^1 (1+x)^n\, dx = \int_0^1 (C_0 + C_1x + C_2x^2 + \ldots + C_nx^n)\, dx$

$\left[\dfrac{(1+x)^{n+1}}{n+1}\right]_0^1$

$= \left[C_0x + C_1\dfrac{x^2}{2} + C_2\dfrac{x^3}{3} + \ldots + C_n\dfrac{x^{n+1}}{n+1}\right]_0^1$

$\Rightarrow \dfrac{1}{n+1}[2^{n+1} - 1] = C_0 + \dfrac{C_1}{2} + \dfrac{C_2}{3} + \ldots + \dfrac{C_n}{n+1}$

$\Rightarrow C_0 + \dfrac{C_1}{2} + \dfrac{C_2}{3} + \ldots + \dfrac{C_n}{n+1} = \dfrac{2^{n+1} - 1}{n+1}$

(ii) Integrating Eq. (i) on both sides within limits -1 to 0, we get

$$\left[\frac{(1+x)^{n+1}}{n+1}\right]_{-1}^{0} = \left[C_0 \cdot x + C_1 \cdot \frac{x^2}{2} + C_2 \frac{x^3}{3} + \ldots + C_n \frac{x^{n+1}}{n+1}\right]_{-1}^{0}$$

$$\Rightarrow \frac{1-0}{n+1} = 0 - \left[-C_0 + \frac{C_1}{2} - \frac{C_2}{3} + \ldots + (-1)^{n+1}\frac{C_n}{n+1}\right]$$

$$\Rightarrow C_0 - \frac{C_1}{2} + \frac{C_2}{3} + \ldots (-1)^n \frac{C_n}{n+1} = \frac{1}{n+1}$$

27. $C_0 \cdot {}^{2n}C_n - C_1 {}^{2n-2}C_n + C_2 \cdot {}^{2n-4}C_n \ldots$

$= \text{Coefficient of } x^n \text{ in } [C_0(1+x)^{2n} - C_1(1+x)^{2n-2} + \ldots]$

$= \text{Coefficient of } x^n \text{ in } [C_0[(1+x)^2]^n - C_1[(1+x)^2]^{n-1} \ldots]$

$= \text{Coefficient of } x^n \text{ in } [(1+x)^2 - 1]^n$

$= \text{Coefficient of } x^n \text{ in } (2x + x^2)^n$

$= \text{Constant term in } (2+x)^n = 2^n$

28. $$\sum_{i=0}^{n}\sum_{j=0}^{n}(C_i + C_j) = \sum_{i=0}^{n}\sum_{j=0}^{n} C_i + \sum_{i=0}^{n}\sum_{j=0}^{n} C_j$$

$$= \sum_{j=0}^{n}\left(\sum_{i=0}^{n} C_i\right) + \sum_{i=0}^{n}\left(\sum_{j=0}^{n} C_j\right)$$

$$= \sum_{j=0}^{n}(2^n) + \sum_{i=0}^{n}(2^n)$$

$$= (n+1)2^n + (n+1)2^n = (n+1)2^{n+1}$$

29. We have, $\dfrac{1}{\sqrt{5+4x}} = \dfrac{1}{\sqrt{5} \times \sqrt{1 + \dfrac{4x}{5}}}$

$$= 5^{-1/2}\left(1 + \frac{4x}{5}\right)^{-1/2}$$

This can be expanded as infinite series, only if

$$\left|\frac{4}{5}x\right| < 1 \text{ or } |x| < \frac{5}{4}$$

30. $\because (1+x)^n = 1 + nx + \dfrac{n(n-1)}{2!}x^2 + \dfrac{n(n-1)(n-2)}{3!}x^3 + \ldots$

Putting $n = \dfrac{1}{2}$ and x as $-x$, we get

$$(1-x)^{1/2} = 1 + \left(\frac{1}{2}\right)(-x) + \frac{\left(\frac{1}{2}\right)\left(\frac{1}{2}-1\right)}{2!}(-x)^2 + \frac{\frac{1}{2}\left(\frac{1}{2}-1\right)\left(\frac{1}{2}-2\right)}{3!}(-x)^3 + \ldots$$

$$= 1 - \frac{x}{2} - \frac{x^2}{8} - \frac{x^3}{16} \ldots \infty$$

31. $\because (1+x)^n = 1 + nx + \dfrac{n(n-1)}{2!}x^2 + \dfrac{n(n-1)(n-2)}{3!}x^3 + \ldots$

$$\therefore (1-2x)^{3/2} = 1 + \frac{3}{2}(-2x) + \frac{\frac{3}{2}\left(\frac{1}{2}\right)}{2!}(-2x)^2 + \frac{\frac{3}{2}\left(\frac{1}{2}\right)\left(\frac{-1}{2}\right)}{3!}(-2x)^3 + \ldots$$

Hence, 4th term $= \dfrac{\frac{3}{2}\left(\frac{1}{2}\right)\left(-\frac{1}{2}\right)}{6} \times (-8x^3) = \dfrac{x^3}{2}$

32. (i) $$\frac{1}{(3+2x)^2} = \frac{1}{9}\left(1 + \frac{2}{3}x\right)^{-2}$$

$$= \frac{1}{9}\left[1 - 2\left(\frac{2}{3}x\right) + 3\left(\frac{2}{3}x\right)^2 - \ldots\right]$$

$$= \frac{1}{9} - \frac{4}{27}x + \frac{4}{27}x^2 \text{ (three terms)}$$

(ii) $$\frac{1}{(2x+3)^2} = \frac{1}{4x^2}\left(1 + \frac{3}{2x}\right)^{-2}$$

$$= \frac{1}{4x^2}\left[1 - 2\left(\frac{3}{2x}\right) + 3\left(\frac{3}{2x}\right)^2 + \ldots\right]$$

$$= \frac{1}{4x^2} - \frac{3}{4x^3} + \frac{27}{16x^4} \text{ (three terms)}$$

33. $$\frac{1+x}{1-x} = (1+x)(1-x)^{-1}$$

$$= (1+x)\left[1 + \frac{(-1)}{1!}(-x) + \frac{(-1)(-1-1)}{2!}(-x)^2 + \frac{(-1)(-1-1)(-1-2)}{3!}(-x)^2 + \ldots\right]$$

$$= (1+x)(1 + x + x^2 + x^3 + x^4 + \ldots \infty)$$

$$= (1 + x + x^2 + x^3 + x^4 + \ldots \infty) + (x + x^2 + x^3 + x^4 + \ldots \infty)$$

$$= 1 + 2x + 2x^2 + 2x^3 + 2x^4 + \ldots \infty$$

$\therefore$ Coefficient of $x^5 = 2$

34. $\because (1+x)^n = 1 + nx + \dfrac{n(n-1)}{2!}x^2 + \ldots \infty; |x| < 1$

On comparing with given, $nx = \dfrac{1}{3}$ and $\dfrac{n(n-1)}{2!}x^2 = \dfrac{1 \cdot 3}{3 \cdot 6}$

$$\Rightarrow \frac{n^2x^2}{n(n-1)} \times \frac{2!}{x^2} = \frac{1}{9}\left(\frac{3 \cdot 6}{1 \cdot 3}\right) \qquad \left[\because (nx)^2 = \frac{1}{9}\right]$$

or $$\frac{2n}{n-1} = \frac{2}{3} \Rightarrow n = -\frac{1}{2}$$

$$\therefore \quad x = \frac{1}{3\left(-\frac{1}{2}\right)} = -\frac{2}{3}$$

Hence, sum of the given series

$$=(1+x)^n=\left(1-\frac{2}{3}\right)^{-1/2}=\left(\frac{1}{3}\right)^{-1/2}=\frac{1}{\left(\frac{1}{\sqrt{3}}\right)}=\sqrt{3}$$

35. Coefficient of x^7 in the expansion of $(1+3x-2x^3)^{10}$ is

$$\sum\frac{10!}{\alpha!\beta!\gamma!}(1)^{\alpha}(3x)^{\beta}(-2x^3)^{\gamma},$$

where $\alpha+\beta+\gamma=10$ and $\beta+3\gamma=7$

The possible values

α	β	γ
3	7	0
5	4	1
7	1	2

$\therefore$ Coefficient of x^7

$$=\frac{10!}{3!7!0!}(1)^3(3)^7(-2)^0+\frac{10!}{5!4!1!}(1)^5(3)^4(-2)^1+\frac{10!}{7!1!2!}(1)^7(3)^1(-2)^2$$

$$=262440-204120+4320=62640$$

36. We have,

$$(1+x^2-x^3)^9=\sum_{\alpha+\beta+\gamma=9}\frac{9!}{\alpha!\beta!\gamma!}1^{\alpha}\cdot(x^2)^{\beta}(-x^3)^{\gamma}$$

$$=\sum_{\alpha+\beta+\gamma=9}\frac{9!}{\alpha!\beta!\gamma!}(-1)^{\gamma}(x)^{2\beta+3\gamma}$$

Now, we must have $2\beta+3\gamma=8$, also $\alpha+\beta+\gamma=9$. It is easy to see that there are only two triplets of values of α,β and γ which can satisfy both equations namely (6, 1, 2) and (5, 4, 0).

Thus, the coefficient of x^8 is

$$\frac{9!}{6!1!2!}(-1)^2+\frac{9!}{5!4!0!}(-1)^0=378$$

37. Here, $a^3b^4c^7=(ab)^x(bc)^y(ca)^z$

$\Rightarrow z+x=3, x+y=4$ and $y+z=7$

$\Rightarrow \quad y=4, z=3$ and $x=0$

Coefficient of $a^3b^4c^7$ in the expansion of $(ab+bc+ca)^8$

= Coefficient of $(ab)^0(bc)^4(ca)^3$ in the expansion of $(bc+ca+ab)^8$

$$=\frac{8!}{0!4!3!}=280$$

38. The general term of the given expression

$$=\frac{10!}{\alpha!\beta!\gamma!}(2)^{\alpha/2}(3)^{\beta/3}(5)^{\gamma/6}$$

where α,β and γ satisfy the following properties.

$0\le\alpha,\beta,\gamma\le10$ and $\alpha+\beta+\gamma=10$

$\therefore\ \alpha=0,2,4,6,8,10; \beta=0,3,6,9$ and $\gamma=0,6$

Hence, possible values of

$$(\alpha,\beta,\gamma)=(4,6,0),(4,0,6),(10,0,0).$$

$\therefore$ There exist three rational terms.

So, sum of rational terms

$$=\frac{10!}{4!6!}2^2\cdot3^2+\frac{10!}{4!6!}2^2\cdot5+\frac{10!}{10!}2^5$$

$$=12632$$

39. The total number of distinct or dissimilar terms in the expansion of $(x+y+z+w)^n$ is

$$={}^{n+4-1}C_{4-1}={}^{n+3}C_3$$

$$=\frac{(n+3)(n+2)(n+1)}{1\cdot2\cdot3}=\frac{(n+1)(n+2)(n+3)}{6}$$

40. Putting $x=1$, we have $(1-2+3)^{50}=2^{50}$

Thus, the sum of coefficients is 2^{50}.

41. We have, $(\alpha^2x^2-3\alpha x+2)^{51}$

Sum of coefficients $=(\alpha^2-3\alpha+2)^{51}$ [putting $x=1$]

$\because\quad(\alpha^2-3\alpha+2)^{51}=0$ [given]

$\Rightarrow\quad \alpha^2-3\alpha+2=0$

$\Rightarrow\quad(\alpha-2)(\alpha-1)=0$

$\Rightarrow\quad \alpha=1,2$

42. $\because$ Sum of coefficients of $(\alpha x^2-2x+1)^{35}$

= Sum of coefficients of $(x-\alpha y)^{35}$

Putting $x=y=1$, we get

$(\alpha-1)^{35}=(1-\alpha)^{35}$

$\Rightarrow\quad(\alpha-1)^{35}=-(\alpha-1)^{35}$

$\Rightarrow\quad 2(\alpha-1)^{35}=0\Rightarrow\alpha-1=0$

$\therefore\quad \alpha=1$

43. Let $p(n):\ \sum_{r=0}^{n}r\cdot{}^nC_r=n\cdot2^{n-1}$ ...(i)

Step I For $n=1$,

LHS of Eq. (i) $=\sum_{r=0}^{1}r\cdot{}^1C_r=0+1\cdot{}^1C_1=1$

and RHS of Eq. (i) $=1\cdot2^{1-1}=1\cdot2^0=1$

$\therefore p(1)$ is true.

Step II Assume that $p(k)$ is true, then

$$p(k):\ \sum_{r=0}^{k}r\cdot{}^kC_r=k\cdot2^{k-1}$$

Step III For $n=k+1$,

$$p(k+1):\ \sum_{r=0}^{k+1}r\cdot{}^{k+1}C_r=(k+1)2^k$$

$\therefore$ LHS $= \sum_{r=0}^{k+1} r \cdot {}^{k+1}C_r = 0 + \sum_{r=1}^{k+1} r \cdot {}^{k+1}C_r$

$= \sum_{r=1}^{k+1} r \cdot {}^{k+1}C_r = \sum_{r=1}^{k} \{r \cdot {}^{k+1}C_r + (k+1) \cdot {}^{k+1}C_{k+1}\}$

$= \sum_{r=1}^{k} \{r \cdot ({}^kC_r + {}^kC_{r-1}) + (k+1)\}$

$= \sum_{r=0}^{k} r \cdot {}^kC_r + \sum_{r=1}^{k} \{r \cdot {}^kC_{r-1} + (k+1)\}$

$= \sum_{r=0}^{k} r \cdot {}^kC_r + \sum_{r=0}^{k+1} r \cdot {}^kC_{r-1}$

$= \sum_{r=0}^{k} r \cdot {}^kC_r + \sum_{r=0}^{k} (r+1)\, {}^kC_r$

$= \sum_{r=0}^{k} r \cdot {}^kC_r + \sum_{r=0}^{k} r \cdot {}^kC_r + \sum_{r=0}^{k} {}^kC_r$

$= p(k) + p(k) + 2^k$ [by step III]

$= k \cdot 2^{k-1} + k \cdot 2^{k-1} + 2^k$

$= (k+1)\, 2^k =$ RHS

$\therefore p(k+1)$ is true.

Hence, by mathematical induction $p(n)$ is true for all $n \in N$.

44. Let $p(n): 5^{2n+1} + 3^{n+2} \cdot 2^{n-1}$

Step I For $n = 1$, $p(1) = 5^{2+1} + 3^{1+2} \cdot 2^{1-1}$

$= 125 + 27 = 152,$

which is divisible by 19.

$\therefore$ The result is true for $n = 1$.

Step II Assume that the result is true for $n = k$

i.e. $p(k): 5^{2k+1} + 3^{k+2} \cdot 2^{k-1}$ is divisible by 19.

$\Rightarrow p(k) = 19r$, where r is an integer.

Step III For $n = k+1$,

$p(k+1) = 5^{2(k+1)+1} + 3^{k+1+2} \cdot 2^{k+1-1}$

$= 5^{2k+3} + 3^{k+3} \cdot 2^k$

$= 25 \cdot 5^{2k+1} + 3 \cdot 3^{k+2} \cdot 2 \cdot 2^{k-1}$

$= 25 \cdot 5^{2k+1} + 6 \cdot 3^{k+2} \cdot 2^{k-1}$

Now,

$\because 25 \cdot 5^{2k+1} + 6 \cdot 3^{k+2} \cdot 2^{k-1}$

$= 25 \cdot 5^{2k+1} + 25 \cdot 3^{k+2} \cdot 2^{k-1} - 19 \cdot 3^{k+2} \cdot 2^{k-1}$

$\therefore 25 \cdot 5^{2k+1} + 6 \cdot 3^{k+2} \cdot 2^{k-1}$

$= 25\,(5^{2k+1} + 3^{k+2} \cdot 2^{k-1}) - 19\,(3^{k+2} \cdot 2^{k-1})$

i.e. $p(k+1) = 25\, p(k) - 19 \cdot 3^{k+2} \cdot 2^{k-1}$

But we know $p(k)$ is divisible by 19.

Also, $19 \cdot 3^{k+2} \cdot 2^{k-1}$ is also divisible by 19.

$\therefore p(k+1)$ is divisible by 19.

Hence, by principle of mathematical induction, the result is true for all $n \in N$.

45. Let $p(n)$:

$$\tan^{-1}\left(\frac{1}{3}\right) + \tan^{-1}\left(\frac{1}{7}\right) + \ldots + \tan^{-1}\left(\frac{1}{n^2+n+1}\right) = \tan^{-1}\left(\frac{n}{n+2}\right)$$

Now, $p(1): \tan^{-1}\left(\frac{1}{3}\right) = \tan^{-1}\left(\frac{1}{3}\right)$

$\therefore p(1)$ is true.

Let $p(k)$ be true

i.e. $\tan^{-1}\left(\frac{1}{3}\right) + \tan^{-1}\left(\frac{1}{7}\right) + \ldots + \tan^{-1}\left(\frac{1}{k^2+k+1}\right) = \tan^{-1}\left(\frac{k}{k+2}\right)$, true

For $n = k+1$,

$\tan^{-1}\left(\frac{1}{3}\right) + \tan^{-1}\left(\frac{1}{7}\right) + \ldots + \tan^{-1}\left(\frac{1}{k^2+k+1}\right) + \tan^{-1}\left(\frac{1}{(k+1)^2 + (k+1) + 1}\right)$

$= \tan^{-1}\left(\frac{k}{k+2}\right) + \tan^{-1}\left[\frac{1}{k^2+3k+3}\right]$

$= \tan^{-1}\left[\dfrac{\frac{k}{k+2} + \frac{1}{k^2+3k+3}}{1 - \left(\frac{k}{k+2}\right)\left(\frac{1}{k^2+3k+3}\right)}\right]$

$= \tan^{-1}\left[\frac{(k+1)(k^2+2k+2)}{(k+3)(k^2+2k+2)}\right] = \tan^{-1}\left(\frac{k+1}{k+3}\right)$

$\Rightarrow \quad p(k+1): \tan^{-1}\left(\frac{k+1}{k+3}\right) = \tan^{-1}\left(\frac{k+1}{k+3}\right)$

$=$ RHS

$\therefore p(k+1)$ is also true.

$\therefore$ By the principle of mathematical induction $p(n)$ is true for all $n \in N$.

46. $\because I_m = \int_0^{\pi} \left(\frac{1 - \cos mx}{1 - \cos x}\right) dx$

Step I For $m = 1$,

$I_1 = \int_0^{\pi} \left(\frac{1 - \cos x}{1 - \cos x}\right) dx = \pi \Rightarrow I_1 = \pi$

For $m=2$,

$$I_2=\int_0^\pi \left(\frac{1-\cos 2x}{1-\cos x}\right)dx$$

$$=\int_0^\pi \frac{2\sin^2 x\,(1+\cos x)}{(1-\cos x)(1+\cos x)}dx$$

$$=\int_0^\pi \frac{2\sin^2 x\,(1+\cos x)}{\sin^2 x}dx$$

$$=2\int_0^\pi (1+\cos x)\,dx=2[x+\sin x]_0^\pi$$

$$=2[(\pi+0)-(0+0)]=2\pi,$$

which is true, therefore I_1 and I_2 are true.

Step II Assume, I_{k-1} and I_k to be true.

$\therefore\quad I_{k-1}=(k-1)\pi$ and $I_k=k\pi$

Step III For $m=k+1$,

$$I_{k+1}=\int_0^\pi\left[\frac{1-\cos(k+1)x}{1-\cos x}\right]dx$$

$$\therefore I_{k+1}-I_k=\int_0^\pi\left[\frac{\cos kx-\cos(k+1)x}{1-\cos x}\right]dx$$

$$=\int_0^\pi \frac{2\sin\left(\frac{2k+1}{2}\right)x\sin\left(\frac{x}{2}\right)}{2\sin^2\left(\frac{x}{2}\right)}dx$$

$$\Rightarrow I_{k+1}-I_k=\int_0^\pi \frac{\sin\left(\frac{2k+1}{2}\right)x}{\sin\left(\frac{x}{2}\right)}dx \quad\text{...(i)}$$

$$\text{Similarly, } I_k-I_{k-1}=\int_0^\pi \frac{\sin\left(\frac{2k-1}{2}\right)x}{\sin\left(\frac{x}{2}\right)}dx \quad\text{...(ii)}$$

On subtracting Eq. (ii) from Eq. (i), we get

$I_{k+1}-2I_k+I_{k-1}$

$$=\int_0^\pi \frac{\sin\left(\frac{2k+1}{2}\right)x-\sin\left(\frac{2k-1}{2}\right)x}{\sin\left(\frac{x}{2}\right)}dx$$

$$=\int_0^\pi \frac{2\cos(kx)\sin\left(\frac{x}{2}\right)}{\sin\left(\frac{x}{2}\right)}dx=2\int_0^\pi \cos(kx)\,dx$$

$$=2\left[\frac{\sin kx}{k}\right]_0^\pi=0\Rightarrow I_{k+1}=2I_k-I_{k-1}$$

$$=2k\pi-(k-1)\pi=k\pi+\pi=(k+1)\pi$$

This shows that the result is true for $m=k+1$.

Hence, by the principle of mathematical induction the result is true for all $m\in N$.

Level 2

1. (a) Given, $(1+ax)^n=1+8x+24x^2+\ldots$

$$\Rightarrow 1+n\cdot ax+\frac{n(n-1)}{1\cdot 2}a^2x^2+\ldots.=1+8x+24x^2+\ldots$$

On comparing the coefficients of x and x^2, we get

$$na=8 \text{ and } \frac{n(n-1)}{1\cdot 2}a^2=24$$

$\Rightarrow\quad na(n-1)a=48$

$\Rightarrow\quad 8(8-a)=48\Rightarrow 8-a=6$

$\Rightarrow\quad a=2\Rightarrow n=4$

2. (a) $\because (x+a)^n={}^nC_0\cdot x^n+{}^nC_1\cdot x^{n-1}a+{}^nC_2\cdot x^{n-2}a^2+\ldots$

or $\quad (x+a)^n=({}^nC_0\cdot x^n+{}^nC_2\cdot x^{n-2}a^2$

$\quad +{}^nC_4\cdot x^{n-4}\cdot a^4+\ldots)$

$\quad +({}^nC_1\cdot x^{n-1}\cdot a+{}^nC_3\cdot x^{n-3}\cdot a^3+\ldots)$

or $\quad (x+a)^n=A+B \quad$ ...(i)

Also, $(x-a)^n=A-B \quad$ ...(ii)

On multiplying Eqs. (i) and (ii), we get

$$(x^2-a^2)^n=A^2-B^2$$

3. (c) Given, $(x+10)^{50}+(x-10)^{50}$

$$=a_0+a_1x+a_2x^2+\ldots+a_{50}x^{50}$$

$\because\ (x+a)^n+(x-a)^n=2(T_1+T_3+T_5+T_7+\ldots)$

$\therefore\ (x+10)^{50}+(x-10)^{50}$

$$=2({}^{50}C_0\cdot x^{50}+{}^{50}C_2\cdot x^{48}\times 10^2+{}^{50}C_4\cdot x^{46}\times 10^4+\ldots)$$

$\Rightarrow a_0=2\times {}^{50}C_0\times 10^{50}$ and $a_2=2\times {}^{50}C_2(10)^{48}$

$$\Rightarrow \frac{a_2}{a_0}=\frac{{}^{50}C_2\times(10)^{48}}{{}^{50}C_0\times(10)^{50}}=\frac{50\times 49}{2\times 100}=\frac{49}{4}=12.25$$

4. (d) Using binomial expansion

$(x+a)^n+(x-a)^n=2(T_1+T_3+T_5+T_7+\ldots)$

$\therefore (x+\sqrt{x^2-1})^6+(x-\sqrt{x^2-1})^6=2(T_1+T_3+T_5+T_7)$

$=2[{}^6C_0x^6+{}^6C_2x^4(x^2-1)+{}^6C_4x^2(x^2-1)^2+{}^6C_6(x^2-1)^3]$

$=2[x^6+15(x^6-x^4)+15x^2(x^4-2x^2+1)+(-1+3x^2-3x^4+x^6)]$

$=2(32x^6-48x^4+18x^2-1)$

$\therefore \alpha=-96$ and $\beta=36$

$\therefore \alpha-\beta=-96-36=-132$

5. (a) We have,

$$\frac{2}{\sqrt{2x^2+1}+\sqrt{2x^2-1}}=\frac{2(\sqrt{2x^2+1}-\sqrt{2x^2-1})}{(2x^2+1)-(2x^2-1)}$$

$$=\sqrt{2x^2+1}-\sqrt{2x^2-1}$$

Thus, the given expression can be written as

$$(\sqrt{2x^2+1}+\sqrt{2x^2-1})^6+(\sqrt{2x^2+1}-\sqrt{2x^2-1})^6$$

But $(a+b)^6+(a-b)^6$

$$=2[a^6+{}^6C_2\cdot a^4\cdot b^2+{}^6C_4\cdot a^2\cdot b^4+b^6]$$

Therefore,

$$(\sqrt{2x^2+1}+\sqrt{2x^2-1})^6+(\sqrt{2x^2+1}-\sqrt{2x^2-1})^6$$

$$=2[(2x^2+1)^3+15(2x^2+1)^2(2x^2-1)$$

$$+15(2x^2+1),(2x^2-1)^2+(2x^2-1)^3]$$

which is a polynomial of degree 6.

6. (b) Putting $x=i$, we get

$$(1+i)^5=(\alpha_0-\alpha_2+\alpha_4)+i(\alpha_1-\alpha_3+\alpha_5)$$

$$\Rightarrow |1+i|^5=|(\alpha_0-\alpha_2+\alpha_4)+i(\alpha_1-\alpha_3+\alpha_5)|$$

$$\Rightarrow (\alpha_0-\alpha_2+\alpha_4)^2+(\alpha_1-\alpha_3+\alpha_5)^2=2^5=32$$

7. (b) $T_7={}^9C_6\left(\dfrac{3}{\sqrt[3]{84}}\right)^3(\sqrt{3}\ln x)^6=729$

$$\Rightarrow \frac{84\times 3^3}{84}\times 3^3\times(\ln x)^6=729$$

$$\Rightarrow (\ln x)^6=1 \Rightarrow x=e$$

8. (a) Given, $T_3=2560$

$$\Rightarrow {}^5C_2(x^{\log_2 x})^2=2560$$

$$\Rightarrow 10(x^{\log_2 x})^2=2560$$

$$\Rightarrow (x^{\log_2 x})^2=256=(\pm 16)^2$$

$$\Rightarrow x^{\log_2 x}=16 \text{ or } x^{\log_2 x}=-16 \text{ (rejected)}$$

$$\Rightarrow x^{\log_2 x}=16$$

$$\Rightarrow (\log_2 x)^2=\log_2 16=\log_2(2)^4=4$$

$$\Rightarrow \log_2 x=\pm 2 \Rightarrow x=2^2 \text{ or } 2^{-2}$$

$$\Rightarrow x=4 \text{ or } \frac{1}{4}$$

9. (c) Coefficient of T_5 is ${}^nC_4$ that of T_6 is ${}^nC_5$ and that of T_7 is ${}^nC_6$.

According to the condition, $2\,{}^nC_5={}^nC_4+{}^nC_6$

$$\therefore 2\left[\frac{n!}{(n-5)!5!}\right]=\left[\frac{n!}{(n-4)!4!}+\frac{n!}{(n-6)!6!}\right]$$

$$\text{or } 2\left[\frac{1}{(n-5)5}\right]=\left[\frac{1}{(n-4)(n-5)}+\frac{1}{6\times 5}\right]$$

After solving, we get $n=7$ or 14.

10. (b) $T_5={}^nC_4\cdot a^{n-4}(-2b)^4$

and $T_6={}^nC_5\cdot a^{n-5}(-2b)^5$

As, $T_5+T_6=0$, we get

$${}^nC_4\cdot 2^4a^{n-4}\cdot b^4={}^nC_5\cdot 2^5\cdot a^{n-5}\cdot b^5$$

$$\text{or } \frac{a^{n-4}\cdot b^4}{a^{n-5}\cdot b^5}=\frac{n!2^5}{5!(n-5)!}\cdot\frac{4!(n-4)!}{n!2^4}$$

$$\text{or } \frac{a}{b}=\frac{2(n-4)}{5}$$

11. (b) We have,

$$T_{r+1}={}^{29}C_r\times 3^{29-r}(7x)^r=({}^{29}C_r\times 3^{29-r}\times 7^r)x^r$$

Coefficient of $(r+1)$th term is ${}^{29}C_r\times 3^{29-r}\times 7^r$

and coefficient of rth term is ${}^{29}C_{r-1}\times 3^{30-r}\times 7^{r-1}$

From given condition,

$${}^{29}C_r\times 3^{29-r}\times 7^r={}^{29}C_{r-1}\times 3^{30-r}\times 7^{r-1}$$

$$\Rightarrow \frac{{}^{29}C_r}{{}^{29}C_{r-1}}=\frac{3}{7}$$

$$\Rightarrow \frac{30-r}{r}=\frac{3}{7} \text{ or } r=21$$

12. (c) It is given that 6th term in the expansion of $\left(\dfrac{1}{x^{8/3}}+x^2\log_{10}x\right)^8$ is 5600, therefore

$${}^8C_5(x^2\log_{10}x)^5\left(\frac{1}{x^{8/3}}\right)^3=5600$$

$$\Rightarrow 56x^{10}(\log_{10}x)^5\left(\frac{1}{x^8}\right)=5600$$

$$\Rightarrow x^2(\log_{10}x)^5=100$$

$$\Rightarrow x^2(\log_{10}x)^5=10^2(\log_{10}10)^5$$

$$\therefore x=10$$

13. (b) Given, $T_4=20\times 8^7$

$$\Rightarrow {}^6C_3\left(\frac{2}{x}\right)^3\times(x^{\log_8 x})^3=20\times 8^7$$

$$\Rightarrow 8\times 20\times\left(\frac{x^{\log_8 x}}{x}\right)^3=20\times 8^7$$

$$\Rightarrow \left(\frac{x^{\log_8 x}}{x}\right)^3=8^6$$

$$\Rightarrow \frac{x^{\log_8 x}}{x}=(8^6)^{1/3}=64$$

Now, take $\log_8$ on both sides, then

$$(\log_8 x)^2-\log_8 x=\log_8 64=\log_8 8^2=2$$

$$\Rightarrow (\log_8 x)^2-\log_8 x=2$$

$$\Rightarrow \log_8 x=-1 \text{ or } \log_8 x=2$$

$$\Rightarrow x=\frac{1}{8}$$

$$\text{or } x=8^2$$

14. (c) The 4th term from the end $= 9-4+2$
$= 7$th term from beginning

$$\therefore \quad T_7 = {}^9C_6\left(\frac{x^3}{2}\right)^3\left(-\frac{2}{x^2}\right)^6 = \frac{672}{x^3}$$

15. (b) Given, $\frac{T_7}{T_{n-7+2}} = \frac{1}{6}$

$$\Rightarrow \quad \frac{T_7}{T_{n-5}} = \frac{1}{6}$$

$$\Rightarrow \quad \frac{{}^nC_6(\sqrt[3]{2})^{n-6}\left(\frac{1}{\sqrt[3]{3}}\right)^6}{{}^nC_{n-6}(\sqrt[3]{2})^6\left(\frac{1}{\sqrt[3]{3}}\right)^{n-6}} = \frac{1}{6}$$

$$\Rightarrow \quad 2^{\frac{n-12}{3}} \cdot 3^{\frac{n-12}{3}} = \frac{1}{6}$$

$$\Rightarrow \quad 6^{\frac{n-12}{3}} = 6^{-1}$$

$$\therefore \quad \frac{n-12}{3} = -1 \Rightarrow n = 9$$

16. (c) 5th term from the beginning,

$$T_5 = T_{4+1} = {}^{10}C_4(2^{1/3})^6\left(\frac{1}{2\cdot 3^{1/3}}\right)^4$$

and 5th term from the end,

$$T_{10-5+2} = T_7 = T_{6+1} = {}^{10}C_6(2^{1/3})^4\left(\frac{1}{2\cdot 3^{1/3}}\right)^6$$

$$\therefore \quad \frac{T_5}{T_7} = \frac{{}^{10}C_4(2^{1/3})^6\left(\frac{1}{2\cdot 3^{1/3}}\right)^4}{{}^{10}C_6(2^{1/3})^4\left(\frac{1}{2\cdot 3^{1/3}}\right)^6} = \frac{(2^{1/3})^2}{\left(\frac{1}{2\cdot 3^{1/3}}\right)^2}$$

$$= \frac{2^{2/3}\times 2^2\times 3^{2/3}}{1} = \frac{4(6)^{2/3}}{1} = \frac{4(36)^{1/3}}{1}$$

17. (c) Middle term in the expansion is $\left(\frac{10}{2}+1\right)$th

i.e. 6th term.

Then, $T_6 = 7\frac{7}{8}$

$$\Rightarrow {}^{10}C_5 \cdot \frac{1}{x^5} \cdot x^5 \cdot \sin^5 x = \frac{63}{8}$$

$$\Rightarrow \quad 252 \cdot \sin^5 x = \frac{63}{8}$$

$$\Rightarrow \quad \sin^5 x = \frac{1}{32}$$

$$\Rightarrow \quad \sin x = \frac{1}{2}$$

$$\therefore \quad x = n\pi + (-1)^n\frac{\pi}{6}$$

18. (c) Middle term of $(1+\alpha x)^4$ is T_3, its coefficient is ${}^4C_2(\alpha)^2 = 6\alpha^2$

Middle term of $(1-\alpha x)^6$ is T_4, its coefficient is ${}^6C_3(-\alpha)^3 = -20\alpha^3$

According to the question,

$$6\alpha^2 = -20\alpha^3$$

$$\Rightarrow \quad 3\alpha^2 + 10\alpha^3 = 0 \Rightarrow \alpha^2(3+10\alpha) = 0$$

$$\therefore \quad \alpha = -\frac{3}{10}$$

19. (c) General term, $T_{r+1} = {}^{10}C_r(\sqrt{x})^{10-r}\left(-\frac{k}{x^2}\right)^r$

$$= {}^{10}C_r(-k)^r(x)^{\frac{10-r}{2}-2r}$$

$$= {}^{10}C_r(-k)^r(x)^{\frac{10-5r}{2}}$$

For a constant term, $\frac{10-5r}{2} = 0 \Rightarrow r = 2$

$$\therefore \quad {}^{10}C_2(-k)^2 = 405$$

$$\Rightarrow \quad k^2 = \frac{405\times 2}{10\times 9} = \frac{81}{9} = 9$$

$$\therefore \quad |k| = 3$$

20. (c) General term, $T_{r+1} = {}^9C_r\left(\frac{3}{2}x^2\right)^{9-r}\left(-\frac{1}{3x}\right)^r$

$$= {}^9C_r\left(\frac{3}{2}\right)^{9-r}\left(-\frac{1}{3}\right)^r x^{18-3r}$$

The term independent of x, then

$$18 - 3r = 0 \Rightarrow r = 6$$

$$\therefore \quad T_7 = {}^9C_6\left(\frac{3}{2}\right)^3\left(-\frac{1}{3}\right)^6$$

$$= {}^9C_3\left(\frac{1}{6}\right)^3$$

$$= \frac{9\times 8\times 7}{3\times 2\times 1}\left(\frac{1}{6}\right)^3 = \frac{7}{18}$$

$$\therefore \quad 18k = 18\times\frac{7}{18} = 7$$

21. (c) $\left[\frac{x+1}{x^{2/3}-x^{1/3}+1} - \frac{x-1}{x-x^{1/2}}\right]^{10}$

$$= \left[\frac{(x^{1/3}+1)(x^{2/3}-x^{1/3}+1)}{(x^{2/3}-x^{1/3}+1)} - \frac{(x^{1/2}-1)(x^{1/2}+1)}{x^{1/2}(x^{1/2}-1)}\right]^{10}$$

$$[\because x+1 = (x^{1/3})^3 + 1^3]$$

$$= [(x^{1/3}+1) - (1+x^{-1/2})]^{10} = [x^{1/3} - x^{-1/2}]^{10}$$

$$\Rightarrow T_{r+1} = {}^{10}C_r(x^{1/3})^{10-r}(-x^{-1/2})^r$$

$$\Rightarrow T_{r+1} = {}^{10}C_r(x)^{\frac{10-r}{3} - \frac{r}{2}} \cdot (-1)^r$$

$$\Rightarrow T_{r+1} = {}^{10}C_r(x)^{\frac{20-5r}{6}} \cdot (-1)^r$$

For constant term,

$$20 - 5r = 0 \Rightarrow r = 4$$

$$\Rightarrow \quad T_5 = {}^{10}C_4 = 210$$

22. (c) $[(t^{-1} - 1)x + (t^{-1} + 1)^{-1}x^{-1}]^8$

$$= \left[\left(\frac{1}{t} - 1\right)x + \left(\frac{1}{t} + 1\right)^{-1}\frac{1}{x}\right]^8$$

Let T_{r+1} be the term independent of x, then

$$T_{r+1} = {}^8C_r\left(\frac{1}{t} - 1\right)^{8-r}(x)^{8-r} \cdot \left(\frac{1}{t} + 1\right)^{-r}\left(\frac{1}{x}\right)^r$$

$$= {}^8C_r\left(\frac{1}{t} - 1\right)^{8-r}\left(\frac{1}{t} + 1\right)^{-r}(x)^{8-2r}$$

$\therefore 8 - 2r = 0 \Rightarrow r = 4$

$\therefore T_5$ is the term independent of x.

$$T_5 = {}^8C_4\left(\frac{1}{t} - 1\right)^4\left(\frac{1}{t} + 1\right)^{-4}$$

$$= {}^8C_4\left(\frac{1-t}{t}\right)^4\left(\frac{1+t}{t}\right)^{-4}$$

$$= {}^8C_4\left(\frac{1-t}{1+t}\right)^4$$

$$= 70\left(\frac{1-t}{1+t}\right)^4$$

23. (d) Coefficient of x^n in $(1+x)^{2n} = {}^{2n}C_n$

and coefficient of x^n in $(1+x)^{2n-1} = {}^{2n-1}C_n$

$$\therefore \text{Required ratio} = \frac{{}^{2n}C_n}{{}^{2n-1}C_n} = \frac{\frac{2n!}{n!\,n!}}{\frac{(2n-1)!}{n!\,(n-1)!}} = \frac{2}{1}$$

24. (c) For $\left\{ax^2 + \left(\frac{1}{bx}\right)\right\}^{11}$

$$T_{r+1} = {}^{11}C_r(ax^2)^{11-r}\left(\frac{1}{bx}\right)^r$$

$$= {}^{11}C_r(a)^{11-r}\left(\frac{1}{b^r}\right)(x)^{22-3r}$$

For x^7, $22 - 3r = 7$ or $3r = 15$ or $r = 5$

$$\Rightarrow \quad T_6 = {}^{11}C_5 \cdot a^6\frac{1}{b^5} \cdot x^7$$

$\Rightarrow$ Coefficient of x^7 is ${}^{11}C_5 \cdot \frac{a^6}{b^5}$

Similarly, coefficient of x^{-7} in $\left\{ax - \left(\frac{1}{bx^2}\right)\right\}^{11}$ is

$${}^{11}C_6 \cdot \frac{a^5}{b^6}$$

Given, ${}^{11}C_5 \cdot \frac{a^6}{b^5} = {}^{11}C_6 \cdot \frac{a^5}{b^6} \Rightarrow a = \frac{1}{b}$ or $ab = 1$

25. (b) General term, $T_{r+1} = {}^{16}C_r\left(\frac{x}{\cos\theta}\right)^{16-r}\left(\frac{1}{x\sin\theta}\right)^r$

$$\Rightarrow T_{r+1} = {}^{16}C_r\left(\frac{1}{\cos\theta}\right)^{16-r}\left(\frac{1}{\sin\theta}\right)^r(x)^{16-r-r}$$

$$= {}^{16}C_r\left(\frac{1}{\cos\theta}\right)^{16-r}\left(\frac{1}{\sin\theta}\right)^r(x)^{16-2r}$$

The term is independent of x, when $16 - 2r = 0 \Rightarrow r = 8$

$$\therefore \quad T_9 = {}^{16}C_8\frac{1}{\cos^8\theta\sin^8\theta}$$

$$= {}^{16}C_8\frac{2^8}{(\sin 2\theta)^8} \qquad [\because \sin 2\theta = 2\sin\theta\cos\theta]$$

When $\theta \in \left[\frac{\pi}{8}, \frac{\pi}{4}\right]$, the least value of the term

$l_1 = {}^{16}C_8 \cdot 2^8$ [minimum value will occur at $\theta = \frac{\pi}{4}$]

When $\theta \in \left[\frac{\pi}{16}, \frac{\pi}{8}\right]$, the least value of term

$l_2 = {}^{16}C_8 \times \dfrac{2^8}{\left(\frac{1}{\sqrt{2}}\right)^8}$ [minimum occur at $\theta = \frac{\pi}{8}$]

$$= {}^{16}C_8 \times 2^8 \times 2^4$$

$$\text{Now, } \frac{l_2}{l_1} = \frac{{}^{16}C_8 \times 2^8 \times 2^4}{{}^{16}C_8 \times 2^8} = 2^4 = 16$$

$$l_2 : l_1 = 16 : 1$$

26. (d) $\because T_{r+1} = {}^{60}C_r(7^{1/5})^{60-r}(-3^{1/10})^r$

$$= {}^{60}C_r\,(7)^{12 - \frac{r}{5}}(-1)^r\,(3)^{\frac{r}{10}}$$

Then, for getting rational terms, r should be multiple of LCM of $(5, 10) = 10$. Then, r can be 0, 10, 20, 30, 40, 50 and 60.

Since, total number of terms = 61

Hence, total irrational terms = 61 − 7 = 54

27. (c) Here, $(3^{1/2} + 5^{1/8})^n$

$$\therefore \quad T_{r+1} = {}^nC_r(3)^{\frac{n-r}{2}}(5)^{\frac{r}{8}}$$

$\because \dfrac{n-r}{2}$ and $\dfrac{r}{8}$ are integer.

So, r must be 0, 8, 16, 24, ...

$\because$ Number of terms = 33

$\Rightarrow n = t_{33} = a + (n-1)d = 0 + 32 \times 8 = 256$

28. (a) $T_{r+1} = {}^{4n-2}C_r (ix)^r$

T_{r+1} is negative, if i^r is negative and real

$i^r = -1$

$\Rightarrow r = 2, 6, 10, \ldots$ which form an AP.

$$0 \le r \le 4n-2$$
$$4n-2 = 2 + (r-1)\,4$$
$$\Rightarrow \qquad r = n$$

The required number of terms is n.

29. (c) We have, $\sum_{r=0}^{n} r\,{}^{n}C_r \cdot x^r \cdot y^{n-r}$

$$= \sum_{r=0}^{n} r \cdot \frac{n}{r} \cdot {}^{n-1}C_{r-1} \cdot x^{r-1} \cdot x^1 \cdot y^{n-r}$$

$$= nx \sum_{r=0}^{n} {}^{n-1}C_{r-1} \cdot x^{r-1} \cdot y^{(n-1)-(r-1)}$$

$$= nx\,(x+y)^{n-1} = nx \qquad [\because x + y = 1]$$

30. (c) Given series,

$(x-3)^{100} + {}^{100}C_1 (x-3)^{99} \cdot 2^1 + {}^{100}C_2 (x-3)^{98} \cdot 2^2 + \ldots + {}^{100}C_{100} \cdot 2^{100}$

$$= (x-3+2)^{100} = (x-1)^{100} = (1-x)^{100}$$

$\therefore$ Coefficient of x^{53} in $(1-x)^{100} = -{}^{100}C_{53}$

31. (c) $(1-x-x^3+x^4)^8 = (1-x)^8 (1-x^3)^8$

$$= (1 - {}^8C_1 x + {}^8C_2 x^2 - {}^8C_3 x^3 + \ldots) \times (1 - {}^8C_1 x^3 + {}^8C_2 x^6 - {}^8C_3 x^9 + \ldots)$$

$\therefore$ Coefficient of $x^7 = -{}^8C_7 - {}^8C_1 \times {}^8C_4 - {}^8C_1 \times {}^8C_2$

$= -792$

32. (d) $(x-2)^5 (x+1)^5$

$$= [{}^5C_0 x^5 - {}^5C_1 x^4 \times 2 + \ldots][{}^5C_0 + {}^5C_1 x + \ldots]$$

$\Rightarrow$ Coefficient of x^5

$= {}^5C_0 {}^5C_5 - {}^5C_1 \times 2 \times {}^5C_4 + {}^5C_2 \times 2^2 \times {}^5C_3 - {}^5C_3 \times 2^3 \times {}^5C_2 + {}^5C_4 \times 2^4 \times {}^5C_1 - {}^5C_5 \times 2^5 \times {}^5C_0$

$= 1 - 5 \times 5 \times 2 + 10 \times 10 \times 4 - 10 \times 10 \times 8 + 5 \times 5 \times 16 - 32$

$= -81$

33. (c) We have, $(1+x)^{101} (1-x+x^2)^{100}$

$$= (1+x)\{(1+x)(1-x+x^2)\}^{100}$$
$$= (1+x)(1+x^3)^{100}$$
$$= (1+x)[C_0 + C_1 x^3 + C_2 x^6 + \ldots + C_{100} x^{300}]$$
$$= (1+x) \sum_{r=0}^{n} {}^nC_r \cdot x^{3r} = \sum_{r=0}^{n} {}^nC_r \cdot x^{3r} + \sum_{r=0}^{n} {}^nC_r \cdot x^{3r+1}$$

where, $n = 100$

Hence, there will be no term containing $3r+2$.

34. (b) $(1+x^3-x^6)^{30}$

$$= \{1 + x^3(1-x^3)\}^{30}$$
$$= {}^{30}C_0 + {}^{30}C_1 x^3 (1-x^3) + {}^{30}C_2 x^6 (1-x^3)^2 + \ldots$$

Obviously, each term will contain x^{3m}, $m \in N$. But 28 is not divisible by 3.

Therefore, there will be no term containing x^{28}.

35. (c) $\left(1 + x + \dfrac{7}{x}\right)^{11} = \left(\dfrac{7 + x + x^2}{x}\right)^{11}$

So, we have find the coefficient of x^{11} in $(7+x+x^2)^{11}$

Now, $(7+x+x^2)^{11}$

$= [(7+x)+x^2]^{11}$

$= {}^{11}C_0 (7+x)^{11} + {}^{11}C_1 (7+x)^{10} x^2 + {}^{11}C_2 (7+x)^9 x^4 + \ldots$

$\therefore$ Required coefficient

$$= {}^{11}C_0 + {}^{11}C_1 \times {}^{10}C_9 \times 7 + {}^{11}C_2 \times {}^9C_7 \times 7^2 + {}^{11}C_3 \times {}^8C_5 \times 7^3 + {}^{11}C_4 \times {}^7C_3 \times 7^4 + {}^{11}C_5 \times {}^6C_1 \times 7^5$$

$$= \sum_{r=0}^{5} {}^{11}C_r \cdot {}^{11-r}C_{11-2r} \cdot 7^r$$

36. (d) $(1-x-x^2+x^3)^6 = (1-x)^6 (1-x^2)^6$

$$= (1 - {}^6C_1 x + {}^6C_2 x^2 - {}^6C_3 x^3 + {}^6C_4 x^4 - {}^6C_5 x^5 + {}^6C_6 x^6)(1 - {}^6C_1 x^2 + {}^6C_2 x^4 - {}^6C_3 x^6 + \ldots)$$

$\Rightarrow$ Coefficient of x^7 is

${}^6C_1 \cdot {}^6C_3 - {}^6C_3 \cdot {}^6C_2 + {}^6C_5 \cdot {}^6C_1$

$= 6 \times 20 - 20 \times 15 + 6 \times 6 = -144$

37. (b) We have,

$(x+3)^{n-1} + (x+3)^{n-2}(x+2) + (x+3)^{n-3} (x+2)^2 + \ldots (x+2)^{n-1} = \dfrac{(x+3)^n - (x+2)^n}{(x+3)-(x+2)}$

$$= (x+3)^n - (x+2)^n$$

$$\left[\because \frac{x^n - a^n}{x - a} = x^{n-1} + x^{n-2} \cdot a^1 + x^{n-3} \cdot a^2 + \ldots a^{n-1}\right]$$

Therefore, coefficient of x^r in the given expression is equal to coefficient of x^r in $[(x+3)^n - (x+2)^n]$ which is given by

$${}^nC_r \cdot 3^{n-r} - {}^nC_r \cdot 2^{n-r} = {}^nC_r (3^{n-r} - 2^{n-r})$$

38. (d) $(1+ax+bx^2)(1-2x)^{18}$
$$=1(1-2x)^{18}+ax(1-2x)^{18}+bx^2(1-2x)^{18}$$
Coefficient of x^3
$$=(-2)^3\,{}^{18}C_3+a(-2)^2\,{}^{18}C_2+b(-2)\,{}^{18}C_1=0$$
$$\Rightarrow \quad \frac{4\times(17\times16)}{(3\times2)}-2a\cdot\frac{17}{2}+b=0 \quad \text{...(i)}$$
Coefficient of $x^4=(-2)^4\,{}^{18}C_4+a(-2)^3\,{}^{18}C_3$
$$+b(-2)^2\,{}^{18}C_2=0$$
$$\Rightarrow \quad (4\times20)-2a\cdot\frac{16}{3}+b=0 \quad \text{...(ii)}$$
From Eqs. (i) and (ii), we get
$$4\left(\frac{17\times8}{3}-20\right)+2a\left(\frac{16}{3}-\frac{17}{2}\right)=0$$
$$\Rightarrow \quad a=16$$
$$\Rightarrow \quad b=\frac{2\times16\times16}{3}-80=\frac{272}{3}$$

39. (a) Consider the coefficient of three consecutive terms of $(1+x)^{n+5}$ be ${}^{n+5}C_r,\ {}^{n+5}C_{r+1},\ {}^{n+5}C_{r+2}$

Given, $\dfrac{{}^{n+5}C_r}{{}^{n+5}C_{r+1}}=\dfrac{5}{10}$
$$\Rightarrow \quad \frac{r+1}{n+5-r}=\frac{1}{2} \quad \left[\because \frac{{}^nC_r}{{}^nC_{r-1}}=\frac{n-r+1}{r}\right]$$
$$\Rightarrow \quad 3r=n+3 \quad \text{...(i)}$$
Also, $\dfrac{{}^{n+5}C_{r+1}}{{}^{n+5}C_{r+2}}=\dfrac{10}{14} \quad \left[\because \dfrac{{}^nC_{r-1}}{{}^nC_r}=\dfrac{r}{n-r+1}\right]$
$$\Rightarrow \quad \frac{r+2}{n+4-r}=\frac{5}{7}$$
$$\Rightarrow \quad 12r=5n+6 \quad \text{...(ii)}$$
On solving Eqs. (i) and (ii), we get
$$r=3 \text{ and } n=6$$
$\therefore$ Largest coefficient in the expansion is ${}^{11}C_5=462$.

40. (b) If n is odd, then numerically greatest coefficient in the expansion of $(1-x)^n$ is ${}^nC_{(n-1)/2}$ or ${}^nC_{(n+1)/2}$.

$\therefore$ $(1-x)^{21}$, the numerically greatest coefficient is ${}^{21}C_{10}$ or ${}^{21}C_{11}$.

So, the numerically greatest term
$$={}^{21}C_{11}\cdot x^{11} \text{ or } {}^{21}C_{10}\cdot x^{10}$$
So, $|{}^{21}C_{11}\cdot x^{11}|>|{}^{21}C_{12}\cdot x^{12}|$

and $|{}^{21}C_{10}\cdot x^{10}|>|{}^{21}C_9\cdot x^9|$
$$\Rightarrow \quad \frac{21!}{10!\,11!}>\frac{21!}{9!\,12!}x \text{ and } \frac{21!}{11!\,10!}x>\frac{21!}{9!\,12!}$$
$$\Rightarrow \quad x<\frac{6}{5} \text{ and } x>\frac{5}{6} \Rightarrow x\in\left(\frac{5}{6},\frac{6}{5}\right)$$

41. (a) We know that

$\dfrac{n-r+1}{r}\left|\dfrac{a}{r}\right|\geq1$ is the greatest term in expansion $(x+a)^n$.
$$\Rightarrow \quad \frac{54-r+1}{r}\left|\frac{3x}{1}\right|\geq1$$
$$\Rightarrow \quad \frac{55-r}{r}\left|\frac{3}{1}\times\frac{1}{3}\right|\geq1 \quad \left[\because x=\frac{1}{3}\right]$$
$$\Rightarrow \quad 55-r\geq r$$
$$\Rightarrow \quad 55\geq2r$$
$$\Rightarrow \quad r\leq\frac{55}{2}\approx27.5$$
$$\therefore \quad r=27$$
$\therefore$ The greatest term $=T_{r+1}=T_{27+1}=T_{28}$

Hence, 28th term is the greatest term.

42. (b) Let $f=(8-3\sqrt{7})^{10}$, here $0<f<1$

$\therefore (8+3\sqrt{7})^{10}+(8-3\sqrt{7})^{10}$ is an integer, hence this is the value of n.

43. (b) Let $F=(6\sqrt{6}-14)^{2n+1}$

Then, $R-F=(6\sqrt{6}+14)^{2n+1}-(6\sqrt{6}-14)^{2n+1}$
$$=2[{}^{2n+1}C_1(6\sqrt{6})^{2n}(14)+\dots]$$
$$=\text{an even integer}$$
$\Rightarrow [R]+f-F=$ an even integer

Also, $0\leq f<1$ and $0<F<1$

$-1<f-F<1$ and $f-F$ is integer.

$\therefore f-F=0\Rightarrow f=F$

So, $Rf=RF=(6\sqrt{6}+14)^{2n+1}(6\sqrt{6}-14)^{2n+1}$
$$=(20)^{2n+1}$$
$\therefore$ The value of $k=20$

44. (b) $\because 2^{403}=2^{400}\times2^3=(2^4)^{100}\times2^3=8(2^4)^{100}$
$$\Rightarrow \quad 2^{403}=8(16)^{100}=8(1+15)^{100}$$
$$\Rightarrow \quad 2^{403}=8\left[1+100\times15+\frac{100\times99}{2}(15)^2+\dots\right]$$
$$\Rightarrow \quad \frac{2^{403}}{15}=\frac{8}{15}+\text{Integer}$$
Hence, fractional part of the number is $\dfrac{8}{15}$.

$\therefore$ The value of k is 8.

45. (d) $\because \dfrac{3^{200}}{8}=\dfrac{1}{8}(9^{100})=\dfrac{1}{8}(1+8)^{100}$
$$\Rightarrow \quad \frac{3^{200}}{8}=\frac{1}{8}\left[1+100\cdot8+\frac{100(100-1)}{2!}8^2+\dots\right]$$
$$\Rightarrow \quad \frac{3^{200}}{8}=\frac{1}{8}+\text{Integer}$$
$$\therefore \quad \left\{\frac{3^{200}}{8}\right\}=\left\{\frac{1}{8}+\text{Integer}\right\}=\frac{1}{8}$$

46. (d) We have, $25^{15} = (26-1)^{15}$

$= {}^{15}C_0(26)^{15} - {}^{15}C_1(26)^{14} + {}^{15}C_2(26)^{13} + \dots - {}^{15}C_{15}$

$= 26k - {}^{15}C_{15}$ [where, $k \in I$]

$= 26k - 1 = 26k - 13 + 12$

$= 13(2k-1) + 12$

$\therefore$ The remainder when 25^{15} is divided by 13 is 12.

47. (a) We have, $7^9 + 9^7 = (8-1)^9 + (8+1)^7$

$= (1+8)^7 - (1-8)^9$

$= [1 + {}^7C_1 8 + {}^7C_2 8^2 + \dots + {}^7C_7 8^7]$

$- [1 - {}^9C_1 8 + {}^9C_2 8^2 \dots - {}^9C_9 8^9]$

$= {}^7C_1 8 + {}^9C_1 8 + [{}^7C_2 + {}^7C_3 8$

$\dots - {}^9C_2 + {}^9C_3 8 \dots] 8^2$

$= 8(7+9) + 64k, k \in I$

$= 8 \cdot 16 + 64k$

$= 64(k+2)$

Thus, $7^9 + 9^7$ is divisible by 64.

48. (a) $\because \left(1+\frac{1}{n}\right)^n < 3$ for $n \in N$

Now, $\frac{(1001)^{999}}{(1000)^{1000}} = \frac{1}{1001}\left(\frac{1001}{1000}\right)^{1000}$

$= \frac{1}{1001}\left(1+\frac{1}{1000}\right)^{1000} < \frac{1}{1001} \times 3 < 1$

$\Rightarrow \quad (1001)^{999} < (1000)^{1000}$

$\therefore \quad B < A$

49. (b) $\sum_{r=1}^{20} r^2 \cdot {}^{20}C_r = 20 \sum_{r=1}^{20} r \cdot {}^{19}C_{r-1}$

$= 20\left[\sum_{r=1}^{20} (r-1)\, {}^{19}C_{r-1} + \sum_{r=1}^{20} {}^{19}C_{r-1}\right]$

$= 20\left[19 \sum_{r=2}^{20} {}^{18}C_{r-2} + 2^{19}\right]$

$= 20[19 \cdot 2^{18} + 2^{19}] = 20 \times 2^{18}[19+2]$

$= 20 \times 2^{18} \times 21 = 420 \times 2^{18}$

On comparing, we get

$a = 420$ and $b = 18$

50. (b) $\sum_{r=0}^{20} (3r+2)\, {}^{20}C_r = 3\sum_{r=0}^{20} r \cdot {}^{20}C_r + 2\sum_{r=0}^{20} {}^{20}C_r$

$= 60 \sum_{r=1}^{20} {}^{19}C_{r-1} + 2 \sum_{r=0}^{20} {}^{20}C_r$

$= 60 \times 2^{19} + 2 \times 2^{20} = 2^{21}[15+1] = 2^{21} \times 2^4 = 2^{25}$

51. (b) Given series is ${}^{20}C_0 + {}^{20}C_1 + {}^{20}C_2 + \dots + {}^{20}C_8$

$= \frac{1}{2}(2 \cdot {}^{20}C_0 + 2 \cdot {}^{20}C_1 + \dots\dots + 2 \cdot {}^{20}C_8)$

$= \frac{1}{2}[({}^{20}C_0 + {}^{20}C_1 + \dots + {}^{20}C_8 + {}^{20}C_9$

$+ {}^{20}C_{10} + {}^{20}C_{11} + \dots + {}^{20}C_{20})$

$- ({}^{20}C_9 + {}^{20}C_{10} + {}^{20}C_{11})]$

$= \frac{1}{2}[2^{20} - 2 \cdot {}^{20}C_9 - {}^{20}C_{10}]$

$= 2^{19} - \frac{(2 \cdot {}^{20}C_9 + {}^{20}C_{10})}{2}$

52. (a) $\sum_{r=0}^{11} (r) \cdot {}^{20}C_r = \sum_{r=1}^{11} 20 \times {}^{19}C_{r-1}$

$= 20({}^{19}C_0 + {}^{19}C_1 + \dots + {}^{19}C_{10})$

$= 20({}^{19}C_0 + {}^{19}C_1 + \dots + {}^{19}C_{10})$

$= 20\left(\frac{1}{2} \times 2^{19} + {}^{19}C_{10}\right) = 20(2^{18} + {}^{19}C_{10})$

53. (a) We have,

$({}^{21}C_1 + {}^{21}C_2 + \dots + {}^{21}C_{10}) - ({}^{10}C_1 + {}^{10}C_2 \dots {}^{10}C_{10})$

$= \frac{1}{2}[{}^{21}C_0 + {}^{21}C_1 + {}^{21}C_2 + \dots + {}^{21}C_{21}] - 2^{10}$

$= 2^{20} - 2^{10}$

54. (d) A_r, B_r and C_r denote respectively, the coefficient of x^r in the expansions of $(1+x)^{10}, (1+x)^{20}$ and $(1+x)^{30}$

$\therefore A_r = {}^{10}C_r, B_r = {}^{20}C_r$ and $C_r = {}^{30}C_r$

$\sum_{r=1}^{10} A_r(B_{10}B_r - C_{10}A_r) = B_{10} \sum_{r=1}^{10} A_r B_r - C_{10} \sum_{r=1}^{10} (A_r)^2$

$= B_{10} \sum_{r=1}^{10} {}^{10}C_r \cdot {}^{20}C_r - C_{10} \sum_{r=1}^{10} ({}^{10}C_r)^2$

$= B_{10} \sum_{r=1}^{10} {}^{10}C_r \cdot {}^{20}C_{20-r} - C_{10} \sum_{r=1}^{10} ({}^{10}C_r)^2$

$= B_{10}\left[\left(\sum_{r=0}^{10} {}^{10}C_r \cdot {}^{20}C_{20-r}\right) - 1\right]$

$- C_{10}\left[\left(\sum_{r=0}^{10} ({}^{10}C_r)^2\right) - 1\right]$

$= B_{10}[{}^{30}C_{20} - 1] - C_{10}[{}^{20}C_{10} - 1]$

$[\because {}^nC_0^2 + {}^nC_1^2 + {}^nC_2^2 + \dots + {}^nC_n^2 = {}^{2n}C_n]$

$= [B_{10}\, {}^{30}C_{20} - C_{10} \cdot {}^{20}C_{10}] + [C_{10} - B_{10}]$

$= [{}^{20}C_{10} \cdot {}^{30}C_{20} - {}^{30}C_{10} \cdot {}^{20}C_{10}] + [C_{10} - B_{10}]$

$= C_{10} - B_{10}$

55. (b) The general term, $T_r = (r-1)^{15}C_r, r = 2, 3, 4, \ldots, 15$

$$\therefore T_r = r \cdot {}^{15}C_r - {}^{15}C_r$$

$$= 15 \cdot {}^{14}C_{r-1} - {}^{15}C_r \quad [\because r \cdot {}^nC_r = n \cdot {}^{n-1}C_{r-1}]$$

$$\therefore \sum_{r=2}^{15} T_r = 15[{}^{14}C_1 + {}^{14}C_2 + \ldots + {}^{14}C_{14}] - [{}^{15}C_2 + {}^{15}C_3 + .. + {}^{15}C_{15}]$$

$$= 15[2^{14} - 1] - [2^{15} - 1 - 15]$$

$$= 2^{14} \cdot 15 - 15 - 2^{15} + 1 + 15$$

$$= 2^{14}(15 - 2) + 1$$

$$= 13 \cdot 2^{14} + 1$$

56. (a) We have, $\dfrac{(1+x)^{4n} + (1-x)^{4n}}{2}$

$$= {}^{4n}C_0 + {}^{4n}C_2 \cdot x^2 + {}^{4n}C_4 \cdot x^4 + \ldots + {}^{4n}C_{4n} \cdot x^{4n}$$

Putting $x = 1$ and $x = i$, we get

$${}^{4n}C_0 + {}^{4n}C_2 + {}^{4n}C_4 + \ldots + {}^{4n}C_{4n} = \frac{1}{2} \cdot 2^{4n}$$

and ${}^{4n}C_0 - {}^{4n}C_2 + {}^{4n}C_4 \ldots + {}^{4n}C_{4n}$

$$= \frac{1}{2}[(1+i)^{4n} + (1-i)^{4n}]$$

$$\therefore 2[{}^{4n}C_0 + {}^{4n}C_4 + {}^{4n}C_8 + \ldots + {}^{4n}C_{4n}]$$

$$= 2^{4n-1} + \frac{1}{2}[(1+i)^{4n} + (1-i)^{4n}]$$

$$= 2^{4n-1} + \frac{1}{2}\left[\sqrt{2}\left(\cos\frac{\pi}{4} + i\sin\frac{\pi}{4}\right)\right]^{4n} + \frac{1}{2}\left[\sqrt{2}\left(\cos\frac{\pi}{4} - i\sin\frac{\pi}{4}\right)\right]^{4n}$$

$$= 2^{4n-1} + \frac{1}{2}[2^{2n}(\cos n\pi + i\sin n\pi + \cos n\pi - i\sin n\pi)]$$

$$= 2^{4n-1} + 2^{2n}\cos n\pi$$

$$= 2^{4n-1} + (-1)^n 2^{2n}$$

$$\Rightarrow {}^{4n}C_0 + {}^{4n}C_4 + \ldots + {}^{4n}C_{4n} = 2^{4n-2} + (-1)^n 2^{2n-1}$$

57. (d) $(1+\omega)^n = {}^nC_0 + {}^nC_1\omega + \ldots$

$$= ({}^nC_0 + {}^nC_3 + \ldots) + ({}^nC_1 + {}^nC_4 + \ldots)\left(\frac{-1+\sqrt{3}i}{2}\right) + ({}^nC_2 + {}^nC_5 + \ldots)\left(\frac{-1-\sqrt{3}i}{2}\right)$$

$$= ({}^nC_0 + {}^nC_3 + \ldots) - \frac{1}{2}({}^nC_1 + {}^nC_2 + {}^nC_4 + {}^nC_5 \ldots) + \frac{i\sqrt{3}}{2}({}^nC_1 - {}^nC_2 + {}^nC_4 - {}^nC_5 + \ldots)$$

Equating for modulus, we get $|(-\omega^2)^n| = 1$

58. (a) $\displaystyle\sum_{r=1}^{n+1}\left(\sum_{k=1}^{n} {}^kC_{r-1}\right) = \sum_{r=1}^{n+1}\left(\sum_{k=1}^{n} {}^{k+1}C_r - {}^kC_r\right)$

$$= \sum_{r=1}^{n+1}({}^{n+1}C_r - {}^1C_r)$$

$$= 2^{n+1} - 2$$

59. (d) The expression is divided into two parts

$$= \left\{1 - \frac{n}{1} \cdot \frac{1}{1+nx} + \frac{n(n-1)}{2} \cdot \frac{1}{(1+nx)^2} - \frac{n(n-1)(n-2)}{3 \cdot 2 \cdot 1} \cdot \frac{1}{(1+nx)^3} + \ldots\right\}$$

$$+ \left\{-\frac{nx}{1+nx} + \frac{n(n-1)}{(1+nx)^2} \cdot x - \frac{n(n-1)(n-2)}{1 \cdot 2} \cdot \frac{x}{(1+nx)^3} + \ldots\right\}$$

$$= \left(1 - \frac{1}{1+nx}\right)^n - \frac{nx}{1+nx} \times \left(1 - \frac{(n-1)}{1(1+nx)} + \frac{(n-1)(n-2)}{1 \cdot 2(1+nx)^2} + \ldots\right)$$

$$= \left(\frac{nx}{1+nx}\right)^n - \frac{nx}{1+nx}\left(1 - \frac{1}{1+nx}\right)^{n-1}$$

$$= \left(\frac{nx}{1+nx}\right)^n - \frac{nx}{1+nx}\left(\frac{nx}{1+nx}\right)^{n-1}$$

$$= \left(\frac{nx}{1+nx}\right)^n - \left(\frac{nx}{1+nx}\right)^n = 0$$

60. (c) $(a-b)^{-n} = a^{-n}\left(1 - \dfrac{b}{a}\right)^{-n}$

$$\left(1 - \frac{b}{a}\right)^{-n} = \left[1 + n\left(\frac{b}{a}\right) + \frac{n(n+1)}{2}\left(\frac{b}{a}\right)^2\right]$$

As, we can ignore the powers greater than or equal to 3.

$$a^{-n}\left(1 - \frac{b}{a}\right)^{-n} = \frac{1}{a^n} + \frac{n.b}{a^{n+1}} + \frac{n(n+1)}{2} \cdot \frac{b^2}{a^{n+2}}$$

$$(a-b)^{-n} = \frac{1}{a^n} + \frac{n.b}{a^{n+1}} + \frac{n(n+1)}{2} \cdot \frac{b^2}{a^{n+2}}$$

When $n = 1$

$$(a-b)^{-1} = \frac{1}{a} + \frac{b}{a^2} + \frac{b^2}{a^3} \quad b \to rb$$

$$\sum_{r=b}^{n}(a-rb)^{-1} = \sum\left[\frac{1}{a} + r\left(\frac{b}{a^2}\right) + r^2\left(\frac{b^2}{a^3}\right)\right]$$

$$S = \frac{n}{a} + \frac{n(n+1)b}{2a^2} + \frac{n(n+1)(2n+1)b^2}{6a^3}$$

$$\text{Coefficient of } n^3 = \frac{2b^2}{6a^3} = \frac{b^2}{3a^3}$$

61. (c) $\because (r+1)$th term in the expansion of $(1+x)^{27/5}$

$$\frac{\frac{27}{5}\left(\frac{27}{5}-1\right)\cdots\left(\frac{27}{5}-r+1\right)}{r!}\cdot x^r$$

Now, this term will be negative, if the last factor in numerator is the only one negative factor.

$$\Rightarrow \quad \frac{27}{5}-r+1<0$$

$$\Rightarrow \quad \frac{32}{5}<r \Rightarrow 6\cdot 4<r$$

$\Rightarrow$ Least value of r is 7.

Thus, first negative term will be 8th.

62. (b) $\left(1+\sqrt{a}+\frac{1}{\sqrt{a}-1}\right)^{-30}$

$$=\left(\frac{a}{\sqrt{a}-1}\right)^{-30}=\left(\frac{\sqrt{a}-1}{a}\right)^{30}$$

$$=\frac{1}{a^{30}}(1-\sqrt{a})^{30}$$

$$=\frac{1}{a^{30}}\{{}^{30}C_0-{}^{30}C_1\sqrt{a}+\ldots+{}^{30}C_{30}(\sqrt{a})^{30}\}$$

There is no term independent of a.

63. (d) The general term in the expansion of $(1-x+y)^{20}$ is

$\frac{20!}{r!s!t!}1^r(-x)^s(y)^t$, where $r+s+t=20$.

For x^2y^3, we have the term

$$\frac{20!}{15!2!3!}1^{15}(-x)^2(y)^3$$

Hence, the coefficient of $x^2\cdot y^3$ is $\frac{20!}{15!2!3!}$

64. (d) As, $L\neq 0$

$\therefore \quad m=2+3+4=9$

$\therefore \quad L=\frac{9!}{2!3!4}$

Now, coefficient of $a^4\cdot b^4\cdot c=\frac{9!}{4!4!1!}=\frac{L}{2}$

65. (d) $(a+b+c)^{12}=[(a+b)+c]^{12}$

$$={}^{12}C_0(a+b)^{12}+{}^{12}C_1(a+b)^{11}c+\ldots+{}^{12}C_{12}\cdot c^{12}$$

Total number of terms $=(13+12+11+..+1)$ terms

$$=\frac{13(13+1)}{2}=91 \text{ terms}$$

Also, number of terms in the expansion of $(a+b+c)^n$ is given by ${}^{n+2}C_2$.

For $n=12$, ${}^{12+2}C_2={}^{14}C_2=\frac{14\times 13}{2}=91$

66. (b) Number of terms in the expansion $={}^{n+3-1}C_{3-1}$

$$={}^{n+2}C_2$$

Given, ${}^{n+2}C_2=45\Rightarrow n=8$

67. (d) Put $x=y=z=1$

The sum of coefficient $=(1+2+3)^{10}=6^{10}$

68. (b) $\therefore$ Required sum of coefficients

$=$ Coefficients of t^{13} in $(1+t)^{11}(1+t^2-t)^{10}$.

$=$ Coefficients of t^{13} in $(1+t)(1+t^3)^{10}$

$=$ Coefficients of t^{13} in $[(1+t^3)^{10}+t(1+t^3)^{10}]$

$=$ Coefficients of t^{12} in $(1+t^3)^{10}={}^{10}C_4$

69. (a) Given statement is '$P(n): n^2-n+41$ is prime'.

Clearly, $P(3): 3^2-3+41=9-3+41=47$, which is a prime number

and $P(5): 5^2-5+41=25-5+41=61$, which is also a prime number.

$\therefore$ Both $P(3)$ and $P(5)$ are true.

70. (b) Given, $S(k)=1+3+5+\ldots+(2k-1)=3+k^2$

Putting $k=1$ in both sides, we get

$$S(1)=1, \text{ but } 3+k^2=3+1=4$$

$\Rightarrow$ LHS $\neq$ RHS for $S(1)$

Putting $(k+1)$ on both sides in the place of k, we get

LHS $=1+3+5+\ldots+(2k-1)+(2k+1)$

RHS $=3+(k+1)^2=3+k^2+2k+1$

Here, LHS = RHS

$[\because 1+3+5+\ldots+(2k-1)+(2k+1)=3+k^2+2k+1$ as $1+3+5+\ldots+(2k-1)=3+k^2]$

If $S(k)$ is true, then $S(k+1)$ is also true.

Hence, $S(k)\Rightarrow S(k+1)$

71. (c) $P(n): n!>2^{n-1}$

The statement $P(1)$ and $P(2)$ are false.

$P(3)=3!>2^2$ is true.

Let it is true for $k\geq 3$

$P(k)=k!>2^{k-1}$

Now, for $n=k+1$,

$P(k+1):(k+1)!=(k+1)\cdot k!>(k+1)2^{k-1}>2^k$ as $k+1>2$

So, $P(k)\Rightarrow P(k+1)$

Hence, the given inequality is true for $n\geq 3$ (using mathematical induction)

72. (13) $T_{r+1}={}^{22}C_r(x^m)^{22-r}\left(\frac{1}{x^2}\right)^r$

$$\Rightarrow \quad T_{r+1}={}^{22}C_r(x)^{22m-mr-2r}$$

$\because \quad 22m-mr-2r=1$

$$m=\frac{2r+1}{(22-r)}$$

$$\because \quad {}^{22}C_3={}^{22}C_{19}=1540$$

$\therefore \quad r = 3$ or 19

$r = 3, m = \frac{7}{19} \notin N$

$r = 19, m = 13$

$\therefore$ Only possible value of $m = 13$

73. (4) $T_2 = {}^nC_1 (a^{1/13})^{n-1} \cdot a\sqrt{a} = 14a^{5/2}$

$\Rightarrow \quad n \cdot a^{\frac{n-1}{13}} = 14a$

$\Rightarrow \quad n \cdot a^{\frac{n-14}{13}} = 14 \Rightarrow \frac{n-14}{13} = 0 \Rightarrow n = 14$

$\Rightarrow \quad \frac{{}^{14}C_3}{{}^{14}C_2} = \frac{14!}{3!\cdot 11!} \cdot \frac{2!\cdot 12!}{14!} = \frac{12}{3} = 4$

74. (15) $3 \times 7^{22} + 2 \times 10^{22} - 44$

$= 3 \times (6+1)^{22} + 2 \times (9+1)^{22} - 44$

Now, $(1+6)^{22} = 1 + {}^{22}C_1 6 + {}^{22}C_2 \cdot 6^2 + ... + {}^{22}C_{22} 6^{22}$

$= (1 + 6\lambda)$

$(1+9)^{22} = 1 + {}^{22}C_1 9 + {}^{22}C_2 \cdot 9^2 + ... + {}^{22}C_{22} 6^{22}$

$= (1 + 9\mu)$

$\therefore \quad = 3(1+6\lambda) + 2(1+9\mu) - 44$

$= 18\lambda + 3 + 18\mu + 2 - 44$

$= 18\delta - 39 = 18\alpha + 15$

$3 \times 7^{22} + 2 \times 10^{22} - 44$, when divided by 18 leaves remainder 15.

75. (1) Given, $\left(5^{2/5} \log_5 \sqrt{4^x + 44} + \frac{1}{5^{\log_5 \sqrt[3]{2^{x-1} + 7}}}\right)^8$

$= \left(\sqrt{(4^x + 44)^{2/5}} + \frac{1}{\sqrt[3]{2^{x-1} + 7}}\right)^8$

$= \left((4^x + 44)^{1/5} + \frac{1}{(2^{x-1} + 7)^{1/3}}\right)^8$

Now, $T_4 = T_{3+1} = {}^8C_3 \{(4^x + 44)^{1/5}\}^{8-3} \frac{1}{\{(2^{x-1} + 7)^{1/3}\}^3}$

Given, $336 = {}^8C_3 \left(\frac{4^x + 44}{2^{x-1} + 7}\right)$

Let $2^x = y \Rightarrow 336 = {}^8C_3 \left(\frac{y^2 + 44}{\left(\frac{y}{2}\right) + 7}\right)$

or $336 = \frac{8 \times 7 \times 6}{3 \times 2 \times 1} \left(\frac{2(y^2 + 44)}{y + 14}\right)$

$\Rightarrow \quad y^2 - 3y + 2 = 0$

$\Rightarrow \quad y = 2$ or $x = 1$

76. (2) The general term in $\left(ax^2 + \frac{1}{bx}\right)^{13}$ is

$T_{r+1} = {}^{13}C_r (ax^2)^{13-r} \left(\frac{1}{bx}\right)^r$

$= {}^{13}C_r (a)^{13-r} (b)^{-r} (x)^{26-3r}$

For coefficient of x^r, put $26 - 3r = 8$

$\Rightarrow \quad r = 6$

$\therefore \quad T_7 = {}^{13}C_6 (a)^{13-6} (b)^{-6} x^8 = {}^{13}C_6 a^7 \cdot b^{-6} \cdot x^8$

Now, in $\left(ax - \frac{1}{bx^2}\right)^{13}$, the general term

$T'_{r+1} = {}^{13}C_r (ax)^{13-r} \left(-\frac{1}{bx^2}\right)^r$

$= {}^{13}C_r (a)^{13-r} (b)^{-r} (x)^{13-3r} (-1)^r$

For coefficient of x^{-r}, put $13 - 3r = -8$

$\Rightarrow \quad r = 7$

$\therefore \quad T'_8 = (-1)^7 \, {}^{13}C_7 (a)^{13-7} (b)^{-7} (x)^{-8}$

$= (-1)^7 \, {}^{13}C_7 (a)^6 (b)^{-7} \cdot x^{-8}$

Given, $T_7 = T'_8 \Rightarrow {}^{13}C_6 \cdot a^7 \cdot b^{-6} = -{}^{13}C_7 \cdot a^6 \cdot b^{-7}$

$\Rightarrow \quad \frac{a^7}{a^6} = -\frac{b^6}{b^7}$

$\Rightarrow \quad a = -\frac{1}{b} \Rightarrow ab + 1 = 0 \Rightarrow ab + 3 = 2$

77. (6) rth term in the expansion of $\left(3x - \frac{2}{x^2}\right)^{15}$ is

$T_r = {}^{15}C_{r-1} (3x)^{15-r+1} \left(\frac{-2}{x^2}\right)^{r-1}$

$= {}^{15}C_{r-1} (3)^{15-r+1} (-2)^{r-1} (x)^{15-3r+3}$

For the independent term

$15 - 3r + 3 = 0 \Rightarrow r = 6$

78. (924) Binomial expansion

$(x+y)^n = {}^nC_0 \cdot x^n + {}^nC_1 \cdot x^{n-1} y + \dots {}^nC_n \cdot y^n$

Putting $x = y = 1$, we get

$2^n = {}^nC_0 + {}^nC_1 + {}^nC_2 + \dots + {}^nC_n$ ...(i)

$\because {}^nC_0 + {}^nC_1 + {}^nC_2 + \dots + {}^nC_n = 4096$ [given] ...(ii)

From Eqs. (i) and (ii), we get

$2^n = 4096 = 2^{12}$

$\Rightarrow \quad n = 12$

$\because n$ is even here, so the greatest term is

${}^nC_{n/2} = {}^{12}C_{12/2} = {}^{12}C_6$

$= \frac{12 \times 11 \times 10 \times 9 \times 8 \times 7}{6 \times 5 \times 4 \times 3 \times 2 \times 1} = 924$

79. (120) Coefficient of x^4 in $(1+x+x^2+x^3)^6$

$$= \text{Coefficient of } x^4 \text{ in } \left(\frac{1-x^4}{1-x}\right)^6$$

$$= \text{Coefficient of } x^4 \text{ in } [(1-x^4)^6 (1-x)^{-6}]$$

$$= \text{Coefficient of } x^4 \text{ in } [(1-6\cdot x^4+15x^8+\ldots$$

$$(1+{}^6C_1x+{}^7C_2x^2+{}^8C_3x^3+{}^9C_4x^4\ldots]$$

$$= {}^9C_4 - 6\times 1 = 126-6 = 120$$

80. (4) We have, b = Coefficient of x^3 in

$\{(1+x+2x^2+3x^3)+4x^4\}^4$

$$= \text{Coefficient of } x^3 \text{ in } [{}^4C_0(1+x+2x^2+3x^3)^4](4x^4)^0 + {}^4C_1(1+x+2x^2+3x^3)^3(4x^4)^1+..]$$

$$= \text{Coefficient of } x^3 \text{ in } (1+x+2x^2+3x^3)^4$$

Hence, $\frac{4a}{b} = 4$

81. (118) Let the three consecutive coefficients be ${}^nC_{r-1}, {}^nC_r$ and ${}^nC_{r+1}$.

Given, $\frac{{}^nC_r}{{}^nC_{r-1}} = \frac{5}{2}$

$$\Rightarrow \frac{n-r+1}{r} = \frac{5}{2}$$

$$\Rightarrow 2n-7r+2=0 \quad \ldots\text{(i)}$$

Also, $\frac{{}^nC_{r+1}}{{}^nC_r} = \frac{12}{5} \Rightarrow \frac{n-r}{r+1} = \frac{12}{5}$

$$\Rightarrow 5n-17r-12=0 \quad \ldots\text{(ii)}$$

On solving Eqs. (i) and (ii), we get

$$n = 118 \text{ and } r = 34$$

82. (8) Let the three consecutive coefficients be

${}^nC_{r-1} = 28$, ${}^nC_r = 56$ and ${}^nC_{r+1} = 70$

So, $\frac{{}^nC_r}{{}^nC_{r-1}} = \frac{n-r+1}{r} = \frac{56}{28} = 2$

and $\frac{{}^nC_{r+1}}{{}^nC_r} = \frac{n-r}{r+1} = \frac{70}{56} = \frac{5}{4}$

This gives $n+1 = 3r$ and $4n-5 = 9r$, therefore

$$\frac{4n-5}{n+1} = 3$$

On solving, we get

$$n = 8$$

83. (35) Given, $6^{83}+8^{83}$

$$= (1+7)^{83}+(7-1)^{83} = (1+7)^{83}-(1-7)^{83}$$

$$= 2[{}^{83}C_1\cdot 7 + {}^{83}C_3\cdot 7^3+\ldots+{}^{83}C_{83}\cdot 7^{83}]$$

$= (2\cdot 7\cdot 83)+49I$, where I is an integer

Now, $14\times 83 = 1162$

$$\therefore \frac{1162}{49} = 23\frac{35}{49}$$

$\therefore$ Remainder is 35.

84. (3) $(1+0.00002)^{50000} = \left(1+\frac{1}{50000}\right)^{50000}$

Now, we know that $2 \le \left(1+\frac{1}{n}\right)^n < 3, \forall\, n \ge 1$

$\Rightarrow$ Least integer is 3.

85. (3) $S_n = {}^{2n}C_n$

$$\Rightarrow \frac{S_{n+1}}{S_n} = \frac{{}^{2n+2}C_{n+1}}{{}^{2n}C_n} = \frac{(2n+2)(2n+1)}{(n+1)(n+1)}$$

$$= \frac{2(2n+1)}{n+1} = 4 - \frac{2}{n+1}$$

$$\therefore \left[\frac{S_{n+1}}{S_n}\right]_{max} = 3$$

86. (211) $\sum_{0\le i<j\le 5} \sum ({}^5C_j)({}^jC_i)$

$$= {}^5C_1\cdot{}^1C_0 + {}^5C_2({}^2C_0+{}^2C_1)+\ldots + {}^5C_5({}^5C_0+{}^5C_1+..+{}^5C_4)$$

$$= {}^5C_1(2^1-1)+{}^5C_2(2^2-1)+..+{}^5C_5(2^5-1)$$

$$= ({}^5C_1\cdot 2^1+{}^5C_2\cdot 2^2+\ldots+{}^5C_5\cdot 2^5) - ({}^5C_1+{}^5C_2+\ldots+{}^5C_5)$$

$$= [(1+2)^5-1]-[2^5-1]$$

$$= 3^5-2^5 = 243-32 = 211$$

87. (1) $\lim_{n\to\infty} \sum_{r=1}^{n} 5^{\frac{1}{n}}\cdot{}^nC_r\left(\sum_{t=0}^{r-1} {}^rC_t\cdot 3^t\right)$

$$= \lim_{n\to\infty} \sum_{r=1}^{n} 5^{\frac{1}{n}}\cdot{}^nC_r(4^r-3^r)$$

$$= \lim_{n\to\infty} 5^{\frac{1}{n}}\left(\sum_{r=1}^{n} {}^nC_r\cdot 4^r - \sum_{r=1}^{n} {}^nC_r\cdot 3^r\right)$$

$$= \lim_{n\to\infty} 5^{\frac{1}{n}}(5^n-4^n) = 1$$

88. (5) $\sum_{r=0}^{n}\left(\frac{r+2}{r+1}\right){}^nC_r = \sum_{r=0}^{n}\left(1+\frac{1}{r+1}\right){}^nC_r$

$$= \sum_{r=0}^{n} {}^nC_r + \sum_{r=0}^{n} \frac{{}^nC_r}{r+1}$$

$$= \sum_{r=0}^{n} {}^nC_r + \frac{1}{(n+1)}\sum_{r=0}^{n} {}^{n+1}C_{r+1}$$

$$= 2^n + \frac{1}{(n+1)}(2^{n+1} - 1)$$

$$= \frac{2^n(n+1) + 2^{n+1} - 1}{n+1}$$

$$= \frac{2^n(n+3) - 1}{n+1}$$

$$\therefore \quad \frac{2^n(n+3)-1}{n+1} = \frac{2^8 - 1}{6}$$

Hence, $n = 5$

89. (2499) $\displaystyle\sum_{r=1}^{49} \frac{2r^2 - 48r + 1}{(50-r)\cdot {}^{50}C_r}$

$$= \sum_{r=1}^{49} \frac{(r+1)^2 - r(50-r)}{(50-r)\cdot {}^{50}C_r}$$

$$= \sum_{r=1}^{49} \left(\frac{(r+1)^2}{(50-r)\cdot {}^{50}C_r} - \frac{r}{{}^{50}C_r} \right)$$

$$= \sum_{r=1}^{49} \left(\frac{r+1}{(50-r)\left(\frac{{}^{50}C_r}{r+1}\right)} - \frac{r}{{}^{50}C_r} \right)$$

$$= \sum_{r=1}^{49} \left(\frac{r+1}{{}^{50}C_{r+1}} - \frac{r}{{}^{50}C_r} \right)$$

$$= \frac{50}{{}^{50}C_{50}} - \frac{1}{{}^{50}C_1}$$

$$= 50 - \frac{1}{50}$$

$$\therefore \; 50 \sum_{r=1}^{49} \frac{2r^2 - 48r + 1}{(50-r)\cdot {}^{50}C_r} = 50\left(50 - \frac{1}{50}\right) = 2499$$

90. (40) We have, $(1 - x - x^2)^{20} = \displaystyle\sum_{r=0}^{40} a_r \, . x^r$

Differentiating both sides w.r.t. x, we get

$$20(1 - x - x^2)^{19}(-1 - 2x) = \sum_{r=0}^{40} r a_r \, . x^{r-1} \quad \text{...(i)}$$

Putting $x = 1$ in Eq. (i), we get

$$20(1 - 1 - 1)^{19}(-1 - 2) = \sum_{r=0}^{40} r a_r$$

$$\Rightarrow \quad 60 = a_1 + 2a_2 + 3a_3 + .. + 40a_{40} \quad \text{...(ii)}$$

Putting $x = -1$ in Eq. (i), we get

$$20(1 + 1 - 1)^{19}(-1 + 2) = \sum_{r=0}^{40} r a_r (-1)^{r-1}$$

$$\Rightarrow \quad 20 = a_1 - 2a_2 + 3a_3 + \ldots - 40a_{40} \quad \text{...(iii)}$$

On adding Eqs. (ii) and (iii), we get

$$80 = 2a_1 + 6a_3 + \ldots + 78a_{39}$$

$$\therefore \quad a_1 + 3a_3 + \ldots + 39a_{39} = 40$$

91. (30) Let $(1 - x + x^2 - x^3 + \ldots x^{2n})(1 + x + x^2 + \ldots + x^{2n})$

$$= a_0 + a_1 x + a_2 x^2 + \ldots$$

Putting $x = 1$, we get

$$1(2n+1) = a_0 + a_1 + a_2 + \ldots + a_{2n} \quad \text{...(i)}$$

Putting $x = -1$, we get

$$(2n+1) \times 1 = a_0 - a_1 + a_2 + \ldots + a_{2n} \quad \text{...(ii)}$$

On adding Eqs. (i) and (ii), we get

$$2(2n+1) = 2(a_0 + a_2 + a_4 + \ldots + a_{2n})$$

$$\Rightarrow a_0 + a_2 + a_4 + \ldots + a_{2n} = 2n + 1$$

Given, $a_0 + a_2 + a_4 + \ldots + a_{2n} = 61$

On comparing $2n + 1 = 61 \Rightarrow 2n = 60$

$$\Rightarrow \quad n = 30$$

92. (60) Coefficient of $a^3 b^2 c = \dfrac{6!}{3!2!1!0!}$

$$= \frac{6 \times 5 \times 4 \times 3!}{3! \times 2 \times 1}$$

$$= 60$$

93. (10) $\because \; (1+x)^n = 1 + nx + \dfrac{n(n-1)}{2!}x^2$

$$+ \frac{n(n-1)(n-2)}{3!}x^3 + \ldots$$

$$\Rightarrow \quad (1-x)^{-3} = 1 + 3x + \left[\frac{-3(-3-1)}{2!}\right]x^2$$

$$+ \frac{(-3)(-3-1)(-3-2)}{3!}(-x)^3$$

$\therefore$ Coefficient of x^3 is

$$(-1)\frac{(-3)(-3-1)(-3-2)}{3!}$$

$$= \frac{3 \times 4 \times 5}{3!} = \frac{3 \times 4 \times 5}{3 \times 2} = 10$$

Level 3

1. (c) $(1 + 2x + x^2)^n = \displaystyle\sum_{r=0}^{2n} a_r x^r$

$$\Rightarrow \quad [(1+x)^2]^n = \sum_{r=0}^{2n} a_r x^r$$

$$\Rightarrow \quad (1+x)^{2n} = \sum_{r=0}^{2n} a_r x^r$$

$$\Rightarrow \quad \sum_{r=0}^{2n} {}^{2n}C_r x^r = \sum_{r=0}^{2n} a_r x^r$$

$$\Rightarrow \quad a_r = {}^{2n}C_r$$

2. (b) $(4x^2+1)^n = \sum_{r=0}^{n} a_r (1+x^2)^{n-r}(x)^{2r}$

$$\Rightarrow \left(1+\frac{3x^2}{1+x^2}\right)^n = \sum_{r=0}^{n} a_r \left(\frac{x^2}{1+x^2}\right)^r$$

$\therefore \quad a_r = {}^nC_r \cdot 3^r$

$$\therefore \sum_{r=0}^{n} a_r = \sum_{r=0}^{n} {}^nC_r \cdot 3^r$$

$$= (1+3)^n = 4^n$$

3. (b) $T_{r+1} = {}^{10}C_r(\sqrt{x})^{10-r}\left(\frac{-k}{x^2}\right)^r = {}^{10}C_r(x)^{5-5r/2}(-k)^r$

For this to be independent of x, r must be 2, so that

$${}^{10}C_2 \cdot k^2 = 405$$

or $\quad k = \pm 3$

4. (c) $T_{r+1} = \sqrt{3}\cdot {}^{20}C_r\left(\frac{1}{\sqrt{3}}\right)^r$

and $T_r = \sqrt{3}\cdot {}^{20}C_{r-1}\left(\frac{1}{\sqrt{3}}\right)^{r-1}$

Now, $\frac{T_{r+1}}{T_r} = \frac{20-r+1}{r}\left(\frac{1}{\sqrt{3}}\right)$

$\because \quad T_{r+1} \geq T_r$

$\Rightarrow 20-r+1 \geq \sqrt{3}r$

$\Rightarrow \quad r \leq \frac{21}{\sqrt{3}+1} \approx 7\cdot 6 \Rightarrow r = 7$

$\therefore$ The greatest term is $T_8 = \sqrt{3}\ {}^{20}C_7\left(\frac{1}{\sqrt{3}}\right)^7$

$\Rightarrow \quad T_8 = \frac{25840}{9}$

5. (c) Given, $\left(\frac{1}{x^2} - x^3\right)^n$

General term $T_{r+1} = \frac{n!}{r!(n-r)!}(-1)^{n-r}(x)^{5r-2n}$

If $5r-2n=5$, then $5r = 2n+5$ or $r = \frac{2n}{5}+1$

If $5r-2n=10$, then $5r = 2n+10$ or $r = \frac{2n}{5}+2$

x^5 and x^{10} terms occurs, if $n = 5k$

Given that sum of x^5 and x^{10} is zero

$$\Rightarrow \frac{5k!}{(2k+1)!(3k-1)!} - \frac{5k!}{(2k+2)!(3k-2)!} = 0$$

$$\Rightarrow \frac{1}{3k-1} - \frac{1}{2k+2} = 0$$

$\Rightarrow \quad k = 3 \Rightarrow n = 15$

6. (c) $\because (1+x^2)^4(1+x^3)^7(1+x^4)^{12}$

$$= (1+{}^4C_1\cdot x^2 + {}^4C_2\cdot x^4 + \ldots + {}^4C_4\cdot x^8)$$
$$(1+{}^7C_1\cdot x^3 + {}^7C_2\cdot x^6+\ldots)(1+{}^{12}C_1\cdot x^4 + {}^{12}C_2\cdot x^8+\ldots)$$

Required coefficient

$$= {}^{12}C_2\times{}^7C_1\times 1 + {}^{12}C_1\times{}^7C_1\times{}^4C_2 + {}^7C_1\times{}^4C_4 + {}^7C_3\times{}^4C_1$$

$$= 462+504+7+140 = 1113$$

7. (b) x^9 can be found in the following ways

$x^9, x^{1+8}, x^{2+7}, x^{3+6}, x^{4+5}, x^{1+2+6}, x^{1+3+5}, x^{2+3+4}$

So, there are 8 cases.

The coefficient of x^9 in each cases is 1.

$\therefore$ Required coefficient $= 8$

8. (c) $(1+x)^{21} + (1+x)^{22} + \ldots + (1+x)^{30}$

$$= (1+x)^{21}\left[\frac{(1+x)^{10}-1}{(1+x)-1}\right]$$

$$= \frac{1}{x}[(1+x)^{31} - (1+x)^{21}]$$

$\Rightarrow$ Coefficient of x^5 in the given expression

$$= \text{Coefficient of } x^5 \text{ in } \left\{\frac{1}{x}[(1+x)^{31}-(1+x)^{21}]\right\}$$

$$= \text{Coefficient of } x^6 \text{ in } [(1+x)^{31}-(1+x)^{21}]$$

$$= {}^{31}C_6 - {}^{21}C_6$$

9. (b) $(\sqrt{3}+\sqrt{2})^6 + (\sqrt{3}-\sqrt{2})^6$

$$= 2\{{}^6C_0(\sqrt{3})^6 + {}^6C_2(\sqrt{3})^4(\sqrt{2})^2 + {}^6C_4(\sqrt{3})^2(\sqrt{2})^4 + {}^6C_6(\sqrt{2})^6\}$$

$$= 970$$

Since, $0 < (\sqrt{3}-\sqrt{2})^6 < 1$, the smallest integer larger than $(\sqrt{3}+\sqrt{2})^6$ is 970.

10. (a) $\frac{2^{4n}}{15} = \frac{(15+1)^n}{15}$

$$= \frac{({}^nC_0\cdot 15^n + {}^nC_1\cdot 15^{n-1} + \ldots + {}^nC_{n-1}\cdot 15 + {}^nC_n)}{15}$$

$$= \text{Integer} + \frac{1}{15}$$

Hence, the fractional part of $\frac{2^{4n}}{15}$ is $\frac{1}{15}$.

11. (a) $3^{256} - 3^{12} = 3^{12}(3^{244}-1)$

$$= (1+8)^6((1+8)^{122}-1)$$

$$= (1+8\lambda)(1+8\mu-1)$$

[where λ and μ integers]

$$= 8\mu(1+8\lambda)$$

So, number is divisible by 8.

12. (b) $8^{2n} - (62)^{2n+1} = (1+63)^n - (63-1)^{2n+1}$
$$= (1+63)^n + (1-63)^{2n+1}$$
$$= [1 + {}^nC_1\,63 + {}^nC_2(63)^2 + \ldots + (63)^n]$$
$$+ [1 - {}^{(2n+1)}C_1\,63 + {}^{(2n+1)}C_2\,(63)^2 + \ldots - 63^{2n+1}]$$
$$= 2 + 63k \quad [\because k \text{ is an integer}]$$
Hence, the remainder is 2.

13. (b) $(1+2)^n - (1-2)^n = 2[{}^nC_1 \times 2 + {}^nC_3 \times 2^3 + {}^nC_5 \times 2^5 + \ldots$
$$\therefore {}^nC_1 \times 2 + {}^nC_3 \times 2^3 + {}^nC_5 \times 2^5 + \ldots = \frac{3^n - (-1)^n}{2}$$

14. (d) We have,
$$S = C_0 + (C_0 + C_1) + (C_0 + C_1 + C_2) + \ldots + (C_0 + C_1 + C_2 + \ldots + C_n)$$
$$= \{C_0 + C_0 + \ldots + (n+1) \text{ times}\} + (C_1 + C_1 + \ldots + n \text{ times}) + \ldots\ldots + (C_{n-1} + C_{n-1}) + C_n$$
$$= (n+1)C_0 + nC_1 + (n-1)C_2 + \ldots + 2C_{n-1} + C_n$$
$$= C_0 + 2C_1 + 3C_2 + \ldots\ldots + (n+1)C_n \quad [\because C_r = C_{n-r}]$$
General term,
$$T_{r+1} = (r+1)C_r$$
$$T_{r+1} = r \cdot {}^nC_r + {}^nC_r = n \cdot {}^{n-1}C_{r-1} + {}^nC_r$$
$$\therefore \quad S = \sum_{r=0}^{n} T_{r+1} = n[{}^{n-1}C_0 + {}^{n-1}C_1 + \ldots + {}^{n-1}C_{n-1}] + [{}^nC_0 + {}^nC_1 + \ldots + {}^nC_n]$$
$$\Rightarrow \quad S = n \cdot 2^{n-1} + 2^n$$
$$\Rightarrow \quad S = (n+2)2^{n-1}$$

15. (c) $\because T_{r+1} = {}^{10}C_r\,(\sqrt{2})^{10-r}\,(\sqrt[5]{3})^r$
$$= {}^{10}C_r \cdot 2^{\frac{10-r}{2}} \cdot 3^{\frac{r}{5}}$$
$r = 0, 5, 10$ (multiple of 5)
and $10 - r = 0, 2, 4, 6, 8, 10$ (multiple of 2)
$\therefore$ Possible values of r are 0 and 10.
$\therefore$ Sum of rational terms
$$= T_1 + T_{11}$$
$$= {}^{10}C_0\,(2)^5(3)^0 + {}^{10}C_{10}\,(2)^0(3)^2$$
$$= 2^5 + 3^2 = 32 + 9 = 41$$

16. (d) Given, $P(n): n^2 + n + 1$
$$\Rightarrow \quad P(1) = 1^2 + 1 + 1 = 3,$$
which is not an even integer.
Thus, $P(1)$ is not true.
Also, $n^2 + n + 1 = n(n+1) + 1$ is always an odd integer.

17. (c) Let $P_n: a_n = \sqrt{7 + \sqrt{7 + \sqrt{7 + \ldots}}}$ (n radical sign)
For $n = 1$, $P(1): a_1 = \sqrt{7} < 4$
Assume that $a_k < 4$, for all natural number, $n = k$
For $n = k + 1$,
$$P(k+1): a_{k+1} = \sqrt{7 + \sqrt{7 + \sqrt{7 + \ldots}}} \quad [(k+1) \text{ radical sign}]$$
$$= \sqrt{7 + a_k} < \sqrt{7 + 4} < 4 \quad [\text{by assumption}]$$
This shows that $a_{k+1} < 4$ i.e. the result is true for $n = k + 1$.
Hence, by principle of mathematical induction,
$$a_n < 4, \forall\, n \geq 1$$

18. (a, b, c) We have,
$$(x+a)^n = {}^nC_0 + {}^nC_1(x)^{n-1}(a) + {}^nC_2(x)^{n-2}(a)^2 + \ldots + {}^nC_n a^n$$
$$= [{}^nC_0(x)^n + {}^nC_2(x)^{n-2}(a)^2 + \ldots] + [{}^nC_1(x)^{n-1}(a) + {}^nC_3(x)^{n-3}(a)^3 + \ldots]$$
$$\Rightarrow \quad (x+a)^n = p + q \quad \ldots(i)$$
Similarly, $(x-a)^n = p - q$ ...(ii)

(a) Multiplying Eqs. (i) and (ii),
$$\Rightarrow \quad p^2 - q^2 = (x^2 - a^2)^n$$
(b) Squaring Eqs. (i) and (ii) and subtracting Eq. (ii) from Eq. (i), we get
$$4pq = (x+a)^{2n} - (x-a)^{2n}$$
(c) Squaring Eqs. (i) and (ii) and adding, we get
$$2(p^2 + q^2) = (x+a)^{2n} + (x-a)^{2n}$$

19. (a, d) It is given that the 4th term in the expansion of $\left(ax + \frac{1}{x}\right)^n$ is $\frac{5}{2}$, therefore
$${}^nC_3(ax)^{n-3}\left(\frac{1}{x}\right)^3 = \frac{5}{2}$$
$$\Rightarrow \quad {}^nC_3(a)^{n-3}(x)^{n-6} = \frac{5}{2} \quad \ldots(i)$$
$$\Rightarrow \quad n = 6$$
$[\because$ RHS is independent of $x]$
Putting $n = 6$ in Eq. (i), we get
$${}^6C_3 \cdot a^3 = \frac{5}{2} \text{ or } a^3 = \frac{1}{8} \text{ or } a = \frac{1}{2}$$

20. (a, c) Inclusion of $\log x$ implies $x > 0$.
Now, 3rd term in the expansion is
$$T_{2+1} = {}^5C_2(x)^{5-2}(x^{\log_{10} x})^2 = 1000000 \quad [\text{given}]$$
or $$x^{3 + 2\log_{10} x} = 10^5$$
Taking log on both sides, we get
$$(3 + 2\log_{10} x)\log_{10} x = 5$$
or $$2y^2 + 3y - 5 = 0,$$

where $\log_{10} x = y$

or $(y-1)(2y+5)=0$

or $y = 1$ or $-\dfrac{5}{2}$

or $\log_{10} x = 1$ or $-\dfrac{5}{2}$

$\therefore \quad x = 10^1 = 10$ or $10^{-5/2}$

21. (a, b, c) We have, ${}^{49}C_{39} \cdot p^{10} \cdot q^{39} = {}^{49}C_{40} \cdot p^9 \cdot q^{40}$

$$\Rightarrow \quad p = \frac{{}^{49}C_{40}}{{}^{49}C_{39}} \cdot q$$

$$\Rightarrow \quad p = \frac{49-40+1}{40} \cdot q$$

$$\Rightarrow \quad p = \frac{1}{4} q$$

22. (c, d) $E = \left[\sqrt{9^{x-1}+7} + \dfrac{1}{(3^{x-1}+1)^{1/5}}\right]^7$

Given, $T_6 = 84$

$$\Rightarrow \quad {}^7C_5 (\sqrt{9^{x-1}+7})^{7-5} \left(\frac{1}{(3^{x-1}+1)^{1/5}}\right)^5 = 84$$

$$\Rightarrow {}^7C_5 (9^{x-1}+7) \cdot \frac{1}{(3^{x-1}+1)} = 84$$

$$\Rightarrow \quad 9^{x-1}+7 = 4(3^{x-1}+1)$$

$$\Rightarrow \quad 3^{2x} - 12 \cdot 3^x + 27 = 0$$

$$\Rightarrow \quad (3^x - 3)(3^x - 9) = 0$$

$$\Rightarrow \quad 3^x = 3, 9$$

$$\Rightarrow \quad x = 1, 2$$

23. (a, d) Middle term is $\left(\dfrac{n}{2}+1\right)$th or $(4+1)$th or T_5

$$\Rightarrow \quad T_5 = {}^8C_4 \left(\frac{x}{2}\right)^4 \times 2^4 = 1120$$

$$\Rightarrow \quad \frac{8 \times 7 \times 6 \times 5}{1 \times 2 \times 3 \times 4} \cdot x^4 = 1120$$

$$\Rightarrow \quad x^4 = \frac{1120}{70} = 16$$

$$\Rightarrow \quad (x^2+4)(x^2-4) = 0$$

$$\Rightarrow \quad x = \pm 2 \qquad [\because x \in R]$$

24. (a, b) We have, $(a+b)^n$

$$\frac{T_{r+1}}{T_r} = \frac{n-r+1}{r} \cdot \frac{b}{a} = 1$$

$\therefore \quad (n+1)b - br = ar$

$$\therefore \quad r = \frac{(n+1)b}{a+b} \text{ is integer.}$$

Also considering, $(b+a)^n$

$$\frac{T_{r+1}}{T_r} = \frac{n-r+1}{r} \cdot \frac{a}{b} = 1$$

$$\therefore \quad r = \frac{(n+1)a}{a+b} \text{ is integer.}$$

25. (a, c) $(1-y)^m (1+y)^n$

$$= (1 - {}^mC_1 y + {}^mC_2 y^2 - \ldots)(1 + {}^nC_1 y + {}^nC_2 y^2 + \ldots)$$

$$= 1 + (n-m)y + \left\{\frac{m(m-1)}{2} + \frac{n(n-1)}{2} - mn\right\} y^2 + \ldots$$

Given, $a_1 = 10$

$$\Rightarrow \quad a_1 = n - m = 10 \qquad \ldots(i)$$

$$a_2 = \frac{m^2+n^2-m-n-2mn}{2} = 10$$

$$\Rightarrow (m-n)^2 - (m+n) = 20$$

$$\Rightarrow \quad m+n = 80 \qquad \ldots(ii)$$

On solving Eqs. (i) and (ii), we get

$$m = 35 \text{ and } n = 45$$

26. (a, b, c, d) Let T_5 be numerically the greatest term in the expansion of $\left(1+\dfrac{x}{3}\right)^{10}$.

Then, $\left|\dfrac{T_5}{T_4}\right| \geq 1$ and $\left|\dfrac{T_6}{T_5}\right| \leq 1$

Now, $\dfrac{T_{r+1}}{T_r} = \dfrac{10-r+1}{r} \cdot \dfrac{x}{3}$

$$\Rightarrow \quad \left|\frac{7}{4} \times \frac{x}{3}\right| \geq 1 \text{ and } \left|\frac{6}{5} \times \frac{x}{3}\right| \leq 1$$

$$\Rightarrow \quad |x| \geq \frac{12}{7} \text{ and } |x| \leq \frac{5}{2} \Rightarrow \frac{12}{7} \leq |x| \leq \frac{5}{2}$$

$$\Rightarrow \quad x \in \left[-\frac{5}{2}, -\frac{12}{7}\right] \cup \left[\frac{12}{7}, \frac{5}{2}\right]$$

27. (a, b, c) $\left(x + \dfrac{1}{x} + 1\right)^6 = \displaystyle\sum_{r=0}^{6} {}^6C_r \left(x + \frac{1}{x}\right)^r$

(a) $a_0 = 1 + {}^6C_2 \cdot {}^2C_1 + {}^6C_4 \cdot {}^4C_2 + {}^6C_6 \cdot {}^6C_3$

$= 1 + 30 + 90 + 20 = 141$

(b) $a_5 = {}^6C_5 \times$ coefficient of x^5 in $\left(x + \dfrac{1}{x}\right)^5$

$= {}^6C_5 \cdot {}^5C_0 = 6$

(c) Putting $x = 1$,

$$a_0 + (a_1+b_1) + (a_2+b_2) + \ldots + (a_6+b_6) = 3^6$$

$$\therefore \quad \sum_{i=1}^{6} (a_i + b_i) = 3^6 - 141 = 588$$

28. (a, b, c, d) To get the sum of coefficients, we put $x = 1$

Then, sum of coefficients in $(1 + ax - 2x^2)^n$ is $(a-1)^n$

Obviously, when $a > 1$, sum is positive for any n.

29. (a, c) $\sum_{r=0}^{n} \frac{r}{^nC_r} = \sum_{r=0}^{n} \frac{n-r}{^nC_{n-r}} = \frac{n}{2} \sum_{r=0}^{n} \frac{1}{^nC_r}$

$\therefore \quad \frac{n}{2} = \frac{n^2 - 3n + 3}{2}$

$\Rightarrow \quad n = 1, 3$

30. (a, c) $\left(x^2 + 1 + \frac{1}{x^2}\right)^n$

$$= {^nC_0} + {^nC_1}\left(x^2 + \frac{1}{x^2}\right) + {^nC_2}\left(x^2 + \frac{1}{x^2}\right)^2 + \dots$$

$${^nC_n}\left(x^2 + \frac{1}{x^2}\right)^n$$

Thus, contains terms having

$x^0, x^2, x^4, \dots, x^{2n}, x^{-2}, x^{-4}, \dots, x^{-2n}$

Coefficient of constant term

$= {^nC_0} + {^nC_2}(2) + ({^nC_4})({^4C_2}) + ({^nC_6})({^6C_3}) + \dots \neq 2^{n-1}$.

Coefficient of $(x)^{2n-2}$ is ${^nC_{n-1}} = n$

Coefficient of x^2 is ${^nC_1} + ({^nC_3})({^3C_1}) + ({^nC_5})({^5C_2}) + \dots\ n$

31. (a, c) $S = \sum_{r=0}^{n} (pr + 2) \cdot {^nC_r}$ ...(i)

$\therefore \quad S = \sum_{r=0}^{n} \{p(n-r) + 2\} \cdot {^nC_r}$ ...(ii) $[\because {^nC_r} = {^nC_{n-r}}]$

Adding Eqs. (i) and (ii), we get

$2S = (pn + 4) \sum_{r=0}^{n} {^nC_r} \Rightarrow S = (pn + 4) \times 2^{n-1}$

Given that the value of $S = (25)(64)$

On comparing, we get

So, $n = 7$ and $7p + 4 = 25$ or $p = 3$.

32. (a, c, d) $f(m) = \sum_{i=0}^{m} \binom{30}{30-i}\binom{20}{m-i} = \sum_{i=0}^{m} \binom{30}{i}\binom{20}{m-i}$

$= {^{50}C_m}$

$f(m)$ is greatest, when $m = 25$.

Also, $f(0) + f(1) + \dots + f(50)$

$= {^{50}C_0} + {^{50}C_1} + {^5C_2} + \dots + {^{50}C_{50}} = 2^{50}$

Also, ${^{50}C_m}$ is not divisible by 50 for all as 50 is not a prime number.

$\sum_{m=0}^{50} \{f(m)\}^2 = ({^{50}C_0})^2 + ({^{50}C_1})^2 + ({^{50}C_2})^2 + \dots + ({^{50}C_{50}})^2$

$= {^{100}C_{50}}$

33. (b, c, d) $\frac{(n-1)(n-2)\dots(n-m+1)}{(m-1)!}$

$= \frac{(n-1)(n-2)\dots(n-m+1)(n-m)\dots 2 \cdot 1}{(n-m)!\,(m-1)!}$

$= {^{n-1}C_{m-1}}$

= Coefficient of x^{m-1} in $(1+x)^{n-1}$

= Coefficient of x^{m-1} in $(1+x)^n(1+x)^{-1}$

Now, $(1+x)^n = C_0 + C_1x + C_2x^2 + \dots + C_{m-1}x^{m-1} + \dots + C_nx^n$ ...(i)

$(1+x)^{-1} = 1 - x + x^2 - x^3 + \dots + (-1)^{m-1}x^{m-1} + \dots$ ...(ii)

Collecting the coefficient of x^{m-1} in the product of Eqs. (i) and (ii), we get

$(-1)^{m-1}C_0 + (-1)^{m-2}C_1 + \dots + C_{m-1}$

= Coefficient of x^{m-1} in $(1+x)^{n-1}$

$= {^{n-1}C_{m-1}}$

$\therefore C_0 - C_1 + C_2 - \dots + (-1)^{m-1}C_{m-1}$

$= {^{n-1}C_{m-1}}(-1)^{m-1}$

$= \frac{(n-1)(n-2)\dots(n-m+1)}{(m-1)!}(-1)^{m-1}$

34. (a, b, d) $(1 + z^2 + z^4)^8 = C_0 + C_1z^2 + C_2z^4 + \dots + C_{16}z^{32}$ ...(i)

Putting $z = i$, where $i = \sqrt{-1}$,

$(1 - 1 + 1)^8 = C_0 - C_1 + C_2 - C_3 + \dots + C_{16}$

or $\quad C_0 - C_1 + C_2 - C_3 + \dots + C_{16} = 1$

Also, putting $z = \omega$,

$(1 + \omega^2 + \omega^4)^8 = C_0 + C_1\omega^2 + C_2\omega^4 + \dots + C_{16}\omega^{32}$

or $\quad C_0 + C_1\omega^2 + C_2\omega + C_3 + \dots + C_{16}\omega^2 = 0$ ...(ii)

Putting $z = \omega^2$,

$(1 + \omega^4 + \omega^8)^8 = C_0 + C_1\omega^4 + C_2\omega^8 + \dots C_{16}\omega^{64}$

or $\quad C_0 + C_1\omega + C_2\omega^2 + \dots C_{16}\omega = 0$ ...(iii)

Putting $z = 1$, $3^8 = C_0 + C_1 + C_2 + \dots + C_{16}$ ...(iv)

Adding Eqs. (ii), (iii) and (iv), we get

$3(C_0 + C_3 + \dots + C_{15}) = 3^8$ or $C_0 + C_3 + \dots + C_{15} = 3^7$

Similarly, first multiplying Eq. (i) by z and then putting $1, \omega, \omega^2$ and adding, we get

$C_1 + C_4 + C_7 + C_{10} + C_{13} + C_{16} = 3^7$

Multiplying Eq. (i) by z^2 and then putting $1, \omega, \omega^2$ and adding, we get

$C_2 + C_5 + C_8 + C_{11} + C_{14} = 3^7$

35. (b, c) Given expression is

$$C_0-(C_0+C_1)+(C_0+C_1+C_2)$$
$$-(C_0+C_1+C_2+C_3)+\ldots(-1)^{n-1}(C_0+C_1+\ldots C_{n-1})$$

For $n=2m$,

$$=C_0-(C_0+C_1)+(C_0+C_1+C_2)$$
$$-(C_0+C_1+C_2+C_3)+\ldots+(C_0+C_1+\ldots.+C_{2m-1})$$
$$=-(C_1+C_3+C_5+\ldots+C_{2m-1})$$
$$=-(C_1+C_3+C_5+\ldots+C_{n-1})=-2^{n-1}$$

36. (a, b) $^nC_0\times{}^{2n}C_r-{}^nC_1\times{}^{2n-2}C_r+{}^nC_2\times{}^{2n-4}C_r+\ldots$

$$=\text{Coefficient of } x^r \text{ in } \{^nC_0(1+x)^{2n}-{}^nC_1(1+x)^{2n-2}+{}^nC_2(1+x)^{2n-4}+\ldots\}$$
$$=\text{Coefficient of } x^r \text{ in } [^nC_0\{(1+x)^2\}^n-{}^nC_1\{(1+x)^2\}^{n-1}+{}^nC_2\{(1+x)^2\}^{n-2}+\ldots]$$
$$=\text{Coefficient of } x^r \text{ in } \{(1+x)^2-1\}^n$$
$$=\text{Coefficient of } x^r \text{ in } (2x+x^2)^n$$

General term, $T_{p+1}={}^nC_p(2x)^{n-p}(x^2)^p$

$$={}^nC_p(2)^{n-p}(x)^{n+p}$$

For $x^r, n+p=r$

So, $p=r-n$

$\therefore$ Coefficient of $x^2={}^nC_{r-n}\times(2)^{2n-r}$

This is non-zero, if $r\geq n$ and zero, if $r<n$.

37. (c, d) $\because a_2=$ Coefficient of x^2 in $(1+x+x^2)^{2n}$

$$\therefore\ (1+x+x^2)^{2n}=\sum_{\alpha+\beta+\gamma=2n}\frac{2n!}{\alpha!\beta!\gamma!}(1)^{\alpha}(x)^{\beta}(x^2)^{\gamma}$$
$$=\sum_{\alpha+\beta+\gamma=2n}\frac{2n!}{\alpha!\beta!\gamma!}(x)^{\beta+2\gamma}$$

For a_2, $\beta+2\gamma=2$

Possible values of α,β,γ are $(2n-2,2,0)$ and $(2n-1,0,1)$

$$\therefore\quad a_2=\frac{2n!}{(2n-2)!\,0!\,2!}+\frac{2n!}{(2n-1)!\,0!\,1!}$$
$$\therefore\quad a_2={}^{2n}C_2+{}^{2n}C_1$$
$$\Rightarrow\quad a_2={}^{2n+1}C_2\qquad[\because {}^nC_r+{}^nC_{r+1}={}^{n+1}C_{r+1}]$$

Sol. (Q.Nos. 38-40)

38. (c) **39.** (a) **40.** (c)

$a_0+a_1x+a_2x^2+\ldots+a_{99}x^{99}+x^{100}=0$ has roots $^{99}C_0, {}^{99}C_1, {}^{99}C_2,\ldots, {}^{99}C_{99}$

or $a_0+a_1x+a_2x^2\ldots+a_{99}x^{99}+x^{100}$

$$=(x-{}^{99}C_0)(x-{}^{99}C_1)(x-{}^{99}C_2)\ldots(x-{}^{99}C_{99})$$

Now, sum of roots is

$$^{99}C_0+{}^{99}C_1+{}^{99}C_2+\ldots.+{}^{99}C_{99}=-\frac{a_{99}}{\text{coefficient of } x^{100}}$$

or $a_{99}=-2^{99}$

Also, sum of product of roots taken two at a time is

$$\frac{a_{98}}{\text{Coefficient of } x^{100}}$$

$$\therefore \sum_{0\leq i<j\leq 99}\sum {}^{99}C_i\cdot{}^{99}C_j$$
$$=\frac{\left(\sum_{i=0}^{99}\sum_{j=0}^{99}{}^{99}C_i\cdot{}^{99}C_j\right)-\sum_{i=0}^{99}({}^{99}C_i)^2}{2}$$
$$=\frac{\left(\sum_{i=0}^{99}{}^{99}C_i\,2^{99}\right)-\sum_{i=0}^{99}({}^{99}C_i)^2}{2}$$
$$=\frac{2^{99}\cdot2^{99}-\sum_{i=0}^{99}({}^{99}C_i)^2}{2}$$
$$=\frac{2^{198}-{}^{198}C_{99}}{2}$$

$$({}^{99}C_0)^2+({}^{99}C_1)^2+\ldots+({}^{99}C_{99})^2$$
$$=({}^{99}C_0+{}^{99}C_1+{}^{99}C_2+..+{}^{99}C_{99})^2-2\sum_{0\leq i<j\leq 99}\sum {}^{99}C_i\cdot{}^{99}C_j$$
$$=(-a_{99})^2-2a_{98}=a_{99}^2-2a_{98}$$

Sol. (Q.Nos. 41-43)

41. (b) **42.** (b) **43.** (c)

In P, general term of the series is

$$T(r)=\frac{^{50+r}C_r(2r-1)}{^{50}C_r(50+r)}$$
$$=\frac{^{50+r}C_r}{^{50}C_r}\left(1-\frac{50-r+1}{50+r}\right)$$
$$=\frac{^{50+r}C_r}{^{50}C_r}-\frac{^{50+r}C_r}{^{50}C_r}\left(\frac{50-r+1}{50+r}\right)$$

Now, $\dfrac{^{50+r}C_r}{^{50}C_r}\left(\dfrac{50-r+1}{50+r}\right)$

$$=\frac{(50-r+1)(50+r)!\,r!\,(50-r)!}{r!\,50!\,(50+r)\,50!}$$
$$=\frac{(50-r+1)!\,(50+r-1)!}{50!\,50!}$$
$$=\frac{^{50+r-1}C_{r-1}}{^{50}C_{r-1}}$$
$$\Rightarrow\ T(r)=\frac{^{50+r}C_r}{^{50}C_r}-\frac{^{50+r-1}C_{r-1}}{^{50}C_{r-1}}=\gamma(r)-\gamma(r-1)$$

where, $\gamma(r)=\dfrac{^{50+r}C_r}{^{50}C_r}$

Now, sum of the given series

$$P = \sum_{r=1}^{50} T(r) = \gamma(50) - \gamma(0)$$

$$= \frac{^{100}C_{50}}{^{50}C_{50}} - \frac{^{50}C_0}{^{50}C_0} = {}^{100}C_{50} - 1$$

Also, $Q = \sum_{r=0}^{50} ({}^{50}C_r)^2 = {}^{50}C_0^2 + {}^{50}C_1^2 + {}^{50}C_2^2 + \ldots + {}^{50}C_{50}^2$

$$= {}^{100}C_{50}$$

$\Rightarrow P - Q = -1$

We know that

$$C_0^2 - C_1^2 + C_2^2 + \ldots + (-1)^n C_n^2 = \begin{cases} (-1)^n \, {}^nC_{n/2}, & \text{if } n \text{ is even} \\ 0, & \text{if } n \text{ is odd} \end{cases}$$

$$\Rightarrow \sum_{r=0}^{100} (-1)^r ({}^{100}C_r)^2 = (-1)^{100} \cdot {}^{100}C_{50} = {}^{100}C_{50}$$

$$\Rightarrow \quad P - R = -1$$

$$\therefore \; Q + R = 2\,{}^{100}C_{50} = 2P + 2$$

44. (i) → C, (ii) → D, (iii) → B, (iv) → A

(i) If $T(r+1)$ is greatest, then $\dfrac{n}{1+\left|\dfrac{a}{bx}\right|} \le r \le \dfrac{n+1}{1+\left|\dfrac{a}{bx}\right|}$

For $x = 1$ and $r = 11$, $\dfrac{n}{1+\dfrac{3}{5}} \le 11 \le \dfrac{n+1}{1+\dfrac{3}{5}}$

$$\Rightarrow \quad 5n \le 88 \le 5(n+1) \Rightarrow n = 17$$

(ii) If $T(r+1)$ is greatest, then

$$\left(\frac{r}{n-r+1}\right)\frac{a}{b} < x < \left(\frac{r+1}{n-r}\right)\frac{a}{b} \text{ for } (a+bx)^n.$$

For $n = 25$ and $r = 20$, $2 < x < \dfrac{63}{25}$

(iii) We have, $\dfrac{T_{r+1}}{T_r} = \dfrac{n-r+1}{r} \cdot \dfrac{bx}{a}$ for $(a+bx)^n$

For $n = 20$ and $r = 10$, $\dfrac{11}{10} \cdot \dfrac{5x}{3} = 1$ or $x = \dfrac{6}{11}$

(iv) We have, $\dfrac{n}{1+\left|\dfrac{a}{bx}\right|} \le r \le \dfrac{n+1}{1+\left|\dfrac{a}{bx}\right|}$ for $(a+bx)^n$,

For $x = 2$ and $r = 8$ (as T_9 will be largest)

$$\frac{n}{1+\dfrac{3}{10}} \le 8 \le \frac{n+1}{1+\dfrac{3}{10}}$$

$$\Rightarrow \quad 10n \le 104 \le 10(n+1)$$

$$\Rightarrow \quad n = 10$$

45. (i) → B, (ii) → A, (iii) → D, (iv) → C

(i) $\displaystyle\sum_{r=0}^{14} \frac{(15-r)}{r+1} \cdot {}^{15}C_r \cdot {}^{20}C_{m-r}$

$$= \sum_{r=0}^{14} \frac{(15-r)}{r+1} \cdot \frac{15!}{r!(15-r)!} \cdot {}^{20}C_{m-r}$$

$$= \sum_{r=0}^{14} \frac{15!}{(r+1)!(14-r)!} \cdot {}^{20}C_{m-r}$$

$$= \sum_{r=0}^{14} {}^{15}C_{r+1} \cdot {}^{20}C_{m-r}$$

$$= {}^{35}C_{m+1},$$

which has maximum value of ${}^{35}C_{18}$.

(ii) ${}^{35}C_0 - {}^{35}C_1 + {}^{35}C_2 - {}^{35}C_3 + \ldots + (-1)^r \cdot {}^{35}C_r + \ldots$

= Coefficient of x^r in

$({}^{35}C_0 - {}^{35}C_1 x + {}^{35}C_2 x^2 - {}^{35}C_3 x^3 + \ldots + (-1)^r \, {}^{35}C_r + \ldots) \times (1 + x + x^2 + x^3 + \ldots + x^r + \ldots)$

= Coefficient of x^r in $(1-x)^{35}(1-x)^{-1}$

= Coefficient of x^r in $(1-x)^{34}$

$= (-1)^r \, {}^{34}C_r$

∴ Required maximum value is ${}^{34}C_{17}$.

(iii) ${}^{31}C_{r-5} + 5 \times {}^{31}C_{r-4} + 10 \times {}^{31}C_{r-3} + 10 \times {}^{31}C_{r-2} + 5 \times {}^{31}C_{r-1} + {}^{31}C_r$

$= {}^5C_5 \times {}^{31}C_{r-5} + {}^5C_4 \times {}^{31}C_{r-4} + {}^5C_3 \times {}^{31}C_{r-3} + {}^5C_2 \times {}^{31}C_{r-2} + {}^5C_1 \times {}^{31}C_{r-1} + {}^5C_0 \times {}^{31}C_r$

= Coefficient of x^r in $(1+x)^5(1+x)^{31}$

= Coefficient of x^r in $(1+x)^{36}$

$= {}^{36}C_r$,

which has maximum value of ${}^{36}C_{18}$.

(iv) $\displaystyle\sum_{r=0}^{k} 2^{k-r} \cdot (-1)^r \, {}^{37}C_r \cdot {}^{37-r}C_{37-k}$

$$= \sum_{r=0}^{k} 2^{k-r} \cdot (-1)^r \cdot \frac{37!}{(37-r)!\,r!} \cdot \frac{(37-r)!}{(37-k)!(k-r)!}$$

$$= \frac{37!}{k!(37-k)!} \sum_{r=0}^{k} 2^{k-r}(-1)^r \cdot \frac{k!}{r!(k-r)!}$$

$$= {}^{37}C_k \sum_{r=0}^{k} {}^kC_r \cdot 2^{k-r} \cdot (-1)^r$$

$$= {}^{37}C_k (2-k)k$$

$$= {}^{37}C_k$$

which has minimum value of ${}^{37}C_{18}$.

46. (i) → D, (ii) → A, (iii) → B, (iv) → C

(i) We know that

$$^mC_r + {}^mC_{r-1} \times {}^nC_1 + {}^mC_{r-2} \times {}^nC_2 + \dots + {}^nC_r = {}^{m+n}C_r$$

So, if $m, n > r$ and $m + n < r$, then

$$^mC_r + {}^mC_{r-1} \times {}^nC_1 + {}^mC_{r-2} \times {}^nC_2 + \dots + {}^nC_r = 0$$

(ii) If $m, n > r$, then

$$^mC_r + {}^mC_{r-1} \times {}^nC_1 + {}^mC_{r-2} \times {}^nC_2 + \dots + {}^nC_r = {}^{m+n}C_r$$

(iii) If $m, n < r < m + n$, then

$$^mC_m \times {}^nC_{r-m} + {}^mC_{m-1} \times {}^nC_{r-m+1} + {}^mC_{m-2} \times {}^nC_{r-m+2} + \dots\dots + {}^mC_{r-m} \times {}^nC_m = {}^{m+n}C_r$$

(iv) If $m < r < n$, then

$$^mC_m \times {}^nC_{r-m} + {}^mC_{m-1} \times {}^nC_{r-m+1} + {}^mC_{m-2} \times {}^nC_{r-m+2} + \dots + {}^nC_r = {}^{m+n}C_r$$

Sol. (Q.Nos. 47-49)

47. (7) **48.** (2) **49.** (128)

The coefficient of the 2nd, 3rd and 4th terms in the expansion are $^mC_1$, $^mC_2$ and $^mC_3$, which are given in AP.

Hence, $2\,{}^mC_2 = {}^mC_1 + {}^mC_3$

$$\Rightarrow \frac{2m(m-1)}{2!} = m + \frac{m(m-1)(m-2)}{3!}$$

$$\Rightarrow m(m^2 - 9m + 14) = 8$$

$$\Rightarrow m(m-2)(m-7) = 0$$

$$\Rightarrow m = 7$$

[$\because m \neq 0$ or 2 as 6th term is given equal to 21]

Now, 6th term in the expansion, when $m = 7$, is

$$^7C_5 \left[\sqrt{\{2^{\log(10-3^x)}\}}\right]^{7-5} \times \left[\sqrt[5]{\{2^{(x-2)\log 3}\}}\right]^5 = 21$$

$$\Rightarrow \frac{7 \times 6}{2!} 2^{\log(10-3^x)} \times 2^{(x-2)\log 3} = 21$$

$$\Rightarrow 2^{\log(10-3^x) + (x-2)\log 3} = 1 = 2^0$$

$$\Rightarrow \log(10-3^x) + (x-2)\log 3 = 0$$

$$\Rightarrow \log(10-3^x)(3)^{(x-2)} = 0$$

$$\Rightarrow (10-3^x) \times 3^x \times 3^{-2} = 1$$

$$\Rightarrow 10 \times 3^x - (3^x)^2 = 9$$

$$\Rightarrow (3^x)^2 - 10 \times 3^x + 9 = 0$$

$$\Rightarrow (3^x - 1)(3^x - 9) = 0$$

$$\Rightarrow 3^x = 1, 9$$

$$\Rightarrow x = 0, 2$$

When $x = 2$,

$$\left[\sqrt{\{2^{\log(10-3^x)}\}} + \sqrt[5]{\{2^{(x-2)\log 3}\}}\right]^m = [1+1]^7 = 128$$

When $x = 0$,

$$\left[\sqrt{\{2^{\log(10-3^x)}\}} + \sqrt[5]{\{2^{(x-2)\log 3}\}}\right]^m$$

$$= \left[\sqrt{\{2^{\log 9}\}} + \sqrt[5]{\{2^{-2\log 3}\}}\right]^7 = \left[2^{\frac{\log 9}{2}} + \frac{1}{2^{\frac{\log 9}{5}}}\right]^7 > 2^7$$

Hence, the minimum value is 128.

Sol. (Q.Nos. 50-52)

50. (– 1) **51.** (32) **52.** (1562)

2nd term is $^nC_1(x)^{n-1}(a) = 240$ ...(i)

3rd term is $^nC_2(x)^{n-2}(a)^2 = 720$ ...(ii)

4th term is $^nC_3(x)^{n-3}(a)^3 = 1080$ ...(iii)

Multiplying Eqs. (i) and (iii) and dividing by the square of Eq. (ii), we get

$$\frac{^nC_1 \times {}^nC_3}{(^nC_2)^2} = \frac{240 \times 1080}{(720)^2}$$

$$\Rightarrow \frac{n \times n(n-1)(n-2)(2!)^2}{n^2(n-1)^2 \times 3!} = \frac{1}{2}$$

$$\Rightarrow 4(n-2) = 3(n-1) \quad [\because n \neq 1]$$

$$\Rightarrow n = 5$$

Putting $n = 5$ in Eqs. (i) and (ii), we get

$$5x^4a = 240 \text{ and } 10x^3a^2 = 720$$

$$\Rightarrow \frac{(5x^4a)^2}{10x^3a^2} = \frac{(240)^2}{720}$$

or $x^5 = 32$ or $x = 2$

$$a = \frac{240}{5x^4} = \frac{48}{2^4} = 3$$

Hence, $x = 2, a = 3$ and $n = 5$

$$(x-a)^n = (2-3)^5 = -1$$

Also, $(2+3)^5 = 2^5 + {}^5C_1 2^4 \times 3 + {}^5C_2 2^3 \times 3^2 + {}^5C_3 2^2 \times 3^3 + {}^5C_4 2^1 \times 3^4 + {}^5C_5 3^5$

$$= 32 + 240 + 720 + 1080 + 810 + 243$$

Hence, least value of the term is 32.

Sum of odd numbered terms is $32 + 720 + 810 = 1562$

53. (6) Let T_{r-1}, T_r and T_{r+1} are three consecutive terms of $(1+x)^{n+5}$.

$$T_{r-1} = {}^{n+5}C_{r-2}(x)^{r-2},$$

$$T_r = {}^{n+5}C_{r-1}(x)^{r-1} \text{ and } T_{r+1} = {}^{n+5}C_r(x)^r,$$

where $^{n+5}C_{r-2} : {}^{n+5}C_{r-1} : {}^{n+5}C_r = 5 : 10 : 14$

So, $\dfrac{^{n+5}C_{r-2}}{5}=\dfrac{^{n+5}C_{r-1}}{10}=\dfrac{^{n+5}C_r}{14}$

Comparing first two results, we have $n-3r=-6$ ...(i)

Comparing last two results, we have

$$5n-12r=-18 \quad ...(ii)$$

From Eqs. (i) and (ii), $n=6$

54. (6) $(1-2x+5x^2-10x^3)[^nC_0+{}^nC_1x+{}^nC_2x^2+\dots]$

$$=1+a_1x+a_2x^2+\dots$$

Given, $a_1^2=2a_2$

$\Rightarrow (n-2)^2=n(n-1)-4n+10$

$\Rightarrow n^2-4n+4=n^2-5n+10$

$\Rightarrow n=6$

55. (646) $X=\sum_{r=1}^{10} r\cdot(^{10}C_r)^2=\sum_{r=1}^{10} r\cdot{}^{10}C_r\cdot{}^{10}C_r$

$$=10\sum_{r=1}^{10} {}^9C_{r-1}\cdot{}^{10}C_{10-r}$$

$X=10\cdot{}^{19}C_9$

Now, $\dfrac{X}{1430}=\dfrac{10\cdot{}^{19}C_9}{1430}=\dfrac{^{19}C_9}{143}=\dfrac{^{19}C_9}{11\times13}$

$$=\frac{19\times17\times16}{8}$$

$$=19\times34=646$$

56. (5) Coefficient of x^2 in expansion

$=1+{}^3C_2+{}^4C_2+{}^5C_2+\dots+{}^{49}C_2+{}^{50}C_2\cdot m^2$

[as $^nC_r+{}^nC_{r-1}={}^{n+1}C_r$]

$=(^3C_3+{}^3C_2)+{}^4C_2+{}^5C_2+\dots+{}^{49}C_2+{}^{50}C_2\cdot m^2$

$=(^4C_3+{}^4C_2)+\dots+{}^{50}C_2m^2$

$={}^5C_3+\dots+{}^{49}C_2+{}^{50}C_2m^2$

$={}^{50}C_3+{}^{50}C_2m^2+{}^{50}C_2-{}^{50}C_2$

$={}^{51}C_3+{}^{50}C_2(m^2-1)$

$=(3n+1)\,{}^{51}C_3$ [given]

$\therefore 3n\cdot\dfrac{51}{3}\cdot{}^{50}C_2={}^{50}C_2(m^2-1)$

$$\frac{m^2-1}{51}=n$$

$\therefore$ Value of n is 5.

57. (0) $E=(1+x+x^2+x^3+x^4)^{199}(x-1)^{201}$

$$=\left(\frac{1-x^5}{1-x}\right)^{199}(x-1)^{201}$$

$$=-(1-x)^2(1-x^5)^{199}$$

$$=-(1-2x+x^2)(1-{}^{199}C_1\cdot x^5+{}^{199}C_2\cdot x^{10}-\dots)$$

$\therefore$ Coefficient x^{103} in $E=0$

58. (495) Given, $\{\log(x^{\log x})-\log_{x^2}100\}^{12}$

$$=\left((\log x)^2-\frac{1}{\log x}\right)^{12}$$

$\therefore$ Constant term $={}^{12}C_4\{(\log x)^2\}^4\left(-\dfrac{1}{\log x}\right)^8$

$$={}^{12}C_4$$

$$=\frac{12\times11\times10\times9}{24}$$

$$=55\times9=495$$

59. (102) Given, $(^{101}C_0-{}^{101}C_1x+{}^{101}C_2x^2-\dots$

$-{}^{101}C_{101}x^{101})(1+x+x^2+\dots+x^{100})^{101}$

$$=(1-x)^{101}\cdot\left(\frac{1-x^{101}}{1-x}\right)^{101}$$

$$=(1-x^{101})^{101}$$

$\therefore$ Number of different terms $=102$.

60. (220) $\sum_{r=0}^{3}{}^8C_r\cdot(^5C_{r+1}-{}^4C_r)=\sum_{r=0}^{3}{}^8C_r\cdot{}^4C_{r+1}$

$={}^8C_0\times{}^4C_1+{}^8C_1\times{}^4C_2+{}^8C_2\times{}^4C_3+{}^8C_3\times{}^4C_4$

$=$ Coefficient of x^3 in $(1+x)^4(1+x)^8$

$=$ Coefficient of x^3 in $(1+x)^{12}$

$={}^{12}C_3=220$

61. (5151) Here, $T_r=\dfrac{r\cdot{}^{101}C_r}{^{101}C_{r-1}}$

$$=\frac{r\cdot(101-r+1)}{r}$$

$$=102-r$$

$\therefore$ Given, sum $=\sum_{r=1}^{101}(102-r)$

$$=101+100+99+\dots+2+1$$

$$=5151$$

62. (9) $f(n)={}^nC_0(a)^{n-1}-{}^nC_1(a)^{n-2}+{}^nC_2(a)^{n-3}$

$+\dots+(-1)^{n-1}\,{}^nC_{n-1}a^0$

$+(-1)^n\cdot{}^nC_n\cdot\left(\dfrac{1}{a}\right)-(-1)^n\cdot{}^nC_n\cdot\left(\dfrac{1}{a}\right)$

$=\dfrac{1}{a}(^nC_0a^n-{}^nC_1a^{n-1}+{}^nC_2a^{n-2}$

$+\dots+(-1)^n\,{}^nC_n)-(-1)^n\cdot{}^nC_n\left(\dfrac{1}{a}\right)$

$$=\frac{1}{a}\{(a-1)^n-(-1)^n\,{}^nC_n\}$$

$$=\frac{1}{a}\{(3^{\frac{n}{223}}-(-1)^n)\}$$

$$f(n)=\frac{3^{\frac{n}{223}}-(-1)^n}{(3^{\frac{1}{223}}+1)} \qquad [\text{since, } a=1+3^{1/223}]$$

$$\Rightarrow f(2007)=\frac{3^{\frac{2007}{223}}+1}{3^{\frac{1}{223}}+1}$$

$$\Rightarrow f(2008)=\frac{3^{\frac{2008}{223}}-1}{3^{\frac{1}{223}}+1}$$

$$\Rightarrow f(2007)+f(2008)=\frac{3^{\frac{2007}{223}}+3^{\frac{2008}{223}}}{3^{\frac{1}{223}}+1}$$

$$=\frac{3^9+3^{9+\frac{1}{223}}}{3^{\frac{1}{223}}+1}=3^9\,\frac{(1+3^{\frac{1}{223}})}{(1+3^{\frac{1}{223}})}=3^9$$

$$\Rightarrow \qquad 3^9=3^k, \text{ then } k=9$$

63. (0) $1+2+2^2+2^3+\ldots+2^{1999}$

$$=\frac{1\,(2^{2000}-1)}{2-1}=2^{2000}-1$$

$$=(1-5)^{1000}-1$$

$$=1-{}^{1000}C_1\cdot 5+{}^{1000}C_2\cdot 5^2+\ldots+{}^{1000}C_{1000}\cdot 5^{1000}-1$$

which is divisible by 5.

64. (576) The general term in the expansion of

$$(1+3x+2x^2)^6=\sum\frac{6!}{r_1!\,r_2!\,r_3!}(1)^{r_1}(3x)^{r_2}(2x^2)^{r_3}$$

where, $r_1+r_2+r_3=6$ ...(i)

For coefficient of x^{11}, we have

$$r_2+2r_3=11 \qquad \text{...(ii)}$$

Now, from Eqs. (i) and (ii), we get

$$r_1=r_3-5$$

For $r_3=5, r_1=0$ and $r_2=1$,

$$\therefore \text{ Coefficient of } x^{11}=\frac{6!}{0!1!5!}(1)^0(3)^1(2)^5$$

$$=6\times 3\times 2^5$$

$$=18\times 32$$

$$=576$$

CHAPTER

07

PROBABILITY

LEVEL 01 \\ Starter Level

7.1 Probability-A Theoretical Approach

1 The probability that it will rain tomorrow is 0.85. What is the probability that it will not rain tomorrow?

2 A die is thrown. Find probability of getting
(i) a prime number
(ii) 2 or 4
(iii) a multiple of 2 or 3
(iv) an even prime number
(v) a number greater than 5
(vi) a number lying between 2 and 6

3 In a lottery of 50 tickets numbered 1 to 50, one ticket is drawn. Find the probability that the drawn ticket bears a prime number.

4 What is the probability that a number selected from the numbers 1, 2, 3, ..., 15 is a multiple of 4?

5 If the probability of winning a game is 0.3, what is the probability of loosing it?

6 Two dice are rolled if both dice have six faces numbered 1, 2, 3, 5, 7 and 11, then find the probability that the sum of the numbers on the top faces is less than or equal to 8. **[JEE Main 2021]**

7 Find the probability that the birth days of 6 different persons will fall in exactly 2 calendar month.

8 Words from the letters of the word PROBABILITY are formed by taking all letters at a time. Find the probability that both B's are not together and both I's are not together.

9 The odds against a certain event are 5 : 2 and the odds in favour of other event, independent of the former, are 6 : 5. Then, find the probability that at least one of the events will happen.

10 A four figure number is formed of the figures 1, 2, 3, 5 with no repetitions. Find the probability that the number is divisible by 5.

11 Let E be an event which is neither a certainly nor an impossibility. If probability is such that $P(E) = 1 + \lambda + \lambda^2$ and $P(E') = (1 + \lambda)^2$ are given in terms of an unknown λ. Then, find $P(E)$.

12 In a given race the odds in favour of 3 horses, A, B, C are $1:3$; $1:4$, $1:5$, respectively. Assuming that dead head is impossible. Find the probability that one of them wins.

13 If the probability that A and B will die within a year are p and q respectively, then find the probability that only one of them will be alive at the end of the year.

14 A die is loaded in such a ways that each odd number is twice all likely to occur as each even number. If E is the event of a number greater than or equal to 4 on single toss of the die, then find $P(E)$.

15 A card is drawn at random from a pack of 52 cards. Find the probability that the drawn card is either a club or an ace of diamond.

16 A herd contains 30 cows numbered from 1 to 30. One cow is selected at random. Find the probability that the number of the selected cow is multiple of 5 or 6.

17 A number was drawn at random from the number 1 to 50. What is the probability that it will be a multiple of 2 or 3 or 10.

7.2 Conditional Probability

18 Two dice are thrown. What is the probability that the sum of the numbers appearing on the two dice is 11, if 5 appears on the 1st?

19 A family has 2 children. What is the probability that both the children are boys given that at least one of them is a boy?

20 In a school, there are 1000 students, out of which 430 are girls. It is known than out of 430, 10% of the girls study in class XII. What is the probability that a student chosen randomly studies in class XII given that the chosen student is a girl?

21 A die is thrown twice and the sum of the numbers appearing is observed to be 6. What is the conditional probability that the number 4 has appeared at least once?

22 If two events A and B are such that $P(A) = 0.3$; $P(B) = 0.4$ and $P(A' \cap B') = 0.5$, then find the value of $P[B \mid (A \cup B')]$.

23 A coin is tossed 3 times, where

(i) A : head on 3rd toss, B : heads on 1st 2 tosses

(ii) A : atleast two heads, B : atmost two heads

(iii) A : atmost two tails, B : atleast one tail

In each case, find $P\left(\frac{A}{B}\right)$.

24 A dice is thrown 3 times and the sum of the thrown numbers is 15. Find the probability for which number 4 appears in 1st throw.

25 Consider the experiment of tossing a coin. If the coin shows head, toss it again but if it shows tails, then throw a dice. Find the conditional probability of the event 'the dice shows a number greater than 4' given that there is at least one tail.

26 A box contains 10 mangoes out of which 4 are rotten. Two mangoes are taken out together. If one of them is found to be good, then find the probability that the others is also good.

27 An urn contains 10 black and 5 white balls. 2 balls are drawn from the urn one after the other without replacement. What is the probability that 1st ball is black and 2nd ball is white?

28 3 cards are drawn successively, without replacement from a pack of 52 well shuffled cards. What is the probability that 1st 2 cards are kings and third card drawn is an ace?

29 3 cards are drawn successively without replacement from a pack of 52 well shuffled cards. What is the probability that 1st, 2nd and 3rd cards are jack, queen and king, respectively?

30 An unbiased die is thrown twice. Let the event A be 'odd number on the 1st throw and B the event 'odd numbers on the 2nd throw'. Check the independence of the events A and B.

31 A fair coin is tossed repeatedly. If tail appears on 1st four tosses, then find the probability of head appearing on 5th toss.

32 A die marked 1, 2, 3 in red and 4, 5, 6 in green is tossed. Let A be the event, 'the number is even' and B be the event, 'the number is red'. Are A and B independent?

33 3 persons work independently on a problem. If the probability that they will solve it are $\frac{1}{3}, \frac{1}{4}$ and $\frac{1}{5}$, then find the probability that none can solve it.

34 The probability of hitting a target by 3 marksmen A, B and C are $\frac{1}{2}, \frac{1}{3}$ and $\frac{1}{4}$, respectively. Find the probability that one and only one of them will hit the target when they fire simultaneously.

35 If A and B are 2 independent events, the probability that both A and B occur is $\frac{1}{8}$ and the probability that neither of them occurs is $\frac{3}{8}$. Find the probability of the occurrence of A.

36 'x' speaks truth in 60% and 'y' in 50% of the cases. Find the probability that they contradict each other narrating the same incident.

.3 Baye's Theorem

37 A bag contains $n+1$ coins. It is known that one of these coins shows heads on both sides, whereas the other coins are fair. One coin is selected at random and tossed. If the probability that toss result in heads is $\frac{7}{12}$, then find the value of n.

38 An urn contains 6 white and 4 black balls. A fair dice is rolled and that number of balls are chosen from the urn. Find the probability that the balls selected are white.

39 A purse contains 2 silver and 4 copper coins. A second purse contains 4 silver and 3 copper coins. If a coin is pulled at random from one of the two purse, what is the probability that it is a silver coin?

40 Bag A contains 5 white and 6 black balls. Another bag B contains 4 white and 3 black balls. A ball is transferred from bag A to the bag B and then a ball is taken out of the second bag. Find the probability of this ball being black.

41 The contents of three bags I, II and III are as follows :

Bag I 1 white, 2 black and 3 red balls
Bag II 2 white, 1 black and 1 red ball
Bag III 4 white, 5 black and 3 red balls

A bag is chosen at random and two balls are drawn. What is the probability that the balls are white and red?

42 A person has undertaken a construction job. The probabilities are 0.65 that there will be strike, 0.80 that the construction job will be completed on time, if there is no strike and 0.32 that the construction job will be completed on time, if there is a strike. Determine the probability that the construction job will be completed on time.

43 In a factory which manufactures bolts, machines A, B and C manufacture respectively 25%, 35% and 40% of the bolts of their outputs 5, 4 and 2 percent are respectively defective bolts. A bolts is drawn at random from the product and is found to be defective. What is the probability that it is manufactured by the machine B?

44 Bag I contains 3 red and 4 black balls while another bag II contains 5 red and 6 black balls. One balls is drawn at random from one of the bags and it is found to be red. Find the probability that it was drawn from Bag II.

7.4 Random Variable Probability Distribution

45 Let x denotes the number of hours you study during a randomly selected school day. The probability that x can take the values x has the following form, where k is some unknown constant

$$P(X=x)=\begin{cases} 0.1, \text{ if } x=0 \\ kx, \text{ if } x=1 \text{ or } 2 \\ k(5-x), \text{ if } x=3 \text{ or } 4 \\ 0, \text{ otherwise} \end{cases}$$

(i) Find the value of k.

(ii) What is the probability that you study atleast two hours, exactly two hours and atmost two hours?

46 2 cards are drawn successively with replacement from a well shuffled deck of 52 cards. Find the probability distribution of the number of aces.

47 2 cards are drawn simultaneously from a well shuffled pack of 52 cards. Find the mean, variance and standard deviation of the number of kings.

7.5 Bernoulli's Trials and Binomial Distribution

48 6 balls are drawn successively from an urn containing 7 red and 9 blacks balls. Tell whether or not the trials of drawing ball are Bernoulli trials when after each draw the ball drawn is

(i) replaced

(ii) not replaced in the urn

49 Find the probability of 4 turning up atleast once in two tosses of a fair die.

50 A die is thrown 7 times. What is the chance that an odd number turns up (i) exactly 4 times (ii) atleast 4 times?

51 Suppose that 90% of people are right handed. What is the probability that at most 6 of a random sample of 10 people are right handed?

52 An experiment succeeds twice as often as its fails. Then, find the probability that in the next 6 trials, there will be atleast 4 successes.

53 What is the probability of guessing correctly at least 8 out of 10 answers on a true-false examination?

54 A rifleman is firing at a distant target and hence, has only 10% chances of hitting it. Find the number of rounds, he must fire in order to have more than 50% chances of hitting it atleast once.

55 A and B throw a die alternatively till one of them gets a 6 and wins the game. Find their respective probabilities of winning, if A starts first.

56 A dice is thrown thrice. A success is 1 or 6 in a throw. Find the mean and variance of the number of success.

57 In eight throws of a die 5 or 6 is considered a success, find the mean number of success and the standard deviation.

58 If a random variable x follows binomial distribution with mean 3 and variance $\frac{3}{2}$, find $P(x \leq 5)$.

LEVEL 02 \ JEE Main Level

Single Choice Correct Questions

7.1 Probability-A Theoretical Approach

1 A single letter is selected at random from the word 'PROBABILITY'. The probability that it is a vowel, is

(a) $\frac{8}{11}$ (b) $\frac{4}{11}$ (c) $\frac{2}{11}$ (d) $\frac{3}{11}$

2 The probability that a leap year selected at random contains 53 sundays, is

(a) $\frac{7}{366}$ (b) $\frac{28}{183}$ (c) $\frac{1}{7}$ (d) $\frac{2}{7}$

3 Three of six vertices of regular hexagon are chosen at random. The probability that the triangle with three vertices is equilateral, is equal to

(a) $\frac{1}{2}$ (b) $\frac{1}{5}$ (c) $\frac{1}{10}$ (d) $\frac{1}{20}$

4 4 ball are drawn at random from a bag containing 5 white, 4 green and 3 black balls. The probability that exactly 2 of them are white, is

(a) $\frac{14}{33}$ (b) $\frac{7}{16}$ (c) $\frac{18}{33}$ (d) $\frac{9}{16}$

5 7 people seat themselves in discriminately at round table. The probability that two distinguished person will be next to each other, is

(a) $\frac{1}{3}$ (b) $\frac{1}{2}$ (c) $\frac{1}{4}$ (d) $\frac{2}{3}$

6 If two cards are drawn simultaneously from the set, then probability that atleast one of them will be the ace of heart, is

(a) $\frac{1}{13}$ (b) $\frac{1}{26}$ (c) $\frac{1}{52}$ (d) $\frac{3}{13}$

7 South African cricket captain lost the toss of a coin 13 times out of 14. The chance of this happening was

(a) $\frac{7}{2^{13}}$ (b) $\frac{1}{2^{13}}$ (c) $\frac{13}{2^{14}}$ (d) $\frac{13}{12^{13}}$

8 Four dice are thrown simultaneously. The probability that 4 and 3 appear on two of the die given that 5 and 6 have appeared on other two die, is

(a) $\frac{1}{6}$ (b) $\frac{1}{36}$ (c) $\frac{12}{151}$ (d) None of these

9 Three houses are available in a locality. Three persons apply for the houses. Each applies for one house without consulting others. The probability that all three apply for the same house is

(a) $\frac{1}{9}$ (b) $\frac{2}{9}$ (c) $\frac{7}{9}$ (d) $\frac{8}{9}$

10 A 3-digit number is selected at random from the set of all 3-digit numbers. The probability that the number selected has all 3 digits same, is

(a) $\frac{1}{9}$ (b) $\frac{1}{10}$ (c) $\frac{1}{50}$ (d) $\frac{1}{100}$

11 Three integers are chosen at random from the set of first 20 natural numbers. The chances that their product is a multiple of 3, is

(a) $\frac{194}{285}$ (b) $\frac{1}{57}$ (c) $\frac{13}{19}$ (d) $\frac{3}{4}$

12 The probability that a random chosen three-digit number has exactly 3 factors, is

(a) $\frac{2}{225}$ (b) $\frac{7}{900}$ (c) $\frac{1}{800}$ (d) None of these

13 If three squares are selected at random from chessboard, then the probability that they form letter "L" is

(a) $\frac{196}{{}^{64}C_3}$ (b) $\frac{49}{{}^{64}C_3}$ (c) $\frac{36}{{}^{64}C_3}$ (d) $\frac{98}{{}^{64}C_3}$

14 Six dice are thrown. The probability that different numbers will it turn up, is

(a) $\frac{129}{1296}$ (b) $\frac{1}{54}$ (c) $\frac{5}{324}$ (d) $\frac{5}{54}$

15 A drawer contains 5 brown socks and 4 blue socks well mixed. A man reaches the drawer and pull out socks at random. What is the probability that they match?

(a) $\frac{4}{9}$ (b) $\frac{5}{8}$ (c) $\frac{5}{9}$ (d) $\frac{7}{12}$

16 A composite number is selected at random from the first 30 natural numbers and it is divided by 5. The probability that there will be a remainder, is

(a) $\frac{14}{19}$ (b) $\frac{5}{19}$ (c) $\frac{5}{6}$ (d) $\frac{7}{15}$

17 One mapping is selected at random from all mappings of the set $S = \{1, 2, 3, \ldots, n\}$ into itself. If the probability that the mapping is one-one is $\frac{3}{32}$, then the value of n is

(a) 1 (b) 3 (c) 4 (d) None of these

18 A draws a card from pack of n cards marked 1, 2, ..., n. The card is replaced in the pack and B draws a card. Then, the probability that A draws a higher card than B, is

(a) $\frac{(n+1)}{2n}$ (b) $\frac{1}{2}$ (c) $\frac{(n-1)}{2n}$ (d) None of these

19 2 numbers a, b are chosen from the set of integers 1, 2, 3, ..., 39. Then, probability that the equation $7a - 9b = 0$ is satisfied, is

(a) $\frac{1}{247}$ (b) $\frac{2}{247}$ (c) $\frac{4}{741}$ (d) $\frac{5}{741}$

20 Dialing a telephone number an old man forgets the last digits remembering only that these are different dialed at random. The probability that the numbers is dialed correctly, is

(a) $\frac{1}{45}$ (b) $\frac{1}{90}$ (c) $\frac{1}{100}$ (d) None of these

21 A natural number is chosen at random from the first 100 natural numbers. The probability that $x + \frac{100}{x} > 50$ is

(a) $\frac{1}{10}$ (b) $\frac{11}{50}$ (c) $\frac{11}{20}$ (d) None of these

22 A fair six faced die is rolled 12 times. The probability each face turns up twice is equal to

(a) $\frac{12!}{6!6!6^{12}}$ (b) $\frac{2^{12}}{2^6 6^{12}}$ (c) $\frac{12!}{2^6 6^{12}}$ (d) $\frac{12!}{6^2 6^{12}}$

23 The probability that in a family of five members, exactly two members have birthday on Sunday, is

(a) $\frac{(12 \times 5^3)}{7^5}$ (b) $\frac{(10 \times 6^2)}{7^5}$ (c) $\frac{2}{5}$ (d) $\frac{(10 \times 6^3)}{7^5}$

24 A dice is thrown six times, it being known that each time a different digit is shown. The probability that a sum of 12 will be obtained in first three throws, is

(a) $\frac{5}{24}$ (b) $\frac{25}{216}$ (c) $\frac{3}{20}$ (d) $\frac{1}{12}$

25 The probability that a randomly selected 2-digit number belongs to the set $\{n \in N : (2^n - 2)$ is a multiple of $3\}$ is equal to **(JEE Main 2021)**

(a) $\frac{1}{6}$ (b) $\frac{2}{3}$ (c) $\frac{1}{2}$ (d) $\frac{1}{3}$

26 If two different numbers are taken from set {0, 1, 2, 3, ..., 10}, then the probability that their sum as well as absolute difference are both multiples of 4, is **(JEE Main 2017)**

(a) $\frac{7}{55}$ (b) $\frac{6}{55}$ (c) $\frac{12}{55}$ (d) $\frac{14}{45}$

27 Out of 11 consecutive natural numbers, if three numbers are selected at random (without repetition), then the probability that they are in AP with positive common, difference is **(JEE Main 2020)**

(a) $\frac{15}{101}$ (b) $\frac{5}{101}$ (c) $\frac{5}{33}$ (d) $\frac{10}{99}$

28 Three randomly chosen non-negative integers x, y and z are found to satisfy the equations $x + y + z = 10$. Then, the probability that z is even, is

(a) $\frac{1}{2}$ (b) $\frac{36}{55}$ (c) $\frac{6}{11}$ (d) $\frac{5}{11}$

29 Two squares are chosen at random on a chessboard (see figure). The probability that they have a side in common is **(JEE Main 2021)**

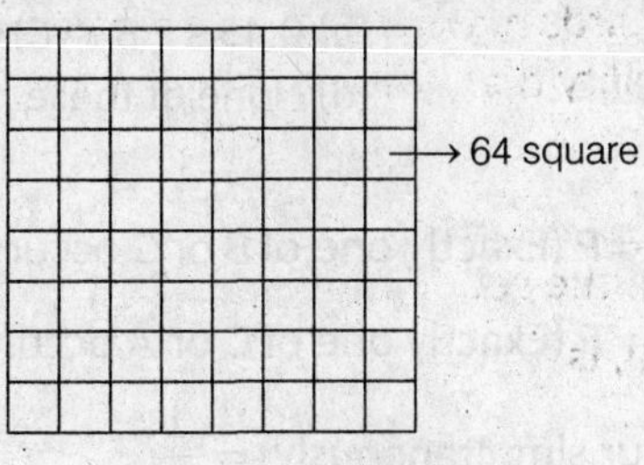

(a) $\frac{2}{7}$ (b) $\frac{1}{18}$ (c) $\frac{1}{7}$ (d) $\frac{1}{9}$

30 A determinant is chosen at random from the set of all determinants of order 2 with elements 0 and 1 only. The probability that the determinant chosen is positive is

(a) $\frac{1}{2}$ (b) $\frac{1}{8}$ (c) $\frac{3}{16}$ (d) $\frac{1}{4}$

31 15 persons, among whom are A and B sit down at random at a round table. The probability that there are 4 persons between A and B is

(a) $\frac{9!}{14!}$ (b) $\frac{10!}{14!}$ (c) $\frac{9!}{15!}$ (d) $\frac{1}{7}$

32 A is a set of 4 elements. A bijection on A is selected at random. The probability that it satisfies the condition $f(x) \neq x$ for all $x \in A$ is

(a) $\frac{17}{24}$ (b) $\frac{15}{24}$ (c) $\frac{11}{24}$ (d) $\frac{9}{24}$

33 A bag has 13 red, 14 green and 15 white balls, p_1 is the probability of drawing exactly 2 white balls when 4 balls are drawn. Then, the number of balls of each colour are doubled. Let P_2 be the probability of drawing 4 white balls when 8 ball are drawn, then

(a) $P_1 = P_2$ (b) $P_1 > P_2$ (c) $P_1 < P_2$ (d) None of these

34 Five different games are to be distributed among 4 children randomly. The probability that each child get at least one game is

(a) $\frac{1}{4}$ (b) $\frac{15}{64}$ (c) $\frac{21}{64}$ (d) None of these

35 If A and B are two events, then which of the following does not represent the probability that exactly one of them occurs?

(a) $P(A) + P(B) - 2P(A \cap B)$ (b) $P(A \cap B') + P(A' \cap B)$

(c) $P(A \cup B) - P(A \cap B)$ (d) $P(A') + P(B') - P(A' \cap B')$

36 Let A and B be two events such that probability that exactly one of them occurs is $\frac{2}{5}$ and the probability that A or B occurs is $\frac{1}{2}$, then the probability of both of them occur together is **(JEE Main 2020)**

(a) 0.02 (b) 0.20 (c) 0.01 (d) 0.10

37 The probabilities of three events, A, B and C are given by $P(A) = 0.6$, $P(B) = 0.4$ and $P(C) = 0.5$. If $P(A \cup B) = 0.8$, $P(A \cap C) = 0.3$, $P(A \cap B \cap C) = 0.2$, $P(B \cap C) = \beta$ and $P(A \cup B \cup C) = \alpha$, where $0.85 \leq \alpha \leq 0.95$, then β lies in the interval. **(JEE Main 2020)**

(a) [0.35, 0.36] (b) [0.25, 0.35] (c) [0.20, 0.25] (d) [0.36, 0.40]

38 Let A, B, C be three events such that $P(A) = 0.3$, $P(B) = 0.4$, $P(C) = 0.8$, $P(A \cap B) = 0.08$, $P(A \cap C) = 0.28$, $P(A \cap B \cap C) = 0.09$. If $P(A \cup B \cup C) \geq 0.75$, then

(a) $0.23 \leq P(B \cap C) \leq 0.48$ (b) $0.45 \leq P(B \cap C) \leq 0.75$

(c) $0.48 \leq P(B \cap C) \leq 0.75$ (d) None of these

39 For three events A, B and C

$$P(\text{Exactly one of } A \text{ or } B \text{ occurs}) = P(\text{Exactly one of } B \text{ or } C \text{ occurs})$$

$$= P(\text{Exactly one of } C \text{ or } A \text{ occurs}) = \frac{1}{4}$$

and $P(\text{All the three events occur simultaneously}) = \frac{1}{16}$

Then, the probability that at least one of the events occurs **[JEE Main 2017]**

(a) $\frac{3}{16}$ (b) $\frac{7}{32}$ (c) $\frac{7}{16}$ (d) $\frac{7}{64}$

7.2 Conditional Probability

40 One ticket is selected at random from 50 tickets numbered 00, 01, 02, ..., 49. Then, the probability that the sum of the digits on the selected ticket is 8, given that the product of these digits is zero equals

(a) $\frac{1}{14}$ (b) $\frac{1}{7}$ (c) $\frac{5}{14}$ (d) $\frac{1}{50}$

41 *3 numbers are chosen at random without replacement from {1, 2, 3, ..., 8}. The probability that their minimum is 3, given that their maximum is 6 is* **[AIEEE 2012]**

(a) $\frac{3}{8}$ (b) $\frac{1}{5}$ (c) $\frac{1}{4}$ (d) $\frac{2}{5}$

42 The probabilities of 4 cricketers A, B, C and D scoring more than 50 runs in a match are $\frac{1}{2}, \frac{1}{3}, \frac{1}{4}$ and $\frac{1}{10}$. It is known that exactly 2 of the players scored more than 50 runs in particular match. The probabilities that these players were A and B is

(a) $\frac{27}{65}$ (b) $\frac{5}{6}$ (c) $\frac{1}{6}$ (d) None of these

43 A bag contains $n+1$ coins. It is known that one of these coins shows heads on both sides, whereas the other coins are fair. One coin is selected at random and tossed. If the probability that toss resulted in head is $\frac{7}{12}$, then the value of n is

(a) 3 (b) 4 (c) 5 (d) None of these

44 A signal which can be green or red with probability $\frac{4}{5}$ and $\frac{1}{5}$ respectively is received by station A and then transmitted to station B. The probability of each station receiving the signal correctly is $\frac{3}{4}$. If the signal received at station B is green, then the probability that the original signal green is

(a) $\frac{3}{5}$ (b) $\frac{6}{7}$ (c) $\frac{20}{23}$ (d) $\frac{9}{20}$

45 Let A and B be two events such that $P(\overline{A \cup B}) = \frac{1}{6}$, $P(A \cap B) = \frac{1}{4}$ and $P(\overline{A}) = \frac{1}{4}$, where $\overline{A}$ stands for the complement of the event A. Then, the events A and B are **[JEE Main 2014]**

(a) mutually exclusive and independent
(b) equally likely but not independent
(c) independent but not equally likely
(d) independent and equally likely

46 Let X and Y be two independent events such that $P(X) = 0.3$ and $P(Y) = 0.7$. Then, $P(X \cup Y)$ is

(a) 0.69 (b) 0.73 (c) 0.79 (d) 0.83

47 Amit and Ankit are playing with cards and in a pack there are 52 cards. Amit drew a card at random with a replacement. Then, he asked Ankit, what is the probability of drawing a queen followed by a king?

(a) $\frac{1}{171}$ (b) $\frac{1}{165}$ (c) $\frac{1}{169}$ (d) $\frac{1}{173}$

7.3 Baye's Theorem

48 An unbiased coin is tossed. If the result is a head, a pair of unbiased dice is rolled and the sum of the numbers obtained is noted. If the results is a tail, a card from a well shuffled pack of eleven cards numbered 2, 3, 4, ..., 12 is picked and the number on the card is noted. The probability that the noted number is either 7 or 8 is

(a) $\frac{191}{792}$ (b) $\frac{193}{792}$ (c) $\frac{195}{792}$ (d) $\frac{197}{792}$

49 A man is known to speak to truth 3 out of 4 times. He throws a die and reports that it is six. The probability that it is actually a six, is

(a) $\frac{3}{8}$ (b) $\frac{1}{5}$ (c) $\frac{3}{4}$ (d) None of these

50 For $k = 1, 2, 3$ the box B_k contains k red balls and $(k + 1)$ white balls. Let $P(B_1) = \frac{1}{2}$, $P(B_2) = \frac{1}{3}$ and $P(B_3) = \frac{1}{6}$. A box is selected at random and a ball is drawn from it. If a red ball is drawn, then the probability that it has come from box B_2 is

(a) $\frac{35}{78}$ (b) $\frac{14}{39}$ (c) $\frac{10}{13}$ (d) $\frac{12}{13}$

51 A bag contains 4 green and 3 red balls and bag *B* contains 4 red and 3 green balls. One bag is taken at random and a ball is drawn and noted to be green. The probability that it comes from bag *B* is

(a) $\frac{2}{7}$ (b) $\frac{2}{3}$ (c) $\frac{3}{7}$ (d) $\frac{1}{3}$

52 Box A contains 2 black and 3 red balls, white box *B* contains 3 black and 4 red balls. Out of these two boxes one is selected at random and the probability of choosing box *A* is double that of box *B*. If a red ball is drawn from the selected box, then the probability that it has come from box *B*, is

(a) $\frac{21}{41}$ (b) $\frac{10}{31}$ (c) $\frac{12}{31}$ (d) $\frac{13}{41}$

7.4 Probability Distribution of Random Variables

53 A sample space consists of 3 sample points with associated probabilities given as $2p, p^2, 4p - 1$. Then, the value of p is

(a) $p = \sqrt{11} - 3$ (b) $\sqrt{10 - 3}$ (c) $\frac{1}{4} < p < \frac{1}{2}$ (d) None of these

54 If m and σ^2 are the mean and variance of the random variable X, whose distribution is given by

X	0	1	2	3
P(X)	$\frac{1}{3}$	$\frac{1}{2}$	0	$\frac{1}{6}$

Then,

(a) $m = \sigma^2 = 1$ (b) $m = \sigma^2 = 2$ (c) $m = 2, \sigma^2 = 1$ (d) $m = 1, \sigma^2 = 2$

55 A fair die is tossed. Let X denoted twice the number appearing on the die. The variance of X is

(a) 11.1 (b) 11.3 (c) 11.5 (d) 11.7

56 A fair die is tossed. Let X denote 1 or 3 according as an odd or an even number appears. The standard deviation of X is

(a) 1 (b) 0.5 (c) 0.1 (d) 0.3

.5 Bernoulli's Trails and Binomial Distribution

57 Consider 5 independent, Bernoulli' is trials each with probability of success p. If the probability of at least one failure is greater than or equal to $\frac{31}{32}$, then p lies in the interval

(a) $\left(\frac{3}{4}, \frac{11}{12}\right]$ (b) $\left[0, \frac{1}{2}\right]$ (c) $\left(\frac{11}{12}, 1\right]$ (d) $\left(\frac{1}{2}, \frac{3}{4}\right]$

58 If 12 identical balls are to be placed in 3 identical boxes, then the probability that one of the boxes contains exactly 3 balls is **(JEE Main 2015)**

(a) $\frac{55}{3}\left(\frac{2}{3}\right)^{11}$ (b) $55\left(\frac{2}{3}\right)^{10}$ (c) $220\left(\frac{1}{3}\right)^{12}$ (d) $22\left(\frac{1}{3}\right)^{11}$

59 In a trial, the probability, the success is twice the probability of failure. In six trials, the probability of at least four success will be

(a) $\frac{496}{729}$ (b) $\frac{400}{729}$ (c) $\frac{500}{729}$ (d) $\frac{600}{729}$

60 If the mean and variance of a Binomial variate X are 8 and 4 respectively, then $P(X < 3)$ equals

(a) $\frac{265}{2^{15}}$ (b) $\frac{137}{2^{14}}$ (c) $\frac{137}{2^{16}}$ (d) $\frac{265}{2^{16}}$

61 A multiple choice examination has 5 questions. Each questions has three alternative answers of which exactly one is correct. The probability that a student will get 4 or more correct answers just by guessing is

(a) $\frac{17}{3^5}$ (b) $\frac{13}{3^5}$ (c) $\frac{11}{3^5}$ (d) $\frac{10}{3^5}$

62 The probability of a man hitting a target is $\frac{1}{4}$. The number of times he must shoot, so that the probability he hits the target at least once is more than 0.9 is

(a) 7 (b) 8 (c) 6 (d) 5

63 One hundred identical coins, each with probability p of showing up heads, one tossed. If $0 < p < 1$ and the probability of heads showing on 50 coins is equal to that of heads showing on 51 coins. The value of p is

(a) $\frac{1}{2}$ (b) $\frac{49}{101}$ (c) $\frac{50}{101}$ (d) $\frac{51}{101}$

64 Suppose X follows a binomial distribution with parameters n and p, where $0 < p < 1$, if $\frac{P(X = r)}{P(X = n - r)}$ is independent of n and r, then

(a) $p = \frac{1}{2}$ (b) $p = \frac{1}{3}$ (c) $p = \frac{1}{4}$ (d) None of these

65 Five fair coins are tossed. If p is the probability that not more than 2 heads appears and q is the probability that not less than 3 heads appears, then

(a) $p > q$ (b) $p = q$ (c) $p < q$ (d) $pq = 1$

66 A bag contains 3 white, 2 black and 4 red balls. 4 balls are drawn at random with replacement, the probability by that the sample contains just one white ball is

(a) $\frac{16}{81}$ (b) $\frac{8}{81}$ (c) $\frac{32}{81}$ (d) $\frac{4}{81}$

67 A fair die is thrown 20 times. The probability that on the tenth throw the 4th six appears is

(a) $\frac{{}^{20}C_{10} \times 5^6}{6^{20}}$ (b) $\frac{120 \times 5^2}{6^{10}}$ (c) $\frac{84 \times 5^6}{6^{10}}$ (d) None of these

68 If X follows a binomial distribution with parameters $n = 8$ and $p = \frac{1}{2}$, then $P(|x - 4| \le 2)$ is

(a) $\frac{119}{128}$ (b) $\frac{119}{228}$ (c) $\frac{19}{128}$ (d) $\frac{18}{128}$

69 A coin is tossed n times. The probability of getting head at least once is greater than 0.8, then the least value of n is

(a) 2 (b) 3 (c) 5 (d) 4

70 In a Binomial distribution $B\left(n, p = \frac{1}{4}\right)$, if the probability of at least one success is greater than or equal to $\frac{9}{10}$, then n is greater than

(a) $\frac{1}{\log_{10} 4 - \log_{10} 3}$ (b) $\frac{1}{\log_{10} 4 + \log_{10} 3}$ (c) $\frac{9}{\log_{10} 4 - \log_{10} 3}$ (d) $\frac{4}{\log_{10} 4 - \log_{10} 3}$

71 If the mean and the standard deviation of a Binomial variate X are 4 and $\sqrt{3}$ respectively, then $P(X \ge 1)$ is equal to

(a) $1 - \left(\frac{1}{4}\right)^{16}$ (b) $1 - \left(\frac{3}{4}\right)^{16}$ (c) $1 - \left(\frac{2}{3}\right)^{16}$ (d) $1 - \left(\frac{1}{3}\right)^{16}$

72 If the mean of a binomial of distribution is 25, then its standard deviation lies in the interval given below

(a) [0, 5) (b) (0, 5] (c) [0, 25) (d) (0, 25]

73 A die is thrown 100 times, getting an even number is considered a success. The variance of the number of success is

(a) 10 (b) 25 (c) 18 (d) 10

74 A box contains 15 green and 10 yellow balls. If 10 balls are randomly drawn, one-by-one, with replacement, then the variance of the number of green balls drawn is **(JEE Main 2017)**

(a) $\frac{6}{25}$ (b) $\frac{12}{5}$ (c) 6 (d) 4

75 If the mean and variance of a binomial variate X are 2 and 1 respectively, then the probability that X takes a value at least one is

(a) $\frac{2}{3}$ (b) $\frac{4}{5}$ (c) $\frac{7}{8}$ (d) $\frac{15}{16}$

Numerical Value Type Questions

76 If the probability of 6-digit number N whose 6 digits are 1, 2, 3, 4, 5, 6 written as random order is divisible by 6 is p, then the value of $\frac{1}{p}$ is ......

77 If the probability that the product of the outcomes of 3 rolls of a fair dice is a prime number is p, then the value of $\frac{1}{(4p)}$ is ....

78 There are 2 red, 2 blue, 2 white and certain number (greater than 0) of green socks in a drawer If 2 socks are taken at random from the drawer without replacement, the probability that they are of the same colour is $\frac{1}{5}$, then the number of green socks are ....

79 If two loaded dice each have the property that 2 or 4 is three times as likely to appear as 1, 3, 5 or 6 on each roll. When 2 such dice are rolled, the probability of obtaining a total of 7 is p, then the value of $\left[\frac{1}{p}\right]$ is, where $[x]$ represents the greatest integer less than or equal to x ....

80 Five different games are to be distributed among four children randomly. The probability that each child get at least one game is p, then the value of $\left[\frac{1}{p}\right]$ is, where $[\cdot]$ represents the greatest integer function, ...

81 An urn contains 3 red balls and n white balls. Mr. A draws two balls together from the urn. The probability that they have the same colour is $\frac{1}{2}$. Mr. B draws one balls from the urn, notes its colour and replaces it. He then draws a 2nd ball from the urn and finds that both balls have the same colour is $\frac{5}{8}$. The possible value of n is ......

82 A bag contains 10 different balls. Five balls are drawn simultaneously and then replaced and then seven balls are drawn. If the probability that exactly 3 balls are common to the 2 draws is p, then the value of $8p$ is ...

83 A dice is weighted such that the probability of rolling the face numbered n is proportional to n^2 $(n = 1, 2, 3, 4, 5, 6)$. The dice is rolled twice, yielding the numbers a and b. The probability that $a < b$ is p, then the value of $\left[\frac{2}{p}\right]$ is ....

84 Suppose A and B are two events with $P(A) = 0.5$ and $P(A \cup B) = 0.8$. Let $P(B) = p$, if A and B are mutually exclusive and $P(B) = q$ if A and B are independent events, then the value of $\frac{q}{p}$ is ....

85 If A and B are 2 events such that $P(A) = 0.6$ and $P(B) = 0.8$, if the greatest value that $P\left(\frac{A}{B}\right)$ can have is p, then the value of $8p$ is ....

86 A fair coin is flipped n times. Let E be the event a head is obtained on the 1st flip' and let F_k be the event 'exactly k heads are obtained'. Then, the value of $\frac{n}{k}$ for which E and F_k are independent is ....

87 A random variable X has the following probability distribution

X	1	2	3	4
$p(X)$	k	$2k$	$3k$	$4k$

The mean of X is ....

88 If X is a Binomial variate with parameter $n = 6$ and $P(X = 2) = 4P(X = 4)$, then the parameter p of X is ....

LEVEL 03 \\ JEE Advanced Level

Single Choice Correct Questions

1 A cube with all six faces coloured is cut into 64 cubical blocks of the same size, which are thoroughly mixed. Then, the probability that the 2 randomly chosen blocks have 2 coloured faces each, is

(a) $\frac{23}{168}$ (b) $\frac{51}{168}$ (c) $\frac{1}{8}$ (d) $\frac{1}{16}$

2 3 boys and 2 girls stand in a queue. The probability that the number of boys ahead of every girls is atleast one more than the number of girls ahead of her, is

(a) $\frac{1}{2}$ (b) $\frac{1}{3}$ (c) $\frac{2}{3}$ (d) $\frac{3}{4}$

3 12 balls are placed in three boxes. The probability that the first box contains three balls, is

(a) $\frac{110}{9}\left(\frac{2}{3}\right)^{10}$ (b) $\frac{9}{110}\left(\frac{2}{3}\right)^{10}$ (c) $\frac{^{12}C_3}{12^3}\times 2^9$ (d) $\frac{^{12}C_3}{3^{12}}$

4 8 players $P_1, P_2, P_3, \ldots, P_8$ play a knockout tournament. It is known that whenever the players P_i and P_j play the player P_i will win, if $i < j$. Assuming that the players are paired at random in each round, then the probability that the player P_4 reaches the final, is

(a) $\frac{6}{35}$ (b) $\frac{8}{35}$ (c) $\frac{4}{35}$ (d) None of these

5 Three married couple sit in a row. Then, the probability that no husband sits with his wife, is

(a) $\frac{1}{6}$ (b) $\frac{1}{3}$ (c) $\frac{2}{3}$ (d) $\frac{5}{12}$

6 Four fair dice D_1, D_2, D_3 and D_4 each having six faces numbered 1, 2, 3, 4, 5 and 6 are rolled simultaneously. The probability that D_4 shows a number appearing on one of D_1, D_2 and D_3 is

(a) $\frac{91}{216}$ (b) $\frac{108}{216}$ (c) $\frac{125}{216}$ (d) $\frac{127}{216}$

7 Consider $f(x) = x^3 + ax^2 + bx + c$. Parameters a, b, c are chosen, respectively by throwing a die three times. Then, the probability that $f(x)$ is an increasing function is

(a) $\frac{5}{36}$ (b) $\frac{1}{4}$ (c) $\frac{4}{9}$ (d) $\frac{1}{3}$

8 If two different numbers are taken from the set {0, 1, 2, 3, ..., 10}, then the probability that their sum as well as absolute difference are both multiples of 4, is

(a) $\frac{7}{35}$ (b) $\frac{6}{55}$ (c) $\frac{12}{55}$ (d) $\frac{14}{45}$

9 A man alternately tosses a coin and throws a die beginning with the coin. The probability that he gets a head in the coin before he gets 5 or 6 in the die, is

(a) $\frac{3}{4}$ (b) $\frac{1}{2}$ (c) $\frac{1}{3}$ (d) None of these

10 A box contains 24 identical balls of which 12 are white and 12 are black. The balls are drawn at random from the box one at a time with replacement. The probability that a white ball is drawn for the 4th time in the 7th draw, is

(a) $\frac{5}{64}$ (b) $\frac{27}{32}$ (c) $\frac{5}{32}$ (d) $\frac{1}{2}$

11 A pair of unbiased dice is rolled together till a sum of either 5 or 7 is obtained. The probability that 5 comes before 7, is

(a) $\frac{2}{5}$ (b) $\frac{3}{5}$ (c) $\frac{4}{5}$ (d) $\frac{3}{10}$

12 Miss Rita has either tea or coffee at morning break. If she has tea one morning, the probability she has tea the next morning is 0.4. If she has coffee one morning, the probability she has coffee next morning is 0.3. Suppose, she has coffee on Monday morning. The probability she has tea on the following Wednesday morning, is

(a) 0.46 (b) 0.49 (c) 0.51 (d) 0.61

13 Suppose families always have one, two or three children, with probabilities $\frac{1}{4}$, $\frac{1}{2}$ and $\frac{1}{4}$, respectively. Assume everyone eventually gets married and has children. The probability of a couple having exactly four grandchildren, is

(a) $\frac{27}{128}$ (b) $\frac{37}{128}$ (c) $\frac{25}{128}$ (d) $\frac{20}{128}$

14 A signal which can be green or red with probability $\frac{2}{3}$ and $\frac{1}{5}$, respectively is received by station A and then transmitted to station B. The probability of each station receiving the signal correctly is $\frac{3}{4}$. If the signal received at station B is green. Then, the probability that the original signal was green, is

(a) $\frac{3}{5}$ (b) $\frac{6}{7}$ (c) $\frac{20}{23}$ (d) $\frac{9}{20}$

15 Four persons independently solve a certain problem correctly with probability $\frac{1}{2}, \frac{3}{4}, \frac{1}{4}, \frac{1}{8}$ respectively. Then, the probability that problem is solved correctly by atleast one of them, is

(a) $\frac{235}{256}$ (b) $\frac{21}{256}$ (c) $\frac{3}{256}$ (d) $\frac{253}{256}$

16 A bag contains some white and some black balls. All combinations of ball being equally likely. The total number of balls in bag is 10. If three balls are drawn at random without replacement and all of them are found to be black. The probability that the bag contains 1 white and 9 black balls, is

(a) $\frac{14}{55}$ (b) $\frac{12}{55}$ (c) $\frac{2}{11}$ (d) $\frac{8}{55}$

17 Each of two persons A and B toss three fair coins. The probability that both get the same number of heads, is

(a) $\frac{3}{8}$ (b) $\frac{1}{9}$ (c) $\frac{5}{16}$ (d) $\frac{7}{16}$

18 Let X be a set containing n elements. Two subsets A and B of X are chosen at random. The probability that $A \cup B = X$, is

(a) $\frac{{}^{2n}C_n}{2^{2n}}$ (b) $\frac{1}{{}^{2n}C_n}$ (c) $\left(\frac{3}{4}\right)^n$ (d) None of these

19 16 players $P_1, P_2, P_3 \dots P_{16}$ play in a tournament. They are divided into eight pairs at random. From each pair a winner is decided on the basis of a game played between the two players of the pair. Assuming that all the players are of equal strength. The probability that exactly one of the two players P_1 and P_2 is among the eight winners, is

(a) $\frac{4}{15}$ (b) $\frac{7}{15}$ (c) $\frac{8}{15}$ (d) $\frac{17}{30}$

20 A and B are two inaccurate arithmeticians whose chance of solving a given question correctly are $\frac{1}{8}$ and $\frac{1}{12}$, respectively. They solve a problem and obtained the same result. If it is 1000 to 1 against their making the same mistake, then chance that the result is correct is

(a) $\frac{13}{14}$ (b) $\frac{1}{14}$ (c) $\frac{1}{7}$ (d) $\frac{6}{7}$

21 If a machine is correctly set up, it produces 90% acceptable items. If it is incorrectly set up, it produces only 40% acceptable items. Past experience shows that 80% of the set up are correctly done. If after a certain set up, the machine produces 2 acceptable items, then the probability that the machine correctly set up, is

(a) 0.85 (b) 0.90 (c) 0.95 (d) 0.75

22 A computer producing factory has only plants T_1 and T_2. Plant T_1 produces 20% and plant T_2 produces 80% of the total computer produced. 7% of computers produced in the factory turn out to be defective. It is known that

P(Computer turns out to be defective given that it is produced in plant T_1) = $10P$(Computer turns out to be defective given that it is produced in plant T_2),

where $P(E)$ denotes the probability of an event E. A computer produced in the factory is randomly selected and it does not turn out to be defective. Then, the probability that it is produced in plant T_2, is **[JEE Advanced 2016]**

(a) $\frac{36}{73}$ (b) $\frac{47}{69}$ (c) $\frac{78}{93}$ (d) $\frac{75}{83}$

23 The standard deviation of the following probability distribution is

x_i	2	3	4
p_i	2.2	0.5	0.3

(a) 0.6 (b) 0.7 (c) 0.8 (d) 0.5

24 Consider three sets $E_1 = \{1, 2, 3\}$, $F_1 = \{1, 3, 4\}$ and $G_1 = \{2, 3, 4, 5\}$. Two elements are chosen at random, without replacement, from the set E_1 and let S_1 denote the set of these chosen elements. Let $E_2 = E_1 - S_1$ and $F_2 = F_1 \cup S_1$. Now, two elements are chosen at random, without replacement, from the set F_2 and let S_2 denote the set of these chosen elements. Let $G_2 = G_1 \cup S_2$. Finally, two elements are chosen at random, without replacement, from the set G_2 and let S_3 denote the set of these chosen elements.

Let $E_3 = E_2 \cup S_3$. Given that $E_1 = E_3$, let p be the conditional probability of the event $S_1 = \{1, 2\}$. Then, the value of p is **[JEE Advanced 2021]**

(a) $\frac{1}{5}$ (b) $\frac{3}{5}$ (c) $\frac{1}{2}$ (d) $\frac{2}{5}$

25 Let C_1 and C_2 be two biased coins such that the probabilities of getting head in a single toss are $\frac{2}{3}$ and $\frac{1}{3}$, respectively. Suppose α is the number of heads that appear when C_1 is tossed twice, independently, and suppose β is the number of heads that appear when C_2 is tossed twice, independently. Then, the probability that the roots of the quadratic polynomial $x^2 - \alpha x + \beta$ are real and equal, is **[JEE Advanced 2020]**

(a) $\frac{40}{81}$ (b) $\frac{20}{81}$ (c) $\frac{1}{2}$ (d) $\frac{1}{4}$

26 If the mean and variance of a binomial distribution are 4 and 3, respectively. Then, the probability of no success, is

(a) $\left(\frac{4}{3}\right)^{16}$ (b) $\left(\frac{3}{4}\right)^{16}$ (c) $1-\left(\frac{4}{3}\right)^{16}$ (d) $1-\left(\frac{3}{4}\right)^{16}$

More than One Correct

27 Let A and B are two events such that $P(A)=\frac{1}{2}$ and $P(B)=\frac{2}{3}$, then

(a) $P(A\cup B)\geq\frac{2}{3}$ (b) $P(A\cap B')\leq\frac{1}{3}$ (c) $\frac{1}{6}\leq P(A\cap B)\leq\frac{1}{2}$ (d) $\frac{1}{6}\leq P(A'\cap B)\leq\frac{1}{2}$

28 Let E, F and G be three events having probabilities $P(E)=\frac{1}{8}$, $P(F)=\frac{1}{6}$ and $P(G)=\frac{1}{4}$, and let $P(E\cap F\cap G)=\frac{1}{10}$. For any event H, if H^C denotes its complement, then which of the following statements is (are) TRUE? [JEE Advanced 2021]

(a) $P(E\cap F\cap G^C)\leq\frac{1}{40}$ (b) $P(E^C\cap F\cap G)\leq\frac{1}{15}$

(c) $P(E\cup F\cup G)\leq\frac{13}{24}$ (d) $P(E^C\cap F^C\cap G^C)\leq\frac{5}{12}$

29 A bag contains b blue balls and r red balls. If two balls are drawn at random, the probability of drawing two red balls is five times the probability of drawing two blue balls. Further more, the probability of drawing one ball of each colour is six times the probability of drawing two blue balls, then

(a) $b+r=9$ (b) $br=18$ (c) $|b-r|=4$ (d) $\frac{b}{r}=2$

30 If p and q are chosen randomly from the set {1, 2, 3, ... 10} with replacement, then the probability that the roots of the equation $x^2+px+q=0$

(a) are real is $\frac{33}{50}$ (b) are imaginary is $\frac{19}{50}$

(c) are real and equal is $\frac{3}{50}$ (d) are real and distinct is $\frac{3}{5}$

31 Let E and F be two independent events. The probability that exactly one of them occurs is $\frac{11}{25}$ and the probability of none of them occurs is $\frac{2}{25}$. If P(T) denotes the probability of occurrence of the event T, then

(a) $P(E)=\frac{4}{5}$ and $P(F)=\frac{3}{5}$ (b) $P(E)=\frac{1}{5}$ and $P(F)=\frac{2}{5}$

(c) $P(E)=\frac{2}{5}$ and $P(F)=\frac{1}{5}$ (d) $P(E)=\frac{3}{5}$ and $P(F)=\frac{4}{5}$

32 Let $0<P(A)\leq 1$, $0<P(B)<1$ and $P(A\cup B)=P(A)+P(B)-P(A)\cdot P(B)$, then which of the following is correct

(a) $P\left(\frac{B}{A}\right)=P(B)-P(A)$ (b) $P(\overline{A}-\overline{B})=P(\overline{A})-P(\overline{B})$

(c) $P(\overline{A\cup B})=P(\overline{A})\,P(\overline{B})$ (d) $P\left(\frac{A}{B}\right)=P(A)$

33 Two buses A and B are scheduled to arrive at a town central bus station at noon. The probability that bus A will be late is $\frac{1}{5}$. The probability of bus B will be late is $\frac{7}{25}$. The probability that the bus B is late given that bus A is late is $\frac{9}{10}$, then

(a) probability that neither bus will be late on a particular day is $\frac{7}{10}$

(b) probability that bus A is late given that bus B is late is $\frac{18}{28}$

(c) probability that atleast one bus is late is $\frac{3}{10}$

(d) probability that atleast one bus is in time is $\frac{4}{5}$

34 Three six faced fair dice are thrown together. The probability that the sum of the numbers appearing on the dice is k, where $3 \le k \le 8$, is

(a) $\frac{(k-1)(k-2)}{432}$ (b) $\frac{^kC_2}{216}$ (c) $\frac{^{k-1}C_2}{216}$ (d) $\frac{k^2}{432}$

35 Let X and Y be two events such that $P(X) = \frac{1}{3}$, $P\left(\frac{X}{Y}\right) = \frac{1}{2}$ and $P\left(\frac{Y}{X}\right) = \frac{2}{5}$, then

(a) $P(Y) = \frac{4}{15}$ (b) $P(X'/Y) = \frac{1}{2}$ (c) $P(X \cup Y) = \frac{2}{5}$ (d) $P(X \cap Y) = \frac{1}{5}$

36 Let n_1 and n_2 be the number of red and black balls, respectively in box I. Let n_3 and n_4 be the number of red and black balls, respectively in box II. A ball is drawn at random from box I and transferred to box II, if the probability of drawing a red ball from box I after this transfer is $\frac{1}{3}$, then the correct option(s) with the possible values of n_1 and n_2 is (are)

(a) $n_1 = 4$ and $n_2 = 6$ (b) $n_1 = 2$ and $n_2 = 3$
(c) $n_1 = 10$ and $n_2 = 20$ (d) $n_1 = 3$ and $n_2 = 6$

37 A bag initially contains 1 red and 2 blue balls. An experiment consisting of selecting a ball at random noting down its colour and replacing it together with an additional ball of the same colour. If three such trials are made, then

(a) probability that atleast one blue ball is drawn, is 0.9
(b) probability that exactly one blue ball is drawn, is 0.2
(c) probability that all the drawn balls are red given that all the drawn balls are of same colour, is 0.2
(d) probability that atleast one red ball is drawn, is 0.6

38 A boy has a collection of blue and green marbles. The number of blue marbles to the set {2, 3, 4, ... , 13}, if two marbles are chosen simultaneously and at random from this collection, then the probability that they have different colour, is $\frac{1}{2}$. Possible number of blue marbles is

(a) 2 (b) 3 (c) 6 (d) 10

Linked Comprehension Type

Paragraph I Directions (Q.Nos. 39-41)

In an objective paper, there are two sections of 10 questions each. For Section I, each question has 5 options and only one option is correct and for section II, each question has 4 options with one or more than one correct and marks in each section is awarded only, if he ticks all correct answers. Marks for each question in section I is one and in section II is three. (there is no negative marking) Based on above information, answer the following questions.

39 If a candidate attempts only two questions, one from Section I and one from Section II. The probability that he scores in both questions, is

(a) $\frac{74}{75}$ (b) $\frac{1}{25}$ (c) $\frac{1}{15}$ (d) $\frac{1}{75}$

40 If a candidate is total attempts 4 questions, then the probability of scoring 10 marks, is

(a) $\frac{1}{5}\left(\frac{1}{15}\right)^3$ (b) $\frac{4}{5}\left(\frac{1}{15}\right)^3$ (c) $\frac{1}{5}\left(\frac{14}{15}\right)^3$ (d) None of these

41 The probability of getting a score less than 40 in this paper, is

(a) $\left(\frac{1}{75}\right)^{10}$ (b) $1-\left(\frac{1}{75}\right)^{10}$ (c) $\left(\frac{74}{75}\right)^{10}$ (d) $\left(\frac{1}{5}\right)^{10}$

Paragraph II Directions (Q.Nos. 42-44)

There are n urns, each containing $(n+1)$ balls such that ith urn contains i white balls and $(N+1-i)$ red balls. Let U_i be the event of selecting ith urn, $i = 1, 2, 3, \ldots, n$ and W denotes the event of a getting white ball.

42 If $P(u_i) \propto i$, where $i = 1, 2, 3, \ldots, n$, then $\lim_{n\to\infty} P(W)$ is equal to

(a) 1 (b) $\frac{2}{3}$ (c) $\frac{3}{4}$ (d) $\frac{1}{4}$

43 If $P(u_i) = c$, where c is a constant, then $P\left(\frac{u_n}{W}\right)$ is equal to

(a) $\frac{2}{n+1}$ (b) $\frac{1}{n+1}$ (c) $\frac{n}{n+1}$ (d) $\frac{1}{2}$

44 If n is even and E denotes the event of choosing even numbered urn $P(u_i) = \frac{1}{n}$, then the value of $P\left(\frac{W}{E}\right)$ is

(a) $\frac{n+2}{2n+1}$ (b) $\frac{n+2}{2(n+1)}$ (c) $\frac{n}{n+1}$ (d) $\frac{1}{n+1}$

Paragraph III Directions (Q.Nos. 45-47)

A JEE aspirant estimates that she will be successful with an 80% chance, if she studies 10 h per day with a 60% chance, if she studies 7 h per day and with a 40% chance, if she studies 4 h per day. She further believes that she will study 10 h, 7 h and 4 h per day with probabilities 0.1,0.2 and 0.7, respectively.

45 The chance she will be successful, is

(a) 0.28 (b) 0.38 (c) 0.48 (d) 0.58

46 Given that she is successful, the chance that she studied for 4 h, is

(a) $\frac{6}{12}$ (b) $\frac{7}{12}$ (c) $\frac{8}{12}$ (d) $\frac{9}{12}$

47 Given that she does not achieve success, the chance she studied for 4 h, is

(a) $\frac{18}{26}$ (b) $\frac{19}{26}$ (c) $\frac{20}{26}$ (d) $\frac{21}{26}$

Paragraph IV Directions (Q.Nos. 48-50)

A box contains 9 slips bearing the numbers – 4, –3,– 2, – 1, 0, 1, 2, 3 and 4. (one number for one slip)

48 A slip is chosen at random, its number is noted and replaced in the box. If the experiment is repeated 5 times, then the probability that exactly 3 numbers appear in the noted 1.37, is

(a) $\frac{{}^9C_3 \times 90}{9^5}$ (b) $\frac{{}^9C_3 \times 243}{9^5}$ (c) $\frac{{}^9C_3 \times 150}{9^5}$ (d) $\frac{{}^9C_3 \times 240}{9^5}$

49 If a slip is chosen at random from the box and the experiment is repeated nine times (with replacement) to form 3×3 matrix. The chosen number are arranged row wise, then the probability that the matrix. So formed is skew-symmetric, is

(a) $\left(\frac{1}{9}\right)^8$ (b) $\left(\frac{1}{9}\right)^6$ (c) $\left(\frac{1}{9}\right)^3$ (d) $\left(\frac{1}{9}\right)^4$

50 A slip is chosen at random from the box, its number is noted and replaced in the box, if the experiment is repeated 5 times, then the probability that the sum of five numbers is zero, is

(a) $\frac{3491}{9^5}$ (b) $\frac{3951}{9^5}$ (c) $\frac{3923}{9^5}$ (d) None of these

Paragraph V Directions (Q.Nos. 51-53)

A player tosses a coin and scores one point for every head and two points for every tail that turns up. He plays on until his score reaches or passes n. P_n denotes the probability of getting a score of exactly n.

51 The value of P_n is equal to

(a) $\frac{1}{2}[P_{n-1} + P_{n-2}]$ (b) $\frac{1}{2}[2P_{n-1} + P_{n-2}]$

(c) $\frac{1}{2}[P_{n-1} + 2P_{n-2}]$ (d) None of these

52 The value of $P_n + \frac{1}{2}P_{n-1}$ is equal to

(a) $\frac{1}{2}$ (b) $\frac{2}{3}$ (c) 1 (d) None of these

53 Which of the following is not true?

(a) $P_{100} > \frac{2}{3}$ (b) $P_{101} < \frac{2}{3}$ (c) $P_{100}, P_{101} > \frac{2}{3}$ (d) None of these

Matrix Match Type Questions

54

	Column-I		Column-II
(i)	In a knockout tournament, 2^n equally skilled player; $S_1, S_2, S_3, \dots S_n$ are participating in each round, players are moves in the next round. If S_2 reaches the semi-final, then the probability that S_1 wins the tournament is $\frac{1}{20}$. The value of n is	(A)	3
(ii)	In a multiple choice question, there are four alternative answers of which one or more than one is correct. A candidate will get marks on the question only, if he ticks all the correct answers. The candidate getting marks on the question is to be greater than or equal to $\frac{1}{3}$. Then, the least number of chances he should be allowed, is	(B)	4
(iii)	All the face cards from a pack of 52 playing cards are removed from the remaining pack, half of the cards are randomly removed without looking at them and then randomly drawn two cards simultaneously from the remaining, if the probability that two cards drawn are both ace, is $\frac{P({}^{38}C_2)}{{}^{40}C_{20}\,{}^{20}C_2}$. Then, the value of P is	(C)	5
(iv)	An unbiased normal coin is tossed n times. Let E_1 : Event that both head and tails are present in n tosses and E_2: Event that the coin shows up head atmost once. The value of n which E_1 and E are independent is	(D)	6

55 A bag contains some white and some black balls, all combinations being equally likely. The total number of balls in the bag is 12. Four balls are drawn at random from the bag at random without replacement

	Column-I		Column-II
(i)	Probability that all the four balls are black, is equal to	(A)	$\frac{14}{33}$
(ii)	If the bag contains 10 black and 2 white balls, then the probability that all four balls are black, is equal to	(B)	$\frac{1}{5}$
(iii)	If all the four balls are black, then the probability that the bag contains 10 black balls, is equal to	(C)	$\frac{70}{429}$
(iv)	Probability that two balls are black and two are white, is	(D)	$\frac{1}{5}$

56 Three distinct numbers a, b and c are chosen at random from the numbers 1, 2, ..., 100. The probability that

	Column-I		Column-II
(i)	a, b, c are in AP, is	(A)	$\frac{53}{161700}$
(ii)	a, b, c are GP, is	(B)	$\frac{1}{66}$
(iii)	$\frac{1}{a}, \frac{1}{b}, \frac{1}{c}$ are in GP, is	(C)	$\frac{53}{161700}$
(iv)	$a+b+c$ is divisible by 2, is	(D)	$\frac{1}{2}$

57 A triangle is formed with vertices of n sided regular polygon. Then, the probability that the triangle

	Column-I		Column-II
(i)	have exactly one side common with the side of the polygon, is	(A)	$\frac{6}{(n-1)(n-2)}$
(ii)	have exactly two sides common with the side of the polygon, is	(B)	$\frac{6(n-3)}{(n-1)(n-2)}$
(iii)	have atleast one side common with the side of the polygon, is	(C)	$\frac{(n-4)(n-5)}{(n-1)(n-2)}$
(iv)	does not have any side common with the sides of the polygon, is	(D)	$\frac{6(n-4)}{(n-1)(n-2)}$

58 Suppose $n(>6)$, people are asked a question successively in a random order and exactly 3 out of n people know the answer. Let P_r denotes the probability that rth person asked in the first to know the answer.

Column I		Column II
(i) Probability first four do not know the answer	(A)	$\dfrac{3(n-3)}{n(n-1)}$
(ii) P_2 is equal to	(B)	$\dfrac{(n-4)(n-5)(n-6)}{n(n-1)(n-2)}$
(iii) P_{n-2} is equal to	(C)	$\dfrac{3(n-4)(n-5)}{n(n-1)(n-2)}$
(iv) P_4 is equal to	(D)	$\dfrac{6}{n(n-1)(n-2)}$

Numerical Value Type Questions

Paragraph V Directions (Q.Nos. 59-61) The answer to each question is a **NUMERICAL VALUE.**

There are four boxes A_1, A_2, A_3 and A_4. Box A_i has i balls and on each ball a number is printed the number are from 1 to i. A box is selected randomly, the probability of selection of box A_i is $\left(\frac{i}{10}\right)$ and then a ball is drawn. Let E_i represents the event that a ball with number i is drawn.

59 The value of $\sum_{i=1}^{4} P(E_i \cap A_i)$ is

60 $P\left(\frac{A_3}{E_2}\right)$ is equal to

61 One ball is drawn from each box randomly. The probability that the sum of four numbers appearing on the balls is less than 5, is

Question Stem (Q.Nos. 62-63) The answer to each question is a **NUMERICAL VALUE.**

Three numbers are chosen at random, one after another with replacement, from the set $S = \{1, 2, 3, \ldots, 100\}$. Let p_1 be the probability that the maximum of chosen numbers is atleast 81 and p_2 be the probability that the minimum of chosen numbers is atmost 40.

[JEE Advanced 2021]

62 The value of $\frac{625}{4} p_1$ is ....

63 The value of $\frac{125}{4} p_2$ is ....

64 2 cards are drawn from a well-shuffled pack of 52 cards. The probability that one is a heart card and the other is a king, is p, then the value of $104p$ is ......

65 A die is thrown 3 times. The chances that the highest number shown on the die, is 4 is p, then the value of $\left[\frac{1}{p}\right]$, where [.] represents greatest integer function, is ......

66 32 players ranked 1 to 32 are playing in a knockout tournament. Assume that in every match between any two players the better ranked player wins, the probability that ranked 1 and ranked 2 players are winner and runner up respectively is p, then the value of $\left[\frac{2}{p}\right]$, where [.] represents the greatest integer function is ....

67 In a knockout tournament, 2^n equally skilled players, $S_1, S_2, \ldots S_n$ are participating. In each round, players are divided in pairs at random and winner from each pair moves to the next round. If S_2 reaches the semi-final, then the probability that S_1 wins the tournament is $\frac{1}{84}$. The value of n is ....

68 Of the 3 independent events E_1, E_2 and E_3, probability that only E_1 occurs is α, only E_2 occurs is β and only E_3 occurs is γ. Let the probability p that none of event E_1, E_2 or E_3 occurs satisfy the equations$(\alpha - 2\beta)p = \alpha\beta$ and $(\beta - 3\gamma)p = 2\beta\gamma$. All the given probabilities are assumed to lie in the interval (0, 1). Then,

$$\frac{\text{Probability of occurence of } E_1}{\text{Probability of occurence of } E_3} \text{ is equal to ....}$$

[JEE Advanced 2013]

69 Let ω be a complex cube root of unity with $\omega \neq 1$. A fair die is thrown 3 times. If r_1, r_2 and r_3 are the number obtained on the die, then the probability that $\omega^{r_1} + \omega^{r_2} + \omega^{r_3} = 0$ is ....

70 The minimum number of times a fair coin needs to be tossed, so that the probability of getting atleast two heads is atleast 0.96, is .... **[JEE Advanced 2015]**

71 An unbiased normal coin is tossed n times.
Let E_1 : Event that both heads and tails are present in n tosses.
E_2 : Event that the coin shows up head atmost once.
The value of n for which E_1 and E_2 are independent, is........

72 A number is chosen at random from the set {1,2,3, ... ,2000}. Let p be the probability that the chosen number is a multiple of 3 or a multiple of 7. Then, the value of $500p$ is ....
[JEE Advanced 2021]

ANSWER KEY

Level 1

1. 0.15
2. (i) 1/2 (ii) 1/3 (iii) 2/3 (iv) 1/6 (v) 1/6 (vi) 1/2
3. 3/10
4. 1/5
5. 0.7
6. $\frac{17}{36}$
7. $\frac{341}{2^5}$
8. $\frac{53}{55}$
9. $\frac{52}{77}$
10. $\frac{1}{4}$
11. $\frac{3}{4}$
12. $\frac{37}{60}$
13. $p+q-2pq$
14. $\frac{4}{9}$
15. $\frac{7}{26}$
16. $\frac{1}{3}$
17. $\frac{33}{50}$
18. $\frac{1}{6}$
19. $\frac{1}{3}$
20. 0.1
21. $\frac{2}{5}$
22. $\frac{1}{4}$
23. (i) 1/2 (ii) 3/7 (iii) 6/7
24. $\frac{1}{18}$
25. $\frac{2}{9}$
26. $\frac{5}{13}$
27. $\frac{5}{21}$
28. $\frac{2}{5525}$
29. $\frac{8}{16575}$
30. Yes
31. $\frac{1}{2}$
32. No
33. $\frac{2}{5}$
34. $\frac{11}{24}$
35. $\frac{1}{2}, \frac{1}{4}$
36. 0.5
37. $n=5$
38. $\frac{1}{5}$
39. $\frac{19}{42}$
40. $\frac{39}{88}$
41. $\frac{118}{495}$
42. 0.488
43. 28/69
44. 35/68
45. (i) 0.15 (ii) 0.55
46.

x	$P(x)$
0	144/169
1	24/169
2	1/169

47. 0.37
48. (i) Yes (ii) No
49. $\frac{11}{36}$
50. (i) $\frac{35}{128}$ (ii) $\frac{1}{2}$
51. $1-\sum_{r=7}^{10} {}^{10}C_r(0.9)^r(0.1)^{10-r}$
52. $\frac{496}{729}$
53. $\frac{7}{128}$
54. $n=7$
55. $\frac{5}{11}$
56. Mean = 1 and Variance = $\frac{2}{3}$
57. Mean = 2.66 and standard deviation = $\frac{4}{3}$
58. $\frac{63}{64}$

Level 2

1. (b)	2. (d)	3. (c)	4. (a)	5. (a)	6. (b)	7. (a)
8. (c)	9. (a)	10. (d)	11. (a)	12. (b)	13. (a)	14. (c)
15. (a)	16. (a)	17. (c)	18. (c)	19. (c)	20. (b)	21. (c)
22. (c)	23. (d)	24. (c)	25. (c)	26. (b)	27. (c)	28. (c)
29. (b)	30. (c)	31. (d)	32. (d)	33. (b)	34. (b)	35. (d)
36. (d)	37. (b)	38. (a)	39. (c)	40. (a)	41. (b)	42. (a)
43. (c)	44. (c)	45. (c)	46. (c)	47. (c)	48. (b)	49. (a)
50. (b)	51. (c)	52. (b)	53. (a)	54. (a)	55. (d)	56. (a)
57. (b)	58. (a)	59. (a)	60. (c)	61. (c)	62. (b)	63. (d)
64. (a)	65. (b)	66. (c)	67. (c)	68. (a)	69. (b)	70. (a)
71. (b)	72. (a)	73. (b)	74. (b)	75. (d)	76. (2)	77. (6)
78. (4)	79. (7)	80. (4)	81. (1)	82. (7)	83. (5)	84. (2)
85. (6)	86. (2)	87. (3)	88. (0.333)			

Level 3

1. (a)	**2.** (a)	**3.** (a)	**4.** (c)	**5.** (b)	**6.** (a)	**7.** (c)
8. (b)	**9.** (a)	**10.** (c)	**11.** (a)	**12.** (b)	**13.** (a)	**14.** (c)
15. (a)	**16.** (a)	**17.** (c)	**18.** (c)	**19.** (c)	**20.** (a)	**21.** (c)
22. (c)	**23.** (b)	**24.** (a)	**25.** (b)	**26.** (b)	**27.** (a, b, c, d)	
28. (a, b, c)	**29.** (a, b)	**30.** (b, c)	**31.** (a, d)	**32.** (c, d)	**33.** (a, b, c)	**34.** (a, c)
35. (a, b)	**36.** (c, d)	**37.** (a, b, c, d)	**38.** (b, c, d)	**39.** (d)	**40.** (a)	**41.** (b)
42. (b)	**43.** (a)	**44.** (b)	**45.** (c)	**46.** (b)	**47.** (d)	**48.** (c)
49. (b)	**50.** (b)	**51.** (a)	**52.** (c)	**53.** (c)		

54. (i)–B, (ii)–C, (iii)–D, (iv)–A

55. (i)–B, D, (ii)–A, (iii)–C, (iv)–B, D

56. (i)–B, (ii)–A, C, (iii)–A, C, (iv)–D

57. (i)–D, (ii)–A, (iii)–B, (iv)–C

58. (i)–B, (ii)–A, (iii)–D, (iv)–C

59. (0.4)	**60.** (0.33)	**61.** (0.041)	**62.** (76.25)	**63.** (24.50)	**64.** (4)	**65.** (5)
66. (3)	**67.** (6)	**68.** (6)	**69.** (2/9)	**70.** (8)	**71.** (3)	**72.** (214)

SOLUTIONS

Level 1

1. Let total possible events $(n) = 1$

Then, favourable outcome are $(m) = 0.85$

$\therefore$ Not favourable outcomes $(n - m) = 1 - 0.85$

$\therefore$ Probability $P(A) = \frac{n - m}{n}$

$= \frac{1 - 0.85}{1}$

$= 0.15$

2. We know that the numbers on a die are 1, 2, 3, 4, 5, 6 total $(n) = 6$

(i) Prime numbers are (m) 2, 3, 5

$\therefore P(A) = \frac{m}{n} = \frac{3}{6} = \frac{1}{2}$

$\left[\because P(A) = \frac{\text{Favourable outcome}}{\text{Total possible outcome}}\right]$

(ii) 2 or 4, here $m = 2$

$\therefore P(A) = \frac{m}{n} = \frac{2}{6} = \frac{1}{3}$

(iii) A multiple of 2 or 3 are $= 2, 3, 4, 6$

$\therefore P(A) = \frac{m}{n} = \frac{4}{6} = \frac{2}{3}$

(iv) An even prime number is 2

$\therefore P(A) = \frac{m}{n} = \frac{1}{6}$

(v) A number more than 5 is 6

$\therefore P(A) = \frac{m}{n} = \frac{1}{6}$

(vi) Probability of getting a number lying between 2 and 6 which are 3, 4, 5

$P(E) = \frac{\text{Number of actual events}}{\text{Numer of possible events}}$

$= \frac{m}{n}$

$= \frac{3}{6} = \frac{1}{2}$

3. Number of lottery tickets $(n) = 50$

$\therefore$ Prime numbers upto 50 are

2, 3, 5, 7, 11, 13, 17, 19, 23, 29, 31, 37, 41, 43, 47 = 15

$\therefore \quad m = 15$

$\therefore \quad P(A) = \frac{m}{n} \quad \left[P(A) = \frac{\text{Favourable outome } (m)}{\text{Total possible outcomes } (n)}\right]$

$= \frac{15}{50}$

$= \frac{3}{10}$

4. Numbers are 1, 2, 3, ..., 15

$\therefore \quad n = 15$

Now, multiple of 4 are 4, 8, 12

$\therefore \quad m = 3$

$\therefore \quad P(A) = \frac{m}{n} \quad \left[P(A) = \frac{\text{Favourable outomes } (m)}{\text{Total possible outcomes } (n)}\right]$

$= \frac{3}{15}$

$= \frac{1}{5}$

5. Let game $(n) = 1$

and winning of game $= 0.3$

$\therefore$ Loosing the game $(m) = 1 - 0.3$

$= 0.7$

$\therefore \quad P(A) = \frac{m}{n} \quad \left[P(A) = \frac{\text{Favourable outome } (m)}{\text{Total possible outcomes } (n)}\right]$

$= \frac{0.7}{1} = 0.7$

6. Six faces have numbers {1, 2, 3, 5, 7, 11}.

Sum of the numbers on top faces of both the sides is less than or equal to 8

= (1, 1), (1, 2), (1, 3), (1, 5), (1, 7) → 5

(2, 1), (2, 2), (2, 3), (2, 5) → 4

(3, 1), (3, 2), (3, 3), (3, 5) → 4

(5, 1), (5, 2), (5, 3) → 3

(7, 1) → 1

$n(S) = 5 + 4 + 4 + 3 + 1 = 17$

$n(T) = 6 \times 6 = 36$

$\therefore$ Probability $= \frac{17}{36}$

7. Exhaustive number of cases $= 12^6$

Favourable cases $= {}^{12}C_2(2^6 - 2)$

$\therefore \quad$ Probability $= \frac{{}^{12}C_2(2^6 - 2)}{12^6}$

$= \frac{341}{12^5}$

8. The total number of cases is $\frac{11!}{(2! \times 2!)}$

The number of favourable cases is $\left[\frac{11!}{(2! \times 2!)}\right] - 9!$

Therefore, the required probability is

$= 1 - \frac{9! \times 4}{11!} = \frac{53}{55}$

9. $P\{\text{1st event does not happen}\} = \frac{5}{5+2} = \frac{5}{7}$

$P\{\text{2nd event does not happen}\} = \frac{5}{5+6} = \frac{5}{11}$

$\therefore$ $P\{\text{Both the events fails to happen}\} = \frac{5}{7} \times \frac{5}{11} = \frac{25}{77}$

Therefore, the probability that at least one of the events will happen is

$1 - P\text{ (none of two happens)} = 1 - \frac{25}{77} = \frac{52}{77}$

10. The total number of ways in which four figure number can be formed is $4! = 24$. A number is divisible by 5 if at unit's place we have 5. Therefore, unit's place can be filled in one way and the remaining 3 places can be filled with the other digits in 3! ways. Hence, total number of number divisible by 5 is $3! = 6$.

So, the required probability is $\frac{6}{24} = \frac{1}{4}$.

11. $P(E) + P(E') = 1$

$\therefore 1 + \lambda + \lambda^2 + (1+\lambda)^2 = 1$

$\Rightarrow \quad 2\lambda^2 + 3\lambda + 1 = 0$

$\Rightarrow \quad (2\lambda + 1)(\lambda + 1) = 0$

$\Rightarrow \quad \lambda = -1, -\frac{1}{2}$

For $\lambda \Rightarrow -1, P(E) = 1 + (-1) + (-1)^2 = 1$ (not possible)

For $\lambda \Rightarrow -\frac{1}{2},\ P(E) = 1 + \left(-\frac{1}{2}\right) + \left(-\frac{1}{2}\right)^2$

$= \frac{1}{2} + \frac{1}{4} = \frac{3}{4}$

12. Suppose E_1, E_2 and E_3 are the events of winning the race by the horses A, B and C respectively.

$\therefore\ P(E_1) = \frac{1}{1+3} = \frac{1}{4}, P(E_2) = \frac{1}{1+4} = \frac{1}{5}, P(E_3) = \frac{1}{1+5} = \frac{1}{6}$

$\therefore$ Probability of winning the race by one of the horses. A, B and $C = P(E_1 \text{ or } E_2 \text{ or } E_3) = P(E_1) + P(E_2) + P(E_3)$

$= \frac{1}{4} + \frac{1}{5} + \frac{1}{6} = \frac{37}{60}$

13. Only one of A and B can be alive in the following mutually exclusive ways.

E_1 A will die and B will be alive

E_2 B will die and A will be alive

So, required probability $= P(E_1) + P(E_2)$

$= p(1-q) + q(1-p)$

$= p + q - 2pq$

14. If a probability p is assigned to each even number, then $2p$ is the probability to be assigned to each odd number which gives $2p \times 3 + p \times 3 = 9p = 1$

$\Rightarrow \quad p = \frac{1}{9}$

$\therefore$ $P(E)$ = Probability of getting 4, 5 or 6

$= \frac{1}{9} + \frac{2}{9} + \frac{1}{9}$

$= \frac{4}{9}$

15. The probability of drawing a card of club $= P(A) = \frac{13}{52}$

The probability of drawing an ace of diamond

$= P(B) = \frac{1}{52}$

$\because$ The events are mutually, exclusive

$\therefore \quad P(A \cup B) = P(A) + P(B)$

$= \frac{13}{52} + \frac{1}{52}$

$= \frac{14}{52} = \frac{7}{26}$

16. Let event A: number of being a multiple of 5

Event B: number of being a multiple of 6

$\therefore \quad P(A) = \frac{6}{30}; P(B) = \frac{5}{30}$

$P(A \cap B) = P(A \text{ and } B) = \frac{1}{30}$ (only 30 is common)

$\therefore\ P(A \cup B) = P(A) + P(B) - P(A \cap B)$

$= \frac{6}{30} + \frac{5}{30} - \frac{1}{30} = \frac{10}{30} = \frac{1}{3}$

17. A : getting a multiple of 2

B : getting a multiple of 3

C : getting a multiple of 10

$\therefore \quad P(A) = \frac{25}{50}; P(B) = \frac{16}{50}; P(C) = \frac{5}{50}$

$P(A \cap B) = \frac{8}{50}; P(B \cap C) = \frac{1}{50}; P(A \cap C) = \frac{5}{50}$

$P(A \cap B \cap C) = \frac{1}{50}$

$\therefore\ P(A \cup B \cup C) = P(A) + P(B) + P(C) - P(A \cap B) - P(B \cap C) - P(A \cap C) + P(A \cap B \cap C)$

$= \frac{25}{50} + \frac{16}{50} + \frac{5}{50} - \frac{8}{50} - \frac{1}{50} - \frac{5}{50} + \frac{1}{50}$

$= \frac{33}{50}$

18. Let event B is '5 appears on the 1st dice'.

$\therefore\ B = \{(5,1), (5,2), (5,3), (5,4), (5,5), (5,6)\}$

Also, let event A 'sum of the number appearing is 11'.

$\therefore\ A = \{(5,6), (6,5)\}$

$\therefore A \cap B = \{(5,6)\}$

Therefore, $P\left(\frac{A}{B}\right) = \frac{P(A \cap B)}{P(B)}$

$= \frac{1}{6}$

19. Let b stand for boy and g for girl. The sample space of the experiment is

$$S = \{(b,b),(g,b),(b,g),(g,g)\}$$

Let E and F denote the following events

E: Both the children are boys

F: 'at least one of the child is a boy'

Then, $E = \{(b,b)\}$ and $F = \{(b,b),(g,b),(b,g)\}$

Now, $E \cap F = \{(b,b)\}$

Thus, $P(F) = \frac{3}{4}$ and $P(E \cap F) = \frac{1}{4}$

Therefore, $P\left(\frac{E}{F}\right) = \frac{P(E \cap F)}{P(F)}$

$$= \frac{\frac{1}{4}}{\frac{3}{4}} = \frac{1}{3}$$

20. Let E denote the event that a student chosen randomly studies in class XII and F be the event that the randomly chosen student is a girl. We have to find $P\left(\frac{E}{F}\right)$.

Now, $P(F) = \frac{430}{1000} = 0.43$

and $P(E \cap F) = \frac{43}{1000} = 0.043$

Then, $P\left(\frac{E}{F}\right) = \frac{P(E \cap F)}{P(F)} = \frac{0.043}{0.43} = 0.1$

21. Let E be the event that 'number 4 appears at least once and F be the event that 'the sum of the numbers appearing is 6.

Then, $E = \{(4, 1), (4, 2), (4, 3), (4, 4), (4, 5), (4, 6), (1, 4), (2, 4), (3, 4), (5, 4), (6, 4)\}$

and $F = \{(1, 5), (2, 4), (3, 3), (4, 2), (5, 1)\}$

We have, $P(E) = \frac{11}{36}$ and $P(F) = \frac{5}{36}$

Also, $E \cap F = \{(2,4),(4,2)\}$

Therefore, $P(E \cap F) = \frac{2}{36}$

Hence, the required probability

$$P\left(\frac{E}{F}\right) = \frac{P(E \cap F)}{P(F)}$$

$$= \frac{\frac{2}{36}}{\frac{5}{56}} = \frac{2}{5}$$

22. $\frac{P(B)}{(A \cup B')} = \frac{P(B \cap (A \cup B'))}{P(A \cup B')}$

$$= \frac{P(B \cap A) \cup (B \cap B')}{P(A) + P(B') - P(A \cap B')}$$

$$= \frac{P(A \cap B)}{P(A) + P(B') - P(A \cap B')}$$

Now, $P(A' \cap B') = 0.5$

$\therefore \quad 1 - P(A \cup B) = 0.5$

$\Rightarrow \quad P(A \cup B) = 0.5$

$\Rightarrow \quad P(A) + P(B) - P(A \cap B) = 0.5$

$\Rightarrow \quad P(A \cap B) = 0.3 + 0.4 - 0.5 = 0.2$

$\Rightarrow \quad P(A \cap B') = P(A) - P(A \cap B)$

$= 0.3 - 0.2 = 0.1$

$$P\left(\frac{B}{(A \cup B')}\right) = \frac{0.2}{0.3 + 0.6 - 0.1}$$

$$= \frac{0.2}{0.8} = \frac{1}{4}$$

23. If a coin is tossed 3 times, then the sample space S is

$$S = \{HHH, HHT, HTH, HTT, THH, THT, TTH, TTT\}$$

$\therefore \quad n(S) = 8$

(i) $A = \{HHH, HTH, THH, TTH\}$

$B = \{HHH, HHT\}$

$\therefore \quad A \cap B = \{HHH\}$

$$P(B) = \frac{2}{8} = \frac{1}{4} \text{ and } P(A \cap B) = \frac{1}{8}$$

$$P\left(\frac{A}{B}\right) = \frac{P(A \cap B)}{P(B)} = \frac{\frac{1}{8}}{\frac{1}{4}} = \frac{4}{8}$$

$$= \frac{1}{2}$$

(ii) $A = \{HHH, HHT, HTH, THH\}$

$B = \{HHT, HTH, HTT, THH, THT, TTH, TTT\}$

$\therefore \quad A \cap B = \{HHT, HTH, THH\}$

Clearly, $P(A \cap B) = \frac{3}{8}$ and $P(B) = \frac{7}{8}$

$$\therefore P\left(\frac{A}{B}\right) = \frac{P(A \cap B)}{P(B)} = \frac{\frac{3}{8}}{\frac{7}{8}} = \frac{3}{7}$$

(iii) $A = \{HHH, HHT, HTT, HTH, THH, THT, TTH\}$

$B = \{HHT, HTT, HTH, THH, THT, TTH, TTT\}$

$\therefore \quad A \cap B = \{HHT, HTT, HTH, THH, THT, TTH\}$

$$P(B) = \frac{7}{8} \text{ and } P(A \cap B) = \frac{6}{8}$$

Therefore, $P\left(\frac{A}{B}\right) = \frac{P(A \cap B)}{P(B)} = \frac{\frac{6}{8}}{\frac{7}{8}} = \frac{6}{7}$

24. We have to find the conditional probability of sum 15 when 4 appears 1st.

Let the event of getting sum of three thrown numbers as 15 be A and the event of appearing 4 in first throw be B.

So, we have to find $P\left(\frac{A}{B}\right)$.

Now, $P\left(\frac{A}{B}\right) = \frac{P(A \cap B)}{P(B)}$

Event $(A \cap B)$ is '4 appears in 1st throw and sum of numbers' is 15.

$\therefore \quad A \cap B = \{(4,5,6),(4,6,5)\}$

$\therefore \quad P\left(\frac{A}{B}\right) = \frac{2}{36} = \frac{1}{18}$

25. According to the question,

$S = \{(H,H),(H,T),(T,1),(T,2),(T,3),(T,4),(T,5),(T,6)\}$

In the above set, there are 8 elementary events, but all are not equally likely.

However, events (H,H) and (H,T) are equally likely. Each event has probability $\frac{1}{4}$.

Events $(T,1),(T,2),...(T,6)$ are equally likely and each event has probability $\frac{1}{12}$.

Let event A be 'the dice shows a number greater than 4' and event B be 'there is at least one tail'.

$\therefore \quad A = \{(T,5),(T,6)\}$

and $B = \{(H,T),(T,1),(T,2),(T,3),(T,4),(T,5),(T,6)\}$

$\therefore A \cap B = \{(T,5),(T,6)\}$

$\therefore$ Required probability,

$$P\left(\frac{A}{B}\right) = \frac{P(A \cap B)}{P(B)}$$

$$= \frac{\frac{1}{12}+\frac{1}{12}}{\frac{1}{4}+\frac{1}{12}+\frac{1}{12}+\frac{1}{12}+\frac{1}{12}+\frac{1}{12}+\frac{1}{12}}$$

$$= \frac{\frac{1}{6}}{\frac{3}{4}} = \frac{2}{9}$$

26. Let A be the event that the first mango is good and B be the event that the 2nd one is good. Then, required probability is

$$P\left(\frac{B}{A}\right) = \frac{P(A \cap B)}{P(A)}$$

Now, probability that both mangoes are good is

$$P(A \cap B) = \frac{{}^6C_2}{{}^{10}C_2}$$

Probability that 1st mangoes is good is

$$P(A) = \frac{{}^6C_2}{{}^{10}C_2} + \frac{{}^6C_1 \times {}^4C_1}{{}^{10}C_2}$$

Hence, $P\left(\frac{B}{A}\right) = \frac{{}^6C_2}{{}^6C_2 + {}^6C_1 \times {}^4C_1}$

$$= \frac{15}{15+24} = \frac{5}{13}$$

27. Let A and B denote, respectively, the events that 1st ball drawn is black and 2nd ball drawn is white.

We have to find $P(A \cap B)$.

$$P(A \cap B) = P(A)P\left(\frac{B}{A}\right)$$

Now, $P(A) = P$ (black ball is 1st draw) $= \frac{10}{15} = \frac{2}{3}$

Also, given that the 1st ball drawn is black, i.e. event A has occurred.

Now, there are 9 black balls and 5 white balls left in the urn.

Therefore, the probability that the 2nd ball drawn is white, given that the ball in the 1st draw is black, is nothing but the conditional probability of B given that A has accurred.

$\therefore \quad P\left(\frac{B}{A}\right) = \frac{5}{14}$

Therefore, $P(A \cap B) = P(A)P\left(\frac{B}{A}\right)$

$$= \frac{2}{3} \times \frac{5}{14} = \frac{5}{21}$$

28. Let K denote the event that the card drawn is king and A be the event that the card drawn is an ace. Clearly, we have to find $P(KKA)$

Now, $P(K) = \frac{4}{52}$

Also, $P\left(\frac{K}{K}\right)$ is the probability of 2nd king with the condition that one king has already been drawn.

Now, there are 3 kings in $(52-1) = 51$ cards.

Therefore, $P\left(\frac{K}{K}\right) = \frac{3}{51}$

Lastly, $P\left(\frac{A}{KK}\right)$ is the probability of 3rd drawn card to be an ace, with the condition that two kings have already been drawn. Now, there are four aces in left 50 cards.

Therefore, $P\left(\frac{A}{KK}\right) = \frac{4}{50}$

By multiplication law of probability, we have

$$P(KKA) = P(K) \cdot P\left(\frac{K}{K}\right) P\left(\frac{A}{KK}\right)$$

$$= \frac{4}{52} \frac{3}{51} \frac{4}{50} = \frac{2}{5525}$$

29. Let events J be first card is jack, Q be 2nd drawn card is queen and K be third drawn card is king.

We have to find the value of $P(J \cap Q \cap K)$

Now, $P(J \cap Q \cap K) = P(J) \times P\left(\frac{Q}{J}\right) \times P\left(\frac{K}{(J \cap Q)}\right)$

$$P(J) = \frac{4}{52} = \frac{1}{13}$$

$P\left(\frac{Q}{J}\right) = P$ (drawing queen card, if jack has been drawn)

$= \frac{4}{51}$

$P\left(\frac{K}{(J \cap Q)}\right) = P$

[drawing king card, if jack and queen card have been drawn]

$= \frac{4}{50} = \frac{2}{25}$

So, $P(J \cap Q \cap K) = P(J) \times P\left(\frac{Q}{J}\right) \times P\left(\frac{K}{(J \cap Q)}\right)$

$= \frac{1}{13} \times \frac{4}{51} \times \frac{2}{25} = \frac{8}{16575}$

30. If all the 36 elementary events of the experiment are considered to be equally likely, we have,

$$P(A) = \frac{18}{36} = \frac{1}{2} \text{ and } P(B) = \frac{18}{36} = \frac{1}{2}$$

Also, $P(A \cap B) = P$ (odd number on both throws)

$= \frac{9}{36} = \frac{1}{4}$

Now, $P(A) \cdot P(B) = \frac{1}{2} \times \frac{1}{2} = \frac{1}{4}$

Clearly, $P(A \cap B) = P(A) \times P(B)$

Thus, A and B are independent events.

31. Since, the trials are independent, so the probability that head appears on the 5th toss does not depend upon previous results of the tosses. Hence, required probability is equal to probability of getting head, i.e. $\frac{1}{2}$.

32. When a die is thrown, the sample space (S) is

$$S = \{1, 2, 3, 4, 5, 6\}$$

Let A : the number is even $= \{2, 4, 6\}$

$P(A) = \frac{3}{6} = \frac{1}{2}$

B : the number is red $= \{1, 2, 3\}$

$P(B) = \frac{3}{6} = \frac{1}{2}$

$A \cap B = \{2\}$

$P(A \cap B) = \frac{1}{6}$

$P(A) \cdot P(B) = \frac{1}{2} \times \frac{1}{2} = \frac{1}{4} \neq \frac{1}{6}$

$\Rightarrow \quad P(A) \cdot P(B) \neq P(A \cap B)$

Therefore, A and B are not independent.

33. Let 3 persons be A, B and C.

Clearly, the probabilities of solving the problem by A, B and C are independent.

Given, $P(A) = \frac{1}{3}, P(B) = \frac{1}{4}$ and $P(C) = \frac{1}{5}$

$\therefore$ P (none can solve the problem)

$$= P(A' \cap B' \cap C')$$
$$= P(A') P(B') P(C')$$
$$= (1 - P(A))(1 - P(B))(1 - P(C))$$
$$= \left(1 - \frac{1}{3}\right)\left(1 - \frac{1}{4}\right)\left(1 - \frac{1}{5}\right)$$
$$= \frac{2}{3} \cdot \frac{3}{4} \cdot \frac{4}{5} = \frac{2}{5}$$

34. We have,

$$P(A) = \frac{1}{2}, P(B) = \frac{1}{3}, P(C) = \frac{1}{4}$$

Clearly, the probabilities of A, B, C hitting the target are independent.

P (Exactly one of them hitting the target)

$= P(A$ is hitting and B, C missing the target)

$+ P(B$ is hitting and A, C missing the target)

$+ P(C$ is hitting and A, B missing the target)

$= P(A \cap B' \cap C') + P(A' \cap B \cap C') + P(A' \cap B' \cap C)$

$= P(A) P(B') P(C') + P(A') \cdot P(B) P(C')$

$+ P(A') P(B') P(C)$

$= \frac{1}{2} \cdot \frac{2}{3} \cdot \frac{3}{4} + \frac{1}{2} \cdot \frac{1}{3} \cdot \frac{3}{4} + \frac{1}{2} \cdot \frac{2}{3} \cdot \frac{1}{4} = \frac{11}{24}$

35. We have,

$$P(A \cap B) = \frac{1}{8} \text{ and } P(\bar{A} \cap \bar{B}) = \frac{3}{8}$$

$\therefore \quad P(A) P(B) = \frac{1}{8}$ and $P(\bar{A}) P(\bar{B}) = \frac{3}{8}$

[$\because$ A and B are independent]

Now, $P(\bar{A} \cap \bar{B}) = \frac{3}{8}$

$\Rightarrow \quad 1 - P(A \cup B) = \frac{3}{8}$

or $\quad 1 - (P(A) + P(B) - P(A \cap B)) = \frac{3}{8}$

or $\quad 1 - (P(A) + P(B)) + \frac{1}{8} = \frac{3}{8}$

or $\quad P(A) + P(B) = \frac{3}{4}$

The quadratic equation whose roots are $P(A)$ and $P(B)$ is

$x^2 - x\{P(A) + P(B)\} + P(A) P(B) = 0$

or $\quad x^2 - \frac{3}{4}x + \frac{1}{8} = 0$

or $\quad 8x^2 - 6x + 1 = 0$

or $\quad x = \frac{1}{2}, \frac{1}{4}$

36. Let event x be that x speaks the truth and event y be that y speaks the truth.

According to the question,

$$P(x)=0.6 \text{ and } P(y)=0.5$$

$\therefore \quad P(x')=0.4$ and $P(y')=0.5$

Let event C be 'x and y contradict each other narrating the same incident.

Clearly, event C occurs when one of x and y speaks the truth and other does not.

$$\therefore \quad P(C)=P(x\cap y')+P(x'\cap y)$$
$$=P(x)P(y')+P(x')P(y)$$
[since x and y are independent]
$$=0.6\times 0.5+0.4\times 0.5$$
$$=0.5$$

37. Let E_1 denote an event when a coin with two heads is selected and E_2 an event when a fair coin is selected. Let A be the event when the toss results in head. Then,

$$P(E_1)=\frac{1}{(n+1)}, P(E_2)=\frac{n}{(n+1)}, P\left(\frac{A}{E_1}\right)=1$$

and $P\left(\frac{A}{E_2}\right)=\frac{1}{2}$

Using total probability theorem, we have

$$P(A)=P(E_1)P\left(\frac{A}{E_1}\right)+P(E_2)P\left(\frac{A}{E_2}\right)$$

or $\quad \dfrac{7}{12}=\dfrac{1}{n+1}\times 1+\dfrac{n}{n+1}\times\dfrac{1}{2}$

$\Rightarrow \quad 12+6n=7n+7$

$\Rightarrow \quad n=5$

38. Let A_i denote the event that the number i appears on the dice and let E denote the event that only white balls are drawn.

Then, $\quad P(A_i)=\dfrac{1}{6}$ for $i=1,2,\ldots,6$

and $\quad P\left(\dfrac{E}{A_i}\right)=\dfrac{^6C_i}{^{10}C_i}, i=1,2,\ldots,6$

Using total probability theorem, required probability

$$P(E)=\sum_{i=1}^{6} P(E\cap A_i)$$
$$=\sum_{i=1}^{6} P(A_i)P\left(\frac{E}{A_i}\right)$$
$$=\frac{1}{6}\left[\frac{6}{10}+\frac{15}{45}+\frac{20}{120}+\frac{15}{210}+\frac{6}{252}+\frac{1}{210}\right]=\frac{1}{5}$$

39. A silver coin can be drawn in two mutually exclusive ways :

(i) Selecting purse 1 and then drawing a silver coin from it.

(ii) Selecting purse 2 and then drawing a silver coin from it.

Let E_1, E_2 and A be the events as defined below

E_1 : Selecting purse 1

E_2 : Selecting purse 2

A : Drawing a silver coin

$$\therefore \quad P(E_1)=\frac{1}{2}, P(E_2)=\frac{1}{2}$$

Now, $P\left(\dfrac{A}{E_1}\right)=\dfrac{2}{6}=\dfrac{1}{3}; P\left(\dfrac{A}{E_2}\right)=\dfrac{4}{7}$

Using the law of total probability, we get

Required probability $=P(A)$

$$=P(E_1)A\left(\frac{A}{E_1}\right)+P(E_2)P\left(\frac{A}{E_2}\right)$$
$$=\frac{1}{2}\times\frac{1}{3}+\frac{1}{2}\times\frac{4}{7}$$
$$=\frac{1}{6}+\frac{2}{7}$$
$$=\frac{7+12}{42}=\frac{19}{42}$$

40. Let E_1, E_2 and A be the events as defined as

E_1 : A white ball is transferred from bag A to bag B

E_2 : A black ball is transferred from bag A to bag B

A : One black ball drawn from bag B

$$P(E_1)=\frac{5}{11}, P(E_2)=\frac{6}{11}$$

Now, $P\left(\dfrac{A}{E_1}\right)=\dfrac{3}{8}, P\left(\dfrac{A}{E_2}\right)=\dfrac{4}{8}$

Using the law of total probabilities, we get

$$P(A)=P(E_1)P\left(\frac{A}{E_1}\right)+P(E_2)P\left(\frac{A}{E_2}\right)$$
$$=\frac{5}{11}\times\frac{3}{8}+\frac{6}{11}\times\frac{4}{8}$$
$$=\frac{15}{88}+\frac{24}{88}=\frac{39}{88}$$

41. Let E_1, E_2, E_3 and A be the events as defined as

E_1 : Selecting bag I

E_2 : Selecting bag II

E_3 : Selecting bag III

A : Drawing a white and a red ball

It is given that one of the bags is selected randomly.

$$\therefore \quad P(E_1)=\frac{1}{3}, P(E_2)=\frac{1}{3}, P(E_3)=\frac{1}{3}$$

Now, $P\left(\dfrac{A}{E_1}\right)=\dfrac{^1C_1\times {^3C_1}}{^6C_2}=\dfrac{3}{15}$

$$P\left(\frac{A}{E_2}\right)=\frac{^2C_1\times {^1C_1}}{^4C_2}=\frac{2}{6}$$

$$P\left(\frac{A}{E_3}\right)=\frac{^4C_1\times {^3C_1}}{^{12}C_2}=\frac{12}{66}$$

Using the law of total probability, we get

$$P(A)=P(E_1)P\left(\frac{A}{E_1}\right)+P(E_2)P\left(\frac{A}{E_2}\right)+P(E_3)P\left(\frac{A}{E_3}\right)$$

$$=\frac{1}{3}\times\frac{3}{15}+\frac{1}{3}\times\frac{2}{6}+\frac{1}{3}\times\frac{12}{66}$$

$$=\frac{1}{15}+\frac{1}{9}+\frac{2}{33}=\frac{33+55+30}{495}$$

$$=\frac{118}{495}$$

42. Let A be the event that the construction job will be completed on time and B be the event that there will be a strike. We have to find $P(A)$. We have,

$$P(B)=0.65, P(\text{no strike})=P(B')=1-P(B)$$

$$=1-0.65=0.35$$

$$P\left(\frac{A}{B}\right)=0.32, P\left(\frac{A}{B'}\right)=0.80$$

Since events, B and B' from a partition of the sample spaces therefore, by theorem on total probability, we have

$$P(A)=P(B)P\left(\frac{A}{B}\right)+P(B')P\left(\frac{A}{B'}\right)$$

$$=0.65\times0.32+0.35\times0.8$$

$$=0.208+0.28$$

$$=0.488$$

43. Let events B_1, B_2, B_3 be the following

B_1 : the bolt is manufactured by machine A

B_2 : the bolt is manufactured by machine B

B_3 : the bolt is manufactured by machine C

Clearly, B_1, B_2, B_3 are mutually exclusive and exhaustive events and hence, they represent a partition of the same space.

Let the event E be 'the bolt is defective'. The event E occurs with B_1 or with B_2 or with B_3, given that $P(B_1)=25\%=0.25, P(B_2)=0.35$ and $P(B_3)=0.40$

Again, $P\left(\frac{E}{B_1}\right)$ = Probability that the bolt drawn is defective given that it is manufactured by machine

$$A=5\%=0.05$$

Similarly, $P\left(\frac{E}{B_2}\right)=0.04, P\left(\frac{E}{B_3}\right)=0.02$

Hence, by the Bayes' theorem, we have

$$P\left(\frac{B_2}{E}\right)=\frac{P(B_2)P\left(\frac{E}{B_2}\right)}{P(B_1)P\left(\frac{E}{B_1}\right)P(B_2)P\left(\frac{E}{B_2}\right)+P(B_3)P\left(\frac{E}{B_3}\right)}$$

$$=\frac{0.35\times0.04}{0.25\times0.05+0.35\times0.04+0.40\times0.02}$$

$$=\frac{0.0140}{0.0345}=\frac{28}{69}$$

44. Let E_1 be the event of choosing the bag I, E_2 the event of choosing the bag II and A be the event of drawing a red ball

Then, $P(E_1)=P(E_2)=\frac{1}{2}$

Also, $P\left(\frac{A}{E_1}\right)=P=\frac{3}{7}$ and $P\left(\frac{A}{E_2}\right)=P=\frac{5}{11}$

Now, the probability of drawing a ball from bag II, being given that it is red, is $P\left(\frac{E_2}{A}\right)$

By using Bayes' theorem, we have

$$P\left(\frac{E_2}{A}\right)=\frac{P(E_2)P\left(\frac{A}{E_2}\right)}{P(E_1)P\left(\frac{A}{E_1}\right)+P(E_2)P\left(\frac{A}{E_2}\right)}$$

$$=\frac{\frac{1}{2}\times\frac{5}{11}}{\frac{1}{2}\times\frac{3}{7}+\frac{1}{2}\times\frac{5}{11}}=\frac{35}{68}$$

45. The probability distribution of x is

x	0	1	2	3	4
$P(x)$	0.1	k	$2k$	$2k$	k

(i) We know that $\sum_{i=1}^{n} P_i = 1$

Therefore, $0.1+k+2k+2k+k=1$

i.e. $k=0.15$

(ii) P(you study at least two hours)

$$=P(x\geq2)$$

$$=P(x=2)+P(x=3)+P(x=4)$$

$$=2k+2k+k=5k=5\times0.15=0.75$$

P(you study exactly two hours)

$$=P(x=2)$$

$$=2k=2\times0.15=0.3$$

P(you study at most two hours)

$$=P(x\leq2)$$

$$=P(x=0)+P(x=1)+P(x=2)$$

$$=0.1+k+2k=0.1+3k$$

$$=0.1+3\times0.15=0.55$$

46. The number of aces is a random variable. Let it be denoted by x. Clearly, x can take the values 0, 1 or 2. Now, since the draws are done with replacement, therefore the two draws form independent experiments.

Therefore, $P(x=0)=P$ (non-ace and non-ace)

$$=P(\text{non-ace})\times P(\text{non-ace})$$

$$=\frac{48}{52}\times\frac{48}{52}=\frac{144}{169}$$

$P(x=1) = P$ (ace and non-ace or non-ace and ace)

$= P(\text{ace and non-ace}) + P(\text{non-ace and ace})$

$= P(\text{ace}) \cdot P(\text{non-ace}) + P(\text{non-ace}) \cdot P(\text{ace})$

$= \frac{4}{52} \times \frac{48}{52} + \frac{48}{52} \times \frac{4}{52} = \frac{24}{169}$

and $P(x=2) = P$ (ace and ace)

$= \frac{4}{52} \cdot \frac{4}{52} = \frac{1}{169}$

Thus, the required probability distribution is

x	0	1	2
P(x)	$\frac{144}{169}$	$\frac{24}{169}$	$\frac{1}{169}$

47. Let x denote the number of kings in a draw of two cards. x is a random variable which can assume the values 0, 1 or 2.

Now, $P(x=0) = P(\text{no king}) = \frac{^{48}C_2}{^{52}C_2} = \frac{\frac{48!}{2!(48-2!)}}{\frac{52!}{2!(52-2!)}}$

$\frac{48\ 47}{52\ 51} = \frac{188}{221}$

$P(x=1) = P$ (one king and one non-king)

$= \frac{^4C_1 \times {^{48}C_1}}{^{52}C_2}$

$= \frac{4 \times 48 \times 2}{52 \times 51} = \frac{32}{221}$

and $P(x=2) = P$ (2 kings) $= \frac{^4C_2}{^{52}C_2} = \frac{4 \times 3}{52 \times 51} = \frac{1}{221}$

Thus, the probability distribution of x is

x	0	1	2
P(x)	$\frac{188}{221}$	$\frac{32}{221}$	$\frac{1}{221}$

Now, mean of $x = E(x) = \sum_{i=1}^{n} x_i P(x_i)$

$= 0 \times \frac{188}{221} + 1 \times \frac{32}{221} + 2 \times \frac{1}{221} = \frac{34}{221}$

Also, $E(x^2) = \sum_{i=1}^{n} x_i^2 p(x_i)$

$= 0^2 \times \frac{188}{221} + 1^2 \times \frac{32}{221} + 2^2 \times \frac{1}{221}$

$= \frac{36}{221}$

Now, $\text{Var}(x) = E(x^2) - [E(x)]^2$

$= \frac{36}{221} - \left(\frac{34}{221}\right)^2 = \frac{6800}{(221)^2}$

Therefore, $\sigma_x = \sqrt{\text{Var}(x)}$

$= \frac{\sqrt{6800}}{221} = 0.37$

48. (i) The numbers of trials is finite. When the drawing is done with replacement, there probability of success is $P = \frac{7}{16}$ which is same for all 6 trials (draws). Hence, the drawing of balls with replacement are Bernoulli trials.

(ii) When the drawing is done without replacement, the probability of success in 1st trials is $\frac{7}{16}$ in 2nd trials is $\frac{6}{15}$, if the 1st ball drawn is red or $\frac{7}{15}$, if the 1st ball drawn is black and so on. Clearly, the probability of success is not same for all trials, hence the trials are not Bernoulli trials.

49. Let p denote the 4 turning up in a toss of a fair die, so

$p = \frac{1}{6}, q = 1 - \frac{1}{6} = \frac{5}{6}$ $[\because p + q = 1]$

Probability of getting 4 at least once in two tosses of a fair die

$= P(x=1) + P(x=2)$

$= 1 - P(x=0)$

$= 1 - {^2C_0}\left(\frac{1}{6}\right)^0 \left(\frac{5}{6}\right)^2$

$= 1 - \frac{25}{36}$

$= \frac{11}{36}$

50. Probability of success is $\frac{3}{6} = \frac{1}{2}$

$\therefore$ $p = \frac{1}{2}$ and $q = \frac{1}{2}$

(i) For exactly 4 successes, the required probability is

${^7C_4}\left(\frac{1}{2}\right)^4 \left(\frac{1}{3}\right)^3 = \frac{35}{128}$

(ii) For at least 4 successes, the required probability is

${^7C_4}\left(\frac{1}{2}\right)^4 \left(\frac{1}{2}\right)^3 + {^7C_5}\left(\frac{1}{2}\right)^5 \left(\frac{1}{2}\right)^2$

$+ {^7C_6}\left(\frac{1}{2}\right)^6 \left(\frac{1}{2}\right)^1 + {^7C_7}\left(\frac{1}{2}\right)^7$

$= \frac{35}{128} + \frac{21}{128} + \frac{7}{128} + \frac{1}{128}$

$= \frac{64}{128}$

$= \frac{1}{2}$

51. A person can be either right handed or left handed. It is given that 90% of the people are right handed

$\therefore \quad p = P \text{ (right handed)} = \frac{9}{10}$

$q = P \text{ (left handed)} = 1 - \frac{9}{10} = \frac{1}{10}$

Using binomial distribution, the probability that at most six people are right handed is given by

$$\sum_{r=0}^{6} {}^{10}C_r p^r q^{n-r} = \sum_{r=0}^{6} {}^{10}C_r \left(\frac{9}{10}\right)^r \left(\frac{1}{10}\right)^{10-r}$$

Therefore, the probability that at most 6 people are right handed $= 1 - \sum_{r=7}^{10} {}^{10}C_r \, (0.9)^r \, (0.1)^{10-r}$

52. Let p be its probability of success and q be the failure.

Then, $p = 2q$.

Also, $p + q = 1$.

It gives $p = \frac{2}{3}$ and $q = \frac{1}{3}$

$P\{4 \text{ successes in 6 trials}\} = {}^6C_4 p^4 q^2$

$= {}^6C_4 \left(\frac{2}{3}\right)^4 \left(\frac{1}{3}\right)^2 \quad \ldots\text{(i)}$

$P\{5 \text{ successes in 6 trials}\} = {}^6C_5 \left(\frac{2}{3}\right)^5 \left(\frac{1}{3}\right) \quad \ldots\text{(ii)}$

$P\{6 \text{ successes in 6 trials}\} = {}^6C_6 \left(\frac{2}{3}\right)^6 \left(\frac{1}{3}\right)^0 \quad \ldots\text{(iii)}$

Therefore, the probability that there are at least 4 successes is

$P\{\text{either 4 or 5 or 6 successes}\}$

$$= {}^6C_4 \left(\frac{2}{3}\right)^4 \left(\frac{1}{3}\right)^2 + {}^6C_5 \left(\frac{2}{3}\right)^5 \left(\frac{1}{3}\right) + {}^6C_6 \left(\frac{2}{3}\right)^6$$

$$= \frac{496}{729}$$

53. $P(x \geq 8)$

$$= {}^{10}C_8 \left(\frac{1}{2}\right)^8 \left(\frac{1}{2}\right)^2 + {}^{10}C_9 \left(\frac{1}{2}\right)^9 \left(\frac{1}{2}\right)^1 + {}^{10}C_{10} \left(\frac{1}{2}\right)^{10}$$

$$= \left(\frac{1}{2}\right)^{10} [{}^{10}C_2 + {}^{10}C_1 + {}^{10}C_0]$$

$$= \left(\frac{1}{2}\right)^{10} [45 + 10 + 1] = \frac{56}{8 \times 2^7} = \frac{7}{128}$$

54. Let a rifleman fires n numbers of rounds.

Probability of hitting the target, $p = \frac{1}{10}$.

$\therefore$ Probability of not hitting the target,

$$q = 1 - \frac{1}{10} = \frac{9}{10}$$

$\therefore$ Probability of hitting the target at least once

$$= 1 - \left(\frac{9}{10}\right)^n$$

Given, $1 - \left(\frac{9}{10}\right)^n > \frac{1}{2}$

$\therefore \quad \left(\frac{9}{10}\right)^n < \frac{1}{2}$

So, the least value of n is 7.

55. Let S denote the success and F denote the failure.

Thus, $P(S) = \frac{1}{6}, P(F) = \frac{5}{6}$

P (A wins in the 1st throw) $= P(S) = \frac{1}{6}$

A get the third throw, when the first throw by A and 2nd throw by B result into failures

Therefore, P(A wins in 3rd throw)

$= P(FFS) = P(F)\,P(F)\,P(S)$

$= \frac{5}{6} \times \frac{5}{6} \times \frac{1}{6} = \left(\frac{5}{6}\right)^2 \times \frac{1}{6}$

P (A wins in the 5th throw) $= P(FF\,FF\,S)$

$= \left(\frac{5}{6}\right)^4 \frac{1}{6}$ and so on

Hence, $P \text{ (A wins)} = \frac{1}{6} + \frac{5^2}{6} \times \frac{1}{6} + \left(\frac{5}{6}\right)^4 \times \frac{1}{6}$

$= \frac{\frac{1}{6}}{1 - \frac{25}{36}} = \frac{6}{11}$

$P(B \text{ wins}) = 1 - P \text{ (A wins)} = 1 - \frac{6}{11} = \frac{5}{11}$

56. Let p be the probability of success, so

$p = \frac{2}{6}; q = 1 - \frac{2}{6} = \frac{2}{3}$

Mean $= np = 3\left(\frac{2}{6}\right) = 1$

Variance $= npq = 3\left(\frac{2}{6}\right)\left(\frac{2}{3}\right) = \frac{2}{3}$

57. $p = \frac{2}{6}, q = 1 - \frac{2}{6} = \frac{2}{3}$

Given, $n = 8$,

Mean $= np = 8\left(\frac{2}{6}\right) = \frac{8}{3} = 2.66$

Variance $= npq = 8\left(\frac{2}{6}\right)\left(\frac{2}{3}\right) = \frac{16}{9}$

$\therefore$ Standard deviation $= \sqrt{\text{Variance}} = \sqrt{\frac{16}{9}} = \frac{4}{3}$

58. Mean $= np = 3$

Variance $= npq = \frac{3}{2}$

$\Rightarrow \quad 3(q) = \frac{3}{2} \Rightarrow q = \frac{1}{2}$

$\therefore \quad p + q = 1$

$p = 1 - q = 1 - \frac{1}{2} = \frac{1}{2}$

$\Rightarrow \quad n = \frac{3}{p} = \frac{3}{1/2} = 6$

Hence, $P(x \le 5) = 1 - P(x = 6)$

$= 1 - {}^6C_6 \left(\frac{1}{2}\right)^6 \left(\frac{1}{2}\right)^0$

$= 1 - \left(\frac{1}{2}\right)^6$

$= 1 - \frac{1}{64} = \frac{63}{64}$

Level 2

1. (b) Total number of letters in the word PROBABILITY $= 11$

Total number of vowels in the word PROBABILITY $= 4\,(O, A, I, I)$

Hence, probability that the single letter to be selected at random is vowel $= \frac{4}{11}$

2. (d) A leap year has 366 days in which 52 weak and 2 days are extra as (Sunday, Monday), (Monday, Tuesday), (Tuesday, Wednesday), (Wednesday, Thursday), (Thursday, Friday), (Friday, Saturday), (Saturday, Sunday) [7 choices]

So, probability that a leap year contains 53 Sunday is $\frac{2}{7}$.

3. (c) In a regular hexagon, there are two equilateral triangles are possible.

$\therefore$ Required probability $= \frac{2}{{}^6C_3} = \frac{2}{20} = \frac{1}{10}$

4. (a) Total number of balls $= 12$

Hence, the required probability

$= \frac{{}^5C_2 \times {}^7C_2}{{}^{12}C_4} = \frac{14}{33}$

5. (a) Seven people can seat themselves at a round table in $6!$ ways. The number of ways in which 2 distinguished person will be next to each other $= 2\,(5)!$

Hence, the required probability $= \frac{2\,(5)!}{6!} = \frac{1}{3}$

6. (b) Total number of cards $= 52$

Total number of ace of heart $= 1$

$\because$ Two cards are drawn simultaneously such that at least one of them is ace of heart.

$\therefore$ Required probability $= \frac{{}^1C_1 \times {}^{51}C_1}{{}^{52}C_2}$

$= \frac{1 \times 51 \times 2}{52 \times 51} = \frac{2}{52} = \frac{1}{26}$

7. (a) L and W can be filled at 14 places in 2^{14} ways.

$\therefore \quad n(S) = 2^{14}$

Now, 13 L's and 1W can be arranged at 14 places in 14 ways.

Hence, $n(A) = 14$

$\therefore \quad p = \frac{14}{2^{14}} = \frac{7}{2^{13}}$

8. (c) Given that 5 and 6 have appeared on two of the dice, the sample space reduces to $6^4 - 2 \times 5^4 + 4$.

Also, the number of favourable cases are $4! = 24$

So, the required probability is

$\frac{24}{302} = \frac{12}{151}$.

9. (a) Required probability

$= \frac{\text{Number of favourable cases}}{\text{Total number of exhaustive cases}}$

$= \frac{3}{3 \times 3 \times 3} = \frac{1}{9}$

10. (d) Three-digits numbers are 100, 101, ..., 999. Total number of such numbers is 900. The 3-digit number are 111, 222, 333, ... 999. Favourable number of cases is 9.

Therefore, the required probability is $\frac{9}{900} = \frac{1}{100}$.

11. (a) The total number of ways of selecting 3 integers from 20 natural numbers is ${}^{20}C_3 = 1140$. Their product is a multiple of 3 means at least one number is divisible by 3. The numbers which are divisible by 3 are 3, 6, 9, 12, 15, 18 and the number of ways of selecting at least one of them is ${}^6C_1 \times {}^{14}C_2 + {}^6C_2 \times {}^{14}C_1 + {}^6C_3 = 776$.

Hence, the required probability is $\frac{776}{1140} = \frac{194}{285}$

12. (b) A number has exactly 3 factors if the number is squares of a prime number. Squares of 11, 13, 17, 19, 23, 29, 31 and 3-digit numbers. Hence, the required probability is $\frac{7}{900}$.

13. (a) We have,

$n(S) = {}^{64}C_3$

Let 'E' be the event of selecting 3 squares which form the letter 'L'.

The number of ways of selecting squares consisting of 4 unit squares is $7 \times 7 = 49$

Each squares with 4 unit squares form 4 L-shaped consisting of 3 unit squares.

Therefore,

$$n(E) = 7 \times 7 \times 4 = 196$$

$$\therefore \quad P(E) = \frac{196}{^{64}C_3}$$

14. (c) The number of ways of getting the different numbers 1, 2, ..., 6 in six dice $= 6!$

Total number of ways $= 6^6$

Hence, required probability $= \frac{6!}{6^6}$

$$= \frac{1 \times 2 \times 3 \times 4 \times 5 \times 6}{6^6} = \frac{5}{324}$$

15. (a) Out of 9 socks, 2 can be drawn in $^9C_2$ ways.

Therefore, the total number of cases in $^9C_2$.

Two socks drawn from the drawer will match, if either both are brown or bere

Hence, the required probability $= \frac{^5C_2 + {^4C_2}}{^9C_2} = \frac{4}{9}$

16. (a) The number of composite number from 1 to 30 is $n(S) = 19$.

The number of composite numbers when divided by 5 leaves a remainder is $n(E) = 14$.

Therefore, the required probability is $\frac{14}{19}$.

17. (c) The total number of mapping is n^n.

The number of one-one mapping is $n!$.

Hence, the probability is

$$\frac{n!}{n^n} = \frac{3}{32} = \frac{4!}{4^4}$$

Comparing, we get $n = 4$.

18. (c) If A draws card higher than B, then the number of favourable cases is $(n-1) + (n-2) + \ldots + 3 + 2 + 1$

Therefore, the required probability is

$$\frac{\frac{n(n-1)}{2}}{n^2} = \frac{n-1}{2n}$$

19. (c) Given, $7a - 9b = 0$ or $b = \frac{7}{9}a$

Hence, number of pairs (a, b) can be (9, 7); (18, 14); (27, 21); (36, 28).

Hence, the required probability is

$$\frac{4}{^{39}C_2} = \frac{4}{741}$$

20. (b) There are 10 digits 0, 1, 2, 3, 4, 5, 6, 7, 8, 9. The last two digits can be dialed in $^{10}P_2 = 90$ ways out of which only one way is favourable.

Thus, the required probability is $\frac{1}{90}$.

21. (c) We have, $x + \frac{100}{x} > 50$

$$\text{or} \quad x^2 + 100 > 50x$$

$$\text{or} \quad (x - 25)^2 > 525$$

$$\Rightarrow \quad x - 25 < -\sqrt{525} \text{ or } x - 25 > \sqrt{525}$$

$$\Rightarrow \quad x < 25 - \sqrt{525} \text{ or } x > 25 + \sqrt{525}$$

As x is a possible integer and $\sqrt{525} = 22.91$, we must have $x \le 2$ or $x \ge 48$.

Thus, favourable number of cases is $2 + 53 = 55$.

Hence, the required probability is

$$\frac{55}{100} = \frac{11}{20}$$

22. (c) Required probability

$$= \frac{^{12}C_2 \times {^{10}C_2} \times {^8C_2} \times {^6C_2} \times {^4C_2} \times {^2C_2}}{6^{12}}$$

$$= \frac{\frac{12!}{10! \times 2!} \times \frac{10!}{8! \times 2!} \times \frac{8!}{6! \times 2!} \times \frac{6!}{4! \times 2!} \times \frac{4!}{2! \times 2!} \times \frac{2!}{2! \times 0!}}{6^{12}}$$

$$= \frac{12!}{2^6 \times 6^{12}}$$

23. (d) A person can have his/her birthday on any one of the seven days of the week. So, 5 person can have their birthday in 7^5 ways. Out of 5, three person can have their birthdays on days other than Sunday in 6^3 ways and other 2 on Sunday. Hence, the required probability is

$$= \frac{^5C_2 \times 6^3}{7^5} = \frac{10 \times 6^3}{7^5}$$

[Note that 2 persons can be selected out of 5 in $^5C_2$ ways.]

24. (c) Since, it is known that each time a different digit appears, total number of cases is $6!$.

The sum in the first three throws is 12, if we get (1, 5, 6), (2, 4, 6) or (3, 4, 5) in any order.

So, number of cases for the first three throws $= 3 \times 3!$

The remaining three throws have $3!$ ways.

$\therefore$ Number of favourable cases $= 3 \times 3! \times 3!$

$\therefore$ Required probability $= \frac{3 \times 3! \times 3!}{6!}$

$$= \frac{3}{20}$$

25. (c) Sample set $S = \{10, 11, 12, ..., 99\}$

$$n(S) = 99 - 10 + 1 = 90$$

n	2^n	$2^n - 2$	Divisible by 3
1	2	0	Yes
2	4	2	No
3	8	6	Yes
4	16	14	No
5	32	30	Yes
6	64	62	No
7	128	126	Yes
8	256	254	No
9	512	510	Yes
10	1024	1022	No
11	2048	2046	Yes
12	4096	4094	No

As we can see, for all the odd values of n, $2^n - 2$ is divisible by 3. So, the event set will be all the two digit odd numbers.

$$E = \{11, 13, 15, ..., 99\}$$

$$n(E) = \left(\frac{99 - 11}{2} + 1\right) = 45$$

$$\therefore \text{Probability} = \frac{n(E)}{n(S)} = \frac{45}{90} = \frac{1}{2}$$

26. (b) We have, $A = \{0, 1, 2, 3, 4, ..., 10\}$

$$n(S) = {}^{11}C_2$$

Let E be the given event

$\therefore \quad E = \{(0, 4), (0, 8), (2, 6), (2, 10), (4, 8), (6, 10)\}$

$\therefore \quad n(E) = 6$

$\therefore \quad P(E) = \dfrac{6}{55}$

27. (c) For an AP, $2b = a + c$ (even), so both (a) and (c) even numbers or odd number from given numbers.

$$\text{Required probability} = \frac{{}^6C_2 + {}^5C_2}{{}^{11}C_3} = \frac{25}{165} = \frac{5}{33}$$

28. (c) $x + y + z = 10$

Total number of non-negative solutions

$$= {}^{10+3-1}C_{3-1} = {}^{12}C_2 = 66$$

Now, let $\quad z = 2n$

$\therefore \quad x + y = 10 - 2n; n \geq 0$

Number of non-negative solutions of above equations is

$$11 + 9 + 7 + 5 + 3 + 1 = 36$$

$$\therefore \text{Required probability} = \frac{36}{66} = \frac{6}{11}$$

29. (b) Total ways $= {}^{64}C_2 = 32 \times 63$

Favourable ways $= 2 \times 7 \times 8$

$$\text{Required probability} = \frac{2 \times 7 \times 8}{32 \times 63} = \frac{1}{18}$$

30. (c) Total number of possibilities $= 2 \times 2 \times 2 \times 2 = 16$

Determinants whose value is positive are

$$\begin{vmatrix} 1 & 0 \\ 0 & 1 \end{vmatrix}, \begin{vmatrix} 1 & 1 \\ 0 & 1 \end{vmatrix} \text{ and } \begin{vmatrix} 1 & 0 \\ 1 & 1 \end{vmatrix}$$

So, the desired probability $= \dfrac{3}{16}$

31. (d) Exhaustive number of cases $= (15 - 1)! = 14!$

Favourable cases ${}^{13}C_4 (10 - 1)! \times 2! \times 4! = 2 \times 13!$

$$\therefore \text{Probability} = \frac{2 \times 13!}{14!} = \frac{1}{7}$$

32. (d) Total number of bijections $= 4! = 24$

Selecting functions satisfying $f(x) \neq x$ for all x is equivalent to derangement

$\therefore$ Favourable number of cases

$$= 4!\left[1 - \frac{1}{1!} + \frac{1}{2!} - \frac{1}{3!} + \frac{1}{4!}\right] = 9$$

$$\therefore \text{Probability} = \frac{9}{24} = \frac{3}{8}$$

33. (b) $p_1 = \dfrac{{}^{15}C_2}{{}^{42}C_4} = \dfrac{15 \times 14 \times 4!}{2! \times 42 \times 41 \times 40 \times 39} = \dfrac{1}{41 \times 26}$

and $p_2 = \dfrac{{}^{30}C_4}{{}^{84}C_8} = \dfrac{15 \times 14 \times 13 \times 12 \times 8!}{4! \times 84 \times 83 \times 82 \times \ldots \times 77}$

$$= \frac{15 \times 14 \times 13 \times 12 \times 8 \times 7 \times 6 \times 5}{84 \times 83 \times 82 \times 81 \times 79 \times 78 \times 77} < p_1$$

$\Rightarrow \quad p_1 > p_2$

34. (b) Total number of ways of distribution is 4^5

$\therefore \quad n(S) = 4^5$

Total number of ways of distribution so that each child get at least one game is

$$4^5 - {}^4c_1 3^5 + {}^4C_2 2^5 - {}^4C_3$$

$$= 1024 - 4 \times 243 + 6 \times 32 - 4 = 240$$

$\therefore \quad n(E) = 240$

Therefore, the required probability is

$$\frac{n(E)}{n(S)} = \frac{240}{4^5} = \frac{15}{64}$$

35. (d) $\because$ P(Exactly one of A and B occurs)

$$= P[(A \cap B') \cup (A' \cap B)]$$

$$= P(A \cap B') + P(A' \cap B) \quad \text{...(i)}$$

$$= P(A) - P(A \cap B) + P(B) - P(A \cap B)$$

$$= P(A) + P(B) - 2P(A \cap B) \quad \text{...(ii)}$$

$$= P(A \cup B) - P(A \cap B) \quad \text{...(iii)}$$

$$= [1 - P(A \cup B)'] - [1 - P(A \cap B)']$$
$$= P(A' \cup B') - P(A' \cap B')$$
[De-Morgan's law]
$$= P(A') + P(B') - 2P(A' \cap B') \quad \text{...(iv)}$$

Clearly, (d) is not correct.

36. (d) $P(\text{Exactly one}) = \frac{2}{5}$

$$\Rightarrow \quad P(A) + P(B) - 2P(A \cap B) = \frac{2}{5} \quad \text{...(i)}$$
$$P(A \text{ or } B) - P(A \cup B) = \frac{1}{2}$$
$$\Rightarrow \quad P(A) + P(B) - P(A \cap B) = \frac{1}{2} \quad \text{...(ii)}$$

From Eqs. (i) and (ii), we get

$$P(A \cap B) = \frac{1}{2} - \frac{2}{5}$$
$$= \frac{5-4}{10} = \frac{1}{10}$$
$$= 0.10$$

37. (b) $\because \; P(A \cap B) = P(A) + P(B) - P(A \cup B)$
$$= 0.6 + 0.4 - 0.8$$
$$= 1 - 0.8 = 0.2$$

$\because P(A \cup B \cup C) = P(A) + P(B) + P(C)$
$$- P(A \cap B) - P(B \cap C) - P(C \cap A) + P(A \cap B \cap C)$$
$$\alpha = 0.6 + 0.4 + 0.5 - 0.2 - \beta - 0.3 + 0.2$$
$\therefore \quad \beta = 1.2 - \alpha$

$\because \quad \alpha \in [0.85, 0.95]$, then $\beta \in [0.25, 0.35]$

38. (a) Since, $P(A \cup B \cup C) \geq 0.75$, therefore
$$0.75 \leq P(A \cup B \cup C) \leq 1$$
$$\Rightarrow \quad 0.75 \leq P(A) + P(B) + P(C) - P(A \cap B) - P(B \cap C) - P(A \cap C) + P(A \cap B \cap C) \leq 1$$
$$\Rightarrow \quad 0.75 \leq 0.3 + 0.4 + 0.8 - 0.08 - P(B \cap C) - 0.28 + 0.09 \leq 1$$
$$\Rightarrow \quad 0.75 \leq 1.23 - P(B \cap C) \leq 1$$
$$\Rightarrow \quad -0.48 \leq -P(B \cap C) \leq -0.23$$
$$\Rightarrow \quad 0.23 \leq P(B \cap C) \leq 0.48$$

39. (c) $P(\text{Exactly one of } A \text{ or } B \text{ occurs})$
$$= P(A) + P(B) - 2P(A \cap B) = \frac{1}{4}$$

$P(\text{Exactly one of } B \text{ or } C \text{ occurs})$
$$= P(B) + P(C) - 2P(B \cap C) = \frac{1}{4}$$

$P(\text{Exactly one of } C \text{ or } A \text{ occurs})$
$$= P(C) + P(A) - 2P(C \cap A) = \frac{1}{4}$$

Adding all, we get

$$P(A) + P(B) + P(C) - P(A \cap B) - P(B \cap C) - P(A \cap C) = \frac{3}{8}$$

$$\therefore \quad \Sigma P(A) - \Sigma P(A \cap B) = \frac{3}{8}$$

Now, $P(A \cup B \cup C) = \Sigma P(A) - \Sigma P(A \cap B) + P(A \cap B \cap C)$
$$= \frac{3}{8} + \frac{1}{16} = \frac{7}{16}$$

40. (a) $S = \{00, 01, 02, \ldots, 49\}$

Let A be the event that sum of the digits on the selected ticket is 8

Then, $A = \{08, 17, 26, 35, 44\}$

Let B be the event that the product of the digits is zero, then

$$B = \{00, 01, 02, 03, \ldots, 10, 20, 30, 40\}$$
$$\Rightarrow \quad A \cap B = \{08\}$$

$\therefore$ Required probability $= P\left(\frac{A}{B}\right) = \frac{P(A \cap B)}{P(B)}$
$$= \frac{\frac{1}{50}}{\frac{14}{50}} = \frac{1}{14}$$

41. (b) Consider the two events A and B as

A: Event for getting minimum as 3

B: Event for getting maximum as 6

We have to find $P(A/B)$

and $P(A/B) = \frac{P(A \cap B)}{P(B)}$

Now, $P(A \cap B) = \frac{{}^2C_1}{{}^8C_3}$

[$\because$ favourable count is ${}^2C_1$ and total count is ${}^8C_3$]

also $P(B) = \frac{{}^5C_2}{{}^8C_3}$

[$\because$ 6 is already chosen as maximum and two more has to be taken from (1, 2, 3, 4, 5)]

$$\therefore \; P(A/B) = \frac{{}^2C_1 / {}^8C_3}{{}^5C_2 / {}^8C_3} = \frac{1}{5}$$

42. (a) Let E_1 be the event that exactly 2 players scored more than 50 runs, then

$$P(E_1) = \frac{1}{2} \times \frac{1}{3} \times \frac{3}{4} \times \frac{9}{10} + \frac{1}{2} \times \frac{2}{3} \times \frac{1}{4} \times \frac{9}{10} + \frac{1}{2} \times \frac{2}{3} \times \frac{3}{4} \times \frac{1}{10} + \frac{1}{2} \times \frac{1}{3} \times \frac{1}{4} \times \frac{9}{10} + \frac{1}{2} \times \frac{2}{3} \times \frac{3}{4} \times \frac{1}{10} + \frac{1}{2} \times \frac{2}{3} \times \frac{1}{4} \times \frac{1}{10}$$
$$= \frac{65}{240}$$

Let E_2 be the event that A and B scored more than 50 runs, then

$$P(E_1 \cap E_2) = \frac{1}{2} \times \frac{1}{3} \times \frac{3}{4} \times \frac{9}{10} = \frac{27}{240}$$

$\therefore$ Desired probability $= P\left(\dfrac{E_2}{E_1}\right)$

$$= \frac{P(E_1 \cap E_2)}{P(E_1)}$$

$$= \frac{27}{65}$$

43. (c) Let A be the event that fair coin is selected and B that biased coin is selected

$$P = P\left(\frac{H}{A}\right) + P\left(\frac{H}{B}\right)$$

$$= \frac{7}{12} = \frac{n}{n+1} \times \frac{1}{2} + \frac{1}{n+1} \times 1$$

$\Rightarrow \quad n = 5$

44. (c)

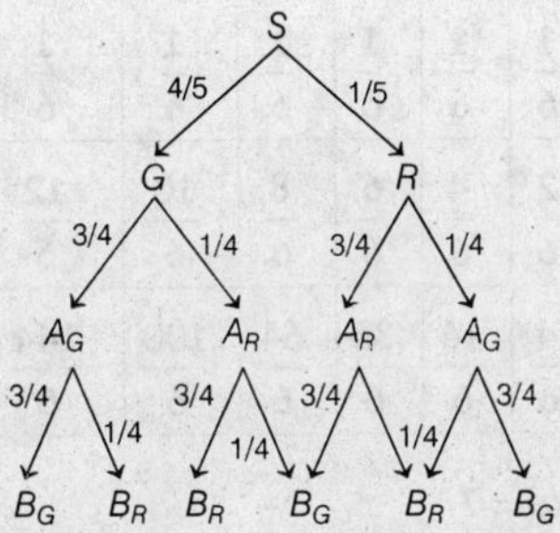

$$P(B_G) = \frac{46}{80}$$

$$P\left(\frac{B_G}{G}\right) = \frac{10}{16} = \frac{5}{8}$$

$$\therefore \quad P(B_G \cap G) = \frac{5}{8} \times \frac{4}{5} = \frac{1}{2}$$

$$P\left(\frac{G}{B_G}\right) = \frac{P(B_G \cap G)}{P(B_G)}$$

$$= \frac{1}{2} \times \frac{80}{46} = 40/46 = 20/23$$

45. (c) $P(\overline{A \cup B}) = \dfrac{1}{6}$

$$P(A \cup B) = \frac{5}{6}, P(A) = \frac{3}{4}$$

$$P(A \cup B) = P(A) + P(B) - P(A \cap B) = \frac{5}{6}$$

$$\Rightarrow \quad P(B) = \frac{5}{6} - \frac{3}{4} + \frac{1}{4} = \frac{1}{3}$$

$$\Rightarrow P(A \cap B) = P(A) \cdot P(B) = \frac{1}{4}$$

46. (c) $P(X) = 0.3$ and $P(Y) = 0.7$

$\because$ X and Y are independent of each other.

$\therefore$ $P(X \cup Y) = P(X) + P(Y) - P(X \cap Y)$

$\because$ $P(X \cap Y) = P(X) \cdot P(Y)$ $\quad [\because X$ and Y are independent$]$

$$= 0.3 \times 0.7 = 0.21$$

$$\Rightarrow P(X \cup Y) = 0.3 + 0.7 - 0.21 = 0.79$$

47. (c) This is the case of independent events.

P(drawing a queen in the first condition) $= \dfrac{4}{52}$

P(drawing a king in the second condition after a queen with replacement) $= \dfrac{4}{52}$

$\therefore$ P(drawing a queen followed by king) $= \dfrac{4}{52} \times \dfrac{4}{52}$

$$= \frac{16}{2704} = \frac{1}{169}$$

48. (b) Let E_1 and E_2 and A be the events as defined below

E_1 : The coin shows a head

E_2 : The coin shows a tail

A : The noted number is 7 or 8

$$P(E_1) = \frac{1}{2}, P(E_2) = \frac{1}{2}$$

$$P\left(\frac{A}{E_1}\right) = P \text{ [pair of dice shows 7 or 8 as sum]}$$

P [(1, 6), (2, 5), (3, 4), (4, 3), (5, 2), (6, 1), (2, 6), (3, 5), (4, 4), (5, 3), (6, 2)]

$$\Rightarrow \quad P\left(\frac{A}{E_1}\right) = \frac{11}{36}$$

Also, $P\left(\dfrac{A}{E_2}\right) = \dfrac{2}{11}$

By law of total probability, we get

$$P(A) = P(E_1) \cdot P\left(\frac{A}{E_1}\right) + P(E_2) \cdot P\left(\frac{A}{E_2}\right)$$

$$= \frac{1}{2} \times \frac{11}{36} + \frac{1}{2} \times \frac{2}{11}$$

$$= \frac{11}{72} + \frac{2}{22} = \frac{121 + 72}{792}$$

$$= \frac{193}{792}$$

49. (a) Let E denotes the event that a six occurs and A the event that the man reports that it is a 6.

Then, $P(E) = \dfrac{1}{6}, P(E') = \dfrac{5}{6}$

Then $P\left(\dfrac{A}{E}\right) = \dfrac{3}{4}$ and $P\left(\dfrac{A}{E'}\right) = \dfrac{1}{4}$

From Baye's theorem

$$P\left(\frac{E}{A}\right) = \frac{P(E) \cdot P\left(\frac{A}{E}\right)}{P(E) \cdot P\left(\frac{A}{E'}\right) + P(E') \cdot P\left(\frac{A}{E'}\right)}$$

$$= \frac{\frac{1}{6} \times \frac{3}{4}}{\frac{1}{6} \times \frac{3}{4} + \frac{5}{6} \times \frac{1}{4}} = \frac{3}{8}$$

50. (b) $P\left(\frac{B_2}{R}\right)=\frac{P(B_2)P\left(\frac{R}{B_2}\right)}{P(B_1)P\left(\frac{R}{B_1}\right)+P(B_2)P\left(\frac{R}{B_2}\right)+P(B_3)P\left(\frac{R}{B_3}\right)}$

$$=\frac{\frac{1}{2}\times\frac{2}{5}}{\frac{1}{2}\times\frac{1}{3}+\frac{1}{3}\times\frac{2}{5}+\frac{1}{6}\times\frac{3}{7}}$$

$$=\frac{\frac{2}{15}}{\frac{1}{6}+\frac{2}{15}+\frac{1}{14}}$$

$$=\frac{14}{39}$$

51. (c) $P\left(\frac{G}{A}\right)=\frac{^4C_1}{^7C_1}=\frac{4}{7},P(A)=\frac{1}{2},P(B)=\frac{1}{2}$

$$P\left(\frac{G}{B}\right)=\frac{^3C_1}{^7C_1}=\frac{3}{7}$$

Now, $P\left(\frac{B}{G}\right)=\frac{P(B)\cdot P\left(\frac{G}{B}\right)}{P(A)\cdot P\left(\frac{G}{A}\right)+P(B)\cdot P\left(\frac{G}{B}\right)}$

$$=\frac{\frac{1}{2}\cdot\frac{3}{7}}{\frac{1}{2}\cdot\frac{4}{7}+\frac{1}{2}\cdot\frac{3}{7}}=\frac{3}{7}$$

52. (b) Let probability of choosing box $B, P(B)=p$

$\therefore\ P(A)=2P(B)=2p$

Now, $p\left(\frac{R}{A}\right)=\frac{^3C_1}{^5C_1}=\frac{3}{5}$ and $P\left(\frac{R}{B}\right)=\frac{^4C_1}{^7C_1}=\frac{4}{7}$

$$\therefore\ P\left(\frac{B}{R}\right)=\frac{P(B)\cdot P\left(\frac{R}{B}\right)}{P(A)\cdot P\left(\frac{R}{A}\right)+P(B)\cdot P\left(\frac{R}{B}\right)}=\frac{p\cdot\frac{4}{7}}{2p\cdot\frac{3}{5}+p\cdot\frac{4}{7}}=\frac{10}{31}$$

53. (a) $p^2+2p+4p-1=1$ (Exhaustive)

$$p^2+6p-2=0$$

or $\quad p=-3\pm\sqrt{11}=\sqrt{11}-3$ (neglecting negative sign)

54. (a) Given distribution is

X	0	1	2	3
P(X)	$\frac{1}{3}$	$\frac{1}{2}$	0	$\frac{1}{6}$

Mean, $m=\sum_{i=1}^{4}p_ix_i=0\times\frac{1}{3}+1\times\frac{1}{2}+2\times0+3\times\frac{1}{6}$

$$=\frac{1}{2}+\frac{1}{2}=1$$

Variance, $\sigma^2=\sum_{i=1}^{4}p_i(x_i-m)^2$

$$=\frac{1}{3}(0-1)^2+\frac{1}{2}(1-1)^2+0(2-1)^2+\frac{1}{6}(3-1)^2$$

$$=\frac{1}{3}+0+0+\frac{4}{6}$$

$$=\frac{1}{3}+\frac{4}{6}$$

$$=1$$

55. (d) Let X denotes twice find the number appearing on the die.

So, $X=2,4,6,8,10,12$

x_i	2	4	6	8	10	12	
$p(x_i)$	$\frac{1}{6}$	$\frac{1}{6}$	$\frac{1}{6}$	$\frac{1}{6}$	$\frac{1}{6}$	$\frac{1}{6}$	
$x_ip(x_i)$	$\frac{2}{6}$	$\frac{4}{6}$	$\frac{6}{6}$	$\frac{8}{6}$	$\frac{10}{6}$	$\frac{12}{6}$	$\Sigma xp=7$
$x_i^2p(x_i)$	$\frac{4}{6}$	$\frac{16}{6}$	$\frac{36}{6}$	$\frac{64}{6}$	$\frac{100}{6}$	$\frac{144}{6}$	$\Sigma x_i^2p=\frac{364}{6}$

Mean $=\Sigma xp=7$

Variance $=\Sigma x_i^2p-(\text{mean})^2$

$$=\left(\frac{364}{6}\right)-(7)^2$$

$$=\frac{364-294}{6}$$

$$=\frac{70}{6}$$

$\Rightarrow$ Variance $=11.7$

56. (a) Probability of even number $=P(E)=\frac{3}{6}=\frac{1}{2}$

and probability of odd number $=\frac{3}{6}=1/2$

x or x_i	P(x) or p_i	x_ip_i	$x_i^2p_i$
1	$\frac{1}{2}$	$\frac{1}{2}$	$\frac{1}{2}$
3	$\frac{1}{2}$	$\frac{3}{2}$	$\frac{9}{2}$
		$\Sigma xp=2$	$\Sigma x^2p=5$

Mean $=\Sigma xp=2$

Variance $=\Sigma x^2p-(\text{mean})^2=5-4$

$$=1$$

$\therefore$ Standard deviation $=\sqrt{\text{Variance}}$

$$=\sqrt{1}$$

$$=1$$

57. (b) Given, $n=5$ and $r\geq 0$

$$P(X=r)={}^nC_r p^{n-r}q^r$$

$$P(X\geq 1)=1-P(X=0)$$

$$=1-{}^5C_0 p^5 q^0$$

$$=1-p^5\geq \frac{31}{32} \quad \text{[given]}$$

$$\Rightarrow \quad p^5\leq 1-\frac{31}{32}=\frac{1}{32}=\left(\frac{1}{2}\right)^5$$

$$\therefore \quad p\leq \frac{1}{2} \text{ and } p\geq 0$$

$$\Rightarrow \quad p\in\left[0,\frac{1}{2}\right]$$

58. (a) Since the 3 boxes are identical, probability of selecting a box is $\frac{1}{3}$.

$$\Rightarrow \quad P=\frac{1}{3}, q=1-\frac{1}{3}=\frac{2}{3}$$

Probability that the selected box should contain exactly 3 ball is the same as probability of exactly 3 success in 12 trials.

$$\therefore \text{Required probability}={}^{12}C_3\left(\frac{1}{3}\right)^3\left(\frac{2}{9}\right)^9$$

$$=\frac{55}{3}\left(\frac{2}{3}\right)^{11}$$

59. (a) Let the probability of success and failure be p and q respectively, then

$$p=2q \text{ and } p+q=1$$

$$\Rightarrow \quad 3q=1\Rightarrow q=\frac{1}{3}$$

$$\therefore \quad p=\frac{2}{3}$$

Required probability

$$={}^6C_4\left(\frac{2}{3}\right)^4\left(\frac{1}{3}\right)^2+{}^6C_5\left(\frac{2}{3}\right)^5\left(\frac{1}{3}\right)+{}^6C_6\left(\frac{2}{3}\right)^6$$

$$=\frac{496}{729}$$

60. (c) Mean of Binomial variable, $np=8$

Variance of Binomial variable, $npq=4$

$$\therefore \quad q=\frac{1}{2} \text{ and } p=1-q=1-\frac{1}{2}=\frac{1}{2}$$

and $n\left(\frac{1}{2}\right)=8\Rightarrow n=16$

$$\therefore P(X<3)=P(X=0)+P(X=1)+P(X=2)$$

$$={}^{16}C_0\left(\frac{1}{2}\right)^0\left(\frac{1}{2}\right)^{16}+{}^{16}C_1\left(\frac{1}{2}\right)\left(\frac{1}{2}\right)^{15}$$

$$+{}^{16}C_2\left(\frac{1}{2}\right)^2\left(\frac{1}{2}\right)^{14}$$

$$=\left(\frac{1}{2}\right)^{16}+16\left(\frac{1}{2}\right)^{16}+120\left(\frac{1}{2}\right)^{16}$$

$$=\frac{137}{2^{16}}$$

61. (c) Here, $p=\frac{1}{3}, q=1-p=1-\frac{1}{3}=\frac{2}{3}$

Probability of guessing a 4 or more correct answer

$$={}^5C_4\left(\frac{1}{3}\right)^4\left(\frac{2}{3}\right)+{}^5C_5\left(\frac{1}{3}\right)^5$$

$$=5\times\frac{2}{3^5}+\frac{1}{3^5}=\frac{11}{3^5}$$

62. (b) Let n denote the required number of shots and x the number of shots that hit the target.

Then, $x\sim B(n,p)$ with $p=\frac{1}{4}$.

Now, $P(x\geq 1)\geq 0.9$

$$\Rightarrow \quad 1-P(x=0)\geq 0.9$$

$$\Rightarrow \quad 1-{}^nC_0\left(\frac{3}{4}\right)^n\geq 0.9$$

$$\Rightarrow \quad \left(\frac{3}{4}\right)^n\leq\frac{1}{10}$$

$$\Rightarrow \quad \left(\frac{4}{3}\right)^n\geq 10$$

$$\Rightarrow \quad n(\log 4-\log 3)\geq 1 \quad \text{[base of logarithm is 10]}$$

$$\Rightarrow n(0.602-0.4777)\geq 1$$

$$\Rightarrow \quad n\geq\frac{1}{0.125}=8$$

63. (d) Let $x\sim B(100,p)$ be the number of coins showing heads and let $q=1-p$.

Then, since $P(x=51)=P(x=50)$, we have

$${}^{100}C_{51}(p^{51})(q^{49})={}^{100}C_{50}(p^{50})(q^{50})$$

$$\Rightarrow \quad \frac{p}{q}=\left(\frac{100!}{50!50!}\right)\left(\frac{51!49!}{100!}\right)$$

$$\Rightarrow \quad \frac{p}{1-p}=\frac{51}{50}$$

$$\Rightarrow \quad 50p=51-51p$$

$$\Rightarrow \quad p=\frac{51}{101}$$

64. (a) $\frac{P(x=r)}{P(x=n-r)}=\frac{{}^nC_r p^r(1-p)^{(n-r)}}{{}^nC_{n-r}\,p^{n-r}(1-p)^r}$

$$=\frac{(1-p)^{n-2r}}{p^{n-2r}}$$

$$=\left(\frac{1-p}{p}\right)^{n-2r}=\left(\frac{1}{p}-1\right)^{n-2r}$$

$\therefore$ Ratio will be independent of n and r, if

$$\left(\frac{1}{p}\right)-1=1$$

$$\Rightarrow \quad p=\frac{1}{2}$$

65. (b) Probability that not more than two heads or 2 tails appears

$$=\left(\frac{1}{5}\right)^5+{}^5C_1\left(\frac{1}{2}\right)\left(\frac{1}{2}\right)^4+{}^5C_2\left(\frac{1}{2}\right)^2\left(\frac{1}{2}\right)^3=\frac{1}{2}$$

$q=$ probability that not less than 3 heads appears
$=$ probability that not more than 2 tails appear
$=p$

Hence, $q=p$.

66. (c) The probability p of getting a white ball in a single draw $\frac{3}{9}=\frac{1}{3}$. If X is the number of white balls drawn, then $x \sim B(n,p)$, with $n=4$. Therefore, the probability of getting exactly one white ball is

$$P(x=1)={}^4C_1pq^3=4\left(\frac{1}{3}\right)\left(\frac{2}{3}\right)^3=\frac{32}{81}$$

67. (c) Desired probability

= Probability of getting 3 sixes in 1st 9 thrown
× getting 6 in the 10 throw

$$={}^9C_3\left(\frac{1}{6}\right)^3\left(\frac{5}{6}\right)^6\times\frac{1}{6}=\frac{84\times5^6}{6^{10}}$$

68. (a) We have,

$$P(|x-4|\le 2)=P(-2\le x-4\le 2)=P(2\le x\le 6)$$
$$=1-[P(x=0)+P(x=1)+P(x=7)+P(x=8)]$$
$$=1-\left[{}^8C_0\left(\frac{1}{2}\right)^8+{}^8C_1\left(\frac{1}{2}\right)^8+{}^8C_7\left(\frac{1}{2}\right)^8+{}^8C_8\left(\frac{1}{2}\right)^8\right]$$
$$=1-\left(\frac{1}{2}\right)^8(1+8+8+1)$$
$$=1-\frac{18}{2^8}$$
$$=\frac{119}{128}$$

69. (b) Let X be the number of heads getting in n tossed. Therefore, X follows Binomial distributions

$n, p=\frac{1}{2}, q=\frac{1}{2}$.

$$\because \quad P(X\ge 1)>0.8$$
$$1-P(X=0)>0.8$$
$$P(X=0)<0.2$$
$${}^nC_0\left(\frac{1}{2}\right)^0\left(\frac{1}{2}\right)^n<0.2$$
$$\Rightarrow \quad \left(\frac{1}{2}\right)^n<\frac{1}{5}$$
$$\Rightarrow \quad 2^n>5$$

Hence, the least value of n is 3.

70. (a) According to the condition,

$$1-\left(\frac{3}{4}\right)^n\ge\frac{9}{10}$$
$$\Rightarrow \quad \left(\frac{3}{4}\right)^n\le 1-\frac{9}{10}=\frac{1}{10}$$
$$\Rightarrow \quad \left(\frac{4}{3}\right)^n\ge 10$$
$$\Rightarrow n[\log_{10}4-\log_{10}3]\ge\log_{10}10=1$$
$$\Rightarrow \quad n\ge\frac{1}{\log_{10}4-\log_{10}3}$$

71. (b) Mean $=np=4$, Variance $=npq=3$

$$\Rightarrow \quad 4(q)=3$$
$$\Rightarrow \quad q=\frac{3}{4}\text{ and }n=16, p=\frac{1}{4}$$

Now, $P(X\ge 1)=1-P(X=0)=1-{}^nC_0p^0q^{n-0}$

$$=1-\left(\frac{3}{4}\right)^{16}$$

72. (a) Standard deviation $\sigma=\sqrt{npq}\ge 0$

Now, mean $=np=25$ and $q<1$

So, $\sigma=\sqrt{npq}<\sqrt{np}=5$

$\therefore \quad 0\le\sigma<5$

73. (b) Let $E=$ Event of getting an even numbers $=\{2,4,6\}$

$$n(E)=3$$

$\therefore$ Probability of success, $p=\frac{3}{6}=\frac{1}{2}$

and probability of failure, $q=\frac{1}{2}$

$\therefore$ Variance $=npq=100\times\frac{1}{2}\times\frac{1}{2}=25$

74. (b) $p=\frac{15}{25}, q=\frac{10}{25}$

$$n=10$$
$$\sigma^2=npq=10\cdot\frac{3}{5}\cdot\frac{2}{5}=\frac{12}{5}$$

75. (d) We have, mean $=np=2$

Variance $=npq=1$

$$\Rightarrow \quad q=\frac{1}{2}\text{ and }p=\frac{1}{2}$$

According to the given,

$$n\cdot\frac{1}{2}=2\Rightarrow n=4$$

Thus, $P(x \geq 1) = 1 - P(x = 0)$

$$= 1 - {}^4C_0\left(\frac{1}{2}\right)^4$$

$$= 1 - \frac{1}{16} = \frac{15}{16}$$

76. (2) Total number of cases $= n(S) = 6!$

Now, sum the given digits is $1+2+3+4+5+6=21$, which is divisible by 3.

Now, we have to form the number which is divisible by 6, then we have to ensure that the digit in unit place is even.

$\Rightarrow$ Favourable cases $= n(A) = 3 \cdot 5!$

Hence, $P(A) = \frac{3 \cdot 5!}{6!} = \frac{1}{2}$

77. (6) Total number of cases $n(S) = 6^3 = 216$

Product is prime only when two outcomes are 1 and the 3rd is prime, i.e. 2, 3, 5.

If it is 2, 1, 1, then the number of cases is 3.

Similarly, 3 cases for 3, 1, 1 and 5, 1, 1 each.

Hence, favourable cases = 9

Hence, required probability $(p) = \frac{1}{24}$

$$\Rightarrow \quad \frac{1}{4p} = 6$$

78. (4) Let the number of green socks be $x > 0$.

E : 2 socks drawn are of the same colour

$$\Rightarrow \quad P(E) = P(RR \text{ or } BB \text{ or } WW \text{ or } GG)$$

$$= \frac{3}{{}^{6+x}C_2} + \frac{{}^xC_2}{{}^{6+x}C_2}$$

$$= \frac{6}{(x+6)(x+5)} + \frac{x(x-1)}{(x+6)(x+5)} = \frac{1}{5} \text{ (given)}$$

$$\Rightarrow \quad 5(x^2 - x + 6) = x^2 + 11x + 30$$

$$\Rightarrow \quad 4x^2 - 16x = 0$$

$$\Rightarrow \quad x = 4$$

79. (7) Let the probability of the faces 1, 3, 5 or 6 be p for each face.

Hence, probability of each of the faces 2 or 4 is $3p$.

Now, according to the question,

$$4p + 6p = 1$$

$$\Rightarrow \quad p = \frac{1}{10}$$

$$\therefore \quad P(1) = P(3) = P(5) = P(6) = \frac{1}{10}$$

$$\text{and} \quad P(2) = P(4) = \frac{3}{10}$$

$\Rightarrow$ Required probability

$$p = P \text{ (Total of 7 with a draw of dice)}$$

$$= P(16, 61, 25, 52, 43, 34)$$

$$= 2\left(\frac{1}{10} \cdot \frac{1}{10}\right) + 2\left(\frac{3}{10} \cdot \frac{1}{10}\right) + 2\left(\frac{3}{10} \cdot \frac{1}{10}\right)$$

$$= \frac{2+6+6}{100}$$

$$= \frac{14}{100} = \frac{7}{50}$$

80. (4) Total ways of distribution $= n(S) = 4^5$

Total ways of distribution so that each child get at least one game

$$n(E) = 4^5 - {}^4C_1 3^5 + {}^4C_2 2^5 - {}^4C_3$$

$$= 1024 - 4 \times 243 + 6 \times 32 - 4$$

$$= 240$$

Required probability $p = \frac{n(E)}{n(S)} = \frac{240}{4^5} = \frac{15}{64}$

81. (1) There are n white balls in the urn.

$\Rightarrow$ Probability of Mr. A to draw 2 balls of same color is

$$\frac{{}^3C_2 + {}^nC_2}{{}^{n+3}C_2} = \frac{1}{2} \text{ (given)}$$

$$\text{or} \quad \frac{6 + n(n-1)}{(n+3)(n+2)} = \frac{1}{2}$$

$$\text{or} \quad n^2 - 7n + 6 = 0$$

$$\Rightarrow \quad n = 1 \text{ or } 6 \quad \text{...(i)}$$

Also, required probability for Mr. B according to the question is

$$\frac{3}{n+3}\frac{3}{n+3} + \frac{n}{n+3}\frac{n}{n+3} = \frac{5}{8} \text{ (given)}$$

Solving, we get $n^2 - 10n + 9 = 0$; $n = 1$ or 9 ...(ii)

From Eqs. (i) and (ii),

$$n = 1$$

82. (7) Number of ways of drawing 7 balls $= {}^{10}C_7$

For each set of 7 balls of the 2nd draw, 3 must be common to the set of 5 balls of the first draw, i.e. 2 other balls can be drawn in ${}^3C_2$ ways.

Thus, for each set of 7 balls of the 2nd draw, there are ${}^7C_3 \times {}^3C_2$ ways of making the 1st draw so that there are 3 balls common.

$\therefore$ Probability of getting 3 balls in common

$$= \frac{{}^7C_3 \times {}^3C_2}{{}^{10}C_7} = \frac{7}{8}$$

83. (5) $P(n) = Kn^2$

Given, $P(1) = K, P(2) = 2^2K, P(3) = 3^2K, P(4) = 4^2K$

$$P(5) = 5^2K, P(6) = 6^2K$$

So, total probability $= 91K$

$$\Rightarrow \quad 1 = 91K$$

$\therefore \quad K = \frac{1}{91}$

Therefore, $P(1) = \frac{1}{91}; P(2) = \frac{4}{91}$ and so on.

Let 3 events A, B and C be defined as

$A : a < b$

$B : a = b$

$C : a > b$

By symmetry, $P(A) = P(C)$

Also, $P(A) + P(B) + P(C) = 1$

Since, $P(B) = \sum_{i=1}^{6} [P(i)]^2$

$$= \left[\frac{1+16+81+256+625+1296}{91 \times 91}\right]$$

$$P(B) = \frac{2275}{91 \times 91} = \frac{25}{91}$$

Now, $P(A) + P(B) + P(C) = 1$

$P(A) + P(B) + P(A) = 1 \quad [\because P(A) = P(C)]$

$$\Rightarrow \quad P(A) = \frac{1}{2}[1 - P(B)]$$

$$= \frac{1}{2}\left[1 - \frac{25}{91}\right]$$

$$= \frac{33}{91}$$

$$\therefore \quad P(A) = \frac{33}{91}$$

$$\therefore \quad \left[\frac{2}{P}\right] = \left[\frac{182}{33}\right] = 5$$

84. (2) When A and B are mutually exclusive

$P(A \cup B) = 0.8$

$\therefore \quad P(A \cup B) = P(A) + P(B) \quad$ ...(i)

$\Rightarrow \quad 0.8 = 0.5 + p$

or $\quad p = 0.3 \quad$ ...(ii)

$P(A \cup B) = P(A) + P(B) - P(A \cup B)$

$= P(A) + P(B) - P(A) P(B)$

or $\quad 0.8 = 0.5 + q - (0.5) q$

or $\quad 0.3 = \frac{q}{2}$

$\Rightarrow \quad q = 0.6 \quad$ ...(iii)

$\Rightarrow \quad \frac{q}{p} = 2$

85. (6) $P\left(\frac{A}{B}\right) = \frac{P(A \cap B)}{P(B)} = \frac{0.6}{0.8}$

(Maximum value of $P(A \cap B) = P(A) = 0.6$)

$= \frac{3}{4}$

86. (2) Here, $P(E) = \frac{1}{2}$ and $P(F_k) = {}^nC_k \cdot \frac{1}{2^n}$

Also, $P(E \cap F_k) = P$ (Exactly K heads are obtained and head obtained in 1st flip)

$$= \frac{1}{2} \cdot {}^{n-1}C_{k-1}\left(\frac{1}{2}\right)^{n-1}$$

Events E and F_k are independent.

Therefore, $P(E \cap F_k) = P(E) \cdot P(F_k)$

or $\quad {}^{n-1}C_{k-1} \times \frac{1}{2^n} = \frac{1}{2} \times {}^nC_k \frac{1}{2^n}$

or $\quad 2 \times {}^{n-1}C_{k-1} = {}^nC_k$

or $\quad n = 2k$

87. (3) We know that sum of probabilities distribution is 2.

$\therefore \; k + 2k + 3k + 4k = 1$

$\Rightarrow \quad k = \frac{1}{10}$

Now, mean of

$\bar{X} = k \times 1 + 2 \times 2k + 3 \times 3k + 4 \times 4k$

$= 30k$

$= 30 \times \frac{1}{10} = 3$

88. (0.333) Here, $n = 6$

$\because \quad P(X = 2) = 4P(X = 4)$

${}^6C_2 p^2 q^4 = 4 \cdot {}^6C_4 p^4 q^2$

$\Rightarrow \quad q^2 = 4p^2$

$\Rightarrow \quad (1 - p)^2 = 4p^2$

$\Rightarrow \quad 3p^2 + 2p - 1 = 0$

$\Rightarrow \quad (p + 1)(3p - 1) = 0$

$\Rightarrow \quad p = \frac{1}{3} = 0.333 \quad [\because p$ cannot be negative$]$

Level 3

1. (a) Let the side of small cube $= x$

$\Rightarrow \quad 64x^3 = a^3$

$\therefore \quad x = \frac{a}{4}$

There are 12 edges and from each edge, we get 2 cubes with only two coloured faces.

$\therefore$ Faces $= 12 \times 2 = 24$

$n(S) = {}^{64}C_2$ and $n(E) = {}^{24}C_2$

$\because$ Required probability

$$P(E) = \frac{n(E)}{n(S)} = \frac{{}^{24}C_2}{{}^{64}C_2}$$

$$= \frac{24 \times 23}{64 \times 63} = \frac{23}{168}$$

2. (a) We have 3 boys and 2 girls

1 B 2 B 3 B 4

Girls can not occupy the 4the position. Either girls can occupy 2 of 1, 2, 3 position or they can be at position of 1 or 2.

Hence, the total number of ways in which a girl can be seated, is

$$({}^3C_2 \times 2! \times 3!) + {}^2C_1 \times 2! \times 3! = 36 + 24$$
$$= 60$$

and $n(S) = 5! = 120$

$\therefore$ Required probability $= \frac{60}{120} = \frac{1}{2}$

3. (a) We have 12 balls and 3 boxes.

Since, each ball can be placed in anyone of the 3 boxes. Therefore, there are 3 ways in which a ball can be placed in anyone of the three boxes. Thus, there are 3^{12} ways in which 12 balls can be placed in 3 boxes. The number of ways in which 3 balls out of 12 can be put in the box is ${}^{12}C_3$, the remaining 9 balls can be placed in 2 boxes in 2^9 ways.

$\therefore$ Required probability $= \frac{{}^{12}C_3}{3^{12}} \cdot 2^9 = \frac{110}{9}\left(\frac{2}{3}\right)^{10}$

4. (c) $P_1, P_2, P_3 \ldots P_8$. Given that if P_i, P_j play with $i < j$, then P_i will win for the first round P_4 should be paired with any from P_5 to P_8. It can be done in ${}^4C_1$ ways. Then, P_4 to be finalist at least one player from P_5 to P_8 should reach in the second round. Therefore, one pair should be from remaining 3 from P_5 to P_8 in ${}^3C_2$. Then, favourable pairing in first round is ${}^4C_1 \cdot {}^3C_2 \cdot {}^3C_2$. Then in the second round, we have four players. Favourable ways is 1.

Now, total possible pairing is

$$\frac{{}^8C_2 \times {}^6C_2 \times {}^4C_2 \times {}^2C_2}{4!} \times \frac{{}^4C_2 \times {}^2C_2}{2!}$$

$\therefore$ Required probability

$$= \frac{{}^4C_1 \times {}^3C_2 \times {}^3C_2 \times 4! \times 2!}{{}^8C_2 \times {}^6C_2 \times {}^4C_2 \times {}^2C_2 \times {}^2C_2 \times {}^4C_2}$$
$$= \frac{4}{35}$$

5. (b) We have three married couples i.e. 6 persons

$\therefore$ Total number of cases $= 6! = 720$

Number of favourable case = Total number of cases

– Cases when atleast one husband sits with his wife

$$= 6! - [{}^3C_1 \times 2! \times 5! - {}^3C_2 \times 2! \times 2! \times 4! + {}^3C_3 \times 2! \times 2! \times 3! \times 2!]$$
$$= 720 - [720 - 288 + 48] = 240$$

$\therefore$ Required probability $= \frac{240}{720} = \frac{1}{3}$

6. (a) Clearly, the number of outcomes in sample space of the given experiment is $6 \times 6 \times 6 \times 6 = 6^4$

Now, let us favourable cases D_4 shows a number, then only 1 of $D_1D_2D_3$ shows some number or only 2 of $D_1D_2D_3$ shows some number or all 3 of $D_1D_2D_3$ shows some number.

$\therefore$ Number of favourable cases

$$= {}^6C_1({}^3C_1 \times 5 \times 5 + {}^3C_2 \times 5 + {}^3C_3)$$
$$= 6(75 + 15 + 1) = 6 \times 91$$

$\therefore$ Required probability $= \frac{6 \times 91}{6^4} = \frac{91}{6^3} = \frac{91}{216}$

7. (c) We have, $f(x) = x^3 + ax^2 + bx + c$

$$f'(x) = 3x^2 + 2ax + b$$

Now, $f(x)$ is increasing function.

$\therefore f'(x) \geq 0 \; \forall \; x \; f'(x) = 0$ should not form an interval.

For $f'(x) \geq 0$

$$D \leq 0$$
$$\therefore \quad 4a^2 - 12b \leq 0$$
$$\Rightarrow \quad a^2 - 3b \leq 0$$

This is true for exactly 16 ordered pairs $(a, b), 1 \leq a, b \leq 6$ are (1, 1,), (1, 2), (1, 3), (1, 4), (1, 5), (1, 6), (2, 2), (2, 3), (2, 4), (2, 5), (2, 6), (3, 3), (3, 4), (3, 5), (3, 6), (4, 6)

$\therefore$ Required probability $= \frac{16}{36} = \frac{4}{9}$

8. (b) We have, $A = \{0, 1, 2, 3, \ldots, 10\}$

$$n(A) = 11$$

and $\quad n(S) =$ Two numbers taken from A

$$= {}^{11}C_2 = 55$$

Let E be given event such that

$$(a + b) = 4n \text{ and } |a - b| = 4n$$

$\therefore$ Outcomes of $E = \{(0, 4), (0, 8), (2, 6), (2, 10), (4, 8), (6, 10)\}$

$$n(E) = 6$$

$\therefore$ Required probability, $P(E) = \frac{n(E)}{n(S)} = \frac{6}{55}$

9. (a) The probability of getting a head in single toss of a coin, $\quad p = \frac{1}{2}$

The probability of getting 5 or 6 in a single throw of a die, $\quad q = \frac{2}{6} = \frac{1}{3}$

$\therefore$ Required probability

$$= p + (1 - p)(1 - q)p + (1 - p)(1 - q)(1 - p)(1 - q)p + \ldots$$
$$= \frac{1}{2} + \left(\frac{1}{2}\right)\left(\frac{2}{3}\right)\left(\frac{1}{2}\right) + \left(\frac{1}{2}\right)\left(\frac{2}{3}\right)\left(\frac{1}{2}\right)\left(\frac{2}{3}\right)\left(\frac{1}{2}\right) + \ldots$$
$$= \frac{1}{2} + \frac{1}{3} \times \frac{1}{2} + \left(\frac{1}{3}\right)^2 \times \frac{1}{2} + \left(\frac{1}{3}\right)^3 \times \frac{1}{2} + \ldots$$

$$=\frac{1}{2}\left(1+\frac{1}{3}+\frac{1}{3^2}+\frac{1}{3^3}+\ldots\right)$$

$$=\frac{1}{2}\times\frac{1}{1-\frac{1}{3}}=\frac{3}{4}$$

10. (c) In any trial P(getting white ball),

$$P(W)=\frac{1}{12} \text{ and } P(B)=\frac{1}{12}$$

The event will occur, if in the first six trial, 3 white balls are drawn in any of the 3 trials from six.

For remaining three trials, black must be drawn.

Three trials in which black balls, 3 is drawn can be selected in ${}^6C_3=20$ ways

∴ Required probability $=P(WWW\ BBB\ldots)\times P(W)$

$$=20\times\left(\frac{1}{2}\right)^3\left(\frac{1}{2}\right)^3\times\frac{1}{2}$$

$$=\frac{5}{32}$$

11. (a) Let A denotes the event that a sum of 5 occurs, B the event that a sum of 7 occurs and C the event that neither a sum of 5 nor a sum of 7 occurs.

We have,

$$P(A)=\frac{1}{9}, P(B)=\frac{1}{6} \text{ and } P(C)=\frac{13}{18}$$

Thus, required probability that A occurs before B is

$P\{A \text{ or } (C\cap A)+(C\cap C\cap A) \text{ or}\ldots\}$

$$=P(A)+P(C\cap A)+P(C\cap C\cap A)\ldots$$

$$=P(A)+P(C)\cdot P(A)+\{P(C)\}^2\cdot P(A)\ldots$$

$$=\frac{1}{9}+\frac{1}{9}\times\frac{13}{18}+\frac{1}{9}\times\left(\frac{13}{18}\right)^2+\ldots.$$

$$=\frac{1}{9}\left(1+\frac{13}{18}+\left(\frac{13}{18}\right)^2+\ldots\right)$$

$$=\frac{1}{9}\times\frac{1}{1-\frac{13}{18}}=\frac{2}{5}$$

12. (b)

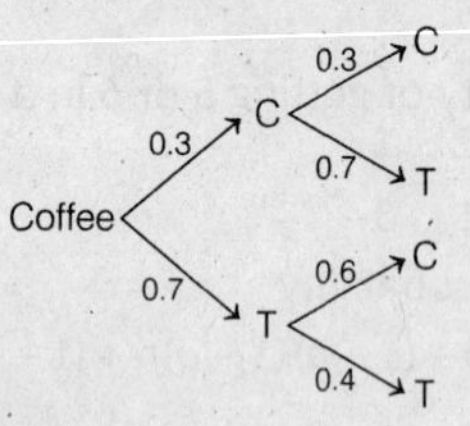

Required probability

$$P(E)=P(CCT) \text{ or } P(CTT)$$

$$=(0.3)(0.7)+(0.7)(0.4)$$

$$=0.21+0.28$$

$$=0.49$$

13. (a) Given, families have one, two or three children. A family have four grand children is possible, if they have two or three children.

Now, two children have three possible cases (2, 2), (1, 3) and (3, 1) and three children have three possible cases (1, 1, 2), (1, 2, 1) and (2, 1, 1).

∴ Required probability

$$=\frac{1}{2}\times\frac{1}{2}\times\frac{1}{2}+\frac{1}{2}\times\frac{1}{4}\times\frac{1}{4}+\frac{1}{2}\times\frac{1}{4}\times\frac{1}{4}$$

$$+\frac{1}{4}\times\frac{1}{4}\times\frac{1}{4}\times\frac{1}{2}+\frac{1}{4}\times\frac{1}{4}\times\frac{1}{2}\times\frac{1}{4}+\frac{1}{4}\times\frac{1}{4}\times\frac{1}{4}\times\frac{1}{4}$$

$$=\frac{1}{8}+\frac{1}{32}+\frac{1}{32}+\frac{1}{128}+\frac{1}{128}+\frac{1}{128}=\frac{27}{128}$$

14. (c) Consider the event

$G=$ Original signal is green

$E_1=A$ receives the signal correct

$E_2=B$ receives the signal correct

$E=$ Signal received is green

$$P(E_1)=\frac{3}{4}, P(E_2)=\frac{3}{4}, P(G)=\frac{4}{5}$$

$$P(E)=P(GE_1E_2)+P(G\bar{E}_1\bar{E}_2)+P(\bar{G}E_1\bar{E}_2)+P(\bar{G}\bar{E}_1E_2)$$

$$=\frac{4}{5}\times\frac{3}{4}\times\frac{3}{4}+\frac{4}{5}\times\frac{1}{4}\times\frac{1}{4}+\frac{1}{5}\times\frac{3}{4}\times\frac{1}{4}+\frac{1}{5}\times\frac{1}{4}\times\frac{3}{4}$$

$$=\frac{36+4+3+3}{80}$$

$$=\frac{46}{80}=\frac{23}{40}$$

∴ Required probability

$$P\left(\frac{G}{E}\right)=\frac{[P(GE_1E_2)+P(G\bar{E}_1\bar{E}_2)]}{P(E)}$$

$$=\frac{\frac{18}{40}+\frac{2}{40}}{\frac{23}{40}}=\frac{20}{23}$$

15. (a) Let $P(A)=\frac{1}{2}, P(B)=\frac{3}{4}, P(C)=\frac{1}{4}$ and $P(D)=\frac{1}{8}$

$$\therefore \quad P(\bar{A})=\frac{1}{2}, P(\bar{B})=\frac{1}{4}, P(\bar{C})=\frac{3}{4} \text{ and } P(\bar{D})=\frac{7}{8}$$

Probability that the problem is solved correctly by atleast one of them

$=1-$ Problem is not solved by all

∴ Required probability

$=1-P$ (problem is not solved)

$$=1-P(\bar{A})\cdot P(\bar{B})\cdot P(\bar{C})\cdot P(\bar{D})$$

$$=1-\frac{1}{2}\times\frac{1}{4}\times\frac{3}{4}\times\frac{7}{8}$$

$$=1-\frac{21}{256}$$

$$=\frac{256-21}{256}=\frac{235}{256}$$

16. (a) Let E_i denotes the event that the bag contains i black and $(10-i)$ white balls, $i = 0, 1, 2, ..., 10$. Let A denotes the event that the three balls drawn at random from the bag are black. We have,

$$P(E_i) = \frac{1}{11} (i = 0, 1, 2, ..., 10)$$

$$P\left(\frac{A}{E_i}\right) = 0 \text{ for } i = 0,\ 1,\ 2, \text{ and } P\left(\frac{A}{E_i}\right) = \frac{{}^iC_3}{{}^{10}C_3} \text{ for } i \geq 3$$

$$\Rightarrow \quad P(A) = \frac{1}{11} \times \frac{1}{{}^{10}C_3} ({}^3C_3 + {}^4C_3 + \ldots {}^{10}C_3)$$

$$\Rightarrow \quad P(A) = \frac{1}{11} \times \frac{1}{{}^{10}C_3} \times {}^{11}C_4$$

$$[\because {}^3C_3 + {}^4C_3 \ldots + {}^{10}C_3 = {}^{11}C_4]$$

$$\Rightarrow \quad P(A) = \frac{{}^{11}C_4}{11 \times {}^{10}C_3} = \frac{1}{4}$$

$$\therefore \text{ Required probability} = \frac{P(E_9) P\left(\frac{A}{E_9}\right)}{P(A)} = \frac{\frac{1}{11} \times \frac{{}^9C_3}{{}^{10}C_3}}{\frac{1}{4}} = \frac{14}{55}$$

17. (c) Let X be the numbers of head obtained by A and Y be the number of head obtained by B. Note that X and Y are binomial variate with parameter $n = 3$ and $P = \frac{1}{2}$.

Probability that both A and B obtain the same number of head is

$$\therefore \quad P(X=0)P(Y=0) + P(X=1)P(Y=1) + P(X=2)P(Y=2) + P(X=3)P(Y=3)$$

$$= \left[{}^3C_0\left(\frac{1}{2}\right)^3\right]^2 + \left[{}^3C_1\left(\frac{1}{2}\right)^3\right]^2 + \left[{}^3C_2\left(\frac{1}{2}\right)^3\right]^2 + \left[{}^3C_3\left(\frac{1}{2}\right)^3\right]^2$$

$$= \left(\frac{1}{2}\right)^6 [1+9+9+1] = \frac{20}{64} = \frac{5}{16}$$

18. (c) Let $X = \{x_1, x_2, x_3, \ldots x_n\}$

For each $x_i \in X (1 \leq i \leq n)$, we have the following four choices

(i) $x_i \in A$ and $x_i \in B$

(ii) $x_i \in A$ and $x_i \notin B$

(iii) $x_i \notin A$ and $x_i \in B$

(iv) $x_i \notin A$ and $x_i \notin B$

Thus, the number of ways of choosing A and B is 4^n

Out of the four choices listed above (iv) is not favourable for the occurrence of $A \cup B = X$.

Therefore, the number of favourable ways is 3^n.

$$\text{Hence, required probability} = \frac{3^n}{4^n} = \left(\frac{3}{4}\right)^n$$

19. (c) Let E_1 and E_2 denote the event that P_1 and P_2 are paired and not paired together and let A denotes the event that one of two players P_1 and P_2 is among the winner.

Since, P_1 can be paired with any of the remaining 15 players.

We have, $P(E_1) = \frac{1}{15}$

and $P(E_2) = 1 - P(E_1) = 1 - \frac{1}{15} = \frac{14}{15}$

In case E_1 occurs, it is certain that one of P_1 and P_2 will among the winners. In case E_2 occurs, the probability that exactly one of P_1 and P_2 is among the winner

$$P[(P_1 \cap \overline{P}_2) \cup (\overline{P}_1 \cap P_2)] = P(P_1 \cap \overline{P}_2) + P(\overline{P}_1 \cap P_2)$$

$$= \frac{1}{2}\left(1 - \frac{1}{2}\right) + \left(1 - \frac{1}{2}\right)\frac{1}{2} = \frac{1}{2}$$

i.e. $P\left(\frac{A}{E_1}\right) = 1$ and $P\left(\frac{A}{E_2}\right) = \frac{1}{2}$

$$\therefore \text{ Required probability } P(A) = P(E_1)P\left(\frac{A}{E_1}\right) + P(E_2)P\left(\frac{A}{E_2}\right)$$

$$= \frac{1}{15} \times 1 + \frac{14}{15} \times \frac{1}{2} = \frac{2+14}{30} = \frac{16}{30} = \frac{8}{15}$$

20. (a) Consider the event

$$E_1 = A \cap \overline{B}; P(E_1) = \frac{1}{8} \times \frac{11}{12}$$

$$E_2 = \overline{A} \cap B; P(E_2) = \frac{7}{8} \times \frac{1}{12}$$

$$E_3 = A \cap B; P(E_3) = \frac{1}{8} \times \frac{1}{12}$$

$$E_4 = \overline{A} \cap \overline{B}; P(E_4) = \frac{7}{8} \times \frac{11}{12}$$

E = They obtained same results

$$P\left(\frac{E}{E_1}\right) = 0, P\left(\frac{E}{E_2}\right) = 0, P\left(\frac{E}{E_3}\right) = 1$$

$$P\left(\frac{E}{E_4}\right) = \frac{1}{1001}$$

Required probability

$$P\left(\frac{E_3}{E}\right)=\frac{P(E_3)\times P\left(\frac{E}{E_3}\right)}{P(E_3)P\left(\frac{E}{E_3}\right)+P\left(\frac{E}{E_4}\right)\times P(E_4)}$$

$$=\frac{\frac{1}{8}\times\frac{1}{12}}{\frac{1}{8}\times\frac{1}{12}+\frac{7}{8}\times\frac{11}{12}\times\frac{1}{1001}}=\frac{13}{14}$$

21. (c) Let A be the event that the machine produces 2 acceptable items.
Also, let B be the event of correct set up and B_2 represents the event of incorrect set up.

$$P(B_1)=0.8 \text{ and } P(B_2)=0.2$$

$$P\left(\frac{A}{B_1}\right)=0.9\times 0.9 \text{ and } P\left(\frac{A}{B_2}\right)=0.4\times 0.4$$

$$\therefore\quad P\left(\frac{B_1}{A}\right)=\frac{P(B_1)\cdot P\left(\frac{A}{B_1}\right)}{P(B_1)\cdot P\left(\frac{A}{B_1}\right)+P(B_2)\cdot P\left(\frac{A}{B_2}\right)}$$

$$=\frac{0.8\times 0.9\times 0.9}{0.8\times 0.9\times 0.9+0.2\times 0.4\times 0.4}$$

$$=\frac{648}{680}=0.95$$

22. (c) $P(T_1)=\frac{20}{100}, P(T_2)=\frac{80}{100}, P(D)=\frac{7}{100}$

$$P\left(\frac{D}{T_2}\right)=P,$$

then $P\left(\frac{D}{T_1}\right)=10P$

By total probability

$$P(D)=P(T_1)\cdot P\left(\frac{D}{T_1}\right)+P(T_2).P\left(\frac{D}{T_2}\right)$$

$$\frac{7}{100}=\frac{20}{100}\times 10P+\frac{80}{100}\times P$$

On solving, $P=\frac{1}{40}$

$$\therefore\quad P\left(\frac{D}{T_1}\right)=\frac{10}{40} \text{ and } P\left(\frac{D}{T_2}\right)=\frac{1}{40}$$

$$\Rightarrow\quad P\left(\frac{\overline{D}}{T_1}\right)=1-\frac{10}{40}=\frac{30}{40}, P\left(\frac{\overline{D}}{T_2}\right)=1-\frac{1}{40}=\frac{39}{40}$$

$$P\left(\frac{T_2}{\overline{D}}\right)=\frac{P\left(\frac{\overline{D}}{T_2}\right)\cdot P(T_2)}{P\left(\frac{\overline{D}}{T_1}\right)\cdot P(T_1)+P\left(\frac{\overline{D}}{T_2}\right)\cdot P(T_2)}$$

$$=\frac{\frac{80}{100}\times\frac{39}{40}}{\frac{20}{100}\times\frac{30}{40}+\frac{80}{100}\times\frac{39}{40}}=\frac{156}{186}=\frac{78}{93}$$

23. (b)

x	p_i	$p_i x_i$	$p_i x_i^2$
2	2.2	0.4	0.8
3	0.5	1.5	4.5
4	0.3	1.2	4.8
		$\Sigma p_i x_i=3.1$	$\Sigma p_i x_i^2=10.1$

Variance $=\Sigma p_i x_i^2-\Sigma p_i x_i=10.1-(3.1)^2=0.49$

$\therefore$ Standard deviation $=\sqrt{\text{Variance}}=\sqrt{0.49}=0.7$

24. (a)

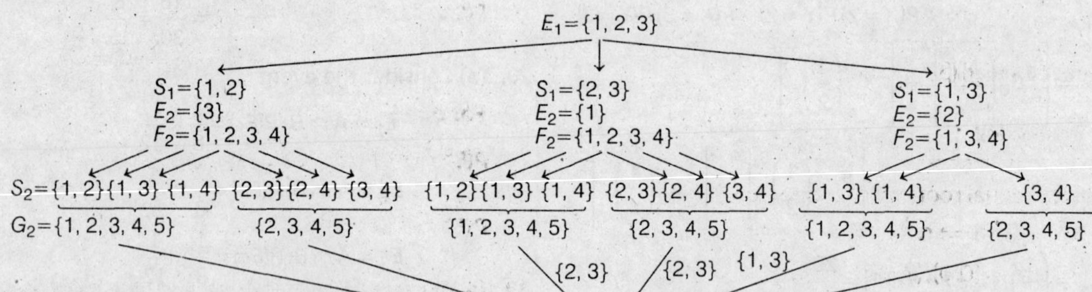

To find : Probability $P=\frac{P(S_1\cap(E_1=E_3))}{P(E_1=E_3)}=\frac{P(A_{1,2})}{P(A)}$,

where $P(A)=P(A_{1,2})+P(A_{1,3})+P(A_{2,3})$

Also, $A_{1,2}$ represents 1, 2 chosen at start and similarly others.

Now, $P(A_{1,2}) = \frac{1}{3} \times \frac{1 \times {}^3C_1}{{}^4C_2} \times \frac{1}{{}^5C_2} = \frac{1}{3} \times \frac{3}{6} \times \frac{1}{10}$

$$= \frac{1}{3} \times \frac{1}{2} \times \frac{1}{10}$$

$$P(A_{1,3}) = \frac{1}{3} \times \frac{1 \times {}^2C_1}{{}^4C_2} \times \frac{1}{{}^5C_2} = \frac{1}{3} \times \frac{2}{3} \times \frac{1}{10}$$

$$P(A_{2,3}) = \frac{1}{3} \times \left[\frac{{}^3C_2 \times 1}{{}^4C_2} \times \frac{1}{{}^4C_2} + \frac{1 \times {}^3C_1}{{}^4C_2} \times \frac{1}{{}^5C_2} \right]$$

$$= \frac{1}{3} \times \left[\frac{3}{6} \times \frac{1}{6} + \frac{3}{6} \times \frac{1}{10} \right] = \frac{1}{3}\left[\frac{1}{2} \times \frac{1}{6} + \frac{1}{2} \times \frac{1}{10} \right]$$

$$P(A) = \frac{1}{3}\left[\begin{array}{l} \frac{1}{2} \times \frac{1}{10} + \frac{1}{2} \times 0 + \frac{1}{2} \times \frac{1}{10} + \frac{1}{2} \times \frac{1}{6} \\ + \frac{2}{3} \times \frac{1}{10} + \frac{1}{3} \times 0 \end{array} \right]$$

$$= \frac{1}{3}\left[\frac{1}{4} \right] = \frac{1}{12}$$

$\therefore$ Required probability $= \dfrac{\frac{1}{3}\left(\frac{1}{2} \times \frac{1}{10}\right)}{\frac{1}{12}} \quad \left[\because P = \frac{P(A_{1,2})}{P(A)} \right]$

$$= \frac{1}{60} \times \frac{12}{1} = \frac{1}{5}$$

25. (b) Probability of getting head on coin $C_1 = P(H) = \frac{2}{3}$

Probability of getting head on coin $C_2 = P(H) = \frac{1}{3}$

For the coin C_1

Number of head (α)	0	1	2
Probability	$\frac{1}{9}$	$\frac{4}{9}$	$\frac{4}{9}$

For the coin C_2

Number of head (β)	0	1	2
Probability	$\frac{4}{9}$	$\frac{4}{9}$	$\frac{1}{9}$

For real and equal roots of the given polynomial,

$$\alpha^2 - 4\beta = 0 \Rightarrow \alpha^2 = 4\beta$$

$\therefore \quad (\alpha, \beta) = (0, 0), (2, 1)$

So, probability $= \frac{1}{9} \times \frac{4}{9} + \frac{4}{9} \times \frac{4}{9} = \frac{20}{81}$

26. (b) Given, mean $= np = 4$ ...(i)

and variance $= npq = 3$ ...(ii)

From Eqs. (i) and (ii), we get

$$q = \frac{3}{4} \Rightarrow p = 1 - q = 1 - \frac{3}{4} = \frac{1}{4}$$

$\because \quad np = 4 \Rightarrow n = 16$

$\therefore$ P(no success) $= P(X = 0)$

$$= {}^{16}C_0 \left(\frac{1}{4}\right)^0 \left(\frac{3}{4}\right)^{16} = \left(\frac{3}{4}\right)^{16}$$

27. (a, b, c, d) We have,

$$P(A) = \frac{1}{2} \text{ and } P(B) = \frac{2}{3}$$

$$P(A \cup B) \geq \max\{P(A), P(B)\}$$

$\therefore \quad P(A \cup B) \geq \frac{2}{3}$

$$P(A \cap B) = P(A) + P(B) - P(A \cup B)$$

$$P(A \cap B) \geq P(A) + P(B) - 1 = \frac{1}{6}$$

and $P(A \cap B) \leq \min\{P(A), P(B)\} = \frac{1}{2}$

$$P(A \cap B') = P(A) - P(A \cap B) \leq \frac{1}{2} - \frac{1}{6} = \frac{1}{3}$$

$$P(A' \cap B) = P(B) - P(A \cap B)$$

Hence, $\frac{2}{3} - \frac{1}{2} \leq P(A' \cap B) \leq \frac{2}{3} - \frac{1}{6}$

$\Rightarrow \quad \frac{1}{6} \leq P(A' \cap B) \leq \frac{1}{2}$

28. (a, b, c)

For option (a)

$$P(E \cap F \cap G^C) = P(E \cap F) - P(E \cap F \cap G)$$

$$\leq P(E) - P(E \cap F \cap G) = \frac{1}{8} - \frac{1}{10} = \frac{1}{40}$$

For option (b)

$$P(E^C \cap F \cap G) = P(F \cap G) - P(E \cap F \cap G) \leq P(F) - P(E \cap F \cap G)$$

$$= \frac{1}{6} - \frac{1}{10} = \frac{1}{15}$$

For option (c)

$$P(E \cup F \cup G) \leq P(E) + P(F) + P(G)$$

$$= \frac{1}{8} + \frac{1}{6} + \frac{1}{4} = \frac{3 + 4 + 6}{24} = \frac{13}{24}$$

For option (d)

$$P(E^C \cap F^C \cap G^C) = 1 - P(E \cap F \cap G) \geq 1 - \frac{13}{24}$$

$$P(E^C \cap F^C \cap G^C) \geq \frac{11}{24} > \frac{5}{12}$$

29. (a, b) Let the number of red and blue balls be r and b, respectively, then the probability of drawing two red balls, is

$$P_1 = \frac{{}^rC_2}{{}^{r+b}C_2} = \frac{r(r-1)}{(r+b)(r+b-1)}$$

The probability of drawing two blue balls is

$$P_2 = \frac{{}^bC_2}{{}^{r+b}C_2} = \frac{b(b-1)}{(r+b)(r+b-1)}$$

The probability of drawing one red and one blue balls is

$$P_3 = \frac{{}^rC_1 \cdot {}^bC_1}{{}^{(r+b)}C_2} = \frac{2br}{(r+b)(r+b-1)}$$

Given, $P_1 = 5P_2$ and $P_3 = 6P_2$

$\therefore r(r-1) = 5(b)(b-1)$ and $2br = 6b(b-1)$

$\Rightarrow \quad r = 6$ and $b = 3$

$\therefore \quad b + r = 9, br = 18, |b - r| = 3$ and $\frac{b}{r} = \frac{1}{2}$

30. (b, c) Roots of $x^2 + px + q = 0$ will be real if $p^2 \geq 4q$

The possible selections are as follows

p	q
1	–
2	1
3	1, 2
4	1, 2, 3, 4
5	1, 2, 3, 4, 5, 6
6	1, 2, 3, ..., 9
7	1, 2, 3, ..., 10
8	1, 2, 3, ..., 10
9	1, 2, 3, ..., 10
10	1, 2, 3, ..., 10

Total = 62

(a) Probability that roots are real $= \frac{62}{100} = \frac{31}{50}$

(b) Probability that roots are not real $= 1 - \frac{31}{50} = \frac{19}{50}$

(c) Roots are equal when $(p, q) = (2, 1), (4, 4), (6, 9)$

$\therefore$ Required probability $= \frac{3}{50}$

(d) Probability that roots are real and distinct

$$= \frac{62-3}{100} = \frac{59}{100}$$

31. (a, d) Probability that exactly one of them occur is $= \frac{11}{25}$

$$\therefore \quad P(E \cup F) - P(E \cap F) = \frac{11}{25} \quad \text{...(i)}$$

Probability of neither occurs is $\frac{2}{25}$

$$\therefore \quad P(\bar{E} \cap \bar{F}) = \frac{2}{25} \quad \text{...(ii)}$$

From Eq. (i),

$$P(E) + P(F) - 2P(E \cap F) = \frac{11}{25} \quad \text{...(iii)}$$

From Eq. (ii),

$$1 - P(E) - P(F) + P(E \cap F) = \frac{2}{25} \quad \text{...(iv)}$$

From Eq. (iii) and (iv),

$$P(E) + P(F) = \frac{7}{5}$$

and $\quad P(E \cap F) = P(E) \cdot P(F) = \frac{12}{25}$

$$\Rightarrow \quad P(E)\left(\frac{7}{5} - P(E)\right) = \frac{12}{25}$$

$$\Rightarrow \quad P(E)^2 - \frac{7}{5}P(E) + \frac{12}{25} = 0$$

$$\Rightarrow \quad \left(P(E) - \frac{3}{5}\right)\left(P(E) - \frac{4}{5}\right) = 0$$

$$\Rightarrow \quad P(E) = \frac{3}{5} \text{ or } \frac{4}{5} \text{ and } P(F) = \frac{4}{5} \text{ or } \frac{3}{5}$$

32. (c, d) We have,

$$P(A \cup B) = P(A) + P(B) - P(A)P(B)$$

$$\therefore \quad P(A \cap B) = P(A) \cdot P(B)$$

(a) $P\left(\frac{B}{A}\right) = \frac{P(A \cap B)}{P(A)} = \frac{P(A) \cdot P(B)}{P(A)}$

$= P(B) \neq P(B) - P(A)$

(b) $P(\bar{A} - \bar{B}) = P(\bar{A} \cap B)$

$= P(B) - (A \cap B)$

$= P(B) - P(A) \cdot P(B)$

(c) $P(\overline{A \cup B}) = 1 - P(A \cup B)$

$= 1 - P(A) - P(B) + P(A)P(B)$

$= 1 - P(A) - P(B)\{1 - P(A)\}$

$= \{1 - P(A)\}\{1 - P(B)\}$

$= P(\bar{A})P(\bar{B})$

(d) $P\left(\frac{A}{B}\right) = \frac{P(A \cap B)}{P(B)}$

$= \frac{P(A)P(B)}{P(B)} = P(A)$

33. (a, b, c) We have,

$$P(A) = \frac{1}{5}, P\left(\frac{B}{A}\right) = \frac{9}{10} \text{ and } P(B) = \frac{7}{25}$$

(a) $P(\bar{A} \cap \bar{B}) = 1 - P(A \cup B)$

$= 1 - P(A) - P(B) + P(A \cap B)$

$= 1 - P(A) - P(B) + P(A)P\left(\frac{B}{A}\right)$

$= 1 - \frac{1}{5} - \frac{7}{25} + \frac{1}{5} \times \frac{9}{10} = \frac{7}{10}$

(b) $P\left(\frac{A}{B}\right) = \frac{P(A \cap B)}{P(B)} = \frac{9/50}{7/25} = \frac{9}{14}$

(c) $P(A \cup B) = P(A) + P(B) - P(A \cap B)$

$= \frac{1}{5} + \frac{7}{25} - \frac{9}{50} = \frac{3}{10}$

(d) $P(A' \cup B') = 1 - P(A \cap B)$

$= 1 - \frac{9}{50} = \frac{41}{50}$

34. (a, c) The total number of cases $= 6^3 = 216$

The number of favourable ways

$= \text{Coefficient of } x^k \text{ in } (x + x^2 + x^3 + x^4 + x^5 + x^6)^3$

$= \text{Coefficient of } x^{k-3} \text{ in } (1 + x + x^2 \ldots x^5)^3$

$= \text{Coefficient of } x^{k-3} \text{ in } (1 - x^6)^3 (1 - x)^{-3}$

$= \text{Coefficient of } x^{k-3} \text{ in } (1 - x)^{-3} \quad [\because 0 \le k - 3 \le 5]$

$= {}^{k-3+2}C_2 = {}^{k-1}C_2$

$= \dfrac{(k-1)(k-2)}{2}$

$\therefore$ Required probability $= \dfrac{{}^{k-1}C_2}{216} = \dfrac{(k-1)(k-2)}{432}$

35. (a, b) We have,

$$P(X) = \frac{1}{3}, P\left(\frac{X}{Y}\right) = \frac{1}{2} \text{ and } P\left(\frac{Y}{X}\right) = \frac{2}{5}$$

$$P\left(\frac{Y}{X}\right) = \frac{P(X \cap Y)}{P(X)}$$

$$P(X \cap Y) = P(X) \cdot P\left(\frac{Y}{X}\right) = \frac{1}{3} \times \frac{2}{5} = \frac{2}{15}$$

$$P\left(\frac{X}{Y}\right) = \frac{P(X \cap Y)}{P(Y)}$$

$$\Rightarrow \quad P(Y) = \frac{P(X \cap Y)}{P\left(\frac{X}{Y}\right)} = \frac{2/15}{1/2} = \frac{4}{15}$$

$$P(X \cup Y) = P(X) + P(Y) - P(X \cap Y)$$

$$= \frac{1}{3} + \frac{4}{15} - \frac{2}{15} = \frac{7}{15}$$

$$P\left(\frac{X'}{Y}\right) = \frac{P(X' \cap Y)}{P(Y)} = \frac{P(Y) - P(X \cap Y)}{P(Y)}$$

$$= 1 - \frac{P(X \cap Y)}{P(Y)} = 1 - P\left(\frac{X}{Y}\right)$$

$$= 1 - \frac{1}{2} = \frac{1}{2}$$

36. (c, d) Given, number of red balls in box I $= n_1$

and number of black balls in box I $= n_2$

Consider the events

R = Red ball is drawn from box I and transferred in box II

B = Black ball is drawn from box I and transferred in box II

Given, $\dfrac{n_1}{n_1 + n_2} \cdot \dfrac{n_1 - 1}{n_1 + n_2 - 1} + \dfrac{n_2}{n_1 + n_2} \cdot \dfrac{n_1}{n_1 + n_2 - 1} = \dfrac{1}{3}$

$\Rightarrow \quad 3(n_1^2 - n_1 + n_1 n_2) = (n_1 + n_2)(n_1 + n_2 - 1)$

$\Rightarrow \quad 3n_1(n_1 + n_2 - 1) = (n_1 + n_2)(n_1 + n_2 - 1)$

$\Rightarrow \quad 3n_1 = n_1 + n_2$

$\Rightarrow \quad 2n_1 = n_2$

$\therefore \quad n_1 = 10, n_2 = 20$

or $\quad n_1 = 3, n_2 = 6$

37. (a, b, c, d)

(a) $P(E_1) = 1 - P(RRR) = 1 - \left[\dfrac{1}{3} \times \dfrac{2}{4} \times \dfrac{3}{5}\right] = 0.9$

(b) $P(E_2) = 3P(BRR) = 3 \times \dfrac{2}{3} \times \dfrac{1}{4} \times \dfrac{2}{5} = 0.2$

(c) $P(E_3) = P\left(\dfrac{RRR}{RRR \cup BBB}\right) = \dfrac{P(RRR)}{P(RRR) + P(BBB)}$

$= \dfrac{0.1}{0.1 + \dfrac{2}{3} \times \dfrac{3}{4} \times \dfrac{4}{5}} = \dfrac{0.1}{0.1 + 0.4} = 0.2$

(d) $P(E_4) = i - P(BBB) = 1 - \dfrac{2}{5} = \dfrac{3}{5} = 0.6$

38. (b, c, d) Let the number of blue marbles is b and the number of green marbles is g.

Accordingly, $\dfrac{bg}{{}^{b+g}C_2} = \dfrac{1}{2}$

$\Rightarrow \quad 4bg = (b + g)(b + g - 1)$

$\Rightarrow \quad (b + g)^2 - (b + g) - 4bg = 0$

$\Rightarrow \quad b^2 + g^2 - 2bg - b - g = 0$

$\Rightarrow \quad g^2 - 2bg - g + b^2 - b = 0$

$\Rightarrow \quad g^2 - (2b + 1)g + (b^2 - b) = 0$

$D = (2b + 1)^2 - 4(b^2 - b)$

$= 4b^2 + 4b + 1 - 4b^2 + 4b$

$D = 8b + 1$ must be perfect square

Hence, possible values of b is 3, 6 and 10.

Sol. (Q.Nos. 39-41)

39. (d) Let P_1 be the probability of being an answer correct of questions attempted from Section I, then

$$P_1 = \frac{1}{5}$$

Again, let P_2 be the probability of being an answer correct of questions attempted from Section II, then

$$P_2 = \frac{1}{15}$$

$\therefore$ Required probability $= P_1 \times P_2 = \dfrac{1}{5} \times \dfrac{1}{15} = \dfrac{1}{75}$

40. (a) Scoring 10 marks from four questions can be obtained as 3 marks + 3 marks + 3 marks + 1 mark = 10 i.e. to answer 3 questions from Section II and one question from Section I correctly.

$\therefore$ Required probability $= \dfrac{1}{5}\left(\dfrac{1}{15}\right)^3$

41. (b) To get 40 marks, candidate has to answer all questions correctly and its probability is $\left(\dfrac{1}{5}\right)^{10}\left(\dfrac{1}{15}\right)^{10}$

Hence, probability of getting a score less than 40

$$= 1 - \left(\frac{1}{5}\right)^{10}\left(\frac{1}{15}\right)^{10} = 1 - \left(\frac{1}{75}\right)^{10}$$

Sol. (Q.Nos. 42-44)

42. (b) Given, $P(u_i) \propto i$, where $i = 1, 2, 3, \ldots, n$

$\Rightarrow \quad P(u_i) = ki$

Since, $\quad \Sigma P(u_i) = 1$

$\Rightarrow \quad k\Sigma i = 1$

$\Rightarrow \quad k = \dfrac{2}{n(n+1)}$

$\therefore \quad P(u_i) = ki = \dfrac{2i}{n(n+1)}$

$$P(W) = \sum_{i=1}^{n} P(u_i) P\left(\frac{W}{u_i}\right)$$

$$= \frac{2i}{n(n+1)} \times \frac{i}{n+1}$$

$$= \sum_{i=1}^{n} \frac{2i^2}{n(n+1)^2} = \frac{2}{n(n+1)^2} \cdot \frac{n(n+1)(2n+1)}{6}$$

$$= \frac{2n+1}{3n+3}$$

$$\therefore \quad \lim_{n \to \infty} P(W) = \frac{2}{3}$$

43. (a) If $P(u_i) = c$, then

$$P\left(\frac{u_n}{W}\right) = \frac{P(u_n) P\left(\dfrac{W}{u_n}\right)}{\sum_{i=1}^{n} P(u_i) P\left(\dfrac{W}{u_i}\right)}$$

$$= \frac{\dfrac{cn}{n+1}}{c\left(\dfrac{1}{n+1} + \dfrac{2}{n+1} + \ldots \dfrac{n}{n+1}\right)}$$

$$= \frac{n}{n+1} \times \frac{n+1}{n\dfrac{(n+1)}{2}} = \frac{2}{n+1}$$

44. (b) $P\left(\dfrac{W}{E}\right) = \dfrac{P(W \cap E)}{P(E)}$

$$= \frac{\dfrac{1}{n} \times \dfrac{2}{n+1} + \dfrac{1}{n} \times \dfrac{4}{n+1} + \dfrac{1}{n} \times \dfrac{6}{n+1} + \ldots \dfrac{n}{n(n+1)}}{\left\{\dfrac{1}{n} + \dfrac{1}{n} + \dfrac{1}{n} + \ldots + \dfrac{1}{n}\right\}\left(\dfrac{n}{2} \text{ times}\right)}$$

$$= \frac{2}{n(n+1)} \cdot \frac{\left(1 + 2 + 3 \ldots \dfrac{n}{2}\right)}{\dfrac{1}{n} \times \dfrac{n}{2}}$$

$$= \frac{4}{n(n+1)} \left(\frac{\dfrac{n}{2}\left(\dfrac{n}{2} + 1\right)}{2}\right) = \frac{n+2}{2(n+1)}$$

Sol. (Q.Nos. 45-47)

45. (c) Consider the events

E_1 = She studies 10 h

E_2 = She studies 7 h

E_3 = She studies 4 h

A = She will be successful

$P(E_1) = 0.1, P(E_2) = 0.2$ and $P(E_3) = 0.7$

$P\left(\dfrac{A}{E_1}\right) = 0.8, P\left(\dfrac{A}{E_2}\right) = 0.6$ and $P\left(\dfrac{A}{E_3}\right) = 0.4$

$\because$ Probability of successful

$$P(A) = P(E_1) P\left(\frac{A}{E_1}\right) + P(E_2) P\left(\frac{A}{E_2}\right) + P(E_3) P\left(\frac{A}{E_3}\right)$$

$$= 0.1 \times 0.8 + 0.2 \times 0.6 + 0.7 \times 0.4$$

$$= 0.08 + 0.12 + 0.28 = 0.48$$

46. (b) $\therefore$ Required probability

$$P\left(\frac{E_3}{A}\right) = \frac{P(E_3) \times P\left(\dfrac{A}{E_3}\right)}{P(A)}$$

$$= \frac{0.7 \times 0.4}{0.48}$$

$$= \frac{0.28}{0.48} = \frac{7}{12}$$

47. (d) $\therefore$ Required probability

$$P\left(\frac{E_3}{\overline{A}}\right) = \frac{P(E_3 \cap \overline{A})}{P(\overline{A})}$$

$$= \frac{P(E_3) - P(E_3 \cap A)}{1 - P(A)}$$

$$= \frac{0.7 - 0.28}{1 - 0.48}$$

$$= \frac{0.42}{0.52} = \frac{21}{26}$$

Sol. (Q.Nos. 48-50)

48. (c) $P(E) = \dfrac{{}^9C_3\left[{}^3C_1 \cdot {}^2C_2 \dfrac{5!}{3!} + {}^3C_2 \cdot {}^1C_1 \cdot \dfrac{5!}{2!}\right]}{9^5}$

$$= \frac{{}^9C_3 (60 + 90)}{9^5} = \frac{{}^9C_3 \times 150}{9^5}$$

49. (b) 3×3 matrices $\begin{bmatrix} a_{11} & a_{12} & a_{13} \\ a_{21} & a_{22} & a_{23} \\ a_{31} & a_{32} & a_{33} \end{bmatrix}$ is

Skew-symmetric if $a_{11} = a_{22} = a_{33} = 0$ and $a_{ij} = -a_{ji}$.

a_{12}, a_{13} and a_{23} can be filled in 9 ways each

a_{21}, a_{31} and a_{32} has one way

$\because n(E) = 9^3$ and $n(S) = 9^9$

$\therefore$ Required probability, $P(E) = \dfrac{n(E)}{n(S)} = \dfrac{9^3}{9^9} = \dfrac{1}{9^6}$

50. (b) Total number of ways of selecting five numbers from $\{-4, -3, -2, -1, 0, 1, 2, 3, 4\}$ with replacement $= 9^5$

Now, $x_1 + x_2 + x_3 + x_4 + x_5 = 0, -4 \le x_i \le 4$

$\therefore$ Favourable case

$= \text{Coefficient of } x^0 \text{ in}$

$(x^{-4} + x^{-3} + x^{-2} + x^{-1} + x^0 + x^1 + x^2 + x^3 + x^4)^5$

$= \text{Coefficient of } x^0 \text{ in } \left(\frac{1 + x + x^2 \ldots + x^8}{x^4}\right)^5$

$= \text{Coefficient of } x^{20} \text{ in } (1 + x + x^2 \ldots + x^8)^5$

$= \text{Coefficient of } x^{20} \text{ in } (1 - x^9)^5 (1 - x)^{-5}$

$= \text{Coefficient of } x^{20} \text{ in } (1 - 5x^9 + 10x^{18} \ldots)$
$(1 + 5x + {}^6C_2x^2 + {}^7C_3x^3 \ldots)$

$= {}^{24}C_4 - 5 \times {}^{15}C_4 + 10 \times {}^6C_4$

$= \frac{24!}{4!20!} - \frac{5 \times 15!}{4!11!} + \frac{10 \times 6!}{4!2!}$

$= 10626 - 6825 + 150 = 3951$

$\therefore$ Required probability $= \frac{3951}{9^5}$

Sol. (Q.Nos. 51-53)

51. (a) The scores of n can be reached in the following two mutually exclusive events

(i) by throwing a head when the score is $(n - 1)$

(ii) by throwing a tail when the score is $(n - 2)$

Hence, $P_n = P_{n-1} \times \frac{1}{2} + P_{n-2} \times \frac{1}{2}$ $\quad \left[\because P(H) = P(T) = \frac{1}{2}\right]$

$\Rightarrow \quad P_n = \frac{1}{2}[P_{n-1} + P_{n-2}]$

52. (c) We have, $P_n = \frac{1}{2}[P_{n-1} + P_{n-2}]$

$\Rightarrow P_n + \frac{1}{2}P_{n-1} = P_{n-1} + \frac{1}{2}P_{n-2} = P_{n-2} + \frac{1}{2}P_{n-3}$

Similarly, $P_n + \frac{1}{2}P_{n-1} = P_2 + \frac{1}{2}P_1$

Now, a score of 1 can be obtained by throwing a head at a single toss

$\therefore \quad P_1 = \frac{1}{2}$

and a score of 2 can be obtained by throwing a tail at a single toss or a head of the first as well as second toss

$\therefore \quad P_2 = \frac{1}{2} + \left(\frac{1}{2}\right)^2 = \frac{3}{4}$

$\therefore P_n + \frac{1}{2}P_{n-1} = P_2 + \frac{1}{2}P_1$

$= \frac{3}{4} + \frac{1}{2} \times \frac{1}{2} = 1$

53. (c) $P_n = 1 - \frac{1}{2}P_{n-1}$

$\Rightarrow P_n - \frac{2}{3} = 1 - \frac{1}{2}P_{n-1} - \frac{2}{3} = -\frac{1}{2}\left(P_{n-1} - \frac{2}{3}\right)$

$\Rightarrow P_n - \frac{2}{3} = \left(-\frac{1}{2}\right)^2 \left(P_{n-2} - \frac{2}{3}\right)$

Similarly, $P_n - \frac{2}{3} = \left(-\frac{1}{2}\right)^{n-1} \left(P_1 - \frac{2}{3}\right)$

$= \left(\frac{-1}{2}\right)^{n-1} \left(\frac{1}{2} - \frac{2}{3}\right)$

$\Rightarrow \quad P_n - \frac{2}{3} = \left(-\frac{1}{2}\right)^n \frac{1}{3} \Rightarrow P_n = \left(-\frac{1}{2}\right)^n \frac{1}{3} + \frac{2}{3}$

Also, $P_{100} = \frac{2}{3} + \frac{1}{3 \cdot 2^{100}} > \frac{2}{3}$

$P_{101} = \frac{2}{3} + \frac{1}{3 \cdot 2^{101}} < \frac{2}{3}$

$\Rightarrow \quad P_{101} < \frac{2}{3} < P_{100}$

54. (i) $\to$ B, (ii) $\to$ C, (iii) $\to$ D, (iv) $\to$ A

(i) Given, S_2 reaches the semi-final, hence all the other players $1 + (2^n - 1)$ are equally likely to wins the final with probability P.

$\therefore (2^n - 1)P + \frac{1}{4} = 1$

$\Rightarrow \quad P = \frac{3}{4(2^n - 1)}$

$\Rightarrow \quad \frac{1}{20} = \frac{3}{4(2^n - 1)}$ $\quad \left[\because P = \frac{1}{20}\right]$

$\Rightarrow \quad 2^n = 16 \Rightarrow n = 4$

(ii) Let n number of chances is allowed. Then, total number of outcomes of ticking one or more alternative out of 4 is

$2^4 - 1 = 15$

Out of these 15 only one is correct. The probability of ticking the alternatives correct at the first trial

$= \frac{1}{15}$

The second trial is $\frac{14}{15} \times \frac{1}{14} = \frac{1}{15}$

and third trial is also $\frac{1}{15}$.

Thus, the probability that the candidate will get marks on the questions, if he is allowed n trials is

$\frac{1}{15} + \frac{1}{15} + \ldots$ to n times $= \frac{n}{15} \ge \frac{1}{3}$

$\Rightarrow \quad n \ge 5$

$\Rightarrow \quad n = 5$

(iii) In 52 cards 12 face and 4 ace cards

$$n(S) = {}^{40}C_{20} \cdot {}^{20}C_2$$

Probability that two cards drawn are both aces is

$$= \frac{{}^4C_2 \cdot {}^{38}C_2}{{}^{40}C_{20} \cdot {}^{20}C_2}$$

$$\therefore \quad P = {}^4C_2 = 6$$

(iv) $E_1 = 1 - 2\left(\frac{1}{2}\right)^n$

$$E_2 = \left(\frac{1}{2}\right)^n (n+1)$$

$$P(E_1 \cap E_2) = P(X = 1)$$

$$= \left(\frac{1}{2}\right)^n \cdot n$$

Now, E_1 and E_2 are independent

$\therefore \; P(E_1)P(E_2) = P(E_1 \cap E_2)$

$$\Rightarrow \left(1 - 2\left(\frac{1}{2}\right)^n\right)\left(\frac{1}{2}\right)^n (n+1) = \left(\frac{1}{2}\right)^n \cdot n$$

$$\Rightarrow \quad \frac{n}{n+1} = 1 - \frac{1}{2^{n-1}}$$

$$\Rightarrow \quad n = 3$$

55. (i) → B, D, (ii) → A, (iii) → C, (iv) → B, D

Let E_i denotes the event that the bag contains i black and $(12 - i)$ white balls $(i = 0, 1, 2, 3, \ldots, 12)$ and A denotes the event that the four balls drawn are black.

Then, $P(E_i) = \frac{1}{13}$ $(i = 0, 1, 2, \ldots, 12)$

$$P\left(\frac{A}{E_i}\right) = 0 \text{ for } i = 0, 1, 2, 3$$

$$P\left(\frac{A}{E_i}\right) = \frac{{}^iC_4}{{}^{12}C_4} \text{ for } i \geq 4$$

(i) $P(A) = \sum_{i=0}^{12} P(E_i)P\left(\frac{A}{E_i}\right)$

$$= \frac{1}{13} \times \frac{({}^4C_4 + {}^5C_4 + \ldots {}^{12}C_4)}{{}^{12}C_4}$$

$$= \frac{1}{13} \times \frac{{}^{13}C_5}{{}^{12}C_4}$$

$$= \frac{1}{5}$$

(ii) Clearly, $P\left(\frac{A}{E_{10}}\right) = \frac{{}^{10}C_4}{{}^{12}C_4}$

$$= \frac{14}{33}$$

(iii) By Baye's theorem,

$$P\left(\frac{E_{10}}{A}\right) = \frac{P(E_{10})P\left(\frac{A}{E_{10}}\right)}{P(A)}$$

$$= \frac{\frac{1}{13} \times \frac{14}{33}}{\frac{1}{5}}$$

$$P\left(\frac{E_{10}}{A}\right) = \frac{70}{429}$$

(iv) Let B denotes the probability of drawing 2 white and 2 black balls, then

$$P\left(\frac{B}{E_i}\right) = 0, i = 0, 1 \text{ or } 11, 12$$

$$P\left(\frac{B}{E_i}\right) = \frac{{}^iC_2 \times {}^{12-i}C_2}{{}^{12}C_4} \text{ for } i = 1, 2, 3, \ldots, 10$$

$$\therefore \quad P(B) = \sum_{i=0}^{12} P(E_i)P\left(\frac{B}{E_i}\right)$$

$$= \frac{1}{13} \times \frac{1}{{}^{12}C_4}[2({}^2C_2{}^{10}C_2 + {}^3C_2{}^9C_2 + \ldots {}^{10}C_2 \times {}^2C_2)]$$

$$= \frac{1}{13} \times \frac{1}{{}^{12}C_4} \times 1287$$

$$= \frac{1 \times 1287}{13 \times 495} = \frac{1}{5}$$

56. (i) → B, (ii) → A, C, (iii) → A, C, (iv) → D

Three distinct numbers a, b and c are selected from $\{1, 2, \ldots, 100\}$

$$\therefore \quad n(S) = {}^{100}C_3 = \frac{100!}{3!97!} = \frac{1}{161700}$$

(i) a, b and c are in AP.

$\because \; 2b = a + c$

This means that a and c are both even or both odd

$$n(E) = {}^{50}C_2 + {}^{50}C_2 = 2({}^{50}C_2)$$

$$\text{Required probability} = \frac{2 \cdot {}^{50}C_2}{{}^{100}C_3}$$

$$= \frac{2 \times 50 \times 49 \times 2 \times 3}{2 \times 100 \times 99 \times 98}$$

$$= \frac{1}{66}$$

(ii) a, b and c are in GP.

Taking $r = 2, 3, 4, \ldots 10$

$\because a, b$ and c are in GP in 53 ways

$$\therefore \text{Required probability} = \frac{53 \times 2 \times 3}{100 \times 99 \times 98}$$

$$= \frac{53}{161700}$$

(iii) $\frac{1}{a}, \frac{1}{b}$ and $\frac{1}{c}$ are in GP $= a, b$ and c are in GP.

$\therefore$ Required probability $= \frac{53}{161700}$

(iv) $a+b+c$ is even

$\therefore$ Required probability $= \frac{^{50}C_3 + {}^{50}C_1 \times {}^{50}C_2}{^{100}C_3} = \frac{1}{2}$

57. (i) $\to$ D, (ii) $\to$ A, (iii) $\to$ B, (iv) $\to$ C

Total number of ways $= {}^nC_3$

$$= \frac{n(n-1)(n-2)}{1\cdot2\cdot3} = \frac{n(n-1)(n-2)}{6}$$

(i) Favourable ways, if side AB and take one vertex out of $(n-4)$ vertices

$$= {}^{n-4}C_1 \times n = n(n-4)$$

$\therefore$ Required probability $= \frac{6(n-4)}{(n-1)(n-2)}$

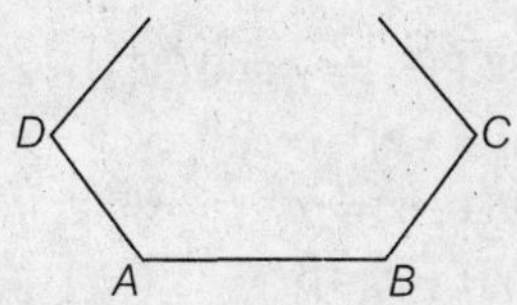

(ii) $A_1, A_2, A_3, A_2, A_3, A_4, \ldots\ldots A$ or $A_nA_1A_2$

Favourable ways $= n$

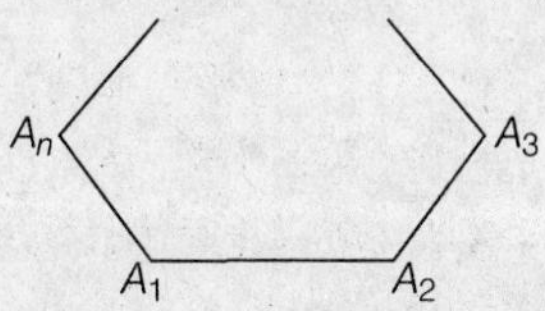

$\therefore$ Required probability $= \frac{n}{^nC_3}$

$$= \frac{6}{(n-1)(n-2)}$$

(iii) Favourable ways, one side common or two sides common $= n(E_1) + n(E_2)$

$$= n(n-4) + n$$
$$= n(n-3)$$

$\therefore$ Required probability $= \frac{n(n-3)}{^nC_3}$

$$= \frac{6(n-3)}{(n-1)(n-2)}$$

(iv) No sides are common

${}^nC_3 - n(n-4) - n$

$$= \frac{n(n-1)(n-2)}{6} - n(n-4) - n$$

$$= \frac{n}{6}(n^2 - 3n + 2 - 6n + 24 - 6)$$

$$= \frac{n(n-4)(n-5)}{6}$$

$\therefore$ Required probability

$$= \frac{n(n-4)(n-5)}{n(n-1)(n-2)}$$

$$= \frac{(n-4)(n-5)}{(n-1)(n-2)}$$

58. (i) $\to$ B, (ii) $\to$ A, (iii) $\to$ D, (iv) $\to$ C

(i) As $(n-3)$ people do not know the the answer probability that the first four do not know the answer is

$$\frac{^{n-3}C_4}{^nC_4} = \frac{(n-4)(n-5)(n-6)}{n(n-1)(n-2)}$$

(ii) $P_2 = \left(\frac{n-3}{n}\right)\left(\frac{3}{n-1}\right) = \frac{3(n-3)}{n(n-1)}$

(iii) The first person to answer the question will be the rth person, if $(r-1)$ persons are chosen from $(n-3)$ persons who do not know the answer and the rth person is chosen from 3 persons who know the answer, the probability of this event is

$$\frac{^{n-3}C_{r-1}}{^nC_{r-1}} \cdot \frac{3}{n-(r-1)}$$

$$= \frac{(n-3)!}{(r-1)!(n-r-2)!} \times \frac{(r-1)!(n-r+1)!}{n!} \times \frac{3}{n-r+1}$$

$$= \frac{3(n-r)(n-r-1)}{n(n-1)(n-2)}$$

$$\therefore P_{n-2} = \frac{3[(n-(n-2)(n-(n-2)-1))]}{n(n-1)(n-2)}$$

$$= \frac{6}{n(n-1)(n-2)}$$

(iv) $P_4 = \frac{3(n-4)(n-5)}{n(n-1)(n-2)}$

Sol. (Q.Nos. 59-61)

59. (0.4) We have, A_1 have one ball {1}

A_2 have two balls {1, 2}; A_3 have three balls {1, 2, 3} and A_4 have four balls {1, 2, 3, 4}

$\therefore$ $P(A_1) = \frac{1}{10}, P(A_2) = \frac{2}{10}, P(A_3) = \frac{3}{10}$ and $P(A_4) = \frac{4}{10}$

$P\left(\frac{E_1}{A_1}\right) = 1, P\left(\frac{E_2}{A_2}\right) = \frac{1}{2}, P\left(\frac{E_3}{A_3}\right) = \frac{1}{3}$ and $P\left(\frac{E_4}{A_4}\right) = \frac{1}{4}$

$$\therefore \sum_{i=1}^{4} P(E_i \cap A_i) = P(A_1)P\left(\frac{E_1}{A_1}\right) + P(A_2)\cdot P\left(\frac{E_2}{A_2}\right) + P(A_3)P\left(\frac{E_3}{A_3}\right) + P(A_4)\times P\left(\frac{E_4}{A_4}\right)$$

$$= \frac{1}{10}\times1 + \frac{2}{10}\times\frac{1}{2} + \frac{3}{10}\times\frac{1}{3} + \frac{4}{10}\times\frac{1}{4}$$

$$= \frac{4}{10} = \frac{2}{5} = 0.4$$

60. (0.33) $P\left(\frac{A_3}{E_2}\right)=\dfrac{P(A_3)\times P\left(\frac{E_2}{A_3}\right)}{\left[P(A_1)\times P\left(\frac{E_2}{A_1}\right)+P(A_2)P\left(\frac{E_2}{A_2}\right)+P(A_3)P\left(\frac{E_2}{A_3}\right)+P(A_4)P\left(\frac{E_2}{A_4}\right)\right]}$

$$=\frac{\frac{3}{10}\times\frac{1}{3}}{\frac{1}{10}\times 0+\frac{2}{10}\times\frac{1}{2}+\frac{3}{10}\times\frac{1}{3}+\frac{4}{10}\times\frac{1}{4}}=\frac{1}{3}=0.33$$

61. (0.041) Sum of four numbers is less than 5

i.e. only one possible cases $(1+1+1+1)=4$

$\therefore$ Required probability $=\dfrac{1}{{}^1C_1\times{}^2C_1\times{}^3C_1\times{}^4C_1}=\dfrac{1}{24}$

$=0.041$

62. (76.25) $p_1=1-p(\text{all 3 numbers are}\le 80)$

$$=1-\left(\frac{80}{100}\right)^3=\frac{125-64}{125}=\frac{61}{125}$$

So, $\dfrac{625p_1}{4}=\dfrac{625}{4}\times\dfrac{61}{125}=\dfrac{5}{4}\times 61=76.25$

63. (24.50) $p_2=1-p$ (all three numbers are > 40)

$$=1-\left(\frac{60}{100}\right)^3=1-\frac{27}{125}=\frac{98}{125}$$

So, $\dfrac{125p_2}{4}=\dfrac{125}{4}\times\dfrac{98}{125}=\dfrac{98}{4}=24.50$

64. (4) Let Event A : Card is of heart but not king (12 cards)

Event B : King but not heart (3 cards)

Event C : Heart and king (1 card)

$\therefore$ Required probability

$$p=P(E)=\frac{{}^{12}C_1\cdot{}^3C_1+{}^3C_1\cdot{}^1C_1+{}^1C_1\cdot{}^{12}C_1}{{}^{52}C_2}$$

$$=\frac{2}{52}$$

$\therefore\quad 104p=4$

65. (5) Highest number in 3 throws is 4, which means at least one of the throws must be equal to 4.

Number of ways when 3 blocks are filled from {1, 2, 3}

$=4^3$

$\therefore$ Number of ways when filled from $\{1, 2, 3\}=3^3$

$\therefore$ Required number of ways $=4^3-3^3$

$\therefore$ Required probability, $P=\dfrac{4^3-3^3}{6^3}=\dfrac{37}{216}$

66. (3) For ranked 1 and 2 players to be winners and runner up, respectively, they should not be paired with each other in any round

$\Rightarrow\quad P=\dfrac{30}{31}\times\dfrac{14}{15}\times\dfrac{6}{7}\times\dfrac{2}{3}=\dfrac{16}{31}$

67. (6) Given, S_2 reaches the semi-finals.

Since, all other players (2^n-1) are equally likely to win the final with probability P.

$\therefore\quad (2^n-1)P+\dfrac{1}{4}=1$

$\Rightarrow\quad (2^n-1)P=\dfrac{3}{4}\Rightarrow P=\dfrac{3}{4(2^n-1)}$

If $P=\dfrac{1}{84}$, then

$\therefore\quad \dfrac{1}{84}=\dfrac{3}{4(2^n-1)}$

$\Rightarrow\ 2^n-1=63$

$\Rightarrow\quad 2^n=64$

$\Rightarrow\quad n=6$

68. (6) Let $P(E_1)=x, P(E_2)=y$ and $P(E_3)=z$, then

$$(1-x)(1-y)(1-z)=p$$
$$x(1-y)(1-z)=\alpha$$
$$(1-x)y(1-z)=\beta$$
$$(1-x)(1-y)(z)=\gamma$$

So, $\dfrac{1-x}{x}=\dfrac{p}{\alpha}\Rightarrow x=\dfrac{\alpha}{\alpha+p}$

Similarly, $z=\dfrac{\gamma}{\gamma+p}$

So, $\dfrac{P(E_1)}{P(E_3)}=\dfrac{\frac{\alpha}{\alpha+p}}{\frac{\gamma}{\gamma+p}}$

$$=\frac{\frac{\gamma+p}{\gamma}}{\frac{\alpha+p}{\alpha}}=\frac{1+\frac{p}{\gamma}}{1+\frac{p}{\alpha}}$$

Also, $\dfrac{\alpha\beta}{\alpha-2\beta}=p=\dfrac{2\beta\gamma}{\beta-3\gamma}$

$\Rightarrow\quad \beta=\dfrac{5\alpha\gamma}{\alpha+4\gamma}$

Substituting in given relations

$$\left(\alpha-2\left(\frac{5\alpha\gamma}{\alpha+4\gamma}\right)\right)p=\frac{\alpha\times 5\alpha\gamma}{\alpha+4\gamma}$$

or $\quad\left(\dfrac{p}{\gamma}+1\right)=6\left(\dfrac{p}{\alpha}+1\right)$

or $\quad\dfrac{\frac{p}{\gamma}+1}{\frac{p}{\alpha}+1}=6$

69. (2/9) $r_1, r_2, r_3 \in \{1, 2, 3, 4, 5, 6\}$

r_1, r_2 and r_3 are of the form $3k, 3k+1$ and $3k+2$

$$\text{Required probability} = \frac{3! \times {}^2C_1 \times {}^2C_1 \times {}^2C_1}{6 \times 6 \times 6}$$

$$= \frac{6 \times 8}{216} = \frac{2}{9}$$

70. (8) Let coin be tossed n times.

P (atleast two head) $= 1 - P$ (no head) $- P$(exactly one head)

$$P(\text{atleast two head}) = 1 - \left(\frac{1}{2}\right)^n - {}^nC_1 \cdot \left(\frac{1}{2}\right)^n \geq 0.96$$

$$\Rightarrow \quad \frac{4}{100} \geq \frac{n+1}{2^n}$$

$$\Rightarrow \quad \frac{2^n}{n+1} \geq 25$$

Therefore, least value of n is 8.

71. (3) $P(E_1) = 1 - [P(\text{all heads}) + P(\text{all tails})]$

$$= 1 - \left[\frac{1}{2^n} + \frac{1}{2^n}\right] = 1 - \frac{1}{2^{n-1}}$$

$P(E_2) = P$ (no head) $+ P$ (exactly one head)

$$= \frac{1}{2^n} + {}^nC_1 \times \frac{1}{2^n} = \frac{n+1}{2^n}$$

$P(E_1 \cap E_2) = \{$exactly one head and $(n-1)$ tail$\}$

$$= {}^nC_1 \times \frac{1}{2} \times \frac{1}{2^{n-1}} = \frac{n}{2^n}$$

If E_1 and E_2 are independent, then

$$\frac{n}{2^n} = \left(1 - \frac{1}{2^{n-1}}\right)\left(\frac{n+1}{2^n}\right)$$

or $$n = \left[1 - \left(\frac{1}{2^{n-1}}\right)\right](n+1)$$

or $$n = n + 1 - \frac{n+1}{2^{n-1}}$$

or $$n + 1 = 2^{n-1}$$

or $$n = 3$$

72. (214) Given, set $= \{1, 2, 3, \ldots, 2000\}$

Let E_1 = Event that it is a multiple of 3

$= \{3, 6, 9, \ldots, 1998\}$

$\therefore \quad n(E_1) = 666$

and E_2 = Event that it is a multiple of 7

$= \{7, 14, \ldots, 1995\}$

$\therefore \quad n(E_2) = 285$

$E_1 \cap E_2$ = multiple of 21 $= \{21, 42, \ldots, 1995\}$

$n(E_1 \cap E_2) = 95$

$\therefore \quad P(E_1 \cup E_2) = P(E_1) + P(E_2) - P(E_1 \cap E_2)$

$$P(E_1 \cup E_2) = \frac{666 + 285 - 95}{2000}$$

$$= \frac{856}{2000} = p \quad \text{(given)}$$

Hence, $500p = 500 \times \frac{856}{2000} = \frac{856}{4} = 214$

CHAPTER

08

MATRICES

LEVEL 01 \\ Starter Level

8.1 Matrices and its Types

1 If a matrix has 8 elements, what are the possible orders it can have?

2 Construct a 3×2 matrix whose elements are given by $a_{ij} = \frac{1}{2}|i - 3j|$.

3 Construct a 2×2 matrix, where

(i) $a_{ij} = \frac{(i-2j)^2}{2}$

(ii) $a_{ij} = |-2i + 3j|$

4 $\begin{bmatrix} x+3 & z+4 & 2y-7 \\ -6 & a-1 & 0 \\ b-3 & -21 & 0 \end{bmatrix} = \begin{bmatrix} 0 & 6 & 3y-2 \\ -6 & -3 & 2c+2 \\ 2b+4 & -21 & 0 \end{bmatrix}$

Find the values of a, b, c, x, y and z.

8.2 Algebra of Matrices, Transpose and Trace

5 Given, $A = \begin{bmatrix} 1 & 3 & 5 \\ -2 & 0 & 2 \\ 0 & 4 & -3 \end{bmatrix}$, $B = \begin{bmatrix} 0 & 3 \\ -2 & 0 \\ 0 & -4 \end{bmatrix}$ and $C = \begin{bmatrix} 4 & 1 & -2 \\ 3 & 2 & 1 \\ 2 & -1 & 7 \end{bmatrix}$.

Find (which ever defined)

(i) $A + B$ (ii) $A + C$

6 If $A = \begin{bmatrix} 1 & 1 & 1 \\ 1 & 1 & 1 \\ 1 & 1 & 1 \end{bmatrix}$ and $B = \begin{bmatrix} 0 & 1 & 0 \\ 0 & 0 & 1 \\ 0 & 1 & 0 \end{bmatrix}$, then find $A - B$.

7 If $A = \begin{bmatrix} 0 & 2 \\ 3 & -4 \end{bmatrix}$ and $kA = \begin{bmatrix} 0 & 3a \\ 2b & 24 \end{bmatrix}$

Then, find the value of $b - a - k$.

8 Find X and Y if $X + Y = \begin{bmatrix} 5 & 2 \\ 0 & 9 \end{bmatrix}$ and $X - Y = \begin{bmatrix} 3 & 6 \\ 0 & -1 \end{bmatrix}$.

9 Solve the following equation for X and Y.

$$2X - Y = \begin{bmatrix} 3 & -3 & 0 \\ 3 & 3 & 2 \end{bmatrix}, 2Y + X = \begin{bmatrix} 4 & 1 & 5 \\ -1 & 4 & -4 \end{bmatrix}$$

10 Find the values of x and y from the following equations

$$2\begin{bmatrix} x & 5 \\ 7 & y-3 \end{bmatrix} + \begin{bmatrix} 3 & -4 \\ 1 & 2 \end{bmatrix} = \begin{bmatrix} 7 & 6 \\ 15 & 14 \end{bmatrix}$$

11 If $A = \begin{bmatrix} 8 & 0 \\ 4 & -2 \\ 3 & 6 \end{bmatrix}$ and $B = \begin{bmatrix} 2 & -2 \\ 4 & 2 \\ -5 & 1 \end{bmatrix}$, then find the matrix X such that $2A + 3X = 5B$.

12 If $\begin{bmatrix} \lambda^2 - 2\lambda + 1 & \lambda - 2 \\ 1 - \lambda^2 + 3\lambda & 1 - \lambda^2 \end{bmatrix} = A\lambda^2 + B\lambda + C$, where A, B and C are matrices then find matrices B and C.

13 Find AB, if $A = \begin{bmatrix} 0 & -1 \\ 0 & 2 \end{bmatrix}$ and $B = \begin{bmatrix} 3 & 5 \\ 0 & 0 \end{bmatrix}$.

14 If $A = \begin{bmatrix} 1 & -2 & 3 \\ -4 & 2 & 5 \end{bmatrix}$ and $B = \begin{bmatrix} 2 & 3 \\ 4 & 5 \\ 2 & 1 \end{bmatrix}$, then find the product AB and BA.

15 The book shop of a particular school has 10 dozen chemistry books, 8 dozen physics books, 10 dozen economic books. Their selling prices are ₹ 80, ₹ 60 and ₹ 40 each respectively. Find the total amount the book shop will receive from selling all the books using matrix algebra.

16 If $A = \begin{bmatrix} 2 & -1 \\ 3 & 4 \end{bmatrix}$, $B = \begin{bmatrix} 5 & 2 \\ 7 & 4 \end{bmatrix}$, $C = \begin{bmatrix} 2 & 5 \\ 3 & 8 \end{bmatrix}$. Find a matrix D such that $CD - AB = 0$.

17 If $A = \begin{bmatrix} 1 & 2 & 3 \\ 3 & -2 & 1 \\ 4 & 2 & 1 \end{bmatrix}$, then evaluate $A^3 - 23A - 40I$.

18 Find the possible square roots of the two rowed unit matrix.

19 If $A = \begin{bmatrix} 1 & 2 \\ -2 & 3 \end{bmatrix}$, $B = \begin{bmatrix} 2 & 1 \\ 2 & 3 \end{bmatrix}$ and $C = \begin{bmatrix} -3 & 1 \\ 2 & 0 \end{bmatrix}$, verify that $(AB)C = A(BC)$ and $A(B + C) = AB + AC$.

20 Find x, so that $[1 \; x \; 1]\begin{bmatrix} 1 & 3 & 2 \\ 0 & 5 & 1 \\ 0 & 3 & 2 \end{bmatrix}\begin{bmatrix} 1 \\ 1 \\ x \end{bmatrix} = 0$.

21 If $A = \begin{bmatrix} 3 & \sqrt{3} & 2 \\ 4 & 2 & 0 \end{bmatrix}$ and $B = \begin{bmatrix} 2 & -1 & 2 \\ 1 & 2 & 4 \end{bmatrix}$, verify that

(i) $(A')' = A$ (ii) $(A + B)' = A' + B'$ (iii) $(kB)' = kB'$, where k is any constant.

22 Let $A = \begin{bmatrix} 2 & -3 \\ -7 & 5 \end{bmatrix}$ and $B = \begin{bmatrix} 1 & 0 \\ 2 & -4 \end{bmatrix}$ verify that

(i) $(2A)^T = 2A^T$ (ii) $(A+B)^T = A^T + B^T$

(iii) $(A-B)^T = A^T - B^T$ (iv) $(AB)^T = B^T A^T$

23 $A = \begin{bmatrix} 3 \\ 5 \\ 2 \end{bmatrix}$ and $B = [1 \quad 0 \quad 4]$, verify that $(AB)^T = B^T A^T$.

24 Find the values of x, y, z if the matrix $A = \begin{bmatrix} 0 & 2y & z \\ x & y & -z \\ x & -y & z \end{bmatrix}$ satisfies the equation $A^T A = I$.

25 Find the trace of the following matrices.

(i) $A = \begin{bmatrix} 1 & 0 & 0 \\ 0 & 1 & 0 \\ 0 & 0 & 1 \end{bmatrix}$ (ii) $B = \begin{bmatrix} 1 & 2 & 3 \\ 4 & 5 & 6 \\ 7 & 8 & 9 \end{bmatrix}$

(iii) $C = \begin{bmatrix} 5 & 2 & 3 \\ 0 & 5 & 0 \\ 0 & 0 & 5 \end{bmatrix}$

26 If $A = \begin{bmatrix} 1 & 2 & 3 \\ 1 & 1 & 2 \\ 2 & 3 & 5 \end{bmatrix}$ and $B = \begin{bmatrix} 1 & 0 & 0 \\ 5 & 0 & 0 \\ 7 & 0 & 7 \end{bmatrix}$, then find

(i) Trace $(A+B)$ (ii) Trace $(3A)$

(iii) Trace (A^T)

27 Let $A + 2B = \begin{bmatrix} 1 & 2 & 0 \\ 6 & -3 & 3 \\ -5 & 3 & 1 \end{bmatrix}$ and $2A - B = \begin{bmatrix} 2 & -1 & 5 \\ 2 & -1 & 6 \\ 0 & 1 & 2 \end{bmatrix}$, then find trace$(A)$ – trace(B).

8.3 Special Kind of Matrices

28 If the matrix $A = \begin{bmatrix} 5 & 2 & x \\ y & z & -3 \\ 4 & t & -7 \end{bmatrix}$, is a symmetric matrix find x, y, z and t.

29 If a square matrix A of order three is defined $A = [a_{ij}]$ where $a_{ij} = \text{sgn}\,(i-j)$, then verify that A is skew-symmetric matrix.

30 If $A = \begin{bmatrix} 3 & -4 \\ 1 & -1 \end{bmatrix}$, verify that $A - A^T$ is skew-symmetric matrix.

31 Express A as the sum of a symmetric and a skew-symmetric matrix, where $A = \begin{bmatrix} 3 & 5 \\ -1 & 2 \end{bmatrix}$.

32 Express the matrix $B = \begin{bmatrix} 2 & -2 & -4 \\ -1 & 3 & 4 \\ 1 & -2 & -3 \end{bmatrix}$ as the sum of a symmetric and a skew-symmetric matrix.

33 $A = \begin{bmatrix} 3 & 2 & 7 \\ 1 & 4 & 3 \\ -2 & 5 & 8 \end{bmatrix}$, find matrices X and Y such that $X + Y = A$, where X is a symmetric and Y is a skew-symmetric matrix.

34 If $\begin{bmatrix} 4 & 1+3i \\ 1-ai & 7 \end{bmatrix}$ is Hermitian. Then, find the value of a.

35 Verify that $A = \begin{bmatrix} 0 & -2-i \\ 2-i & i \end{bmatrix}$ is skew-hermitian.

36 Express A as the sum of a Hermitian and skew-hermitian matrix, where $A = \begin{bmatrix} 2+3i & 7 \\ 1-i & 2i \end{bmatrix}, i = \sqrt{-7}$.

37 Verify that the given matrix $\begin{bmatrix} \frac{b^2-a^2}{a^2+b^2} & \frac{-2ab}{a^2+b^2} \\ \frac{-2ab}{a^2+b^2} & \frac{a^2-b^2}{a^2+b^2} \end{bmatrix}$ is orthogonal.

38 If $\begin{bmatrix} 0 & 2\beta & \gamma \\ \alpha & \beta & -\gamma \\ \alpha & -\beta & \gamma \end{bmatrix}$ is orthogonal, then find the value of $2\alpha^2 + 6\beta^2 + 3\gamma^2$.

39 Verify the matrix $\frac{1}{\sqrt{3}}\begin{bmatrix} 1 & 1+i \\ 1-i & -1 \end{bmatrix}$ is unitary matrix, where $i = \sqrt{-1}$.

40 Verify that matrix

$$A = \begin{bmatrix} 2 & -2 & -4 \\ -1 & 3 & 4 \\ 1 & -2 & -3 \end{bmatrix}$$ is idempotent.

41 Verify that $\begin{bmatrix} 1 & 1 & 3 \\ 5 & 2 & 6 \\ -2 & -1 & -3 \end{bmatrix}$ is nilpotent matrix of order 3.

42 Show that the matrix

$$A = \begin{bmatrix} -5 & -8 & 0 \\ 3 & 5 & 0 \\ 1 & 2 & -1 \end{bmatrix}$$ is involutory. Justify your answer.

8.4 Invertible Matrices, Solution of System of Simultaneous Linear Equations and Reflection of Matrix

43 Verify that the matrices

$$A = \begin{bmatrix} 1 & -1 & 0 \\ 2 & 1 & 1 \end{bmatrix} \text{ and } B = \begin{bmatrix} 3 & 0 & 1 \\ 0 & 3 & 1 \end{bmatrix}$$ are row equivalent.

44 Using elementary operations. Find the inverse of the matrix

$$A = \begin{bmatrix} 1 & 2 \\ 2 & -1 \end{bmatrix}$$

45 Find P^{-1}, if it exists, given $P = \begin{bmatrix} 10 & -2 \\ -5 & 1 \end{bmatrix}$.

46 Let A be a 3×3 matrix such that

$$A \begin{bmatrix} 1 & 2 & 3 \\ 0 & 2 & 3 \\ 0 & 1 & 1 \end{bmatrix} = \begin{bmatrix} 0 & 0 & 1 \\ 1 & 0 & 0 \\ 0 & 1 & 0 \end{bmatrix}, \text{ then find } A^{-1}.$$

47 Obtain the inverse of the following matrix using elementary operations

(i) $A = \begin{bmatrix} 0 & 1 & 2 \\ 1 & 2 & 3 \\ 3 & 1 & 1 \end{bmatrix}$ (ii) $\begin{bmatrix} 2 & 0 & -1 \\ 5 & 1 & 0 \\ 0 & 1 & 3 \end{bmatrix}$

48 Matrices A and B satisfy $AB = B^{-1}$, where $B = \begin{bmatrix} 2 & -2 \\ -1 & 0 \end{bmatrix}$, find the value of λ for which $\lambda A - 2B^{-1} + I = 0$ without finding B^{-1}.

49 Find the rank of $\begin{bmatrix} 3 & -1 & 2 \\ -3 & 1 & 2 \\ -6 & 2 & 4 \end{bmatrix}$.

50 Determine for what values of λ and μ the following system of equations

$x + y + z = 6$

$x + 2y + 3z = 10$

and $x + 2y + \lambda z = \mu$ have

(i) no solution ?

(ii) a unique solution ?

(iii) an infinite number of solutions?

51 Find the image of the point $(-2, -7)$ under the transformations

$$(x, y) \to (x - 2y, -3x + y)$$

52 The image of the point $A(2, 3)$ by the line mirror $y = x$ is the point B and the image of B by the line mirror $y = 0$ is the point (α, β). Find α and β.

53 Find the image of the point $(-\sqrt{2}, \sqrt{2})$ by the line mirror $y = x \tan\left(\frac{\pi}{8}\right)$.

LEVEL 02 \ JEE Main Level

Single Choice Correct Questions

8.1 Matrices and Its Types

1 Total number of possible matrices of order 3×3 with each entry 2 or 0 is

(a) 9 (b) 27 (c) 81 (d) 512

2 If $A = [a_{ij}]_{2 \times 2}$ where $a_{ij} = 2^i \cdot \omega^{i+j}$, where ω is cube root of unity, then A is equal to

(a) $\begin{bmatrix} 2\omega & 2\omega^2 \\ 4 & 4 \end{bmatrix}$ (b) $\begin{bmatrix} 2\omega^2 & 2 \\ 4 & 4\omega \end{bmatrix}$ (c) $\begin{bmatrix} 2 & 2\omega \\ 4 & 4 \end{bmatrix}$ (d) $\begin{bmatrix} 2 & 2 \\ 4\omega & 4\omega^2 \end{bmatrix}$

3 The values of x, y and z from the following equation of $\begin{bmatrix} 4 & x-z \\ 2+y & xz \end{bmatrix} = \begin{bmatrix} 4 & 3 \\ -1 & 10 \end{bmatrix}$, are

(a) $x=-5, y=3, z=2$ (b) $x=5, y=-3, z=2$
(c) $x=5, y=3, z=-2$ (d) $x=5, y=-3, z=-2$

8.2 Algebra on Matrices, Transpose and Trace

4 If $A = \begin{bmatrix} {}^5C_2 & {}^6C_2 \\ {}^4C_2 & {}^3C_2 \end{bmatrix}$ and $B = \begin{bmatrix} i^4 & i^6 \\ i^8 & i^{10} \end{bmatrix}$, where $i=\sqrt{-1}$.

Then, find $A+B$.

(a) $\begin{bmatrix} 11 & 15+i \\ 6-i & 3 \end{bmatrix}$ (b) $\begin{bmatrix} 11+i & 15 \\ 6 & 3-i \end{bmatrix}$ (c) $\begin{bmatrix} 11 & 14 \\ 7 & 2 \end{bmatrix}$ (d) $\begin{bmatrix} 11 & 15-i \\ 7 & 3-i \end{bmatrix}$

5 If $A = \begin{bmatrix} x & 1 \\ 1 & 0 \end{bmatrix}$ and A^2 is the identity matrix, then x is equal to

(a) −1 (b) 0 (c) 1 (d) 2

6 If the matrix $A = \begin{pmatrix} 0 & 2 \\ K & -1 \end{pmatrix}$ satisfies $A(A^3+3I)=2I$, then the value of K is

(a) $\frac{1}{2}$ (b) $-\frac{1}{2}$ (c) −1 (d) 1

7 If $A = \begin{bmatrix} 1 & 0 & 0 \\ 0 & 1 & 0 \\ a & b & -1 \end{bmatrix}$ and I is the unit matrix of order 3, then $A^2+2A^4+4A^6$ is equal to

(a) $7A^8$ (b) $7A^7$ (c) $8I$ (d) $6I$

8 If $A = \begin{bmatrix} 3 & 1 \\ -1 & 2 \end{bmatrix}$, then

(a) $A^2+7A-5I=0$ (b) $A^2-7A+5I=0$
(c) $A^2+5A-7I=0$ (d) $A^2-5A+7I=0$

9 If $A = \begin{bmatrix} \alpha & 0 \\ 1 & 1 \end{bmatrix}$ and $B = \begin{bmatrix} 1 & 0 \\ 5 & 1 \end{bmatrix}$, then value of α for which $A^2=B$ is

(a) 1 (b) −1 (c) 4 (d) No real values

10 Let $A = \begin{bmatrix} 1 & 2 \\ 3 & 4 \end{bmatrix}$ and $B = \begin{bmatrix} a & 0 \\ 0 & b \end{bmatrix}$, $a, b \in N$, then

(a) there cannot exist any B such that AB = BA
(b) there exist more than one but finite number
(c) there exist exactly one B such that AB = BA
(d) there exist infinitely many B's such that AB = BA

11 If $A = \begin{bmatrix} a & b \\ b & a \end{bmatrix}$ and $A^2 = \begin{bmatrix} \alpha & \beta \\ \beta & \alpha \end{bmatrix}$, then

(a) $\alpha=2ab, \beta=a^2+b^2$ (b) $\alpha=a^2+b^2, \beta=ab$
(c) $\alpha=a^2+b^2, \beta=2ab$ (d) $\alpha=a^2+b^2, \beta=a^2-b^2$

12 If AB = A and BA = B, then B^2 is equal to

(a) B (b) A (c) −B (d) B^2

13 If A and B are square matrices of size $n \times n$ such that $A^2 - B^2 = (A - B)(A + B)$, then which of the following will be always true

(a) $A = B$
(b) $AB = BA$
(c) either of A or B is a zero matrix
(d) either of A or B is identity matrix

14 Let $A = \begin{bmatrix} \cos\alpha & -\sin\alpha \\ \sin\alpha & \cos\alpha \end{bmatrix}, \alpha \in R$ such that $A^{32} = \begin{bmatrix} 0 & -1 \\ 1 & 0 \end{bmatrix}$. Then, a value of α is **[JEE Main 2019]**

(a) $\frac{\pi}{22}$
(b) 0
(c) $\frac{\pi}{64}$
(d) $\frac{\pi}{16}$

15 If $A = \begin{bmatrix} \cos\theta & i\sin\theta \\ i\sin\theta & \cos\theta \end{bmatrix}; \theta = \frac{\pi}{24}$ and $A^5 = \begin{bmatrix} a & b \\ c & d \end{bmatrix}$, where $i = \sqrt{-1}$, then which of the following is not true? **[JEE Main 2020]**

(a) $0 \le a^2 + b^2 \le 1$
(b) $a^2 - d^2 = 0$
(c) $a^2 - c^2 = 1$
(d) $a^2 - b^2 = \frac{1}{2}$

16 Let α be a root of the equation $x^2 + x + 1 = 0$ and the matrix $A = \frac{1}{\sqrt{3}} \begin{bmatrix} 1 & 1 & 1 \\ 1 & \alpha & \alpha^2 \\ 1 & \alpha^2 & \alpha^4 \end{bmatrix}$, then the matrix A^{31} is equal to **[JEE Main 2020]**

(a) A
(b) I_3
(c) A^2
(d) A^3

17 Let $P = \begin{bmatrix} 1 & 0 & 0 \\ 3 & 1 & 0 \\ 9 & 3 & 1 \end{bmatrix}$ and $Q = [a_{ij}]$ be two 3×3 matrices such that $Q - P^5 = I_3$.

Then, $\frac{q_{21} + q_{31}}{q_{32}}$ is equal to

(a) 10
(b) 135
(c) 15
(d) 9

18 If A is 3×4 matrix and B is a matrix such that $A'B$ and BA' are both defined, then B is of the type

(a) 4×4
(b) 3×4
(c) 4×3
(d) 3×3

19 Let $A = \begin{bmatrix} 0 & 2q & r \\ p & q & -r \\ p & -q & r \end{bmatrix}$. If $AA^T = I_3$, then $|P|$ is **[JEE Main 2019]**

(a) $\frac{1}{\sqrt{2}}$
(b) $-\frac{1}{\sqrt{2}}$
(c) $\frac{2}{\sqrt{2}}$
(d) $\frac{1}{\sqrt{3}}$

20 If $A = \begin{bmatrix} 1 & 2 & 2 \\ 2 & 1 & -2 \\ a & 2 & b \end{bmatrix}$ is a matrix satisfying the equation $AA^T = 9I$, where I is 3×3 identity matrix, then the ordered pair (a, b) is equal to

(a) $(2, -1)$
(b) $(-2, 1)$
(c) $(2, 1)$
(d) $(-2, -1)$

21 The total number of matrices $A = \begin{bmatrix} 0 & 2y & 1 \\ 2x & y & -1 \\ 2x & -y & 1 \end{bmatrix}, (x, y \in R, x \ne y)$ for which $A^T A = 3I_3$ is **[JEE Main 2019]**

(a) 2
(b) 4
(c) 3
(d) 6

22 If $A = \begin{bmatrix} 3 & -4 \\ 1 & -1 \end{bmatrix}$, then trace (A^{2012}) is

(a) 1
(b) 2
(c) -2
(d) -1

8.3 Special Kind of Matrices

23 If $A = \begin{bmatrix} 3 & x-1 \\ 2x+3 & x+2 \end{bmatrix}$, is a symmetric matrix, then the value of x is

(a) 4 (b) 3 (c) −4 (d) −3

24 If A is a square matrix then,

(a) $A + A^T$ is symmetric (b) AA^T is skew-symmetric
(c) $A^T + A$ is skew-symmetric (d) $A^T A$ is skew-symmetric

25 If A and B are two symmetric matrices of same order. Then, the matrix $AB - BA$ is equal to

(a) a symmetric matrix (b) a skew-symmetric matrix
(c) a null matrix (d) the identity matrix

26 The symmetric part of the matrix

$$A = \begin{bmatrix} 1 & 2 & 4 \\ 6 & 8 & 2 \\ 2 & -2 & 7 \end{bmatrix}$$ is equal to

(a) $\begin{bmatrix} 0 & -2 & -1 \\ -2 & 0 & -2 \\ -1 & -2 & 0 \end{bmatrix}$ (b) $\begin{bmatrix} 1 & 4 & 3 \\ 2 & 8 & 0 \\ 3 & 0 & 7 \end{bmatrix}$ (c) $\begin{bmatrix} 0 & -2 & 1 \\ 2 & 0 & 2 \\ -1 & 2 & 0 \end{bmatrix}$ (d) $\begin{bmatrix} 1 & 4 & 3 \\ 4 & 8 & 0 \\ 3 & 0 & 7 \end{bmatrix}$

27 If A and B are symmetric matrices of the same order and $X = AB + BA$ and $Y = AB - BA$, then $(XY)^T$ is equal to

(a) XY (b) YX (c) $-YX$ (d) None of these

28 Let A and B be two square matrices of the same size such that $AB^T + BA^T = 0$. If A is a skew-symmetric matrix, then BA is

(a) a symmetric matrix (b) a skew-symmetric matrix
(c) an orthogonal matrix (d) an invertible matrix

29 If A is symmetric matrix and B is a skew-symmetric matrix such that $A + B = \begin{bmatrix} 2 & 3 \\ 5 & -1 \end{bmatrix}$, then AB is equal to **[JEE Main 2019]**

(a) $\begin{bmatrix} -4 & -1 \\ -1 & 4 \end{bmatrix}$ (b) $\begin{bmatrix} 4 & -2 \\ -1 & -4 \end{bmatrix}$ (c) $\begin{bmatrix} 4 & -2 \\ 1 & -4 \end{bmatrix}$ (d) $\begin{bmatrix} -4 & 2 \\ 1 & 4 \end{bmatrix}$

30 Let $A = \begin{bmatrix} 2 & 3 \\ a & 0 \end{bmatrix}$, $a \in R$ be written as $P + Q$, where P is a symmetric matrix and Q is skew symmetric matrix. If $\det(Q) = 9$, then the modulus of the sum of all possible values of determinant of P is equal to

(a) 36 (b) 24 (c) 45 (d) 18

31 If both $A - \frac{1}{2}I$ and $A + \frac{1}{2}I$ are orthogonal matrices, then

(a) A is orthogonal (b) A is skew-symmetric of even order
(c) $A^2 = -\frac{3}{4}I$ (d) None of these

32 If $A, B, A + I, A + B$ are idempotent matrices, then AB is equal to

(a) BA (b) $-BA$ (c) I (d) 0

8.4 Invertible Matrices, System of Simultaneous Linear Equations and Reflection Matrix

33 If $A^2 - A + I = 0$, then the inverse of A is

(a) $A + I$ (b) A (c) $A - I$ (d) $I - A$

34 Which one of the following is correct always for any two non-singular matrices A and B of same order?

(a) $AB = BA$ (b) $(AB)^T = A^T B^T$ (c) $(A + B)(A - B) = A^2 - B^2$ (d) $(AB)^{-1} = B^{-1}A^{-1}$

35 If A and B are two non-singular matrices of the same order such that $B^r = I$, for some positive integer $r > 1$, then $A^{-1}B^{r-1}A - A^{-1}B^{-1}A$ is equal to

(a) I (b) $2I$ (c) 0 (d) $-I$

36 If $A = \begin{bmatrix} 0 & 1 & 2 \\ 1 & 2 & 3 \\ 3 & a & 1 \end{bmatrix}$ and $A^{-1} = \begin{bmatrix} 1/2 & -1/2 & 1/2 \\ -4 & 3 & c \\ 5/2 & -3/2 & 1/2 \end{bmatrix}$, then the values of a and c are equal to

(a) 1, 1 (b) 1, −1 (c) 1, 2 (d) −1, 1

37 If A and B are two square matrices such that $B = -A^{-1}BA$, then $(A + B)^2$ is equal to

(a) $A^2 + B^2$ (b) 0 (c) $A^2 + 2AB + B^2$ (d) $A + B$

38 If A is non-singular and $(A - 2I)(A - 4I) = 0$, then $\frac{1}{6}A + \frac{4}{3}A^{-1}$ is equal to

(a) 0 (b) I (c) $2I$ (d) $6I$

39 Let $P = \begin{bmatrix} 1 & 2 & 1 \\ 0 & 1 & -1 \\ 3 & 1 & 1 \end{bmatrix}$, if the product PQ has inverse $R = \begin{bmatrix} -1 & 0 & 1 \\ 1 & 1 & 3 \\ 2 & 0 & 2 \end{bmatrix}$, then Q^{-1} equals

(a) $\begin{bmatrix} 3 & 2 & 9 \\ -1 & 1 & 1 \\ 0 & 1 & 8 \end{bmatrix}$ (b) $\begin{bmatrix} 2 & 2 & 9 \\ -1 & 1 & 1 \\ 0 & 1 & 7 \end{bmatrix}$ (c) $\begin{bmatrix} 2 & -1 & 0 \\ 10 & 6 & 3 \\ 8 & 6 & 4 \end{bmatrix}$ (d) None of these

40 If A and B are square matrices of the same order such that $(A + B)(A - B) = A^2 - B^2$, then $(ABA^{-1})^2$ is equal to

(a) B^2 (b) I (c) A^2B^2 (d) A^2

41 If A and B are square matrices of the same order and $AB = 3I$, then A^{-1} is equal to

(a) $3B$ (b) $\frac{1}{3}B$ (c) $3B^{-1}$ (d) $\frac{1}{3}B^{-1}$

42 Let $A = \begin{bmatrix} 1 & 2 \\ -1 & 4 \end{bmatrix}$. If $A^{-1} = \alpha I + \beta A$, $\alpha, \beta \in R$, I is a 2×2 identity matrix, then $4(\alpha - \beta)$ is equal to

(a) 5 (b) 8/3 (c) 2 (d) 4

43 If A is a non-singular matrix such that $AA^T = A^TA$ and $B = A^{-1}A^T$, then matrix B is

(a) involuntary (b) orthogonal (c) idempotent (d) None of these

44 If A and B are square matrices of the same order and A is non-singular, then for a positive integer n, $(A^{-1}BA)^n$ is equal to

(a) $A^{-n}B^nA^n$ (b) $A^nB^nA^{-n}$ (c) $A^{-1}B^nA$ (d) $n(A^{-1}BA)$

45 If P is an orthogonal matrix and $Q = PAP^T$ and $X = P^TQ^{1000}P$, then X^{-1} is where A is involuntary matrix

(a) A (b) I (c) A^{1000} (d) None of these

46 Let M and N be two 3×3 non-singular skew-symmetric matrices such that $MN = NM$. If P^T denotes the transpose, then $M^2N^2\ (M^TN)^{-1}(MN^{-1})^T$ is equal to

(a) M^2 (b) $-N^2$ (c) $-M^2$ (d) MN

47 If the matrix $A = \begin{bmatrix} 1 & 2 & 3 & 0 \\ 2 & 4 & 3 & 2 \\ 3 & 2 & 1 & 3 \\ 6 & 8 & 7 & \alpha \end{bmatrix}$ is of rank 3, then α equals

(a) -5 (b) 5 (c) 4 (d) 1

48 The number of 3×3 matrices A, whose entries are either 0 or 1 and for which the system $A\begin{bmatrix} x \\ y \\ z \end{bmatrix} = \begin{bmatrix} 1 \\ 0 \\ 0 \end{bmatrix}$ has exactly two distinct solution is **[IIT 2010]**

(a) 0 (b) $2^9 - 1$ (c) 168 (d) 2

49 The system of linear equation $x - y - 2z = 6, -x + y + z = \mu$ and $\lambda x + y + z = 3$ has

(a) infinite number of soludions, for $\lambda \neq -1$ and all μ

(b) infinite number of solution, for $\lambda = -1$ and $\mu = 3$

(c) no solution, for $\lambda \neq -1$

(d) unique solution for $\lambda = -1$ and $\mu = -3$

50 Consider the system of linear equations

$x_1 + 2x_2 + x_3 = 3, 2x_1 + 3x_2 + x_3 = 3$ and $3x_1 + 5x_2 + 2x_3 = 1$. The system has

(a) infinite number of solutions (b) exactly 3 solution

(c) a unique solution (d) no solution

51 The value of λ and μ for which the system of equation $x + y + z = 6, x + 2y + 3z = 10$ and $x + 2y + \lambda z = \mu$ have infinite number of solutions are

(a) $\lambda = 3, \mu = 10$ (b) $\lambda = 3, \mu \neq 10$ (c) $\lambda \neq 3, \mu = 10$ (d) $\lambda \neq 3, \mu \neq 10$

52 If the system of linear equations $x + y + z = 5, x + 2y + 2z = 6, x + 3y + \lambda z = \mu, (\lambda, \mu \in R)$ has infinitely many solutions, then the value of $\lambda + \mu$ is **[JEE Main 2019]**

(a) 7 (b) 3 (c) 10 (d) 8

53 The image of the point (1, 2) about the Y-axis.

(a) (1, 2) (b) (−1, 2) (c) (1, −2) (d) (−1, −2)

54 The image of the point (−1, 2) about origin.

(a) (−1, −2) (b) (−1, 2) (c) (1, −2) (d) (1, 2)

Numerical Value Type Questions

55 A square matrix M of order 3 satisfies $M^2 = I - M$, where I is an identity matrix of order 3. If $M^n = 5I - 8M$, then n is equal to ....

56 $A = \begin{bmatrix} 0 & 1 \\ 3 & 0 \end{bmatrix}$ and $(A^8 + A^6 + A^4 + A^2 + I) \cdot V = \begin{bmatrix} 0 \\ 11 \end{bmatrix}$ (where I is the 2×2 identity matrix), then the product of all elements of matrix V is ....

57 Let $A = \begin{bmatrix} 0 & \alpha \\ 0 & 0 \end{bmatrix}$ and $(A + I)^{70} - 70A = \begin{bmatrix} a-1 & b-1 \\ c-1 & d-1 \end{bmatrix}$, then the value of $a + b + c + d$ is ....

58 Let M be a 3×3 matrix satisfying

$$M\begin{bmatrix}0\\1\\0\end{bmatrix}=\begin{bmatrix}-1\\2\\3\end{bmatrix}, M\begin{bmatrix}1\\-1\\0\end{bmatrix}=\begin{bmatrix}1\\1\\-1\end{bmatrix}, \text{ and } M\begin{bmatrix}1\\1\\1\end{bmatrix}=\begin{bmatrix}0\\0\\12\end{bmatrix}$$, then the sum of the diagonal entries of M is ………

59 Let $A=\begin{bmatrix}1&0&0\\1&0&1\\0&1&0\end{bmatrix}$ satisfies

$A^n=A^{n-2}+A^2-I$, then trace of A^{50} equals ………

60 If a, b and c are integers, then the number of matrices

$A=\begin{bmatrix}a&b&c\\b&c&a\\c&a&b\end{bmatrix}$ which are possible such that $AA^T=I$ is ………

61 If $\begin{bmatrix}a&b\\c&1-a\end{bmatrix}$ is an idempotent matrix and $f(x)=x-x^2$ and $bc=\frac{1}{4}$, then the value of $f(a)$ is ………

62 If A is an idempotent matrix satisfying, $[I-0.4A]^{-1}=I-\alpha A$, where i is the unit matrix of the same order as that of A, then the value of $|9\alpha|$ is equal to ………

63 The rank of the matrix $\begin{bmatrix}1&0&2\\0&0&1\\0&0&1\end{bmatrix}$ is ………

64 Let A be a 3×3 matrix given by $A=[a_{ij}]$. If for every column vector X, $X^TAX=0$ and $a_{23}=-1008$, the sum of the digits of a_{32} is ………

LEVEL 03 \ JEE Advanced Level

Single Choice Correct Questions

1 If $\begin{bmatrix}\alpha&\beta\\\gamma&-\alpha\end{bmatrix}$ is to be the square root of two rowed unit matrix, then α, β and γ should satisfy the relation

(a) $1-\alpha^2+\beta\gamma=0$ (b) $\alpha^2+\beta\gamma-1=0$ (c) $1+\alpha^2+\beta\gamma=0$ (d) $1+\alpha^2-\beta\gamma=0$

2 If $A=\begin{bmatrix}i&-i\\-i&i\end{bmatrix}$ and $B=\begin{bmatrix}1&-1\\-1&1\end{bmatrix}$, then A^8 equals

(a) $4B$ (b) $128B$ (c) $-128B$ (d) $-64B$

3 The number of diagonal matrix A of order n for which $A^3=A$ is

(a) 1 (b) 0 (c) 2^n (d) 3^n

4 Matrix A such that $A^2=2A-I$, where I is the identity matrix, then for $n\ge2$, A^n is equal to

(a) $2^{n-1}A-(n-1)I$ (b) $2^{n-1}A-I$ (c) $nA-(n-1)I$ (d) $nA-I$

5 If A and B are square matrices of order n, then $A-\lambda I$ and $B-\lambda I$. Commute for every scalar λ, only if

(a) $AB=BA$ (b) $AB+BA=0$ (c) $A=-B$ (d) None of these

6 If $\begin{bmatrix} 2 & -1 \\ 1 & 0 \\ -3 & 4 \end{bmatrix} A = \begin{bmatrix} -1 & -8 & -10 \\ 1 & -2 & -5 \\ 9 & 22 & 15 \end{bmatrix}$, then sum of all the elements of matrix A is

(a) 0 (b) 1 (c) 2 (d) -3

7 Let $P = \begin{bmatrix} 1 & 0 & 0 \\ 4 & 1 & 0 \\ 16 & 4 & 1 \end{bmatrix}$ and I be the identity matrix of order 3. If $P = [a_{ij}]$ is a matrix such that $P^{50} - Q = I$, then $\dfrac{q_{31} + q_{32}}{q_{21}}$ equals **[JEE Advanced 2016]**

(a) 52 (b) 103 (c) 201 (d) 205

8 If $\begin{bmatrix} \cos\dfrac{2\pi}{7} & -\sin\dfrac{2\pi}{7} \\ \sin\dfrac{2\pi}{7} & \cos\dfrac{2\pi}{7} \end{bmatrix}^k = \begin{bmatrix} 1 & 0 \\ 0 & 1 \end{bmatrix}$, then the least positive integral value of k is

(a) 3 (b) 6 (c) 7 (d) 14

9 A is a 2×2 matrix such that $A \begin{bmatrix} 1 \\ -1 \end{bmatrix} = \begin{bmatrix} -1 \\ 2 \end{bmatrix}$ and $A^2 \begin{bmatrix} 1 \\ -1 \end{bmatrix} = \begin{bmatrix} 1 \\ 0 \end{bmatrix}$ the sum of the elements of A is

(a) -1 (b) 0 (c) 2 (d) 5

10 If $A = \begin{bmatrix} a & b \\ 0 & a \end{bmatrix}$ is nth root of I (identity matrix of order 2×2), then choose the correct statements:

I. if n is odd, $a = 1, b = 0$
II. if n is odd, $a = -1, b = 0$
III. if n is even, $a = 1, b = 0$
IV. if n is even, $a = -1, b = 0$

(a) I, II, III (b) II, III, IV (c) I, II, III, IV (d) I, III, IV

11 For each real x, $-1 < x < 1$, let $A(x)$ be the matrix $(1-x)^{-1} \begin{bmatrix} 1 & -x \\ -x & 1 \end{bmatrix}$ and $z = \dfrac{x+y}{1+xy}$, then

(a) $A(z) = A(x) \cdot A(y)$ (b) $A(z) = A(x) - A(y)$
(c) $A(z) = A(x) + A(y)$ (d) $A(z) = A(x)[A(y)]^{-1}$

12 If $P = \begin{bmatrix} \dfrac{\sqrt{3}}{2} & \dfrac{1}{2} \\ \dfrac{-1}{2} & \dfrac{\sqrt{3}}{2} \end{bmatrix}$; $A = \begin{bmatrix} 1 & 1 \\ 0 & 1 \end{bmatrix}$ and $Q = PAP^T$, then $P^T Q^{2005} P$ is equal to

(a) $\begin{bmatrix} 1 & 2005 \\ 0 & 1 \end{bmatrix}$ (b) $\begin{bmatrix} 1 & 2005 \\ 2005 & 1 \end{bmatrix}$ (c) $\begin{bmatrix} 1 & 0 \\ 2005 & 1 \end{bmatrix}$ (d) $\begin{bmatrix} 1 & 0 \\ 0 & 1 \end{bmatrix}$

13 How many 2×3 matrices M with entries from {0, 1, 2} are there for which the sum of the diagonal entries of $M^T M$ is 5 **[JEE Advanced 2017]**

(a) 198 (b) 126 (c) 135 (d) 162

14 Elements of a matrix A order 10×10 are defined as $a_{ij} = \omega^{i+j}$ (where ω is imaginary cube root of unity, then trace (A) of the matrix is

(a) 0 (b) 1 (c) 3 (d) None of these

15 Consider three matrices $A=\begin{bmatrix}2 & 1\\ 4 & 1\end{bmatrix}$, $B=\begin{bmatrix}3 & 4\\ 2 & 3\end{bmatrix}$ and $C=\begin{bmatrix}3 & -4\\ -2 & 3\end{bmatrix}$, then the value of sum

$tr(A)+tr\left(\frac{ABC}{2}\right)+tr\left(\frac{A(BC)^2}{4}\right)+tr\left(\frac{A(BC)^3}{8}\right)+......+\infty$ is

(a) 6 (b) 9 (c) 12 (d) None of these

16 If $A_1, A_3, ..., A_{2n-1}$ are n skew-symmetric matrices of same order, then $B=\sum_{r=1}^{n}(2r-1)(A_{2r-1})^{2r-1}$ will be

(a) symmetric
(b) skew-symmetric
(c) neither symmetric nor skew-symmetric
(d) data not adequate

17 If A is symmetric as well as skew-symmetric matrix, then A is

(a) diagonal matrix (b) null matrix
(c) triangular matrix (d) None of these

18 The matrix $A=\begin{bmatrix}\frac{1}{\sqrt{2}} & \frac{1}{\sqrt{2}}\\ \frac{-1}{\sqrt{2}} & \frac{-1}{\sqrt{2}}\end{bmatrix}$ is

(a) unitary (b) orthogonal (c) nilpotent (d) involuntary

19 If B is an invertible matrix and A is a matrix, then

(a) rank $(BA)=$ rank (A) (b) rank $(BA)\geq$ rank (B)
(c) rank $(BA)>$ rank (A) (d) rank $(BA)>$ rank (B)

20 If A is a non-singular matrix such that $A^3=A+I$, then the inverse of $B=A^6-A^5$ is

(a) A (b) A^{-1} (c) $-A$ (d) $-A^{-1}$

More than One Correct

21 Let $C_k={}^nC_k$ for $0\leq k\leq n$ and $A_k=\begin{bmatrix}C_{k-1}^2 & 0\\ 0 & C_k^2\end{bmatrix}$ for $k\geq 1$

and $A_1+A_2+A_3+....+A_n=\begin{bmatrix}k_1 & 0\\ 0 & k_2\end{bmatrix}$, then

(a) $k_1=k_2$ (b) $k_1+k_2=2$ (c) $k_1={}^{2n}C_n-1$ (d) $k_2={}^{2n}C_{n-1}$

22 If $A=\begin{bmatrix}1 & 1 & 1\\ 1 & 1 & 1\\ 1 & 1 & 1\end{bmatrix}$, then

(a) $A^3=9A$ (b) $A^3=27A$ (c) $A+A=A^2$ (d) $A+A+A=A^2$

23 If $A=\begin{bmatrix}1 & -1\\ 2 & 1\end{bmatrix}$, $B=\begin{bmatrix}a & 1\\ b & -1\end{bmatrix}$ and $(A+B)^2=A^2+B^2+2AB$, then

(a) $a=-1$ (b) $a=1$ (c) $b=2$ (d) $b=-2$

24 If D_1 and D_2 are two 3×3 diagonal matrices where none of the diagonal element is zero, then

(a) D_1D_2 is a diagonal matrix (b) $D_1D_2=D_2D_1$
(c) $D_1^2+D_2^2$ is diagonal matrix (d) None of these

25 Let $A = \begin{bmatrix} 1 & 0 & 0 \\ 2 & 1 & 0 \\ 4 & 2 & 1 \end{bmatrix}$ and $B = [b_{ij}]$ is a 3×3 matrix such that $A^{100} - B = I$, then

(a) $b_{32} = b_{21}$ (b) B is lower triangular matrix
(c) B is upper triangular matrix (d) $b_{21} = 100$

26 Let A and B are two square matrices such that $AB = A$, $BA = B$, then $(A^2 - AB + B^2)$ is always equal to

(a) A (b) B (c) A^2 (d) B^2

27 Let A and B are two matrices such that $AB = BA$, then for every $n \in N$

(a) $A^nB = BA^n$ (b) $(AB)^n = A^nB^n$
(c) $(A+B)^n = {}^nC_0A^n + {}^nC_1A^{n-1}B + \dots + {}^nC_nB^n$
(d) $A^{2n} - B^{2n} = (A^n - B^n)(A^n + B^n)$

28 The product of matrices

$A = \begin{bmatrix} \cos^2\theta & \cos\theta\sin\theta \\ \cos\theta\sin\theta & \sin^2\theta \end{bmatrix}$ and $B = \begin{bmatrix} \cos^2\phi & \cos\phi\sin\phi \\ \cos\phi\sin\phi & \sin^2\phi \end{bmatrix}$ is a null matrix, if $\theta - \phi$ is equal to

(a) $2n\pi, n \in z$ (b) $n\frac{\pi}{2}, n \in z$ (c) $(2n+1)\frac{\pi}{2}, n \in z$ (d) $n\pi + \frac{\pi}{2}, n \in z$

29 Let A and B be two non-singular square matrices, A^T and B^T are the transpose matrices of A and B, respectively, then which of the following are correct?

(a) B^TAB is symmetric matrix if A is symmetric
(b) B^TAB is symmetric matrix if B is symmetric
(c) B^TAB is skew-symmetric matrix for every matrix A
(d) B^TAB is skew-symmetric matrix if A is skew-symmetric

30 If P and Q are symmetric non-singular matrices and $PQ = QP$, then

(a) PQ is symmetric (b) $P^{-1}Q$ is symmetric (c) $QP^{-1} = P^{-1}Q$ (d) $Q^{-1}P$ is symmetric

31 Let x and y be two arbitrary 3×3, non-zero, skew-symmetric matrices and z be an arbitrary, 3×3, non-zero, symmetric matrix. Then, which of the following matrices is (are) skew-symmetric? **[JEE Advanced 2015]**

(a) $y^3z^4 - z^4y^3$ (b) $x^{44} + y^{44}$ (c) $x^4z^3 - z^3x^4$ (d) $x^{23} + y^{23}$

32 If $AB = A$ and $BA = B$, then which of the following is/are true?

(a) A is idempotent (b) B is idempotent (c) A^T is idempotent (d) None of these

33 If B is an idempotent matrix and $A = I - B$, then

(a) $A^2 = A$ (b) $A^2 = I$ (c) $AB = 0$ (d) $BA = 0$

34 If A and B are square matrices such that $AB = B$ and $BA = A$, then

(a) A is an idempotent matrix (b) $A^3 + B^2 = A^2 + B^3$
(c) $A^3 + B^3 = A + B$ (d) A is an involutary matrix

35 If the matrix A and B are of 3×3 and $(I - AB)$ is invertible, then which of the following statement is/are correct?

(a) $I - BA$ is not invertible (b) $I - BA$ is invertible
(c) $I - BA$ has for its inverse $I + B(I - AB)^{-1}A$ (d) $I - BA$ has for its inverse $I + A(I - BA)^{-1}B$

36 The rank of the matrix $\begin{bmatrix} -1 & 2 & 5 \\ 2 & -4 & a-4 \\ 1 & -2 & a+1 \end{bmatrix}$ is

(a) 2, if $a = -6$ (b) 2, if $a = 1$ (c) 1, if $a = 2$ (d) 1, if $a = -6$

Linked Comprehension Type

Paragraph I Directions (Q.Nos. 37-39)

If A is symmetric and B is skew-symmetric matrix and $A+B$ is non-singular and $C=(A+B)^{-1}(A-B)$.

37 $C^T(A+B)C$ equals

(a) $A+B$ (b) $A-B$ (c) A (d) B

38 $C^T(A-B)C$ equals

(a) $A+B$ (b) $A-B$ (c) A (d) B

39 C^TAC equals

(a) $A+B$ (b) $A-B$ (c) A (d) B

Paragraph II Directions (Q.Nos. 40-42)

Let $A=\begin{bmatrix} a & b & c \\ b & c & a \\ c & a & b \end{bmatrix}$ is an orthogonal matrix and $abc=\lambda(<0)$.

40 The value of $a^2b^2+b^2c^2+c^2a^2$ is

(a) 2λ (b) -2λ (c) λ (d) $-\lambda$

41 The value of $a^3+b^3+c^3$ is

(a) λ (b) 2λ (c) 3λ (d) None of these

42 The equation whose roots are a, b, c is

(a) $x^3-2x^2+\lambda=0$ (b) $x^3-\lambda x^2+\lambda x+\lambda=0$
(c) $x^3-2x^2+2\lambda x+\lambda=0$ (d) $x^3\pm x^2-\lambda=0$

Paragraph III Directions (Q.Nos. 43-45)

Suppose A and B be two non-singular matrices such that $AB=BA^m$, $B^n=I$ and $A^p=I$, where I is an identity matrix.

43 If $m=2$ and $n=5$, then p is equal to

(a) 30 (b) 31 (c) 33 (d) 81

44 The relation between m, n and p is

(a) $p=mn^2$ (b) $p=m^n-1$ (c) $p=n^m-1$ (d) $p=m$

45 Which of the following ordered triplet(m, n, p) is false?

(a) (3, 4, 80) (b) (6, 3, 215) (c) (8, 3, 50) (d) (2, 8, 255)

Matrix Match Type Questions

46 Let A and B are two square matrices of same order such that $AB+BA=0$, then match the following list.

List-I		List-II	
(i)	$(A-B)(A^2+AB+B^2)$	A.	A^3-B^3
(ii)	$(A+B)(A^2-AB-B^2)$	B.	$A^3-2AB^2-B^3$
(iii)	$(A+B)(A^2+AB-B^2)$	C.	$-B$
(iv)	$A^{-1}BA$	D.	A^3+B^3

47 Match the following lists.

	List I		List II
(i)	$(I-A)^n$ is if A is idempotent	A.	$2^{n-1}(I-A)$
(ii)	$(I-A)^n$ is if A is involutory	B.	$I-nA$
(iii)	$(I-A)^n$ is if A is nilpotent of index 2	C.	A
(iv)	If A is orthogonal then $(A^T)^{-1}$	D.	$I-A$

48 Consider a matrix $A=[a_{ij}]$ of order 3×3 such that $a_{ij}=(k)^{i+j}$ where $k\in I$

Match List I with List II.

	List-I		List-II
(i)	A is singular, if	A.	$k\in\{0\}$
(ii)	A is null matrix, if	B.	$k\in\phi$
(iii)	A is skew-symmetric which is not null matrix, if	C.	$k\in I$
(iv)	$A^2=3A$, if	D.	$k\in\{-1,0,1\}$

Numerical Value Type Questions

Question Stem Directions (Q.Nos. 49-50) The answer to each question is a **NUMERICAL VALUE.**

Let A be a matrix of order 2×2 such that $A^2=0$

49 $A^2-(a+d)A+(ad-bc)$ is equal to ....

50 Trace (A) is equal to ....

51 Let A, B, C, D be real matrices such that $A^T=BCD$; $B^T=CDA$; $C^T=DAB$ and $D^T=ABC$ for the matrix $S=ABCD$, the least value of k such that $S^k=S$ is ....

52 Let x be the solution set of the equation $A^n=I$, where $A=\begin{bmatrix}0&1&-1\\4&-3&4\\3&-3&4\end{bmatrix}$ and I is the corresponding unit matrices $x\in N$, the minimum value of $\Sigma(\cos^n\theta+\sin^n\theta)$, $\theta\in R-\left\{\frac{n\pi}{2}, n\in I\right\}$ is ....

53 If A is a square matrix of order 2×2 and $\det A=10$, then $\text{trace}(A)^2-\text{trace}(A^2)$ is equal to ....

54 Let A be the set of all 3×3 skew-symmetric matrices whose entries are either -1, 0 or 1. If there are exactly three 0's, three 1's and three $(-1)^1$ s, then the number of such matrices is ....

55 If the matrix $A=\begin{bmatrix}\lambda_1^2&\lambda_1\lambda_2&\lambda_1\lambda_3\\\lambda_2\lambda_1&\lambda_2^2&\lambda_2\lambda_3\\\lambda_3\lambda_1&\lambda_3\lambda_2&\lambda_3^2\end{bmatrix}$ is idempotent, then the value of $\lambda_1^2+\lambda_2^2+\lambda_3^2$ is ....

56 If A is an idempotent matrix and I is an identity matrix of the same order, then the value of n, such that $(A+I)^n=I+127\,A$ is ....

57 The rank of the matrix $A=\begin{bmatrix}1&2&3\\2&1&4\\3&0&5\end{bmatrix}$ is ....

ANSWER KEY

Level 1

1. $1\times 8, 8\times 1, 4\times 2, 2\times 4$ **2.** $A=\begin{bmatrix}1 & 5/2\\ 1/2 & 2\\ 0 & 2\end{bmatrix}$ **3.** (i) $A=\begin{bmatrix}1/2 & 9/2\\ 0 & 2\end{bmatrix}$ (ii) $A=\begin{bmatrix}1 & 4\\ 1 & 2\end{bmatrix}$

4. $a=-2, b=-7, c=-1, x=-3, y=-5$ and $z=2$ **5.** (i) not defined (ii) $\begin{bmatrix}5 & 4 & 3\\ 1 & 2 & 3\\ 2 & 3 & 4\end{bmatrix}$

6. $\begin{bmatrix}1 & 0 & 1\\ 1 & 1 & 1\\ 1 & 0 & 1\end{bmatrix}$ **7.** 1 **8.** $X=\begin{bmatrix}4 & 4\\ 0 & 4\end{bmatrix}$ and $Y=\begin{bmatrix}1 & -2\\ 0 & 5\end{bmatrix}$

9. $X=\begin{bmatrix}2 & -1 & 1\\ 1 & 2 & 0\end{bmatrix}$ and $Y=\begin{bmatrix}1 & 1 & 2\\ -1 & 1 & -2\end{bmatrix}$ **10.** $x=2$ and $y=9$ **11.** $\begin{bmatrix}-2 & -10/3\\ 4 & 14/3\\ -31/3 & -7/3\end{bmatrix}$

12. $B=\begin{bmatrix}-2 & 1\\ 3 & 0\end{bmatrix}$ and $C=\begin{bmatrix}1 & -2\\ 1 & 1\end{bmatrix}$ **13.** 0 (zero) **14.** $AB=\begin{bmatrix}0 & -4\\ 10 & 3\end{bmatrix}$ and $B=\begin{bmatrix}-10 & 2 & 21\\ -16 & 2 & 37\\ -2 & -2 & 11\end{bmatrix}$

15. 20160 **16.** $D=\begin{bmatrix}-191 & -110\\ 77 & 44\end{bmatrix}$ **17.** 0 (zero) **18.** $\begin{bmatrix}1 & 0\\ 0 & 1\end{bmatrix}$ or $\begin{bmatrix}-1 & 0\\ 0 & -1\end{bmatrix}$

20. $x=\dfrac{-9\pm\sqrt{53}}{2}$ **24.** $x=\pm\dfrac{1}{\sqrt{2}}, y=\pm\dfrac{1}{\sqrt{6}}$ and $z=\pm\dfrac{1}{\sqrt{3}}$

25. (i) 3 (ii) 15 (iii) 15 **26.** (i) 15 (ii) 21 (iii) 7 **27.** 2

28. $x=4, y=2, t=-3$

33. $X=\begin{bmatrix}3 & 3/2 & 5/2\\ 3/2 & 4 & 4\\ 5/2 & 4 & 8\end{bmatrix}$ and $Y=\begin{bmatrix}0 & 1/2 & 9/2\\ -1/2 & 0 & -1\\ -9/2 & 1 & 0\end{bmatrix}$

34. $a=3$ **38.** 3 **44.** $A^{-1}=\begin{bmatrix}1/5 & 2/5\\ 2/5 & -1/5\end{bmatrix}$ **45.** do not exist **46.** $A^{-1}=\begin{bmatrix}3 & 1 & 2\\ 3 & 0 & 2\\ 1 & 0 & 1\end{bmatrix}$

47. (i) $A^{-1}=\begin{bmatrix}1/2 & -1/2 & 1/2\\ -4 & 3 & -1\\ 5/2 & -3/2 & 1/2\end{bmatrix}$ (ii) $A^{-1}=\begin{bmatrix}3 & -1 & 1\\ -15 & 6 & -5\\ 5 & -2 & 2\end{bmatrix}$ **48.** $\lambda=-2$ **49.** 2

50. (i) $\lambda=3$ and $\mu\neq 10$ (ii) $\lambda\neq 3$ and $\mu\in R$ (iii) $\lambda=3$ and $\mu=10$

51. $(12, -1)$ **52.** $\alpha=3$ and $\beta=-2$ **53.** $(0, -2)$

Level 2

1. (d)	**2.** (b)	**3.** (b)	**4.** (c)	**5.** (b)	**6.** (a)	**7.** (a)
8. (d)	**9.** (d)	**10.** (c)	**11.** (c)	**12.** (a)	**13.** (b)	**14.** (c)
15. (d)	**16.** (d)	**17.** (a)	**18.** (b)	**19.** (a)	**20.** (d)	**21.** (b)
22. (b)	**23.** (c)	**24.** (a)	**25.** (b)	**26.** (d)	**27.** (c)	**28.** (b)
29. (b)	**30.** (a)	**31.** (c)	**32.** (b)	**33.** (d)	**34.** (d)	**35.** (c)
36. (a)	**37.** (a)	**38.** (b)	**39.** (c)	**40.** (a)	**41.** (b)	**42.** (d)
43. (b)	**44.** (c)	**45.** (b)	**46.** (c)	**47.** (b)	**48.** (a)	**49.** (b)
50. (d)	**51.** (a)	**52.** (c)	**53.** (b)	**54.** (c)	**55.** (6)	**56.** (0)
57. (6)	**58.** (9)	**59.** (3)	**60.** (6)	**61.** (1/4)	**62.** (6)	**63.** (2)
64. (9)						

Level 3

1. (b)	**2.** (b)	**3.** (d)	**4.** (c)	**5.** (a)	**6.** (b)	**7.** (b)
8. (c)	**9.** (d)	**10.** (d)	**11.** (a)	**12.** (a)	**13.** (a)	**14.** (d)
15. (a)	**16.** (b)	**17.** (b)	**18.** (c)	**19.** (a)	**20.** (b)	**21.** (a, d)
22. (a,d)	**23.** (a, d)	**24.** (a, b, c)	**25.** (a, b)	**26.** (b, d)	**27.** (a, c, d)	**28.** (c, d)
29. (a, d)	**30.** (a, b, c, d)	**31.** (c, d)	**32.** (a, b, c)	**33.** (a, c, d)	**34.** (a, b, c)	**35.** (b, c)
36. (b, d)	**37.** (a)	**38.** (b)	**39.** (c)	**40.** (b)	**41.** (d)	**42.** (d)
43. (b)	**44.** (b)	**45.** (c)				

46. (i)–B, (ii)–A, (iii)–D, (iv)–C

47. (i)–D, (ii)–A, (iii)–B, (iv)–C

48. (i)–C, (ii)–A, (iii)–B, (iv)–D

49. 0	**50.** 0	**51.** (3)	**52.** (2)	**53.** (20)	**54.** (8)	**55.** (1)
56. (7)	**57.** (2)					

SOLUTIONS

Level 1

1. We know that if a matrix is of order $m \times n$, then it has mn elements. Thus, to find all possible orders of a matrix with 8 element, we will find all ordered pairs of natural numbers, whose product is 8.
Thus, all possible ordered pairs are
(1, 8), (8, 1), (4, 2), (2, 4).
Hence, possible orders are $1 \times 8, 8 \times 1, 4 \times 2, 2 \times 4$

2. In general a 3×2 matrix is given by

$$A = \begin{bmatrix} a_{11} & a_{12} \\ a_{21} & a_{22} \\ a_{31} & a_{32} \end{bmatrix}$$

Now, $a_{ij} = \frac{1}{2}|i - 3j|, i = 1, 2, 3$ and $j = 1, 2$

Therefore,

$$a_{11} = \frac{1}{2}|1 - 3 \times 1| = 1$$

$$a_{12} = \frac{1}{2}|1 - 3 \times 2| = \frac{5}{2}$$

$$a_{21} = \frac{1}{2}|2 - 3 \times 1| = \frac{1}{2}$$

$$a_{22} = \frac{1}{2}|2 - 3 \times 2| = 2$$

$$a_{31} = \frac{1}{2}|3 - 3 \times 1| = 0$$

$$a_{32} = \frac{1}{2}|3 - 3 \times 2| = \frac{3}{2}$$

Hence, the required matrix is given by

$$A = \begin{bmatrix} 1 & 5/2 \\ 1/2 & 2 \\ 0 & 3/2 \end{bmatrix}$$

3. (i) We have, $A = [a_{ij}]_{2 \times 2}$
such that

$$a_{ij} = \frac{(i - 2j)^2}{2};$$

where $1 \le i \le 2; 1 \le j \le 2$

$$a_{11} = \frac{(1-2)^2}{2} = \frac{1}{2}$$

$$a_{12} = \frac{(1 - 2 \times 2)^2}{2} = \frac{9}{2}$$

$$a_{21} = \frac{(2 - 2 \times 1)^2}{2} = 0$$

$$a_{22} = \frac{(2 - 2 \times 2)^2}{2} = 2$$

So, $A = \begin{bmatrix} \frac{1}{2} & \frac{9}{2} \\ 0 & 2 \end{bmatrix}$

(ii) We have,

$$A = [a_{ij}]_{2 \times 2}$$

such that

$$a_{ij} = |-2i + 3j|, \text{ where } 1 \le i \le 2; 1 \le j \le 2$$

$$a_{11} = |-2 \times 1 + 3 \times 1| = 1$$

$$a_{12} = |-2 \times 1 + 3 \times 2| = 4$$

$$a_{21} = |-2 \times 2 + 3 \times 1| = 1$$

$$a_{22} = |-2 \times 2 + 3 \times 2| = 2$$

$$\therefore \quad A = \begin{bmatrix} 1 & 4 \\ 1 & 2 \end{bmatrix}$$

4. As the given matrices are equal therefore, their corresponding elements must be equal.
Comparing the corresponding elements, we get

$$x + 3 = 0$$
$$a - 1 = -3$$
$$z + 4 = 6$$
$$0 = 2c + 2$$
$$2y - 7 = 3y - 2$$
$$b - 3 = 2b + 4$$

Simplifying, we get

$$a = -2, b = -7, c = -1, x = -3, y = -5, z = 2$$

5. (i) Given, A is a matrix of the type 3×3 and B is a matrix of the type 3×2.
Since, A and B are not the same type.
$\therefore$ Sum $A + B$ is not defined.

(ii) As A and C are two matrices of the same type.
Therefore, the sum $A + C$ is defined

$$\therefore A + C = \begin{bmatrix} 1 & 3 & 5 \\ -2 & 0 & 2 \\ 0 & 4 & -3 \end{bmatrix} + \begin{bmatrix} 4 & 1 & -2 \\ 3 & 2 & 1 \\ 2 & -1 & 7 \end{bmatrix}$$

$$= \begin{bmatrix} 1+4 & 3+1 & 5-2 \\ -2+3 & 0+2 & 2+1 \\ 0+2 & 4-1 & -3+7 \end{bmatrix}$$

$$= \begin{bmatrix} 5 & 4 & 3 \\ 1 & 2 & 3 \\ 2 & 3 & 4 \end{bmatrix}$$

6. $A - B = \begin{bmatrix} 1 & 1 & 1 \\ 1 & 1 & 1 \\ 1 & 1 & 1 \end{bmatrix} - \begin{bmatrix} 0 & 1 & 0 \\ 0 & 0 & 1 \\ 0 & 1 & 0 \end{bmatrix} = \begin{bmatrix} 1 & 0 & 1 \\ 1 & 1 & 0 \\ 1 & 0 & 1 \end{bmatrix}$

7. We have, $A=\begin{bmatrix}0 & 2\\ 3 & -4\end{bmatrix}$

$$\Rightarrow \quad kA=\begin{bmatrix}0 & 2k\\ 3k & -4k\end{bmatrix}$$

But $$kA=\begin{bmatrix}0 & 3a\\ 2b & 24\end{bmatrix}$$

$$\therefore \begin{bmatrix}0 & 2k\\ 3k & -4k\end{bmatrix}=\begin{bmatrix}0 & 3a\\ 2b & 24\end{bmatrix}$$

$\Rightarrow$ $2k=3a, 3k=2b, -4k=24$

$\Rightarrow$ $k=-6, a=-4, b=-9$

Hence, $b-a-k=-9-(-4)-(-6)=-9+10=1$

8. We have,

$$(X+Y)+(X-Y)=\begin{bmatrix}5 & 2\\ 0 & 9\end{bmatrix}+\begin{bmatrix}3 & 6\\ 0 & -1\end{bmatrix}$$

$$(X+X)=\begin{bmatrix}8 & 8\\ 0 & 8\end{bmatrix}$$

$$\Rightarrow \quad 2X=\begin{bmatrix}8 & 8\\ 0 & 8\end{bmatrix}$$

$$X=\frac{1}{2}\begin{bmatrix}8 & 8\\ 0 & 8\end{bmatrix}=\begin{bmatrix}4 & 4\\ 0 & 4\end{bmatrix}$$

$$(X+Y)-(X-Y)=\begin{bmatrix}5 & 2\\ 0 & 9\end{bmatrix}\begin{bmatrix}3 & 6\\ 0 & -1\end{bmatrix}$$

$$(Y+Y)=\begin{bmatrix}5-3 & 2-6\\ 0 & 9+1\end{bmatrix}$$

$$\Rightarrow \quad 2Y=\begin{bmatrix}2 & -4\\ 0 & 10\end{bmatrix}$$

$$Y=\frac{1}{2}\begin{bmatrix}2 & -4\\ 0 & 10\end{bmatrix}$$

$$=\begin{bmatrix}1 & -2\\ 0 & 5\end{bmatrix}$$

9. Given, $2X-Y=\begin{bmatrix}3 & -3 & 0\\ 3 & 3 & 2\end{bmatrix}$

On multiplying both side by 2, we get

$$4X-2Y=2\begin{bmatrix}3 & -3 & 0\\ 3 & 3 & 2\end{bmatrix};$$

$$4X-2Y=\begin{bmatrix}6 & -6 & 0\\ 6 & 6 & 4\end{bmatrix} \quad \text{...(i)}$$

Also, $$X+2Y=\begin{bmatrix}4 & 1 & 5\\ -1 & 4 & -4\end{bmatrix} \quad \text{...(ii)}$$

On adding Eqs. (i) and (ii), we get

$$5X=\begin{bmatrix}6 & -6 & 0\\ 6 & 6 & 4\end{bmatrix}+\begin{bmatrix}4 & 1 & 5\\ -1 & 4 & -4\end{bmatrix}$$

$$=\begin{bmatrix}6+4 & -6+1 & 0+5\\ 6-1 & 6+4 & 4-4\end{bmatrix}$$

$$=\begin{bmatrix}10 & -5 & 5\\ 5 & 10 & 0\end{bmatrix}$$

$$X=\frac{1}{5}\begin{bmatrix}10 & -5 & 5\\ 5 & 10 & 0\end{bmatrix}$$

$$\Rightarrow \quad X=\begin{bmatrix}2 & -1 & 1\\ 1 & 2 & 0\end{bmatrix}$$

Putting the value of X in Eq. (ii), we get

$$\begin{bmatrix}2 & -1 & 1\\ 1 & 2 & 0\end{bmatrix}+2Y=\begin{bmatrix}4 & 1 & 5\\ -1 & 4 & -4\end{bmatrix}$$

$$\Rightarrow \quad 2Y=\begin{bmatrix}4 & 1 & 5\\ -1 & 4 & -4\end{bmatrix}-\begin{bmatrix}2 & -1 & 1\\ 1 & 2 & 0\end{bmatrix}$$

$$=\begin{bmatrix}4-2 & 1+1 & 5-1\\ -1-1 & 4-2 & -4-0\end{bmatrix}$$

$$=\begin{bmatrix}2 & 2 & 4\\ -2 & 2 & -4\end{bmatrix}$$

$$Y=\begin{bmatrix}1 & 1 & 2\\ -1 & 1 & -2\end{bmatrix}$$

Hence, $X=\begin{bmatrix}2 & -1 & 1\\ 1 & 2 & 0\end{bmatrix}$ and $Y=\begin{bmatrix}1 & 1 & 2\\ -1 & 1 & -2\end{bmatrix}$.

10. We have,

$$2\begin{bmatrix}x & 5\\ 7 & y-3\end{bmatrix}+\begin{bmatrix}3 & -4\\ 1 & 2\end{bmatrix}=\begin{bmatrix}7 & 6\\ 15 & 14\end{bmatrix}$$

$$\Rightarrow \quad \begin{bmatrix}2x & 10\\ 14 & 2y-6\end{bmatrix}+\begin{bmatrix}3 & -4\\ 1 & 2\end{bmatrix}=\begin{bmatrix}7 & 6\\ 15 & 14\end{bmatrix}$$

or $$\begin{bmatrix}2x+3 & 10-4\\ 14+1 & 2y-6+2\end{bmatrix}=\begin{bmatrix}7 & 6\\ 15 & 14\end{bmatrix}$$

$$\Rightarrow \quad \begin{bmatrix}2x+3 & 6\\ 15 & 2y-4\end{bmatrix}=\begin{bmatrix}7 & 6\\ 15 & 14\end{bmatrix}$$

or $2x+3=7$ and $2y-4=14$

$2x=7-3$ and $2y=18$

or $x=\frac{4}{2}$ and $y=\frac{18}{2}$

i.e $x=2$ and $y=9$

11. We have, $2A+3X=5B$

or $2A+3X-2A=5B-2A$

or $2A-2A+3X=5B-2A$

[matrix addition is commutative]

or $0+3X=5B-2A$

[$-2A$ is the additive inverse of $2A$]

$3X=5B-2A$ [0 is the additive identity]

or $X=\frac{1}{3}(5B-2A)$

or

$$X=\frac{1}{3}\left(5\begin{bmatrix}2 & -2\\ 4 & 2\\ -5 & 1\end{bmatrix}-2\begin{bmatrix}8 & 0\\ 4 & -2\\ 3 & 6\end{bmatrix}\right)$$

$$=\frac{1}{3}\left(\begin{bmatrix}10 & -10\\ 20 & 10\\ -25 & 5\end{bmatrix}+\begin{bmatrix}-16 & 0\\ -8 & 4\\ -6 & -12\end{bmatrix}\right)$$

$$=\frac{1}{3}\begin{bmatrix}10-16 & -10+0\\ 20-8 & 10+4\\ -25-6 & 5-12\end{bmatrix}$$

$$=\frac{1}{3}\begin{bmatrix}-6 & -10\\ 12 & 14\\ -31 & -7\end{bmatrix}$$

$$=\begin{bmatrix}-2 & \frac{-10}{3}\\ 4 & \frac{14}{3}\\ \frac{-31}{3} & \frac{-7}{3}\end{bmatrix}$$

12. We have, $\begin{bmatrix}\lambda^2-2\lambda+1 & \lambda-2\\ 1-\lambda^2+3\lambda & 1-\lambda^2\end{bmatrix}$

$$=A\lambda^2+B\lambda+C$$

Putting $\lambda=0$, we get

$$C=\begin{bmatrix}1 & -2\\ 1 & 1\end{bmatrix}$$

Putting $\lambda=1$, we get

$$A+B+C=\begin{bmatrix}0 & -1\\ 3 & 0\end{bmatrix} \quad \text{...(i)}$$

Putting $\lambda=-1$, we get

$$A-B+C=\begin{bmatrix}4 & -3\\ -3 & 0\end{bmatrix} \quad \text{...(ii)}$$

Subtracting Eq. (ii) from Eq. (i), we get

$$2B=\begin{bmatrix}0 & -1\\ 3 & 0\end{bmatrix}-\begin{bmatrix}4 & -3\\ -3 & 0\end{bmatrix}$$

$$=\begin{bmatrix}-4 & 2\\ 6 & 0\end{bmatrix}$$

$$\therefore \quad B=\begin{bmatrix}-2 & 1\\ 3 & 0\end{bmatrix}$$

13. We have, $AB=\begin{bmatrix}0 & -1\\ 0 & 2\end{bmatrix}\begin{bmatrix}3 & 5\\ 0 & 0\end{bmatrix}$

$$=\begin{bmatrix}0 & 0\\ 0 & 0\end{bmatrix}$$

Thus, if the product of two matrices is a zero matrix, it is not necessary that one of the matrices is a zero matrix.

14. $AB=\begin{bmatrix}1 & -2 & 3\\ -4 & 2 & 5\end{bmatrix}\begin{bmatrix}2 & 3\\ 4 & 5\\ 2 & 1\end{bmatrix}$

$$=\begin{bmatrix}2-8+6 & 3-10+3\\ -8+8+10 & -12+10+5\end{bmatrix}=\begin{bmatrix}0 & -4\\ 10 & 3\end{bmatrix}$$

$$BA=\begin{bmatrix}2 & 3\\ 4 & 5\\ 2 & 1\end{bmatrix}\begin{bmatrix}1 & -2 & 3\\ -4 & 2 & 5\end{bmatrix}=\begin{bmatrix}2-12 & -4+6 & 6+15\\ 4-20 & -8+10 & 12+25\\ 2-4 & -4+2 & 6+5\end{bmatrix}$$

$$=\begin{bmatrix}-10 & 2 & 21\\ -16 & 2 & 37\\ -2 & -2 & 11\end{bmatrix}$$

15. Number of chemistry books $=10$ dozen $=10\times12=120$

Number of physics books $=8$ dozen $=8\times12=96$

Number of economics books $=10$ dozen $=10\times12=120$

Let matrix A denote the number of books per subject

Chemistry Physics Economics

$$A=[120 \quad 96 \quad 120]$$

Let matrix B denote the selling price of books per subject

$$B=\begin{bmatrix}80\\ 60\\ 40\end{bmatrix}\begin{matrix}\text{Chemistry}\\ \text{Physics}\\ \text{Economics}\end{matrix}$$

Now, total amount the shopkeeper receives

$=$ Number of books × selling price

$=A\times B$

$$=[120 \quad 96 \quad 120]_{1\times3}\begin{bmatrix}80\\ 60\\ 40\end{bmatrix}_{3\times1}$$

$$=[120\times80+96\times60+120\times40]_{1\times1}$$

$$=9600+5760+4800$$

$$=20160$$

16. Since A, B and C are all square matrices of order 2, and $CD-AB$ is well defined D must be a square matrix of order 2.

Let $D=\begin{bmatrix}a & b\\ c & d\end{bmatrix}$, then $CD-AB=0$ given

or $\begin{bmatrix}2 & 5\\ 3 & 8\end{bmatrix}\begin{bmatrix}a & b\\ c & d\end{bmatrix}-\begin{bmatrix}2 & -1\\ 3 & 4\end{bmatrix}\begin{bmatrix}5 & 2\\ 7 & 4\end{bmatrix}=0$

or $\begin{bmatrix}2a+5c & 2b+5d\\ 3a+8c & 3b+8d\end{bmatrix}-\begin{bmatrix}3 & 0\\ 43 & 22\end{bmatrix}=\begin{bmatrix}0 & 0\\ 0 & 0\end{bmatrix}$

or $\begin{bmatrix}2a+5c-3 & 2b+5d\\ 3a+8c-43 & 3b+8d-22\end{bmatrix}=\begin{bmatrix}0 & 0\\ 0 & 0\end{bmatrix}$

By equality of matrices, we get

$$2a+5c-3=0 \quad \text{...(i)}$$

$$3a+8c-43=0 \quad \text{...(ii)}$$

$$2b+5d=0 \quad \text{...(iii)}$$
$$3b+8d-22=0 \quad \text{...(iv)}$$

Solving Eqs. (i) and (ii), we get

$a=-191, c=77$

Solving Eqs. (iii) and (iv) we get

$b=-110, d=44$

Therefore, $D=\begin{bmatrix} a & b \\ c & d \end{bmatrix}=\begin{bmatrix} -191 & -110 \\ 77 & 44 \end{bmatrix}$

17. We have,

$$A^2 = A\cdot A = \begin{bmatrix} 1 & 2 & 3 \\ 3 & -2 & 1 \\ 4 & 2 & 1 \end{bmatrix}\begin{bmatrix} 1 & 2 & 3 \\ 3 & -2 & 1 \\ 4 & 2 & 1 \end{bmatrix}$$
$$= \begin{bmatrix} 19 & 4 & 8 \\ 1 & 12 & 8 \\ 14 & 6 & 15 \end{bmatrix}$$
$$A^3 = AA^2 = \begin{bmatrix} 1 & 2 & 3 \\ 3 & -2 & 1 \\ 4 & 2 & 1 \end{bmatrix}\begin{bmatrix} 19 & 4 & 8 \\ 1 & 12 & 8 \\ 14 & 6 & 15 \end{bmatrix}$$
$$= \begin{bmatrix} 63 & 46 & 69 \\ 69 & -6 & 23 \\ 92 & 46 & 63 \end{bmatrix}$$

Now, $A^3-23A-40I = \begin{bmatrix} 63 & 46 & 69 \\ 69 & -6 & 23 \\ 92 & 46 & 63 \end{bmatrix} - 23\begin{bmatrix} 1 & 2 & 3 \\ 3 & -2 & 1 \\ 4 & 2 & 1 \end{bmatrix} - 40\begin{bmatrix} 1 & 0 & 0 \\ 0 & 1 & 0 \\ 0 & 0 & 1 \end{bmatrix}$

$$= \begin{bmatrix} 63 & 46 & 69 \\ 69 & -6 & 23 \\ 92 & 46 & 63 \end{bmatrix} + \begin{bmatrix} -23 & -46 & -69 \\ -69 & 46 & -23 \\ -92 & -46 & -23 \end{bmatrix} + \begin{bmatrix} -40 & 0 & 0 \\ 0 & -40 & 0 \\ 0 & 0 & -40 \end{bmatrix}$$
$$= \begin{bmatrix} 63-23-40 & 46-46+0 & 69-69+0 \\ 69-69+0 & -6+46-40 & 23-23-0 \\ 92-92+0 & 46-46+0 & 63-23-40 \end{bmatrix}$$
$$= \begin{bmatrix} 0 & 0 & 0 \\ 0 & 0 & 0 \\ 0 & 0 & 0 \end{bmatrix} = 0$$

18. Let $A=\begin{bmatrix} a & b \\ c & d \end{bmatrix}$ be a square roots of matrix I.

Then, $A^2=I$, where $I=\begin{bmatrix} 1 & 0 \\ 0 & 1 \end{bmatrix}$

$$\Rightarrow \begin{bmatrix} a & b \\ c & d \end{bmatrix}\begin{bmatrix} a & b \\ c & d \end{bmatrix}=\begin{bmatrix} 1 & 0 \\ 0 & 1 \end{bmatrix}$$
$$\Rightarrow \begin{bmatrix} a^2+bc & ab+bd \\ ac+cd & cb+d^2 \end{bmatrix}=\begin{bmatrix} 1 & 0 \\ 0 & 1 \end{bmatrix}$$
$$\Rightarrow a^2+bc=1 \quad \text{...(i)}$$
$$ab+bd=0 \quad \text{...(ii)}$$
$$ac+cd=0 \quad \text{...(iii)}$$
$$cb+d^2=1 \quad \text{...(iv)}$$

Case I $a+d=0$

The above four equations hold simultaneously, if $d=-a$ and $a^2+bc=1$

Hence, one possible square roots of I is $A=\begin{bmatrix} \alpha & \beta \\ \gamma & -\alpha \end{bmatrix}$

Where α, β, γ are the 3 numbers related by the condition $\alpha^2+\beta\gamma=1$

Case II $a+d\neq 0$

The above four equations hold simultaneously, if $b=0, c=0, a=1, d=1$ or if $b=0, c=0, a=-1, d=-1$

Hence, $\begin{bmatrix} 1 & 0 \\ 0 & 1 \end{bmatrix}$ or $\begin{bmatrix} -1 & 0 \\ 0 & -1 \end{bmatrix}$ i.e. $\pm I$ are other possible square roots of I.

19. We have,

$$AB=\begin{bmatrix} 1 & 2 \\ -2 & 3 \end{bmatrix}\times\begin{bmatrix} 2 & 1 \\ 2 & 3 \end{bmatrix}$$
$$=\begin{bmatrix} 1\cdot2+2\cdot2 & 1\cdot1+2\cdot3 \\ (-2)\cdot2+3\cdot2 & (-2)\cdot1+3\cdot3 \end{bmatrix}$$
$$=\begin{bmatrix} 6 & 7 \\ 2 & 7 \end{bmatrix}$$
$$BC=\begin{bmatrix} 2 & 1 \\ 2 & 3 \end{bmatrix}\times\begin{bmatrix} -3 & 1 \\ 2 & 0 \end{bmatrix}$$
$$=\begin{bmatrix} 2\cdot(-3)+1\cdot2 & 2\cdot1+1\cdot0 \\ 2\cdot(-3)+3\cdot2 & 2\cdot1+3\cdot0 \end{bmatrix}$$
$$=\begin{bmatrix} -6+2 & 2+0 \\ -6+6 & 2+0 \end{bmatrix}$$
$$=\begin{bmatrix} -4 & 2 \\ 0 & 2 \end{bmatrix}$$
$$AC=\begin{bmatrix} 1 & 2 \\ -2 & 3 \end{bmatrix}\times\begin{bmatrix} -3 & 1 \\ 2 & 0 \end{bmatrix}$$
$$=\begin{bmatrix} 1\cdot(-3)+2\cdot2 & 1\cdot1+2\cdot0 \\ (-2)\cdot(-3)+3\cdot2 & (-2)\cdot1+3\cdot0 \end{bmatrix}$$
$$=\begin{bmatrix} -3+4 & 1+0 \\ 6+6 & -2+0 \end{bmatrix}=\begin{bmatrix} 1 & 1 \\ 12 & -2 \end{bmatrix}$$
$$B+C=\begin{bmatrix} 2 & 1 \\ 2 & 3 \end{bmatrix}+\begin{bmatrix} -3 & 1 \\ 2 & 0 \end{bmatrix}$$

$$=\begin{bmatrix}2-3 & 1+1\\2+2 & 3+0\end{bmatrix}=\begin{bmatrix}-1 & 2\\4 & 3\end{bmatrix}$$

$$\text{Now, } (AB)C=\begin{bmatrix}6 & 7\\2 & 7\end{bmatrix}\times\begin{bmatrix}-3 & 1\\2 & 0\end{bmatrix}$$

$$=\begin{bmatrix}-18+14 & 6+0\\-6+14 & 2+0\end{bmatrix}$$

$$=\begin{bmatrix}-4 & 6\\8 & 2\end{bmatrix} \quad \text{...(i)}$$

$$A(BC)=\begin{bmatrix}1 & 2\\-2 & 3\end{bmatrix}\times\begin{bmatrix}-4 & 2\\0 & 2\end{bmatrix}$$

$$=\begin{bmatrix}-4+0 & 2+4\\8+0 & -4+6\end{bmatrix}$$

$$=\begin{bmatrix}-4 & 6\\8 & 2\end{bmatrix} \quad \text{...(ii)}$$

Thus, from Eqs. (i) and (ii), we get

$$(AB)C=A(BC)$$

$$\text{Now, } A(B+C)=\begin{bmatrix}1 & 2\\-2 & 3\end{bmatrix}\times\begin{bmatrix}-1 & 2\\4 & 3\end{bmatrix}$$

$$=\begin{bmatrix}-1+8 & 2+6\\2+12 & -4+9\end{bmatrix}$$

$$=\begin{bmatrix}7 & 8\\14 & 5\end{bmatrix} \quad \text{...(iii)}$$

$$\text{and } AB+AC=\begin{bmatrix}6 & 7\\2 & 7\end{bmatrix}+\begin{bmatrix}1 & 1\\12 & -2\end{bmatrix}$$

$$=\begin{bmatrix}6+1 & 7+1\\2+12 & 7-2\end{bmatrix}$$

$$=\begin{bmatrix}7 & 8\\14 & 5\end{bmatrix} \quad \text{...(iv)}$$

Thus, from Eqs. (iii) and (iv), we get

$$A(B+C)=AB+AC$$

20. We have,

$$[1 \text{ x } 1]\begin{bmatrix}1 & 3 & 2\\0 & 5 & 1\\0 & 3 & 2\end{bmatrix}\begin{bmatrix}1\\1\\x\end{bmatrix}=0$$

$$\Rightarrow \quad [1 \quad 5x+6 \quad x+4]\begin{bmatrix}1\\1\\x\end{bmatrix}=0$$

$$\Rightarrow \quad [1+5x+6+x^2+4x]=0$$

$$\text{or,} \quad x^2+9x+7=0$$

$$\therefore \quad x=\frac{-9\pm\sqrt{81-28}}{2}$$

$$\Rightarrow \quad x=\frac{-9\pm\sqrt{53}}{2}$$

21. (i) We have,

$$A=\begin{bmatrix}3 & \sqrt{3} & 2\\4 & 2 & 0\end{bmatrix}$$

$$\Rightarrow \quad A'=\begin{bmatrix}3 & 4\\\sqrt{3} & 2\\2 & 0\end{bmatrix}$$

$$\Rightarrow \quad (A')'=\begin{bmatrix}3 & \sqrt{3} & 2\\4 & 2 & 0\end{bmatrix}=A$$

Thus, $(A')'=A$

(ii) We have,

$$A=\begin{bmatrix}3 & \sqrt{3} & 2\\4 & 2 & 0\end{bmatrix}, B=\begin{bmatrix}2 & -1 & 2\\1 & 2 & 4\end{bmatrix},$$

$$\Rightarrow A+B=\begin{bmatrix}5 & \sqrt{3}-1 & 4\\5 & 4 & 4\end{bmatrix}$$

$$\text{Therefore, } (A+B)'=\begin{bmatrix}5 & 5\\\sqrt{3}-1 & 4\\4 & 4\end{bmatrix}$$

$$\text{Now, } A'=\begin{bmatrix}3 & 4\\\sqrt{3} & 2\\2 & 0\end{bmatrix}$$

$$B'=\begin{bmatrix}2 & 1\\-1 & 2\\2 & 4\end{bmatrix}$$

$$\text{So, } A'+B'=\begin{bmatrix}5 & 5\\\sqrt{3}-1 & 4\\4 & 4\end{bmatrix}$$

Thus, $(A+B)'=A'+B'$

(iii) We have,

$$kB=k\begin{bmatrix}2 & -1 & 2\\1 & 2 & 4\end{bmatrix}=\begin{bmatrix}2k & -k & 2k\\k & 2k & 4k\end{bmatrix}$$

$$\text{Then, } (kB)'=\begin{bmatrix}2k & k\\-k & 2k\\2k & 4k\end{bmatrix}=k\begin{bmatrix}2 & 1\\-1 & 2\\2 & 4\end{bmatrix}=kB'$$

Thus, $(kB)'=kB'$

22. (i) Given, $A=\begin{bmatrix}2 & -3\\-7 & 5\end{bmatrix}$ and $B=\begin{bmatrix}1 & 0\\2 & -4\end{bmatrix}$

Consider $(2A)^T=2A^T$

Put the value of A

$$\left(2\begin{bmatrix}2 & -3\\-7 & 5\end{bmatrix}\right)^T=2\begin{bmatrix}2 & -3\\-7 & 5\end{bmatrix}^T$$

$$\Rightarrow \quad \begin{bmatrix}4 & -6\\-14 & 10\end{bmatrix}^T=2\begin{bmatrix}2 & -7\\-3 & 5\end{bmatrix}$$

$$\Rightarrow \quad \begin{bmatrix}4 & -14\\-6 & 10\end{bmatrix}=\begin{bmatrix}4 & -14\\-6 & 10\end{bmatrix}$$

LHS = RHS

(ii) Given, $A=\begin{bmatrix}2 & -3\\ -7 & 5\end{bmatrix}$

and $\quad B=\begin{bmatrix}1 & 0\\ 2 & -4\end{bmatrix}$

Consider, $(A+B)^T = A^T + B^T$

$$\Rightarrow \left(\begin{bmatrix}2 & -3\\ -7 & 5\end{bmatrix}+\begin{bmatrix}1 & 0\\ 2 & -4\end{bmatrix}\right)^T = \begin{bmatrix}2 & -3\\ -7 & 5\end{bmatrix}^T+\begin{bmatrix}1 & 0\\ 2 & -4\end{bmatrix}^T$$

$$\Rightarrow \begin{bmatrix}2+1 & -3+0\\ -7+2 & 5-4\end{bmatrix}^T = \begin{bmatrix}2 & -7\\ -3 & 5\end{bmatrix}+\begin{bmatrix}1 & 2\\ 0 & -4\end{bmatrix}$$

$$\Rightarrow \begin{bmatrix}3 & -3\\ -5 & 1\end{bmatrix}^T = \begin{bmatrix}3 & -5\\ -3 & 1\end{bmatrix}$$

$$\Rightarrow \begin{bmatrix}3 & -5\\ -3 & 1\end{bmatrix} = \begin{bmatrix}3 & -5\\ -3 & 1\end{bmatrix}$$

LHS = RHS

(iii) Given, $A=\begin{bmatrix}2 & -3\\ -7 & 5\end{bmatrix}$ and $B=\begin{bmatrix}1 & 0\\ 2 & -4\end{bmatrix}$

Consider, $(A-B)^T = A^T - B^T$

$$\Rightarrow \left(\begin{bmatrix}2 & -3\\ -7 & 5\end{bmatrix}-\begin{bmatrix}1 & 0\\ 2 & -4\end{bmatrix}\right)^T = \begin{bmatrix}2 & -3\\ -7 & 5\end{bmatrix}^T-\begin{bmatrix}1 & 0\\ 2 & -4\end{bmatrix}^T$$

$$\Rightarrow \begin{bmatrix}2-1 & -3-0\\ -7-2 & 5+4\end{bmatrix}^T = \begin{bmatrix}2 & -7\\ -3 & 5\end{bmatrix}-\begin{bmatrix}1 & 2\\ 0 & -4\end{bmatrix}$$

$$\Rightarrow \begin{bmatrix}1 & -3\\ -9 & 9\end{bmatrix}^T = \begin{bmatrix}1 & -9\\ -3 & 9\end{bmatrix}$$

$$\Rightarrow \begin{bmatrix}1 & -9\\ -3 & 9\end{bmatrix} = \begin{bmatrix}1 & -9\\ -3 & 9\end{bmatrix}$$

LHS = RHS

(iv) Given, $A=\begin{bmatrix}2 & -3\\ -7 & 5\end{bmatrix}$ and $B=\begin{bmatrix}1 & 0\\ 2 & -4\end{bmatrix}$

Consider, $(AB)^T = B^T A^T$

$$\Rightarrow \left(\begin{bmatrix}2 & -3\\ -7 & 5\end{bmatrix}\begin{bmatrix}1 & 0\\ 2 & -4\end{bmatrix}\right)^T = \begin{bmatrix}1 & 0\\ 2 & -4\end{bmatrix}^T\begin{bmatrix}2 & -3\\ -7 & 5\end{bmatrix}^T$$

$$\Rightarrow \begin{bmatrix}2-6 & 0+12\\ -7+10 & 0-20\end{bmatrix}^T = \begin{bmatrix}1 & 2\\ 0 & -4\end{bmatrix}\begin{bmatrix}2 & -7\\ -3 & 5\end{bmatrix}$$

$$\Rightarrow \begin{bmatrix}-4 & 12\\ 3 & -20\end{bmatrix}^T = \begin{bmatrix}2-6 & -7+10\\ 0+12 & 0-20\end{bmatrix}$$

$$\Rightarrow \begin{bmatrix}-4 & 3\\ 12 & -20\end{bmatrix} = \begin{bmatrix}-4 & 3\\ 12 & -20\end{bmatrix}$$

So, $(AB)^T = B^T A^T$

23. $A=\begin{bmatrix}3\\ 5\\ 2\end{bmatrix}$ and $B=[1 \;\; 0 \;\; 4]$

Consider, $(AB)^T = B^T A^T$

$$\Rightarrow \left(\begin{bmatrix}3\\ 5\\ 2\end{bmatrix}[1 \;\; 0 \;\; 4]\right)^T = [1 \;\; 0 \;\; 4]^T\begin{bmatrix}3\\ 5\\ 2\end{bmatrix}^T$$

$$\Rightarrow \begin{bmatrix}3 & 0 & 12\\ 5 & 0 & 20\\ 2 & 0 & 8\end{bmatrix}^T = \begin{bmatrix}1\\ 0\\ 4\end{bmatrix}[3 \;\; 5 \;\; 2]$$

$$\Rightarrow \begin{bmatrix}3 & 5 & 2\\ 0 & 0 & 0\\ 12 & 20 & 8\end{bmatrix} = \begin{bmatrix}3 & 5 & 2\\ 0 & 0 & 0\\ 12 & 20 & 8\end{bmatrix}$$

LHS = RHS

So, $(AB)^T = B^T A^T$

24. It is given that

$$A=\begin{bmatrix}0 & 2y & z\\ x & y & -z\\ x & -y & z\end{bmatrix}$$

$$\therefore \quad A^T=\begin{bmatrix}0 & x & x\\ 2y & y & -y\\ z & -z & z\end{bmatrix}$$

Now, $A^T A = I$

$$\begin{bmatrix}0 & x & x\\ 2y & y & -y\\ z & -z & z\end{bmatrix}\begin{bmatrix}0 & 2y & z\\ x & y & -z\\ x & -y & z\end{bmatrix} = \begin{bmatrix}1 & 0 & 0\\ 0 & 1 & 0\\ 0 & 0 & 1\end{bmatrix}$$

$$\text{or} \begin{bmatrix}0+x^2+x^2 & 0+xy-xy & 0-xz+xz\\ 0+xy-xy & 4y^2+y^2+y^2 & 2yz-yz-yz\\ 0-xz+zx & 2yz-yz-yz & z^2+z^2+z^2\end{bmatrix} = \begin{bmatrix}1 & 0 & 0\\ 0 & 1 & 0\\ 0 & 0 & 1\end{bmatrix}$$

$$\Rightarrow \begin{bmatrix} 2x^2 & 0 & 0 \\ 0 & 6y^2 & 0 \\ 0 & 0 & 3z^2 \end{bmatrix} = \begin{bmatrix} 1 & 0 & 0 \\ 0 & 1 & 0 \\ 0 & 0 & 1 \end{bmatrix}$$

On comparing the corresponding elements, we have

$$2x^2 = 1 \text{ or } x = \pm\frac{1}{\sqrt{2}}$$

$$6y^2 = 1 \text{ or } y = \pm\frac{1}{\sqrt{6}}$$

$$3z^2 = 1 \text{ or } z = \pm\frac{1}{\sqrt{3}}$$

25. $\because$ Trace of a matrix = sum of all diagonal elements

(i) Trace $(A) = 1 + 1 + 1 = 3$

(ii) Trace $(B) = 1 + 5 + 9 = 15$

(iii) Trace $(C) = 5 + 5 + 5 = 15$

26. (i) $\because$ Trace $(A + B) =$ Trace (A) + Trace (B)

[$\because$ A and B are 3×3 matrices]

$$= (1 + 1 + 5) + (1 + 0 + 7) = 7 + 8 = 15$$

(ii) Trace $(3A) = 3$ Trace $(A) = 3(1 + 1 + 5) = 21$

(iii) Trace$(A^T) =$ Trace $(A) = 1 + 1 + 5 = 7$

27. Here to find value of Trace (A) – Trace (B), we need not to find the matrices A and B.

We can find Trace (A) – Trace (B) using the properties of trace of matrix i.e.

$$A + 2B = \begin{bmatrix} 1 & 2 & 0 \\ 6 & -3 & 3 \\ -5 & 3 & 1 \end{bmatrix}$$

$\Rightarrow \quad tr(A + 2B) = -1$

or, $\quad tr(A) + 2tr(B) = -1 \quad$ ...(i)

$$2A - B = \begin{bmatrix} 2 & -1 & 5 \\ 2 & -1 & 6 \\ 0 & 1 & 2 \end{bmatrix}$$

$\Rightarrow \quad tr(2A - B) = 3$

or $\quad 2tr(A) - tr(B) = 3 \quad$ ...(ii)

On solving Eqs. (i) and (ii), we get

$tr(A) = 1$ and $tr(B) = -1$

$\Rightarrow \quad tr(A) - tr(B) = 2$

28. Given, $A = \begin{bmatrix} 5 & 2 & x \\ y & z & -3 \\ 4 & t & -7 \end{bmatrix}$ is a symmetric matrix.

We know that, $A = [A_{ij}]_{m \times n}$ is a symmetric matrix, if

$$a_{ij} = a_{ji}$$

So, $\quad x = a_{13} = a_{31} = 4$

$y = a_{21} = a_{12} = 2$

$z = a_{22} = a_{22} = z$

$t = a_{32} = a_{23} = -3$

Hence, $x = 4$, $y = 2$, $t = -3$ and z can have any value.

29. We know that,

$$\text{sgn}(x) = \begin{cases} 1 & \text{if } x > 0 \\ -1 & \text{if } x < 0 \\ 0 & \text{if } x = 0 \end{cases}$$

$$\therefore \quad a_{ij} = \text{sgn}(i - j) = \begin{cases} 1, & \text{if } i > j \\ -1, & \text{if } i < j \\ 0, & \text{if } i = j \end{cases}$$

$$\therefore \quad A = \begin{bmatrix} 0 & -1 & -1 \\ 1 & 0 & -1 \\ 1 & 1 & 0 \end{bmatrix}$$

Hence, matrix A is skew-symmetric. $\quad [\because A^T = -A]$

30. Given, $A = \begin{bmatrix} 3 & -4 \\ 1 & -1 \end{bmatrix}$

Consider, $(A - A^T) = \begin{bmatrix} 0 & -5 \\ 5 & 0 \end{bmatrix} \quad$ ...(i)

or $\quad -(A - A^T)^T = -\begin{bmatrix} 0 & -5 \\ 5 & 0 \end{bmatrix}^T$

$$= -\begin{bmatrix} 0 & 5 \\ -5 & 0 \end{bmatrix}$$

$\Rightarrow \quad -(A - A^T)^T = \begin{bmatrix} 0 & -5 \\ 5 & 0 \end{bmatrix} \quad$ ...(ii)

From Eqs. (i) and (ii), we can see that $A - A^T$ is a skew-symmetric matrix.

31. We have,

$$A = \begin{bmatrix} 3 & 5 \\ -1 & 2 \end{bmatrix}, \text{ then}$$

$$A' = \begin{bmatrix} 3 & -1 \\ 5 & 2 \end{bmatrix}$$

Let $\quad P = \frac{1}{2}(A + A') = \frac{1}{2}\begin{bmatrix} 6 & 4 \\ 4 & 4 \end{bmatrix} = \begin{bmatrix} 3 & 2 \\ 2 & 2 \end{bmatrix}$

Now, $P' = \begin{bmatrix} 3 & 2 \\ 2 & 2 \end{bmatrix}' = \begin{bmatrix} 3 & 2 \\ 2 & 2 \end{bmatrix} = P$

Thus, $P = \frac{1}{2}(A + A')$ is a symmetric matrix.

Also, let $Q = \frac{1}{2}(A - A') = \frac{1}{2}\begin{bmatrix} 0 & 6 \\ -6 & 0 \end{bmatrix} = \begin{bmatrix} 0 & 3 \\ -3 & 0 \end{bmatrix}$

Then, $Q' = \begin{bmatrix} 0 & -3 \\ 3 & 0 \end{bmatrix} = -\begin{bmatrix} 0 & 3 \\ -3 & 0 \end{bmatrix} = -Q$

Thus, $Q = \frac{1}{2}(A - A')$ is a skew-symmetric matrix.

Now, $P + Q = \begin{bmatrix} 3 & 2 \\ 2 & 2 \end{bmatrix} + \begin{bmatrix} 0 & 3 \\ -3 & 0 \end{bmatrix} = \begin{bmatrix} 3 & 5 \\ -1 & 2 \end{bmatrix} = A$

Hence, A is represented as the sum of a symmetric and a skew-symmetric matrix.

32. Here, $B=\begin{bmatrix}2 & -2 & -4\\ -1 & 3 & 4\\ 1 & -2 & -3\end{bmatrix}$

Then, $B'=\begin{bmatrix}2 & -1 & 1\\ -2 & 3 & -2\\ -4 & 4 & -3\end{bmatrix}$

Let $P=\frac{1}{2}(B+B')=\frac{1}{2}\begin{bmatrix}4 & -3 & -3\\ -3 & 6 & 2\\ -3 & 2 & -6\end{bmatrix}$

$$=\begin{bmatrix}2 & \frac{-3}{2} & \frac{-3}{2}\\ \frac{-3}{2} & 3 & 1\\ \frac{-3}{2} & 1 & -3\end{bmatrix}$$

Now, $P'=\begin{bmatrix}2 & \frac{-3}{2} & \frac{-3}{2}\\ \frac{-3}{2} & 3 & 1\\ \frac{-3}{2} & 1 & -3\end{bmatrix}=P$

Thus, $P=\frac{1}{2}(B+B')$ is a symmetric matrix.

Also, let

$$Q=\frac{1}{2}(B-B')=\frac{1}{2}\begin{bmatrix}0 & -1 & -5\\ 1 & 0 & 6\\ 5 & -6 & 0\end{bmatrix}=\begin{bmatrix}0 & \frac{-1}{2} & \frac{-5}{2}\\ \frac{1}{2} & 0 & 3\\ \frac{5}{2} & -3 & 0\end{bmatrix}$$

Then, $Q'=\begin{bmatrix}0 & \frac{1}{2} & \frac{5}{3}\\ \frac{-1}{2} & 0 & -3\\ \frac{-5}{2} & 3 & 0\end{bmatrix}=-Q$

Thus, $Q=\frac{1}{2}(B-B')$ is a skew-symmetric matrix.

$$P+Q=\begin{bmatrix}2 & \frac{-3}{2} & \frac{-3}{2}\\ \frac{-3}{2} & 3 & 1\\ \frac{-3}{2} & 1 & -3\end{bmatrix}+\begin{bmatrix}0 & \frac{-1}{2} & \frac{-5}{2}\\ \frac{1}{2} & 0 & 3\\ \frac{5}{2} & -3 & 0\end{bmatrix}$$

$$=\begin{bmatrix}2 & -2 & -4\\ -1 & 3 & 4\\ 1 & -2 & -3\end{bmatrix}=B$$

Thus, B is represented as the sum of a symmetric and skew-symmetric matrix.

33. Given $A=\begin{bmatrix}3 & 2 & 7\\ 1 & 4 & 3\\ -2 & 5 & 8\end{bmatrix}$, then $A^T=\begin{bmatrix}3 & 1 & -2\\ 2 & 4 & 5\\ 7 & 3 & 8\end{bmatrix}$

$$X=\frac{1}{2}(A+A^T)$$

$$=\frac{1}{2}\left(\begin{bmatrix}3 & 2 & 7\\ 1 & 4 & 3\\ -2 & 5 & 8\end{bmatrix}+\begin{bmatrix}3 & 1 & -2\\ 2 & 4 & 5\\ 7 & 3 & 8\end{bmatrix}\right)$$

$$=\frac{1}{2}\begin{bmatrix}3+3 & 2+1 & 7-2\\ 1+2 & 4+4 & 3+5\\ -2+7 & 5+3 & 8+8\end{bmatrix}$$

$$=\frac{1}{2}\begin{bmatrix}6 & 3 & 5\\ 3 & 8 & 8\\ 5 & 8 & 16\end{bmatrix}$$

$$X=\begin{bmatrix}3 & \frac{3}{2} & \frac{5}{2}\\ \frac{3}{2} & 4 & 4\\ \frac{5}{2} & 4 & 8\end{bmatrix}$$

Now, $X^T=\begin{bmatrix}3 & \frac{3}{2} & \frac{5}{2}\\ \frac{3}{2} & 4 & 4\\ \frac{5}{2} & 4 & 8\end{bmatrix}^T=\begin{bmatrix}3 & \frac{3}{2} & \frac{5}{2}\\ \frac{3}{2} & 4 & 4\\ \frac{5}{2} & 4 & 8\end{bmatrix}=X$

$\Rightarrow X$ is a symmetric matrix

Now, $Y=\frac{1}{2}(A-A^T)$

$$=\frac{1}{2}\left(\begin{bmatrix}3 & 2 & 7\\ 1 & 4 & 3\\ -2 & 5 & 8\end{bmatrix}-\begin{bmatrix}3 & 1 & -2\\ 2 & 4 & 5\\ 7 & 3 & 8\end{bmatrix}\right)$$

$$=\frac{1}{2}\begin{bmatrix}3-3 & 2-1 & 7+2\\ 1-2 & 4-4 & 3-5\\ -2-7 & 5-3 & 8-8\end{bmatrix}=\frac{1}{2}\begin{bmatrix}0 & 1 & 9\\ -1 & 0 & -2\\ -9 & 2 & 0\end{bmatrix}$$

$$Y=\begin{bmatrix}0 & \frac{1}{2} & \frac{9}{2}\\ \frac{-1}{2} & 0 & -1\\ \frac{-9}{2} & 1 & 0\end{bmatrix}$$

Now, $-Y^T=-\begin{bmatrix}0 & \frac{1}{2} & \frac{9}{2}\\ \frac{-1}{2} & 0 & -1\\ \frac{-9}{2} & 1 & 0\end{bmatrix}^T=\begin{bmatrix}0 & \frac{-1}{2} & \frac{-9}{2}\\ \frac{1}{2} & 0 & 1\\ \frac{9}{2} & -1 & 0\end{bmatrix}^T$

$$-Y^T = \begin{bmatrix} 0 & \frac{1}{2} & \frac{9}{2} \\ \frac{-1}{2} & 0 & -1 \\ \frac{-9}{2} & 1 & 0 \end{bmatrix}$$

$$-Y^T = Y$$

$\Rightarrow Y$ is a skew-symmetric matrix.

$$X+Y = \begin{bmatrix} 3 & \frac{3}{2} & \frac{5}{2} \\ \frac{3}{2} & 4 & 4 \\ \frac{5}{2} & 4 & 8 \end{bmatrix} + \begin{bmatrix} 0 & \frac{1}{2} & \frac{9}{2} \\ \frac{-1}{2} & 0 & -1 \\ \frac{-9}{2} & 1 & 0 \end{bmatrix}$$

$$= \begin{bmatrix} 3+0 & \frac{3}{2}+\frac{1}{2} & \frac{5}{2}+\frac{9}{2} \\ \frac{3}{2}-\frac{1}{2} & 4+0 & 4-1 \\ \frac{5}{2}-\frac{9}{2} & 4+1 & 8+0 \end{bmatrix}$$

$$= \begin{bmatrix} 3 & 2 & 7 \\ 1 & 4 & 3 \\ -2 & 5 & 8 \end{bmatrix} = A$$

Hence, $X+Y=A$

34. Let $A = \begin{bmatrix} 4 & 1+3i \\ 1-ai & 7 \end{bmatrix}$

$$\bar{A} = \begin{bmatrix} 4 & 1-3i \\ 1+ai & 7 \end{bmatrix}$$

$$(\bar{A})^T = \begin{bmatrix} 4 & 1+ai \\ 1-3i & 7 \end{bmatrix}$$

Given, that A is Hermitian

$\therefore \quad (\bar{A})^T = A$

$$\Rightarrow \begin{bmatrix} 4 & 1+ai \\ 1-3i & 7 \end{bmatrix} = \begin{bmatrix} 4 & 1+3i \\ 1-ai & 7 \end{bmatrix}$$

On comparing, we have

$$1+ai = 1+3i \Rightarrow a = 3$$

35. Let $A = \begin{bmatrix} 0 & -2-i \\ 2-i & i \end{bmatrix}$

$$\bar{A} = \begin{bmatrix} 0 & -2+i \\ 2+i & -i \end{bmatrix}$$

$$(\bar{A})^T = \begin{bmatrix} 0 & 2+i \\ -2+i & -i \end{bmatrix} \quad \ldots(i)$$

For skew Hermitian, we must have

$$(\bar{A})^T = -A$$

From Eq. (i),

$$(\bar{A})^T = \begin{bmatrix} 0 & 2+i \\ -2+i & -i \end{bmatrix} \quad \ldots(ii)$$

and $\quad A = \begin{bmatrix} 0 & -2-i \\ 2-i & i \end{bmatrix}$

$$-A = -\begin{bmatrix} 0 & -2-i \\ 2-i & i \end{bmatrix} = \begin{bmatrix} 0 & 2+i \\ -2+i & -i \end{bmatrix} \quad \ldots(iii)$$

From Eqs. (ii) and (iii), it is clear that A is skew Hermitian.

36. We have, $A = \begin{bmatrix} 2+3i & 7 \\ 1-i & 2i \end{bmatrix}$, then $A^\theta = (\bar{A}')$

$$= \begin{bmatrix} 2-3i & 1+i \\ 7 & -2i \end{bmatrix}$$

Let $P = \frac{1}{2}(A+A^\theta) = \frac{1}{2}\begin{bmatrix} 4 & 8+i \\ 8-i & 0 \end{bmatrix}$

$$= \begin{bmatrix} 2 & 4+\frac{i}{2} \\ 4-\frac{i}{2} & 0 \end{bmatrix} = P^\theta$$

Thus, $P = \frac{1}{2}(A+A^\theta)$ is a Hermitian matrix.

Also, let $Q = \frac{1}{2}(A-A^\theta) = \frac{1}{2}\begin{bmatrix} 6i & 6-i \\ -6-i & 4i \end{bmatrix}$

$$= \begin{bmatrix} 3i & 3-\frac{i}{2} \\ -3-\frac{i}{2} & 2i \end{bmatrix} = -\begin{bmatrix} -3i & -3+\frac{i}{2} \\ 3+\frac{i}{2} & -2i \end{bmatrix} = -Q^\theta$$

Thus, $Q = \frac{1}{2}(A+A^\theta)$ is a skew-hermitian matrix.

Now, $P+Q = \begin{bmatrix} 2 & 4+\frac{i}{2} \\ 4-\frac{i}{2} & 0 \end{bmatrix} + \begin{bmatrix} 3i & 3-\frac{i}{2} \\ -3-\frac{i}{2} & 2i \end{bmatrix}$

$$= \begin{bmatrix} 2+3i & 7 \\ 1-i & 2i \end{bmatrix} = A$$

Hence, A is represented as the sum of a hermitian and a skew-hermitian matrix.

37. Let $\quad A = \begin{bmatrix} \frac{b^2-a^2}{a^2+b^2} & \frac{-2ab}{a^2+b^2} \\ \frac{-2ab}{a^2+b^2} & \frac{a^2-b^2}{a^2+b^2} \end{bmatrix}$

$$= \begin{bmatrix} \dfrac{1-\frac{a^2}{b^2}}{1+\frac{a^2}{b^2}} & \dfrac{-2\frac{a}{b}}{1+\frac{a^2}{b^2}} \\ \dfrac{-2\frac{a}{b}}{1+\frac{a^2}{b^2}} & -\dfrac{1-\frac{a^2}{b^2}}{1+\frac{a^2}{b^2}} \end{bmatrix}$$

$$=\begin{bmatrix}\cos 2\theta & -\sin 2\theta\\ -\sin 2\theta & -\cos 2\theta\end{bmatrix}, \text{ where } \frac{a}{b}=\tan\theta$$

$$\therefore\ AA^T=\begin{bmatrix}\cos 2\theta & -\sin 2\theta\\ -\sin 2\theta & -\cos 2\theta\end{bmatrix}\begin{bmatrix}\cos 2\theta & -\sin 2\theta\\ -\sin 2\theta & -\cos 2\theta\end{bmatrix}$$

$$=\begin{bmatrix}1 & 0\\ 0 & 1\end{bmatrix}$$

Thus, A is orthogonal.

38. Let $A=\begin{bmatrix}0 & 2\beta & \gamma\\ \alpha & \beta & -\gamma\\ \alpha & -\beta & \gamma\end{bmatrix}$, then $A'=\begin{bmatrix}0 & \alpha & \alpha\\ 2\beta & \beta & -\beta\\ \gamma & -\gamma & \gamma\end{bmatrix}$

Since, A is orthogonal.

$\therefore\quad AA'=I$

$$\Rightarrow\begin{bmatrix}0 & 2\beta & \gamma\\ \alpha & \beta & -\gamma\\ \alpha & -\beta & \gamma\end{bmatrix}\begin{bmatrix}0 & \alpha & \alpha\\ 2\beta & \beta & -\beta\\ \gamma & -\gamma & \gamma\end{bmatrix}=\begin{bmatrix}1 & 0 & 0\\ 0 & 1 & 0\\ 0 & 0 & 1\end{bmatrix}$$

$$\Rightarrow\begin{bmatrix}4\beta^2+\gamma^2 & 2\beta^2-\gamma^2 & -2\beta^2+\gamma^2\\ 2\beta^2-\gamma^2 & \alpha^2+\beta^2+\gamma^2 & \alpha^2-\beta^2-\gamma^2\\ -2\beta^2+\gamma^2 & \alpha-\beta^2-\gamma^2 & \alpha^2+\beta^2+\gamma^2\end{bmatrix}$$

$$=\begin{bmatrix}1 & 0 & 0\\ 0 & 1 & 0\\ 0 & 0 & 1\end{bmatrix}$$

Equating the corresponding elements, we get

$$4\beta^2+\gamma^2=1 \quad \text{...(i)}$$

$$2\beta^2-\gamma^2=0 \quad \text{...(ii)}$$

$$\alpha^2+\beta^2+\gamma^2=1 \quad \text{...(iii)}$$

From Eqs. (i) and (ii), we get

$$\beta^2=\frac{1}{6} \text{ and } \gamma^2=\frac{1}{3}$$

Now, $\alpha^2=1-\beta^2-\gamma^2$

$$=1-\frac{1}{6}-\frac{1}{3}=\frac{1}{2}$$

Hence, $2\alpha^2+6\beta^2+3\gamma^2$

$$=2\times\frac{1}{2}+6\times\frac{1}{6}+3\times\frac{1}{3}=3$$

39. Let $A=\frac{1}{\sqrt{3}}\begin{bmatrix}1 & 1+i\\ 1-i & -1\end{bmatrix}$ and $A^\theta=\frac{1}{\sqrt{3}}\begin{bmatrix}1 & 1+i\\ 1-i & -1\end{bmatrix}$

$$\therefore AA^\theta=\frac{1}{\sqrt{3}}\begin{bmatrix}1 & 1+i\\ 1-i & -1\end{bmatrix}\times\frac{1}{\sqrt{3}}\begin{bmatrix}1 & 1+i\\ 1-i & -1\end{bmatrix}$$

$$=\frac{1}{3}\begin{bmatrix}3 & 0\\ 0 & 3\end{bmatrix}=\begin{bmatrix}1 & 0\\ 0 & 1\end{bmatrix}=I$$

Hence, A is unitary matrix.

40. $A^2=A\cdot A=\begin{bmatrix}2 & -2 & -4\\ -1 & 3 & 4\\ 1 & -2 & -3\end{bmatrix}\times\begin{bmatrix}2 & -2 & -4\\ -1 & 3 & 4\\ 1 & -2 & -3\end{bmatrix}$

$$=\begin{bmatrix}2\cdot 2+(-2)\cdot(-1)+(-4)\cdot 1 & 2\cdot(-2)+(-2)\cdot 3+(-4)(-2) & 2\cdot(-4)+(-2)\cdot 4+(-4)\cdot(-3)\\ (-1)\cdot 2+3\cdot(-1)+4\cdot 1 & (-1)\cdot(-2)+3\cdot 3+4\cdot(-2) & (-1)\cdot(-4)+(3)\cdot 4+(4)\cdot(-3)\\ 1\cdot 2+(-2)\cdot(-1)+(-3)\cdot 1 & 1\cdot(-2)+(-2)\cdot 3+(-3)\cdot(-2) & 1\cdot(-4)+(-2)\cdot(4)+(-3)\cdot(-3)\end{bmatrix}$$

$$=\begin{bmatrix}2 & -2 & -4\\ -1 & 3 & 4\\ 1 & -2 & -3\end{bmatrix}=A$$

Hence, the matrix A is idempotent.

41. Let $A=\begin{bmatrix}1 & 1 & 3\\ 5 & 2 & 6\\ -2 & -1 & -3\end{bmatrix}$

$$A^2=A\cdot A=\begin{bmatrix}1 & 1 & 3\\ 5 & 2 & 6\\ -2 & -1 & -3\end{bmatrix}\times\begin{bmatrix}1 & 1 & 3\\ 5 & 2 & 6\\ -2 & -1 & -3\end{bmatrix}$$

$$=\begin{bmatrix}1+5-6 & 1+2-3 & 3+6-9\\ 5+10-12 & 5+4-6 & 15+12-18\\ -2-5+6 & -2-2+3 & -6-6+9\end{bmatrix}$$

$$=\begin{bmatrix}0 & 0 & 0\\ 3 & 3 & 9\\ -1 & -1 & -3\end{bmatrix}$$

$$\therefore\ A^3=A^2A=\begin{bmatrix}0 & 0 & 0\\ 3 & 3 & 9\\ -1 & -1 & -3\end{bmatrix}\times\begin{bmatrix}1 & 1 & 3\\ 5 & 2 & 6\\ -2 & -1 & -3\end{bmatrix}$$

$$=\begin{bmatrix}0+0+0 & 0+0+0 & 0+0+0\\ 3+15-18 & 3+6-9 & 9+18-27\\ -1-5+6 & -1-2+3 & -3-6+9\end{bmatrix}$$

$$=\begin{bmatrix}0 & 0 & 0\\ 0 & 0 & 0\\ 0 & 0 & 0\end{bmatrix}=0$$

$A^3=0$, i.e. $A^K=0$

Here, $K=3$

Hence, the matrix A is nilpotent of order 3.

42. $A^2=A\cdot A=\begin{bmatrix}-5 & -8 & 0\\ 3 & 5 & 0\\ 1 & 2 & -1\end{bmatrix}\times\begin{bmatrix}-5 & -8 & 0\\ 3 & 5 & 0\\ 1 & 2 & -1\end{bmatrix}$

$$=\begin{bmatrix}25-24+0 & 40-40+0 & 0+0+0\\ -15+15+0 & -24+25+0 & 0+0+0\\ -5+6-1 & -8+10-2 & 0+0+1\end{bmatrix}$$

$$=\begin{bmatrix}1 & 0 & 0\\ 0 & 1 & 0\\ 0 & 0 & 1\end{bmatrix}=1$$

Hence, the given matrix A is involutory.

43. In matrix A, $A=\begin{bmatrix}1 & -1 & 0\\ 2 & 1 & 1\end{bmatrix}$

Applying $R_1 \to R_1+R_2$, we get

$$E_1=\begin{bmatrix}3 & 0 & 1\\ 2 & 1 & 1\end{bmatrix}$$

In E_1; Applying $R_2 \to 3R_2$, we get

$$E_2=\begin{bmatrix}3 & 0 & 1\\ 6 & 3 & 3\end{bmatrix}$$

In E_2, Applying $R_2 \to R_2-2R_1$, we get

$$E_3=\begin{bmatrix}3 & 0 & 1\\ 0 & 3 & 1\end{bmatrix}=B$$

Thus, A and B are row equivalent.

44. In order to use elementary row operations we may write

$$A=IA$$

or, $\begin{bmatrix}1 & 2\\ 2 & -1\end{bmatrix}=\begin{bmatrix}1 & 0\\ 0 & 1\end{bmatrix}A$, then $\begin{bmatrix}1 & 2\\ 0 & -5\end{bmatrix}=\begin{bmatrix}1 & 0\\ -2 & 1\end{bmatrix}A$

[applying $R_2 \to R_2-2R_1$]

or, $\begin{bmatrix}1 & 2\\ 0 & 1\end{bmatrix}=\begin{bmatrix}1 & 0\\ \frac{2}{5} & \frac{-1}{5}\end{bmatrix}A$ [applying $R_2 \to -\frac{1}{5}R_2$]

or, $\begin{bmatrix}1 & 0\\ 0 & 1\end{bmatrix}=\begin{bmatrix}\frac{1}{5} & \frac{2}{5}\\ \frac{2}{5} & \frac{-1}{5}\end{bmatrix}A$ [applying $R_1 \to R_1-2R_2$]

Thus, $A^{-1}=\begin{bmatrix}\frac{1}{5} & \frac{2}{5}\\ \frac{2}{5} & \frac{-1}{5}\end{bmatrix}$

45. We have,

$P=IP$ i.e.

$$\begin{bmatrix}10 & -2\\ -5 & 1\end{bmatrix}=\begin{bmatrix}1 & 0\\ 0 & 1\end{bmatrix}P$$

Or, $\begin{bmatrix}1 & \frac{-1}{5}\\ -5 & 1\end{bmatrix}=\begin{bmatrix}\frac{1}{10} & 0\\ 0 & 1\end{bmatrix}P$ [applying $R_1 \to \frac{1}{10}R_1$]

or, $\begin{bmatrix}1 & \frac{-1}{5}\\ 0 & 0\end{bmatrix}=\begin{bmatrix}\frac{1}{10} & 0\\ \frac{1}{2} & 1\end{bmatrix}P$ [applying $R_2 \to R_2+5R_1$]

We have all zeroes in the second row of the left hand side matrix of the above equation.
Therefore, P^{-1} does not exist.

46. We have, $A\begin{bmatrix}1 & 2 & 3\\ 0 & 2 & 3\\ 0 & 1 & 1\end{bmatrix}=\begin{bmatrix}0 & 0 & 1\\ 1 & 0 & 0\\ 0 & 1 & 0\end{bmatrix}$

To get A^{-1} we have to transform this equation to $AB=I$

So, we will be using elementary column transformation.

Applying $C_1 \leftrightarrow C_3$ we get

$$A\begin{bmatrix}3 & 2 & 1\\ 3 & 2 & 0\\ 1 & 1 & 0\end{bmatrix}=\begin{bmatrix}1 & 0 & 0\\ 0 & 0 & 1\\ 0 & 1 & 0\end{bmatrix}$$

Applying $C_2 \leftrightarrow C_3$ we get

$$A\begin{bmatrix}3 & 1 & 2\\ 3 & 0 & 2\\ 1 & 0 & 1\end{bmatrix}=\begin{bmatrix}1 & 0 & 0\\ 0 & 1 & 0\\ 0 & 0 & 1\end{bmatrix} \Rightarrow A^{-1}=\begin{bmatrix}3 & 1 & 2\\ 3 & 0 & 2\\ 1 & 0 & 1\end{bmatrix}$$

47. (i) Write $A=IA$,

i.e. $\begin{bmatrix}0 & 1 & 2\\ 1 & 2 & 3\\ 3 & 1 & 1\end{bmatrix}=\begin{bmatrix}1 & 0 & 0\\ 0 & 1 & 0\\ 0 & 0 & 1\end{bmatrix}A$

or $\begin{bmatrix}1 & 2 & 3\\ 0 & 1 & 2\\ 3 & 1 & 1\end{bmatrix}=\begin{bmatrix}0 & 1 & 0\\ 1 & 0 & 0\\ 0 & 0 & 1\end{bmatrix}A$

[applying $R_1 \leftrightarrow R_2$]

or, $\begin{bmatrix}1 & 2 & 3\\ 0 & 1 & 2\\ 0 & -5 & -8\end{bmatrix}=\begin{bmatrix}0 & 1 & 0\\ 1 & 0 & 0\\ 0 & -3 & 1\end{bmatrix}A$

[applying $R_3 \to R_3-3R_1$]

or, $\begin{bmatrix}1 & 0 & -1\\ 0 & 1 & 2\\ 0 & -5 & -8\end{bmatrix}=\begin{bmatrix}-2 & 1 & 0\\ 1 & 0 & 0\\ 0 & -3 & 1\end{bmatrix}A$

[applying $R_1 \to R_1-2R_2$]

or $\begin{bmatrix}1 & 0 & -1\\ 0 & 1 & 2\\ 0 & 0 & 2\end{bmatrix}=\begin{bmatrix}-2 & 1 & 0\\ 1 & 0 & 0\\ 5 & -3 & 1\end{bmatrix}A$

[applying $R_3 \to R_3+5R_2$]

or, $\begin{bmatrix}1 & 0 & -1\\ 0 & 1 & 2\\ 0 & 0 & 1\end{bmatrix}=\begin{bmatrix}-2 & 1 & 0\\ 1 & 0 & 0\\ \frac{5}{2} & \frac{-3}{2} & \frac{1}{2}\end{bmatrix}A$

[applying $R_3 \to \frac{1}{2}R_3$]

or, $\begin{bmatrix}1 & 0 & 0\\ 0 & 1 & 2\\ 0 & 0 & 1\end{bmatrix}=\begin{bmatrix}\frac{1}{2} & \frac{-1}{2} & \frac{1}{2}\\ 1 & 0 & 0\\ \frac{5}{2} & \frac{-3}{2} & \frac{1}{2}\end{bmatrix}A$

[applying $R_1 \to R_1+R_3$]

or, $\begin{bmatrix}1 & 0 & 0\\ 0 & 1 & 0\\ 0 & 0 & 1\end{bmatrix}=\begin{bmatrix}1/2 & -1/2 & 1/2\\ -4 & 3 & -1\\ 5/2 & -3/2 & 1/2\end{bmatrix}A$

[applying $R_2 \to R_2-2R_3$]

$$\therefore\ A^{-1} = \begin{bmatrix} 1/2 & -1/2 & 1/2 \\ -4 & 3 & -1 \\ 5/2 & -3/2 & 1/2 \end{bmatrix}$$

(ii) Let $A = \begin{bmatrix} 2 & 0 & -1 \\ 5 & 1 & 0 \\ 0 & 1 & 3 \end{bmatrix}$

We know that $A = IA$. Therefore,

$$\begin{bmatrix} 2 & 0 & -1 \\ 5 & 1 & 0 \\ 0 & 1 & 3 \end{bmatrix} = \begin{bmatrix} 1 & 0 & 0 \\ 0 & 1 & 0 \\ 0 & 0 & 1 \end{bmatrix} A$$

Applying $R_1 \to \frac{1}{2}R_1$, we have

$$\begin{bmatrix} 1 & 0 & \frac{-1}{2} \\ 5 & 1 & 0 \\ 0 & 1 & 3 \end{bmatrix} = \begin{bmatrix} \frac{1}{2} & 0 & 0 \\ 0 & 1 & 0 \\ 0 & 0 & 1 \end{bmatrix} A$$

Applying $R_2 \to R_2 - 5R_1$, we have

$$\begin{bmatrix} 1 & 0 & \frac{-1}{2} \\ 0 & 1 & \frac{5}{2} \\ 0 & 1 & 3 \end{bmatrix} = \begin{bmatrix} \frac{1}{2} & 0 & 0 \\ \frac{-5}{2} & 1 & 0 \\ 0 & 0 & 1 \end{bmatrix} A$$

Applying $R_3 \to R_3 - R_2$, we have

$$\begin{bmatrix} 1 & 0 & \frac{-1}{2} \\ 0 & 1 & \frac{5}{2} \\ 0 & 0 & \frac{1}{2} \end{bmatrix} = \begin{bmatrix} \frac{1}{2} & 0 & 0 \\ \frac{-5}{2} & 1 & 0 \\ \frac{5}{2} & -1 & 1 \end{bmatrix} A$$

Applying $R_3 \to 2R_3$, we have

$$\begin{bmatrix} 1 & 0 & \frac{-1}{2} \\ 0 & 1 & \frac{5}{2} \\ 0 & 0 & 1 \end{bmatrix} = \begin{bmatrix} \frac{1}{2} & 0 & 0 \\ \frac{-5}{2} & 1 & 0 \\ 5 & -2 & 2 \end{bmatrix} A$$

Applying $R_1 \to R_1 + \frac{1}{2}R_3$ and $R_2 \to R_2 - \frac{5}{2}R_3$, we have

$$\begin{bmatrix} 1 & 0 & 0 \\ 0 & 1 & 0 \\ 0 & 0 & 1 \end{bmatrix} = \begin{bmatrix} 3 & -1 & 1 \\ -15 & 6 & -5 \\ 5 & -2 & 2 \end{bmatrix} A$$

$$\therefore\ A^{-1} = \begin{bmatrix} 3 & -1 & 1 \\ -15 & 6 & -5 \\ 5 & -2 & 2 \end{bmatrix}$$

48. $\because AB = B^{-1}$ or $AB^2 = I$

Now, $\lambda A - 2B^{-1} + I = 0$

$\Rightarrow \lambda AB - 2B^{-1}B + IB = 0$ [post multiplying by B]

$\Rightarrow \lambda AB - 2I + B = 0$

$\Rightarrow \lambda AB^2 - 2IB + B^2 = 0$ [again post multiplying by B]

$\Rightarrow \lambda AB^2 - 2B + B^2 = 0$

$\Rightarrow \lambda I - 2B + B^2 = 0$ [$\because AB^2 = I$]

$$\Rightarrow \lambda\begin{bmatrix} 1 & 0 \\ 0 & 1 \end{bmatrix} - 2\begin{bmatrix} 2 & -2 \\ -1 & 0 \end{bmatrix} + \begin{bmatrix} 2 & -2 \\ -1 & 0 \end{bmatrix}^2 = \begin{bmatrix} 0 & 0 \\ 0 & 0 \end{bmatrix}$$

$$\Rightarrow \begin{bmatrix} \lambda & 0 \\ 0 & \lambda \end{bmatrix} - \begin{bmatrix} 4 & -4 \\ -2 & 0 \end{bmatrix} + \begin{bmatrix} 6 & -4 \\ -2 & 2 \end{bmatrix} = \begin{bmatrix} 0 & 0 \\ 0 & 0 \end{bmatrix}$$

$$\Rightarrow \begin{bmatrix} \lambda+2 & 0 \\ 0 & \lambda+2 \end{bmatrix} = \begin{bmatrix} 0 & 0 \\ 0 & 0 \end{bmatrix}$$

$\Rightarrow \lambda + 2 = 0$

$\therefore \lambda = -2$

49. Let $A = \begin{bmatrix} 3 & -1 & 2 \\ -3 & 1 & 2 \\ -6 & 2 & 4 \end{bmatrix}$

Applying $R_2 \to R_2 + R_1$ and $R_3 \to R_3 + 2R_1$, we get

$$A = \begin{bmatrix} 3 & -1 & 2 \\ 0 & 0 & 4 \\ 0 & 0 & 8 \end{bmatrix}$$

Applying $R_3 \to R_3 - 2R_2$, we get

$$A = \begin{bmatrix} 3 & -1 & 2 \\ 0 & 0 & 4 \\ 0 & 0 & 0 \end{bmatrix}$$

Applying $R_1 \to \left(\frac{1}{3}\right)R_1$ and $R_2 \to \left(\frac{1}{4}\right)R_2$, then

$$A = \begin{bmatrix} 1 & \frac{-1}{3} & \frac{2}{3} \\ 0 & 0 & 1 \\ 0 & 0 & 0 \end{bmatrix}$$

This is Echelon form of matrix A.

$\therefore$ Rank = Number of non-zero rows (in echelon form)

$\Rightarrow$ Rank of A is 2.

50. We can write the above system of equation in the matrix form

$$\begin{bmatrix} 1 & 1 & 1 \\ 1 & 2 & 3 \\ 1 & 2 & \lambda \end{bmatrix}\begin{bmatrix} x \\ y \\ z \end{bmatrix} = \begin{bmatrix} 6 \\ 10 \\ \mu \end{bmatrix}$$

$\Rightarrow AX = B$

Where $A = \begin{bmatrix} 1 & 1 & 1 \\ 1 & 2 & 3 \\ 1 & 2 & \lambda \end{bmatrix}$, $X = \begin{bmatrix} x \\ y \\ z \end{bmatrix}$ and $B = \begin{bmatrix} 6 \\ 10 \\ \mu \end{bmatrix}$

The augmented matrix

$$C=[A:B]=\begin{bmatrix}1 & 1 & 1: & 6\\ 1 & 2 & 3: & 10\\ 1 & 2 & \lambda: & \mu\end{bmatrix}$$

Applying $R_2 \to R_2 - R_1$ and $R_3 \to R_3 - R_1$, we get

$$C=\begin{bmatrix}1 & 1 & 1 & : & 6\\ 0 & 1 & 2 & : & 4\\ 0 & 1 & \lambda-1 & : & \mu-6\end{bmatrix}$$

Applying $R_3 \to R_3 - R_2$, we get

$$\begin{bmatrix}1 & 1 & 1 & : & 6\\ 0 & 1 & 2 & : & 4\\ 0 & 0 & \lambda-3 & : & \mu-10\end{bmatrix}$$

(i) No solution $\rho(A) \neq \rho(C)$
i.e. $\lambda - 3 = 0$ and $\mu - 10 \neq 0$
$\therefore \lambda = 3$ and $\mu \neq 10$

(ii) A unique solution $\rho(A) = \rho(C) = 3$
i.e. $\lambda - 3 \neq 0$ and $\mu \in R$
$\therefore \lambda \neq 3$ and $\mu \in R$

(iii) Infinite number of solutions
$\rho(A) = \rho(C)(< 3)$
i.e. $\lambda - 3 = 0$ and $\mu - 10 = 0$
$\therefore \lambda = 3$ and $\mu = 10$

51. Let (x_1, y_1) be the image of the point (x, y) under the given transformation, then

$$x_1 = x - 2y \quad = 1 \cdot x + (-2) \cdot y$$

and $\quad y_1 = -3x + y \; = (-3) \cdot x + 1 \cdot y$

$$\Rightarrow \begin{bmatrix}x_1\\ y_1\end{bmatrix} = \begin{bmatrix}1 & -2\\ -3 & 1\end{bmatrix}\begin{bmatrix}x\\ y\end{bmatrix}$$

$$= \begin{bmatrix}1 & -2\\ -3 & 1\end{bmatrix}\begin{bmatrix}-2\\ -7\end{bmatrix} = \begin{bmatrix}-2+14\\ 6-7\end{bmatrix}$$

$$= \begin{bmatrix}12\\ -1\end{bmatrix}$$

Therefore, the required image is $(12, -1)$.

52. Let $B(x_1, y_1)$ be the image of the point $A(2, 3)$ about the line $y = x$, then

$$\begin{bmatrix}x_1\\ y_1\end{bmatrix} = \begin{bmatrix}0 & 1\\ 1 & 0\end{bmatrix}\begin{bmatrix}2\\ 3\end{bmatrix} = \begin{bmatrix}3\\ 2\end{bmatrix}$$

Therefore, the image of $A\,(2, 3)$ by the line mirror $y = x$ is $B\,(3, 2)$.

Given, image of B by the line mirror $y = 0$ (X-axis) is (α, β), then

$$\begin{bmatrix}\alpha\\ \beta\end{bmatrix} = \begin{bmatrix}1 & 0\\ 0 & -1\end{bmatrix}\begin{bmatrix}3\\ 2\end{bmatrix} = \begin{bmatrix}3\\ -2\end{bmatrix}$$

On comparing, we get

$$\alpha = 3 \text{ and } \beta = -2$$

53. Let (x_1, y_1) be the image of $(-\sqrt{2}, \sqrt{2})$ about the line

$$y = x \tan\left(\frac{\pi}{8}\right)$$

On comparing,

$$y = x \tan\left(\frac{\pi}{8}\right) \text{ by } y = x \tan\theta$$

$$\therefore \quad \theta = \frac{\pi}{8}$$

Now, $\begin{bmatrix}x_1\\ y_1\end{bmatrix} = \begin{bmatrix}\cos 2\theta & \sin 2\theta\\ \sin 2\theta & -\cos 2\theta\end{bmatrix}\begin{bmatrix}-\sqrt{2}\\ \sqrt{2}\end{bmatrix}$

$$= \begin{bmatrix}\frac{1}{\sqrt{2}} & \frac{1}{\sqrt{2}}\\ \frac{1}{\sqrt{2}} & -\frac{1}{\sqrt{2}}\end{bmatrix}\begin{bmatrix}-\sqrt{2}\\ \sqrt{2}\end{bmatrix}$$

$$= \begin{bmatrix}0\\ -2\end{bmatrix}$$

On comparing,

$$x_1 = 0 \text{ and } y_1 = -2$$

Therefore, the required image is $(0, -2)$.

Level 2

1. (d) Since each element a_{ij} can be filled in two ways (with either 2 or 0).

Total number of matrices is 2^9 i.e. 512.

2. (b) $A = \begin{bmatrix}a_{11} & a_{12}\\ a_{21} & a_{22}\end{bmatrix}$

$$= \begin{bmatrix}2^1 \cdot \omega^{1+1} & 2^1 \cdot \omega^{1+2}\\ 2^2 \omega^{2+1} & 2^2 \omega^{2+2}\end{bmatrix}$$

$$= \begin{bmatrix}2 \cdot \omega^2 & 2 \cdot \omega^3\\ 4\omega^3 & 4\omega^4\end{bmatrix}$$

$$= \begin{bmatrix}2\omega^2 & 2\\ 4 & 4\omega\end{bmatrix} \quad [\because \omega^3 = 1]$$

3. (b) Given, $\begin{bmatrix}4 & x-3\\ 2+y & xz\end{bmatrix} = \begin{bmatrix}4 & 3\\ -1 & 10\end{bmatrix}$

On comparing the corresponding elements, we get

$$x - z = 3 \quad \text{...(i)}$$

$$2 + y = -1$$

$$\Rightarrow \quad y = -1 - 2 = -3 \quad \text{...(ii)}$$

and $\quad xz = 10 \quad$ ...(iii)

Now, $(x + z)^2 = (x - z)^2 + 4xz$

$$= (3)^2 + 4 \times 10 = 9 + 40$$

$$\Rightarrow \quad (x + z)^2 = 49$$

$$\Rightarrow \quad x + z = 7 \quad \text{...(iv)}$$

On adding Eqs. (i) and (iv), we get

$$2x = 10 \Rightarrow x = 5$$

From Eq. (i),

$$5 - z = 3 \Rightarrow -z = 3 - 5$$

$$\therefore \quad z = 2$$

Hence, $x = 5, y = -3$

and $\quad z = 2$

4. (c) $A + B = \begin{bmatrix} {}^5C_2 & {}^6C_2 \\ {}^4C_2 & {}^3C_2 \end{bmatrix} + \begin{bmatrix} i^4 & i^6 \\ i^8 & i^{10} \end{bmatrix}$

$$= \begin{bmatrix} 10 & 15 \\ 6 & 3 \end{bmatrix} + \begin{bmatrix} 1 & -1 \\ 1 & -1 \end{bmatrix}$$

$$= \begin{bmatrix} 10+1 & 15-1 \\ 6+1 & 3-1 \end{bmatrix}$$

$$= \begin{bmatrix} 11 & 14 \\ 7 & 2 \end{bmatrix}$$

5. (b) Given, $A = \begin{bmatrix} x & 1 \\ 1 & 0 \end{bmatrix}$

Now, $\quad A^2 = \begin{bmatrix} x & 1 \\ 1 & 0 \end{bmatrix}\begin{bmatrix} x & 1 \\ 1 & 0 \end{bmatrix}$

$$\Rightarrow \begin{bmatrix} x^2+1 & x \\ x & 1 \end{bmatrix} = \begin{bmatrix} 1 & 0 \\ 0 & 1 \end{bmatrix} \qquad [\text{given } A^2 = 1]$$

On comparing, we get

$$x = 0$$

6. (a) Given matrix,

$$A = \begin{bmatrix} 0 & 2 \\ K & -1 \end{bmatrix}$$

Characteristic equation of A is

$$|A - \lambda I| = 0$$

$$\Rightarrow \begin{vmatrix} -\lambda & 2 \\ K & -1-\lambda \end{vmatrix} = 0$$

$$\Rightarrow \lambda(1+\lambda) - 2k = 0$$

$$\Rightarrow \lambda^2 + \lambda - 2k = 0$$

$\because$ Every square matrix satisfied its own characteristic equation.

$$\therefore \quad A^2 + A - 2KI = 0$$

$$\Rightarrow \quad A^2 = 2KI - A$$

$$\Rightarrow \quad A^4 = 4K^2I + A^2 - 2(2KI)(A)$$

$$\Rightarrow \quad A^4 = 4K^2I + 2KI - A - 4KA$$

$$\Rightarrow \quad A^4 = (4K^2 + 2K)I - (1 + 4K)A \qquad \text{...(i)}$$

Now, $A(A^3 + 3I) = 2I$

$$\Rightarrow \quad A^4 = 2I - 3A \qquad \text{.... (ii)}$$

Comparing Eqs. (i) and (ii), we get

$$1 + 4K = 3$$

$$\Rightarrow \quad K = \frac{1}{2}$$

7. (a) $A^2 = \begin{bmatrix} 1 & 0 & 0 \\ 0 & 1 & 0 \\ a & b & -1 \end{bmatrix}\begin{bmatrix} 1 & 0 & 0 \\ 0 & 1 & 0 \\ a & b & -1 \end{bmatrix}$

$$\Rightarrow \quad A^2 = \begin{bmatrix} 1 & 0 & 0 \\ 0 & 1 & 0 \\ 0 & 0 & 1 \end{bmatrix} = I_3$$

$$\Rightarrow \quad A^4 = A^2A^2 = I_3 \cdot I_3 = I_3$$

$$\Rightarrow \quad A^6 = A^2 \cdot A^4 = I_3 \cdot I_3 = I_3$$

$$\therefore \quad A^2 + 2A^4 + 4A^6$$

$$= I_3 + 2I_3 + 4I_3$$

$$= 7I_3 = 7(I_3)^4$$

$$= 7(A^2)^4 = 7A^8$$

8. (d) We have, $A = \begin{bmatrix} 3 & 1 \\ -1 & 2 \end{bmatrix}$

$$\therefore \quad A^2 = \begin{bmatrix} 3 & 1 \\ -1 & 2 \end{bmatrix}\begin{bmatrix} 3 & 1 \\ -1 & 2 \end{bmatrix} = \begin{bmatrix} 8 & 5 \\ -5 & 3 \end{bmatrix}$$

and $\quad 5A = \begin{bmatrix} 15 & 5 \\ -5 & 10 \end{bmatrix}$

Now, $\quad A^2 - 5A + 7I$

$$= \begin{bmatrix} 8 & 5 \\ -5 & 3 \end{bmatrix} - \begin{bmatrix} 15 & 5 \\ -5 & 10 \end{bmatrix} + \begin{bmatrix} 7 & 0 \\ 0 & 7 \end{bmatrix}$$

$$= \begin{bmatrix} 0 & 0 \\ 0 & 0 \end{bmatrix} = 0$$

9. (d) Given, $A = \begin{bmatrix} \alpha & 0 \\ 1 & 1 \end{bmatrix}, B = \begin{bmatrix} 1 & 0 \\ 5 & 1 \end{bmatrix}$ and $A^2 = B$

$$\Rightarrow \begin{bmatrix} \alpha & 0 \\ 1 & 1 \end{bmatrix}\begin{bmatrix} \alpha & 0 \\ 1 & 1 \end{bmatrix} = \begin{bmatrix} 1 & 0 \\ 5 & 1 \end{bmatrix}$$

$$\Rightarrow \begin{bmatrix} \alpha^2 & 0 \\ \alpha+1 & 1 \end{bmatrix} = \begin{bmatrix} 1 & 0 \\ 5 & 1 \end{bmatrix}$$

$$\Rightarrow \quad \alpha^2 = 1 \text{ and } \alpha + 1 = 5$$

$$\Rightarrow \quad \alpha = \pm 1 \text{ and } \alpha = 4$$

$\because$ There is no common value.

$\therefore$ There is no real value of α for which $A^2 = B$.

10. (c) $A = \begin{bmatrix} 1 & 2 \\ 3 & 4 \end{bmatrix}$ and $B = \begin{bmatrix} a & 0 \\ 0 & b \end{bmatrix}$

$$AB = \begin{bmatrix} a & 2b \\ 3a & 4b \end{bmatrix}$$

$$BA = \begin{bmatrix} a & 0 \\ 0 & b \end{bmatrix}\begin{bmatrix} 1 & 2 \\ 3 & 4 \end{bmatrix}$$

$$= \begin{bmatrix} a & 2a \\ 3b & 4b \end{bmatrix}$$

Hence, $AB = BA$ only when $a = b$

11. (c) $A^2 = A \cdot A = \begin{bmatrix} a & b \\ b & a \end{bmatrix}\begin{bmatrix} a & b \\ b & a \end{bmatrix}$

$$A^2 = \begin{bmatrix} \alpha & \beta \\ \beta & \alpha \end{bmatrix} = \begin{bmatrix} a^2+b^2 & 2ab \\ 2ab & a^2+b^2 \end{bmatrix}$$

$$\alpha = a^2+b^2, \beta = 2ab$$

12. (a) Given, $AB = A$ and $BA = B$

Now, $BA = B$

$\Rightarrow$ $BAB = B^2$ [post multiplying both sides by B]

$\Rightarrow$ $B(A) = B^2$ [$\because AB = A$]

$\Rightarrow$ $B = B^2$ [$\because BA = B$]

13. (b) $A^2 - B^2 = (A-B)(A+B)$

$$A^2 - B^2 = A^2 + AB - BA - B^2$$

$\Rightarrow$ $AB = BA$

14. (c) $A = \begin{bmatrix} \cos\alpha & -\sin\alpha \\ \sin\alpha & \cos\alpha \end{bmatrix}$

$$A^2 = \begin{bmatrix} \cos\alpha & -\sin\alpha \\ \sin\alpha & \cos\alpha \end{bmatrix}\begin{bmatrix} \cos\alpha & -\sin\alpha \\ \sin\alpha & \cos\alpha \end{bmatrix}$$

$$= \begin{bmatrix} \cos^2\alpha - \sin^2\alpha & -\cos\alpha\sin\alpha - \sin\alpha\cos\alpha \\ \sin\alpha\cos\alpha + \cos\alpha\sin\alpha & -\sin^2\alpha + \cos^2\alpha \end{bmatrix}$$

$$= \begin{bmatrix} \cos 2\alpha & -\sin 2\alpha \\ \sin 2\alpha & \cos 2\alpha \end{bmatrix}$$

Similarly, $A^4 = A^2 \cdot A^2 = \begin{bmatrix} \cos 4\alpha & -\sin 4\alpha \\ \sin 4\alpha & \cos 4\alpha \end{bmatrix}$

Similarly, $A^{32} = \begin{bmatrix} \cos 32\alpha & -\sin 32\alpha \\ \sin 32\alpha & \cos 32\alpha \end{bmatrix} = \begin{bmatrix} 0 & -1 \\ 1 & 0 \end{bmatrix}$ [given]

On comparing, we get

$$\sin 32\alpha = 1, \cos 32\alpha = 0$$

$\Rightarrow$ $32\alpha = n\pi + (-1)^n \frac{\pi}{2}$ and $32\alpha = 2n\pi + \frac{\pi}{2}$

$\Rightarrow$ $\alpha = \frac{n\pi}{32} + (-1)^n \frac{\pi}{64}$ and $\alpha = \frac{n\pi}{16} + \frac{\pi}{64}$

where $n \in Z$

Put $n = 0$; $\alpha = \frac{\pi}{64}$

15. (d) $\because$ $A = \begin{bmatrix} \cos\theta & i\sin\theta \\ i\sin\theta & \cos\theta \end{bmatrix}$

$\therefore$ $A^n = \begin{bmatrix} \cos n\theta & i\sin n\theta \\ i\sin n\theta & \cos n\theta \end{bmatrix}; n \in N$

$$A^5 = \begin{bmatrix} a & b \\ c & d \end{bmatrix}$$

$\Rightarrow \begin{bmatrix} \cos 5\theta & i\sin 5\theta \\ i\sin 5\theta & \cos 5\theta \end{bmatrix} = \begin{bmatrix} a & b \\ c & d \end{bmatrix}$

$\therefore$ $a = \cos 5\theta, b = i\sin 5\theta, c = i\sin 5\theta, d = \cos 5\theta$

$$a^2 - b^2 = \cos^2 5\theta + \sin^2 5\theta = 1$$

$$a^2 - c^2 = \cos^2 5\theta + \sin^2 5\theta = 1$$

$$a^2 - d^2 = \cos^2 5\theta - \cos^2 5\theta = 0$$

$$a^2 + b^2 = \cos^2 5\theta - \sin^2 5\theta$$

$$= \cos 10\theta = \cos\frac{10\pi}{24}$$

and $0 \le \cos\frac{5\pi}{12} \le 1$

$\Rightarrow$ $0 \le a^2 + b^2 \le 1$

Thus, option (d) is not correct as per all given options.

16. (d) Solution of $x^2 + x + 1 = 0$ is ω and ω^2

So, $\alpha = \omega$ and $\omega^4 = \omega^3 \cdot \omega = \omega$

$$A^2 = \frac{1}{3}\begin{bmatrix} 1 & 1 & 1 \\ 1 & \omega & \omega^2 \\ 1 & \omega^2 & \omega \end{bmatrix}\begin{bmatrix} 1 & 1 & 1 \\ 1 & \omega & \omega^2 \\ 1 & \omega^2 & \omega \end{bmatrix}$$

$$A^2 = \begin{bmatrix} 1 & 0 & 0 \\ 0 & 0 & 1 \\ 0 & 1 & 0 \end{bmatrix}$$

$$A^4 = A^2 \cdot A^2 = \begin{bmatrix} 1 & 0 & 0 \\ 0 & 0 & 1 \\ 0 & 1 & 0 \end{bmatrix}\begin{bmatrix} 1 & 0 & 0 \\ 0 & 0 & 1 \\ 0 & 1 & 0 \end{bmatrix} = I$$

$\therefore$ $A^{31} = A^{28} \cdot A^3 = I \cdot A^3 = A^3$

17. (a) $P^2 = \begin{bmatrix} 1 & 0 & 0 \\ 3 & 1 & 0 \\ 9 & 3 & 1 \end{bmatrix}\begin{bmatrix} 1 & 0 & 0 \\ 3 & 1 & 0 \\ 9 & 3 & 1 \end{bmatrix}$

$$P^2 = \begin{bmatrix} 1 & 0 & 0 \\ 6 & 1 & 0 \\ 27 & 6 & 1 \end{bmatrix}$$

Similarly, $P^4 = \begin{bmatrix} 1 & 0 & 0 \\ 12 & 1 & 0 \\ 90 & 12 & 1 \end{bmatrix}$

$$P^5 = P^4 \cdot P = \begin{bmatrix} 1 & 0 & 0 \\ 15 & 1 & 0 \\ 135 & 15 & 1 \end{bmatrix}$$

Given, $Q - P^5 = I_3$

$\therefore$ $Q = I_3 + P^5$

$$= \begin{bmatrix} 1 & 0 & 0 \\ 0 & 1 & 0 \\ 0 & 0 & 1 \end{bmatrix} + \begin{bmatrix} 1 & 0 & 0 \\ 15 & 1 & 0 \\ 135 & 15 & 1 \end{bmatrix}$$

$$Q = \begin{bmatrix} 2 & 0 & 0 \\ 15 & 2 & 0 \\ 135 & 15 & 2 \end{bmatrix}$$

$\therefore$ $\frac{q_{21} + q_{31}}{q_{32}} = \frac{15 + 135}{15} = \frac{150}{15} = 10$

18. (b) Given, A is 3×4 matrix. So, A' is 4×3 matrix.

Since $A'B$ is defined, therefore B should be 3×4 Also, BA' is defined. So, B should be 3×4. Hence, B is of the type 3×4.

19. (a) Given, $AA^T = I$

$$\Rightarrow \begin{bmatrix} 0 & 2q & r \\ p & q & -r \\ p & -q & r \end{bmatrix} \begin{bmatrix} 0 & p & p \\ 2q & q & -q \\ r & -r & r \end{bmatrix} = \begin{bmatrix} 1 & 0 & 0 \\ 0 & 1 & 0 \\ 0 & 0 & 1 \end{bmatrix}$$

$$\Rightarrow \begin{bmatrix} 0+4q^2+r^2 & 0+2q^2-r^2 & 0-2q^2+r^2 \\ 0+2q^2-r^2 & p^2+q^2+r^2 & p^2-q^2-r^2 \\ 0-2q^2+r^2 & p^2-q^2-r^2 & p^2+q^2+r^2 \end{bmatrix} = \begin{bmatrix} 1 & 0 & 0 \\ 0 & 1 & 0 \\ 0 & 0 & 1 \end{bmatrix}$$

We know that if two matrices are equal, then corresponding elements are also equal, so

$$4q^2 + r^2 = 1 = p^2 + q^2 + r^2 \quad \text{...(i)}$$

$$2q^2 - r^2 = 0 \Rightarrow r^2 = 2q^2 \quad \text{...(ii)}$$

and $p^2 - q^2 - r^2 = 0$ ...(iii)

Using Eqs. (ii) and (iii), we get

$$p^2 = 3q^2 \quad \text{...(iv)}$$

Using Eq. (ii) in Eq. (i), we get

$$4q^2 + 2q^2 = 1$$

$$\Rightarrow 6q^2 = 1$$

$$\Rightarrow 2p^2 = 1 \quad \text{[using Eq. (iv)]}$$

$$p^2 = \frac{1}{2}$$

$$\Rightarrow |p| = \frac{1}{\sqrt{2}}$$

20. (d) $AA^T = 9I$

$$\Rightarrow \begin{bmatrix} 1 & 2 & 2 \\ 2 & 1 & -2 \\ a & 2 & b \end{bmatrix} \begin{bmatrix} 1 & 2 & a \\ 2 & 1 & 2 \\ 2 & -2 & b \end{bmatrix} = 9 \begin{bmatrix} 1 & 0 & 0 \\ 0 & 1 & 0 \\ 0 & 0 & 1 \end{bmatrix}$$

$$\Rightarrow \begin{bmatrix} 9 & 0 & a+4+2b \\ 0 & 9 & 2a+2-2b \\ a+4+2b & 2a+2-2b & a^2+4+b^2 \end{bmatrix} = \begin{bmatrix} 9 & 0 & 0 \\ 0 & 9 & 0 \\ 0 & 0 & 9 \end{bmatrix}$$

$$\Rightarrow a + 4 + 2b = 0, 2a + 2 - 2b = 0, a^2 + 4 + b^2 = 9$$

$$\Rightarrow a = -2, b = -1$$

21. (b) Given matrix

$$A = \begin{bmatrix} 0 & 2y & 1 \\ 2x & y & -1 \\ 2x & -y & 1 \end{bmatrix}, (x, y \in R, x \neq y)$$

For which $A^T A = 3I_3$

$$\Rightarrow \begin{bmatrix} 0 & 2x & 2x \\ 2y & y & -y \\ 1 & -1 & 1 \end{bmatrix} \begin{bmatrix} 0 & 2y & 1 \\ 2x & y & -1 \\ 2x & -y & 1 \end{bmatrix} = \begin{bmatrix} 3 & 0 & 0 \\ 0 & 3 & 0 \\ 0 & 0 & 3 \end{bmatrix}$$

$$\Rightarrow \begin{bmatrix} 8x^2 & 0 & 0 \\ 0 & 6y^2 & 0 \\ 0 & 0 & 3 \end{bmatrix} = \begin{bmatrix} 3 & 0 & 0 \\ 0 & 3 & 0 \\ 0 & 0 & 3 \end{bmatrix}$$

Here, two matrices are equal, therefore equating the corresponding element, we get

$$8x^2 = 3 \text{ and } 6y^2 = 3$$

$$\Rightarrow x = \pm\sqrt{\frac{3}{8}} \text{ and } y = \pm\frac{1}{\sqrt{2}}$$

$\therefore$ There are 2 different value x and y each.

So, 4 matrices are possible such that

$$A^T A = 3I_3$$

22. (b)
$$A = \begin{bmatrix} 3 & -4 \\ 1 & -1 \end{bmatrix} = \begin{bmatrix} 1 & 0 \\ 0 & 1 \end{bmatrix} + \begin{bmatrix} 2 & -4 \\ 1 & -2 \end{bmatrix}$$

$$\Rightarrow A = I + B$$

where, $B = \begin{bmatrix} 2 & -4 \\ 1 & -2 \end{bmatrix}$

$$B^2 = \begin{bmatrix} 2 & -4 \\ 1 & -2 \end{bmatrix} \begin{bmatrix} 2 & -4 \\ 1 & -2 \end{bmatrix} = \begin{bmatrix} 0 & 0 \\ 0 & 0 \end{bmatrix} = 0$$

$$\Rightarrow B^3 = B^4 = \dots = 0$$

$$\therefore A^{2012} = (I + B)^{2012}$$

$$= I^{2012} + {}^{2012}C_1 B + {}^{2012}C_2 B^2 + \dots$$

$$= I + 2012B + 0 + 0 + \dots$$

$$\Rightarrow A^{2012} = I + 2012B$$

$$\Rightarrow A^{2012} = \begin{bmatrix} 1 & 0 \\ 0 & 1 \end{bmatrix} + 2012 \begin{bmatrix} 2 & -4 \\ 1 & -2 \end{bmatrix}$$

$\therefore$ Trace $(A^{2012}) = (1 + 1) + 2012\,(2 - 2) = 2$

23. (c) We have, $A = \begin{bmatrix} 3 & x-1 \\ 2x+3 & x+2 \end{bmatrix}$ is a symmetric matrix

i.e. $A^T = A$

$$\Rightarrow \begin{bmatrix} 3 & 2x+3 \\ x-1 & x+2 \end{bmatrix} = \begin{bmatrix} 3 & x-1 \\ 2x+3 & x+2 \end{bmatrix}$$

$$\Rightarrow 2x + 3 = x - 1$$

$$x = -4$$

24. (a) $(A + A^T)^T = A^T + (A^T)^T = A^T + A$

Hence, $A + A^T$ is symmetric matrix.

25. (b) Given, $A = A', B = B'$

Now, $(AB - BA)' = (AB)' - (BA)' = B'A' - A'B'$

$$= BA - AB$$

$$= -(AB - BA)$$

26. (d) Given, matrix $A=\begin{bmatrix}1 & 2 & 4\\6 & 8 & 2\\2 & -2 & 7\end{bmatrix}$

Symmetric part of $A=\frac{1}{2}[A+A']$

$$=\frac{1}{2}\left\{\begin{bmatrix}1 & 2 & 4\\6 & 8 & 2\\2 & -2 & 7\end{bmatrix}+\begin{bmatrix}1 & 6 & 2\\2 & 8 & -2\\4 & 2 & 7\end{bmatrix}\right\}$$

$$=\frac{1}{2}\left\{\begin{bmatrix}2 & 8 & 6\\8 & 16 & 0\\6 & 0 & 14\end{bmatrix}\right\}=\begin{bmatrix}1 & 4 & 3\\4 & 8 & 0\\3 & 0 & 7\end{bmatrix}$$

27. (c) Given, $X=AB+BA$

$\Rightarrow \quad X=X^T \quad [\because A=A^T \text{ and } B=B^T]$

and $\quad Y=AB-BA$

$\Rightarrow \quad -Y=Y^T$

Now, $\quad (XY)^T=Y^TX^T=-YX$

28. (b) Let $C=BA$, then

$C^T=(BA)^T=A^TB^T$

$=-AB^T$ [as A is skew-symmetric]

$=BA^T$ [as $AB^T+BA^T=0$]

$=B(-A)$

$=-BA=-C$

Therefore, C is skew-symmetric.

29. (b) Given matrix A is a symmetric and matrix B is a skew-symmetric.

$\therefore \quad A^T=A$ and $B^T=-B$

Since, $A+B=\begin{bmatrix}2 & 3\\5 & -1\end{bmatrix}$ ...(i)

On taking transpose both sides, we get

$$(A+B)^T=\begin{bmatrix}2 & 3\\5 & -1\end{bmatrix}^T$$

$\Rightarrow A^T+B^T=\begin{bmatrix}2 & 5\\3 & -1\end{bmatrix}$ ...(ii)

Given, $\quad A^T=A$ and $B^T=-B$

$\Rightarrow \quad A-B=\begin{bmatrix}2 & 5\\3 & -1\end{bmatrix}$

On solving Eqs. (i) and (ii), we get

$$A=\begin{bmatrix}2 & 4\\4 & -1\end{bmatrix} \text{ and } B=\begin{bmatrix}0 & -1\\1 & 0\end{bmatrix}$$

So, $\quad AB=\begin{bmatrix}2 & 4\\4 & -1\end{bmatrix}\begin{bmatrix}0 & -1\\1 & 0\end{bmatrix}=\begin{bmatrix}4 & -2\\-1 & -4\end{bmatrix}$

30. (a) $\quad A=\begin{bmatrix}2 & 3\\a & 0\end{bmatrix}, A^T=\begin{bmatrix}2 & a\\3 & 0\end{bmatrix}$

$$A=\frac{A+A^T}{2}+\frac{A-A^T}{2}$$

Let $\quad P=\frac{A+A^T}{2}$ and $Q=\frac{A-A^T}{2}$

$$Q=\begin{pmatrix}0 & \frac{3-a}{2}\\\frac{a-3}{2} & 0\end{pmatrix}$$

Det $(Q)=9$

$$0-\left(\frac{3-a}{2}\right)\left(\frac{a-3}{2}\right)=9$$

$$\Rightarrow \left(\frac{a-3}{2}\right)^2=9 \Rightarrow (a-3)^2=36$$

$$a-3=\pm 6 \Rightarrow a=9,-3$$

$$P=\begin{bmatrix}2 & \frac{a+3}{2}\\\frac{a+3}{2} & 0\end{bmatrix}$$

$$P=\begin{bmatrix}2 & 6\\6 & 0\end{bmatrix} \text{ or } \begin{bmatrix}2 & 0\\0 & 0\end{bmatrix}$$

$$|P|=-36 \text{ or } 0$$

$\therefore \quad |-36+0|=36$

31. (c) $\left(A-\frac{1}{2}I\right)\left(A-\frac{1}{2}I\right)^T=I$ [given]

$\Rightarrow \quad \left(A-\frac{1}{2}I\right)\left(A^T-\frac{1}{2}I\right)=I$ ...(i)

Also, $\left(A+\frac{1}{2}I\right)\left(A+\frac{1}{2}I\right)^T=I$

$\Rightarrow \quad \left(A+\frac{1}{2}I\right)\left(A^T+\frac{1}{2}I\right)=I$ ...(ii)

On subtracting Eqs. (i) and (ii), we get

$$\left(AA^T-\frac{1}{2}A-\frac{1}{2}A^T+\frac{1}{4}I\right)-\left(AA^T+\frac{1}{2}A+\frac{1}{2}A^T+\frac{1}{4}I\right)$$

$$=I-I$$

$\Rightarrow \quad -A-A^T=0$

$\Rightarrow \quad A^T=-A$

$\therefore$ A is skew-symmetric matrix.

$$\left(A-\frac{1}{2}I\right)\left(A-\frac{1}{2}I\right)^T=I$$

$\Rightarrow \quad \left(A-\frac{1}{2}I\right)\left(A^T-\frac{1}{2}I\right)=I$

$\Rightarrow \quad AA^T-\frac{1}{2}A-\frac{1}{2}A^T+\frac{1}{4}I=I.$

$\Rightarrow \quad -A\cdot A-\frac{1}{2}A-\frac{1}{2}(-A)=I-\frac{I}{4}$ $[\because A^T=-A]$

$\Rightarrow \quad -A^2-\frac{1}{2}A+\frac{1}{2}A=\frac{3I}{4}$

$\Rightarrow \quad A^2=\frac{-3I}{4}$

32. (b) Given the matrices $A, B, A+I$ and $A+B$ are idempotent.

Hence, $A^2 = A, B^2 = B, (A+I)^2 = A+I$ and $(A+B)^2 = A+B$

$\Rightarrow \quad A^2 + B^2 + AB + BA = A + B$

$\Rightarrow \quad A + B + AB + BA = A + B$

$\Rightarrow \quad AB + BA = 0$

33. (d) Given, $A^2 - A + I = 0$

On multiplying A^{-1} both sides, we get

$A^{-1}A^2 - A^{-1}A + A^{-1} \cdot I = A^{-1} \cdot 0$

$\Rightarrow A - I + A^{-1} = 0$ or $A^{-1} = I - A$

34. (d) $(AB)^{-1} = B^{-1}A^{-1}$

35. (c) Given, $B^r = I$ or $B^rB^{-1} = IB^{-1}$ or $B^{r-1} = B^{-1}$

$\Rightarrow A^{-1}B^{r-1}A - A^{-1}B^{-1}A = A^{-1}B^{-1}A - A^{-1}B^{-1}A = 0$

36. (b) We have,

$$I = AA^{-1} = \frac{1}{2}\begin{bmatrix} 0 & 1 & 2 \\ 1 & 2 & 3 \\ 3 & a & 1 \end{bmatrix}\begin{bmatrix} 1 & -1 & 1 \\ -8 & 6 & 2c \\ 5 & -3 & 1 \end{bmatrix}$$

$$= \begin{bmatrix} 1 & 0 & c+1 \\ 0 & 1 & 2(c+1) \\ 4(1-a) & 3(a-1) & 2+ac \end{bmatrix}$$

Comparing the elements of AA^{-1} with corresponding elements of I, we have

$c + 1 = 0$ or $c = -1$

$a - 1 = 0$ or $a = 1$

37. (a) As, $B = -A^{-1}BA$, we get

$AB = -BA$ or $AB + BA = 0$

Now, $(A+B)^2 = (A+B)(A+B)$

$= A^2 + AB + BA + B^2$

$= A^2 + 0 + B^2 = A^2 + B^2$

38. (b) We have, $(A - 2I)(A - 4I) = 0$

or $\quad A^2 - 2A - 4A + 8I = 0$

or $\quad A^2 - 6A + 8I = 0$

or $\quad A^{-1}(A^2 - 6A + 8I) = A^{-1}0$

or $\quad A - 6I + 8A^{-1} = 0$

or $\quad A + 8A^{-1} = 6I$

or $\quad \frac{1}{6}A + \frac{4}{3}A^{-1} = I$

39. (c) $(PQ)^{-1} = R \Rightarrow Q^{-1} \times P^{-1} = R$

$$\Rightarrow \quad Q^{-1} = RP = \begin{bmatrix} -1 & 0 & 1 \\ 1 & 1 & 3 \\ 2 & 0 & 2 \end{bmatrix}\begin{bmatrix} 1 & 2 & 1 \\ 0 & 1 & -1 \\ 3 & 1 & 1 \end{bmatrix}$$

$$= \begin{bmatrix} 2 & -1 & 0 \\ 10 & 6 & 3 \\ 8 & 6 & 4 \end{bmatrix}$$

40. (a) Given, $(A+B)(A-B) = A^2 - B^2$

$\Rightarrow A^2 - AB + BA - B^2 = A^2 - B^2$

$\Rightarrow \quad AB = BA$

Now, $\quad (ABA^{-1})^2 = (BAA^{-1})^2 = (BI)^2 = B^2$

41. (b) Since, $AB = 3I$

$\Rightarrow A^{-1}AB = A^{-1}(3I) = 3(A^{-1}I) \Rightarrow B = 3A^{-1}$

$\therefore \quad A^{-1} = \frac{B}{3}$

42. (d) $\quad A = \begin{bmatrix} 1 & 2 \\ -1 & 4 \end{bmatrix}$

$A^{-1} = \alpha I + \beta A, \alpha, \beta \in R$

$A^{-1} = \frac{\text{adj} A}{|A|}$

If $\quad X = \begin{bmatrix} a & b \\ c & d \end{bmatrix}$ adj$X = \begin{bmatrix} d & -b \\ -c & a \end{bmatrix}$

Similarly, adj $A = \begin{bmatrix} 4 & -2 \\ 1 & 1 \end{bmatrix}$

$|A| = 4 - (-2) = 6$

$\therefore \quad A^{-1} = \frac{1}{6}\begin{bmatrix} 4 & -2 \\ 1 & 1 \end{bmatrix}$

Now, $\quad A^{-1} = \alpha I + \beta A$

$$= \begin{bmatrix} \alpha & 0 \\ 0 & \alpha \end{bmatrix} + \begin{bmatrix} \beta & 2\beta \\ -\beta & 4\beta \end{bmatrix}$$

$$= \begin{bmatrix} \alpha+\beta & 2\beta \\ -\beta & \alpha+4\beta \end{bmatrix}$$

$$\begin{bmatrix} \alpha+\beta & 2\beta \\ -\beta & \alpha+4\beta \end{bmatrix} = \begin{bmatrix} \frac{2}{3} & -\frac{1}{3} \\ \frac{1}{6} & \frac{1}{6} \end{bmatrix}$$

$\therefore \quad \beta = -\frac{1}{6} \Rightarrow \alpha + \beta = \frac{2}{3}$

$\Rightarrow \quad \alpha = \frac{4}{6} + \frac{1}{6} = \frac{5}{6}$

$\therefore \quad 4(\alpha - \beta) = 4\left(\frac{5}{6} + \frac{1}{6}\right) = 4$

43. (b) Given, $B = A^{-1}A^T$

$\Rightarrow \quad B^T = (A^{-1}A^T)^T = A \times (A^{-1})^T$

$\Rightarrow B \times B^T = A^{-1}A^T \times A \times (A^{-1})^T$

$= A^{-1} \times (A^T \times A)(A^{-1})^T$

$= A^{-1}(A \times A^T)(A^{-1})^T$

$= (A^{-1}A)(A^{-1}A)^T = I(I^T) = I$

44. (c) $(A^{-1}BA)^2 = (A^{-1}BA)(A^{-1}BA)$

$= A^{-1}B(AA^{-1})BA$

$= A^{-1}BIBA = A^{-1}B^2A$

$(A^{-1}BA)^3 = (A^{-1}B^2A)(A^{-1}BA)$

$= A^{-1}B^2(AA^{-1})BA$

$= A^{-1}B^3A$ and so on

$\therefore \; (A^{-1}BA)^n = A^{-1}B^nA$

45. (b) $Q = PAP^T$

$\therefore\; X = P^T Q^{1000} P = P^T (PAP^T)^{1000} P$

$= P^T PAP^T (PAP^T)^{999} P$

$= IAP^T \cdot PAP^T (PAP^T)^{998} P$

$= AIAP^T (PAP^T)^{998} P$

$= A^2 P^T PAP^T (PAP^T)^{997} P$

$= A^3 P^T (PAP^T)^{997} P$

$= A^{1000} = I$ [$\because$ A is involuntary]

Hence, $X^{-1} = I$

46. (c) Given, $M^T = -M, N^T = -N$ and $MN = NM$ ...(i)

$\therefore M^2N^2(M^TN)^{-1}(MN^{-1})^T = M^2N^2N^{-1}(M^T)^{-1}(N^{-1})^T M^T$

$= M^2N(NN^{-1})(-M)^{-1}(-N)^{-1}(-M)$

$= -M^2NM^{-1}N^{-1}M$

$= -M\cdot(MN)M^{-1}N^{-1}M$

$= -M(NM)M^{-1}N^{-1}M$

$= -MN(MM^{-1})N^{-1}M$

$= -M(NN^{-1})M = -M^2$

47. (b) Given, $A = \begin{bmatrix} 1 & 2 & 3 & 0 \\ 2 & 4 & 3 & 2 \\ 3 & 2 & 1 & 3 \\ 6 & 8 & 7 & \alpha \end{bmatrix}$

On applying $R_2 \to R_2 - 2R_1, R_3 \to R_3 - 3R_1$ and $R_4 \to R_4 - 2R_3$, we get

$$A = \begin{bmatrix} 1 & 2 & 3 & 0 \\ 0 & 0 & -3 & 2 \\ 0 & -4 & -8 & 3 \\ 0 & 4 & 5 & \alpha - 6 \end{bmatrix}$$

On applying $R_4 \to R_4 + R_3$, we get

$$A = \begin{bmatrix} 1 & 2 & 3 & 0 \\ 0 & 0 & -3 & 2 \\ 0 & -4 & -8 & 3 \\ 0 & 0 & -3 & \alpha - 3 \end{bmatrix}$$

On applying $R_4 \to R_4 - R_2$, we get

$$A = \begin{bmatrix} 1 & 2 & 3 & 0 \\ 0 & 0 & -3 & 2 \\ 0 & -4 & -8 & 3 \\ 0 & 0 & 0 & \alpha - 5 \end{bmatrix}$$

Since, the matrix A is of rank 3.

$\therefore\; \alpha - 5 = 0 \Rightarrow \alpha = 5$

48. (a) Since, $A\begin{bmatrix} x \\ y \\ z \end{bmatrix} = \begin{bmatrix} 1 \\ 0 \\ 0 \end{bmatrix}$ is a system of linear equation in three variables and that could have unique, no solution or infinitely many solution.

So, it is not possible to have exactly two solution.

Hence, the number of matrices A is zero.

49. (b) Augmented matrix

$$[A:B] = \begin{bmatrix} 1 & -1 & -2 & : & 6 \\ -1 & 1 & 1 & : & \mu \\ \lambda & 1 & 1 & : & 3 \end{bmatrix}$$

$$\begin{bmatrix} 1 & -1 & -2 & : & 6 \\ 0 & 0 & -1 & : & \mu + 6 \\ \lambda + 1 & 0 & -1 & : & 9 \end{bmatrix}$$

[using $R_2 \to R_2 + R_1, R_3 \to R_3 + R_1$]

$$\begin{bmatrix} 1 & -1 & -2 & 6 \\ 0 & 0 & -1 & \mu + 6 \\ \lambda + 1 & 0 & 0 & 3 - \mu \end{bmatrix}$$ [using $R_3 - R_2$]

So, infinite number of solution for $\lambda = -1$ and $\mu = 3$

50. (d) On subtracting the addition of first two equations from third equation, we get $0 = -5$, which is an absurd result.

51. (a) Given system of equations are

$x + y + z = 6, x + 2y + 3z = 10$ and $x + 2y + \lambda z = \mu$

The given system of equations has infinite number of solution, if any two equations will be same i.e. the last two equation will be same, if $\lambda = 3$ and $\mu = 10$

52. (c) Given, system of linear equations

$x + y + z = 5$ ...(i)

$x + 2y + 2z = 6$ ...(ii)

$x + 3y + \lambda z = \mu$ ...(iii)

$(\lambda, \mu \in R)$

The above given system has infinitely many solutions, then the plane represented by these equations intersect each other at a line, means

$(x + 3y + \lambda z - \mu)$

$= p(x + y + z - 5) + q(x + 2y + 2z - 6)$

$= (p + q)x + (p + 2q)y + (p + 2q)z - (5p + 6q)$

On comparing, we get

$p + q = 1, p + 2q = 3, p + 2q = \lambda$ and $5p + 6q = \mu$

So, $(p, q) = (-1, 2)$

$\Rightarrow\; \lambda = 3$ and $\mu = 7$

$\Rightarrow \lambda + \mu = 3 + 7 = 10$

53. (b) Reflection of (x, y) about Y-axis is $(x_1, y_1) = (-x, y)$

$\therefore\; x_1 = (-1)\cdot x + 0\cdot y$

$y_1 = 0\cdot x + y$

$\therefore \begin{bmatrix} x_1 \\ y_1 \end{bmatrix} = \begin{bmatrix} -1 & 0 \\ 0 & 1 \end{bmatrix}\begin{bmatrix} x \\ y \end{bmatrix}$

$= \begin{bmatrix} -1 & 0 \\ 0 & 1 \end{bmatrix}\begin{bmatrix} 1 \\ 2 \end{bmatrix}$

$\begin{bmatrix} x_1 \\ y_1 \end{bmatrix} = \begin{bmatrix} -1 \\ 2 \end{bmatrix}$

The image is $(-1, 2)$.

54. (c) Reflection of (x, y) about origin is $(x_1, y_1) = (-x, -y)$

$\therefore \quad x_1 = -x + 0y$

$y_1 = 0x - y$

$$\begin{bmatrix} x_1 \\ y_1 \end{bmatrix} = \begin{bmatrix} -1 & 0 \\ 0 & -1 \end{bmatrix}\begin{bmatrix} x \\ y \end{bmatrix}$$

$$= \begin{bmatrix} -1 & 0 \\ 0 & -1 \end{bmatrix}\begin{bmatrix} -1 \\ 2 \end{bmatrix} = \begin{bmatrix} 1 \\ -2 \end{bmatrix}$$

55. (6) $M^3 = M^2M = (I - M)M = M - M^2 = 2M - I$

$[\because M^2 = I - M]$

Similarly, $M^4 = 2I - 3M$

$M^5 = 5M - 3I$

$M^6 = 5I - 8M$

$n = 6$

56. (0) $A = \begin{bmatrix} 0 & 1 \\ 3 & 0 \end{bmatrix}$

$$\Rightarrow A^2 = A \cdot A = \begin{bmatrix} 0 & 1 \\ 3 & 0 \end{bmatrix}\begin{bmatrix} 0 & 1 \\ 3 & 0 \end{bmatrix} = \begin{bmatrix} 3 & 0 \\ 0 & 3 \end{bmatrix}$$

$$\Rightarrow A^4 = A^2 \cdot A^2 = \begin{bmatrix} 3 & 0 \\ 0 & 3 \end{bmatrix}\begin{bmatrix} 3 & 0 \\ 0 & 3 \end{bmatrix} = \begin{bmatrix} 3^2 & 0 \\ 0 & 3^2 \end{bmatrix}$$

$$\Rightarrow A^8 = \begin{bmatrix} 3^4 & 0 \\ 0 & 3^4 \end{bmatrix}$$

and $A^6 = A^4 \cdot A^2 = \begin{bmatrix} 3^2 & 0 \\ 0 & 3^2 \end{bmatrix}\begin{bmatrix} 3 & 0 \\ 0 & 3 \end{bmatrix} = \begin{bmatrix} 3^3 & 0 \\ 0 & 3^3 \end{bmatrix}$

Let $V = \begin{bmatrix} x \\ y \end{bmatrix}$

$A^8 + A^6 + A^4 + A^2 + I$

$$= \begin{bmatrix} 81 & 0 \\ 0 & 81 \end{bmatrix} + \begin{bmatrix} 27 & 0 \\ 0 & 27 \end{bmatrix} + \begin{bmatrix} 9 & 0 \\ 0 & 9 \end{bmatrix} + \begin{bmatrix} 3 & 0 \\ 0 & 3 \end{bmatrix} + \begin{bmatrix} 1 & 0 \\ 0 & 1 \end{bmatrix}$$

$$= \begin{bmatrix} 121 & 0 \\ 0 & 121 \end{bmatrix}$$

$$(A^8 + A^6 + A^4 + A^2 + I)V = \begin{bmatrix} 0 \\ 11 \end{bmatrix}$$

or $\begin{bmatrix} 121 & 0 \\ 0 & 121 \end{bmatrix}\begin{bmatrix} x \\ y \end{bmatrix} = \begin{bmatrix} 0 \\ 11 \end{bmatrix}$

or $\begin{bmatrix} 121x \\ 121y \end{bmatrix} = \begin{bmatrix} 0 \\ 11 \end{bmatrix}$

$\Rightarrow \quad x = 0$ and $y = \dfrac{1}{11}$

$\Rightarrow \quad V = \begin{bmatrix} x \\ y \end{bmatrix} = \begin{bmatrix} 0 \\ 1/11 \end{bmatrix}$

57. (6) $A = \begin{bmatrix} 0 & \alpha \\ 0 & 0 \end{bmatrix}$

$$\therefore A^2 = A \cdot A = \begin{bmatrix} 0 & \alpha \\ 0 & 0 \end{bmatrix}\begin{bmatrix} 0 & \alpha \\ 0 & 0 \end{bmatrix}\begin{bmatrix} 0 & 0 \\ 0 & 0 \end{bmatrix} = 0$$

$\Rightarrow A^2 = A^3 = A^4 = A^5 = \dots = 0$

Now, $(A + I)^{70} = (I + A)^{70}$

$= I + {}^{70}C_1A + {}^{70}C_2A^2 + {}^{70}C_3A^3 + \dots + {}^{70}C_{70}A^{70}$

$= I + 70A + 0 + 0 + \dots = I + 70A$

$$\Rightarrow (A + I)^{70} - 70A = I = \begin{bmatrix} 1 & 0 \\ 0 & 1 \end{bmatrix} = \begin{bmatrix} a-1 & b-1 \\ c-1 & d-1 \end{bmatrix} \text{given}$$

$\therefore \; a - 1 = 1, b - 1 = 0, c - 1 = 0, d - 1 = 1$

$\Rightarrow \quad a = 2, b = 1, c = 1$ and $d = 2$

Hence, $a + b + c + d = 6$.

58. (9) Let $M = \begin{bmatrix} a & b & c \\ d & e & f \\ g & h & i \end{bmatrix}$

$$M\begin{bmatrix} 0 \\ 1 \\ 0 \end{bmatrix} = \begin{bmatrix} -1 \\ 2 \\ 3 \end{bmatrix}$$

$\Rightarrow \quad b = -1, e = 2, h = 3$

$$M\begin{bmatrix} 1 \\ -1 \\ 0 \end{bmatrix} = \begin{bmatrix} 1 \\ 1 \\ -1 \end{bmatrix}$$

$\Rightarrow \quad a = 0, d = 3, g = 2$

$$M\begin{bmatrix} 1 \\ 1 \\ 1 \end{bmatrix} = \begin{bmatrix} 0 \\ 0 \\ 12 \end{bmatrix} \Rightarrow g + h + i = 12 \Rightarrow i = 7$$

Therefore, sum of diagonal elements = 9

59. (3) $A^2 = \begin{bmatrix} 1 & 0 & 0 \\ 1 & 0 & 1 \\ 0 & 1 & 0 \end{bmatrix}\begin{bmatrix} 1 & 0 & 0 \\ 1 & 0 & 1 \\ 0 & 1 & 0 \end{bmatrix} = \begin{bmatrix} 1 & 0 & 0 \\ 1 & 1 & 0 \\ 1 & 0 & 1 \end{bmatrix}$

Given, $A^n = A^{n-2} + A^2 - I$

$\Rightarrow \quad A^{50} = A^{48} + A^2 - I$

Further $A^{48} = A^{46} + A^2 - I$

$A^{46} = A^{44} + A^2 - I$

$\vdots \quad \vdots \quad \vdots \quad \vdots$

$A^4 = A^2 + A^2 - I$

On adding all

$A^{50} = 25A^2 - 24I$

$$\Rightarrow \quad A^{50} = 25\begin{bmatrix} 1 & 0 & 0 \\ 1 & 1 & 0 \\ 1 & 0 & 1 \end{bmatrix} - 24\begin{bmatrix} 1 & 0 & 0 \\ 0 & 1 & 0 \\ 0 & 0 & 1 \end{bmatrix}$$

$$\Rightarrow \quad A^{50} = \begin{bmatrix} 1 & 0 & 0 \\ 25 & 1 & 0 \\ 25 & 0 & 1 \end{bmatrix}$$

Hence, trace of $A^{50} = 1 + 1 + 1 = 3$

60. (6) $AA^T = 1$

$a^2 + b^2 + c^2 = 1$

and $ab + bc + ca = 0$

This is possible when

(i) $a=\pm 1, b=c=0$

(ii) $a=0, b=\pm 1, c=0$

(iii) $a=0, b=0, c=\pm 1$

So, there are six such matrices A.

61. (1/4) $\begin{bmatrix} a & b \\ c & 1-a \end{bmatrix}$ is an idempotent matrix

$$\Rightarrow \begin{bmatrix} a & b \\ c & 1-a \end{bmatrix}^2 = \begin{bmatrix} a & b \\ c & 1-a \end{bmatrix}$$

or $\begin{bmatrix} a & b \\ c & 1-a \end{bmatrix}\begin{bmatrix} a & b \\ c & 1-a \end{bmatrix} = \begin{bmatrix} a & b \\ c & 1-a \end{bmatrix}$

or $\begin{bmatrix} a^2+bc & ab+b-ab \\ ac+c-ac & bc+(1-a)^2 \end{bmatrix} = \begin{bmatrix} a & b \\ c & 1-a \end{bmatrix}$

or $\begin{bmatrix} a^2+bc & b \\ c & bc+(1-a)^2 \end{bmatrix} = \begin{bmatrix} a & b \\ c & 1-a \end{bmatrix}$

or $a^2+bc=a$

$$a-a^2=bc=\frac{1}{4}$$

$$\therefore \quad f(a)=\frac{1}{4}$$

62. (6) Given, $A^2=A$

Now, $I=(I-0.4A)(I-\alpha A)$

$=I-\alpha A-0.4A+0.4\alpha A^2$

$=I-A\alpha-0.4A+0.4\alpha A$

$=I-A(0.4+\alpha)+0.4\alpha A$

$\Rightarrow 0.4\alpha=0.4+\alpha$

$\Rightarrow \quad \alpha=\frac{-2}{3} \Rightarrow |9\alpha|=6$

63. (2) Let $A=\begin{bmatrix} 1 & 0 & 2 \\ 0 & 0 & 1 \\ 0 & 0 & 1 \end{bmatrix}$

$$=\begin{bmatrix} 1 & 0 & 2 \\ 0 & 0 & 1 \\ 0 & 0 & 0 \end{bmatrix} \quad [\text{using } R_3 \to R_3 - R_2]$$

Here, R_1 and R_2 are non-zero rows and R_3 is a zero row.

The matrix is in Echelon form.

Hence, the rank is 2.

64. (9) Let $X=\begin{bmatrix} x_1 \\ x_2 \\ x_3 \end{bmatrix}$ and given, $X^TAX=0$

$$\Rightarrow [x_1\ x_2\ x_3]\begin{bmatrix} a_{11} & a_{12} & a_{13} \\ a_{21} & a_{22} & a_{23} \\ a_{31} & a_{32} & a_{33} \end{bmatrix}\begin{bmatrix} x_1 \\ x_2 \\ x_3 \end{bmatrix}=0$$

$$\Rightarrow [x_1\ x_2\ x_3]\begin{bmatrix} a_{11}x_1 & a_{12}x_2 & a_{13}x_3 \\ a_{21}x_1 & a_{22}x_2 & a_{23}x_3 \\ a_{31}x_1 & a_{32}x_2 & a_{33}x_3 \end{bmatrix}=0$$

$$\Rightarrow a_{11}x_1^2+a_{12}x_1x_2+a_{13}x_1x_3+a_{22}x_2^2+a_{21}x_1x_2+a_{23}x_2x_3 + a_{31}x_1x_3+a_{32}x_2x_3+a_{33}x_3^2=0$$

$$\Rightarrow a_{11}x_1^2+a_{22}x_2^2+a_{33}x_3^2+(a_{12}+a_{21}) x_1x_2+(a_{23}+a_{32})x_2x_3+(a_{31}+a_{13})x_1x_3=0$$

It is true for every x_1, x_2, x_3, then

$$a_{11}=a_{22}=a_{33}=0$$

and $\quad a_{21}=-a_{12}; a_{23}=-a_{32}, a_{13}=-a_{31}$

Now, as $a_{23}=-1008$

$\Rightarrow \quad a_{32}=1008$

$\therefore$ Sum of digits $=1+0+0+8=9$

Level 3

1. (b) We have, $\begin{bmatrix} \alpha & \beta \\ \gamma & -\alpha \end{bmatrix}\begin{bmatrix} \alpha & \beta \\ \gamma & -\alpha \end{bmatrix}=\begin{bmatrix} 1 & 0 \\ 0 & 1 \end{bmatrix}$

or $\quad \begin{bmatrix} \alpha^2+\beta\gamma & 0 \\ 0 & \alpha^2+\beta\gamma \end{bmatrix}=\begin{bmatrix} 1 & 0 \\ 0 & 1 \end{bmatrix}$

$\therefore \quad \alpha^2+\beta\gamma-1=0$

2. (b) We have, $A=iB$

or $\quad A^2=(iB)^2=i^2B^2=-B^2$

$$=-\begin{bmatrix} 2 & -2 \\ -2 & 2 \end{bmatrix}=-2B$$

or $\quad A^4=(-2B)^2=4B^2=4(2B)=8B$

or $\quad (A^4)^2=(8B)^2$

or $\quad A^8=64B^2=128B$

3. (d) $A=diag(d_1, d_2, \ldots d_n)$

Given, $A^3=A$

$\Rightarrow \text{diag}(d_1^3, d_2^3, \ldots d_n^3)=\text{diag}(d_1, d_2, \ldots d_n)$

$\Rightarrow d_1^3=d_1, d_2^3=d_2, \ldots, d_n^3=d_n$

Hence, all $d_1, d_2, d_3 \ldots d_n$ have three possible values $\pm 1, 0$. Each diagonal element can be selected in three ways. Hence. the number of different matrices is 3^n.

4. (c) Given, $A^2=2A-I$

Now, $\quad A^3=A(A^2)$

$=A(2A-I)$

$=2A^2-A$

$=2(2A-I)-A$

$=3A-2I$

$A^4=A(A^3)$

$=A(3A-2I)$

$=3A^2-2A$

$=3(2A-I)-2A$

$=4A-3I$

Following this, we can say $A^n=nA-(n-1)I$

5. (a) $(A-\lambda I)(B-\lambda I)=(B-\lambda I)(A-\lambda I)$

or $AB-\lambda(A+B)I+\lambda^2I^2$

$$=BA-\lambda(B+A)I+\lambda^2I^2$$

or $$AB=BA$$

6. (b) Since the product matrix is 3×3 matrix and the pre multiple of A is a 3×2 matrix, therefore A is 2×3 matrix. Let

$A=\begin{bmatrix} l & m & n \\ x & y & z \end{bmatrix}$, then the given equation becomes

$$\begin{bmatrix} 2 & -1 \\ 1 & 0 \\ -3 & 4 \end{bmatrix}\begin{bmatrix} l & m & n \\ x & y & z \end{bmatrix}=\begin{bmatrix} -1 & -8 & -10 \\ 1 & -2 & -5 \\ 9 & 22 & 15 \end{bmatrix}$$

or $\begin{bmatrix} 2l-x & 2m-y & 2n-z \\ l & m & n \\ -3l+4x & -3m+4y & -3n+4z \end{bmatrix}$

$$=\begin{bmatrix} -1 & -8 & -10 \\ 1 & -2 & -5 \\ 9 & 22 & 15 \end{bmatrix}$$

or $2l-x=-1, 2m-y=-8, 2n-z=-10$

$l=1, m=-2, n=-5$

$\Rightarrow$ $x=3, y=4, z=0, l=1, m=-2, n=-5$

$\Rightarrow$ $A=\begin{bmatrix} l & m & n \\ x & y & z \end{bmatrix}=\begin{bmatrix} 1 & -2 & -5 \\ 3 & 4 & 0 \end{bmatrix}$

$\therefore$ Sum of all the elements of matrix A is

$1-2-5+3+4+0=1$

7. (b) $P^2=\begin{bmatrix} 1 & 0 & 0 \\ 4 & 1 & 0 \\ 16 & 4 & 1 \end{bmatrix}\begin{bmatrix} 1 & 0 & 0 \\ 4 & 1 & 0 \\ 16 & 4 & 1 \end{bmatrix}=\begin{bmatrix} 1 & 0 & 0 \\ 8 & 1 & 0 \\ 48 & 8 & 1 \end{bmatrix}$

$P^3=\begin{bmatrix} 1 & 0 & 0 \\ 8 & 1 & 0 \\ 48 & 8 & 1 \end{bmatrix}\begin{bmatrix} 1 & 0 & 0 \\ 4 & 1 & 0 \\ 16 & 4 & 1 \end{bmatrix}=\begin{bmatrix} 1 & 0 & 0 \\ 12 & 1 & 0 \\ 96 & 12 & 1 \end{bmatrix}$

$\therefore$ $P^n=\begin{bmatrix} 1 & 0 & 0 \\ 4n & 1 & 0 \\ 8(n^2+n) & 4n & 1 \end{bmatrix}$

$\therefore$ $P^{50}=\begin{bmatrix} 1 & 0 & 0 \\ 200 & 1 & 0 \\ 8\times50(51) & 200 & 1 \end{bmatrix}$

$P^{50}-Q=I$

Equating we get

$200-q_{21}=0$

$\Rightarrow$ $q_{21}=200$

$400\times51-q_{31}=0$

$\Rightarrow$ $q_{31}=400\times51$

$200-q_{32}=0$

$\Rightarrow$ $q_{32}=200$

$\Rightarrow$ $\dfrac{q_{31}+q_{32}}{q_{21}}=\dfrac{400\times51+200}{200}$

$=2(51)+1=103$

8. (c) We know that,

$$\begin{bmatrix} \cos\theta & -\sin\theta \\ \sin\theta & \cos\theta \end{bmatrix}^n=\begin{bmatrix} \cos n\theta & -\sin n\theta \\ \sin n\theta & \cos n\theta \end{bmatrix}$$

$$\begin{bmatrix} \cos\frac{2\pi}{7} & -\sin\frac{2\pi}{7} \\ \sin\frac{2\pi}{7} & \cos\frac{2\pi}{7} \end{bmatrix}^k=\begin{bmatrix} \cos\frac{2k\pi}{7} & -\sin\frac{2k\pi}{7} \\ \sin\frac{2k\pi}{7} & \cos\frac{2k\pi}{7} \end{bmatrix}$$

$=\begin{bmatrix} 1 & 0 \\ 0 & 1 \end{bmatrix}$, if $k=7m$, where $m\in N$

Hence, the least value of k is 7.

9. (d) $A\begin{bmatrix} 1 \\ -1 \end{bmatrix}=\begin{bmatrix} -1 \\ 2 \end{bmatrix}$ ...(i)

and $A^2\begin{bmatrix} 1 \\ -1 \end{bmatrix}=\begin{bmatrix} 1 \\ 0 \end{bmatrix}$ ...(ii)

Let A be given by $A=\begin{bmatrix} a & b \\ c & d \end{bmatrix}$ the first equation gives

$a-b=-1$ ...(iii)

$c-d=2$ ...(iv)

For the second equation

$$A^2\begin{bmatrix} 1 \\ -1 \end{bmatrix}=A\left(A\begin{bmatrix} 1 \\ -1 \end{bmatrix}\right)=A\left(\begin{bmatrix} -1 \\ 2 \end{bmatrix}\right)=\begin{bmatrix} 1 \\ 0 \end{bmatrix}$$

This gives

$-a+2b=1$ ...(v)

$-c+2d=0$ ...(vi)

On adding Eqs. (iii) and (v), we get

$b=0$ and $a=-1$

On adding Eqs. (iv) and (vi), we get

$d=2$ and $c=4$

So, the sum of $a+b+c+d=5$

10. (d) If A is nth root of I_2, then

$A^n=I_2$.

Now, $A^2=\begin{bmatrix} a & b \\ 0 & a \end{bmatrix}\begin{bmatrix} a & b \\ 0 & a \end{bmatrix}=\begin{bmatrix} a^2 & 2ab \\ 0 & a^2 \end{bmatrix}$

$A^3=A^2A=\begin{bmatrix} a^2 & 2ab \\ 0 & a^2 \end{bmatrix}\begin{bmatrix} a & b \\ 0 & a \end{bmatrix}=\begin{bmatrix} a^3 & 3a^2b \\ 0 & a^3 \end{bmatrix}$

Thus, $A^n=\begin{bmatrix} a^n & na^{n-1}b \\ 0 & a^n \end{bmatrix}$

Now, $A^n=I$

$\Rightarrow\begin{bmatrix} a^n & na^{n-1}b \\ 0 & a^n \end{bmatrix}=\begin{bmatrix} 1 & 0 \\ 0 & 1 \end{bmatrix}$

$\Rightarrow$ $a^n=1, b=0$

11. (a) $A(x)A(y)=(1-x)^{-1}(1-y)^{-1}\begin{bmatrix}1 & -x\\ -x & 1\end{bmatrix}\begin{bmatrix}1 & -y\\ -y & 1\end{bmatrix}$

$$=(1+xy-(x+y))^{-1}\begin{bmatrix}1+xy & -(x+y)\\ -(x+y) & 1+xy\end{bmatrix}$$

$$=\left(1-\frac{x+y}{1+xy}\right)^{-1}\begin{bmatrix}1 & -\frac{x+y}{1+xy}\\ -\frac{x+y}{1+xy} & 1\end{bmatrix}=A(z)$$

12. (a) $PP^T=\begin{bmatrix}\frac{\sqrt{3}}{2} & \frac{1}{2}\\ \frac{-1}{2} & \frac{\sqrt{3}}{2}\end{bmatrix}\begin{bmatrix}\frac{\sqrt{3}}{2} & -\frac{1}{2}\\ \frac{1}{2} & \frac{\sqrt{3}}{2}\end{bmatrix}$

$$=\begin{bmatrix}1 & 0\\ 0 & 1\end{bmatrix}=I$$

Also, $Q=PAP^T$

$\therefore P^TQ^{2005}P=P^T[(PAP^T)(PAP^T)....2005 \text{ times}]P$

$$=(P^TP)A(P^TP)A(P^TP)\ldots(P^TP)$$

$$=I\cdot A^{2005}$$

$$=A^{2005}$$

$\because\quad A=\begin{bmatrix}1 & 1\\ 0 & 1\end{bmatrix};$

$$A^2=\begin{bmatrix}1 & 1\\ 0 & 1\end{bmatrix}\begin{bmatrix}1 & 1\\ 0 & 1\end{bmatrix}=\begin{bmatrix}1 & 2\\ 0 & 1\end{bmatrix}$$

$\therefore\quad A^3=\begin{bmatrix}1 & 2\\ 0 & 1\end{bmatrix}\begin{bmatrix}1 & 1\\ 0 & 1\end{bmatrix}=\begin{bmatrix}1 & 3\\ 0 & 1\end{bmatrix}$

Similarly, $A^{2005}=\begin{bmatrix}1 & 2005\\ 0 & 1\end{bmatrix}$

$\therefore\quad P^TQ^{2005}P=\begin{bmatrix}1 & 2005\\ 0 & 1\end{bmatrix}$

13. (a) Let $M=\begin{bmatrix}a & b & c\\ d & e & f\\ g & h & i\end{bmatrix}$

$\therefore tr(M^TM)=a^2+b^2+c^2+d^2+e^2+f^2+g^2+h^2+i^2$

$= 5$, where entries are $\{0, 1, 2\}$

only two cases are possible

Case I Five entries are 1 and other four are 0.

$\therefore$ Number of matrices $={}^9C_5\times 1!=126$

Case II One entry is 2, one entry is 1 and others are 0.

$\therefore$ Number of matrices $={}^9C_2\times 2!=72$

So, total number of matrices $=126+72=198$

14. (d) $tr(A)=\sum_{i=j} a_{ij}$

$$=(a_{11}+a_{22}+a_{33}+\ldots+a_{1010})$$

$$=(\omega^2+\omega^4+\omega^6+\ldots+\omega^{20})$$

$$=\omega^2(1+\omega^2+\omega^4+\ldots+\omega^{18})$$

$$=\omega^2[(1+\omega+\omega^2)+....+(1+\omega+\omega^2)+1]$$

$$=\omega^2\times 1=\omega^2$$

15. (a) $BC=\begin{bmatrix}3 & 4\\ 2 & 3\end{bmatrix}\begin{bmatrix}3 & -4\\ -2 & 3\end{bmatrix}$

$\Rightarrow\quad BC=\begin{bmatrix}1 & 0\\ 0 & 1\end{bmatrix}$

$$\Rightarrow tr(A)+tr\left(\frac{ABC}{2}\right)+tr\left(\frac{A(BC)^2}{4}\right)+tr\left(\frac{A(BC)^3}{8}\right)+\ldots+\infty$$

$$=tr(A)+tr\left(\frac{A}{2}\right)+tr\left(\frac{A}{2^2}\right)+\ldots$$

$$=tr(A)+\frac{1}{2}tr(A)+\frac{1}{2^2}tr(A)+....$$

$$=\frac{tr(A)}{1-(1\div 2)}=2tr(A)=2(2+1)=6$$

16. (b) $B=A_1+3A_3^3+....+(2n-1)(A_{2n-1})^{2n-1}$.

$B^T=-[A_1+3A_3^3+....+(2n-1)(A_{2r-1})^{2r-1}]=-B$

Hence, B is skew-symmetric.

17. (b) Let $A=[a_{ij}]$, since A is skew-symmetric, we have

$a_{ii}=0$ and $a_{ij}=-a_{ji}(i\neq j)$

A is symmetric as well so $a_{ij}=a_{ji}$ for all i and j.

$\therefore\quad a_{ij}=0$ for all $i\neq j$

Hence, $a_{ij}=0$ for all i and j, i.e. A is a null matrix.

18. (c) $A^2=\begin{bmatrix}\frac{1}{\sqrt{2}} & \frac{1}{\sqrt{2}}\\ \frac{-1}{\sqrt{2}} & \frac{-1}{\sqrt{2}}\end{bmatrix}\begin{bmatrix}\frac{1}{\sqrt{2}} & \frac{1}{\sqrt{2}}\\ \frac{-1}{\sqrt{2}} & \frac{-1}{\sqrt{2}}\end{bmatrix}$

$$=\begin{bmatrix}\frac{1}{2}-\frac{1}{2} & \frac{1}{2}-\frac{1}{2}\\ -\frac{1}{2}+\frac{1}{2} & -\frac{1}{2}+\frac{1}{2}\end{bmatrix}=\begin{bmatrix}0 & 0\\ 0 & 0\end{bmatrix}$$

$\because\quad A\neq 0;\ A^2=0$

A is nilpotent matrix.

19. (a) Since B is invertible, therefore B^{-1} exists.

Now, rank (A) = rank $[(BA)B^{-1}]\leq$ rank (BA)

But rank $(BA)\leq$ rank (A)

$\therefore$ rank (BA) = rank (A)

20. (b) Given, $B=A^6-A^5$, Where $A^3=A+I$

$\Rightarrow\quad B=(A^3)^2-A^3A^2=(A+I)^2-(A+I)A^2$

$$=A^2+i^2+2AI-A^3-IA^2=I+2A-(A+I)$$

$\Rightarrow\quad B=A$

$\therefore$ Inverse of $B=A^{-1}$

21. (a, d) $A_1+A_2+A_3+\ldots+A_n$

$$=\begin{bmatrix}C_0^2 & 0\\ 0 & C_1^2\end{bmatrix}+\begin{bmatrix}C_1^2 & 0\\ 0 & C_2^2\end{bmatrix}+\begin{bmatrix}C_2^2 & 0\\ 0 & C_3^2\end{bmatrix}+\ldots+\begin{bmatrix}C_{n-1}^2 & 0\\ 0 & C_2^2\end{bmatrix}$$

$$=\text{diagonal}[C_0^2+C_1^2+C_2^2+....+C_{n-1}^2, C_1^2+C_2^2+C_3^2+\ldots+C_2^2]$$

$$=\begin{bmatrix}{}^{2n}C_{n-1} & 0\\ 0 & {}^{2n}C_{n-1}\end{bmatrix}=\begin{bmatrix}k_1 & 0\\ 0 & k_2\end{bmatrix}\quad\text{[given]}$$

$\therefore k_1=k_2={}^{2n}C_{n-1}$

22. (a, d) $A = \begin{bmatrix} 1 & 1 & 1 \\ 1 & 1 & 1 \\ 1 & 1 & 1 \end{bmatrix}$

$$A^2 = A \cdot A = \begin{bmatrix} 1 & 1 & 1 \\ 1 & 1 & 1 \\ 1 & 1 & 1 \end{bmatrix}\begin{bmatrix} 1 & 1 & 1 \\ 1 & 1 & 1 \\ 1 & 1 & 1 \end{bmatrix}$$

$$= \begin{bmatrix} 3 & 3 & 3 \\ 3 & 3 & 3 \\ 3 & 3 & 3 \end{bmatrix} = 3A$$

$$A^3 = A^2 \cdot A = (3A)A = 3A^2 = 3(3A) = 9A$$

$$A^4 = A^3 \cdot A = (9A)A = 9A^2$$

$$= 9(3A) = 27A$$

23. (a, d) Given, $(A+B)^2 = A^2 + B^2 + 2AB$

or $\quad (A+B)(A+B) = A^2 + B^2 + 2AB$

or $A^2 + AB + BA + B^2 = A^2 + B^2 + 2AB$

or $\quad BA = AB$

or $\begin{bmatrix} a & 1 \\ b & -1 \end{bmatrix}\begin{bmatrix} 1 & -1 \\ 2 & 1 \end{bmatrix} = \begin{bmatrix} 1 & -1 \\ 2 & 1 \end{bmatrix}\begin{bmatrix} a & 1 \\ b & -1 \end{bmatrix}$

or $\begin{bmatrix} a+2 & -a+1 \\ b-2 & -b-1 \end{bmatrix} = \begin{bmatrix} a-b & 1+1 \\ 2a+b & 2-1 \end{bmatrix}$

The corresponding elements of equal matrices are equal

$$a+2 = a-b, -a+1 = 2$$

$\Rightarrow \quad a = -1$

$$b-2 = 2a+b, -b-1 = 1$$

$\Rightarrow \quad b = -2$

$\Rightarrow \quad a = -1, b = -2$

24. (a, b, c) Let $D_1 = \begin{bmatrix} d_1 & 0 & 0 \\ 0 & d_2 & 0 \\ 0 & 0 & d_3 \end{bmatrix}$ and $D_2 = \begin{bmatrix} d_4 & 0 & 0 \\ 0 & d_5 & 0 \\ 0 & 0 & d_6 \end{bmatrix}$

$\therefore \quad D_1D_2 = \begin{bmatrix} d_1d_4 & 0 & 0 \\ 0 & d_2d_5 & 0 \\ 0 & 0 & d_3d_6 \end{bmatrix} = D_2D_1$

and $\quad D_1^2 + D_2^2 = \begin{bmatrix} d_1^2 & 0 & 0 \\ 0 & d_2^2 & 0 \\ 0 & 0 & d_3^2 \end{bmatrix} + \begin{bmatrix} d_4^2 & 0 & 0 \\ 0 & d_5^2 & 0 \\ 0 & 0 & d_6^2 \end{bmatrix}$

$$= \begin{bmatrix} d_1^2 + d_4^2 & 0 & 0 \\ 0 & d_2^2 + d_5^2 & 0 \\ 0 & 0 & d_3^2 + d_6^2 \end{bmatrix}$$

25. (a, b) Let $D = A - I$

$$= \begin{bmatrix} 1 & 0 & 0 \\ 2 & 1 & 0 \\ 4 & 2 & 1 \end{bmatrix} - \begin{bmatrix} 1 & 0 & 0 \\ 0 & 1 & 0 \\ 0 & 0 & 1 \end{bmatrix}$$

$$= \begin{bmatrix} 0 & 0 & 0 \\ 2 & 0 & 0 \\ 4 & 2 & 0 \end{bmatrix}$$

$$(I+D)^{100} = I + {}^{100}C_1 \cdot D + {}^{100}C_2 D^2 + {}^{100}C_3 D^3 + \dots + {}^{100}C_{100} D^{100}$$

$$D^2 = \begin{bmatrix} 0 & 0 & 0 \\ 2 & 0 & 0 \\ 4 & 2 & 0 \end{bmatrix}\begin{bmatrix} 0 & 0 & 0 \\ 2 & 0 & 0 \\ 4 & 2 & 0 \end{bmatrix}$$

$$= \begin{bmatrix} 0 & 0 & 0 \\ 0 & 0 & 0 \\ 4 & 0 & 0 \end{bmatrix}$$

Similarly, $D^3 = D^2 \cdot D = \begin{bmatrix} 0 & 0 & 0 \\ 0 & 0 & 0 \\ 4 & 0 & 0 \end{bmatrix}\begin{bmatrix} 0 & 0 & 0 \\ 2 & 0 & 0 \\ 4 & 2 & 0 \end{bmatrix}$

$\Rightarrow \quad D^3 = \begin{bmatrix} 0 & 0 & 0 \\ 0 & 0 & 0 \\ 0 & 0 & 0 \end{bmatrix}$

$\Rightarrow \quad A = I + D$

$\Rightarrow \quad A^{100} = (I+D)^{100} = I + 100 \cdot D + 4950D^2 + 0 + \dots + 0$

$$= I + 100D + 4950D^2$$

$\Rightarrow \quad B = A^{100} - I = 100D + 4950D^2$

$$= 100\begin{bmatrix} 0 & 0 & 0 \\ 2 & 0 & 0 \\ 4 & 2 & 0 \end{bmatrix} + 4950\begin{bmatrix} 0 & 0 & 0 \\ 0 & 0 & 0 \\ 4 & 0 & 0 \end{bmatrix}$$

$$= \begin{bmatrix} 0 & 0 & 0 \\ 200 & 0 & 0 \\ 20200 & 200 & 0 \end{bmatrix}$$

26. (b, d) $\quad AB = A$

$\Rightarrow \quad ABA = A^2$

$\Rightarrow \quad AB = A^2 \quad [\because BA = B]$

Also, $\quad BA = B$

$\Rightarrow \quad BAB = B^2$

$\Rightarrow \quad BA = B^2 \quad [\because AB = A]$

$\Rightarrow \quad B = B^2 \quad [\because BA = B]$

$\therefore \quad A^2 - AB + B^2 = AB - AB + B^2 \quad [\because AB = A^2]$

$\Rightarrow \quad B^2 = B \quad [\because B^2 = B]$

27. (a, c, d) $A^2B = A(AB) = A(BA) = (AB)A = (BA)A = BA^2$

Similarly, $A^3B = BA^3$.

In general, $A^nB = BA^n, \forall n \geq 1$

and $(A+B)^n = {}^nC_0A^n + {}^nC_1A^{n-1}B + {}^nC_2A^{n-2}B^2 + \dots + {}^nC_nB^n$

Also, $(A^n - B^n)(A^n + B^n)$

$$= A^nA^n + A^nB^n - B^nA^n - B^nB^n$$

$$= A^{2n} - B^{2n} \quad [\because AB = BA]$$

28. (c,d) $AB = \begin{bmatrix} \cos^2\theta & \cos\theta\sin\theta \\ \cos\theta\sin\theta & \sin^2\theta \end{bmatrix}$

$$\begin{bmatrix} \cos^2\phi & \cos\phi\sin\phi \\ \cos\phi\sin\phi & \sin^2\phi \end{bmatrix}$$

$$= \begin{bmatrix} \cos\theta \cos\phi \cos(\theta-\phi) & \cos\theta \sin\phi \cos(\theta-\phi) \\ \sin\theta \cos\phi \cos(\theta-\phi) & \sin\theta \sin\phi \cos(\theta-\phi) \end{bmatrix}$$

$$= \cos(\theta-\phi) \begin{bmatrix} \cos\theta \cos\phi & \cos\theta \sin\phi \\ \sin\theta \cos\phi & \sin\theta \sin\phi \end{bmatrix}$$

Now, $AB = 0 \Rightarrow \cos(\theta - \phi) = 0$

$\Rightarrow \theta - \phi = (2n+1)\dfrac{\pi}{2}, n \in Z$

29. (a, d) $(B^T AB)^T = B^T A^T (B^T)^T = B^T A^T B = B^T AB$, if A is symmetric therefore, $B^T AB$ is symmetric, if A is symmetric.

Also, $(B^T AB)^T = B^T A^T B = B^T(-A)B = -(B^T AB)$

Therefore, $B^T AB$, is skew-symmetric, if A is skew-symmetric.

30. (a, b, c, d) $\because PQ = QP$

$\Rightarrow (PQ)^T = Q^T P^T = QP = PQ$ $\quad [\because QP = PQ]$ [symmetric]

PQ is symmetric

$\because \quad PQ = QP$

$\Rightarrow \quad P^{-1}PQ = P^{-1}QP$

$\Rightarrow \quad Q = P^{-1}QP$

$\Rightarrow \quad QP^{-1} = P^{-1}Q$

$\Rightarrow \quad (P^{-1}Q)^T = Q^T(P^{-1})^T = Q^T P^{-1}$

$QP^{-1} = P^{-1}Q$ [symmetric]

Also, $(Q^{-1}P)^T = P^T(Q^{-1})^T = PQ^{-1} = Q^{-1}P$ [symmetric]

31. (c, d)

(a) $(y^3z^4 - z^4y^3)^T = (y^3z^4)^T - (z^4y^3)^T$
$= (z^T)^4(y^T)^3 - (y^T)^3(z^T)^4$
$= -z^4y^3 + y^3z^4$

$\Rightarrow y^3z^4 - z^4y^3$ is symmetric but not skew-symmetric.

(b) $(x^{44} + y^{44})^T = (x^T)^{44} + (y^T)^{44}$
$= (-x)^{44} + (-y)^{44}$
$= x^{44} + y^{44}$

$\Rightarrow x^{44} + y^{44}$ is symmetric but not skew-symmetric.

(c) $(x^4z^3 - z^3x^4)^T = (x^4z^3)^T - (z^3x^4)^T$
$= (z^T)^3(x^T)^4 - (x^T)^4(z^T)^3$
$= z^3x^4 - x^4z^3$
$= -(x^4z^3 - z^3x^4)$

$\Rightarrow x^4z^3 - z^3x^4$ = skew-symmetric

(d) $(x^{23} + y^{23})^T = (x^T)^{23} + (y^T)^{23}$
$= (-x)^{23} + (-y)^{23}$
$= -x^{23} - y^{23}$

$\Rightarrow x^{23} + y^{23}$ is skew-symmetric

32. (a, b, c) Given, $AB = A$, $BA = B$

$\Rightarrow \quad B \times AB = B \times A$

or $\quad (BA)B = B$

or $\quad B^2 = B$

Also, $A \times BA = AB$

$\Rightarrow \quad (AB)A = A$

$A^2 = A$

Now, $(A^T)^2 = (A^T \times A^T) = (A \times A)^T = (A^2)^T = A^T$

Similarly, $(B^T)^2 = B^T$

Hence, A^T and B^T are idempotent.

33. (a, c, d) $A = I - B$

$\Rightarrow \quad A^2 = I^2 + B^2 - 2B = I - B = A$ $\quad [\because B$ is idempotent]

and $\quad AB = B - B^2 = B - B = 0$ [null matrix]

and $\quad BA = B - B^2 = B - B = 0$ [null matrix]

34. (a, b, c) $\because \quad AB = B$

$\Rightarrow \quad BAB = B^2$

$\Rightarrow \quad AB = B^2$ $\quad [\because BA = A]$

$\Rightarrow \quad B = B^2$ $\quad [\because AB = B]$

$\Rightarrow \quad B^2 = B^3$ ...(i)

Also, $\quad BA = A$

$\Rightarrow \quad ABA = A^2$ $\quad [\because AB = B]$

$\Rightarrow \quad BA = A^2$

$\Rightarrow \quad A = A^2$

$\Rightarrow \quad A^2 = A^3$ ...(ii)

From Eqs. (i) and (ii),

$A^3 + B^3 = A^2 + B^2 = A + B$

Also, $\quad A^3 + B^2 = A^2 + B^3$

$\because A^2 = A$; A is an idempotent matrix.

35. (b, c) Let $(I - AB)^{-1} = P$

$\Rightarrow \quad P(I - AB) = I$

$\Rightarrow \quad P - PAB = I$

$\Rightarrow \quad PB^{-1} - PA = B^{-1}$

$\Rightarrow \quad BPB^{-1} - BPA = I$

$\Rightarrow \quad BPB^{-1} = I + BPA$

Now, $BPB^{-1} = B(I - AB)^{-1}B^{-1}$
$= B(B(I - AB))^{-1}$
$= (B^{-1})^{-1}(B(I - AB))^{-1}$
$= (B(I - AB)B^{-1})^{-1}$
$= ((B - BAB)B^{-1})^{-1}$
$= (I - BA)^{-1}$

36. (b, d) Let $A = \begin{bmatrix} -1 & 2 & 5 \\ 2 & -4 & a-4 \\ 1 & -2 & a+1 \end{bmatrix}$

Applying $R_2 \to R_2 + 2R_1$ and $R_3 \to R_3 + R_1$, then

$$A = \begin{bmatrix} -1 & 2 & 5 \\ 0 & 0 & a+6 \\ 0 & 0 & a+6 \end{bmatrix}$$

Applying $R_3 \to R_3 - R_2$, then

$$A = \begin{bmatrix} -1 & 2 & 5 \\ 0 & 0 & a+6 \\ 0 & 0 & 0 \end{bmatrix}$$

For $a = -6$, $\rho(A) = 1$

For $a = 1, 2$, $\rho(A) = 2$

Sol. (Q.Nos. 37-39)

Given, $A^T = A, B^T = -B$, $\det(A+B) \neq 0$

and $C = (A+B)^{-1}(A-B)$

$\Rightarrow (A+B)C = A-B$ ...(i)

Also, $(A+B)^T = A-B$ ...(ii)

and $(A-B)^T = A+B$ ...(iii)

$\Rightarrow C^T(A+B)C = C^T[(A+B)C]$

$= C^T(A-B)$ [from Eq. (i)]

$= C^T(A+B)^T$ [from Eq. (ii)]

$= [(A+B)C]^T$

$= (A-B)^T$ [from Eq. (i)]

$= A+B$ [from Eq. (iii)]

$C^T(A-B)C = [C^T(A+B)^T]C$ [from Eq. (ii)]

$= [(A+B)C]^T C$

$= (A-B)^T C$ [from Eq. (i)]

$= (A+B)C$ [from Eq. (ii)]

$= A-B$ [from Eq. (i)]

$$C^TAC = C^T\left(\frac{A+B+A-B}{2}\right)C$$

$$= \frac{1}{2}C^T(A+B)C + \frac{1}{2}C^T(A-B)C$$

$$= \frac{1}{2}(A+B) + \frac{1}{2}(A-B)$$

$$= A$$

37. (a) **38.** (b) **39.** (c)

Sol. (Q.Nos. 40-42)

A is an orthogonal matrix

$\therefore AA^T = I$

$$\begin{bmatrix} a & b & c \\ b & c & a \\ c & a & b \end{bmatrix}\begin{bmatrix} a & b & c \\ b & c & a \\ c & a & b \end{bmatrix} = \begin{bmatrix} 1 & 0 & 0 \\ 0 & 1 & 0 \\ 0 & 0 & 1 \end{bmatrix}$$

$$\begin{bmatrix} a^2+b^2+c^2 & ab+bc+ca & ab+bc+ca \\ ab+bc+ca & a^2+b^2+c^2 & ab+bc+ca \\ ab+bc+ca & ab+bc+ca & a^2+b^2+c^2 \end{bmatrix}$$

$$= \begin{bmatrix} 1 & 0 & 0 \\ 0 & 1 & 0 \\ 0 & 0 & 1 \end{bmatrix}$$

By equality of matrices, we get

$a^2+b^2+c^2 = 1$ ...(i)

$ab+bc+ca = 0$ ...(ii)

$(a+b+c)^2 = a^2+b^2+c^2+2(ab+bc+ca)$

$= 1+0 = 1$

$\therefore a+b+c = \pm 1$ ...(iii)

$\because a^2b^2+b^2c^2+c^2a^2 = (ab+bc+ca)^2 - 2abc(a+b+c)$

$= 0 - 2abc(\pm 1) = \mp 2\lambda$ [$\because abc = \lambda$]

$= -2\lambda$

$\because a^3+b^3+c^3-3abc$

$= (a+b+c)(a^2+b^2+c^2-ab-bc-ca)$

$\Rightarrow a^3+b^3+c^3-3\lambda = (\pm 1)(1-0)$

[from Eqs. (i), (ii) and (iii) and $abc = \lambda$]

$\Rightarrow a^3+b^3+c^3 = 3\lambda \pm 1$

Equation whose roots are a, b, c is

$x^3 - (a+b+c)x^2 + (ab+bc+ca)x - abc = 0$

$\Rightarrow x^3 - (\pm 1)x^2 + 0 - \lambda = 0$

$\therefore x^3 \pm x^2 - \lambda = 0$

40. (b) **41.** (d) **42.** (d)

Sol. (Q.Nos. 43-45)

$AB = BA^m$

$\Rightarrow B = A^{-1}BA^m$

$$\therefore B^n = \frac{(A^{-1}BA^m)(A^{-1}BA^m)\ldots(A^{-1}BA^m)}{n \text{ times}}$$

$$= \frac{A^{-1}BA^{m-1}BA^{m-1}\ldots BA^{m-1}BA^{m-1}A}{n \text{ times}} \quad \text{...(i)}$$

Given, $AB = BA^m$

$\Rightarrow AAB = ABA^m = BA^{2m}$

$\Rightarrow AAAB = BA^{3m}$

Similarly, $A^xB = BA^{mx} \ \forall \ m \in N$

From Eq. (i), we get

$$B^n = A^{-1}\frac{BA^{m-1}BA^{m-1}BA^{m-1}\ldots BA^{m-1}BA^{m-1}A}{(n-1)\text{times}}$$

$$= A^{-1}B(A^{m-1}B)A^{m-1}$$

$$\frac{BA^{m-1}\ldots BA^{m-1}BA^{m-1}}{(n-2)\text{times}}\cdot A$$

$$A^{-1}BBA^{(m-1)m}A^{m-1}$$

$$\frac{BA^{m-1}\ldots BA^{m-1}BA^{m-1}}{(n-2)\text{times}}A \quad \text{[using Eq. (i)]}$$

$$= A^{-1}B^2A^{(m^2-1)}BA^{m-1}\ldots\frac{BA^{m-1}BA^{m-1}}{(n-2)\text{times}}A$$

$$= A^{-1}B^n(A)^{(m^n-1)}A$$

$I = A^{-1}IA^{(m^n-1)}A$ [$\because B^n = I$]

$I = A^{-1}A^{(m^n-1)}A$

$= A^{-1}A^{m^n}$

$\Rightarrow I = A^{(m^n-1)}$

$\therefore P = m^n - 1$ [$\because A^P = I$] ...(ii)

$\Rightarrow$ Put $m = 2$, $n = 5$ in Eq. (ii), we get

$P = 2^5 - 1 = 31$

From Eq. (ii), we get

$P = m^n - 1$

From Eq. (ii), we get

$50 \neq 8^3 - 1$

43. (b) **44.** (b) **45.** (c)

46. (i) $\rightarrow$ B, (ii) $\rightarrow$ A, (iii) $\rightarrow$ D, (iv) $\rightarrow$ C

(i) $(A-B)(A^2+AB+B^2)$

$$= A^3 + AAB + ABB - BAA - BAB - B^3$$
$$= A^3 - 2AB^2 - B^3$$

(ii) $(A+B)(A^2-AB-B^2)$

$$= A^3 - AAB - ABB + BAA - BAB - B^3$$
$$= A^3 + ABA + BAA - B^3 \quad [\because AB = -BA]$$
$$= A^3 - BAA + BAA - B^3 \quad [\because AB = -BA]$$
$$= A^3 - B^3$$

(iii) $(A-B)(A^2+AB-B^2)$

$$= A^3 + AAB - ABB - BAA - BAB + B^3$$
$$= A^3 + BAA + BAB - BAA - BAB + B^3$$
$$= A^3 + B^3$$

(iv) $AB + BA = 0$

$$AB = -BA$$
$$\Rightarrow \quad B = -A^{-1}BA$$

47. (i) $\rightarrow$ D, (ii) $\rightarrow$ A, (iii) $\rightarrow$ B, (iv) $\rightarrow$ C

(i) Since, A is idempotent, we have

$$A^2 = A$$
$$\Rightarrow A^3 = AA^2 = AA = A^2 = A,\ A^4 = AA^3 = AA = A^2 = A$$
$$\Rightarrow A^n = A$$
$$\Rightarrow (I-A)^n = {}^nC_0 I - {}^nC_1 A + {}^nC_2 A^2 - {}^nC_3 A^3 + \dots$$
$$= I + (-{}^nC_1 + {}^nC_2 - {}^nC_3 + \dots)A$$
$$= I + [({}^nC_0 - {}^nC_1 + {}^nC_2 - {}^nC_3 + \dots) - {}^nC_0]A$$
$$= I - A$$

(ii) A is involuntary.

Hence, $A^2 = I$

$$\Rightarrow \quad A^3 = A^5 = A$$

and $\quad A^2 = A^4 = A^6 = I$

$$\Rightarrow (I-A)^n = {}^nC_0 I - {}^nC_1 A + {}^nC_2 A^2 - {}^nC_3 A^3 + \dots$$
$$= {}^nC_0 I - {}^nC_1 A + {}^nC_2 I - {}^nC_3 A + {}^nC_4 I -$$
$$= ({}^nC_0 + {}^nC_2 + {}^nC_4 + \dots)I$$
$$-({}^nC_1 A + {}^nC_3 + {}^nC_5 + \dots)A$$
$$= 2^{n-1}(I-A)$$

(iii) If A is nilpotent of index 2, then

$$A^2 = A^3 = A^4 = A^n = 0$$
$$\Rightarrow \quad (I-A)^n = {}^nC_0 I - {}^nC_1 A + {}^nC_2 A^2 - {}^nC_3 A^3 +$$
$$= I - nA + 0 + 0$$
$$= I - nA$$

(iv) For orthogonal matrix $(A^T)^{-1} = A$.

48. (i) $\rightarrow$ C, (ii) $\rightarrow$ A, (iii) $\rightarrow$ B, (iv) $\rightarrow$ D

Let $A = [A_{ij}]_{3\times 3}$, where $a_{ij} = (k)^{i+j}$

$$\text{So, } A = \begin{bmatrix} k^2 & k^3 & k^4 \\ k^3 & k^4 & k^5 \\ k^4 & k^5 & k^6 \end{bmatrix}$$

(i) If A is singular, then $|A| = 0$

$$\Rightarrow k^2 \cdot k^3 \cdot k^4 \begin{vmatrix} 1 & 1 & 1 \\ k & k & k \\ k^2 & k^2 & k^2 \end{vmatrix} = 0$$
$$\Rightarrow \quad k \in I$$

(ii) If A is null matrix, then $k \in \{0\}$

(iii) There is no value of k for A to be skew-symmetric matrix which is not null matrix.

$$\therefore \quad k \in \phi$$

(iv) If $A^2 = 3A$, then $\begin{bmatrix} k^2 & k^3 & k^4 \\ k^3 & k^4 & k^5 \\ k^4 & k^5 & k^6 \end{bmatrix}\begin{bmatrix} k^2 & k^3 & k^4 \\ k^3 & k^4 & k^5 \\ k^4 & k^5 & k^6 \end{bmatrix}$

$$= \begin{bmatrix} 3k^2 & 3k^3 & 3k^4 \\ 3k^3 & 3k^4 & 3k^5 \\ 3k^4 & 3k^5 & 3k^6 \end{bmatrix}$$

$$\Rightarrow \begin{bmatrix} k^4+k^6+k^8 & k^5+k^7+k^9 & k^6+k^8+k^{10} \\ k^5+k^7+k^9 & k^6+k^8+k^{10} & k^7+k^9+k^{11} \\ k^6+k^8+k^{10} & k^7+k^9+k^{11} & k^8+k^{10}+k^{12} \end{bmatrix}$$

$$= \begin{bmatrix} 3k^2 & 3k^3 & 3k^4 \\ 3k^3 & 3k^4 & 3k^5 \\ 3k^4 & 3k^5 & 3k^6 \end{bmatrix}$$

$$\Rightarrow k \in \{-1, 0, 1\}$$

Sol. (Q.Nos. 49-50)

Let $A = \begin{bmatrix} a & b \\ c & d \end{bmatrix}$

49. (0) $A^2 - (a+d)A + (ad-bc)I$

$$= \begin{bmatrix} a & b \\ c & d \end{bmatrix}^2 - (a+d)\begin{bmatrix} a & b \\ c & d \end{bmatrix} + (ad-bc)\begin{bmatrix} 1 & 0 \\ 0 & 1 \end{bmatrix}$$

$$= \begin{bmatrix} a^2+bc & ab+bd \\ ac+cd & bc+d^2 \end{bmatrix} - \begin{bmatrix} a^2+ad & ab+bd \\ ac+cd & ad+d^2 \end{bmatrix} + \begin{bmatrix} ad-bc & 0 \\ 0 & ad-bc \end{bmatrix}$$

$$= 0$$

50. (0) If $A = 0, tr(A) = 0$, suppose $A \neq 0$ and $A = \begin{bmatrix} a & b \\ c & d \end{bmatrix}$,

then $A^2 - (a+d)A + (ad-bc)I = 0$

$$\Rightarrow \quad a + d = 0 = tr(A)$$

51. (3) $S = ABCD = A(BCD) = AA^T$ ...(i)

$$S^3 = (ABCD)(ABCD)(ABCD)$$
$$= (ABC)(DAB)(CDA)(BCD)$$
$$= D^T C^T B^T A^T = (BCD)^T A^T$$
$$= (A^T)^T A^T = AA^T = S \quad \text{[from Eq. (i)]}$$
$$\Rightarrow S^3 = S$$

52. (2) $\because A=\begin{bmatrix}0 & 1 & -1\\ 4 & -3 & 4\\ 3 & -3 & 4\end{bmatrix}$

$$\therefore \quad A^2 = A\cdot A = \begin{bmatrix}0 & 1 & -1\\ 4 & -3 & 4\\ 3 & -3 & 4\end{bmatrix}\begin{bmatrix}0 & 1 & -1\\ 4 & -3 & 4\\ 3 & -3 & 4\end{bmatrix}$$

$$= \begin{bmatrix}1 & 0 & 0\\ 0 & 1 & 0\\ 0 & 0 & 1\end{bmatrix} = I$$

$\Rightarrow \quad A^2 = I$

$\Rightarrow \quad A^4 = A^6 = A^8 = \ldots = I$

Now, $\quad A^n = 1$

$\Rightarrow \quad n = 2, 4, 6, 8 \ldots\ldots$

$$\therefore \Sigma(\cos^n\theta + \sin^n\theta) = (\cos^2\theta + \sin^2\theta) + (\cos^4\theta + \sin^4\theta) + (\cos^6\theta + \sin^6\theta) + \ldots\ldots$$

$$= (\cos^2\theta + \cos^4\theta + \cos^6\theta \ldots\ldots) + (\sin^2\theta + \sin^4\theta + \sin^4\theta + \ldots)$$

$$= \frac{\cos^2\theta}{1-\cos^2\theta} + \frac{\sin^2\theta}{1-\sin^2\theta}$$

$$= \cot^2\theta + \tan^2\theta$$

Since, $\cot^2\theta + \tan^2\theta \geq 2$

Hence, minimum value of $\Sigma(\cos^n\theta + \sin^n\theta)$ is 2.

53. (20) Let $A = \begin{bmatrix}a & b\\ c & d\end{bmatrix}$

$$A^2 = \begin{bmatrix}a & b\\ c & d\end{bmatrix}\begin{bmatrix}a & b\\ c & d\end{bmatrix}$$

$$= \begin{bmatrix}a^2+bc & ab+bd\\ ac+cd & bc+d^2\end{bmatrix}$$

$$\therefore \quad (tr\cdot A)^2 - (tr\cdot A^2) = (a+d)^2 - (a^2+bc+bc+d^2)$$

$$= 2(ad-bc)$$

$$= 2 \text{ determinant of } (A) = 2\times 10 = 20$$

54. (8) In a skew-symmetric matrix, diagonal elements are zero.

Also, $a_{ij} + a_{ji} = 0$

Hence, number of matrices $= 2\times 2\times 2 = 8$

55. (1) Given, $A = \begin{bmatrix}\lambda_1^2 & \lambda_1\lambda_2 & \lambda_1\lambda_3\\ \lambda_2\lambda_1 & \lambda_2^2 & \lambda_2\lambda_3\\ \lambda_3\lambda_1 & \lambda_3\lambda_2 & \lambda_3^2\end{bmatrix}$

$$\therefore A^2 = A\cdot A = \begin{bmatrix}\lambda_1^2 & \lambda_1\lambda_2 & \lambda_1\lambda_3\\ \lambda_2\lambda_1 & \lambda_2^2 & \lambda_2\lambda_3\\ \lambda_3\lambda_1 & \lambda_3\lambda_2 & \lambda_3^2\end{bmatrix}\begin{bmatrix}\lambda_1^2 & \lambda_1\lambda_2 & \lambda_1\lambda_3\\ \lambda_2\lambda_1 & \lambda_2^2 & \lambda_2\lambda_3\\ \lambda_3\lambda_1 & \lambda_3\lambda_2 & \lambda_3^2\end{bmatrix}$$

$$= \begin{bmatrix}\lambda_1^2(\lambda_1^2+\lambda_2^2+\lambda_3^2) & \lambda_1\lambda_2(\lambda_1^2+\lambda_2^2+\lambda_3^2) & \lambda_1\lambda_3(\lambda_1^2+\lambda_2^2+\lambda_3^2)\\ \lambda_1\lambda_2(\lambda_1^2+\lambda_2^2+\lambda_3^2) & \lambda_2^2(\lambda_1^2+\lambda_2^2+\lambda_3^2) & \lambda_2\lambda_3(\lambda_1^2+\lambda_2^2+\lambda_3^2)\\ \lambda_1\lambda_3(\lambda_1^2+\lambda_2^2+\lambda_3^2) & \lambda_3\lambda_2(\lambda_1^2+\lambda_2^2+\lambda_3^2) & \lambda_3^2(\lambda_1^2+\lambda_2^2+\lambda_3^2)\end{bmatrix}$$

$$= (\lambda_1^2+\lambda_2^2+\lambda_3^2)A$$

Given, A is idempotent

$$A^2 = A$$

$$\therefore \lambda_1^2+\lambda_2^2+\lambda_3^2 = 1$$

56. (7) A is idempotent matrix

$$A^2 = A$$

$\Rightarrow \quad A = A^2 = A^3 = A^4 = A^5 = \ldots\ldots\ldots \quad$ ...(i)

Now, $(A+I)^n = (I+A)^n$

$$= I + {}^nC_1A + {}^nC_2A^2 + {}^nC_3A^3 + \ldots. + {}^nC_nA^n$$

$$= I + ({}^nC_1 + {}^nC_2 + {}^nC_3 + \ldots + {}^nC_n)A$$

[from Eq. (i)]

$\Rightarrow \quad (A+I)^n = I + (2^n-1)A \quad$ ...(ii)

Given, $\quad (A+I)^n = I + 127\,A \quad$ ...(iii)

From Eqs. (ii) and (iii), we get

$$2^n - 1 = 127$$

$\Rightarrow \quad 2^n = 128 = 2^7$

$\therefore \quad n = 7$

57. (2) Let $A = \begin{bmatrix}1 & 2 & 3\\ 2 & 1 & 4\\ 3 & 0 & 5\end{bmatrix}$

$$A = \begin{bmatrix}1 & 2 & 3\\ 0 & -3 & -2\\ 0 & -6 & -4\end{bmatrix} R_2 \to R_2 - 2R_1, R_3 \to R_3 - 3R_1$$

$$A = \begin{bmatrix}1 & 2 & 3\\ 0 & -3 & -2\\ 0 & 0 & 0\end{bmatrix} R_3 \to R_3 - 2R_2$$

The above matrix, in row echelon form. Number of non-zero rows = 2

Hence, the rank of the matrix is 2.

CHAPTER

09

DETERMINANTS

LEVEL 01 \\ Starter Level

9.1 Determinants

1 Evaluate the following determinants

(i) $\begin{vmatrix} \cos\theta & -\sin\theta \\ \sin\theta & \cos\theta \end{vmatrix}$ (ii) $\begin{vmatrix} \cos 15^\circ & \sin 15^\circ \\ \sin 75^\circ & \cos 75^\circ \end{vmatrix}$

2 Find values of x for which $\begin{vmatrix} 3 & x \\ x & 1 \end{vmatrix} = \begin{vmatrix} 3 & 2 \\ 4 & 1 \end{vmatrix}$.

3 Evaluate

(i) $\Delta = \begin{vmatrix} 0 & \sin\alpha & -\cos\alpha \\ -\sin\alpha & 0 & \sin\beta \\ \cos\alpha & -\sin\beta & 0 \end{vmatrix}$ (ii) $\begin{vmatrix} 1 & 2 & 4 \\ -1 & 3 & 0 \\ 4 & 1 & 0 \end{vmatrix}$

4 If $a, b, c \in R$, then find the number of real roots of the equation

$$\Delta = \begin{vmatrix} x & c & -b \\ -c & x & a \\ b & -a & x \end{vmatrix} = 0$$

9.2 Properties of Determinants

5 Evaluate the following determinants

(i) $\begin{vmatrix} 102 & 18 & 36 \\ 1 & 3 & 4 \\ 17 & 3 & 6 \end{vmatrix}$ (ii) $\begin{vmatrix} 41 & 1 & 5 \\ 79 & 7 & 9 \\ 29 & 5 & 3 \end{vmatrix}$

(iii) $\begin{vmatrix} 6 & -3 & -2 \\ 2 & -1 & 2 \\ -10 & 5 & 2 \end{vmatrix}$ (iv) $\begin{vmatrix} \sqrt{13}+\sqrt{3} & 2\sqrt{5} & \sqrt{5} \\ \sqrt{15}+\sqrt{26} & 5 & \sqrt{10} \\ 3+\sqrt{65} & \sqrt{15} & 5 \end{vmatrix}$

(v) $\begin{vmatrix} 1 & 1 & 1 & 1 \\ 1 & 2 & 3 & 4 \\ 1 & 3 & 6 & 10 \\ 1 & 4 & 10 & 20 \end{vmatrix}$ (vi) $\begin{vmatrix} 1 & 3 & 9 & 27 \\ 3 & 9 & 27 & 1 \\ 9 & 27 & 1 & 3 \\ 27 & 1 & 3 & 9 \end{vmatrix}$

6 Without expanding, verify that $\Delta = \begin{vmatrix} x+y & y+z & z+x \\ z & x & y \\ 1 & 1 & 1 \end{vmatrix} = 0$.

7 If $\Delta = \begin{vmatrix} 1 & x & x^2 \\ 1 & y & y^2 \\ 1 & z & z^2 \end{vmatrix}$ and $\Delta_1 = \begin{vmatrix} 1 & 1 & 1 \\ yz & zx & xy \\ x & y & z \end{vmatrix}$, then find the value of $\Delta + \Delta_1$.

8 Let $p\lambda^4 + q\lambda^3 + r\lambda^2 + s\lambda + t = \begin{vmatrix} \lambda^2+3\lambda & \lambda-1 & \lambda+3 \\ \lambda+1 & -2\lambda & \lambda-4 \\ \lambda-3 & \lambda+4 & 3\lambda \end{vmatrix}$ be an identity in λ, where p, q, r, s and t are constants. Then, find the value of t.

9 Verify that $\begin{vmatrix} b+c & a & a \\ b & c+a & b \\ c & c & a+b \end{vmatrix} = 4abc$.

10 Verify the following

$$\begin{vmatrix} a & b & c \\ a-b & b-c & c-a \\ b+c & c+a & a+b \end{vmatrix} = a^3 + b^3 + c^3 - 3abc$$

11 Verify that

$$\begin{vmatrix} 1+a & 1 & 1 \\ 1 & 1+b & 1 \\ 1 & 1 & 1+c \end{vmatrix} = abc\left(1 + \frac{1}{a} + \frac{1}{b} + \frac{1}{c}\right) = abc + bc + ca + ab.$$

12 Evaluate $\begin{vmatrix} 1 & 2 & 3 \\ -2 & 3 & 2 \\ 3 & 4 & -4 \end{vmatrix} \times \begin{vmatrix} -2 & 1 & 3 \\ 3 & -2 & 1 \\ 2 & 1 & -2 \end{vmatrix}$, using the concept of multiplication of determinants.

9.3 Adjoint and Inverse of a Matrix

13 Find the minor of element 6 in the determinant $\Delta = \begin{vmatrix} 1 & 2 & 3 \\ 4 & 5 & 6 \\ 7 & 8 & 9 \end{vmatrix}$.

14 Write the minors and cofactors of each element of the first column of the following matrices and hence, evaluate the determinant in each case.

(i) $A = \begin{bmatrix} 5 & 20 \\ 0 & -1 \end{bmatrix}$ (ii) $A = \begin{bmatrix} 1 & -3 & 2 \\ 4 & -1 & 2 \\ 3 & 5 & 2 \end{bmatrix}$

(iii) $A = \begin{bmatrix} 1 & a & bc \\ 1 & b & ca \\ 1 & c & ab \end{bmatrix}$

15 Find the minors and cofactors of the elements of the determinant $\begin{vmatrix} 2 & -3 & 5 \\ 6 & 0 & 4 \\ 1 & 5 & -7 \end{vmatrix}$, verify that

$a_{11}A_{31} + a_{12}7A_{32} + a_{13}A_{33} = 0$.

16 Find the adjoint of the matrices

(i) $A = \begin{bmatrix} 2 & 3 \\ 4 & 1 \end{bmatrix}$ (ii) $A = \begin{bmatrix} 1 & 1 & 1 \\ 2 & 1 & -3 \\ -1 & 2 & 3 \end{bmatrix}$

17 Evaluate $\begin{vmatrix} 2 & 3 & 7 \\ 13 & 17 & 5 \\ 15 & 20 & 12 \end{vmatrix}^2$.

18 If det $(A) = 3$, det $(B) = 5$ and A and B are 3×3 matrices, then find

(i) $\det(A^T)$ (ii) $\det(AB)$

(iii) $\det(A^T B)$ (iv) $\det(A^T B^T)$

(v) $\det(2A)$ (vi) $\det(3B)$

19 If $A = \begin{bmatrix} -1 & 1 & 1 \\ 1 & -1 & 1 \\ 1 & 1 & -1 \end{bmatrix}$, then find the values of

(i) $|A|\,|\text{adj } A|$ (ii) $|\text{adj }(3A)|$

(iii) adj (adj A)

20 Let A be a square matrix such that $A \cdot (\text{adj } A) = \begin{bmatrix} 4 & 0 & 0 \\ 0 & 4 & 0 \\ 0 & 0 & 4 \end{bmatrix}$, then find the values of

(i) $|\text{adj } A|$ (ii) $|\text{adj }(\text{adj } A)|$

(iii) $|\text{adj }(3A)|$

21 For a matrix $\begin{bmatrix} 4 & -1 & 0 \\ 2 & 3 & 5 \\ -1 & 7 & 2 \end{bmatrix}$, find if the matrix is singular or a non-singular matrix.

22 Find the inverse of the following matrix

(i) $A = \begin{bmatrix} -3 & 4 \\ 2 & 5 \end{bmatrix}$ (ii) $A = \begin{bmatrix} 4 & -2 & 1 \\ 5 & 0 & 3 \\ -1 & 2 & 6 \end{bmatrix}$

23 Find the inverse of the matrix $A = \begin{bmatrix} 5 & 0 & 0 \\ 0 & 6 & 0 \\ 0 & 0 & 7 \end{bmatrix}$.

24 Find the eigen values of the matrices.

(i) $A = \begin{bmatrix} 2 & 0 \\ 0 & 3 \end{bmatrix}$ (ii) $A = \begin{bmatrix} 0 & -2 \\ 3 & 4 \end{bmatrix}$

25 If 3 and 5 are eigen values of A, then find the eigen values of

(i) A^T (ii) A^{-1} (iii) A^2

26 Find the characteristics equation of the matrix $A = \begin{bmatrix} 2 & 1 \\ 3 & 2 \end{bmatrix}$ and hence, find its inverse using Cayley Hamilton theorem.

9.4 Application of Determinants, Differentiation and Integration of Determinants

27 Find the area of the triangle with vertices at the points

(i) (3, 8), (– 4, 2) and (5, – 1) (ii) (2, 7), (1, 1) and (10, 8)

28 Find the values of a for which the lines $2x + y - 1 = 0$, $ax + 3y - 3 = 0$ and $3x + 2y - 2 = 0$ are concurrent.

29 Find the values of a and b, if the system of equations $a^2x - by = a^2 - b$ and $bx - b^2y = 2 + 4b$

(i) possess unique solution (ii) infinite solution

30 If $2ax - 2y + 3z = 0$, $x + ay + 2z = 0$ and $2x + az = 0$ have a non-trivial solution, find the value of a.

31 Solve the equations $2x + 5y = 1$ and $3x + 2y = 7$.

32 Solve the following system of equations by matrix method

$$3x - 2y + 3z = 8,\ 2x + y - z = 1 \text{ and } 4x - 3y + 2z = 4.$$

33 Solve the following system of equations, using matrix method $x + 2y + z = 7$, $x + 3z = 11$ and $2x - 3y = 1$.

34 Using matrix method, solve the system of equations,
$3x + 2y - 2z = 3$, $x + 2y + 3z = 6$ and $2x - y + z = 2$.

35 If $A = \begin{bmatrix} 1 & 2 & 0 \\ -2 & -1 & -2 \\ 0 & -1 & 1 \end{bmatrix}$, then find A^{-1} and using A^{-1}, solve the system of linear equations, $x - 2y = 10$, $2x - y - z = 8$ and $-2y + z = 7$.

36 Given, $A = \begin{bmatrix} 2 & 2 & -4 \\ -4 & 2 & -4 \\ 2 & -1 & 5 \end{bmatrix}$ and $B = \begin{bmatrix} 1 & -1 & 0 \\ 2 & 3 & 4 \\ 0 & 1 & 2 \end{bmatrix}$, then find BA and use this to solve the system of equations $y + 2z = 7$, $x - y = 3$ and $2x + 3y + 4z = 17$.

37 The sum of three numbers is 6. If we multiply third number by 3 and add second number to it, we get 11. By adding first and third numbers, we get double of the second number. Represent it algebraically and find the numbers, using matrix method.

38 Solve the following system of linear equations by Cramer's rule

$$x - 2y = 4 \text{ and}$$
$$-3x + 5y = -7$$

39 Solve by Cramer's rule

$$x + y + z = 6$$
$$x - y + z = 2$$
$$3x + 2y - 4z = -5$$

40 For what values of p and q, the system of equations $2x + py + 6z = 8$, $x + 2y + qz = 5$ and $x + y + 3z = 4$ has

(i) no solution (ii) a unique solution

(iii) infinitely many solutions

41 If $f(x) = \begin{vmatrix} x^n & n! & 2 \\ \cos x & \cos \frac{n\pi}{2} & 4 \\ \sin x & \sin \frac{n\pi}{2} & 8 \end{vmatrix}$, then find the value of $\frac{d^n}{dx^n}[f(x)]_{x=0} (n \in Z)$.

42 If $\Delta(x) = \begin{vmatrix} 1 & x^2 & x^2 \\ 6 & 4x & 3 \\ 9 & x & -7 \end{vmatrix}$, then find the value of $\int_0^1 \Delta(x)\, dx$ without expanding $\Delta(x)$.

LEVEL 02 \\ JEE Main Level

Single Choice Correct Questions

9.1 Determinants

1 If $P = \begin{bmatrix} 1 & 2 & 1 \\ 1 & 3 & 1 \end{bmatrix}$ and $Q = PP^T$, then the value of determinant Q is

(a) 2 (b) -2 (c) 1 (d) 0

2 If $A = \begin{bmatrix} a & b \\ c & d \end{bmatrix}$ is such that $|A| = 0$ and $A^2 - (a + d) A + kI = O$, then k is equal to

(a) $b + c$ (b) $a + d$ (c) $ab + cd$ (d) 0

3 If $f(x) = a + bx + cx^2$ and α, β, γ are the roots of the equation $x^3 = 1$, then $\begin{vmatrix} a & b & c \\ b & c & a \\ c & a & b \end{vmatrix}$ is equal to

(a) $f(\alpha) + f(\beta) + f(\gamma)$ (b) $f(\alpha) \cdot f(\beta) + f(\beta) \cdot f(\gamma) + f(\gamma) \cdot f(\alpha)$

(c) $f(\alpha) \cdot f(\beta) \cdot f(\gamma)$ (d) $-f(\alpha) \cdot f(\beta) \cdot f(\gamma)$

4 When the determinant $\begin{vmatrix} \cos 2x & \sin^2 x & \cos 4x \\ \sin^2 x & \cos 2x & \cos^2 x \\ \cos 4x & \cos^2 x & \cos 2x \end{vmatrix}$ is expanded in power of $\sin x$. Then, the constant term in that expansion is

(a) 1 (b) 0 (c) -1 (d) 2

9.2 Properties of Determinants

5 The solutions of the equation

$$\begin{vmatrix} 1+\sin^2 x & \sin^2 x & \sin^2 x \\ \cos^2 x & 1+\cos^2 x & \cos^2 x \\ 4\sin 2x & 4\sin 2x & 1+4\sin 2x \end{vmatrix} = 0, (0 < x < \pi), \text{ are}$$ **[JEE Main 2021]**

(a) $\frac{\pi}{12}, \frac{\pi}{6}$ (b) $\frac{\pi}{6}, \frac{5\pi}{6}$

(c) $\frac{5\pi}{12}, \frac{7\pi}{12}$ (d) $\frac{7\pi}{12}, \frac{11\pi}{12}$

6 Let $P = [a_{ij}]$ be a 3×3 matrix and $Q = [b_{ij}]$, where $b_{ij} = 2^{i+j} \cdot a_{ij}$ for $1 \le i, j \le 3$. If the determinant of P is 2, then the determinant of the matrix Q is

(a) 2^{10} (b) 2^{11} (c) 2^{12} (d) 2^{13}

7 If $k > 1$ and the determinant of the matrix A^2, where $A = \begin{bmatrix} k & k\alpha & \alpha \\ 0 & \alpha & k\alpha \\ 0 & 0 & k \end{bmatrix}$ is k^2, then $|\alpha|$ is equal to

(a) $\frac{1}{k^2}$ (b) k (c) k^2 (d) $\frac{1}{k}$

8 Let $A = \begin{pmatrix} [x+1] & [x+2] & [x+3] \\ [x] & [x+3] & [x+3] \\ [x] & [x+2] & [x+4] \end{pmatrix}$, where $[t]$ denotes the greatest integer less than or equal to t.

If $\det(A) = 192$, then the set of values of x is the interval **[JEE Main 2021]**

(a) [68, 69) (b) [62, 63) (c) [65, 66) (d) [60, 61)

9 If $\theta \in R$, then maximum value of $\Delta = \begin{vmatrix} 1 & 1 & 1 \\ 1 & 1+\sin\theta & 1 \\ 1 & 1 & 1+\cos\theta \end{vmatrix}$ is

(a) $\sqrt{3/2}$ (b) 1/2

(c) $1/\sqrt{2}$ (d) None of these

10 Let ω be a complex number such that $2\omega + 1 = z$, where $z = \sqrt{-3}$. **[JEE Main 2017]**

If $\begin{vmatrix} 1 & 1 & 1 \\ 1 & -\omega^2 - 1 & \omega^2 \\ 1 & \omega^2 & \omega \end{vmatrix} = 3k$, then k is equal to

(a) 1 (b) $-z$ (c) z (d) -1

11 If $\begin{vmatrix} x-4 & 2x & 2x \\ 2x & x-4 & 2x \\ 2x & 2x & x-4 \end{vmatrix} = (A+Bx)(x-A)^2$, then the ordered pair (A, B) is equal to **[JEE Main 2018]**

(a) (4, 5) (b) (– 4, – 5) (c) (– 4, 3) (d) (– 4, 5)

12 The value of the determinant $\begin{vmatrix} 15! & 16! & 17! \\ 16! & 17! & 18! \\ 17! & 18! & 19! \end{vmatrix}$ is equal to

(a) (15! · 16! · 17!) (b) 2(15! · 16! · 17!) (c) 18! (d) 15! · 16!

13 Let a, b and c be such that $b(a+c) \neq 0$

If $\begin{vmatrix} a & a+1 & a-1 \\ -b & b+1 & b-1 \\ c & c-1 & c+1 \end{vmatrix} + \begin{vmatrix} a+1 & b+1 & c-1 \\ a-1 & b-1 & c+1 \\ (-1)^{n+2}a & (-1)^{n+1}b & (-1)^n c \end{vmatrix} = 0$, then the value of n is

(a) zero
(b) an even integer
(c) an odd integer
(d) an integer

14 Let $M = \left[A = \begin{pmatrix} a & b \\ c & d \end{pmatrix} : a, b, c, d \in \{\pm 3, \pm 2, \pm 1, 0\} \right]$. Define $f : M \to Z$, as $f(A) = \det(A)$, for all $A \in M$, where Z is set of all integers. Then, the number of $A \in M$ such that $f(A) = 15$ is equal to **[JEE Main 2021]**

(a) 16
(b) 19
(c) 13
(d) 11

15 If $\alpha, \beta \neq 0$, $f(n) = \alpha^n + \beta^n$ and $\begin{vmatrix} 3 & 1+f(1) & 1+f(2) \\ 1+f(1) & 1+f(2) & 1+f(3) \\ 1+f(2) & 1+f(3) & 1+f(4) \end{vmatrix} = k(1-\alpha)^2 (1-\beta)^2 (\alpha-\beta)^2$, then k is equal to **[JEE Main 2014]**

(a) $\alpha\beta$
(b) $\dfrac{1}{\alpha\beta}$
(c) 1
(d) -1

9.3 Adjoint and Inverse of a Matrix

16 The minors of $-4, 9$ and the cofactors of $-4, 9$ in matrix $\begin{bmatrix} -1 & -2 & 3 \\ -4 & -5 & -6 \\ -7 & 8 & 9 \end{bmatrix}$ are respectively,

(a) 42, 3 and -42, 3
(b) -42, -3 and 42, -3
(c) 42, 3 and -42, -3
(d) 42, 3 and 42, 3

17 If $A = \begin{bmatrix} 1 & x \\ x^2 & 4y \end{bmatrix}$; $B = \begin{bmatrix} -3 & 1 \\ 1 & 0 \end{bmatrix}$ and $\text{adj}(A) + B = \begin{bmatrix} 1 & 0 \\ 0 & 1 \end{bmatrix}$, then the values of x and y are

(a) 1 and 1
(b) ± 1 and 1
(c) 1 and 0
(d) None of these

18 If matrix $\begin{bmatrix} 0 & 1 & -2 \\ -1 & 0 & 3 \\ \lambda & -3 & 0 \end{bmatrix}$ is singular, then λ is equal to

(a) -2
(b) -1
(c) 1
(d) 2

19 If $A = \begin{bmatrix} 2-k & 2 \\ 1 & 3-k \end{bmatrix}$ is a singular matrix, then the value of $5k - k^2$ is

(a) 0
(b) 6
(c) -6
(d) 4

20 Let $\omega \neq 1$ be a cube root of unity and S be the set of all non-singular matrices of the form $\begin{bmatrix} 1 & a & b \\ \omega & 1 & c \\ \omega^2 & \omega & 1 \end{bmatrix}$, where each of a, b and c is either ω or ω^2. The number of distinct matrices in the set S is **[IIT 2012]**

(a) 2
(b) 6
(c) 4
(d) 8

21 The values of x for which the given matrix $\begin{bmatrix} -x & x & 2 \\ 2 & x & -x \\ x & -2 & -x \end{bmatrix}$ will be non-singular, are

(a) $-2 \le x \le 2$ (b) for all x other than 2 and -2
(c) $x \ge 2$ (d) $x \le -2$

22 If the determinant of a 3×3 matrix A be 6, then B is a matrix defined by $B = 5A^2$. The determinant of B is

(a) 180 (b) 100 (c) 80 (d) None of these

23 Let P and Q be 3×3 matrices, $P \ne Q$. If $P^3 = Q^3$ and $P^2Q = Q^2P$, then determinant of $(P^2 + Q^2)$ is equal to

(a) -2 (b) 1 (c) 0 (d) -1

24 If $A = \begin{bmatrix} 1 & 2 \\ 3 & 5 \end{bmatrix}$, then the value of the determinant $|A^{2009} - 5A^{2008}|$ is

(a) 6 (b) -6 (c) 5 (d) -5

25 If A is a matrix of order 3 such that $A\,(\text{adj } A) = 10I$, then $|\text{adj } A|$ is equal to

(a) 1 (b) 10 (c) 100 (d) $10I$

26 If a square matrix A such that $AA^T = I = A^TA$, then $|A|$ is equal to

(a) 0 (b) ± 1 (c) ± 2 (d) None of these

27 If $A = \begin{bmatrix} a & 0 & 0 \\ 0 & a & 0 \\ 0 & 0 & a \end{bmatrix}$, then $|A|\,|\text{adj } A|$ is equal to

(a) a^{3n} (b) a^{-3n} (c) $-a^{3n}$ (d) $2a^{3n}$

28 If $P = \begin{bmatrix} 1 & \alpha & 3 \\ 1 & 3 & 3 \\ 2 & 4 & 4 \end{bmatrix}$ is the adjoint of a 3×3 matrix A and $|A| = 4$, then α is equal to

(a) 4 (b) 11 (c) 5 (d) 0

29 If the adjoint of a 3×3 matrix P is $\begin{bmatrix} 1 & 4 & 4 \\ 2 & 1 & 7 \\ 1 & 1 & 3 \end{bmatrix}$, then the possible value of the determinant of P is

(a) -2 (b) -1 (c) 1 (d) 3

30 If nth order square matrix A is a orthogonal, then $|\text{adj (adj } A)|$ is

(a) always (-1), if n is even (b) always 1, if n is odd
(c) always 1 (d) None of these

31 If $A = [a_{ij}]_{4 \times 4}$, such that $a_{ij} = \begin{cases} 2, & \text{when } i = j \\ 0, & \text{when } i \ne j \end{cases}$, then $\left\{ \dfrac{\det(\text{adj}(\text{adj } A))}{7} \right\}$ is, where $\{\cdot\}$ represents fractional part function

(a) 1/7 (b) 2/7 (c) 3/7 (d) None of these

32 If A is a square matrix of a order 5 and $2A^{-1} = A^T$, then the remainder when $|\text{adj} \cdot (\text{adj} A)|$ is divided by 7, is

(a) 2 (b) 3 (c) 4 (d) 5

33 If $A=\begin{bmatrix}2 & 3\\5 & -2\end{bmatrix}$ be such that $A^{-1}=kA$, then k is equal to

(a) 19 (b) $\frac{1}{19}$ (c) -19 (d) $-\frac{1}{19}$

34 If $A=\begin{bmatrix}1 & 2\\-5 & 1\end{bmatrix}$ and $A^{-1}=xA+yI$, then the values of x and y are

(a) $x=\frac{-1}{11}$ and $y=\frac{2}{11}$ (b) $x=\frac{-1}{11}$ and $y=\frac{-2}{11}$ (c) $x=\frac{1}{11}$ and $y=\frac{2}{11}$ (d) $x=\frac{1}{11}$ and $y=\frac{-2}{11}$

35 The matrix A satisfying the equation $\begin{bmatrix}1 & 3\\0 & 1\end{bmatrix}A=\begin{bmatrix}1 & 1\\0 & -1\end{bmatrix}$ is

(a) $\begin{bmatrix}1 & 4\\-1 & 0\end{bmatrix}$ (b) $\begin{bmatrix}1 & -4\\1 & 0\end{bmatrix}$ (c) $\begin{bmatrix}1 & 4\\0 & -1\end{bmatrix}$ (d) None of these

36 If $A=\begin{bmatrix}0 & 1 & 2\\1 & 2 & 3\\3 & a & 1\end{bmatrix}$ and $A^{-1}=\begin{bmatrix}1/2 & -1/2 & 1/2\\-4 & 3 & c\\5/2 & -3/2 & 1/2\end{bmatrix}$, then the values of a and c are equal to

(a) 1 and 1 (b) 1 and -1 (c) 1 and 2 (d) -1 and 1

37 If adj $B=A, |P|=|Q|=1$, then adj $(Q^{-1}BP^{-1})$ is

(a) PQ (b) QAP (c) PAQ (d) $PA^{-1}Q$

38 If the eigen values of a 3×3 matrix are 1, 2 and 3, then the eigen values of the matrix A^T is

(a) 1, 2, 3 (b) $1, \frac{1}{2}, \frac{1}{3}$ (c) 1, 4, 9 (d) 1, 8, 2

9.4 Application of Determinants, Differentiation and Integration of Determinants

39 The area of the triangle with vertices (2, 3), (0, 1) and (1, 2) is

(a) $\frac{1}{2}$ sq unit (b) 0 (c) 2 sq units (d) $2\frac{1}{2}$ sq units

40 Let $A=[a_{ij}]$ be a 3×3 matrix, where

$$a_{ij}=\begin{cases}1, & \text{if } i=j\\-x, & \text{if } |i-j|=1\\2x+1, & \text{otherwise}\end{cases}$$

Let a function $f: R\to R$ be defined as $f(x)=\det(A)$. Then, the sum of maximum and minimum values of f on R is equal to **[JEE Main 2021]**

(a) $-\frac{20}{27}$ (b) $\frac{88}{27}$ (c) $\frac{20}{27}$ (d) $-\frac{88}{27}$

41 The area of the triangle whose vertices are (1, 1), (4, 2) and (3, 5) is

(a) 6 sq units (b) 5 sq units (c) 4 sq units (d) 3 sq units

42 The set of all values of λ for which the system of linear equations $2x_1-2x_2+x_3=\lambda x_1$, $2x_1-3x_2+2x_3=\lambda x_2$ and $-x_1+2x_2=\lambda x_3$ has a non-trivial solution, **[JEE Main 2015]**

(a) is an empty set (b) is a singleton set
(c) contains two elements (d) contains more than two elements

43 If the system of equations $x + ky - z = 0, 3x - ky - z = 0$ and $x - 3y + z = 0$ has a non-zero solution, then k is equal to

(a) -1 (b) 0 (c) 1 (d) 2

44 The number of values of k for which the linear equations

$$4x + ky + 2z = 0,\ kx + 4y + z = 0 \text{ and } 2x + 2y + z = 0$$

process a non-zero solution, is **[AIEEE 2011]**

(a) 0 (b) 3 (c) 2 (d) 1

45 The number of values of k, for which the system of equations $(k+1)x + 8y = 4k$ and $kx + (k+3)y = 3k - 1$ has no solution, is

(a) 1 (b) 2 (c) 3 (d) None of these

46 The existence of the unique solution of the system of equations $x + y + z = \beta, 5x - y + \alpha z = 10$ and $2x + 3y - z = 6$ depends on

(a) α only (b) β only (c) both α and β (d) neither α nor β

47 The value of k such that the lines $2x - 3y + k = 0, 3x - 4y - 13 = 0$ and $8x - 11y - 33 = 0$ are concurrent, is

(a) 20 (b) -7 (c) 7 (d) -20

48 Consider the system of equations in x, y and z as

$$x \sin 3\theta - y + z = 0,\ x \cos 2\theta + 4y + 3z = 0 \text{ and } 2x + 7y + 7z = 0.$$

If the system has a non-trivial solution, then for integer x, values of θ are given by

(a) $\pi\left(n + \frac{(-1)^n}{3}\right)$ (b) $\pi\left(n + \frac{(-1)^n}{4}\right)$

(c) $\pi\left(n + \frac{(-1)^n}{6}\right)$ (d) $\frac{n\pi}{2}$

49 If the system of equations $ax + ay - z = 0, bx - y + bz = 0$ and $-x + cy + cz = 0$ has a non-trivial solution, then the value of $\frac{1}{1+a} + \frac{1}{1+b} + \frac{1}{1+c}$ is

(a) 0 (b) 1 (c) 2 (d) 3

50 If S is the set of distinct values of b for which the following system of linear equations

$$x + y + z = 1,\ x + ay + z = 1 \text{ and } ax + by + z = 0$$

has no solution, then S is **[JEE Main 2017]**

(a) a singleton set (b) an empty set

(c) an infinite set (d) a finite set containing two or more elements

51 Consider the system of linear equations

$$x_1 + 2x_2 + x_3 = 3,\ 2x_1 + 3x_2 + x_3 = 3 \text{ and } 3x_1 + 5x_2 + 2x_3 = 1.$$

The system has

(a) no solution (b) infinite number of solutions

(c) exactly three solutions (d) a unique solution

52 The number of values of k, for which the system of equations

$$(k+1)x + 8y = 4k \text{ and } kx + (k+3)y = 3k - 1$$

has no solution, is **[JEE Main 2013]**

(a) infinite (b) 1 (c) 2 (d) 3

53 Let $f(x)=\begin{vmatrix} x^3 & \sin x & \cos x \\ 6 & -1 & 0 \\ p & p^2 & p^3 \end{vmatrix}$ when p is a constant, then the value of $\frac{d}{dx}[f(x)]$ at $x=0$ is

(a) p^3 (b) $1+p+p^2$

(c) $-6p^3$ (d) p^2-6p^3

54 If $\Delta(x)=\begin{vmatrix} \sin^2 x & \log\cos x & \log\tan x \\ x^2 & 2x-1 & 2x+1 \\ 1 & -2\log 2 & 0 \end{vmatrix}$, then $\int_0^{\pi/2} \Delta(x)\,dx$ is

(a) 0 (b) 1 (c) 2 (d) 4

Numerical Value Type Questions

55 If $(1+ax+bx^2)^4=a_0+a_1x+a_2x^2+\ldots+a_8x^8$, where $a, b, a_0, a_1, \ldots, a_8 \in R$ such that $a_0+a_1+a_2 \neq 0$ and $\begin{vmatrix} a_0 & a_1 & a_2 \\ a_1 & a_2 & a_0 \\ a_2 & a_0 & a_1 \end{vmatrix}=0$, then the value of $5\frac{a}{b}$ is ....

56 Let $a+b+c=S$ and $\begin{vmatrix} S+c & a & b \\ c & S+a & b \\ c & a & S+b \end{vmatrix}=532$, the value of S is ....

57 Let $a, b, c \in R$ not all are equal and $\Delta_1=\begin{vmatrix} a & b & c \\ b & c & a \\ c & a & b \end{vmatrix}$ and $\Delta_2=\begin{vmatrix} a+2b & b+3c & c+4a \\ b+2c & c+3a & a+4b \\ c+2a & a+3b & b+4c \end{vmatrix}$, then $\frac{\Delta_2}{\Delta_1}$ is equal to ....

58 If $a_1, a_2, a_3, \ldots, a_{12}$ are in AP and $\Delta_1=\begin{vmatrix} a_1a_5 & a_1 & a_2 \\ a_2a_6 & a_2 & a_3 \\ a_3a_7 & a_3 & a_4 \end{vmatrix}$ and $\Delta_2=\begin{vmatrix} a_2a_{10} & a_2 & a_3 \\ a_3a_{11} & a_3 & a_4 \\ a_4a_{12} & a_4 & a_5 \end{vmatrix}$, then $\Delta_1:\Delta_2$ is equal to ....

59 If three distinct points $P(3u^2, 2u^3)$, $Q(3v^2, 2v^3)$ and $R(3w^2, 2w^3)$ are collinear, then $uv+vw+wu$ is equal to ....

60 If $\Delta_r=\begin{vmatrix} r & 612 & 915 \\ 101r^2 & 2r & 3r \\ r & \frac{1}{r} & \frac{1}{r^2} \end{vmatrix}$, then the value of $\lim_{n\to\infty} \frac{1}{n^3}\sum_{r=1}^{n} \Delta_r$ is ....

61 $A=\begin{bmatrix} 1 & \tan x \\ -\tan x & 1 \end{bmatrix}$ and $f(x)$ is defined as $f(x)=\det(A^TA^{-1})$, then the value of $\underbrace{f(f(f(\ldots f(x)))))}_{n \text{ times}}$ is $(n \geq 2)$ ....

62 Let A and B are two square matrices of order 3 such that $\det(A)=3$ and $\det(B)=2$, then the value of $\det\{(\text{adj}(B^{-1}A^{-1}))^{-1}\}$ is equal to ....

63 Let P, Q and R be invertible matrices of order 3 such that $A = PQ^{-1}, B = QR^{-1}$ and $C = RP^{-1}$, then the value of det $(ABC + BCA + CAB)$ is equal to ....

64 If A is an idempotent matrix satisfying $(1 - 0.4A)^{-1} = I - \alpha A$, where I is the unit matrix of the same order as that of A, then the value of $|9\alpha|$ is equal to ....

65 The equation $\begin{bmatrix} 1 & 2 & 2 \\ 1 & 3 & 4 \\ 3 & 4 & k \end{bmatrix}\begin{bmatrix} x \\ y \\ z \end{bmatrix} = \begin{bmatrix} 0 \\ 0 \\ 0 \end{bmatrix}$ has a solution for (x, y, z) besides $(0, 0, 0)$, then the value of k is ....

66 The sum of values of p for which the equation $x + y + z = 1$, $x + 2y + 4z = p$ and $x + 4y + 10z = p^2$ have a solution, is ....

LEVEL 03 \ JEE Advanced Level

Single Choice Correct Questions

1 If [ ] denotes the greatest integer less than or equal to the real number under considerations, and $-1 \le x < 0, 0 \le y < 1, 1 \le z < 2$, then the value of the determinant

$$\begin{vmatrix} [x]+1 & [y] & [z] \\ [x] & [y]+1 & [z] \\ [x] & [y] & [z]+1 \end{vmatrix}$$ is

(a) $[x]$ (b) $[y]$
(c) $[z]$ (d) None of these

2 If $a = \cos\theta + i\sin\theta, b = \cos 2\theta - i\sin 2\theta, c = \cos 3\theta + i\sin 3\theta$ and if $\begin{vmatrix} a & b & c \\ b & c & a \\ c & a & b \end{vmatrix} = 0$, then

(a) $\theta = 2k\pi, k \in Z$ (b) $\theta = (2k+1)\pi, k \in Z$
(c) $\theta = (4k+1)\pi, k \in Z$ (d) None of these

3 The value of $\begin{vmatrix} \log_5 729 & \log_3 5 \\ \log_5 27 & \log_9 25 \end{vmatrix}\begin{vmatrix} \log_3 5 & \log_{27} 5 \\ \log_5 9 & \log_5 9 \end{vmatrix}$ is

(a) 4 (b) $\frac{1}{4}$ (c) 5 (d) $\frac{1}{5}$

4 If A is square matrix of order 4 and I is a unit matrix, then it is true that
(a) det (2A) = 2 det (A) (b) det (2A) = 16 det (A)
(c) det (– A) = – det (A) (d) det (A + I) = det (A) + I

5 If α, β and γ are the roots of $px^3 + qx^2 + r = 0$, then the value of the determinant

$$\begin{vmatrix} \alpha\beta & \beta\gamma & \gamma\alpha \\ \beta\gamma & \gamma\alpha & \alpha\beta \\ \gamma\alpha & \alpha\beta & \beta\gamma \end{vmatrix}$$ is

(a) p (b) q (c) 0 (d) r

6 If $A = \begin{bmatrix} 5a & -b \\ 3 & 2 \end{bmatrix}$ and $A \cdot \text{adj } A = AA^T$, then $5a + b$ is equal to

(a) 5 (b) 4 (c) 13 (d) -1

7 Let A be an nth-order square matrix and B be its adjoint, then $|AB + k\,I_n|$ is equal to (where k is a scalar units)

(a) $(|A| + k)^{n-2}$ (b) $(|A| + k)^n$

(c) $(|A| + k)^{n-1}$ (d) None of these

8 If $A = \begin{bmatrix} 0 & x \\ y & 0 \end{bmatrix}$ and $A^3 + A = O$, then sum of possible values of xy is

(a) 0 (b) -1 (c) 1 (d) 2

9 If A and B are two non-zero square matrices of the same order such that the product $AB = 0$, then

(a) both A and B must be singular

(b) exactly one of them must be singular

(c) both of them are non-singular

(d) None of the above

10 The number of 3×3 non-singular matrices with four entries as 1 and all other entries as 0 is

(a) atleast 7 (b) less than 4 (c) 5 (d) 6

11 If A is a square matrix of order less than 4 such that $\det(A - A^T) \neq 0$ and $B = \text{adj}(A)$, then $\text{adj}(B^2 A^{-1} B^{-1} A)$ is equal to

(a) A (b) B (c) $|A|A$ (d) $|B|B$

12 If A is order of 3 square matrix such that $|A| = 2$, then $|\text{adj}\{\text{adj}(\text{adj } A)\}|$, is equal to

(a) 512 (b) 256

(c) 64 (d) None of these

13 If $A = \begin{bmatrix} a & b & c \\ x & y & z \\ p & q & r \end{bmatrix}$, $B = \begin{bmatrix} q & -b & y \\ -p & a & -x \\ r & -c & z \end{bmatrix}$ and if A is invertible, then which of the following is not true?

(a) $|A| = |B|$

(b) $|A| = -|B|$

(c) $|\text{adj } A| = |\text{adj } B|$

(d) A is invertible, if and only if B is invertible

14 If $A = \begin{bmatrix} a+ib & c+id \\ -c+id & a-ib \end{bmatrix}$ and $a^2 + b^2 + c^2 + d^2 = 1$, then A^{-1} is equal to

(a) $\begin{bmatrix} a-ib & -c-id \\ c-id & a+ib \end{bmatrix}$ (b) $\begin{bmatrix} a+ib & -c+id \\ -c+id & a-ib \end{bmatrix}$

(c) $\begin{bmatrix} a-ib & -c-id \\ -c-id & a+ib \end{bmatrix}$ (d) None of these

15 If $A(\alpha, \beta) = \begin{bmatrix} \cos\alpha & \sin\alpha & 0 \\ -\sin\alpha & \cos\alpha & 0 \\ 0 & 0 & e^{\beta} \end{bmatrix}$, then $A(\alpha, \beta)^{-1}$ is equal to

(a) $A(-\alpha, -\beta)$ (b) $A(-\alpha, \beta)$ (c) $A(\alpha, -\beta)$ (d) $A(\alpha, \beta)$

16 If $\begin{bmatrix} 1/25 & 0 \\ x & 1/25 \end{bmatrix} = \left(\begin{bmatrix} 5 & 0 \\ -a & 5 \end{bmatrix}^{-1} \right)^2$, then the value of x is

(a) $\frac{a}{125}$ (b) $\frac{2a}{125}$ (c) $\frac{2a}{25}$ (d) None of these

17 If $A = \begin{bmatrix} 1 & \tan x \\ -\tan x & 1 \end{bmatrix}$, then $A^T A^{-1}$ is equal to

(a) $\begin{bmatrix} -\cos 2x & \sin 2x \\ -\sin 2x & \cos 2x \end{bmatrix}$ (b) $\begin{bmatrix} \cos 2x & -\sin 2x \\ \sin 2x & \cos 2x \end{bmatrix}$

(c) $\begin{bmatrix} \cos 2x & \cos 2x \\ \cos 2x & \sin 2x \end{bmatrix}$ (d) None of these

18 Let $A = \begin{bmatrix} 1 & 0 & 0 \\ 2 & 1 & 0 \\ 3 & 2 & 1 \end{bmatrix}$. If u_1 and u_2 are column matrices such that $Au_1 = \begin{bmatrix} 1 \\ 0 \\ 0 \end{bmatrix}$ and $Au_2 = \begin{bmatrix} 0 \\ 1 \\ 0 \end{bmatrix}$, then $u_1 + u_2$ is equal to

(a) $\begin{bmatrix} -1 \\ 1 \\ 0 \end{bmatrix}$ (b) $\begin{bmatrix} -1 \\ 1 \\ -1 \end{bmatrix}$ (c) $\begin{bmatrix} -1 \\ -1 \\ 0 \end{bmatrix}$ (d) $\begin{bmatrix} 1 \\ -1 \\ -1 \end{bmatrix}$

19 If the eigen values of a matrix A are 2 and 3, then the eigen values of the matrix A^3 are

(a) 6 and 9 (b) 8 and 27 (c) 4 and 9 (d) 8 and 9

20 The system of linear equations $3x + y - z = 2$, $x - z = 1$ and $2x + 2y + az = 5$ has unique solution, when

(a) $a \neq 3$ (b) $a \neq 4$ (c) $a \neq 2$ (d) $a \neq 1$

21 The system of equations $-2x + y + z = a$, $x - 2y + z = b$ and $x + y - 2z = c$ is consistent, if

(a) $a + b - c = 0$ (b) $a - b + c = 0$

(c) $a + b + c \neq 0$ (d) $a + b + c = 0$

22 The system of linear equations

$$x + \lambda y - z = 0,\ \lambda x - y - z = 0 \text{ and } x + y - \lambda z = 0$$

has a non-trivial solution for **[JEE Main 2016]**

(a) exactly one value of λ (b) exactly two values of λ

(c) exactly three values of λ (d) infinitely many values of λ

23 If $f(x) = \begin{vmatrix} x^n & \sin x & -\cos x \\ n! & \sin\left(\frac{n\pi}{2}\right) & \cos\left(\frac{n\pi}{2}\right) \\ a & a^2 & a^3 \end{vmatrix}$, where a is a constant, then $\frac{d^n}{dx^n}\{f(x)\}$ at $x = 0$ for $n = 2m + 1$, is

(a) 0 (b) -1 (c) 1 (d) None of these

24 Let $f(x) = \begin{vmatrix} \sec x & \cos x & \sec^2 x + \cot x \operatorname{cosec} x \\ \cos^2 x & \cos^2 x & \operatorname{cosec}^2 x \\ 1 & \cos^2 x & \cos^2 x \end{vmatrix}$, then $\int_0^{\pi/2} f(x)\,dx$ is equal to

(a) $-\left(\frac{\pi}{4} + \frac{8}{15}\right)$ (b) $-\frac{\pi}{4} + \frac{8}{15}$ (c) $\frac{\pi}{4} + \frac{2}{15}$ (d) $-\left(\frac{\pi}{4} + \frac{2}{15}\right)$

More than One Correct

25 For any 3×3 matrix M, let $|M|$ denote the determinant of M. Let

$$E = \begin{bmatrix} 1 & 2 & 3 \\ 2 & 3 & 4 \\ 8 & 13 & 18 \end{bmatrix}, P = \begin{bmatrix} 1 & 0 & 0 \\ 0 & 0 & 1 \\ 0 & 1 & 0 \end{bmatrix} \text{ and } F = \begin{bmatrix} 1 & 3 & 2 \\ 8 & 18 & 13 \\ 2 & 4 & 3 \end{bmatrix}$$

If Q is a non-singular matrix of order 3×3, then which of the following statements is (are) true? **[JEE Advanced 2021]**

(a) $F = PEP$ and $P^2 = \begin{bmatrix} 1 & 0 & 0 \\ 0 & 1 & 0 \\ 0 & 0 & 1 \end{bmatrix}$

(b) $|EQ + PFQ^{-1}| = |EQ| + |PFQ^{-1}|$

(c) $|(EF)^3| > |EF|^2$

(d) Sum of the diagonal entries of $P^{-1}EP + F$ is equal to the sum of diagonal entries of $E + P^{-1}FP$

26 For any 3×3 matrix M, let $|M|$ denote the determinant of M. Let I be the 3×3 identity matrix. Let E and F be two 3×3 matrices such that $(I - EF)$ is invertible. If $G = (I - EF)^{-1}$, then which of the following statements is (are) true? **[JEE Advanced 2021]**

(a) $|FE| = |I - FE||FGE|$

(b) $(I - FE)(I + FGE) = I$

(c) $EFG = GEF$

(d) $(I - FE)(I - FGE) = I$

27 If $f(\alpha, \beta) = \begin{vmatrix} \cos \alpha & -\sin \alpha & 1 \\ \sin \alpha & \cos \alpha & 1 \\ \cos(\alpha + \beta) & -\sin(\alpha + \beta) & 1 \end{vmatrix}$, then

(a) $f(300, 200) = f(400, 200)$

(b) $f(200, 400) = f(200, 600)$

(c) $f(100, 200) = f(200, 200)$

(d) None of these

28 Which of the following has/have value equal to zero?

(a) $\begin{vmatrix} 8 & 2 & 7 \\ 12 & 3 & 5 \\ 16 & 4 & 3 \end{vmatrix}$

(b) $\begin{vmatrix} 1/a & a^2 & bc \\ 1/b & b^2 & ac \\ 1/c & c^2 & ab \end{vmatrix}$

(c) $\begin{vmatrix} a+b & 2a+b & 3a+b \\ 2a+b & 3a+b & 4a+b \\ 4a+b & 5a+b & 6a+b \end{vmatrix}$

(d) $\begin{vmatrix} 2 & 43 & 6 \\ 7 & 35 & 4 \\ 3 & 17 & 2 \end{vmatrix}$

29 If $f(x) = \begin{vmatrix} a & -1 & 0 \\ ax & a & -1 \\ ax^2 & ax & a \end{vmatrix}$, then $f(2x) - f(x)$ is divisible by

(a) x

(b) a

(c) $2a + 3x$

(d) x^2

30 If $\Delta = \begin{vmatrix} -x & a & b \\ b & -x & a \\ a & b & -x \end{vmatrix}$, then a factor of Δ is

(a) $a + b + x$

(b) $x^2 - (a - b)x + a^2 + b^2 + ab$

(c) $x^2 + (a + b)x + a^2 + b^2 - ab$

(d) $a + b - x$

31 The determinant $\Delta = \begin{vmatrix} a^2+x & ab & ac \\ ab & b^2+x & bc \\ ac & bc & c^2+x \end{vmatrix}$ is divisible by

(a) x (b) x^2 (c) x^3 (d) None of these

32 If $f(x) = \begin{vmatrix} 3 & 3x & 3x^2+2a^2 \\ 3x & 3x^2+2a^2 & 3x^3+6a^2x \\ 3x^2+2a^2 & 3x^3+6a^2x & 3x^4+12a^2x^2+2a^4 \end{vmatrix}$, then

(a) $f'(x) = 0$ (b) $y = f(x)$ is a straight line parallel to X-axis
(c) $f(x)$ is a parabola (d) None of these

33 Let $f(x) = \begin{vmatrix} n & n+1 & n+2 \\ {}^nP_n & {}^{n+1}P_{n+1} & {}^{n+2}P_{n+2} \\ {}^nC_n & {}^{n+1}C_{n+1} & {}^{n+2}C_{n+2} \end{vmatrix}$, where the symbols have their usual meanings.

Then, $f(x)$ is divisible by

(a) n^2+n+1 (b) $(n+1)!$ (c) $n!$ (d) None of these

34 Which of the following values of α satisfy the equations

$$\begin{vmatrix} (1+\alpha)^2 & (1+2\alpha)^2 & (1+3\alpha)^2 \\ (2+\alpha)^2 & (2+2\alpha)^2 & (2+3\alpha)^2 \\ (3+\alpha)^2 & (3+2\alpha)^2 & (3+3\alpha)^2 \end{vmatrix} = -648\alpha \text{ ?}$$

[JEE Advanced 2015]

(a) -4 (b) 9 (c) -9 (d) 4

35 If A and B are two square matrices of same orders such that $A+B=AB$ and I is an identity matrix of order same as that of A and B, then

(a) $AB = BA$ (b) $|A-I| = 0$
(c) $|B-I| \neq 0$ (d) $|A-B| = 0$

36 If $A = \begin{bmatrix} 1 & 0 & 0 \\ 1 & 0 & 1 \\ 0 & 1 & 0 \end{bmatrix}$, then

(a) $A^3 - A^2 = A - I$ (b) $|A^{2010} - I| = 0$

(c) $A^{50} = \begin{bmatrix} 1 & 0 & 0 \\ 25 & 1 & 0 \\ 25 & 0 & 1 \end{bmatrix}$ (d) $A^{50} = \begin{bmatrix} 1 & 1 & 0 \\ 25 & 1 & 0 \\ 25 & 0 & 1 \end{bmatrix}$

37 If $A = \begin{bmatrix} 3 & -3 & 4 \\ 2 & -3 & 4 \\ 0 & -1 & 1 \end{bmatrix}$, then

(a) adj (adj A) = A (b) |adj (adj A)| = 1
(c) |adj A| = 1 (d) None of these

38 For 3×3 matrices M and N, which of the following statement(s) is (are) not correct?

(a) $N^T MN$ is symmetric or skew-symmetric, according as M is symmetric or skew-symmetric
(b) $MN - NM$ is skew-symmetric for all symmetric matrices M and N
(c) MN is symmetric for all symmetric matrices M and N
(d) (adj M) (adj N) = adj (MN) for all invertible matrices M and N

39 If $A^{-1} = \begin{bmatrix} 1 & -1 & 2 \\ 0 & 3 & 1 \\ 0 & 0 & -1/3 \end{bmatrix}$, then

(a) $|A| = -1$ (b) adj $A = \begin{bmatrix} -1 & 1 & -2 \\ 0 & -3 & -1 \\ 0 & 0 & 1/3 \end{bmatrix}$

(c) $A = \begin{bmatrix} 1 & 1/3 & 7 \\ 0 & 1/3 & 1 \\ 0 & 0 & -3 \end{bmatrix}$ (d) $A = \begin{bmatrix} 1 & -1/3 & -7 \\ 0 & -3 & 0 \\ 0 & 0 & 1 \end{bmatrix}$

40 Let M be a 2×2 symmetric matrix with integer entries, then M is invertible, if

(a) the first column of M is the transpose of the second row of M
(b) the second row of M is the transpose of the first column of M
(c) M is a diagonal matrix with non-zero entries in the main diagonal
(d) the product of the entries in the main diagonal of M is not the square of an integer

41 If $A(\theta) = \begin{bmatrix} \sin\theta & i\cos\theta \\ i\cos\theta & \sin\theta \end{bmatrix}$, then which of the following is (are) not true?

(a) $A(\theta)^{-1} = A(\pi - \theta)$ (b) $A(\theta) = A(\pi + \theta)$ is a null matrix
(c) $A(\theta)$ is invertible for all $\theta \in R$ (d) $A(\theta)^{-1} = A(-\theta)$

42 If A is a matrix such that $A^2 + A + 2I = 0$, then which of the following is/are true?

(a) A is non-singular (b) A is symmetric (c) $A \neq 0$ (d) $A^{-1} = -\frac{1}{2}(A + I)$

43 Let $\alpha, \lambda, \mu \in R$. Consider the system of linear equations

$$\alpha x + 2y = \lambda \text{ and } 3x - 2y = \mu$$

Which of the following statement(s) is (are) correct? **[JEE Advanced 2016]**

(a) If $\alpha = -3$, then the system has infinitely many solutions for all values of λ and μ
(b) If $\alpha \neq -3$, then the system has a unique solution for all values of λ and μ
(c) If $\lambda + \mu = 0$, then the system has infinitely many solutions for $\alpha = -3$
(d) If $\lambda + \mu \neq 0$, then the system has no solution for $\alpha = -3$

44 If $A = \begin{bmatrix} 8 & -6 & 2 \\ -6 & 7 & -4 \\ 2 & -4 & 3 \end{bmatrix}$ and X is a non-zero column matrix such that $AX = \lambda X$, where λ is a scalar, then values of λ can be

(a) 3 (b) 6 (c) 12 (d) 15

45 The values of $k \in R$ for which the system of equations $x + ky + 3z = 0$, $kx + 2y + 2z = 0$ and $2x + 3y + 4z = 0$ has non-trivial solution, are

(a) 2 (b) 5/2 (c) 3 (d) 5/4

46 The system of equations $-2x + y + z = a$, $x - 2y + z = b$ and $x + y - 2z = c$ has

(a) no solution if $a + b + c \neq 0$ (b) unique solution if $a + b + c = 0$
(c) infinite number of solutions if $a + b + c = 0$ (d) None of these

47 Let α, β and γ be the roots of equation $x^3 + ax^2 + bx + c = 0$ $(a \neq 0)$. If the system of equations $\alpha x + \beta y + \gamma z = 0$, $\beta x + \gamma y + \alpha z = 0$ and $\gamma x + \alpha y + \beta z = 0$ has non-trivial solution, then

(a) $a^2 = 3b$ (b) $a^3 = 27c$ (c) $b^3 = 27c^2$ (d) $\alpha + \beta + \gamma = 0$

Linked Comprehension Type

Paragraph I Directions (Q.Nos. 48-50)

$$f(x)=\begin{vmatrix} x+c_1 & x+a & x+a \\ x+b & x+c_2 & x+a \\ x+b & x+b & x+c_3 \end{vmatrix} \text{ and } g(x)=(c_1-x)(c_2-x)(c_3-x), \text{ then}$$

48 coefficient of x in $f(x)$ is

(a) $\dfrac{g(a)-f(b)}{b-a}$ (b) $\dfrac{g(-a)-g(-b)}{b-a}$ (c) $\dfrac{g(a)-g(b)}{b-a}$ (d) None of these

49 Which of the following is not a constant term in $f(x)$?

(a) $\dfrac{bg(a)-ag(b)}{(b-a)}$ (b) $\dfrac{bg(a)-af(-b)}{(b-a)}$

(c) $\dfrac{bf(-a)-ag(b)}{(b-a)}$ (d) None of these

50 Which of the following is not true?

(a) $f(-a)=g(a)$ (b) $f(-a)=g(-a)$

(c) $f(-b)=g(b)$ (d) None of these

Paragraph II Directions (Q.Nos. 51-52)

Consider an arbitrary 3×3 non-singular matrix $A=[a_{ij}]$. A matrix $B=[b_{ij}]$ is formed such that b_{ij} is the sum of all the elements except a_{ij} in the ith row of A. Answer the following questions.

51 If there exists a matrix X with constant elements such that $AX=B$, then X is

(a) skew-symmetric (b) null matrix

(c) diagonal matrix (d) None of these

52 The value of $|B|$ is equal to

(a) $|A|$ (b) $\dfrac{|A|}{2}$ (c) $2|A|$ (d) None of these

Paragraph III Directions (Q.Nos. 53-55)

Let A be the set of all 3×3 symmetric matrices all of whose entries are either 0 or 1. Five of these entries are 1 and four of them are 0.

53 The number of matrices in A is

(a) 12 (b) 6 (c) 9 (d) 3

54 The number of matrices A in A for which the system of linear equations

$$A\begin{bmatrix} x \\ y \\ z \end{bmatrix}=\begin{bmatrix} 1 \\ 0 \\ 0 \end{bmatrix} \text{ has a unique solution, is}$$

(a) less than 4 (b) atleast 4 but less than 7

(c) atleast 7 but less than 10 (d) atleast 10

55 The number of matrices A in A for which the system of linear equations

$$A\begin{bmatrix} x \\ y \\ z \end{bmatrix}=\begin{bmatrix} 1 \\ 0 \\ 0 \end{bmatrix} \text{ is inconsistent, is}$$

(a) 0 (b) more than 2 (c) 2 (d) 1

Matrix Match Type Questions

56 Match the following lists

	List I		List II
(i)	Coefficient of x in $f(x) = \begin{vmatrix} x & (1+\sin x)^3 & \cos x \\ 1 & \log(1+x) & 2 \\ x^2 & 1+x^2 & 0 \end{vmatrix}$ When each of $\sin x, \cos x$ and $\log(1+x)$ is expanded in power of x.	A.	10
(ii)	Value of $\begin{vmatrix} 1 & 3\cos\theta & 1 \\ \sin\theta & 1 & 3\cos\theta \\ 1 & \sin\theta & 1 \end{vmatrix}$ is	B.	0
(iii)	If a, b, c are in AP and $f(x) = \begin{vmatrix} x+a & x^2+1 & 1 \\ x+b & 2x^2-1 & 1 \\ x+c & 3x^2-2 & 1 \end{vmatrix}$, then $f(0)$ is	C.	-12
(iv)	If $\begin{vmatrix} x & 2 & x \\ 1 & x & 6 \\ x & x & x+1 \end{vmatrix} = a_4x^4 + a_3x^3 + a_2x^2 + a_1x + a_0$, then a_0 is	D.	-2

57 Match the following lists

	List I		List II
(i)	$\begin{vmatrix} \frac{1}{c} & \frac{1}{c} & -\frac{(a+b)}{c^2} \\ -\frac{(b+c)}{a^2} & \frac{1}{a} & \frac{1}{a} \\ -\frac{b(b+c)}{a^2c} & \frac{(a+2b+c)}{ac} & -\frac{b(a+b)}{ac^2} \end{vmatrix}$ is independent of	A.	a, b
(ii)	$\begin{vmatrix} \sin a\cos b & \sin a\sin b & \cos a \\ \cos a\cos b & \cos a\sin b & -\sin a \\ -\sin a\sin b & \sin a\cos b & 0 \end{vmatrix}$ is independent of	B.	a, b, c
(iii)	$\begin{vmatrix} \frac{1}{\sin^2 a\cos b} & \frac{1}{\sin a\sin b} & \frac{1}{\cos a} \\ \frac{-\cos a}{\sin^2 a\cos b} & \frac{-\cos a}{\sin^2 a\sin b} & \frac{\sin a}{\cos^2 a} \\ \frac{\sin b}{\sin a\cos^2 b} & \frac{-\cos b}{\sin a\sin^2 b} & 0 \end{vmatrix}$ is dependent on	C.	b
(iv)	Let a, b and c be the sides of a triangle. If A, B and C are the angles opposite to a, b and c respectively, then $\Delta = \begin{vmatrix} a^2 & b\sin A & c\sin A \\ b\sin A & 1 & \cos A \\ c\sin A & \cos A & 1 \end{vmatrix}$ is independent of	D.	a

58 Match the following lists

	List I		List II
(i)	If A is an idempotent matrix and I is an identity matrix of the same order, then the value of n such that $(A+I)^n = I + 127A$, is	A.	9
(ii)	If $(I-A)^{-1} = I + A + A^2 + \ldots + A^7$, then $A^n = 0$, where n is	B.	10
(iii)	If A is matrix such that $a_{ij} = (i+j)(i-j)$, then A is singular, if order of matrix is	C.	7
(iv)	If a non-singular matrix A is symmetric, show that A^{-1} is also symmetric, then order of A can be	D.	8

59 Consider the system of equations

$$\pi x + y + z = 1,\ x + \pi y + z = \pi \text{ and } x + y + \pi z = \pi^2$$

Now, match the following lists

	List I		List II
(i)	$\pi = 1$	A.	unique solution
(ii)	$\pi \neq 1$	B.	infinite solution
(iii)	$\pi \neq 1, \pi \neq -2$	C.	no solution
(iv)	$\pi = -2$	D.	inconsistent

Numerical Value Type Questions

Question Stem Directions (Q.Nos. 60-62) The answer to each question is a **NUMERICAL VALUE.**

If $x > m, y > n, z > r$, where $(x, y, z > 0)$ such that $\begin{vmatrix} x & n & r \\ m & y & r \\ m & n & z \end{vmatrix} = 0$.

60 The value of $\dfrac{x}{x-m} + \dfrac{y}{y-n} + \dfrac{z}{z-r}$ is ....

61 The value of $\dfrac{m}{x-m} + \dfrac{n}{y-n} + \dfrac{r}{z-r}$ is ....

62 The greatest value of $\dfrac{xyz}{(x-m)(y-n)(z-r)}$ is ....

63 If $a_1, a_2, a_3, 5, 4, a_6, a_7, a_8, a_9$ are in HP and $D = \begin{vmatrix} a_1 & a_2 & a_3 \\ 5 & 4 & a_6 \\ a_7 & a_8 & a_9 \end{vmatrix}$, then the value of $[D]$ is (where $[\cdot]$ represents the greatest integer function) ....

64 If $f(x)=\begin{vmatrix} 1 & x & x+1 \\ 2x & x(x-1) & (x+1)x \\ 3x(x-1) & x(x-1)(x-2) & (x+1)x(x-1) \end{vmatrix}$, then the value of $f(500)$ is ....

65 If $\begin{vmatrix} x & x+y & x+y+z \\ 2x & 3x+2y & 4x+3y+2z \\ 3x & 6x+3y & 10x+6y+3z \end{vmatrix}=64$, then the value of x is ....

66 Let $D_1=\begin{vmatrix} a & b & a+b \\ c & d & c+d \\ a & b & a-b \end{vmatrix}$ and $D_2=\begin{vmatrix} a & c & a+c \\ b & d & b+d \\ a & c & a+b+c \end{vmatrix}$, then the value of $\left|\dfrac{D_1}{D_2}\right|$, where $b\neq 0$ and $ad\neq bc$, is ....

67 The sum of roots of the equation $\begin{vmatrix} x+2 & 2x+3 & 3x+4 \\ 2x+3 & 3x+4 & 4x+5 \\ 3x+5 & 5x+8 & 10x+17 \end{vmatrix}=0$ is ....

68 If $x=31$, $y=32$ and $z=33$, then the value of $\begin{vmatrix} (x^2+1)^2 & (xy+1)^2 & (xz+1)^2 \\ (xy+1)^2 & (y^2+1)^2 & (yz+1)^2 \\ (xz+1)^2 & (yz+1)^2 & (z^2+1)^2 \end{vmatrix}$ is ....

69 Let $A=[a_{ij}]_{3\times 3}, B=[b_{ij}]_{3\times 3}$ and $C=[c_{ij}]_{3\times 3}$ be any three matrices, where $b_{ij}=3^{i-j}\cdot a_{ij}$ and $c_{ij}=4^{i-j}\cdot b_{ij}$. If det $A=2$, then det (BC) is equal to ....

70 Let k be a positive real number and $A=\begin{bmatrix} 2k-1 & 2\sqrt{k} & 2\sqrt{k} \\ 2\sqrt{k} & 1 & -2k \\ -2\sqrt{k} & 2k & -1 \end{bmatrix}$ and $B=\begin{bmatrix} 0 & 2k-1 & \sqrt{k} \\ 1-2k & 0 & 2 \\ -\sqrt{k} & -2\sqrt{k} & 0 \end{bmatrix}$. If det (adj A) + det (adj B) $=10^6$, then $[k]$ is equal to ....

[**Note** adj M denotes the adjoint of a square matrix M and $[k]$ denotes the largest integer less than or equal to k]

71 For a real number α, if the system $\begin{bmatrix} 1 & \alpha & \alpha^2 \\ \alpha & 1 & \alpha \\ \alpha^2 & \alpha & 1 \end{bmatrix}\begin{bmatrix} x \\ y \\ z \end{bmatrix}=\begin{bmatrix} 1 \\ -1 \\ 1 \end{bmatrix}$ of linear equations has infinitely many solutions, then $1+\alpha+\alpha^2$ is equal to .... **[JEE Advanced 2017]**

72 Let P be a matrix of order 3×3 such that all the entries in P are from the set $\{-1, 0, 1\}$. Then, the maximum possible values of the determinant of P is .... **[JEE Advanced 2018]**

ANSWER KEY

Level 1

1. (i) 1 (ii) 0 **2.** $\pm 2\sqrt{2}$ **3.** (i) 0 (ii) -52 **4.** 0

5. (i) 0 (ii) 0 (iii) 0 (iv) $-5\sqrt{3}(5-\sqrt{6})$ (v) 1 (vi) 5120000

7. 0 **8.** 0 **12.** -1725 **13.** -6

14. (i) $M_{11}=-1, M_{21}=20; C_{11}=-1, C_{21}=-20; -5$

(ii) $M_{11}=-12, M_{21}=-16, M_{31}=-4; C_{11}=-12, C_{21}=16, C_{31}=-4; 40$

(iii) $M_{11}=ab^2-ac^2, M_{21}=a^2b-c^2b, M_{31}=a^2c-b^2c; C_{11}=ab^2-ac^2, C_{21}=c^2b-a^2b, C_{31}=a^2c-b^2c;$
$ab^2-ac^2+c^2b-a^2b+a^2c-b^2c$

15. $M_{11}=-20, M_{12}=46, M_{13}=30, M_{21}=-4, M_{22}=-19, M_{23}=13, M_{31}=-12, M_{32}=-22, M_{33}=18$
and $A_{11}=-20, A_{12}=46, A_{13}=30, A_{21}=4, A_{22}=-19, A_{23}=-13, A_{31}=-12, A_{32}=22, A_{33}=18$

16. (i) $\begin{bmatrix} 1 & -3 \\ -4 & 2 \end{bmatrix}$ (ii) $\begin{bmatrix} 9 & -1 & -4 \\ -3 & 4 & 5 \\ 5 & -3 & -1 \end{bmatrix}$ **17.** 0

18. (i) 3 (ii) 15 (iii) 15 (iv) 15 (v) 24 (vi) 135

19. (i) 64 (ii) 11664 (iii) $16\begin{bmatrix} -1 & 1 & 1 \\ 1 & -1 & 1 \\ 1 & 1 & 1 \end{bmatrix}$

20. (i) 16 (ii) 256 (iii) 11664 **21.** Non-singular

22. (i) $-\frac{1}{23}\begin{bmatrix} 5 & -4 \\ -2 & -3 \end{bmatrix}$ (ii) $\frac{1}{52}\begin{bmatrix} -6 & 14 & -6 \\ -33 & 25 & -7 \\ 10 & -6 & -10 \end{bmatrix}$ **23.** $\begin{bmatrix} \frac{1}{5} & 0 & 0 \\ 0 & \frac{1}{6} & 0 \\ 0 & 0 & \frac{1}{7} \end{bmatrix}$

24. (i) 2, 3 (ii) $2 \pm \sqrt{2}i$ **25.** (i) 3, 5 (ii) $\frac{1}{3}, \frac{1}{5}$ (iii) 9, 25 **26.** $\lambda^2-4\lambda+1=0$ and $\begin{bmatrix} 2 & -1 \\ -3 & 2 \end{bmatrix}$

27. (i) $\frac{75}{2}$ sq units (ii) $\frac{47}{2}$ sq units **28.** Infinite values of a

29. (i) $a \neq \pm 1, b \neq 0$ (ii) $(1,-2), (1,-1), (-1,-2), (-1,-1)$

30. 2 **31.** $x=3, y=1$ **32.** $x=1, y=2, z=3$ **33.** $x=2, y=1, z=3$ **34.** $x=1, y=1, z=1$

35. $\begin{bmatrix} -3 & -2 & -4 \\ 2 & 1 & 2 \\ 2 & 1 & 3 \end{bmatrix}$ and $x=0, y=-5, z=-3$ **36.** $\begin{bmatrix} 6 & 0 & 0 \\ 0 & 6 & 0 \\ 0 & 0 & 6 \end{bmatrix}$ and $x=2, y=-1, z=4$

37. 1, 2 and 3 **38.** $x=-6, y=-5$ **39.** $x=1, y=2, z=3$

40. (i) $p \neq 2, q=3$ (ii) $p \neq 2, q \neq 3$ (iii) $p=2, q \in R$ **41.** 0 **42.** 0

Level 2

1. (a)	**2.** (d)	**3.** (d)	**4.** (c)	**5.** (d)	**6.** (d)	**7.** (d)
8. (b)	**9.** (b)	**10.** (b)	**11.** (d)	**12.** (b)	**13.** (c)	**14.** (a)
15. (c)	**16.** (d)	**17.** (a)	**18.** (d)	**19.** (d)	**20.** (a)	**21.** (b)
22. (d)	**23.** (c)	**24.** (b)	**25.** (c)	**26.** (b)	**27.** (a)	**28.** (b)
29. (a)	**30.** (b)	**31.** (a)	**32.** (a)	**33.** (b)	**34.** (a)	**35.** (c)
36. (b)	**37.** (c)	**38.** (a)	**39.** (b)	**40.** (d)	**41.** (b)	**42.** (c)
43. (c)	**44.** (c)	**45.** (a)	**46.** (a)	**47.** (a)	**48.** (c)	**49.** (c)
50. (a)	**51.** (a)	**52.** (b)	**53.** (c)	**54.** (a)	**55.** (8)	**56.** (6)
57. (25)	**58.** (1)	**59.** (0)	**60.** (2)	**61.** (1)	**62.** (36)	**63.** (27)
64. (6)	**65.** (2)	**66.** (3)				

Level 3

1. (c) **2.** (a) **3.** (a) **4.** (b) **5.** (c) **6.** (a) **7.** (b)
8. (b) **9.** (a) **10.** (a) **11.** (a) **12.** (b) **13.** (a) **14.** (a)
15. (a) **16.** (b) **17.** (b) **18.** (d) **19.** (b) **20.** (c) **21.** (d)
22. (c) **23.** (a) **24.** (a) **25.** (a, b, d) **26.** (a, b,c) **27.** (a, c) **28.** (a, b, c)
29. (a, b, c) **30.** (c, d) **31.** (a, b) **32.** (a, b) **33.** (a, c) **34.** (b, c) **35.** (a, c)
36. (a, b, c) **37.** (a, b, c) **38.** (c,d) **39.** (a, b, c) **40.** (c, d) **41.** (b, d) **42.** (a, c, d)
43. (b, c, d) **44.** (a, d) **45.** (a, d) **46.** (a, c) **47.** (a, b, c) **48.** (c) **49.** (d)
50. (b) **51.** (d) **52.** (c) **53.** (a) **54.** (b) **55.** (b)
56. (i)–D, (ii)–A, (iii)–B, (iv)–D
57. (i)–A, B, C, D (ii)–C, (iii)–A, C, D, (iv)–A, B, C, D
58. (i)–C, (ii)–D, (iii)–A, C (iv)–A, B, C, D
59. (i)–B, (ii)–A, (iii)–A, (iv)–D
60. (2) **61.** (−1) **62.** (0.296) **63.** (2) **64.** (0) **65.** (4) **66.** (2)
67. (−4) **68.** (8) **69.** (4) **70.** (4) **71.** (1) **72.** (4)

SOLUTIONS

Level I

1. (i) Let $A = \begin{vmatrix} \cos\theta & -\sin\theta \\ \sin\theta & \cos\theta \end{vmatrix}$

$= \cos\theta \times \cos\theta - (-\sin\theta) \times \sin\theta$

$= \cos^2\theta + \sin^2\theta$

We know that

$\cos^2\theta + \sin^2\theta = 1$

$\therefore \quad |A| = 1$

(ii) Let $A = \begin{vmatrix} \cos 15^\circ & \sin 15^\circ \\ \sin 75^\circ & \cos 75^\circ \end{vmatrix}$

$= \cos 15^\circ \times \cos 75^\circ - \sin 15^\circ \times \sin 75^\circ$

We know that

$\cos(A+B) = \cos A \cos B - \sin A \sin B$

By substituting this, we get

$|A| = \cos(75^\circ + 15^\circ)$

$\Rightarrow \quad |A| = \cos 90^\circ$

$\Rightarrow \quad |A| = 0$

2. We have, $\begin{vmatrix} 3 & x \\ x & 1 \end{vmatrix} = \begin{vmatrix} 3 & 2 \\ 4 & 1 \end{vmatrix}$

i.e. $\quad 3 - x^2 = 3 - 8$

i.e. $\quad x^2 = 8$

Hence, $x = \pm 2\sqrt{2}$

3. (i) Given, $\Delta = \begin{vmatrix} 0 & \sin\alpha & -\cos\alpha \\ -\sin\alpha & 0 & \sin\beta \\ \cos\alpha & -\sin\beta & 0 \end{vmatrix}$

Expanding along the first row,

$$|A| = 0\begin{vmatrix} 0 & \sin\beta \\ -\sin\beta & 0 \end{vmatrix} - \sin\alpha\begin{vmatrix} -\sin\alpha & \sin\beta \\ \cos\alpha & 0 \end{vmatrix} - \cos\alpha\begin{vmatrix} -\sin\alpha & 0 \\ \cos\alpha & -\sin\beta \end{vmatrix}$$

$\Rightarrow |A| = 0 - \sin\alpha(0 - \sin\beta\cos\alpha) - \cos\alpha\{(-\sin\alpha)(-\sin\beta) - 0\}$

$\Rightarrow |A| = \sin\alpha\sin\beta\cos\alpha - \cos\alpha\sin\alpha\sin\beta$

$\Rightarrow |A| = 0$

(ii) Here in the third column, two entries are zero. So, expanding along the third column, we get

$$\Delta = 4\begin{vmatrix} -1 & 3 \\ 4 & 1 \end{vmatrix} - 0\begin{vmatrix} 1 & 2 \\ 4 & 1 \end{vmatrix} + 0\begin{vmatrix} 1 & 2 \\ -1 & 3 \end{vmatrix}$$

$= 4(-1-12) - 0 + 0 = -52$

4. From the symmetry of the determinant, it is easier to expand by Sarrus rule.

$\Delta = x^3 + abc - abc + (b^2x + a^2x + c^2x) = 0$

or $\quad x^3 + x(a^2 + b^2 + c^2) = 0$

$\Rightarrow \quad x = 0$

or $\quad x^2 = -(a^2 + b^2 + c^2)$

$\Rightarrow \quad x = 0$

or $\quad x = \pm i\sqrt{a^2 + b^2 + c^2}$

5. (i) $\begin{vmatrix} 102 & 18 & 36 \\ 1 & 3 & 4 \\ 17 & 3 & 6 \end{vmatrix} = \begin{vmatrix} 6(17) & 6(3) & 6(6) \\ 1 & 3 & 4 \\ 17 & 3 & 6 \end{vmatrix}$

$= 6\begin{vmatrix} 17 & 3 & 6 \\ 1 & 3 & 4 \\ 17 & 3 & 6 \end{vmatrix} = 0 \quad [\because R_1 \text{ and } R_3 \text{ are identical}]$

(ii) Let $\Delta = \begin{vmatrix} 41 & 1 & 5 \\ 79 & 7 & 9 \\ 29 & 5 & 3 \end{vmatrix}$

Applying $C_1 \to C_1 - 8C_3$, we get

$\Delta = \begin{vmatrix} 1 & 1 & 5 \\ 7 & 7 & 9 \\ 5 & 5 & 3 \end{vmatrix}$

$\therefore \Delta = 0 \quad [\because C_1 \text{ and } C_2 \text{ are identical}]$

(iii) Given, $\begin{vmatrix} 6 & -3 & 2 \\ 2 & -1 & 2 \\ -10 & 5 & 2 \end{vmatrix}$

Let $\Delta = \begin{vmatrix} 6 & -3 & 2 \\ 2 & -1 & 2 \\ -10 & 5 & 2 \end{vmatrix}$

Applying row operations,

$R_1 \to R_1 - 3R_2$ and $R_3 \to R_3 + 5R_2$, we get

$\Rightarrow \quad \Delta = \begin{vmatrix} 0 & 0 & -4 \\ 2 & -1 & 2 \\ 0 & 0 & 12 \end{vmatrix} = 0$

(iv) We have,

$\Delta = \begin{vmatrix} \sqrt{13}+\sqrt{3} & 2\sqrt{5} & \sqrt{5} \\ \sqrt{15}+\sqrt{26} & 5 & \sqrt{10} \\ 3+\sqrt{65} & \sqrt{15} & 5 \end{vmatrix}$

Taking $\sqrt{5}$ common from C_2 and C_3, we get

$\Delta = (\sqrt{5})^2\begin{vmatrix} \sqrt{13}+\sqrt{3} & 2 & 1 \\ \sqrt{15}+\sqrt{26} & \sqrt{5} & \sqrt{2} \\ 3+\sqrt{65} & \sqrt{3} & \sqrt{5} \end{vmatrix}$

Applying $C_1 \to C_2 - \sqrt{3}C_2 - \sqrt{13}C_3$, we get

$$\Delta = 5\begin{vmatrix} -\sqrt{3} & 2 & 1 \\ 0 & \sqrt{5} & \sqrt{2} \\ 0 & \sqrt{3} & \sqrt{5} \end{vmatrix} = -5\sqrt{3}\,(5-\sqrt{6})$$

[expanding along C_1]

(v) We have, $\begin{vmatrix} 1 & 1 & 1 & 1 \\ 1 & 2 & 3 & 4 \\ 1 & 3 & 6 & 10 \\ 1 & 4 & 10 & 20 \end{vmatrix} = \begin{vmatrix} 1 & 1 & 1 & 1 \\ 0 & 1 & 2 & 3 \\ 0 & 2 & 5 & 9 \\ 0 & 3 & 9 & 19 \end{vmatrix}$

[applying $R_2 \to R_2 - R_1, R_3 \to R_3 - R_1, R_4 \to R_4 - R_1$]

$$= \begin{vmatrix} 1 & 2 & 3 \\ 2 & 5 & 9 \\ 3 & 9 & 19 \end{vmatrix}$$

$$= \begin{vmatrix} 1 & 2 & 3 \\ 0 & 1 & 3 \\ 0 & 3 & 10 \end{vmatrix}$$

[applying $R_2 \to R_2 - 2R_1, R_3 \to R_3 - 3R_1$]

$= (10-9) = 1$

(vi) Let $\Delta = \begin{vmatrix} 1 & 3 & 9 & 27 \\ 3 & 9 & 27 & 1 \\ 9 & 27 & 1 & 3 \\ 27 & 1 & 3 & 9 \end{vmatrix}$

$$\Rightarrow \Delta = \begin{vmatrix} 1 & 3 & 3^2 & 3^3 \\ 3 & 3^2 & 3^3 & 1 \\ 3^2 & 3^3 & 1 & 3 \\ 3^3 & 1 & 3 & 3^2 \end{vmatrix}$$

Applying $C_1 \to C_1 + C_2 + C_3 + C_4$, we get

$$\begin{vmatrix} 1+3+3^2+3^3 & 3 & 3^2 & 3^3 \\ 1+3+3^2+3^3 & 3^2 & 3^3 & 1 \\ 1+3+3^2+3^3 & 3^3 & 1 & 3 \\ 1+3+3^2+3^3 & 1 & 3 & 3^2 \end{vmatrix}$$

$$\Rightarrow \Delta = (1+3+3^2+3^3)\begin{vmatrix} 1 & 3 & 3^2 & 3^3 \\ 1 & 3^2 & 3^3 & 1 \\ 1 & 3^3 & 1 & 3 \\ 1 & 1 & 3 & 3^2 \end{vmatrix}$$

Now, applying $R_2 \to R_2 - R_1, R_3 \to R_3 - R_1$ and $R_4 \to R_4 - R_1$, we get

$$\Delta = (1+3+3^2+3^3)\begin{vmatrix} 1 & 3 & 3^2 & 3^3 \\ 0 & 3^2-3 & 3^3-3^2 & 1-3^3 \\ 0 & 3^3-3 & 1-3^2 & 3-3^3 \\ 0 & 1-3 & 3-3^2 & 3^2-3^3 \end{vmatrix}$$

$$\Rightarrow \Delta = (1+3+3^2+3^3)\begin{vmatrix} 6 & 18 & -26 \\ 24 & -8 & -24 \\ -2 & -6 & -18 \end{vmatrix}$$

[on expanding with first column]

$$\Rightarrow \Delta = (1+3+3^2+3^3)2^3\begin{vmatrix} 3 & -9 & 13 \\ 12 & 4 & 12 \\ -1 & 3 & 9 \end{vmatrix}$$

Now, applying $R_1 \to R_1 + 3R_3$, we get

$$\Rightarrow \Delta = (1+3+3^2+3^3)2^3\begin{vmatrix} 0 & 0 & 40 \\ 12 & 4 & 12 \\ -1 & 3 & 9 \end{vmatrix}$$

$= (1+3+3^2+3^3)2^3\,[40\{36-(-4)\}]$

$= (40)(8)(40)(40)$

$= 512000$

So, $\Delta = 5120000$

6. Applying $R_1 \to R_1 + R_2$ to Δ, we get

$$\Delta = \begin{vmatrix} x+y+z & x+y+z & x+y+z \\ z & x & y \\ 1 & 1 & 1 \end{vmatrix}$$

Since, the element of R_1 and R_3 are proportional, $\Delta = 0$

7. Let $A = \begin{vmatrix} 1 & x & x^2 \\ 1 & y & y^2 \\ 1 & z & z^2 \end{vmatrix} + \begin{vmatrix} 1 & 1 & 1 \\ yz & zx & xy \\ x & y & z \end{vmatrix}$

$$\Rightarrow A = \begin{vmatrix} 1 & x & x^2 \\ 1 & y & y^2 \\ 1 & z & z^2 \end{vmatrix} + \begin{vmatrix} 1 & yz & x \\ 1 & zx & y \\ 1 & xy & z \end{vmatrix} \qquad [\because |\Delta_1| = |\Delta_1|^T]$$

If any two rows or columns of the determinant are interchanged, then determinant changes its sign

$$\Rightarrow A = \begin{vmatrix} 1 & x & x^2 \\ 1 & y & y^2 \\ 1 & z & z^2 \end{vmatrix} - \begin{vmatrix} 1 & x & yz \\ 1 & y & zx \\ 1 & z & xy \end{vmatrix}$$

$$\Rightarrow A = \begin{vmatrix} 0 & 0 & x^2-yz \\ 0 & 0 & y^2-zx \\ 0 & 0 & z^2-xy \end{vmatrix} = 0$$

So, $A = \Delta + \Delta_1 = 0$

8. Since, given relation is an identity, it is true for all real values of λ.

So, putting $\lambda = 0$ on both sides, we get

$$t = \begin{vmatrix} 0 & -1 & 3 \\ 1 & 0 & -4 \\ -3 & 4 & 0 \end{vmatrix}$$

which is skew-symmetric determinant.

So, $t = 0$

9. Let $\Delta = \begin{vmatrix} b+c & a & a \\ b & c+a & b \\ c & c & a+b \end{vmatrix}$

Applying $R_1 \rightarrow R_1 - R_2 - R_3$, we get

$$\Delta = \begin{vmatrix} 0 & -2c & -2b \\ b & c+a & b \\ c & c & a+b \end{vmatrix}$$

Expanding along R_1, we obtain

$$\Delta = 0\begin{vmatrix} c+a & b \\ c & a+b \end{vmatrix} - (-2c)\begin{vmatrix} b & b \\ c & a+b \end{vmatrix} + (-2b)\begin{vmatrix} b & c+a \\ c & c \end{vmatrix}$$

$$= 2c(ab + b^2 - bc) - 2b(bc - c^2 - ac)$$

$$= 2abc + 2cb^2 - 2c^2b - 2b^2c + 2bc^2 + 2abc$$

$$= 4abc$$

10. Given, $\begin{vmatrix} a & b & c \\ a-b & b-c & c-a \\ b+c & c+a & a+b \end{vmatrix}$

$$\text{LHS} = \begin{vmatrix} a & b & c \\ a-b & b-c & c-a \\ b+c & c+a & a+b \end{vmatrix}$$

Applying $C_1 \rightarrow C_1 + C_2 + C_3$, we get

$$= \begin{vmatrix} a+b+c & b & c \\ 0 & b-c & c-a \\ 2(a+b+c) & c+a & a+b \end{vmatrix}$$

Taking $(a+b+c)$ common from C_1, we get

$$= (a+b+c)\begin{vmatrix} 1 & b & c \\ 0 & b-c & c-a \\ 2 & c+a & a+b \end{vmatrix}$$

Applying $R_3 \rightarrow R_3 - 2R_1$, we get

$$= (a+b+c)\begin{vmatrix} 1 & b & c \\ 0 & b-c & c-a \\ 0 & c+a-2b & a+b-2c \end{vmatrix}$$

$$= (a+b+c)[(b-c)(a+b-2c) - (c-a)(c+a-2b)]$$

$$= a^3 + b^3 + c^3 - 3abc$$

As, LHS = RHS

11. Taking out factors a, b and c common from R_1, R_2 and R_3, we get

$$\text{LHS} = abc\begin{vmatrix} \frac{1}{a}+1 & \frac{1}{a} & \frac{1}{a} \\ \frac{1}{b} & \frac{1}{b}+1 & \frac{1}{b} \\ \frac{1}{c} & \frac{1}{c} & \frac{1}{c}+1 \end{vmatrix}$$

Applying $R_1 \rightarrow R_1 + R_2 + R_3$, we get

$$\Delta = abc\begin{vmatrix} 1+\frac{1}{a}+\frac{1}{b}+\frac{1}{c} & 1+\frac{1}{a}+\frac{1}{b}+\frac{1}{c} & 1+\frac{1}{a}+\frac{1}{b}+\frac{1}{c} \\ \frac{1}{b} & \frac{1}{b}+1 & \frac{1}{b} \\ \frac{1}{c} & \frac{1}{c} & \frac{1}{c}+1 \end{vmatrix}$$

$$= abc\left(1+\frac{1}{a}+\frac{1}{b}+\frac{1}{c}\right)\begin{vmatrix} 1 & 1 & 1 \\ \frac{1}{b} & \frac{1}{b}+1 & \frac{1}{b} \\ \frac{1}{c} & \frac{1}{c} & \frac{1}{c}+1 \end{vmatrix}$$

Now, applying $C_2 \rightarrow C_2 - C_1, C_3 \rightarrow C_3 - C_1$, we get

$$\Delta = abc\left(1+\frac{1}{a}+\frac{1}{b}+\frac{1}{c}\right)\begin{vmatrix} 1 & 0 & 0 \\ \frac{1}{b} & 1 & 0 \\ \frac{1}{c} & 0 & 1 \end{vmatrix}$$

$$= abc\left(1+\frac{1}{a}+\frac{1}{b}+\frac{1}{c}\right)[1(1-0)]$$

$$= abc\left(1+\frac{1}{a}+\frac{1}{b}+\frac{1}{c}\right)$$

$$= abc + bc + ca + ab = \text{RHS}$$

12. Let $\Delta = \begin{vmatrix} 1 & 2 & 3 \\ -2 & 3 & 2 \\ 3 & 4 & -4 \end{vmatrix} \times \begin{vmatrix} -2 & 1 & 3 \\ 3 & -2 & 1 \\ 2 & 1 & -2 \end{vmatrix}$

$$= \begin{vmatrix} -2+2+9 & 3-4+3 & 2+2-6 \\ 4+3+6 & -6-6+2 & -4+3-4 \\ -6+4-12 & 9-8-4 & 6+4+8 \end{vmatrix}$$

$$= \begin{vmatrix} 9 & 2 & -2 \\ 13 & -10 & -5 \\ -14 & -3 & 18 \end{vmatrix}$$

Applying $C_1 \rightarrow C_1 + C_3$ and $C_2 \rightarrow C_2 + C_3$, we get

$$= \begin{vmatrix} 7 & 0 & -2 \\ 8 & -15 & -5 \\ 4 & 15 & 18 \end{vmatrix}$$

Applying $R_2 \rightarrow R_2 + R_3$, then

$$\Delta = \begin{vmatrix} 7 & 0 & -2 \\ 12 & 0 & 13 \\ 4 & 15 & 18 \end{vmatrix}$$

Expanding along C_2, we get

$$\Delta = -15\begin{vmatrix} 7 & -2 \\ 12 & 13 \end{vmatrix}$$

$$= -15(91 + 24) = -15 \times 115$$

$\therefore \quad \Delta = -1725$

13. Since, 6 lies in the second row and third column, its minor M_{23} is given by

$$M_{23} = \begin{vmatrix} 1 & 2 \\ 7 & 8 \end{vmatrix}$$

[obtained by deleting R_2 and C_3 in Δ]

$$= 8 - 14 = -6$$

14. (i) Given, $A=\begin{bmatrix}5 & 20\\0 & -1\end{bmatrix}$

From the given matrix, we have

$M_{11}=-1$

and $M_{21}=20$

$C_{11}=(-1)^{1+1}\times M_{11}=1\times(-1)=-1$

and $C_{21}=(-1)^{2+1}\times M_{21}=(-1)\times 20=-20$

Now, expanding along C_1, we get

$|A|=a_{11}\times C_{11}+a_{21}\times C_{21}$

$=5\times(-1)+0\times(-20)=-5$

(ii) Given, $A=\begin{bmatrix}1 & -3 & 2\\4 & -1 & 2\\3 & 5 & 2\end{bmatrix}$

From given matrix, we have

$M_{11}=\begin{vmatrix}-1 & 2\\5 & 2\end{vmatrix}$

$\Rightarrow M_{11}=-1\times 2-5\times 2$

$\Rightarrow M_{11}=-12$

$M_{21}=\begin{vmatrix}-3 & 2\\5 & 2\end{vmatrix}$

$\Rightarrow M_{21}=-3\times 2-5\times 2$

$\Rightarrow M_{21}=-16$

and $M_{31}=\begin{vmatrix}-3 & 2\\-1 & 2\end{vmatrix}$

$\Rightarrow M_{31}=-3\times 2-(-1)\times 2$

$\Rightarrow M_{31}=-4$

$C_{11}=(-1)^{1+1}\times M_{11}$

$=1\times(-12)=-12$

$C_{21}=(-1)^{2+1}\times M_{21}$

$=-1\times(-16)=16$

and $C_{31}=(-1)^{3+1}\times M_{31}$

$=1\times(-4)=-4$

Now, expanding along C_1, we get

$|A|=a_{11}\times C_{11}+a_{21}\times C_{21}+a_{31}\times C_{31}$

$=1\times(-12)+4\times 16+3\times(-4)$

$=-12+64-12=40$

(iii) Given, $A=\begin{bmatrix}1 & a & bc\\1 & b & ca\\1 & c & ab\end{bmatrix}$

$M_{11}=\begin{bmatrix}b & ca\\c & ab\end{bmatrix}$

$\Rightarrow M_{11}=b\times ab-c\times ca$

$\Rightarrow M_{11}=ab^2-ac^2$

$M_{21}=\begin{bmatrix}a & bc\\c & ab\end{bmatrix}$

$\Rightarrow M_{21}=a\times ab-c\times bc$

$\Rightarrow M_{21}=a^2b-c^2b$

$M_{31}=\begin{bmatrix}a & bc\\b & ca\end{bmatrix}$

$\Rightarrow M_{31}=a\times ca-b\times bc$

$\Rightarrow M_{31}=a^2c-b^2c$

$C_{11}=(-1)^{1+1}\times M_{11}$

$\Rightarrow C_{11}=1\times(ab^2-ac^2)$

$\Rightarrow C_{11}=ab^2-ac^2$

$C_{21}=(-1)^{2+1}\times M_{21}$

$\Rightarrow C_{21}=-1\times(a^2b-c^2b)$

$\Rightarrow C_{21}=c^2b-a^2b$

and $C_{31}=(-1)^{3+1}\times M_{31}$

$\Rightarrow C_{31}=1\times(a^2c-b^2c)$

$\Rightarrow C_{31}=a^2c-b^2c$

Now, expanding along C_1, we get

$|A|=a_{11}\times C_{11}+a_{21}\times C_{21}+a_{31}\times C_{31}$

$=1\times(ab^2-ac^2)+1\times(c^2b-a^2b)+1\times(a^2c-b^2c)$

$=ab^2-ac^2+c^2b-a^2b+a^2c-b^2c$

15. Let $\Delta=\begin{vmatrix}2 & -3 & 5\\6 & 0 & 4\\1 & 5 & -7\end{vmatrix}$

$M_{11}=\begin{vmatrix}0 & 4\\5 & -7\end{vmatrix}=0-20=-20;\ A_{11}=(-1)^{1+1}(-20)=-20$

$M_{12}=\begin{vmatrix}6 & 4\\1 & -7\end{vmatrix}=-42-4=-46;\ A_{12}=(-1)^{1+2}(-46)=46$

$M_{13}=\begin{vmatrix}6 & 0\\1 & 5\end{vmatrix}=30-0=30;\ A_{13}=(-1)^{1+3}(30)=30$

$M_{21}=\begin{vmatrix}-3 & 5\\5 & -7\end{vmatrix}=21-25=-4;\ A_{21}=(-1)^{2+1}(-4)=4$

$M_{22}=\begin{vmatrix}2 & 5\\1 & -7\end{vmatrix}=-14-5=-19;\ A_{22}=(-1)^{2+2}(-19)=-19$

$M_{23}=\begin{vmatrix}2 & -3\\1 & 5\end{vmatrix}=10+3=13;\ A_{23}=(-1)^{2+3}(13)=-13$

$M_{31}=\begin{vmatrix}-3 & 5\\0 & 4\end{vmatrix}=-12-0=-12;\ A_{31}=(-1)^{3+1}(-12)=-12$

$M_{32}=\begin{vmatrix}2 & 5\\6 & 4\end{vmatrix}=8-30=-22;\ A_{32}=(-1)^{3+2}(-22)=22$

$M_{33}=\begin{vmatrix}2 & -3\\6 & 0\end{vmatrix}=0+18=18;\ A_{33}=(-1)^{3+3}(18)=18$

Now, $a_{11}=2$, $a_{12}=-3$, $a_{13}=5$, $A_{31}=-12$, $A_{32}=22$ and $A_{33}=18$

So, $a_{11}A_{31}+a_{12}A_{32}+a_{13}A_{33}$

$=2(-12)+(-3)(22)+(5)(18)$

$=-24-66+90=0$

16. (i) $A=\begin{bmatrix}2 & 3\\ 4 & 1\end{bmatrix}$

$C_{11}=1, C_{12}=-4, C_{21}=-3$ and $C_{22}=2$

$$\text{adj } A=\begin{bmatrix}C_{11} & C_{12}\\ C_{21} & C_{22}\end{bmatrix}^T=\begin{bmatrix}1 & -4\\ -3 & 2\end{bmatrix}^T$$

$$\therefore \text{ adj }(A)=\begin{bmatrix}1 & -3\\ -4 & 2\end{bmatrix}$$

(ii) $A=\begin{bmatrix}1 & 1 & 1\\ 2 & 1 & -3\\ -1 & 2 & 3\end{bmatrix}$

$$C_{11}=\begin{vmatrix}1 & -3\\ 2 & 3\end{vmatrix}=9, C_{12}=-\begin{vmatrix}2 & -3\\ -1 & 3\end{vmatrix}=-3,$$

$$C_{13}=\begin{vmatrix}2 & 1\\ -1 & 2\end{vmatrix}=5, C_{21}=-\begin{vmatrix}1 & 1\\ 2 & 3\end{vmatrix}=-1,$$

$$C_{22}=\begin{vmatrix}1 & 1\\ -1 & 3\end{vmatrix}=-4,$$

$$C_{23}=-\begin{vmatrix}1 & 1\\ -1 & 2\end{vmatrix}=-3, C_{31}=\begin{vmatrix}1 & 1\\ 1 & -3\end{vmatrix}=-4$$

$$C_{32}=-\begin{vmatrix}1 & 1\\ 2 & -3\end{vmatrix}=5 \text{ and } C_{33}=\begin{vmatrix}1 & 1\\ 2 & 1\end{vmatrix}=-1$$

$$\text{adj }(A)=\begin{vmatrix}9 & -3 & 5\\ -1 & 4 & -3\\ -4 & 5 & -1\end{vmatrix}^T=\begin{bmatrix}9 & -1 & -4\\ -3 & 4 & 5\\ 5 & -3 & -1\end{bmatrix}$$

17. Let $|A|=\begin{vmatrix}2 & 3 & 7\\ 13 & 17 & 5\\ 15 & 20 & 12\end{vmatrix}$

$$|A|=2\begin{vmatrix}17 & 5\\ 20 & 12\end{vmatrix}-3\begin{vmatrix}13 & 5\\ 15 & 12\end{vmatrix}+7\begin{vmatrix}13 & 17\\ 15 & 20\end{vmatrix}$$

$$=2(17\times 12-5\times 20)-3(13\times 12-5\times 15)+7(13\times 20-15\times 17)$$

$$=2(204-100)-3(156-75)+7(260-255)$$

$$=2\times 104-3\times 81+7\times 5$$

$$=208-243+35=0$$

Now, $|A|^2=|A|\times|A|$

$$|A|^2=0$$

18. (i) Since, the determinant of A is equal to the determinant of A^T.

$\therefore \det(A^T)=\det(A)=3$

(ii) $\det(AB)=\det(A)\cdot\det(B)=3\times 5=15$

(iii) $\det(A^TB)=\det(A^T)\cdot\det(B)=\det(A)\cdot\det(B)$

$=3\times 5=15$

(iv) $\det(A^TB^T)=\det(A^T)\cdot\det(B^T)=\det(A)\cdot\det(B)$

$=3\times 5=15$

(v) $\det(2A)=2^3\det(A)$

[$\because \det(kA)=k^n\det(A)$, where n-order of matrix A]

$=8\times 3=24$

(vi) $\det(3B)=3^3\det(B)=3^3\times 5=135$

19. $|A|=-1(1-1)-1(-1-1)+1(1+1)=4\neq 0$

(i) $|A||\text{adj }A|=|A||A|^{3-1}=|A||A|^2=|A|^3=4^3=64$

(ii) $|\text{adj}(3A)|=|3^2\text{ adj }A|=(3^2)^3|\text{adj }A|$

$=3^6|A|^2=729\times 4^2=11664$

(iii) $\text{adj}(\text{adj }A)=|A|^2A=16A=16\begin{bmatrix}-1 & 1 & 1\\ 1 & -1 & 1\\ 1 & 1 & 1\end{bmatrix}$

20. $\because$ We know that

$$A\cdot\text{adj}(A)=|A|I_n$$

$\therefore \qquad |A|=4$

(i) $|\text{adj }A|=|A|^{3-1}=|A|^2=(4)^2=16$

[$\because |\text{adj }A|=|A|^{n-1}$, where n-order of the matrix]

(ii) $|\text{adj}(\text{adj }A)|=|A|^{(n-1)^2}=(4)^{(2)^2}=(4)^4=256$

(iii) $|\text{adj}(3A)|=|3^{3-1}\text{ adj}(A)|=|3^2\text{ adj}(A)|=|9\text{ adj}(A)|$

$=9^3|A|^{3-1}=9^3\times|A|^2=9^3\times 16=11664$

21. Let $A=\begin{bmatrix}4 & -1 & 0\\ 2 & 3 & 5\\ -1 & 7 & 2\end{bmatrix}$

$$|A|=4\begin{vmatrix}3 & 5\\ 7 & 2\end{vmatrix}+1\begin{vmatrix}2 & 5\\ -1 & 2\end{vmatrix}+0\begin{vmatrix}2 & 3\\ -1 & 7\end{vmatrix}$$

$$=4(6-35)+1(4+5)+0$$

$$=4(-29)+9=-116+9=-107$$

$\because |A|\neq 0$

$\therefore$ The given matrix is non-singular.

22. (i) $A=\begin{bmatrix}-3 & 4\\ 2 & 5\end{bmatrix}$

$$|A|=-3\times 5-(4\times 2)$$

$$=-15-8=-23$$

$$\text{adj}(A)=\begin{bmatrix}C_{11} & C_{12}\\ C_{21} & C_{22}\end{bmatrix}^T$$

$C_{11}=(-1)^{1+1}\cdot 5, C_{12}=(-1)^{1+2}\cdot 2=-2,$

$C_{21}=(-1)^3(4)=-4$

and $C_{22}=(-1)^{2+2}(-3)=-3$

$$\text{adj}(A)=\begin{bmatrix}5 & -2\\ -4 & -3\end{bmatrix}^T=\begin{bmatrix}5 & -4\\ -2 & -3\end{bmatrix}$$

$$\therefore A^{-1}=\frac{1}{-23}\begin{bmatrix}5 & -4\\ -2 & -3\end{bmatrix} \qquad \left[\because A^{-1}=\frac{\text{adj }A}{|A|}\right]$$

(ii) $A=\begin{bmatrix}4 & -2 & 1\\ 5 & 0 & 3\\ -1 & 2 & 6\end{bmatrix}$

$$|A|=4\begin{vmatrix}0 & 3\\ 2 & 6\end{vmatrix}-(-2)\begin{vmatrix}5 & 3\\ -1 & 6\end{vmatrix}+1\begin{vmatrix}5 & 0\\ -1 & 2\end{vmatrix}$$

$$=4(0-6)+2(30+3)+1(10-0)$$

$$=-24+66+10=52$$

$$\text{adj}(A)=\begin{bmatrix}-6 & -33 & 10\\ 14 & 25 & -6\\ -6 & -7 & 10\end{bmatrix}^T$$

$$\text{adj}(A)=\begin{bmatrix}-6 & 14 & -6\\ -33 & 25 & -7\\ 10 & -6 & 10\end{bmatrix}$$

$$\therefore A^{-1}=\frac{1}{52}\begin{bmatrix}-6 & 14 & -6\\ -33 & 25 & -7\\ 10 & -6 & 10\end{bmatrix}$$

23. $\because$ A is a diagonal matrix.

If $A=\begin{bmatrix}d_1 & 0 & 0\\ 0 & d_2 & 0\\ 0 & 0 & d_3\end{bmatrix}$, then $A^{-1}=\begin{bmatrix}\frac{1}{d_1} & 0 & 0\\ 0 & \frac{1}{d_2} & 0\\ 0 & 0 & \frac{1}{d_3}\end{bmatrix}$.

$$\Rightarrow A^{-1}=\begin{bmatrix}\frac{1}{5} & 0 & 0\\ 0 & \frac{1}{6} & 0\\ 0 & 0 & \frac{1}{7}\end{bmatrix}$$

24. (i) For eigen values,

$$|A-\lambda I|=0$$

$$\Rightarrow \begin{vmatrix}2-\lambda & 0\\ 0 & 3-\lambda\end{vmatrix}=0$$

$$\Rightarrow (2-\lambda)(3-\lambda)=0 \Rightarrow \lambda=2,3$$

(ii) $A=\begin{bmatrix}0 & -2\\ 3 & 4\end{bmatrix}$

$|A-\lambda I|=0$ [for eigen values]

$$\Rightarrow \left|\begin{bmatrix}0 & -2\\ 3 & 4\end{bmatrix}-\lambda\begin{bmatrix}1 & 0\\ 0 & 1\end{bmatrix}\right|=0$$

$$\Rightarrow \begin{vmatrix}-\lambda & -2\\ 3 & 4-\lambda\end{vmatrix}=0$$

$$\Rightarrow -\lambda(4-\lambda)+6=0$$

$$\Rightarrow -4\lambda+\lambda^2+6=0$$

$$\Rightarrow \lambda^2-4\lambda+6=0$$

$$\Rightarrow \lambda=2\pm\sqrt{2}i$$

25. If λ_1 and λ_2 are eigen values of A, then

(i) the eigen values of A^T be λ_1 and λ_2.

(ii) the eigen values of A^{-1} be $\frac{1}{\lambda_1}$ and $\frac{1}{\lambda_2}$.

(iii) the eigen values of A^2 be λ_1^2 and λ_2^2.

$\therefore$ Eigen values of A^T be 3 and 5, the eigen values of A^{-1} be $\frac{1}{3}$ and $\frac{1}{5}$ and the eigen values of A^2 be $(3)^2=9$ and $(5)^2=25$

26. Characteristics equation is

$$|A-\lambda I|=0$$

$$\Rightarrow \begin{vmatrix}2-\lambda & 1\\ 3 & 2-\lambda\end{vmatrix}=0$$

$$\Rightarrow (2-\lambda^2)-3=0$$

$$\Rightarrow \lambda^2-4\lambda+1=0$$

$\therefore$ By Cayley Hamilton theorem, every square matrix satisfies its characteristics equation.

$$\Rightarrow A^2-4A+I=0$$

$$\Rightarrow A^{-1}A^2-4A^{-1}A+A^{-1}I=0 \quad \text{[pre multiplying by } A^{-1}\text{]}$$

$$\Rightarrow A-4I+A^{-1}=0$$

$$\Rightarrow A^{-1}=4I-A=\begin{bmatrix}4 & 0\\ 0 & 4\end{bmatrix}-\begin{bmatrix}2 & 1\\ 3 & 2\end{bmatrix}$$

$$\Rightarrow A^{-1}=\begin{bmatrix}2 & -1\\ -3 & 2\end{bmatrix}$$

27. (i) $\Delta=\frac{1}{2}\left\|\begin{matrix}x_1 & y_1 & 1\\ x_2 & y_2 & 1\\ x_3 & y_3 & 1\end{matrix}\right\|$

Now, substituting given value in above formula,

$$\Delta=\frac{1}{2}\left\|\begin{matrix}3 & 8 & 1\\ -4 & 2 & 1\\ 5 & -1 & 1\end{matrix}\right\|$$

Expanding along R_1,

$$=\frac{1}{2}\left|\left[3\begin{vmatrix}2 & 1\\ -1 & 1\end{vmatrix}-8\begin{vmatrix}-4 & 1\\ 5 & 1\end{vmatrix}+1\begin{vmatrix}-4 & 2\\ 5 & -1\end{vmatrix}\right]\right|$$

$$=1|[3(3)-8(-9)+1(-6)]|$$

$$=\frac{1}{2}|[9+72-6]|$$

$$=\frac{75}{2} \text{ sq units.}$$

(ii) $\Delta=\frac{1}{2}\left\|\begin{matrix}x_1 & y_1 & 1\\ x_2 & y_2 & 1\\ x_3 & y_3 & 1\end{matrix}\right\|$

Now, substituting given value in above formula,

$$\Delta = \frac{1}{2}\left\|\begin{matrix} 2 & 7 & 1 \\ 1 & 1 & 1 \\ 10 & 8 & 1 \end{matrix}\right\|$$

Expanding along R_1,

$$= \frac{1}{2}\left|\left[2\begin{vmatrix} 1 & 1 \\ 8 & 1 \end{vmatrix} - 7\begin{vmatrix} 1 & 1 \\ 10 & 1 \end{vmatrix} + 1\begin{vmatrix} 1 & 1 \\ 10 & 8 \end{vmatrix}\right]\right|$$

$$= \frac{1}{2}|[2(-7) - 7(-9) + 1(-2)]|$$

$$= \frac{1}{2}|[-14 + 63 - 2]|$$

$$= \frac{47}{2} \text{ sq units}$$

28. Lines are concurrent if $\begin{vmatrix} 2 & 1 & -1 \\ a & 3 & -3 \\ 3 & 2 & -2 \end{vmatrix} = 0$

Since, C_2 and C_3 are proportional, thus lines are concurrent for infinite values of a.

29. $\because$ System of equation is

$a^2x - by = a^2 - b$ and $bx - b^2y = 2 + 4b$

(i) If system has unique solution, then lines must be non-parallel or $\Delta \neq 0$.

Hence, $\begin{vmatrix} a^2 & -b \\ b & -b^2 \end{vmatrix} \neq 0$

$\Rightarrow \quad -a^2b^2 + b^2 \neq 0$

or $\quad b^2(1 - a^2) \neq 0$

$\Rightarrow \quad b \neq 0$ and $a \neq \pm 1$

(ii) If system has infinite solutions, then lines must be coincident.

Hence, $\dfrac{a^2}{b} = \dfrac{b}{b^2} = \dfrac{a^2 - b}{2 + 4b}$

$\Rightarrow \quad b = 0$ or $a = \pm 1 \quad [b = 0$ is not possible$]$

If $a = 1$, then $2 + 4b = b - b^2$

or $\quad b^2 + 3b + 2 = 0$

$\Rightarrow \quad b = -2$ or -1

If $a = -1$, then $b = -2$ or -1

Thus, ordered pairs (a, b) for which system has infinite solutions are

$(1, -2), (1, -1), (-1, -2), (-1, -1)$.

30. $\because$ For non-trivial solution, we must have

$$\begin{vmatrix} 2a & -2 & 3 \\ 1 & a & 2 \\ 2 & 0 & a \end{vmatrix} = 0$$

$\Rightarrow \quad 2a(a^2 - 0) + 2(a - 4) + 3(0 - 2a) = 0$

$\Rightarrow \quad 2a^3 + 2a - 8 + 0 - 6a = 0$

$\Rightarrow \quad 2a^3 - 4a - 8 = 0$

$\Rightarrow \quad a^3 - 2a - 4 = 0$

$\Rightarrow \quad a^3 - 2a^2 + 2a^2 - 4a + 2a - 4 = 0$

$\Rightarrow \quad a^2(a - 2) + 2a(a - 2) + 2(a - 2) = 0$

$\Rightarrow \quad (a - 2)(a^2 + 2a + 2) = 0$

$\Rightarrow \quad a = 2$

31. The system of equations can be written in the form $AX = B$, where

$$A = \begin{bmatrix} 2 & 5 \\ 3 & 2 \end{bmatrix}, X = \begin{bmatrix} x \\ y \end{bmatrix} \text{ and } B = \begin{bmatrix} 1 \\ 7 \end{bmatrix}$$

Now, $|A| = -11 \neq 0$.

Hence, A is non-singular matrix and so has a unique solution.

Note that $A^{-1} = -\dfrac{1}{11}\begin{bmatrix} 2 & -5 \\ -3 & 2 \end{bmatrix}$

Therefore, $X = A^{-1}B = -\dfrac{1}{11}\begin{bmatrix} 2 & -5 \\ -3 & 2 \end{bmatrix}\begin{bmatrix} 1 \\ 7 \end{bmatrix}$

i.e. $\begin{bmatrix} x \\ y \end{bmatrix} = \dfrac{1}{11}\begin{bmatrix} 33 \\ -11 \end{bmatrix}$

Hence, $x = 3$ and $y = -1$

32. The system of equation can be written in the form $AX = B$, where

$$A = \begin{bmatrix} 3 & -2 & 3 \\ 2 & 1 & -1 \\ 4 & -3 & 2 \end{bmatrix}, X = \begin{bmatrix} x \\ y \\ z \end{bmatrix} \text{ and } B = \begin{bmatrix} 8 \\ 1 \\ 4 \end{bmatrix}$$

We see that

$$|A| = 3(2 - 3) + 2(4 + 4) + 3(-6 - 4) = -17 \neq 0$$

Hence, A is non-singular and so its inverse exists.

Now,

$$A_{11} = -1, A_{12} = -8, A_{13} = -10$$
$$A_{21} = -5, A_{22} = -6, A_{23} = 1$$
$$A_{31} = -1, A_{32} = 9 \text{ and } A_{33} = 7$$

Therefore, $A^{-1} = -\dfrac{1}{17}\begin{bmatrix} -1 & -5 & -1 \\ -8 & -6 & 9 \\ -10 & 1 & 7 \end{bmatrix}$

So, $X = A^{-1}B = -\dfrac{1}{17}\begin{bmatrix} -1 & -5 & -1 \\ -8 & -6 & 9 \\ -10 & 1 & 7 \end{bmatrix}\begin{bmatrix} 8 \\ 1 \\ 4 \end{bmatrix}$

i.e. $\begin{bmatrix} x \\ y \\ z \end{bmatrix} = -\dfrac{1}{17}\begin{bmatrix} -17 \\ -34 \\ -51 \end{bmatrix} = \begin{bmatrix} 1 \\ 2 \\ 3 \end{bmatrix}$

Hence, $x = 1, y = 2$ and $z = 3$

33. The given system of equation is

$x + 2y + z = 7;\ x + 0y + 3z = 11$ and $2x - 3y + 0z = 1$

or $\begin{bmatrix} 1 & 2 & 1 \\ 1 & 0 & 3 \\ 2 & -3 & 0 \end{bmatrix}\begin{bmatrix} x \\ y \\ z \end{bmatrix} = \begin{bmatrix} 7 \\ 11 \\ 1 \end{bmatrix}$

or $AX = B$, where

$$A = \begin{bmatrix} 1 & 2 & 1 \\ 1 & 0 & 3 \\ 2 & -3 & 0 \end{bmatrix}, X = \begin{bmatrix} x \\ y \\ z \end{bmatrix} \text{ and } B = \begin{bmatrix} 7 \\ 11 \\ 1 \end{bmatrix}$$

$$\text{Now, } |A| = \begin{vmatrix} 1 & 2 & 1 \\ 1 & 0 & 3 \\ 2 & -3 & 0 \end{vmatrix} = 18$$

So, the given system of equations has a unique solution given by $X = A^{-1}B$. Therefore,

$$\text{adj } A = \begin{bmatrix} 9 & 6 & -3 \\ -3 & -2 & 7 \\ 6 & -2 & -2 \end{bmatrix}^T = \begin{bmatrix} 9 & -3 & 6 \\ 6 & -2 & -2 \\ -3 & 7 & -2 \end{bmatrix}$$

$$\Rightarrow A^{-1} = \frac{1}{|A|} \text{adj } A = \frac{1}{18} \begin{bmatrix} 9 & -3 & 6 \\ 6 & -2 & -2 \\ -3 & 7 & -2 \end{bmatrix}$$

Now, $X = A^{-1} B$

$$\Rightarrow \quad X = \frac{1}{18} \begin{bmatrix} 9 & -3 & 6 \\ 6 & -2 & -2 \\ -3 & 7 & -2 \end{bmatrix} \begin{bmatrix} 7 \\ 11 \\ 1 \end{bmatrix} = \frac{1}{18} \begin{bmatrix} 63-33+6 \\ 42-22-2 \\ -21+77-2 \end{bmatrix}$$

$$\Rightarrow \begin{bmatrix} x \\ y \\ z \end{bmatrix} = \frac{1}{18} \begin{bmatrix} 36 \\ 18 \\ 54 \end{bmatrix} = \begin{bmatrix} 2 \\ 1 \\ 3 \end{bmatrix}$$

$\Rightarrow \quad x = 2, y = 1$ and $z = 3$

34. Given system of equations is

$3x + 2y - 2z = 3$; $x + 2y + 3z = 6$ and $2x - y + z = 2$

or $\qquad AX = B$

$$\text{i.e.} \quad \begin{bmatrix} 3 & 2 & -2 \\ 1 & 2 & 3 \\ 2 & -1 & 1 \end{bmatrix} \begin{bmatrix} x \\ y \\ z \end{bmatrix} = \begin{bmatrix} 3 \\ 6 \\ 2 \end{bmatrix}$$

$\therefore \qquad X = A^{-1}B$

For A^{-1}, cofactors are

$A_{11} = 5, A_{12} = 5, A_{13} = -5,$

$A_{21} = 0, A_{22} = 7, A_{23} = 7,$

$A_{31} = 10, A_{32} = -11$ and $A_{33} = 4$

$$\therefore \text{adj } A = \begin{bmatrix} 5 & 5 & -5 \\ 0 & 7 & 7 \\ 10 & -11 & 4 \end{bmatrix}^T = \begin{bmatrix} 5 & 0 & 10 \\ 5 & 7 & -11 \\ -5 & 7 & 4 \end{bmatrix}$$

$\Rightarrow |A| = 3(5) + 2(5) + (-2)(-5) = 35$

$$\therefore A^{-1} = \frac{\text{adj } A}{|A|} = \frac{1}{35} \begin{bmatrix} 5 & 0 & 10 \\ 5 & 7 & -11 \\ -5 & 7 & 4 \end{bmatrix}$$

Now, $X = A^{-1}B$

$$\begin{bmatrix} x \\ y \\ z \end{bmatrix} = \frac{1}{35} \begin{bmatrix} 5 & 0 & 10 \\ 5 & 7 & -11 \\ -5 & 7 & 4 \end{bmatrix} \begin{bmatrix} 3 \\ 6 \\ 2 \end{bmatrix}$$

$$= \frac{1}{35} \begin{bmatrix} 15+0+20 \\ 15+42-22 \\ -15+42+8 \end{bmatrix} = \frac{1}{35} \begin{bmatrix} 35 \\ 35 \\ 35 \end{bmatrix} = \begin{bmatrix} 1 \\ 1 \\ 1 \end{bmatrix}$$

$\therefore x = 1, y = 1$ and $z = 1$

35. We have, $A = \begin{bmatrix} 1 & 2 & 0 \\ -2 & -1 & -2 \\ 0 & -1 & 1 \end{bmatrix}$ ...(i)

Cofactors are

$A_{11} = -3, A_{12} = 2, A_{13} = 2,$

$A_{21} = -2, A_{22} = 1, A_{23} = 1,$

$A_{31} = -4, A_{32} = 2$ and $A_{33} = 3$

$$\text{adj } A = \begin{bmatrix} -3 & 2 & 2 \\ -2 & 1 & 1 \\ -4 & 2 & 3 \end{bmatrix}^T = \begin{bmatrix} -3 & -2 & -4 \\ 2 & 1 & 2 \\ 2 & 1 & 3 \end{bmatrix}$$

$\Rightarrow \qquad |A| = 1(-3) + 2(2) + 0 = 1$

$$\therefore \qquad A^{-1} = \frac{\text{adj } A}{|A|} = \begin{bmatrix} -3 & -2 & -4 \\ 2 & 1 & 2 \\ 2 & 1 & 3 \end{bmatrix}$$

Now, the system of linear equations is

$x - 2y = 10$, $2x - y - z = 8$ and $-2y + z = 7$

which can be written in matrix form as

$$\begin{bmatrix} 1 & -2 & 0 \\ 2 & -1 & -1 \\ 0 & -2 & 1 \end{bmatrix} \begin{bmatrix} x \\ y \\ z \end{bmatrix} = \begin{bmatrix} 10 \\ 8 \\ 7 \end{bmatrix}$$

i.e. $\quad A^T X = B,$

$$\text{where } A^T = \begin{bmatrix} 1 & -2 & 0 \\ 2 & -1 & -1 \\ 0 & -2 & 1 \end{bmatrix}, X = \begin{bmatrix} x \\ y \\ z \end{bmatrix} \text{ and } B = \begin{bmatrix} 10 \\ 8 \\ 7 \end{bmatrix}$$

Now, $|A^T| = |A| = 1 \neq 0$

$X = (A^T)^{-1}B = (A^{-1})^T B$

$$\Rightarrow \begin{bmatrix} x \\ y \\ z \end{bmatrix} = \begin{bmatrix} -3 & 2 & 2 \\ -2 & 1 & 1 \\ -4 & 2 & 3 \end{bmatrix} \begin{bmatrix} 10 \\ 8 \\ 7 \end{bmatrix} = \begin{bmatrix} -30+16+14 \\ -20+8+7 \\ -40+16+21 \end{bmatrix} = \begin{bmatrix} 0 \\ -5 \\ 3 \end{bmatrix}$$

$\therefore \qquad x = 0, y = -5$ and $z = -3$

36. We have,

$$A = \begin{bmatrix} 2 & 2 & -4 \\ -4 & 2 & -4 \\ 2 & -1 & 5 \end{bmatrix} \text{ and } B = \begin{bmatrix} 1 & -1 & 0 \\ 2 & 3 & 4 \\ 0 & 1 & 2 \end{bmatrix}$$

$$\therefore BA = \begin{bmatrix} 1 & -1 & 0 \\ 2 & 3 & 4 \\ 0 & 1 & 2 \end{bmatrix} \begin{bmatrix} 2 & 2 & -4 \\ -4 & 2 & -4 \\ 2 & -1 & 5 \end{bmatrix} = \begin{bmatrix} 6 & 0 & 0 \\ 0 & 6 & 0 \\ 0 & 0 & 6 \end{bmatrix} = 6I$$

$$\therefore \quad B^{-1}=\frac{A}{6}=\frac{1}{6}\begin{bmatrix}2 & 2 & -4\\ -4 & 2 & -4\\ 2 & -1 & 5\end{bmatrix} \quad \text{...(i)}$$

Given, system of equations is

$x-y=3, 2x+3y+4z=17$ and $y+2z=7$

$$\text{or } \begin{bmatrix}1 & -1 & 0\\ 2 & 3 & 4\\ 0 & 1 & 2\end{bmatrix}\begin{bmatrix}x\\ y\\ z\end{bmatrix}=\begin{bmatrix}3\\ 17\\ 7\end{bmatrix}$$

$$\therefore \begin{bmatrix}x\\ y\\ z\end{bmatrix}=\begin{bmatrix}1 & -1 & 0\\ 2 & 3 & 4\\ 0 & 1 & 2\end{bmatrix}^{-1}\begin{bmatrix}3\\ 17\\ 7\end{bmatrix}=\frac{1}{6}\begin{bmatrix}2 & 2 & -4\\ -4 & 2 & -4\\ 2 & -1 & 5\end{bmatrix}\begin{bmatrix}3\\ 17\\ 7\end{bmatrix}$$

$$=\frac{1}{6}\begin{bmatrix}6+34-28\\ -12+34-28\\ 6-17+35\end{bmatrix}=\frac{1}{6}\begin{bmatrix}12\\ -6\\ 24\end{bmatrix}=\begin{bmatrix}2\\ -1\\ 4\end{bmatrix}$$

$\therefore x=2, y=-1$ and $z=4$

37. Let first, second and third numbers be denoted by x, y and z, respectively. Then, according to given condition, we have

$$x+y+z=6$$
$$y+3z=11$$
and $\quad x+z=2y$ or $x-2y+z=0$

The system can be written as $AX=B$, where

$$A=\begin{bmatrix}1 & 1 & 1\\ 0 & 1 & 3\\ 1 & -2 & 1\end{bmatrix}, X=\begin{bmatrix}x\\ y\\ z\end{bmatrix} \text{ and } B=\begin{bmatrix}6\\ 11\\ 0\end{bmatrix}$$

Here, $|A|=1(1+6)-0+1(3-1)$

$|A|=7+2=9$

Now, we find (adj A)

$A_{11}=1(1+6)=7, A_{12}=-(0-3)=3, A_{13}=-1,$

$A_{21}=-(1+2)=-3, A_{22}=0, A_{23}=-(-2-1)=3,$

$A_{31}=(3-1)=2, A_{32}=-(3-0)=-3$ and $A_{33}=(1-0)=1$

$$\text{Hence, adj } A=\begin{bmatrix}7 & -3 & 2\\ 3 & 0 & -3\\ -1 & 3 & 1\end{bmatrix}$$

$$\text{Thus, } A^{-1}=\frac{1}{|A|}\text{adj}(A)=\frac{1}{9}\begin{bmatrix}7 & -3 & 2\\ 3 & 0 & -3\\ -1 & 3 & 1\end{bmatrix}$$

Since, $X=A^{-1}B$

$$X=\frac{1}{9}\begin{bmatrix}7 & -3 & 2\\ 3 & 0 & -3\\ -1 & 3 & 1\end{bmatrix}\begin{bmatrix}6\\ 11\\ 0\end{bmatrix}$$

$$\text{or } \begin{bmatrix}x\\ y\\ z\end{bmatrix}=\frac{1}{9}\begin{bmatrix}42-33+0\\ 18+0+0\\ -6+33+0\end{bmatrix}=\frac{1}{9}\begin{bmatrix}9\\ 18\\ 27\end{bmatrix}=\begin{bmatrix}1\\ 2\\ 3\end{bmatrix}$$

Thus, $x=1, y=2$ and $z=3$

38. Given, $x-2y=4$

and $-3x+5y=-7$

Now, here we have

$$D=\begin{vmatrix}1 & -2\\ -3 & 5\end{vmatrix}$$

Solving the determinant, we get

$$D=5(1)-(-3)(-2)$$
$$\Rightarrow \quad D=5-6$$
$$\Rightarrow \quad D=-1$$

$$\text{Again, } D_1=\begin{vmatrix}4 & -2\\ -7 & 5\end{vmatrix}$$

Solving the determinant, we get

$$D_1=5(4)-(-7)(-2)$$
$$\Rightarrow \quad D_1=20-14$$
$$\Rightarrow \quad D_1=6$$

$$\text{and} \quad D_2=\begin{vmatrix}1 & 4\\ -3 & -7\end{vmatrix}$$

Solving the determinant, we get

$$D_2=1(-7)-(-3)(4)$$
$$\Rightarrow \quad D_2=-7+12=5$$

Thus, by Cramer's rule, we have

$$x=\frac{D_1}{D}$$
$$\Rightarrow \quad x=\frac{6}{-1}$$
$$\Rightarrow \quad x=-6$$
$$\text{and} \quad y=\frac{D_2}{D}$$
$$\Rightarrow \quad y=\frac{5}{-1}$$
$$\Rightarrow \quad y=-5$$

39. $$\Delta=\begin{vmatrix}1 & 1 & 1\\ 1 & -1 & 1\\ 3 & 2 & -4\end{vmatrix}=\begin{vmatrix}1 & 0 & 0\\ 1 & -2 & 0\\ 3 & -1 & -7\end{vmatrix}=14$$

$$\Delta_x=\begin{vmatrix}6 & 1 & 1\\ 2 & -1 & 1\\ -5 & 2 & -4\end{vmatrix}=\begin{vmatrix}6 & 1 & 1\\ -4 & -2 & 0\\ 19 & 6 & 0\end{vmatrix}=14$$

$$\Delta_y=\begin{vmatrix}1 & 6 & 1\\ 1 & 2 & 1\\ 3 & -5 & -4\end{vmatrix}=\begin{vmatrix}1 & 6 & 1\\ 0 & -4 & 0\\ 0 & -23 & -7\end{vmatrix}=28$$

$$\text{and } \Delta_z=\begin{vmatrix}1 & 1 & 6\\ 1 & -1 & 2\\ 3 & 2 & -5\end{vmatrix}=\begin{vmatrix}1 & 1 & 6\\ 0 & -2 & -4\\ 0 & -1 & -23\end{vmatrix}=42$$

Hence, by Cramer's rule

$$x=\frac{\Delta_x}{\Delta}=1, y=\frac{\Delta_y}{\Delta}=2 \text{ and } z=\frac{\Delta_z}{\Delta}=3$$

40. The given system of equations is

$$2x + py + 6z = 8$$
$$x + 2y + qz = 5$$
and
$$x + y + 3z = 4$$

$$\Delta = \begin{vmatrix} 2 & p & 6 \\ 1 & 2 & q \\ 1 & 1 & 3 \end{vmatrix}$$

On expanding along R_1, we get

$$\Delta = (2-p)(3-q)$$

By Cramer's rule, if $\Delta \neq 0$ i.e. $p \neq 2, q \neq 3$, then system has unique solution. If $p = 2$ or $q = 3$, $\Delta = 0$, then if $\Delta_x = \Delta_y = \Delta_z = 0$, the system has infinite solutions and if atleast one of $\Delta_x, \Delta_y, \Delta_z \neq 0$ system has no solution.

Now,

$$\Delta_x = \begin{vmatrix} 8 & p & 6 \\ 5 & 2 & q \\ 4 & 1 & 3 \end{vmatrix}$$
$$= 30 - 8q - 15p + 4pq$$
$$= (p-2)(4q-15)$$

$$\Delta_y = \begin{vmatrix} 2 & 8 & 6 \\ 1 & 5 & q \\ 1 & 4 & 3 \end{vmatrix} = -8q + 8q = 0$$

$$\Delta_z = \begin{vmatrix} 2 & p & 8 \\ 1 & 2 & 5 \\ 1 & 1 & 4 \end{vmatrix} = p - 2$$

Thus, if $p = 2, \Delta_x = \Delta_y = \Delta_z = 0 \,\forall\, q \in R$

So, the system has infinite solutions.

and if $p \neq 2, q = 3, \Delta_x, \Delta_z \neq 0$, the system has no solution.

Hence, the system has

(i) no solution, if $p \neq 2, q = 3$

(ii) a unique solution, if $p \neq 2, q \neq 3$

(iii) infinitely many solution, $p = 2, q \in R$.

41. $$\frac{d^n}{dx^n}[f(x)] = \begin{vmatrix} \frac{d^n}{dx^n}(x^n) & n! & 2 \\ \frac{d^n}{dx^n}(\cos x) & \cos\frac{n\pi}{2} & 4 \\ \frac{d^n}{dx^n}(\sin x) & \sin\frac{n\pi}{2} & 8 \end{vmatrix}$$

$$= \begin{vmatrix} n! & n! & 2 \\ \cos\left(x + \frac{n\pi}{2}\right) & \cos\frac{n\pi}{2} & 4 \\ \sin\left(x + \frac{n\pi}{2}\right) & \sin\frac{n\pi}{2} & 8 \end{vmatrix}, n \in Z$$

$$\Rightarrow \frac{d^n}{dx^n}[f(x)]_{x=0} = 0 \qquad [\because C_1 \text{ and } C_2 \text{ are same}]$$

42. Taking x common from C_2, then multiplying each elements of R_1 with x, we get

$$\Delta(x) = \begin{vmatrix} x & x^2 & x^3 \\ 6 & 4 & 3 \\ 9 & 1 & -7 \end{vmatrix}$$

$$\therefore \int_0^1 \Delta(x)\,dx = \begin{vmatrix} \int_0^1 x\,dx & \int_0^1 x^2\,dx & \int_0^1 x^3\,dx \\ 6 & 4 & 3 \\ 9 & 1 & -7 \end{vmatrix}$$

$$= \begin{vmatrix} \left[\frac{x^2}{2}\right]_0^1 & \left[\frac{x^3}{3}\right]_0^1 & \left[\frac{x^4}{4}\right]_0^1 \\ 6 & 4 & 3 \\ 9 & 1 & -7 \end{vmatrix}$$

$$= \begin{vmatrix} \frac{1}{2} & \frac{1}{3} & \frac{1}{4} \\ 6 & 4 & 3 \\ 9 & 1 & -7 \end{vmatrix}$$

$$= \frac{1}{12}\begin{vmatrix} 6 & 4 & 3 \\ 6 & 4 & 3 \\ 9 & 1 & -7 \end{vmatrix}$$

[multiplying each elements of R_1 with 12]

$= 0$ [R_1 and R_2 are identical]

Level II

1. (a) $\because P = \begin{bmatrix} 1 & 2 & 1 \\ 1 & 3 & 1 \end{bmatrix}$

$$Q = PP^T = \begin{bmatrix} 1 & 2 & 1 \\ 1 & 3 & 1 \end{bmatrix}\begin{bmatrix} 1 & 1 \\ 2 & 3 \\ 1 & 1 \end{bmatrix} = \begin{bmatrix} 6 & 8 \\ 8 & 11 \end{bmatrix}$$

$$\therefore \quad |Q| = \begin{vmatrix} 6 & 8 \\ 8 & 11 \end{vmatrix} = 66 - 64 = 2$$

2. (d) Given, $A = \begin{bmatrix} a & b \\ c & d \end{bmatrix}$

$$A^2 = \begin{bmatrix} a & b \\ c & d \end{bmatrix}\begin{bmatrix} a & b \\ c & d \end{bmatrix} = \begin{bmatrix} a^2 + bc & ab + bd \\ ac + dc & bc + d^2 \end{bmatrix}$$

$$\therefore A^2 - (a+d)A + kI = O$$

$$\begin{bmatrix} a^2 + bc & ab + bd \\ ac + dc & bc + d^2 \end{bmatrix} - (a+d)\begin{bmatrix} a & b \\ c & d \end{bmatrix} + k\begin{bmatrix} 1 & 0 \\ 0 & 1 \end{bmatrix} = \begin{bmatrix} 0 & 0 \\ 0 & 0 \end{bmatrix}$$

$$\Rightarrow \begin{bmatrix} a^2+bc-a^2-ad+k & ab+bd-ab-bd+0 \\ ac+dc-ac-dc & bc+d^2-ad-d^2+k \end{bmatrix} = \begin{bmatrix} 0 & 0 \\ 0 & 0 \end{bmatrix}$$

$$\Rightarrow \begin{bmatrix} bc-ad+k & 0 \\ 0 & bc-ad+k \end{bmatrix} = \begin{bmatrix} 0 & 0 \\ 0 & 0 \end{bmatrix}$$

$$\Rightarrow \quad bc-ad+k=0$$

$$\Rightarrow \quad k = ad-bc \quad \text{...(i)}$$

Also, given $\begin{vmatrix} a & b \\ c & d \end{vmatrix} = 0$

$$\Rightarrow \quad ad-bc=0 \quad \text{...(ii)}$$

From Eqs. (i) and (ii), we get

$$k=0$$

3. (d) Given, $f(x)=a+bx+cx^2$

and α, β, γ are $1, \omega, \omega^2$ and we know $\omega^3=1$.

Also, $1+\omega+\omega^2=0$

So, $f(\alpha)=f(1)=a+b+c$

$f(\beta)=f(\omega)=a+b\omega+c\omega^2$

and $f(\gamma)=f(\omega^2)=a+b\omega^2+c\omega$

Now, $\begin{vmatrix} a & b & c \\ b & c & a \\ c & a & b \end{vmatrix}$

$$= 3abc-a^3-b^3-c^3$$

$$= -(a+b+c)(a^2+b^2+c^2-ab-bc-ca)$$

$$= -(a+b+c)\{a^2+b^2+c^2+(-1)ab+(-1)bc+(-1)ca\}$$

$$= -(a+b+c)\{a^2+b^2+c^2+(\omega+\omega^2)ab+(\omega+\omega^2)bc + (\omega+\omega^2)ca\}$$

$$= -(a+b+c)(a+b\omega+c\omega^2)(a+b\omega^2+c\omega)$$

$$= -f(1)\,f(\omega)\,f(\omega^2)$$

$$= -f(\alpha)\,f(\beta)\,f(\gamma)$$

4. (c) $f(x)=\begin{vmatrix} 1-2\sin^2 x & \sin^2 x & 1-8\sin^2 x(1-\sin^2 x) \\ \sin^2 x & 1-2\sin^2 x & 1-\sin^2 x \\ 1-8\sin^2 x(1-\sin^2 x) & 1-\sin^2 x & 1-2\sin^2 x \end{vmatrix}$

The required constant term is

$$f(0)=\begin{vmatrix} 1 & 0 & 1 \\ 0 & 1 & 1 \\ 1 & 1 & 1 \end{vmatrix}$$

$$= \begin{vmatrix} 1 & 0 & 0 \\ 0 & 1 & 1 \\ 1 & 1 & 0 \end{vmatrix} = 1(0-1) = -1$$

5. (d) Given, $\begin{vmatrix} 1+\sin^2 x & \sin^2 x & \sin^2 x \\ \cos^2 x & 1+\cos^2 x & \cos^2 x \\ 4\sin 2x & 4\sin 2x & 1+4\sin 2x \end{vmatrix} = 0$

$[0 < x < \pi]$

Applying $R_1 \to R_1+R_2$,

$$\begin{vmatrix} 2 & 2 & 1 \\ \cos^2 x & 1+\cos^2 x & \cos^2 x \\ 4\sin 2x & 4\sin 2x & 1+4\sin 2x \end{vmatrix} = 0$$

Applying $C_1 \to C_1 - C_2$,

$$\begin{vmatrix} 0 & 2 & 1 \\ -1 & 1+\cos^2 x & \cos^2 x \\ 0 & 4\sin 2x & 1+4\sin 2x \end{vmatrix} = 0$$

$$\Rightarrow \quad 2+8\sin 2x-4\sin 2x=0 \quad [\text{expanding along } C_1]$$

$$\Rightarrow \quad 4\sin 2x=-2$$

$$\Rightarrow \quad \sin 2x = -\frac{1}{2}$$

$$\Rightarrow \quad 2x = \pi+\frac{\pi}{6}, 2\pi-\frac{\pi}{6}$$

$$\Rightarrow \quad 2x=\frac{7\pi}{6}, \frac{11\pi}{6} \Rightarrow x=\frac{7\pi}{12}, \frac{11\pi}{12}$$

[**Note** You can also solve by applying $C_1 \to C_1 - C_3$ and $C_2 \to C_2 - C_3$]

6. (d) $P=[a_{ij}]=\begin{bmatrix} a_{11} & a_{12} & a_{13} \\ a_{21} & a_{22} & a_{23} \\ a_{31} & a_{32} & a_{33} \end{bmatrix}$

and $Q=[b_{ij}]=\begin{bmatrix} 4a_{11} & 8a_{12} & 16a_{13} \\ 8a_{21} & 16a_{22} & 32a_{23} \\ 16a_{31} & 32a_{32} & 64a_{33} \end{bmatrix}$ $[\because b_{ij}=2^{i+j}\cdot a_{ij}]$

$$\therefore \quad |Q| = 2^9 \times 2 \times 4 \begin{vmatrix} a_{11} & a_{12} & a_{13} \\ a_{21} & a_{22} & a_{23} \\ a_{31} & a_{32} & a_{33} \end{vmatrix}$$

$$= 2^{12}\cdot|P| = 2^{12}\times 2 = 2^{13}$$

7. (d) Given, $A=\begin{bmatrix} k & k\alpha & \alpha \\ 0 & \alpha & k\alpha \\ 0 & 0 & k \end{bmatrix}$

$$\therefore \quad |A| = \alpha k^2$$

Now, $|A^2|=|A|^2$ $[\because |A^n|=|A|^n]$

$$= (\alpha k^2)^2 = \alpha^2 k^4$$

According to given condition,

$$|A^2| = k^2$$

$$\Rightarrow \quad \alpha^2 k^4 = k^2$$

$$\Rightarrow \quad \alpha^2 = \frac{1}{k^2}$$

$$\Rightarrow \quad |\alpha| = \frac{1}{k}$$

8. (b) Given, $A=\begin{bmatrix}[x+1] & [x+2] & [x+3]\\ [x] & [x+3] & [x+3]\\ [x] & [x+2] & [x+4]\end{bmatrix}$

$$A=\begin{bmatrix}[x]+1 & [x]+2 & [x]+3\\ [x] & [x]+3 & [x]+3\\ [x] & [x]+2 & [x]+4\end{bmatrix}$$

$[\because [x+n]=n+[x], n\in I]$

Applying $R_1\to R_1-R_3, R_2\to R_2-R_3$

$$A=\begin{bmatrix}1 & 0 & -1\\ 0 & 1 & -1\\ [x] & [x]+2 & [x]+4\end{bmatrix}$$

$$\det(A)=1([x]+4+[x]+2)-1(-[x])$$
$$=3[x]+6$$

$\because \quad \det(A)=192$

$\Rightarrow \quad 3[x]+6=192$

$\Rightarrow \quad [x]=62$

$\Rightarrow \quad 62\le x<63$

$\Rightarrow \quad x\in[62,63)$

9. (b) Applying $C_2\to C_2-C_1$ and $C_3\to C_3-C_1$, we get

$$\Delta=\begin{vmatrix}1 & 0 & 0\\ 1 & \sin\theta & 0\\ 1 & 0 & \cos\theta\end{vmatrix}=\sin\theta\cos\theta=(\sin 2\theta)/2$$

So, maximum value of Δ is 1/2.

10. (b) Here, ω is complete cube root of unity.

Applying $R_1\to R_1+R_2+R_3$, the given matrix reduces to

$$\begin{vmatrix}3 & 0 & 0\\ 1 & -\omega^2-1 & \omega^2\\ 1 & \omega^2 & \omega\end{vmatrix}=3(-1-\omega-\omega)=-3z$$

[given, $z=1+2\omega$]

$\Rightarrow \quad k=-z$

11. (d) $\Delta=\begin{vmatrix}x-4 & 2x & 2x\\ 2x & x-4 & 2x\\ 2x & 2x & x-4\end{vmatrix}$

Applying $C_1\to C_1+C_2+C_3$, we get

$$\Delta=\begin{vmatrix}5x-4 & 2x & 2x\\ 5x-4 & x-4 & 2x\\ 5x-4 & 2x & x-4\end{vmatrix}$$

Applying $R_1\to R_1-R_2$ and $R_2\to R_2-R_3$, we get

$$\Delta=\begin{vmatrix}0 & x+4 & 0\\ 0 & -x-4 & x+4\\ 5x-4 & 2x & x-4\end{vmatrix}$$

Expanding along C_1, we get

$$\Delta=(5x-4)(x+4)^2$$

$\therefore \quad A=-4$ and $B=5$

12. (b) On applying $C_2\to C_2-C_1$ and $C_3\to C_3-C_2$, we get

$$\begin{vmatrix}15! & 15\times15! & 16\times16!\\ 16! & 16\times16! & 17\times17!\\ 17! & 17\times17! & 18\times18!\end{vmatrix}$$

$$=(15!)(16!)(17!)\begin{vmatrix}1 & 15 & 16\times16\\ 1 & 16 & 17\times17\\ 1 & 17 & 18\times18\end{vmatrix}$$

On applying $R_1\to R_1-R_2$ and $R_2\to R_2-R_3$, we get

$$=(15!)(16!)(17!)\begin{vmatrix}0 & -1 & -33\\ 0 & -1 & -35\\ 1 & 17 & 18\times18\end{vmatrix}$$

$$=2(15!)(16!)(17!)$$

13. (c) $\begin{vmatrix}a & a+1 & a-1\\ -b & b+1 & b-1\\ c & c-1 & c+1\end{vmatrix}+(-1)^n\begin{vmatrix}a+1 & b+1 & c-1\\ a-1 & b-1 & c+1\\ a & -b & c\end{vmatrix}$

$$=\begin{vmatrix}a & a+1 & a-1\\ -b & b+1 & b-1\\ c & c-1 & c+1\end{vmatrix}+(-1)^n\begin{vmatrix}a+1 & a-1 & a\\ b+1 & b-1 & -b\\ c-1 & c+1 & c\end{vmatrix}$$

[since, $\det(A)=\det(A^T)$]

$$=\begin{vmatrix}a & a+1 & a-1\\ -b & b+1 & b-1\\ c & c-1 & c+1\end{vmatrix}+(-1)^{n+2}\begin{vmatrix}a & a+1 & a-1\\ -b & b+1 & b-1\\ c & c-1 & c+1\end{vmatrix}$$

This is equal to zero only, if $n+2$ is odd i.e. n is an odd integer.

14. (a) $A=\begin{pmatrix}a & b\\ c & d\end{pmatrix}, |A|=ad-bc$

$a,b,c,d=\{-3,-2,-1,0,1,2,3\}$

Case I $ad=9$ and $bc=-6$

$$\begin{cases}a=3, & d=3\\ a=-3, & d=-3\end{cases} \text{ and } \begin{cases}b=2, & c=-3\\ b=-2, & c=3\\ b=3, & c=-2\\ b=-3, & c=2\end{cases}$$

Total possible pairs $=4\times2=8$

Case II $ad=6$ and $bc=-9$

$$\begin{cases}a=2, & d=3\\ a=-2, & d=-3\\ a=3, & d=2\\ a=-3, & d=-2\end{cases}$$

and $\begin{cases}b-3, & c=3\\ b=3, & c=-3\end{cases}$

Total possible pairs $=4\times2=8$

Case I + Case II $=8+8=16$

15. (c) $\begin{vmatrix} 1+1+1 & 1+\alpha+\beta & 1+\alpha^2+\beta^2 \\ 1+\alpha+\beta & 1+\alpha^2+\beta^2 & 1+\alpha^3+\beta^3 \\ 1+\alpha^2+\beta^2 & 1+\alpha^3+\beta^3 & 1+\alpha^4+\beta^4 \end{vmatrix}$

$$=\begin{vmatrix} 1 & 1 & 1 \\ 1 & \alpha & \beta \\ 1 & \alpha^2 & \beta^2 \end{vmatrix}\begin{vmatrix} 1 & 1 & 1 \\ 1 & \alpha & \alpha^2 \\ 1 & \beta & \beta^2 \end{vmatrix}=(1-\alpha)^2(1-\beta)^2(\alpha-\beta)^2$$

$\Rightarrow \quad k=1$

16. (b) Minor of $(-4)=\begin{vmatrix} -2 & 3 \\ 8 & 9 \end{vmatrix}=-42$

Minor of $9=\begin{vmatrix} -1 & -2 \\ -4 & -5 \end{vmatrix}=-3$

Cofactor of $(-4)=(-1)^{2+1}\begin{vmatrix} -2 & 3 \\ 8 & 9 \end{vmatrix}=42$

and cofactor of $9=(-1)^{3+3}\begin{vmatrix} -1 & -2 \\ -4 & -5 \end{vmatrix}=-3$

17. (a) $\text{adj}(A)=\begin{bmatrix} 4y & -x \\ -x^2 & 1 \end{bmatrix}$

$\therefore \text{adj}(A)+B=\begin{bmatrix} 4y & -x \\ -x^2 & 1 \end{bmatrix}+\begin{bmatrix} -3 & 1 \\ 1 & 0 \end{bmatrix}$

$\Rightarrow \begin{bmatrix} 1 & 0 \\ 0 & 1 \end{bmatrix}=\begin{bmatrix} 4y-3 & -x+1 \\ -x^2+1 & 1+0 \end{bmatrix}$

$\Rightarrow \quad 4y-3=1 \Rightarrow y=1$ and $-x+1=0 \Rightarrow x=1$

18. (d) Given, matrix is singular.

$\therefore \begin{vmatrix} 0 & 1 & -2 \\ -1 & 0 & 3 \\ \lambda & -3 & 0 \end{vmatrix}=0$

$\Rightarrow \quad 1(0-6)+3\lambda=0$

$\Rightarrow \quad -6+3\lambda=0 \Rightarrow \lambda=2$

19. (d) Given, $A=\begin{bmatrix} 2-k & 2 \\ 1 & 3-k \end{bmatrix}$

Since, the matrix A is singular.

$\therefore \quad |A|=0$

$\Rightarrow \begin{vmatrix} 2-k & 2 \\ 1 & 3-k \end{vmatrix}=0$

$\Rightarrow \quad (2-k)(3-k)-2=0 \Rightarrow 6-5k+k^2-2=0$

$\Rightarrow \quad 4-5k+k^2=0 \Rightarrow 5k-k^2=4$

20. (a) For the given matrix to be non-singular,

$\begin{vmatrix} 1 & a & b \\ \omega & 1 & c \\ \omega^2 & \omega & 1 \end{vmatrix}\neq 0$

$\Rightarrow 1-(a+c)\omega+ac\omega^2 \neq 0$

$\Rightarrow \quad (1-a\omega)(1-c\omega)\neq 0$

$\Rightarrow \quad a\neq \frac{1}{\omega}, c\neq \frac{1}{\omega}$

$\Rightarrow a=\omega$ and $c=\omega$, where ω is complex cube root of unity.

As a, b and c are complex cube roots of unity.

$\therefore$ a and c can take only one value i.e. ω

While, b can take two values i.e. ω and ω^2.

$\therefore$ Total number of distinct matrices $=1\times 1\times 2=2$

21. (b) For singular matrix, $\begin{vmatrix} -x & x & 2 \\ 2 & x & -x \\ x & -2 & -x \end{vmatrix}=0$

$C_2 \to C_2+C_1$ and $C_3 \to C_3+C_1$, we get

$\begin{vmatrix} -x & 0 & 2-x \\ 2 & 2+x & 2-x \\ x & x-2 & 0 \end{vmatrix}=0$

$\Rightarrow \quad (2-x)\begin{vmatrix} -x & 0 & 1 \\ 2 & 2+x & 1 \\ x & x-2 & 0 \end{vmatrix}=0$

$\Rightarrow (2-x)[-x\{0-(x-2)\}+1(2x-4)-x(2+x)]=0$

$\Rightarrow \quad (2-x)[x^2-2x+2x-4-2x-x^2]=0$

$\Rightarrow \quad (2-x)(-2x-4)=0$

$\Rightarrow \quad -2(2-x)(x+2)=0$

$\Rightarrow \quad x=2,-2$

22. (d) Given, $B=5A^2$

$\det(B)=\det(5A^2)$

$=5^3\det(A^2)$

$=125[\det(A)]^2$

$=125(6)^2 \quad [\because \det(A)=6]$

$=4500$

23. (c) $\because P^3=Q^3$ ...(i)

and $P^2Q=Q^2P$ ...(ii)

On subtracting Eq. (ii) from Eq. (i), we get

$P^3-P^2Q=Q^3-Q^2P$

$\Rightarrow \quad P^2(P-Q)=Q^2(Q-P)$

$\Rightarrow \quad (P-Q)(P^2+Q^2)=0$

$\Rightarrow |(P-Q)(P^2+Q^2)|=0$

$\Rightarrow \quad |P-Q||P^2+Q^2|=0 \quad [\because |AB|=|A||B|]$

$\Rightarrow \quad |P^2+Q^2|=0 \quad [\because P\neq Q]$

24. (b) Given, $A=\begin{bmatrix} 1 & 2 \\ 3 & 5 \end{bmatrix}$

$\Rightarrow \quad |A|=5-6=-1$

$\therefore |A^{2009}-5A^{2008}|=|A^{2008}||A-5I| \quad [\because |AB|=|A|\cdot|B|]$

$=(-1)^{2008}\left|\begin{bmatrix} 1 & 2 \\ 3 & 5 \end{bmatrix}-\begin{bmatrix} 5 & 0 \\ 0 & 5 \end{bmatrix}\right| \quad [|A^n|=|A|^n]$

$=\left|\begin{bmatrix} -4 & 2 \\ 3 & 0 \end{bmatrix}\right|=-6$

25. (c) Given, $A\,(\text{adj}\,A) = 10I$

$\because \quad A \cdot \text{adj}\,(A) = |A|\,I$

$\therefore \quad |A| = 10$

$\Rightarrow |\text{adj}\,A| = |A|^{3-1} = |A|^2 = (10)^2 = 100$

26. (b) $\because AA^T = I$

$\Rightarrow \quad |AA^T| = |I|$

$\Rightarrow |A||A^T| = 1 \Rightarrow |A|\,|A| = 1$

$\Rightarrow \quad |A|^2 = 1 \Rightarrow |A| = \pm 1$

27. (a) Given, $A = \begin{bmatrix} a & 0 & 0 \\ 0 & a & 0 \\ 0 & 0 & a \end{bmatrix}$

Now, $\quad |A| = a^3$

$\therefore |A||\text{adj}\,A| = |A||A|^{n-1} = |A|^n = (a^3)^n = a^{3n}$

28. (b) $|P| = 1(12-12) - \alpha\,(4-6) + 3(4-6) = 2\alpha - 6$

$\because \quad P = \text{adj}\,(A)$

$\therefore \quad |P| = |\text{adj}\,A| = |A|^2 = 16$

$\Rightarrow \quad 2\alpha - 6 = 16$

$\Rightarrow \quad 2\alpha = 22 \Rightarrow \alpha = 11$

29. (a) $\text{adj}\,P = \begin{bmatrix} 1 & 4 & 4 \\ 2 & 1 & 7 \\ 1 & 1 & 3 \end{bmatrix}$

$\Rightarrow |\text{adj}\,P| = \begin{vmatrix} 1 & 4 & 4 \\ 2 & 1 & 7 \\ 1 & 1 & 3 \end{vmatrix}$

$= 1\,(3-7) - 4(6-7) + 4(2-1) = -4 + 4 + 4 = 4$

$\Rightarrow \quad |P|^2 = 4$

$\Rightarrow \quad |P| = \pm 2$

30. (b) Since, A is orthogonal, we have

$A \cdot A^T = I$

or $\quad |AA^T| = 1$

or $\quad |A^2| = 1$

or $\quad |A| = \pm 1$

Now, $|\text{adj} \cdot (\text{adj}\,A)| = |A|^{(n-1)^2}$

31. (a) From given data, $|A| = 2^4$

$\Rightarrow |\text{adj} \cdot (\text{adj} \cdot A)| = (2^4)^9 = 2^{36}$

$\Rightarrow \left\{ \dfrac{\det \cdot \{\text{adj} \cdot (\text{adj} \cdot A)\}}{7} \right\} = \left\{ \dfrac{2^{36}}{7} \right\} = \left\{ \dfrac{(7+1)^{12}}{7} \right\} = \dfrac{1}{7}$

32. (a) We have, $2A^{-1} = A^T$

$\Rightarrow \quad |2A^{-1}| = |A^T|$

$\Rightarrow \quad 2^5 \cdot \dfrac{1}{|A|} = |A| \Rightarrow |A|^2 = 2^5$

$\therefore \quad |\text{adj}\,\{\text{adj}\,(\text{adj}\,A)\}| = |A|^{(n-1)^3} = |A|^{4^3} \qquad [\because n = 5]$

$= |A|^{64} = (|A|^2)^{32} = (2^5)^{32} = 2^{160}$

$= \dfrac{2\,(2^3)^{53}}{7} = \dfrac{2\,(7+1)^{53}}{7} = m + \dfrac{2}{7}$

where m is integer.

Thus, required remainder is 2.

33. (b) $|A| = -4 - (15) = -19$

$\therefore A^{-1} = \dfrac{1}{-19}\begin{bmatrix} -2 & -3 \\ -5 & 2 \end{bmatrix} = \dfrac{1}{19}\begin{bmatrix} 2 & 3 \\ 5 & -2 \end{bmatrix}$

$A^{-1} = \dfrac{1}{19} A$

$\therefore \quad k = \dfrac{1}{19}$

34. (a) $A^{-1} = \dfrac{1}{1+10}\begin{bmatrix} 1 & -2 \\ 5 & 1 \end{bmatrix} = \dfrac{1}{11}\begin{bmatrix} 1 & -2 \\ 5 & 1 \end{bmatrix}$

$\because \quad A^{-1} = xA + yI$

$\Rightarrow \quad \dfrac{1}{11}\begin{bmatrix} 1 & -2 \\ 5 & 1 \end{bmatrix} = \begin{bmatrix} x & 2x \\ -5x & x \end{bmatrix} + \begin{bmatrix} y & 0 \\ 0 & y \end{bmatrix}$

$\Rightarrow \quad x + y = \dfrac{1}{11};\, 2x = \dfrac{-2}{11} \Rightarrow x = \dfrac{-1}{11}$

$\therefore \quad y = \dfrac{1}{11} - x = \dfrac{1}{11} + \dfrac{1}{11} = \dfrac{2}{11}$

35. (c) We have, $\begin{bmatrix} 1 & 3 \\ 0 & 1 \end{bmatrix} A = \begin{bmatrix} 1 & 1 \\ 0 & -1 \end{bmatrix}$

$\Rightarrow \quad A = \begin{bmatrix} 1 & 3 \\ 0 & 1 \end{bmatrix}^{-1} \begin{bmatrix} 1 & 1 \\ 0 & 1 \end{bmatrix}$

$\Rightarrow \quad A = \begin{bmatrix} 1 & -3 \\ 0 & 1 \end{bmatrix} \begin{bmatrix} 1 & 1 \\ 0 & -1 \end{bmatrix}$

$= \begin{bmatrix} 1 & 4 \\ 0 & -1 \end{bmatrix}$

36. (b) We have, $I = AA^{-1} = \dfrac{1}{2}\begin{bmatrix} 0 & 1 & 2 \\ 1 & 2 & 3 \\ 3 & a & 1 \end{bmatrix} \begin{bmatrix} 1 & -1 & 1 \\ -8 & 6 & 2c \\ 5 & -3 & 1 \end{bmatrix}$

$= \begin{bmatrix} 1 & 0 & c+1 \\ 0 & 1 & 2(c+1) \\ 4(1-a) & 3(a-1) & 2+ac \end{bmatrix}$

Comparing the elements of AA^{-1} with those of I, we have

$c + 1 = 0$ or $c = -1$

$a - 1 = 0$ or $a = 1$

37. (c) $\text{adj} \cdot (Q^{-1} BP^{-1}) = \text{adj} \cdot (P^{-1})\,\text{adj}\,(B)\,\text{adj}\,(Q^{-1})$

$= \dfrac{P}{|P|} A \dfrac{Q}{|Q|} = PAQ$

38. (a) If eigen values of A are λ_1, λ_2 and λ_3.

$\therefore$ Eigen values of its transpose will be same.

$\therefore$ Eigen values of A^T = Eigen values of $A = 1, 2, 3$.

39. (b) $\therefore$ Area of triangle $= \frac{1}{2}\left\|\begin{matrix} 2 & 3 & 1 \\ 0 & 1 & 1 \\ 1 & 2 & 1 \end{matrix}\right\|$

$= \frac{1}{2}||2(1-2) - 3(0-1) + 1(0-1)||$

$= \frac{1}{2}|-2+3-1| = 0$

40. (d) $a_{ij} = \begin{cases} 1 & , i = j \\ -x & , |i-j| = 1 \\ 2x+1, & \text{otherwise} \end{cases}$

$A = [a_{ij}]$ be 3×3 matrix.

$\therefore \quad A = \begin{bmatrix} 1 & -x & 2x+1 \\ -x & 1 & -x \\ 2x+1 & -x & 1 \end{bmatrix}$

$|A| = (1-x^2) - x(-2x^2 - x + x) + (2x+1)(x^2 - 2x - 1)$

$|A| = 4x^3 - 4x^2 - 4x$

$\Rightarrow f'(x) = 12x^2 - 8x - 4 \Rightarrow f'(x) = 0$

$\Rightarrow \quad 4(3x^2 - 2x - 1) = 0$

$\Rightarrow \quad (x-1)(3x+1) = 0$

$+ \quad -1/3 \quad - \quad 1 \quad +$

$\Rightarrow \quad x = 1, -\frac{1}{3}$

At $f(1) \rightarrow$ minimum

$f\left(-\frac{1}{3}\right) \rightarrow$ maximum

$f(1) = 4(1^3 - 1^2 - 1) = -4$

$f\left(-\frac{1}{3}\right) = \frac{20}{27}$

$\therefore \quad \text{Sum} = -4 + \frac{20}{27} = \frac{-88}{27}$

41. (b) $\therefore$ Area of triangle $= \frac{1}{2}\left\|\begin{matrix} x_1 & y_1 & 1 \\ x_2 & y_2 & 1 \\ x_3 & y_3 & 1 \end{matrix}\right\|$

$= \frac{1}{2}\left\|\begin{matrix} 1 & 1 & 1 \\ 4 & 2 & 1 \\ 3 & 5 & 1 \end{matrix}\right\|$

$= \frac{1}{2}|1(2-5) - 4(1-5) + 3(1-2)|$

$= \frac{1}{2}|-3+16-3| = \frac{10}{2} = 5$

42. (c) Given, system has non-trivial solution.

$\therefore \quad \begin{vmatrix} \lambda-2 & 2 & -1 \\ -2 & (\lambda+3) & -2 \\ 1 & -2 & \lambda \end{vmatrix} = 0$

$\Rightarrow (\lambda-2)(3\lambda + \lambda^2 - 4) - 2(-2\lambda + 2) - 1(4 - 3 - \lambda) = 0$

$\Rightarrow (\lambda-2)(\lambda-1)(\lambda+4) + 4(\lambda-1) + (\lambda-1) = 0$

$\Rightarrow \quad (\lambda-1)(\lambda^2 + 2\lambda - 8 + 5) = 0$

$\Rightarrow \quad (\lambda-1)(\lambda^2 + 2\lambda - 3) = 0$

$\Rightarrow \quad (\lambda-1)^2(\lambda+3) = 0$

43. (c) The system has non-zero solution if

$\begin{vmatrix} 1 & k & -1 \\ 3 & -k & -1 \\ 1 & -3 & 1 \end{vmatrix} = 0$

$\Rightarrow 1(-k-3) - k(3+1) - 1(-9+k) = 0$

$\Rightarrow \quad -6k + 6 = 0 \Rightarrow k = 1$

44. (c) For non-trivial solution of the given system of linear equations

$\begin{vmatrix} 4 & k & 2 \\ k & 4 & 1 \\ 2 & 2 & 1 \end{vmatrix} = 0$

$\Rightarrow \; 8 + k(2-k) + 2(2k-8) = 0$

$\Rightarrow \quad -k^2 + 6k - 8 = 0$

$\Rightarrow \quad k^2 - 6k + 8 = 0$

$\Rightarrow \quad k = 2, 4$

Clearly, there exist two values of k.

45. (a) For no solution, $|A| = 0$

and adj $(A)B \neq 0$

Now, $|A| = \begin{vmatrix} k+1 & 8 \\ k & k+3 \end{vmatrix} = 0$

$\Rightarrow \quad k = 1, 3$

$\therefore$ adj $(A) = \begin{bmatrix} k+3 & -8 \\ -k & k+1 \end{bmatrix}$

Now, adj $(A)B = \begin{bmatrix} k+3 & -8 \\ -k & k+1 \end{bmatrix}\begin{bmatrix} 4k \\ 3k-1 \end{bmatrix}$

$= \begin{bmatrix} 4k^2 - 12k + 8 \\ -k^2 + 2k - 1 \end{bmatrix}$

Put $k = 1$, (adj $A)B = \begin{bmatrix} 0 \\ 0 \end{bmatrix}$; not true

Put $k = 3$, (adj $A) B = \begin{bmatrix} 8 \\ -4 \end{bmatrix} \neq 0$, true.

46. (a) For unique solution,

$\begin{vmatrix} 1 & 1 & 1 \\ 5 & -1 & \alpha \\ 2 & 3 & -1 \end{vmatrix} \neq 0$

$\Rightarrow 1(1 - 3\alpha) - 1(-5 - 2\alpha) + 1(15 + 2) \neq 0$

$\Rightarrow \quad 1 - 3\alpha + 5 + 2\alpha + 17 \neq 0$

$\Rightarrow \quad -\alpha + 23 \neq 0$

$\Rightarrow \quad \alpha \neq 23$

Hence, depends on only α.

47. (b) Given, lines are concurrent.

$$\therefore \begin{vmatrix} 2 & -3 & k \\ 3 & -4 & -13 \\ 8 & -11 & -33 \end{vmatrix} = 0$$

$\Rightarrow 2(132-143)+3(-99+104)+k(-33+32)=0$

$\Rightarrow \quad -22+15-k=0$

$\Rightarrow \quad k=-7$

48. (c) Given, system has non-trivial solution, if

$$\begin{vmatrix} \sin 3\theta & -1 & 1 \\ \cos 2\theta & 4 & 3 \\ 2 & 7 & 7 \end{vmatrix} = 0$$

$\Rightarrow 7\sin 3\theta + 14\cos 2\theta - 14 = 0$

$\Rightarrow 7(3\sin\theta - 4\sin^3\theta) + 14(1-2\sin^2\theta) - 14 = 0$

$\Rightarrow \quad -28\sin^2\theta - 28\sin^3\theta + 21\sin\theta = 0$

$\Rightarrow \quad -7\sin\theta(4\sin^2\theta + 4\sin\theta - 3) = 0$

$\Rightarrow \quad \sin\theta(2\sin\theta+3)(2\sin\theta-1)=0$

$\Rightarrow \quad \sin\theta = 0, \sin\theta = \frac{1}{2} \qquad \left[\because \sin\theta \neq -\frac{3}{2}\right]$

$\therefore \quad \theta = n\pi, n\pi + (-1)^n \frac{\pi}{6}$

49. (c) Given, $\begin{vmatrix} a & a & -1 \\ b & -1 & b \\ -1 & c & c \end{vmatrix} = 0$

Applying $C_2 \to C_2 - C_1$ and $C_3 \to C_3 - C_1$, we get

$$\begin{vmatrix} a & 0 & -(1+a) \\ b & -(1+b) & 0 \\ -1 & c+1 & c+1 \end{vmatrix} = 0$$

$\Rightarrow -a(1+b)(1+c) - b$

$[(1+a)(1+c)] - 1[-(1+a)(1+b)] = 0$

$\Rightarrow \quad \frac{-a}{1+a} - \frac{b}{1+b} + \frac{1}{1+c} = 0$

Adding 2 on both sides, we get

$$\frac{1}{1+a} + \frac{1}{1+b} + \frac{1}{1+c} = 2$$

50. (a) Since, system of equation has no solution.

$$\therefore \quad \Delta = \begin{vmatrix} 1 & 1 & 1 \\ 1 & a & 1 \\ a & b & 1 \end{vmatrix} = 0$$

$\Rightarrow \quad a = 1$

Also at $a=1$,

$\Delta_1 = \Delta_2 = \Delta_3 = 0$

But when $a=1$ and $b=1$, system of equations becomes

$x+y+z=1$

and $\quad x+y+z=0$

This system has no solution.

$\therefore \quad b = \{1\}$

51. (a) $\Delta = \begin{vmatrix} 1 & 2 & 1 \\ 2 & 3 & 1 \\ 3 & 5 & 2 \end{vmatrix} = 0$

$$\Delta_1 = \begin{vmatrix} 3 & 2 & 1 \\ 3 & 3 & 1 \\ 1 & 5 & 2 \end{vmatrix} \neq 0$$

Hence, the given system does not have any solution.

52. (b) $\because \Delta = \begin{vmatrix} k+1 & 8 \\ k & k+3 \end{vmatrix} = k^2+4k+3-8k$

$= k^2 - 4k + 3 = (k-3)(k-1)$

$\Delta_1 = \begin{vmatrix} 4k & 8 \\ 3k-1 & k+3 \end{vmatrix} = 4k^2 + 12k - 24k + 8$

$= 4k^2 - 12k + 8$

$= 4(k^2 - 3k + 2)$

$= 4(k-2)(k-1)$

$\Delta_2 = \begin{vmatrix} k+1 & 4k \\ k & 3k-1 \end{vmatrix} = 3k^2 + 2k - 1 - 4k^2$

$= -k^2 + 2k - 1 = -(k-1)^2$

As given no solution.

$\Rightarrow k=3$, for which $\Delta = 0$ and $\Delta_1, \Delta_2 \neq 0$

53. (c) $\frac{d}{dx}f(x) = \begin{vmatrix} \frac{d}{dx}(x^3) & \frac{d}{dx}(\sin x) & \frac{d}{dx}(\cos x) \\ 6 & -1 & 0 \\ p & p^2 & p^3 \end{vmatrix} + 0 + 0$

$$= \begin{vmatrix} 3x^2 & \cos x & -\sin x \\ 6 & -1 & 0 \\ p & p^2 & p^3 \end{vmatrix}$$

At $x=0$,

$$\frac{d}{dx}f(x) = \begin{vmatrix} 0 & 1 & 0 \\ 6 & -1 & 0 \\ p & p^2 & p^3 \end{vmatrix} = 0 - 1(6p^3) + 0 = -6p^3$$

54. (a) $\int_0^{\pi/2} \Delta(x)\,dx$

$$= \begin{vmatrix} \int_0^{\pi/2} \sin^2 x\,dx & \int_0^{\pi/2} \log\cos x\,dx & \int_0^{\pi/2} \log\tan x\,dx \\ x^2 & 2x-1 & 2x+1 \\ 1 & -2\log 2 & 0 \end{vmatrix}$$

$$= \begin{vmatrix} \pi/4 & -\frac{\pi}{2}\log 2 & 0 \\ x^2 & 2x-1 & 2x+1 \\ 1 & -2\log 2 & 0 \end{vmatrix}$$

$= \frac{\pi}{2} 2x\log 2 + \frac{\pi}{2}\log 2 - \frac{\pi}{2} 2x \log 2 - \frac{\pi}{2}\log 2$

$= 0$

55. (8) Putting $x=0, a_0=1$

$$(1+ax+bx^2)^4=(1+ax+bx^2)(1+ax+bx^2)(1+ax+bx^2)(1+ax+bx^2)$$

Clearly, $a_0=1, a_1=$ Coefficient of $x=a+a+a+a=4a$

and $a_2=$ Coefficient of $x^2=4b+6a^2$

Now, $\Delta=-(a_0^3+a_1^3+a_2^3-3a_0a_1a_2)$

$\because \quad a_0+a_1+a_2\neq 0$

$\therefore \quad a_0=a_1=a_2$

$1=4a=6a^2+4b$

$\Rightarrow \quad a=\frac{1}{4}$ and $b=\frac{5}{32}$

56. (6) Given that

$$\begin{vmatrix} S+c & a & b \\ c & S+a & b \\ c & a & S+b \end{vmatrix}=532$$

Applying $C_1\to C_1+C_2+C_3$, we get

$$(a+b+c+S)\begin{vmatrix} 1 & a & b \\ 1 & S+a & b \\ 1 & a & S+b \end{vmatrix}=532$$

$$\Rightarrow \quad 2S\begin{vmatrix} 0 & -S & 0 \\ 0 & S & -S \\ 1 & a & S+b \end{vmatrix}=532$$

$$\Rightarrow \quad 2S^3=532$$

$$\Rightarrow \quad S^3=216\Rightarrow S=6$$

57. (25) $$\begin{vmatrix} a & b+3c & c+4a \\ b & c+3a & a+4b \\ c & a+3b & b+4c \end{vmatrix}+2\begin{vmatrix} b & b+3c & c+4a \\ c & c+3a & a+4b \\ a & a+3b & b+4c \end{vmatrix}$$

Δ' $\qquad\qquad$ Δ''

Applying $C_3\to C_3-4C_1$ in Δ' and $C_2\to C_2-C_1$ in Δ'', we get

$$\Delta_2=\begin{vmatrix} a & b+3c & c \\ b & c+3a & a \\ c & a+3b & b \end{vmatrix}+2\times 3\begin{vmatrix} b & c & c+4a \\ c & a & a+4b \\ a & b & b+4c \end{vmatrix}$$

Applying $C_2\to C_2-3C_3$ in Δ' and $C_3\to C_3-C_2$ in Δ'', we get

$$=\begin{vmatrix} a & b & c \\ b & c & a \\ c & a & b \end{vmatrix}+6\begin{vmatrix} b & c & 4a \\ c & a & 4b \\ a & b & 4c \end{vmatrix}$$

$$=\begin{vmatrix} a & b & c \\ b & c & a \\ c & a & b \end{vmatrix}+24\begin{vmatrix} a & b & c \\ b & c & a \\ c & a & b \end{vmatrix}$$

$$=25\begin{vmatrix} a & b & c \\ b & c & a \\ c & a & b \end{vmatrix}=25\Delta_1 \Rightarrow \frac{\Delta_2}{\Delta_1}=25$$

58. (1) $$\Delta_1=\begin{vmatrix} a_1^2+4a_1d & a_1 & d \\ a_2^2+4a_2d & a_2 & d \\ a_3^2+4a_3d & a_3 & d \end{vmatrix} \quad [C_3\to C_3-C_2]$$

where d is the common difference of AP.

$$=d\begin{vmatrix} a_1^2 & a_1 & 1 \\ a_2^2 & a_2 & 1 \\ a_3^2 & a_3 & 1 \end{vmatrix}+4d\begin{vmatrix} a_1 & a_1 & d \\ a_2 & a_2 & d \\ a_3 & a_3 & d \end{vmatrix}$$

$$=d(a_1-a_2)(a_2-a_3)(a_3-a_1)$$

$$=-2d^4$$

Similarly, $\Delta_2=-2d^4$

59. (0) $$\begin{vmatrix} 3u^2 & 2u^3 & 1 \\ 3v^2 & 2v^3 & 1 \\ 3w^2 & 2w^3 & 1 \end{vmatrix}=0$$

Applying $R_1\to R_1-R_2$ and $R_2\to R_2-R_3$, we get

$$\begin{vmatrix} u^2-v^2 & u^3-v^3 & 0 \\ v^2-w^2 & v^3-w^3 & 0 \\ w^2 & w^3 & 1 \end{vmatrix}=0$$

or $$\begin{vmatrix} u+v & u^2+v^2+vu & 0 \\ v+w & v^2+w^2+vw & 0 \\ w^2 & w^3 & 1 \end{vmatrix}=0$$

or $$\begin{vmatrix} 1 & u+w+v & 0 \\ v+w & v^2+w^2+vw & 0 \\ w^2 & w^3 & 1 \end{vmatrix}=0$$

or $(v^2+w^2+vw)-(v+w)[(v+w)+u]=0$

or $v^2+w^2+vw-(v+w)^2-u(v+w)=0$

or $\qquad uv+vw+wu=0$

60. (2) $$\Delta=\begin{vmatrix} r & 612 & 915 \\ 101r^2 & 2r & 3r \\ r & \frac{1}{r} & \frac{1}{r^2} \end{vmatrix}$$

$$=r^2\begin{vmatrix} 1 & 612 & 925 \\ 101 & 2r & 3r \\ 1 & \frac{1}{r} & \frac{1}{r^2} \end{vmatrix}$$

[taking r common from C_1 and R_2]

$$=\begin{vmatrix} 1 & 612 & 915 \\ 101 & 2 & 3 \\ r^2 & r & 1 \end{vmatrix}$$

$$\therefore \lim_{n\to\infty}\frac{1}{n^3}\sum_{r=1}^{n}\begin{vmatrix} 1 & 612 & 915 \\ 101 & 2 & 3 \\ r^2 & r & 1 \end{vmatrix}$$

$$= \lim_{n\to\infty} \frac{1}{n^3} \begin{vmatrix} 1 & 612 & 915 \\ 101 & 2 & 3 \\ \frac{n(n+1)(2n+1)}{6} & \frac{n(n+1)}{2} & n \end{vmatrix}$$

$$= \lim_{n\to\infty} \begin{vmatrix} 1 & 612 & 915 \\ 101 & 2 & 3 \\ \frac{(n+1)(2n+1)}{6n^2} & \frac{n+1}{2n^2} & \frac{1}{n^2} \end{vmatrix}$$

$$= \begin{vmatrix} 1 & 612 & 915 \\ 101 & 2 & 3 \\ \frac{1}{3} & 0 & 0 \end{vmatrix}$$

$$= \frac{1}{3}(3\times 612 - 2\times 915) = 2$$

61. (1) $A = \begin{bmatrix} 1 & \tan x \\ -\tan x & 1 \end{bmatrix}$

Hence, $\det A = \sec^2 x$

$\therefore \quad \det A^T = \sec^2 x$

Now, $\quad f(x) = \det(A^T A^{-1}) = (\det A^T)(\det A^{-1})$

$$= (\det A^T)(\det A)^{-1}$$

$$= \frac{\det(A^T)}{\det(A)} = 1$$

Hence, $f(x) = 1$

62. (36) $\{\text{adj}(B^{-1}A^{-1})\}^{-1} = \{\text{adj}(AB)^{-1}\}^{-1}$

$$= \text{adj}\cdot(AB) \quad [\because \text{adj}\cdot A^{-1} = (\text{adj}\cdot A)^{-1}]$$

$\therefore \quad |\{\text{adj}\cdot(AB)\}| = |AB|^2 = |A|^2|B|^2 = 9\times 4 = 36$

63. (27) We have, $A = PQ^{-1}, B = QR^{-1}$ and $C = RP^{-1}$

$\therefore \quad ABC = I, BCA = I$ and $CAB = I$

$\therefore \det(ABC + BCA + CAB) = \det(3I)$

$$= 3^3 \times \det(I) = 27$$

64. (6) Given, $A^2 = A$

Now, $\quad I = (I - 0.4A)(I - \alpha A)$

$$= I - [\alpha A + 0.4A] + 0.4\alpha A^2$$

$$= I - A(0.4 + \alpha) + 0.4\alpha A$$

$\Rightarrow 0.4\alpha = 0.4 + \alpha$

$\Rightarrow \quad \alpha = -\frac{2}{3}$

$\Rightarrow |9\alpha| = 6$

65. (2) $\begin{vmatrix} 1 & 2 & 2 \\ 1 & 3 & 4 \\ 3 & 4 & k \end{vmatrix} = 0$

$\Rightarrow 1(3k - 16) - 2(k - 12) + 2(4 - 9) = 0$

$\Rightarrow \quad 3k - 16 - 2k + 24 - 10 = 0$

$\Rightarrow \quad k = 2$

66. (3) Given, $x + y + z = 1$ ...(i)

$x + 2y + 4z = p$ ...(ii)

and $\quad x + 4y + 10z = p^2$ ...(iii)

$$\Delta = \begin{vmatrix} 1 & 1 & 1 \\ 1 & 2 & 4 \\ 1 & 4 & 10 \end{vmatrix}$$

Applying $R_1 \to R_1 - R_2$ and $R_2 \to R_2 - R_3$, we get

$$\begin{vmatrix} 0 & -1 & -3 \\ 0 & -2 & -6 \\ 1 & 4 & 10 \end{vmatrix} = 0$$

Since, $\Delta = 0$ solution is not unique.

The system will have infinite solutions

If $\Delta_1 = 0, \Delta_2 = 0, \Delta_3 = 0$

$$\Delta_1 = \begin{vmatrix} 1 & 1 & 1 \\ p & 2 & 4 \\ p^2 & 4 & 10 \end{vmatrix} = 0$$

Applying $C_3 \to C_3 - C_2$, we get

$$\Delta_1 = \begin{vmatrix} 1 & 1 & 0 \\ p & 2 & 2 \\ p^2 & 4 & 6 \end{vmatrix} = 0$$

$\Rightarrow 1(12 - 8) - 1(6p - 2p^2) = 0$

$\Rightarrow \quad 4 - 6p + 2p^2 = 0$

$\Rightarrow \quad 2(p^2 - 3p + 2) = 0 \Rightarrow p^2 - 3p + 2 = 0$

$\Rightarrow \quad p = 1$ or 2

Level 3

1. (c) $\because -1 \le x < 0 \quad \therefore \quad [x] = -1$

$0 \le y < 1 \quad \therefore \quad [y] = 0$

$1 \le z < 2 \quad \therefore [z] = 1$

Hence, the given determinant is

$$\begin{vmatrix} 0 & 0 & 1 \\ -1 & 1 & 1 \\ -1 & 0 & 2 \end{vmatrix} = 1 = [z]$$

2. (a) $\Delta = \begin{vmatrix} a & b & c \\ b & c & a \\ c & a & b \end{vmatrix}$

$$= -(a + b + c)(a^2 + b^2 + c^2 - ab - bc - ca)$$

$$= -\frac{1}{2}(a + b + c)[(a - b)^2 + (b - c)^2 + (c - a)^2] = 0$$

$\Rightarrow a + b + c = 0$ or $a = b = c$

If $a + b + c = 0$, we have

$\cos\theta + \cos 2\theta + \cos 3\theta = 0$ and $\sin\theta - \sin 2\theta + \sin 3\theta = 0$

or $\cos 2\theta(2\cos\theta + 1) = 0$ and $\sin 2\theta(1 - 2\cos\theta) = 0$ ...(i)

which is not possible as $\cos 2\theta = 0$ gives $\sin 2\theta \ne 0$,

$\cos\theta \ne \frac{1}{2}$ and $\cos\theta = -\frac{1}{2}$ gives $\sin 2\theta \ne 0, \cos\theta \ne \frac{1}{2}$.

Therefore, Eq. (i) does not hold simultaneously.
Therefore,

$$a+b+c \neq 0 \Rightarrow a=b=c$$

$$\therefore \quad e^{i\theta} = e^{-2i\theta} = e^{3i\theta}$$

which is satisfied only $e^{i\theta} = 1$ i.e. $\cos\theta = 1, \sin\theta = 0$.

So, $\theta = 2k\pi, k \in Z$.

3. (a) Let $\Delta_1 = \begin{vmatrix} \log_5 729 & \log_3 5 \\ \log_5 27 & \log_9 25 \end{vmatrix} = \begin{vmatrix} \log_5 3^6 & \log_3 5 \\ \log_5 3^3 & \log_{3^2} 5^2 \end{vmatrix}$

$$= \begin{vmatrix} 6\log_5 3 & \log_3 5 \\ 3\log_5 3 & \frac{2}{2}\log_3 5 \end{vmatrix}$$

$$= 6\log_5 3 \cdot \log_3 5 - 3\log_5 3 \cdot \log_3 5 = 6 - 3 = 3$$

and $\Delta_2 = \begin{vmatrix} \log_3 5 & \log_{27} 5 \\ \log_5 9 & \log_5 9 \end{vmatrix} = \begin{vmatrix} \log_3 5 & \log_{3^3} 5 \\ \log_5 3^2 & \log_5 3^2 \end{vmatrix}$

$$= \begin{vmatrix} \log_3 5 & \frac{1}{3}\log_3 5 \\ 2\log_5 3 & 2\log_5 3 \end{vmatrix}$$

$$= 2\log_5 3 \cdot \log_3 5 - \frac{2}{3}\log_5 3 \cdot \log_3 5$$

$$= 2 - \frac{2}{3} = \frac{4}{3}$$

$$\therefore \quad \Delta_1\Delta_2 = 3 \times \frac{4}{3} = 4$$

4. (b) $\because$ $\det(2A) = 2^4 \cdot \det(A) = 16\det(A)$

5. (c) Given, $\begin{vmatrix} \alpha\beta & \beta\gamma & \gamma\alpha \\ \beta\gamma & \gamma\alpha & \alpha\beta \\ \gamma\alpha & \alpha\beta & \beta\gamma \end{vmatrix}$

Applying $C_1 \to C_1 + C_2 + C_3$, we get

$$= (\alpha\beta + \beta\gamma + \gamma\alpha)\begin{vmatrix} 1 & \beta\gamma & \gamma\alpha \\ 1 & \gamma\alpha & \alpha\beta \\ 1 & \alpha\beta & \beta\gamma \end{vmatrix}$$

From the given equations, $\alpha\beta + \beta y + y\alpha = 0$.
So, the value of determinant is 0.

6. (a) $A \cdot \text{adj} A = AA^T$

$$\Rightarrow \quad |A|I = AA^T$$

$$\Rightarrow (10a+3b)\begin{bmatrix} 1 & 0 \\ 0 & 1 \end{bmatrix} = \begin{bmatrix} 5a & -b \\ 3 & 2 \end{bmatrix}\begin{bmatrix} 5a & 3 \\ -b & 2 \end{bmatrix}$$

$$\Rightarrow \; 25a^2 + b^2 = 10 + 3b, 15a - 2b = 0 \text{ and } 10a + 3b = 13$$

$$\Rightarrow \quad 10a + \frac{3 \cdot 15a}{2} = 13$$

$$\Rightarrow \quad 65a = 2 \times 13$$

$$\Rightarrow \quad a = \frac{2}{5}$$

$$\Rightarrow \quad 5a = 2$$

$$\Rightarrow \quad b = 3$$

$$\therefore \quad 5a + b = 5$$

7. (b) We have,

$$AB = A\,(\text{adj}\,A) = |A|\,I_n$$

$$\therefore \quad AB + kI_n = |A|I_n + kI_n$$

$$= (|A| + k)I_n$$

$$\Rightarrow |AB + kI_n| = |(|A| + k)I_n|$$

$$= (|A| + k)^n \qquad [\because |\alpha I_n| = \alpha^n]$$

8. (b) $A^3 + A = O$

$$\Rightarrow \quad A(A^2 + I) = O$$

$$\Rightarrow \quad |A(A^2 + I)| = O$$

$$\Rightarrow |A| = 0 \text{ or } |A^2 + I| = O$$

If $|A| = 0$, then $xy = 0$

$$A^2 + I = \begin{bmatrix} 0 & x \\ y & 0 \end{bmatrix}\begin{bmatrix} 0 & x \\ y & 0 \end{bmatrix} + \begin{bmatrix} 1 & 0 \\ 0 & 1 \end{bmatrix}$$

$$= \begin{bmatrix} xy & 0 \\ 0 & xy \end{bmatrix} + \begin{bmatrix} 1 & 0 \\ 0 & 1 \end{bmatrix}$$

$$\Rightarrow \quad |A^2 + I| = 0$$

$$\Rightarrow \quad (xy + 1)^2 = 0 \Rightarrow xy = -1$$

So, sum of possible values of xy is -1.

9. (a) If possible assume that A is non-singular, then A^{-1} exists. Thus,

$$AB = 0$$

$$\Rightarrow \quad A^{-1}(AB) = (A^{-1}A)B = 0$$

$$\Rightarrow \quad IB = 0 \text{ or } B = 0 \qquad \text{(contradiction)}$$

Hence, both A and B must be singular.

10. (a) Let $A = \begin{bmatrix} a_1 & a_2 & a_3 \\ b_1 & b_2 & b_3 \\ c_1 & c_2 & c_3 \end{bmatrix}$

$$\therefore \; \det(A) = a_1b_2c_3 + a_2b_3c_1 + b_1c_2a_3 - a_3b_2c_1 - a_2b_1c_3 - a_1b_3c_2$$

If any of the terms is non-zero, then det (A) will be non-zero and all the elements of that term will be unity.

Now, there are 6 elements remaining out of which any one can be unity.

Hence, number of non-singular matrices
= (number of ways of choosing any one triplet)
× (number of ways of choosing any one element)

11. (a) We know that
$A - A^T$ is skew symmetric.

Now, it is given that $|A - A^T| \neq 0$.

So, order of A can not be odd.

Thus, the order of A is 2.

Now, $\quad B = \text{Adj}(A)$

$$\therefore \quad AB = BA = |A|I$$

$$\Rightarrow \quad (AB)^{-1} = (BA)^{-1}$$

$$\Rightarrow \quad B^{-1}A^{-1} = A^{-1}B^{-1} \qquad [\because (AB)^{-1} = B^{-1}A^{-1}]$$

$\therefore \text{adj}(B^2A^{-1}B^{-1}A) = \text{adj}(B^2B^{-1}A^{-1}A)$

$= \text{adj}(B \cdot I)$ $\quad [\because A^{-1} \cdot A = I]$

$= \text{adj}(B)$

$= \text{adj}(\text{adj} A)$

$= |A|^{2-2} \cdot A$

$= A$

12. (b) We know that

$$|\text{adj}(\text{adj} A)| = |A|^{(n-1)^2}$$

$$\Rightarrow |\text{adj}\{\text{adj}(\text{adj} A)\}| = |\text{adj} A|^{(n-1)^2}$$

$$= |A|^{(n-1)^3}$$

$$= 2^8 = 256$$

13. (a) $|B| = \begin{vmatrix} q & -b & y \\ -p & a & -x \\ r & -c & z \end{vmatrix}$

$$= -\begin{vmatrix} q & -b & y \\ p & -a & x \\ r & -c & z \end{vmatrix} \quad \text{[multiplying } R_2 \text{ by } -1\text{]}$$

$$= \begin{vmatrix} q & b & y \\ p & a & x \\ r & c & z \end{vmatrix} \quad \text{[multiplying } C_2 \text{ by } -1\text{]}$$

$$= -\begin{vmatrix} p & a & x \\ q & b & y \\ r & c & z \end{vmatrix} \quad \text{[changing } R_1 \text{ with } R_2\text{]}$$

$$= -\begin{vmatrix} a & x & p \\ b & y & q \\ c & z & r \end{vmatrix}$$

Hence, $|A| = -|B|$, obviously when $|A| \neq 0, |B| \neq 0$.

Also, $|\text{adj} B| = |B|^2 = (-|A|)^2 = |A|^2$

14. (a) We have, $|A| = (a+ib)(a-ib) - (-c+id)(c+id)$

$$= a^2 + b^2 + c^2 + d^2 = 1$$

and $\quad \text{adj}(A) = \begin{bmatrix} a-ib & -c-id \\ c-id & a+ib \end{bmatrix}$,

Then, $\quad A^{-1} = \begin{bmatrix} a-ib & -c-id \\ c-id & a+ib \end{bmatrix}$

15. (a) We have,

$$A = (\alpha, \beta)^{-1}$$

$$= \frac{1}{e^{\beta}} \begin{bmatrix} e^{\beta}\cos\alpha & -e^{\beta}\sin\alpha & 0 \\ e^{\beta}\sin\alpha & e^{\beta}\cos\alpha & 0 \\ 0 & 0 & 1 \end{bmatrix}$$

$$= A(-\alpha, -\beta)$$

16. (b) Let $A = \begin{bmatrix} 5 & 0 \\ -a & 5 \end{bmatrix}$

$\therefore \quad |A| = 25$

$\Rightarrow \quad \text{adj}(A) = \begin{bmatrix} 5 & 0 \\ a & 5 \end{bmatrix}$

$\Rightarrow \quad A^{-1} = \frac{1}{|A|}\begin{bmatrix} 5 & 0 \\ a & 5 \end{bmatrix} = \frac{1}{25}\begin{bmatrix} 5 & 0 \\ a & 5 \end{bmatrix}$

$\Rightarrow \quad A^{-2} = (A^{-1})^2 = \frac{1}{25}\begin{bmatrix} 5 & 0 \\ a & 5 \end{bmatrix}\frac{1}{25}\begin{bmatrix} 5 & 0 \\ a & 5 \end{bmatrix}$

$$= \frac{1}{625}\begin{bmatrix} 25 & 0 \\ 10a & 25 \end{bmatrix}$$

$$= \begin{bmatrix} \frac{1}{25} & 0 \\ \frac{2a}{125} & \frac{1}{25} \end{bmatrix}$$

Now, $\begin{bmatrix} \frac{1}{25} & 0 \\ x & \frac{1}{25} \end{bmatrix} = \begin{bmatrix} \frac{1}{25} & 0 \\ \frac{2a}{125} & \frac{1}{25} \end{bmatrix}$

$\Rightarrow \quad x = \frac{2a}{125}$

17. (b) $|A| = \begin{vmatrix} 1 & \tan x \\ -\tan x & 1 \end{vmatrix} = 1 + \tan^2 x \neq 0$

So, A is invertible.

Also, $\text{adj} A = \begin{bmatrix} 1 & \tan x \\ -\tan x & 1 \end{bmatrix}^T = \begin{bmatrix} 1 & -\tan x \\ \tan x & 1 \end{bmatrix}$

Now, $A^{-1} = \frac{1}{|A|}\text{adj} A$

$$= \frac{1}{(1+\tan^2 x)}\begin{bmatrix} 1 & -\tan x \\ \tan x & 1 \end{bmatrix}$$

$$= \begin{bmatrix} \frac{1}{1+\tan^2 x} & \frac{-\tan x}{1+\tan^2 x} \\ \frac{\tan x}{1+\tan^2 x} & \frac{1}{1+\tan^2 x} \end{bmatrix}$$

$$\therefore A^T A^{-1} = \begin{bmatrix} 1 & -\tan x \\ \tan x & 1 \end{bmatrix}\begin{bmatrix} \frac{1}{1+\tan^2 x} & \frac{-\tan x}{1+\tan^2 x} \\ \frac{\tan x}{1+\tan^2 x} & \frac{1}{1+\tan^2 x} \end{bmatrix}$$

$$= \begin{bmatrix} \frac{1-\tan^2 x}{1+\tan^2 x} & \frac{-2\tan x}{1+\tan^2 x} \\ \frac{2\tan x}{1+\tan^2 x} & \frac{1-\tan^2 x}{1+\tan^2 x} \end{bmatrix}$$

$$= \begin{bmatrix} \cos 2x & -\sin 2x \\ \sin 2x & \cos 2x \end{bmatrix}$$

18. (d) $A(u_1+u_2)=\begin{bmatrix}1\\1\\0\end{bmatrix}$

Now, $|A|=1$

$\therefore \quad A^{-1}=\frac{1}{|A|}\text{adj}\,A$

$$u_1+u_2=A^{-1}\begin{bmatrix}1\\1\\0\end{bmatrix}=\begin{bmatrix}1&0&0\\-2&1&0\\1&-2&1\end{bmatrix}\begin{bmatrix}1\\1\\0\end{bmatrix}=\begin{bmatrix}1\\-1\\-1\end{bmatrix}$$

19. (b) If eigen values of A are λ_1 and λ_2.

$\therefore$ Eigen values of A^m are λ_1^m and λ_2^m.

$\Rightarrow$ Eigen values of $A^3=(2)^3$ and $(3)^3=8$ and 27

20. (c) Given, system of equations are

$3x+y-z=2;\ x-z=1$ and $2x+2y+az=5$

Condition for unique solution,

$$\begin{vmatrix}3&1&-1\\1&0&-1\\2&2&a\end{vmatrix}\neq 0$$

$$\Rightarrow 3(0+2)-1(a+2)-1(2-0)\neq 0$$

$$\Rightarrow \quad 6-a-2-2\neq 0$$

$$\Rightarrow \quad a\neq 2$$

21. (d) $|A|=\begin{vmatrix}-2&1&1\\1&-2&1\\1&1&-2\end{vmatrix}=0$

$$\Rightarrow \text{adj}(A)=\begin{vmatrix}3&3&3\\3&3&3\\3&3&3\end{vmatrix}$$

For consistent, $(\text{adj}\,A)B=0$

$$\text{Now, } (\text{adj}\,A)B=\begin{bmatrix}3&3&3\\3&3&3\\3&3&3\end{bmatrix}\begin{bmatrix}a\\b\\c\end{bmatrix}=0$$

$$\Rightarrow 3\begin{bmatrix}a+b+c\\a+b+c\\a+b+c\end{bmatrix}=\begin{bmatrix}0\\0\\0\end{bmatrix}$$

$$\Rightarrow \quad a+b+c=0$$

22. (c) System has non-trivial solution.

$$\therefore \quad \begin{vmatrix}1&\lambda&-1\\\lambda&-1&-1\\1&1&-\lambda\end{vmatrix}=0$$

$$\Rightarrow 1(\lambda+1)-\lambda(-\lambda^2+1)-1(\lambda+1)=0$$

$$\Rightarrow \quad \lambda+1+\lambda^3-\lambda-\lambda-1=0$$

$$\Rightarrow \quad \lambda^3-\lambda=0$$

$$\Rightarrow \quad \lambda=0,\pm 1$$

23. (a) $\frac{d^n}{dx^n}f(x)=\begin{vmatrix}n! & \sin\left(x+\frac{n\pi}{2}\right) & -\cos\left(x+\frac{n\pi}{2}\right)\\ n! & \sin\left(\frac{n\pi}{2}\right) & \cos\frac{n\pi}{2}\\ a & a^2 & a^3\end{vmatrix}$

$$\Rightarrow \frac{d^n}{dx^n}f(x)=\begin{vmatrix}n! & \sin\left(\frac{n\pi}{2}\right) & -\cos\left(\frac{n\pi}{2}\right)\\ n! & \sin\frac{n\pi}{2} & \cos\frac{n\pi}{2}\\ a & a^2 & a^3\end{vmatrix} \quad [\text{at } x=0]$$

$$=\begin{vmatrix}n! & \sin\left(\frac{n\pi}{2}\right) & 0\\ n! & \sin\left(\frac{n\pi}{2}\right) & 0\\ a & a^2 & a^3\end{vmatrix}=0$$

24. (a) Applying $R_1\to R_1-\sec x\,R_3$, we get

$$f(x)=\begin{vmatrix}0 & 0 & \sec^2 x+\cot x\,\text{cosec}x-\cos x\\ \cos^2 x & \cos^2 x & \text{cosec}^2 x\\ 1 & \cos^2 x & \cos^2 x\end{vmatrix}$$

$$=(\sec^2 x+\cot x\,\text{cosec}\,x-\cos x)(\cos^4 x-\cos^2 x)$$

$$=\left(1+\frac{\cos^3 x}{\sin^2 x}-\cos^3 x\right)\cdot(\cos^2 x-1)$$

$$=-\sin^2 x\left(\frac{\sin^2 x+\cos^3 x-\cos^3 x\sin^2 x}{\sin^2 x}\right)$$

$$=-(\sin^2 x+\cos^5 x)$$

$$\Rightarrow \int_0^{\pi/2} f(x)\,dx=-\int_0^{\pi/2}(\sin^2 x+\cos^5 x)\,dx$$

$$=-\left(\frac{1}{2}\cdot\frac{\pi}{2}+\frac{4\cdot 2}{5\cdot 3\cdot 1}\right)=-\left(\frac{\pi}{4}+\frac{8}{15}\right)$$

25. (a, b, d)

For option (a)

$$PEP=\begin{bmatrix}1&0&0\\0&0&1\\0&1&0\end{bmatrix}\begin{bmatrix}1&2&3\\2&3&4\\8&13&18\end{bmatrix}\begin{bmatrix}1&0&0\\0&0&1\\0&1&0\end{bmatrix}$$

$$=\begin{bmatrix}1&2&3\\8&13&18\\2&3&4\end{bmatrix}\begin{bmatrix}1&0&0\\0&0&1\\0&1&0\end{bmatrix}=\begin{bmatrix}1&3&2\\8&18&13\\2&4&3\end{bmatrix}=F$$

$$\text{and} \quad P^2=\begin{bmatrix}1&0&0\\0&0&1\\0&1&0\end{bmatrix}\begin{bmatrix}1&0&0\\0&0&1\\0&1&0\end{bmatrix}=\begin{bmatrix}1&0&0\\0&1&0\\0&0&1\end{bmatrix}$$

For option (b)

$$|EQ+PFQ^{-1}|=|EQ|+|PFQ^{-1}| \quad \text{... (i)}$$

$\because$ $|E|=0$ and $|F|=0$ and $|Q|\neq 0$

$\therefore$ $|EQ|=|E||Q|=0$

and $|PFQ^{-1}|=\dfrac{|P||F|}{|Q|}=0$

Let $R=EQ+PFQ^{-1}$ ...(ii)

$\Rightarrow RQ=EQ^2+PF=EQ^2+P^2EP=EQ^2+EP$ $[\because P^2=I]$

$=E(Q^2+P)$

$\Rightarrow$ $|RQ|=|E(Q^2+P)|$

$\Rightarrow$ $|R||Q|=|E||Q^2+P|=0$ $[\because |E|=0]$

$\Rightarrow$ $|R|=0$ (as $|Q|\neq 0$) ... (iii)

From Eqs. (ii) and (iii), we get Eq. (i) is true.

For option (c)

$|(EF)^3|>|EF|^2$

i.e. $0>0$ which is false.

For option (d)

$\because$ $P^2=I\Rightarrow P^{-1}=P$

$\therefore$ $P^{-1}FP=PFP=PPEPP=E$

So, $E+P^{-1}FP=E+E=2E$

$\Rightarrow$ $Tr(E+P^{-1}FP)=Tr(2E)=2Tr(E)$ ... (iv)

and $P^{-1}EP+F$

$\Rightarrow$ $PEP+F=2PEP$ $[\because F=PEP]$

$\therefore$ $Tr(2PEP)=2Tr(PEP)=2\,Tr(F)=2Tr(E)$ ... (v)

From Eqs. (iv) and (v), option (d) is correct.

26. (a, b, c)

$\because$ $I-EF=G^{-1} \Rightarrow G-GEF=I$ ... (i)

and $G-EFG=I$ ... (ii)

Clearly, $GEF=EFG\rightarrow$ option (c) is correct.

Also, $(I-FE)(I+FGE)$

$=I-FE+FGE-FEFGE$

$=I-FE+FGE-F(G-I)E$

$=I-FE+FGE-FGE+FE$

$=I\rightarrow$ option (b) is correct but option (d) is incorrect.

$\because (I-FE)(I-FGE)=I-FE-FGE+F(G-I)E$

$=I-2FE$

Now, $(I-FE)(-FGE)=-FE$

$\Rightarrow$ $|I-FE||FGE|=|FE|$

$\rightarrow$ option (a) is correct.

27. (a,c) $\begin{vmatrix}\cos\alpha & -\sin\alpha & 1\\ \sin\alpha & \cos\alpha & 1\\ \cos(\alpha+\beta) & -\sin(\alpha+\beta) & 1\end{vmatrix}$

$=\begin{vmatrix}\cos\alpha & -\sin\alpha & 1\\ \sin\alpha & \cos\alpha & 1\\ 0 & 0 & 1+\sin\beta-\cos\beta\end{vmatrix}$

[applying $R_3\rightarrow R_3-R_1(\cos\beta)+R_2(\sin\beta)$]

$=(1+\sin\beta-\cos\beta)(\cos^2\alpha+\sin^2\alpha)$

$=1+\sin\beta-\cos\beta$, which is independent of α.

28. (a,b,c) (a) $\begin{vmatrix}8 & 2 & 7\\ 12 & 3 & 5\\ 16 & 4 & 3\end{vmatrix}$

Applying $R_3\rightarrow R_3-R_2$ and $R_2\rightarrow R_2-R_1$, we get

$=\begin{vmatrix}8 & 2 & 7\\ 4 & 1 & -2\\ 4 & 1 & -2\end{vmatrix}=0$

(b) $\begin{vmatrix}1/a & a^2 & bc\\ 1/b & b^2 & ac\\ 1/c & c^2 & ab\end{vmatrix}=\dfrac{1}{abc}\begin{vmatrix}1 & a^3 & abc\\ 1 & b^3 & abc\\ 1 & c^3 & abc\end{vmatrix}$

$=\begin{vmatrix}1 & a^3 & 1\\ 1 & b^3 & 1\\ 1 & c^3 & 1\end{vmatrix}$

[taking abc common from C_3]

$=0$

(c) $\begin{vmatrix}a+b & 2a+b & 3a+b\\ 2a+b & 3a+b & 4a+b\\ 4a+b & 5a+b & 6a+b\end{vmatrix}$

Applying $R_3\rightarrow R_3-R_2$ and $R_2\rightarrow R_2-R_1$, we get

$=\begin{vmatrix}a+b & 2a+b & 3a+b\\ a & a & a\\ 2a & 2a & 2a\end{vmatrix}=0$

(d) $\begin{vmatrix}2 & 43 & 6\\ 7 & 35 & 4\\ 3 & 17 & 2\end{vmatrix}=\begin{vmatrix}2 & 1 & 6\\ 7 & 7 & 4\\ 3 & 3 & 2\end{vmatrix}$ $[\because C_2\rightarrow C_2-7C_3]$

$=\begin{vmatrix}1 & 1 & 6\\ 0 & 7 & 4\\ 0 & 3 & 2\end{vmatrix}$ $[\because C_1\rightarrow C_1-C_2]$

$=2$

29. (a,b,c) Applying $R_3\rightarrow R_3-xR_2$ and $R_2\rightarrow R_2-xR_1$, we get

$$f(x)=\begin{vmatrix}a & -1 & 0\\ 0 & a+x & -1\\ 0 & 0 & a+x\end{vmatrix}=a(a+x)^2$$

Hence, $f(2x)-f(x)=a[(a+2x)^2-(a+x)^2]$

$=a(a+2x-a-x)(a+2x+a+x)$

$=ax(2a+3x)$

30. (c,d) Applying $C_1 \to C_1 + C_2 + C_3$, we get

$$\Delta = \begin{vmatrix} a+b-x & a & b \\ a+b-x & -x & a \\ a+b-x & b & -x \end{vmatrix}$$

$$= (a+b-x)\begin{vmatrix} 1 & a & b \\ 1 & -x & a \\ 1 & b & -x \end{vmatrix}$$

$$= (a+b-x)\begin{vmatrix} 1 & a & b \\ 0 & -x-a & a-b \\ 0 & b-a & -x-b \end{vmatrix}$$

[applying $R_2 \to R_2 - R_1$ and $R_3 \to R_3 - R_1$]

$$= (a+b-x)[(x+a)(x+b)+(a-b)^2]$$

[expanding along C_1]

$$= (a+b-x)[x^2+(a+b)x+a^2+b^2-ab]$$

31. (a, b) $\Delta = \dfrac{1}{a}\begin{vmatrix} a^3+ax & ab & ac \\ a^2b & b^2+x & bc \\ a^2c & bc & c^2+x \end{vmatrix}$

Applying $C_1 \to C_1 + bC_2 + cC_3$ and taking $a^2+b^2+c^2+x$ common, we get

$$\Delta = \frac{1}{a}(a^2+b^2+c^2+x)\begin{vmatrix} a & ab & ac \\ b & b^2+x & bc \\ c & bc & c^2+x \end{vmatrix}$$

Applying $C_2 \to C_2 - bC_1$ and $C_3 \to C_3 - cC_1$, we get

$$\Delta = \frac{1}{a}(a^2+b^2+c^2+x)\begin{vmatrix} a & 0 & 0 \\ b & x & 0 \\ c & 0 & x \end{vmatrix}$$

$$= x^2(a^2+b^2+c^2+x)$$

Thus, Δ is divisible by x and x^2.

32. (a,b) Applying $C_3 \to C_3 - xC_2$ and $C_2 \to C_2 - xC_1$, we get

$$f(x) = \begin{vmatrix} 3 & 0 & 2a^2 \\ 3x & 2a^2 & 4a^2x \\ 3x^2+2a^2 & 4a^2x & 6a^2x^2+2a^4 \end{vmatrix}$$

$$= 4a^4\begin{vmatrix} 3 & 0 & 1 \\ 3x & 1 & 2x \\ 3x^2+2a^2 & 2x & 3x^2+a^2 \end{vmatrix}$$

Applying $C_3 \to C_3 - xC_2$, we get

$$f(x) = 4a^4\begin{vmatrix} 3 & 0 & 1 \\ 3x & 1 & x \\ 3x^2+2a^2 & 2x & x^2+a^2 \end{vmatrix}$$

Applying $C_1 \to C_1 - 3C_3$, we get

$$f(x) = 4a^4\begin{vmatrix} 0 & 0 & 1 \\ 0 & 1 & x \\ -4a^2 & 2x & x^2+a^2 \end{vmatrix} = 16a^6$$

33. (a,c) $f(x) = \begin{vmatrix} n & n+1 & n+2 \\ n! & (n+1)! & (n+2)! \\ 1 & 1 & 1 \end{vmatrix}$

$$= \begin{vmatrix} n & 1 & 1 \\ n! & nn! & (n+1)(n+1)! \\ 1 & 0 & 0 \end{vmatrix}$$

[applying $C_3 \to C_3 - C_2$ and $C_2 \to C_2 - C_1$]

$$= (n+1)(n+1)! - nn! = n![(n+1)^2 - n]$$

$$= n!(n^2+n+1)$$

Thus, $f(x)$ is divisible by $n!$ and n^2+n+1

34. (b,c) We have,

$$\begin{vmatrix} (1+\alpha^2) & (1+2\alpha)^2 & (1+3\alpha)^2 \\ (2+\alpha)^2 & (2+2\alpha)^2 & (2+3\alpha)^2 \\ (3+\alpha)^2 & (3+2\alpha)^2 & (3+3\alpha)^2 \end{vmatrix} = -648\alpha$$

Applying $R_3 \to R_3 - R_2$ and $R_2 \to R_2 - R_1$, we get

$$\begin{vmatrix} (1+\alpha)^2 & (1+2\alpha)^2 & (1+3\alpha)^2 \\ 3+2\alpha & 3+4\alpha & 3+6\alpha \\ 5+2\alpha & 5+4\alpha & 5+6\alpha \end{vmatrix} = -648\alpha$$

Applying $R_3 \to R_3 - R_2$, we get

$$\begin{vmatrix} (1+\alpha)^2 & (1+2\alpha)^2 & (1+3\alpha)^2 \\ 3+2\alpha & 3+4\alpha & 3+6\alpha \\ 2 & 2 & 2 \end{vmatrix} = -648\alpha$$

Applying $C_3 \to C_3 - C_2$ and $C_2 \to C_2 - C_1$, we get

$$\begin{vmatrix} (1+\alpha)^2 & \alpha(2+3\alpha) & \alpha(2+5\alpha) \\ 3+2\alpha & 2\alpha & 2\alpha \\ 2 & 0 & 0 \end{vmatrix} = -648\alpha$$

$\Rightarrow \quad 2\alpha^2(2+3\alpha) - 2\alpha^2(2+5\alpha) = -324\alpha$

$\Rightarrow \quad -4\alpha^3 = -324\alpha$

$\Rightarrow \quad \alpha = 0, \pm 9$

35. (a,c) $\because A + B = AB$

$\Rightarrow I - (A + B - AB) = I$

$\Rightarrow \quad (I - A)(I - B) = I \Rightarrow |I - A||I - B| = I$

$\Rightarrow |I - A| \cdot |I - B|$ are non-zero.

Also, $(I - B)(I - A) = I \Rightarrow \; I - B - A + BA = I$

$\Rightarrow \quad BA = B + A \Rightarrow BA = AB$

36. (a,b,c) $A^2 = \begin{bmatrix} 1 & 0 & 0 \\ 1 & 0 & 1 \\ 0 & 1 & 0 \end{bmatrix}\begin{bmatrix} 1 & 0 & 0 \\ 1 & 0 & 1 \\ 0 & 1 & 0 \end{bmatrix} = \begin{bmatrix} 1 & 0 & 0 \\ 1 & 1 & 0 \\ 1 & 0 & 1 \end{bmatrix}$

$$A^3 = A^2 \cdot A = \begin{bmatrix} 1 & 0 & 0 \\ 1 & 1 & 0 \\ 1 & 0 & 1 \end{bmatrix}\begin{bmatrix} 1 & 0 & 0 \\ 1 & 0 & 1 \\ 0 & 1 & 0 \end{bmatrix}$$

$$= \begin{bmatrix} 1 & 0 & 0 \\ 2 & 0 & 1 \\ 1 & 1 & 0 \end{bmatrix}$$

$$A^3 - A^2 = \begin{bmatrix} 0 & 0 & 0 \\ 1 & -1 & 1 \\ 0 & 1 & -1 \end{bmatrix} \text{ and } A - I = \begin{bmatrix} 0 & 0 & 0 \\ 1 & -1 & 1 \\ 0 & 1 & -1 \end{bmatrix}$$

$$\Rightarrow A^3 - A^2 = A - I \text{ and } |A - I| = 0$$

$$\Rightarrow |A^n - I| = |(A - I)(I + A + A^2 + \ldots + A^{n-1})|$$

$$= |A - I||I + A + A^2 + \ldots + A^{n-1}| = 0$$

$$\because A^3 - A^2 = A - I \Rightarrow A^4 - A^3 = A^2 - A \quad \ldots(i)$$

$$\Rightarrow A^5 - A^4 = A^3 - A^2 = A - I \quad \ldots(ii) \; [\because A^3 - A^2 = A - I]$$

If n is even, $A^n - A^{n-1} = A^2 - A$ [from Eq. (i)]

If n is odd, $A^n - A^{n-1} = A - I$ [from Eq. (ii)]

Considering, n is even.

$$\therefore A^n - A^{n-1} = A^2 - A$$

$$A^{n-1} - A^{n-2} = A - I$$

On adding, we get

$$A^n - A^{n-2} = A^2 - I$$

$$\Rightarrow A^n = A^{n-2} + A^2 - I$$

$$= (A^{n-4} + A^2 - I) + A^2 - I$$

$$= (A^{n-6} + A^2 - I) + 2(A^2 - I)$$

In similar way, we get

$$\Rightarrow A^n = A^2 + \frac{n-2}{2}(A^2 - I)$$

$$\therefore A^n = \left(\frac{n}{2}\right)A^2 - \left(\frac{n-2}{2}\right)I$$

$$\therefore A^{50} = 25A^2 - 24I$$

$$= 25\begin{bmatrix} 1 & 0 & 0 \\ 1 & 1 & 0 \\ 1 & 0 & 1 \end{bmatrix} - 24\begin{bmatrix} 1 & 0 & 0 \\ 0 & 1 & 0 \\ 0 & 0 & 1 \end{bmatrix}$$

$$= \begin{bmatrix} 1 & 0 & 0 \\ 25 & 1 & 0 \\ 25 & 0 & 1 \end{bmatrix}$$

37. (a,b,c) $\because |A| = \begin{vmatrix} 3 & -3 & 4 \\ 2 & -3 & 4 \\ 0 & -1 & 1 \end{vmatrix}$

$$= 3(-3 + 4) + 3(2 - 0) + 4(-2 - 0) = 1$$

$\therefore \text{adj}(\text{adj } A) = |A|^{3-2} A = A$

and $|\text{adj}(\text{adj } A)| = |A| = 1$

Also, $|\text{adj } A| = |A|^{3-1} = |A|^2 = 1^2 = 1$

38. (c,d)

(a) $(N^T MN)^T = N^T M^T N = N^T MN$, if M is symmetric and is $-N^T MN$, if M is skew-symmetric.

(b) $(MN - NM)^T = N^T M^T - M^T N^T$

$$= NM - MN = -(MN - NM)$$

So, $(MN - NM)$ is skew-symmetric.

(c) $(MN)^T = N^T M^T = NM \neq MN$, if M and N are symmetric.

So, MN is not symmetric

(d) $(\text{adj } M)(\text{adj } N) = \text{adj}(NM) \neq \text{adj}(MN)$

39. (a,b,c) $|A^{-1}| = -1 \Rightarrow |A| = -1$

Now, use $\text{adj } A = |A| A^{-1}$ and $A = (A^{-1})^{-1}$

40. (c,d) Let $M = \begin{bmatrix} a & c \\ c & b \end{bmatrix}$ where $a, b, c \in I$

(a) If the first column of M is transpose of the second row of M, then

$$\begin{bmatrix} a \\ c \end{bmatrix} = \begin{bmatrix} c \\ b \end{bmatrix}$$

$$\therefore \quad a = b = c$$

Thus, $\det(M) = ab - c^2 = 0$

Hence, M is not invertible.

(b) If the second row of M is the transpose of the first column of M, then $[c\, b] = [a\, c]$

Therefore, $a = b = c$

Thus, $\det(M) = ab - c^2 = 0$

Hence, M is not invertible.

(c) If $M = \begin{bmatrix} a & 0 \\ 0 & b \end{bmatrix}$ with $a, b \neq 0$, then $\det(M) = ab \neq 0$

Hence, M is invertible.

(d) If product of elements in main diagonal, which (ab) is not perfect square. Then,

$$\det \cdot (M) = ab - c^2 \neq 0$$

Hence, M is invertible.

41. (b,d) We have, $|A(\theta)| = 1$

Hence, A is invertible

$$A(\pi + \theta) = \begin{bmatrix} \sin(\pi + \theta) & i\cos(\pi + \theta) \\ i\cos(\pi + \theta) & \sin(\pi + \theta) \end{bmatrix}$$

$$= \begin{bmatrix} -\sin\theta & -i\cos\theta \\ -i\cos\theta & -\sin\theta \end{bmatrix} = -A(\theta)$$

$$\text{adj}\{A(\theta)\} = \begin{bmatrix} \sin\theta & -i\cos\theta \\ -i\cos\theta & \sin\theta \end{bmatrix}$$

$$\Rightarrow A(\theta)^{-1} = \begin{bmatrix} \sin\theta & -i\cos\theta \\ -i\cos\theta & \sin\theta \end{bmatrix} = A(\pi - \theta)$$

42. (a,c,d) Given, $A^2 + A + 2I = 0$

$$\Rightarrow \quad A^2 + A = -2I \Rightarrow |A^2 + A| = |-2I|$$

$$\Rightarrow \quad |A||A + I| = (-2)^n \Rightarrow |A| \neq 0$$

Therefore, A is non-singular, hence its inverse exists. Also, multiplying the given equation both sides with A^{-1}, we get

$$A^{-1} = -\frac{1}{2}(A + I)$$

43. (b,c,d) $\alpha x + 2y = \lambda$ and $3x - 2y = \mu$

$$\Delta = \begin{vmatrix} \alpha & 2 \\ 3 & -2 \end{vmatrix} = -2\alpha - 6$$

$\therefore \quad \Delta = 0 \Rightarrow \alpha = -3$

$$\therefore \quad \Delta_1 = \begin{vmatrix} \lambda & 2 \\ \mu & -2 \end{vmatrix} = -2\lambda - 2\mu = -2(\lambda + \mu)$$

$$\Delta_2 = \begin{vmatrix} -3 & \lambda \\ 3 & \mu \end{vmatrix} = -3\mu - 3\lambda = -3(\lambda + \mu)$$

44. (a,d) Let $X = \begin{bmatrix} a \\ b \\ c \end{bmatrix}$

$\therefore \quad AX = \lambda X$

$$\Rightarrow \begin{bmatrix} 8 & -6 & 2 \\ -6 & 7 & -4 \\ 2 & -4 & 3 \end{bmatrix} \begin{bmatrix} a \\ b \\ c \end{bmatrix} = \begin{bmatrix} \lambda a \\ \lambda b \\ \lambda c \end{bmatrix}$$

$\Rightarrow \quad (8 - \lambda)a - 6b + 2c = 0,$

$-6a + (7 - \lambda)b - 4c = 0$

and $\quad 2a - 4b + (3 - \lambda)c = 0$

For non-zero solution,

$$\begin{vmatrix} 8-\lambda & -6 & 2 \\ -6 & 7-\lambda & -4 \\ 2 & -4 & 3-\lambda \end{vmatrix} = 0$$

$\Rightarrow \quad \lambda = 0, 3, 15$

45. (a,d) $\begin{vmatrix} 1 & k & 3 \\ k & 2 & 2 \\ 2 & 3 & 4 \end{vmatrix} = 0$

$\Rightarrow 8 + 4k + 9k - 12 - 4k^2 - 6 = 0$

$\Rightarrow \quad 4k^2 - 13k + 10 = 0$

$\Rightarrow \quad 4k^2 - 8k - 5k + 10 = 0$

$\Rightarrow \quad (4k - 5)(k - 2) = 0$

$\Rightarrow \quad k = \dfrac{5}{4}, 2$

46. (a, c) $\Delta = \begin{vmatrix} -2 & 1 & 1 \\ 1 & -2 & 1 \\ 1 & 1 & -2 \end{vmatrix} = 0$

$$\Delta_x = \begin{vmatrix} a & 1 & 1 \\ b & -2 & 1 \\ c & 1 & -2 \end{vmatrix} = 3a + 3b + 3c$$

$$\Delta_y = \begin{vmatrix} -2 & a & 1 \\ 1 & b & 1 \\ 1 & c & -2 \end{vmatrix} = 3a + 3b + 3c$$

$$\Delta_z = \begin{vmatrix} -2 & 1 & a \\ 1 & -2 & b \\ 1 & 1 & c \end{vmatrix} = 3a + 3b + 3c$$

47. (a,b,c) Given, system of equations has non-trivial solutions.

$$\therefore \quad \begin{vmatrix} \alpha & \beta & \gamma \\ \beta & \gamma & \alpha \\ \gamma & \alpha & \beta \end{vmatrix} = 0$$

$\Rightarrow \quad \alpha^3 + \beta^3 + \gamma^3 = 3\alpha\beta\gamma$

$\Rightarrow \quad \alpha = \beta = \gamma \quad [\because \alpha + \beta + \gamma \neq 0]$

Now, $\alpha + \beta + \gamma = -a,$

$\alpha\beta + \beta\gamma + \alpha\gamma = b$ and $\alpha\beta\gamma = -c$

$\Rightarrow 3\alpha = -a, 3\alpha^2 = b$ and $\alpha^3 = -c$

$\Rightarrow \alpha = -\dfrac{a}{3}, \alpha^2 = \dfrac{b}{3}$ and $\alpha^3 = -c$

$\Rightarrow a^2 = 3b, a^3 = 27c$ and $b^3 = 27c^2$

Sol. (Q.Nos. 48-50)

48. (c) **49.** (d) **50.** (b)

In given determinant, applying $C_2 \to C_2 - C_1$ and $C_3 \to C_3 - C_2$, we get

$$f(x) = \begin{vmatrix} x + c_1 & a - c_1 & 0 \\ x + b & c_2 - b & a - c_2 \\ x + b & 0 & c_3 - b \end{vmatrix}$$

$$= x \begin{vmatrix} 1 & a - c_1 & 0 \\ 1 & c_2 - b & a - c_2 \\ 1 & 0 & c_3 - b \end{vmatrix} + \begin{vmatrix} c_1 & a - c_1 & 0 \\ b & c_2 - b & a - c_2 \\ b & 0 & c_3 - b \end{vmatrix}$$

So, $f(x)$ is linear. Let $f(x) = Px + Q$

Then, $f(-a) = -aP + Q$ and $f(-b) = -bP + Q$

Then, $\quad f(0) = 0 \times P + Q$

$$\Rightarrow \quad Q = \frac{bf(-a) - af(-b)}{(b - a)} = f(0) \quad \text{...(i)}$$

Also, $f(-a) = \begin{vmatrix} c_1 - a & 0 & 0 \\ b - a & c_2 - a & 0 \\ b - a & b - a & c_3 - a \end{vmatrix}$

$= (c_1 - a)(c_2 - a)(c_3 - a)$

Similarly, $f(-b) = (c_1 - b)(c_2 - b)(c_3 - b)$

and $\quad g(x) = (c_1 - x)(c_2 - x)(c_3 - x)$

$\Rightarrow \quad g(a) = f(-a)$ and $g(b) = f(-b)$

Now, from Eq. (i), we get

$$f(0) = \frac{bg(a) - ag(b)}{b - a}$$

Also, $f(-a) - f(-b) = (b - a)P$

$$\therefore \quad P = \frac{g(a) - g(b)}{b - a}$$

Sol. (Q.Nos. 51-52)

51. (d) **52.** (c)

$$A = \begin{bmatrix} a_{11} & a_{12} & a_{13} \\ a_{21} & a_{22} & a_{23} \\ a_{31} & a_{32} & a_{33} \end{bmatrix}$$

$$\Rightarrow B = \begin{bmatrix} a_{12}+a_{13} & a_{11}+a_{13} & a_{11}+a_{12} \\ a_{22}+a_{23} & a_{21}+a_{23} & a_{21}+a_{22} \\ a_{32}+a_{33} & a_{31}+a_{33} & a_{31}+a_{32} \end{bmatrix}$$

$$\Rightarrow X = A^{-1}B$$

$$= \frac{1}{|A|}\begin{bmatrix} C_{11} & C_{21} & C_{31} \\ C_{12} & C_{22} & C_{32} \\ C_{13} & C_{23} & C_{33} \end{bmatrix} \begin{bmatrix} a_{12}+a_{13} & a_{11}+a_{13} & a_{11}+a_{12} \\ a_{22}+a_{23} & a_{21}+a_{23} & a_{21}+a_{22} \\ a_{32}+a_{33} & a_{31}+a_{33} & a_{31}+a_{32} \end{bmatrix}$$

$$= \frac{1}{|A|}\begin{bmatrix} 0 & |A| & |A| \\ |A| & 0 & |A| \\ |A| & |A| & 0 \end{bmatrix}$$

$$= \begin{bmatrix} 0 & 1 & 1 \\ 1 & 0 & 1 \\ 1 & 1 & 0 \end{bmatrix}$$

$\Rightarrow \quad |A^{-1}B| = 2$

or $\quad |A^{-1}||B| = 2$

or $\quad |B| = 2|A|$

Sol. (Q.Nos. 53-55)

53. (a) Let the matrix be $\begin{bmatrix} a & b & c \\ p & q & r \\ x & y & z \end{bmatrix}$

We have five entries as 1 and remaining four entries as 0. Since, matrix is symmetric, we must have even number of zeroes for $i \neq j$, we have two cases.

(i) Two entries in diagonal are zero we can select two places from three (in diagonal) in ${}^3C_2$ ways.

Now, we have to select elements for upper triangle. For upper triangle, we have three places of which one entry is 0 and two are 1. One place from three can be selected in ${}^3C_1$ ways.

Hence, the number of matrices is ${}^3C_2 \times {}^3C_1 = 9$.

(ii) If all the entries in the principal diagonal are 1, we have two 0 and one 1 in upper triangle.

Hence, the number of matrices is 3. Therefore, total number of matrices is 12.

54. (b) $\begin{bmatrix} 0 & a & b \\ a & 0 & c \\ b & c & 1 \end{bmatrix}$ either $b = 0$ or $c = 0 \Rightarrow |A| \neq 0$

$\Rightarrow$ Two matrices.

$A = \begin{bmatrix} 0 & a & b \\ a & 1 & c \\ b & c & 0 \end{bmatrix}$ either $a = 0$ or $c = 0 \Rightarrow |A| \neq 0$

$\Rightarrow$ Two matrices

$A = \begin{bmatrix} 1 & a & b \\ a & 0 & c \\ b & c & 0 \end{bmatrix}$ either $a = 0$ or $b = 0 \Rightarrow |A| \neq 0$

$\Rightarrow$ Two matrices

$A = \begin{bmatrix} 1 & a & b \\ a & 1 & c \\ b & c & 1 \end{bmatrix}$

$a = b = 0 \Rightarrow |A| = 0$

$a = c = 0 \Rightarrow |A| = 0$

$b = c = 0 \Rightarrow |A| = 0$

Therefore, there will be only six matrices.

55. (b) The six matrices A for which $|A| = 0$ are

$\begin{bmatrix} 0 & 0 & 1 \\ 0 & 0 & 1 \\ 1 & 1 & 1 \end{bmatrix}$ (inconsistent) $\begin{bmatrix} 0 & 1 & 0 \\ 1 & 1 & 1 \\ 0 & 1 & 0 \end{bmatrix}$ (inconsistent)

$\begin{bmatrix} 1 & 1 & 1 \\ 1 & 0 & 0 \\ 1 & 0 & 0 \end{bmatrix}$ (infinite solutions) $\begin{bmatrix} 1 & 1 & 0 \\ 1 & 1 & 0 \\ 0 & 0 & 1 \end{bmatrix}$ (inconsistent)

$\begin{bmatrix} 1 & 1 & 0 \\ 1 & 1 & 0 \\ 0 & 0 & 1 \end{bmatrix}$ (inconsistent) $\begin{bmatrix} 1 & 0 & 1 \\ 0 & 1 & 0 \\ 1 & 0 & 1 \end{bmatrix}$ (inconsistent)

$\begin{bmatrix} 1 & 0 & 0 \\ 0 & 1 & 1 \\ 0 & 1 & 1 \end{bmatrix}$ (inconsistent)

56. (i) $\rightarrow$ D, (ii) $\rightarrow$ A, (iii) $\rightarrow$ B, (iv) $\rightarrow$ D

(i) Coefficient of x in $f(x)$ is coefficient of x in

$$\begin{vmatrix} x & 1 & 1 \\ 1 & x & 2 \\ x^2 & 1 & 0 \end{vmatrix}$$

Therefore, coefficient of x is -2.

(ii) Let $D = \begin{vmatrix} 1 & 3\cos\theta & 1 \\ \sin\theta & 1 & 3\cos\theta \\ 1 & \sin\theta & 1 \end{vmatrix}$

$= (3\cos\theta - \sin\theta)^2$

$\Delta_{max} = 10$

(iii) $f'(x) = 0$

$\Rightarrow f'(0) = 0$

(iv) $a_0 = \begin{vmatrix} 0 & 2 & 0 \\ 1 & 0 & 6 \\ 0 & 0 & 1 \end{vmatrix}$

$= -2(1) = -2$

57. (i) $\rightarrow$ A, B, C, D, (ii) $\rightarrow$ C, (iii) $\rightarrow$ A, C, D, (iv) $\rightarrow$ A, B, C, D

(i) We have,

$$\Delta = \begin{vmatrix} \frac{1}{c} & \frac{1}{c} & -\frac{(a+b)}{c^2} \\ -\frac{(b+c)}{a^2} & \frac{1}{a} & \frac{1}{a} \\ \frac{-b(b+c)}{a^2 c} & \frac{(a+2b+c)}{ac} & \frac{-b(a+b)}{ac^2} \end{vmatrix}$$

Multiplying C_1 by a, C_2 by b and C_3 by c, we get

$$\Delta = \frac{1}{abc}\begin{vmatrix} \frac{a}{c} & \frac{b}{c} & -\frac{a+b}{c} \\ -\frac{b+c}{a} & \frac{b}{a} & \frac{c}{a} \\ -\frac{b(b+c)}{ac} & \frac{b(a+2b+c)}{ac} & -\frac{b(a+b)}{ac} \end{vmatrix}$$

Applying $C_1 \rightarrow C_1 + C_2 + C_3$, we get

$$\Delta = \frac{1}{abc}\begin{vmatrix} 0 & \frac{b}{c} & \frac{-a+b}{c} \\ 0 & \frac{b}{a} & \frac{c}{a} \\ 0 & \frac{b(a+2b+c)}{ac} & -\frac{b(a+b)}{ac} \end{vmatrix} = 0$$

This shows that Δ is independent of a, b and c.

(ii) We have, $\Delta = \begin{vmatrix} \sin a \cos b & \sin a \sin b & \cos a \\ \cos a \cos b & \cos a \sin b & -\sin a \\ -\sin a \cdot \sin b & \sin a \cos b & 0 \end{vmatrix}$

Applying $C_1 \rightarrow C_1 - (\cot b) C_2$, we get

$$\Delta = \begin{vmatrix} 0 & \sin a \sin b & \cos a \\ 0 & \cos a \sin b & -\sin a \\ -\sin a / \sin b & \sin a \cos b & 0 \end{vmatrix}$$

$$= -\frac{\sin a}{\sin b}[-\sin b \sin^2 a - \cos^2 a \sin b]$$

[expanding along C_1]

$= \sin a$

This shows that Δ is independent of b.

(iii) We have, $\Delta = \begin{vmatrix} \frac{1}{\sin a \cos b} & \frac{1}{\sin a \sin b} & \frac{1}{\cos a} \\ \frac{-\cos a}{\sin^2 a \cos b} & \frac{-\cos a}{\sin^2 a \sin b} & \frac{\sin a}{\cos^2 a} \\ \frac{\sin b}{\sin a \cos^2 b} & \frac{-\cos b}{\sin a \sin^2 b} & 0 \end{vmatrix}$

Multiplying C_1 by $\sin a \cos b, C_2$ by $\sin a \sin b$ and C_3 by $\cos a$, we get

$$\Delta = \frac{1}{\sin^2 a \cos a \sin b \cos b}\begin{vmatrix} 1 & 1 & 1 \\ -\cot a & -\cot a & \tan a \\ \tan b & -\cot b & 0 \end{vmatrix}$$

Applying $C_1 \rightarrow C_1 - C_2$, we get

$$\Delta = \frac{1}{\sin^2 a \cos a \sin b \cos b}\begin{vmatrix} 0 & 1 & 1 \\ 0 & -\cot a & \tan a \\ \frac{1}{\sin b \cos b} & -\cot b & 0 \end{vmatrix}$$

$$= \frac{1}{\sin^2 a \cos a \sin^2 b \cos^2 b}[\tan a + \cot a]$$

$$= \frac{1}{\sin^3 a \cos^2 a \sin^2 b \cos^2 b}$$

This shows that Δ is dependent on a and b.

(iv) We have,

$$\Delta = \begin{vmatrix} a^2 & b \sin A & c \sin A \\ b \sin A & 1 & \cos A \\ c \sin A & \cos A & 1 \end{vmatrix}$$

$$= \begin{vmatrix} a^2 & a \sin B & a \sin C \\ a \sin B & 1 & \cos A \\ a \sin C & \cos A & 1 \end{vmatrix} \quad \text{[using sine rule]}$$

$$= a^2 \begin{vmatrix} 1 & \sin B & \sin C \\ \sin B & 1 & \cos A \\ \sin C & \cos A & 1 \end{vmatrix}$$

Applying $C_2 \rightarrow C_2 - (\sin B) C_1$ and $C_3 \rightarrow C_3 - (\sin C) C_1$, we get

$$= a^2 \begin{vmatrix} 1 & 0 & 0 \\ \sin B & 1 - \sin^2 B & \cos A - \sin B \sin C \\ \sin C & \cos A - \sin B \sin C & 1 - \sin^2 C \end{vmatrix}$$

$$= a^2[\cos^2 B \cos^2 C - (\cos A - \sin B \sin C)^2]$$

$$= a^2[\cos^2 B \cos^2 C - \{\cos(B+C) + \sin B \sin C\}^2]$$

$$= a^2[\cos^2 B \cos^2 C - \cos^2 B \cos^2 C]$$

$$= 0$$

This shows that Δ is independent of a, b and c.

58. (i) $\rightarrow$ C, (ii) $\rightarrow$ D, (iii) $\rightarrow$ A, C, (iv) $\rightarrow$ A, B, C, D

(i) Since, A is idempotent $A^2 = A^3 = A^4 = A$.

Now,

$$(A + I)^n = I + {}^nC_1 A + {}^nC_2 A^2 + \dots + {}^nC_n A^n$$

$$= I + {}^nC_1 A + {}^nC_2 A + \dots + {}^nC_n A$$

$$= I + ({}^nC_1 + {}^nC_2 + \dots + {}^nC_n) A$$

$$= I + (2^n - 1) A$$

$$\Rightarrow 2^n - 1 = 127 \Rightarrow n = 7$$

(ii) We have,

$$(I - A)(I + A + A^2 + \dots + A^7)$$

$$= I + A + A^2 + \dots + A^7 + (-A - A^2 - A^3 - A^4 - A^8)$$

$= I - A^8 = I$ [if $A^8 = 0$]

(iii) Here, matrix A is skew-symmetric and since, $|A| |-A^T| = (-1)^n |A|$, so $|A|(1-(-1)^n = 0$

As n is odd, hence $|A| = 0$.

Hence, A is singular.

(iv) If A is symmetric, A^{-1} is also symmetric for matrix of any order.

59. (i) → B, (ii) → A, (iii) → A, (iv) → D

$$\Delta = (\pi - 1)^2 (\pi + 2)$$

Now, $\Delta = 0$, if $\pi = 1$ or $\pi = -2$

If $\pi \neq 1, \pi \neq -2$, then the system has a unique solution.

If $\pi = 1$, the system becomes $x + y + z = 1$, which has infinite solution.

If $\pi = -2$, the system is inconsistent.

Sol. (Q.Nos. 60-62)

60. (2) **61.** (–1) **62.** (0.296)

$$\begin{vmatrix} x & n & r \\ m & y & r \\ m & n & z \end{vmatrix} = 0$$

Applying $R_1 \to R_1 - R_2$ and $R_2 \to R_2 - R_3$, we get

$$\begin{vmatrix} x-m & n-y & 0 \\ 0 & y-n & r-z \\ m & n & z \end{vmatrix} = 0$$

$$\Rightarrow (x-m)(y-n)z + (n-y)(r-z)m - n(r-z)(x-m) = 0$$

Dividing by $(x-m)(y-n)(z-r)$, we have

$$\frac{z}{z-r} + \frac{m}{x-m} + \frac{n}{y-n} = 0$$

$$\Rightarrow \frac{z}{z-r} + \frac{m}{x-m} + 1 + \frac{n}{y-n} + 1 = 2$$

$$\Rightarrow \frac{z}{z-r} + \frac{x}{x-m} + \frac{y}{y-n} = 2$$

$$\Rightarrow \frac{z}{z-r} - 1 + \frac{x}{x-m} - 1 + \frac{y}{y-n} - 1 = -1$$

$$\Rightarrow \frac{m}{x-m} + \frac{n}{y-n} + \frac{r}{z-r} = -1$$

Now, AM ≥ GM

$$\Rightarrow \frac{\frac{z}{z-r} + \frac{x}{x-m} + \frac{y}{y-n}}{3} \geq \left(\frac{z}{(z-r)} \frac{x}{(x-m)} \frac{y}{(y-n)} \right)^{1/3}$$

$$\Rightarrow \frac{z}{z-r} \frac{x}{x-m} \frac{y}{y-n} \leq \frac{8}{27}$$

$$\Rightarrow \frac{z}{z-r} \cdot \frac{x}{x-m} \cdot \frac{y}{y-n} \leq 0.296$$

63. (2) We have, $D = \begin{vmatrix} a_1 & a_2 & a_3 \\ 5 & 4 & a_6 \\ a_7 & a_8 & a_9 \end{vmatrix}$

Since, $a_n = \frac{20}{n}$ and $d = \frac{1}{20}$

$$\text{Hence, } D = \begin{vmatrix} 20 & \frac{20}{2} & \frac{20}{3} \\ \frac{20}{4} & \frac{20}{5} & \frac{20}{6} \\ \frac{20}{7} & \frac{20}{8} & \frac{20}{9} \end{vmatrix} = \frac{(20)^3}{4 \times 7} \begin{vmatrix} 1 & \frac{1}{2} & \frac{1}{3} \\ 1 & \frac{4}{5} & \frac{2}{3} \\ 1 & \frac{7}{8} & \frac{7}{9} \end{vmatrix}$$

Applying $R_1 \to R_1 - R_2$ and $R_2 \to R_2 - R_3$, we get

$$= \frac{(20)^3}{4 \times 7} \begin{vmatrix} 0 & \frac{-3}{10} & \frac{-1}{3} \\ 0 & \frac{-3}{40} & \frac{-1}{9} \\ 1 & \frac{7}{8} & \frac{7}{9} \end{vmatrix} = 50/21$$

$$\Rightarrow [D] = 2$$

64. (0) Taking x common from R_2 and $x(x-1)$ common from R_3, we get

$$f(x) = x^2(x-1) \begin{vmatrix} 1 & x & x+1 \\ 2 & x-1 & x+1 \\ 3 & x-2 & x+1 \end{vmatrix}$$

Applying $C_3 \to C_3 - C_2$, we get

$$f(x) = x^2(x-1) \begin{vmatrix} 1 & x & 1 \\ 2 & x-1 & 2 \\ 3 & x-2 & 3 \end{vmatrix} = 0$$

Thus, $f(500) = 0$

65. (4) $\Delta = \begin{vmatrix} x & x+y & x+y+z \\ 2x & 3x+2y & 4x+3y+2z \\ 3x & 6x+3y & 10x+6y+3z \end{vmatrix}$

$$= \begin{vmatrix} x & x & x \\ 2x & 3x & 4x \\ 3x & 6x & 10x \end{vmatrix}$$

$$= x^3 \begin{vmatrix} 1 & 1 & 1 \\ 2 & 3 & 4 \\ 3 & 6 & 10 \end{vmatrix} + 0$$

On solving, we get

$$= x^3(6 - 8 + 3) = 64 \quad \text{[given]}$$

$$\Rightarrow x^3 = 64 \text{ or } x = 4$$

66. (2) Using $C_3 \to C_3 - (C_1 - C_2)$ in D_1 and D_2, we have

$$\frac{D_1}{D_2} = \frac{-2b(ad-bc)}{b(ad-bc)} = -(2)$$

67. (– 4) $\Delta = \begin{vmatrix} x+2 & 2x+3 & 3x+4 \\ 2x+3 & 3x+4 & 4x+5 \\ 3x+5 & 5x+8 & 10x+17 \end{vmatrix} = 0$

Applying $R_3 \to R_3 - R_2$ and $R_2 \to R_2 - R_1$, we get

$$\Delta = \begin{vmatrix} x+2 & 2x+3 & 3x+4 \\ x+1 & x+1 & x+1 \\ x+2 & 2(x+2) & 6(x+2) \end{vmatrix} = 0$$

$$\therefore \quad \Delta = (x+1)(x+2)\begin{vmatrix} x+2 & 2x+3 & 3x+4 \\ 1 & 1 & 1 \\ 1 & 2 & 6 \end{vmatrix} = 0$$

$$\therefore \quad \Delta = (x+1)(x+2)[(x+2)\cdot 4 - (2x+3)\cdot 5 + (3x+4)\cdot 1] = 0$$

$$\Delta = (x+1)(x+2)(-3x-3) = 0$$

or $$(x+1)^2(x+2) = 0$$

$\therefore x = -1, -1, -2$

68. (8) $\begin{vmatrix} (x^2+1)^2 & (xy+1)^2 & (xz+1)^2 \\ (xy+1)^2 & (y^2+1)^2 & (yz+1)^2 \\ (xz+1)^2 & (yz+1)^2 & (z^2+1)^2 \end{vmatrix}$

$$= \begin{vmatrix} 1 & x & x^2 \\ 1 & y & y^2 \\ 1 & z & z^2 \end{vmatrix} \begin{vmatrix} 1 & 1 & 1 \\ 2x & 2y & 2z \\ x^2 & y^2 & z^2 \end{vmatrix}$$

$$= 2(x-y)^2(y-z)^2(z-x)^2$$

Putting $x = 31, y = 32$ and $z = 33$, we get

$$= 2(1)^2(1)^2(2)^2 = 8$$

69. (4) $\det B = \begin{vmatrix} a_{11} & \frac{a_{12}}{3} & \frac{a_{13}}{3^2} \\ 3a_{21} & a_{22} & \frac{a_{23}}{3} \\ 9a_{31} & 3a_{32} & a_{33} \end{vmatrix}$

$$= 3 \times 9 \begin{vmatrix} a_{11} & \frac{a_{12}}{3} & \frac{a_{13}}{3^2} \\ a_{21} & \frac{a_{22}}{3} & \frac{a_{23}}{9} \\ a_{31} & \frac{a_{32}}{3} & \frac{a_{33}}{9} \end{vmatrix}$$

[dividing C_2 by 3 and C_3 by 9]

$$= \begin{vmatrix} a_{11} & a_{12} & a_{13} \\ a_{21} & a_{22} & a_{23} \\ a_{31} & a_{32} & a_{33} \end{vmatrix}$$

$$= \det A = 2$$

Similarly, $\det C = \det A = 2$

$\therefore \quad \det(BC) = (\det B)(\det C) = 2 \times 2 = 4$

70. (4) $|A| = (2k+1)^3$ and $|B| = 0$

[since, B is a skew-symmetric matrix of order 3]

$\Rightarrow \det(\text{adj } A) = |A|^{3-1}$

$$= \{(2k+1)^3\}^2 = 10^6$$

$\Rightarrow \quad 2k+1 = 10$

or $\quad 2k = 9$

$\Rightarrow \quad [k] = [4.5] = 4$

71. (1) Since, the system of equations has infinitely many solutions.

$$\begin{vmatrix} 1 & \alpha & \alpha^2 \\ \alpha & 1 & \alpha \\ \alpha^2 & \alpha & 1 \end{vmatrix} = 0$$

$\Rightarrow 1(1-\alpha^2) - \alpha(\alpha - \alpha^3) + \alpha^2(\alpha^2 - \alpha^2) = 0$

$\Rightarrow \quad (1-\alpha^2) - \alpha^2 + \alpha^4 = 0$

$\Rightarrow \quad (\alpha^2 - 1)^2 = 0$

$\Rightarrow \quad \alpha = \pm 1$

For $\alpha = 1$, we get

$$x + y + z = 1$$

and $\quad x + y + z = -1.$

Hence, system has no solution.

For $\alpha = -1$, all three equations become $x - y + z = 1$, which represents coincident planes.

$\therefore \quad 1 + \alpha + \alpha^2 = 1$

72. (4) $\det(P) = \begin{vmatrix} a_1 & a_2 & a_3 \\ b_1 & b_2 & b_3 \\ c_1 & c_2 & c_3 \end{vmatrix}$

$$= (a_1b_2c_3 + a_2b_3c_1 + a_3b_1c_2) - (a_3b_2c_1 + a_1b_3c_2 + a_2b_1c_3)$$

$$= x - y$$

where, $x = a_1b_2c_3 + a_2b_3c_1 + a_3b_1c_2$

and $\quad y = a_3b_2c_1 + a_1b_3c_2 + a_2b_1c_3$

Now, $\quad x \le 3$ and $y \ge -3$

If we consider $\det(P) = 6$, then

$$a_1b_2c_3 = a_2b_3c_1 = a_3b_1c_2 = 1$$

and $a_3b_2c_1 = a_1b_3c_2 = a_2b_1c_3 = -1$,

which is contradiction.

So, next possibility is 5 for which one of the terms must be zero but that will make two terms zero.

Hence, greatest value of determinant is 4.

CHAPTER

10

MATHEMATICAL REASONING

LEVEL 01 \\ Starter Level

10.1 Statements

1 Which of the following sentences are statements? Justify your answers.

(i) 18 is less than 10.

(ii) Every set is a finite set.

(iii) Mathematics is fun.

(iv) How far is Chennai from here?

(v) $15+8>23$

(vi) Every square is a rectangle.

(vii) $\sin^2 x+\cos^2 x=0$

2 Write the negation of the following statements.

(i) $\sqrt{7}$ is rational.

(ii) Australia is a continent.

(iii) There does not exist a quadrilateral which has all its sides equal.

(iv) 2 is not a prime number.

(v) All similar triangles are congruent.

(vi) Area of a circle is same as the perimeter of the circle.

3 Find the component statements of the following compound statements.

(i) The sky is blue and the grass is green.

(ii) 0 is a positive number or a negative number.

(iii) 24 is a multiple of 4 and 6.

(iv) 2 is an even number and a prime number.

4 For each of the following statements, determine whether an inclusive 'or' or exclusive 'or' is used. Give reasons for your answer.
(i) To enter a country, you need a passport or a voter registration card.
(ii) Two lines intersect at a point or are parallel.
(iii) Students can take French or Sanskrit as their third language.

10.2 Elementary Operations of Logic

5 Write the truth table for following
(i) $p \wedge q$ (ii) $p \vee q$

6 Translate the following statements into symbolic form.
(i) Rahul passed in Hindi and English.
(ii) x and y are even integers.
(iii) A number is either divisible by 2 or 3.
(iv) Students can take Hindi or English as an optional paper.

7 Form the truth table of $\sim p \wedge q$ and $p \vee \sim q$.

8 Find the truth values of
(i) $\sim (p \vee \sim q)$ (ii) $\sim (\sim p \wedge \sim q)$

9 Find the truth values of the following compound statements.
(i) $p \wedge (q \wedge r)$ (ii) $(p \vee q) \vee r$

10 Write negation of statements '$2+3=5$ and $8<10$'.

11 Write down the negation of following compound statements.
(i) All rational numbers are real and complex.
(ii) A triangle has either 3 sides or 4 sides.
(iii) $|x|$ is either x or $-x$

12 Identify the quantifiers in the following statements.
(i) There exist a triangle which is not equilateral.
(ii) For all real numbers x and y, $xy = yx$.
(iii) For every natural number x, $x+1$ is also a natural number.
(iv) There exist a triangle which is not an isosceles triangle.

13 Find the truth values of
(i) $\sim p \rightarrow q$ (ii) $\sim (p \rightarrow q)$

14 Write the negation of the compound proposition. 'If the examination is difficult, then I shall pass, if I study hard'.

15 Form the biconditional statement $p \leftrightarrow q$, where
(i) p: The unit digits of an integer is zero. q: It is divisible by 5.
(ii) p: A natural number is odd. q: Natural number is not divisible by 2.

16 Find the truth values of
(i) $\sim p \leftrightarrow q$ (ii) $\sim (p \leftrightarrow q)$

10.3 Converse, Inverse and Contrapositive of Implications

17 Write down the converse of the following statements.
(i) If today is Monday, then tomorrow is Tuesday.
(ii) If you go to Agra, then you must visit Taj Mahal.
(iii) If all three angles of a triangle are equal, then the triangle is equilateral.
(iv) If $x:y = 3:2$, then $2x = 3y$.

18 For the statements. 'If two angles are congruent, then they have the same measure'. Write down the converse, inverse and contrapositive statements.

19 For the statements. 'If a quadrilateral is a rectangle, then it has two pairs of parallel sides'. Write the converse, inverse and contra positive statements.

20 Write down the contrapositive of the following statements.

(i) If $x = y$ and $y = 3$, then $x = 3$

(ii) If n is a natural number, then n is an integer.

(iii) If natural number n is divisible by 6, then n is divisible by 2 and 3.

21 Using contrapositive methods verify that if n^2 is even integer, then n is also an even integer.

10.4 Tautology, Contradiction and Algebra of Statements

22 Verify that

(i) $p \to (p \vee q)$ is tautology.

(ii) $(p \vee q) \wedge (\sim p \wedge \sim q)$ is a contradiction.

23 Verify that $(p \vee q) \to r \equiv (p \to r) \wedge (q \to r)$.

24 Verify that $[(p \vee q) \vee r] \leftrightarrow (p \vee (q \vee r)]$ is a tautology.

25 Verify that the compound statements $(p \vee q) \wedge \sim p$ and $\sim p \wedge q$ are logical equivalent.

26 If p is any statement, then verify that

(i) $p \vee p \equiv p$ (ii) $p \wedge p \equiv p$

27 If p is any statement, then verify that

(i) $p \vee \sim p = t$ (ii) $\sim\sim p = p$

(iii) $\sim t = f$ (iv) $\sim f = t$

Here, t and f are respectively, some tautology and fallacy in terms of the statement p.

28 If p and q be any two statements, then verify that

(i) $\sim (p \vee q) = \sim p \wedge \sim q$ (ii) $\sim (p \wedge q) = \sim p \vee \sim q$

29 Verify that the negation of the compound proposition is $p \vee (\sim p \vee q)$ is equivalent to $\sim p \wedge [p \wedge \sim q]$

30 Verify that the negation of $\sim s \vee (\sim r \wedge s)$ is equivalent to $s \wedge r$.

LEVEL 02 \ JEE Main Level

Single Choice Correct Questions

10.1 Statements

1 Which of the following is the negation of the statement "for all $M > 0$, there exists $x \in S$ such that $x \geq M$" ? **[JEE Main 2021]**

(a) there exists $M > 0$, such that $x < M$ for all $x \in S$

(b) there exists $M > 0$, there exists $x \in S$ such that $x \geq M$

(c) there exists $M > 0$, there exists $x \in S$ such that $x < M$

(d) there exists $M > 0$, such that $x \geq M$ for all $x \in S$

2 Consider the statement "The match will be played only if the weather is good and ground is not wet". **[JEE Main 2021]**

Select the correct negation from the following
(a) The match will not be played and weather is not good and ground is wet.
(b) If the match will not be played, then either weather is not good or ground is wet.
(c) The match will be played and weather is not good or ground is wet.
(d) The match will not be played or weather is good and ground is not wet.

3 Negation of the statement : $\sqrt{5}$ is an integer or 5 is irrational is **[JEE Main 2020]**
(a) $\sqrt{5}$ is not an integer or 5 is not irrational (b) $\sqrt{5}$ is not an integer and 5 is not irrational
(c) $\sqrt{5}$ is irrational or 5 is an integer (d) $\sqrt{5}$ is an integer and 5 is irrational

0.2 Elementary Operations of Logic

4 Consider the following two statements **[JEE Main 2018]**

Statement *p* The value of sin 120° can be divided by taking $\theta = 240°$ in the equation

$$2\sin\frac{\theta}{2} = \sqrt{1+\sin\theta} - \sqrt{1-\sin\theta}$$

Statement *q* The angles *A*, *B*, *C* and *D* of any quadrilateral *ABCD* satisfy the equation

$$\cos\left(\frac{1}{2}(A+C)\right) + \cos\left(\frac{1}{2}(B+D)\right) = 0$$

Then, the truth value of *p* and *q* are respectively,
(a) F, T (b) T, T (c) F, F (d) T, F

5 If p : 'Ram is tall' and q : 'Ram is intelligent', then the statement $\sim p \vee q$ is
(a) Ram is not tall or he is intelligent
(b) Ram is tall or he is intelligent
(c) Ram is not tall and he is intelligent
(d) Ram is not tall, then he is intelligent

6 The proposition $(\sim p) \vee (p \wedge \sim q)$ is equivalent to **[JEE Main 2017]**
(a) $p \rightarrow \sim q$ (b) $p \wedge (\sim q)$ (c) $q \rightarrow p$ (d) $p \vee (\sim q)$

7 The negation of $q \vee \sim (p \wedge r)$ is
(a) $\sim q \wedge (p \wedge r)$ (b) $\sim q \wedge \sim (p \wedge r)$
(c) $\sim q \vee (\sim p \wedge \sim q)$ (d) None of these

8 Negation of 'Paris is in France and London is in England' is
(a) Paris is in England and London is in France
(b) Paris is not in France or London is not in England
(c) Paris is in England or London is in France
(d) None of the above

9 If *p*, *q* and *r* are simple propositions such that $(p \wedge q) \wedge (q \wedge r)$ is true, then
(a) *p*, *q* and *r* are all false (b) *p*, *q* and *r* are all true
(c) *p*, *q* are true and *r* is false (d) *p* is true and *q*, *r* are false

10 If the statement $(p \wedge \sim r) \rightarrow (q \vee r)$, *q* and *r* are all false, then *p*
(a) is true (b) is false
(c) may be true or false (d) data is insufficient

11 If $(p \wedge \sim r) \rightarrow (\sim p \vee q)$ is false, then the truth values of p, q and r are respectively,
(a) T,T,T (b) T,F,T (c) T,F,F (d) F,T,T

12 The proposition $p \rightarrow \sim (p \wedge \sim q)$ is equivalent to **[JEE Main 2020]**
(a) q (b) $(\sim p) \vee q$ (c) $(\sim p) \wedge q$ (d) $(\sim p) \vee (\sim q)$

13 If $p \rightarrow (\sim p \vee q)$ is false, then truth values of p and q are respectively,

(a) F,T (b) F, F (c) T, T (d) T, F

14 If $p \rightarrow (\sim p \vee \sim q)$ is false, then the truth values of p and q are respectively, **[JEE Main 2018]**

(a) T, F (b) F, F (c) F, T (d) T, T

15 If $p \rightarrow (p \wedge \sim q)$ is false, then the truth values of p and q are respectively, **[JEE Main 2020]**

(a) F, F (b) T, F (c) T, T (d) F, T

16 Consider the following three statements,

(A) If $3+3=7$, then $4+3=8$

(B) If $5+3=8$, then earth is flat

(C) If both (A) and (B) are true, then $5+6=17$

Then, which of the following statements is correct **[JEE Main 2021]**

(a) (A) is false, but (B) and (C) are true

(b) (A) and (C) are true while (B) is false

(c) (A) is true while (B) and (C) are false

(d) (A) and (B) are false while (C) is true

17 If $p \rightarrow (q \vee r)$ is false, then the truth values of p, q and r are respectively.

(a) T, T, F (b) F, F, F (c) F, T, T (d) T, F, F

18 The statement $\sim (p \leftrightarrow \sim q)$ is

(a) equivalent to $\sim p \leftrightarrow \sim q$

(b) equivalent to $p \leftrightarrow q$

(c) equivalent to $p \leftrightarrow \sim q$

(d) None of these

19 In the truth table for the statement $(p \rightarrow q) \leftrightarrow (\sim p \vee q)$, the last column has the truth value in the following order of (p, q) as (T, T), (T, F), (F, T) and (F, F) is

(a) TTTT

(b) FTFT

(c) TTFF

(d) FFFF

10.3 Converse, Inverse and Contrapositive of Implications

20 What is converse for statement, 'if p, then q'?

(a) If not q, then not p

(b) If q, then p

(c) If not p, then q

(d) If not p, then not q

21 Which of the following is converse of 'If x is divisible by 4, then x must be divisible by 2'.

(a) If x is divisible by 2, then x must be divisible by 4

(b) If x is not divisible by 2, then x must be divisible by 4

(c) If x is not divisible by 2, then x is not divisible by 4

(d) If x is divisible by 4, then x is not divisible by 2

22 Which of the following is the inverse of the proposition 'If a number is a prime, then it is odd'?

(a) If a number is not odd, then it is not a prime

(b) If a number is a prime, then it is odd

(c) If a number is not odd, then it is a prime

(d) If a number is not a prime, then it is not odd

23 The contrapositive of $(p \vee q) \rightarrow r$ is

(a) $r \rightarrow (p \vee q)$

(b) $\sim r \rightarrow (p \vee q)$

(c) $\sim r \rightarrow (\sim p \wedge \sim q)$

(d) $p \rightarrow (q \vee r)$

24 The contrapositive of the statement; "If you will work, you will earn money" is **[JEE Main 2021]**
(a) to earn money, you need to work
(b) you will earn money, if you will not work
(c) if you will not earn money, you will not work
(d) if you will earn money, you will work

25 The contrapositive of the statement 'If I am not feeling well, then I will go to the doctor', is
(a) If I am feeling well, then I will not go to the doctor
(b) If I will go to the doctor, then I am feeling well
(c) If I will not go to the doctor, then I am feeling well
(d) If I will go to the doctor, then I am not feeling well

26 Contrapositive of the statement. 'If two numbers are not equal, then their squares are not equal', is **[JEE Main 2017]**
(a) If the squares of two numbers are equal, then the numbers are equal
(b) If the squares of two number are not equal, then the numbers are not equal
(c) If the squares of two numbers are equal, then the numbers are not equal
(d) If the squares of two numbers are not equal, then the numbers are equal

27 The contrapositive of the following statement 'if the side of a square doubles, then its area increases four times' is
(a) If the area of a square increases four times, then its side is not doubled.
(b) If the area of a square does not increase four times, then its side is not doubled
(c) If the area of square increases four times, then its side is doubled
(d) If the side of a square is not doubled, then its area does not increase four times

28 Contrapositive of statement 'if a function f is differentiable at a, then it is also continuous at a', is **[JEE Main 2020]**
(a) If a function f is continuous at a, then it is not differentiable at a
(b) If a function f is not continous at a, then it is not differentiable at a
(c) If a function f is not continuous at a, then it is differentiable at a
(d) If a function f is continuous at a, then it is differentiable at a

29 The contrapositive of the statement 'If I reach the station in time, then I will catch the train', is **[JEE Main 2020]**
(a) If I do not reach the station in time, then I will catch the train
(b) If I do not reach the station in time, then I will not catch the train
(c) If I will catch the train, then I reach the station in time
(d) If I will not catch the train, then I do not reach the station in time

30 Consider the statement; 'For an integer n, if $n^3 - 1$ is even, then n is odd'. The contrapositive statement of this statement, is **[JEE Main 2020]**
(a) For an integer n, if n is even, then $n^3 - 1$ is odd
(b) For an integer n, if $n^3 - 1$ is not even, then n is not odd
(c) For an integer n, if n is even, then $n^3 - 1$ is even
(d) For an integer n, if n is odd, then $n^3 - 1$ is even

10.4 Tautology, Contradiction and Algebra of Statements

31 The conditional statement $(p \wedge q) \rightarrow p$ is
(a) a tautology
(b) a fallacy
(c) neither tautology nor fallacy
(d) None of these

32 $(p \wedge \sim q) \wedge (\sim p \wedge q)$ is
(a) a tautology
(b) a contradiction
(c) neither (a) nor (b)
(d) None of these

33 The proposition $(p \to \sim p) \wedge (\sim p \to p)$ is a
(a) tautology and contradiction
(b) neither tautology nor contradiction
(c) contradiction
(d) tautology

34 $(p \wedge \sim q) \wedge (\sim p \vee q)$ is
(a) contradiction
(b) tautology
(c) either (a) or (b)
(d) neither (a) or (b)

35 The proposition $p \to \sim (p \wedge \sim q)$ is
(a) a contradiction
(b) a tautology
(c) either (a) or (b)
(d) neither (a) nor (b)

36 The statement $\{p \to (q \to p)\} \to \{p \to (p \vee q)\}$ is [JEE Main 2020]
(a) equivalent to $(p \wedge q) \vee (\sim q)$
(b) a contradiction
(c) equivalent to $(p \vee q) \wedge (\sim p)$
(d) a tautology

37 The following statement $(p \to q) \to [(\sim p \to q) \to q]$ is [JEE Main 2017]
(a) a fallacy
(b) a tautology
(c) equivalent to $\sim p \to q$
(d) equivalent to $p \to \sim q$

38 The logical statement $(p \Rightarrow q) \wedge (q \Rightarrow \sim p)$ is equivalent to [JEE Main 2020]
(a) p (b) q (c) $\sim p$ (d) $\sim q$

39 Given the following two statements
$(S_1): (q \vee p) \to (p \leftrightarrow \sim q)$ is a tautology and $(S_2): \sim q \wedge (\sim p \leftrightarrow q)$ is a fallacy. [JEE Main 2020]
(a) Only (S_1) is correct
(b) Only (S_2) is correct
(c) Both are correct
(d) Both are not correct

40 $\sim [\{(\sim (\sim p))\} \wedge q]$ is equal to
(a) $\sim p \wedge q$ (b) $\sim p \vee \sim q$ (c) $p \wedge \sim q$ (d) $\sim p \wedge \sim q$

41 $\sim \{p \vee (\sim q)\}$ is equal to
(a) $\sim p \vee q$ (b) $(\sim p) \wedge q$ (c) $\sim p \vee \sim p$ (d) $\sim p \wedge \sim q$

42 Which of the following is logically equivalent to $\sim (\sim p \to q)$?
(a) $p \wedge q$ (b) $p \wedge \sim q$ (c) $\sim p \wedge q$ (d) $\sim p \wedge \sim q$

43 $\sim \{p \vee (\sim p \vee q)\}$ is equal to
(a) $\sim p \wedge (p \wedge \sim q)$ (b) $(p \wedge \sim q) \vee \sim p$ (c) $(p \vee \sim q) \vee \sim p$ (d) None of these

44 Let p, q and r be three statements such that the truth values of $(p \wedge q) \to (\sim q \vee r)$ is F. Then, the truth values of p, q and r are respectively, [JEE Main 2020]
(a) T,F,T (b) T,T,T (c) F,T,F (d) T,T,F

45 The Boolean expression $\sim (p \vee q) \vee (\sim p \wedge q)$ is equivalent to [JEE Main 2018]
(a) p (b) q (c) $\sim q$ (d) $\sim p$

46 The negation of the Boolean expression $p \vee (\sim p \wedge q)$ is equivalent to [JEE Main 2020]
(a) $p \wedge \sim q$ (b) $\sim p \wedge \sim q$ (c) $\sim p \vee \sim q$ (d) $\sim p \vee q$

47 The Boolean expression $(p \wedge \sim q) \vee q \vee (\sim p \wedge q)$ is equivalent to [JEE Main 2017]
(a) $p \vee q$ (b) $p \vee \sim q$ (c) $\sim p \wedge q$ (d) $p \wedge q$

48 The negation of the Boolean expression $x \leftrightarrow \sim y$ is equivalent to [JEE Main 2020]
(a) $(x \wedge y) \vee (\sim x \wedge \sim y)$
(b) $(x \wedge y) \wedge (\sim x \vee \sim y)$
(c) $(x \wedge \sim y) \vee (\sim x \wedge y)$
(d) $(\sim x \wedge y) \vee (\sim x \wedge \sim y)$

49 The Boolean expression $\{(p \wedge q) \vee (p \vee \sim q)\} \wedge (\sim p \wedge \sim q)$ is equivalent to [JEE Main 2019]
(a) $p \wedge q$ (b) $p \wedge (-q)$ (c) $(\sim p) \wedge (\sim q)$ (d) $p \vee (\sim q)$

ANSWER KEY

Level 1

1. (i) yes (ii) yes (iii) not
   (iv) not (v) yes (vi) yes
   (vii) yes

2. (i) $\sqrt{7}$ is not rational.
   (ii) It is false that Australia is a continent.
   (iii) there exists a quadrilateral which has all its sides equal.
   (iv) 2 is a prime number.
   (v) There exist similar triangles which are not congruent.
   (vi) Area of the circle is not same as the perimeter of the circle

3. (i) and (ii) or
   (iii) and (iv) and

4. (i) inclusive (ii) exclusive (iii) exclusive

5.

p	q	$p \wedge q$	$p \vee q$
T	T	T	T
T	F	F	T
F	T	F	T
F	F	F	F

6. (i) $p \wedge q$ (ii) $p \wedge q$
   (iii) $p \vee q$ (iv) $p \vee q$

7.

p	q	$\sim p$	$\sim q$	$\sim p \wedge q$	$p \vee \sim q$
T	T	F	F	F	T
T	F	F	T	F	T
F	T	T	F	T	F
F	F	T	T	F	T

8. (i)

p	q	$\sim q$	$p \vee \sim q$	$\sim(p \vee \sim q)$
T	T	F	T	F
T	F	T	T	F
F	T	F	F	T
F	F	T	T	F

(ii)

p	q	$\sim p$	$\sim q$	$\sim p \wedge \sim q$	$\sim(\sim p \wedge \sim q)$
T	T	F	F	F	T
T	F	F	T	F	T
F	T	T	F	F	T
F	F	T	T	T	F

9.

p	q	r	$p \wedge (q \wedge r)$	$(p \vee q) \vee r$
T	T	T	T	T
T	T	F	F	T
T	F	T	F	T
F	T	T	F	T
T	F	F	F	T
F	T	F	F	T
F	F	T	F	T
F	F	F	F	F

10. $2 + 3 \neq 5$ or $8 \nless 10$

11. (i) all rational numbers are not real or not complex.
    (ii) a triangle has neither 3 sides nor 4 sides.
    (iii) $|x|$ is neither equal to x nor $-x$.

12. (i) there exist (ii) for all
    (iii) for every (iv) there exist

13. (i)

p	q	$\sim p$	$\sim p \rightarrow q$
T	T	F	T
T	F	F	T
F	T	T	T
F	F	T	F

(ii)

p	q	$p \rightarrow q$	$\sim(p \rightarrow q)$
T	T	T	F
T	F	F	T
F	T	T	F
F	F	T	F

14. The examination is difficult and I study hard but I shall not pass.

15. (i) $p \leftrightarrow q$: The unit digit of an integer is zero, if and only if it is divisible by 5.
    (ii) $p \leftrightarrow q$: A natural number is odd, if and only if it is not divisible by 2.

16. (i)

p	q	$\sim p$	$\sim p \leftrightarrow q$
T	T	F	F
T	F	F	T
F	T	T	T
F	F	T	F

(ii)

p	q	$p \leftrightarrow q$	$\sim(p \leftrightarrow q)$
T	T	T	F
T	F	F	T
F	T	F	T
F	F	T	F

17. (i) If tomorrow is Tuesday, then today is Monday.

(ii) If you visit Taj Mahal, then you go to Agra.

(iii) If the triangles is equilateral, then all three angles of triangle are equal.

(iv) If $2x = 3y$, then $x:y = 3:2$

18. Converse If two angles have the same measure, then they are congruent.

Inverse If two angles are not congruent, then they do not have the same measure.

Contrapositive If two angles do not have the same measure, then they are not congruent.

19. Converse If a quadrilateral has two pairs of parallel sides, then it is a rectangle.

Inverse If a quadrilateral is not a rectangle, then it does not have two pairs of parallel sides.

Contrapositive If a quadrilateral does not have two pairs of parallel sides, then it is not a rectangle.

Here, both converse and inverse are false. Since, the given statement is true, its contrapositive is true.

20. (i) If $x \neq 3$, then, $x \neq y$ or $y \neq 3$

(ii) If n is not an integer, then it is not a natural number.

(iii) If natural number n is not divisible by 2 or 3, then n is not divisible by 6.

Level 2

1. (a)	**2.** (c)	**3.** (b)	**4.** (a)	**5.** (a)	**6.** (a)	**7.** (a)
8. (b)	**9.** (b)	**10.** (a)	**11.** (c)	**12.** (b)	**13.** (d)	**14.** (d)
15. (c)	**16.** (b)	**17.** (d)	**18.** (b)	**19.** (a)	**20.** (b)	**21.** (a)
22. (d)	**23.** (c)	**24.** (c)	**25.** (c)	**26.** (a)	**27.** (b)	**28.** (b)
29. (d)	**30.** (a)	**31.** (a)	**32.** (b)	**33.** (c)	**34.** (a)	**35.** (d)
36. (b)	**37.** (b)	**38.** (c)	**39.** (d)	**40.** (b)	**41.** (b)	**42.** (d)
43. (a)	**44.** (d)	**45.** (d)	**46.** (b)	**47.** (a)	**48.** (a)	**49.** (c)

SOLUTIONS

Level 1

1. (i) The sentence is false because 18 is greater than 10. Hence, it is a statement.
   (ii) This sentence is false, since there are sets which are not finite. Hence, it is a statement.
   (iii) The sentence is subjective in the sense that for those who like Mathematics, it may be fun but for others it may not be. Hence, it is not a statement.
   (iv) This is a question which also contains the word 'Here'. Hence, it is not a statement.
   (v) The sentence is false. Hence, it is a statement.
   (vi) The sentence is true. Hence, it is a statement.
   (vii) The sentence is false. Hence, it is a statement.
2. (i) $\sqrt{7}$ is not rational.
   (ii) It is false that Australia is a continent.
   (iii) there exists a quadrilateral which has all its sides equal.
   (iv) 2 is a prime number.
   (v) There exist similar triangles which are not congruent.
   (vi) Area of the circle is not same as the perimeter of the circle
3. (i) The component statements are
   p : The sky is blue
   q : The grass is green
   The connecting word is 'and'.
   (ii) p : 0 is a positive number.
   q : 0 is a negative number.
   The connecting word is 'or'.
   (iii) p : 24 is a multiple of 4.
   q : 24 is a multiple of 6.
   (iv) p : 2 is an even number.
   q : 2 is a prime number.
4. (i) Here, 'or' is inclusive, since a person can have both a passport and a voter registration card to enter a country.
   (ii) Here, 'or' is exclusive because it is not possible for two lines to intesect and parallel together.
   (iii) Here, 'or' is exclusive because a student can not take both French and Sanskrit.
5.

p	q	$p \wedge q$	$p \vee q$
T	T	T	T
T	F	F	T
F	T	F	T
F	F	F	F

6. (i) p : Rahul passed in Hindi.
   q : Rahul passed in English.
   $\therefore p \wedge q$: Rahul passed in Hindi and English.
   (ii) p : x is an even integer.
   q : y is an even integer.
   $\therefore p \wedge q$: x and y are even integers.
   (iii) p : A number is divisible by 2.
   q : A number is divisible by 3.
   $\therefore p \vee q$: A number is either divisible by 2 or 3.
   (iv) p : Students can take Hindi as an optional paper.
   q : Students can take English as an optional paper.
   $\therefore p \vee q$: Students can take Hindi or English as on optional paper.
7.

p	q	$\sim p$	$\sim q$	$\sim p \wedge q$	$p \vee \sim q$
T	T	F	F	F	T
T	F	F	T	F	T
F	T	T	F	T	F
F	F	T	T	F	T

8. (i)

p	q	$\sim q$	$p \vee \sim q$	$\sim(p \vee \sim q)$
T	T	F	T	F
T	F	T	T	F
F	T	F	F	T
F	F	T	T	F

(ii)

p	q	$\sim p$	$\sim q$	$\sim p \wedge \sim q$	$\sim(\sim p \wedge \sim q)$
T	T	F	F	F	T
T	F	F	T	F	T
F	T	T	F	F	T
F	F	T	T	T	F

9.

p	q	r	$p \wedge (q \wedge r)$	$(p \vee q) \vee r$
T	T	T	T	T
T	T	F	F	T
T	F	T	F	T
F	T	T	F	T
T	F	F	F	T
F	T	F	F	T
F	F	T	F	T
F	F	F	F	F

10. Let $p: 2+3=5$ and $q: 8<10$

Given, proposition is : $p \wedge q$

Its negation is $\sim(p \wedge q) \equiv \sim p \vee \sim q$

Therefore, the required negation is '$2+3 \neq 5$ or $8 \nless 10$'.

11. (i) Let p : All rational numbers are real.

q : All rational numbers are complex.

$\sim p$: All rational numbers are not real.

$\sim q$: All rational numbers are not complex.

Then, negation of the given compound statement, is

$\sim(p \wedge q)$: All rational numbers are not real or not complex.

$\because \sim(p \wedge q) = \sim p \vee \sim q$

(ii) Let p : A triangle has 3-sides.

q : A triangle has 4-sides.

$\sim(p \vee q)$: A triangle has neither 3 sides nor 4 sides.

(iii) $p : |x|$ is equal to x

$q : |x|$ is equal to $-x$

$\sim(p \vee q) : |x|$ is neither equal to x nor $-x$

12. (i) There exist

(ii) For all

(iii) For every

(iv) There exist

13. (i)

p	q	$\sim p$	$\sim p \rightarrow q$
T	T	F	T
T	F	F	T
F	T	T	T
F	F	T	F

(ii)

p	q	$p \rightarrow q$	$\sim(p \rightarrow q)$
T	T	T	F
T	F	F	T
F	T	T	F
F	F	T	F

14. p: Examination is difficult.

q : I shall pass.

r : I study hard.

Given, result is : $p \rightarrow (r \rightarrow q)$

Now, $\sim(r \rightarrow q) \equiv r \wedge \sim q$

$\therefore \sim(p \rightarrow (r \rightarrow q)) \equiv p \wedge (r \wedge \sim q)$

Therefore, required negation is, 'The examination is difficult and I study hard but I shall not pass'.

15. (i) $p \leftrightarrow q$: The unit digit of an integer is zero, if and only if it is divisible by 5.

(ii) $p \leftrightarrow q$: A natural number is odd, if and only if it is not divisible by 2.

16. (i)

p	q	$\sim p$	$\sim p \leftrightarrow q$
T	T	F	F
T	F	F	T
F	T	T	T
F	F	T	F

(ii)

p	q	$p \leftrightarrow q$	$\sim(p \leftrightarrow q)$
T	T	T	F
T	F	F	T
F	T	F	T
F	F	T	F

17. (i) If tomorrow is Tuesday, then today is Monday.

(ii) If you visit Taj Mahal, then you go to Agra.

(iii) If the triangles is equilateral, then all three angles of triangle are equal.

(iv) If $2x = 3y$, then $x : y = 3 : 2$

18. **Converse** If two angles have the same measure, then they are congruent.

Inverse If two angles are not congruent, then they do not have the same measure.

Contrapositive If two angles do not have the same measure, then they are not congruent.

19. **Converse** If a quadrilateral has two pairs of parallel sides, then it is a rectangle.

Inverse If a quadrilateral is not a rectangle, then it does not have two pairs of parallel sides.

Contrapositive If a quadrilateral does not have two pairs of parallel sides, then it is not a rectangle.

Here, both converse and inverse are false. Since, the given statement is true, its contrapositive is true.

20. (i) If $x \neq 3$, then, $x \neq y$ or $y \neq 3$

(ii) If n is not an integer, then it is not a natural number.

(iii) If natural number n is not divisible by 2 or 3, then n is not divisible by 6.

21. Let $p: n^2$ is an even integer.

$q: n$ is also an even integer.

$\sim p$ is true i.e. n is not an even integer.

$\Rightarrow n^2$ is not an even integer.

$\because$ Square of an odd integer is odd.

$\Rightarrow \sim p$ is true.

$\therefore \sim q$ is true which provides that $\sim p$ is true.

22. (i) Truth values of $p \to (p \vee q)$

p	q	$p \vee q$	$p \to (p \vee q)$
T	T	T	T
T	F	T	T
F	T	T	T
F	F	F	T

Thus, for all possible truth values of p and q, the compound statement $p \to (p \vee q)$ is true.
Hence, $p \to (p \vee q)$ is tautology.

(ii) Truth values of $(p \vee q) \wedge (\sim p \wedge \sim q)$

p	q	$\sim p$	$\sim q$	$p \vee q$	$\sim p \wedge \sim q$	$(p \vee q) \wedge (\sim p \wedge \sim q)$
T	T	F	F	T	F	F
T	F	F	T	T	F	F
F	T	T	F	T	F	F
F	F	T	T	F	T	F

Thus, for all possible truth values of p and q, the compound statement $(p \vee q) \wedge (\sim p \wedge \sim q)$ is false.
Hence, $(p \vee q) \wedge (\sim p \wedge \sim q)$ is a contradiction.

23. Truth table for $(p \vee q) \to r$ and $(p \to r) \wedge (q \to r)$

p	q	r	$p \vee q$	$p \to r$	$q \to r$	$(p \vee q) \to r$	$(p \to r) \wedge (q \to r)$
T	T	T	T	T	T	T	T
T	T	F	T	F	F	F	F
T	F	T	T	T	T	T	T
T	F	F	T	F	T	F	F
F	T	T	T	T	T	T	T
F	T	F	T	T	F	F	F
F	F	T	F	T	T	T	T
F	F	F	F	T	T	T	T

It is evident from the last two columns that
$(p \vee q) \to r \equiv (p \to r) \wedge (q \to r)$

24. Truth value of $[(p \vee q) \vee r] \leftrightarrow [p \vee (q \vee r)]$

p	q	r	$p \vee q$	$(p \vee q) \vee r$	$q \vee r$	$p \vee (q \vee r)$	$[(p \vee q) \vee r] \leftrightarrow p \vee (q \vee r)$
T	T	T	T	T	T	T	T
T	T	F	T	T	T	T	T
T	F	T	T	T	T	T	T
F	T	T	T	T	T	T	T
T	F	F	T	T	F	T	T
F	T	F	T	T	T	T	T
F	F	T	F	T	T	T	T
F	F	F	F	F	F	F	T

Thus, for all possible truth values of p, q and r the compound statement
$[(p \vee q) \vee r] \leftrightarrow [p \vee (q \vee r)]$ is true.
$\therefore [(p \vee q) \vee r] \leftrightarrow [p \vee (q \vee r)]$ is a tautology.

25. Truth values of $(p \vee q) \wedge \sim p$ and $\sim p \vee q$ are as follows

p	q	$\sim p$	$p \vee q$	$(p \vee q) \wedge \sim p$	$\sim p \wedge q$
T	T	F	T	F	F
T	F	F	T	F	F
F	T	T	T	T	T
F	F	T	F	F	F

Clearly, $(p \vee q) \wedge \sim p$ and $\sim p \wedge q$ have identical truth values.
$\therefore (p \vee q) \wedge \sim p \equiv \sim p \wedge q.$

26. Truth values $p \vee p$ and $p \wedge p$

p	$p \vee p$	$p \wedge p$
T	T	T
F	F	F

27.

p	$\sim p$	$p \vee \sim p$	t	f	$\sim t$	$\sim f$	$\sim\sim p$
T	F	T	T	F	F	T	T
F	T	T	T	F	F	T	F

28. (i)

p	q	$\sim p$	$\sim q$	$(p \vee q)$	$\sim(p \vee q)$	$\sim p \wedge \sim q$
T	T	F	F	T	F	F
T	F	F	T	T	F	F
F	T	T	F	T	F	F
F	F	T	T	F	T	T

(ii)

p	q	$\sim p$	$\sim q$	$p \wedge q$	$\sim(p \wedge q)$	$\sim p \vee \sim q$
T	T	F	F	T	F	F
T	F	F	T	F	T	T
F	T	T	F	F	T	T
F	F	T	T	F	T	T

29. $\because s : \sim \{p \vee (\sim p \vee q)\}$
$\Rightarrow \sim p \wedge \sim (\sim p \vee q)$ [De Morgan's law]
$\Rightarrow s : \sim p \wedge [p \wedge \sim q]$ [De Morgan's law]

30. $\sim[\sim s \vee (\sim r \wedge s)] \equiv s \wedge \{\sim(\sim r \wedge s)\} = s \wedge (r \vee \sim s)$
$\equiv (s \wedge r) \vee (s \wedge \sim s) \equiv (s \wedge r) \vee F$
[$\because s \wedge \sim s$ is false]
$= s \wedge r$

Level 2

1. (a) Let P : for all $M > 0$, there exists $x \in S$ such that $x \geq M$
$\sim P$: there exists $M > 0$, for all $x \in S$ such that $x < M$
($\because$ negation of 'there exists' is 'for all')

2. (c) Let p: The match will be played
q: weather is good
r: ground is not wet
$\sim[p \to (q \wedge r)] = p \wedge \sim(q \wedge r) = p \wedge (\sim q \vee \sim r)$
The match will be played and weather is not good or ground is wet.

3. (b) Let p and q be the statements such that $p = \sqrt{5}$ is an integer and $q = 5$ is an irrational number.
Then, negation of the given statement is $\sqrt{5}$ is not an integer and 5 is not an irrational number
$$\sim(p \vee q) = \sim p \wedge \sim q$$

4. (a) **Statement p**
$$\sin 120° = \cos 30° = \frac{\sqrt{3}}{2} \Rightarrow 2\sin 120° = \sqrt{3}$$
So, $\sqrt{1 + \sin 240°} - \sqrt{1 - \sin 240°}$
$$= \sqrt{\frac{2 - \sqrt{3}}{2}} - \sqrt{\frac{2 + \sqrt{3}}{2}} \neq \sqrt{3}$$
Statement q
So, $A + B + C + D = 2\pi$
$$\Rightarrow \frac{A + C}{2} + \frac{B + D}{2} = \pi$$
$$\Rightarrow \cos\left(\frac{A + C}{2}\right) + \cos\left(\frac{B + D}{2}\right)$$
$$\Rightarrow \cos\left(\frac{A + C}{2}\right) - \cos\left(\frac{A + C}{2}\right) = 0$$
Therefore, statement p is false and statement q is true.

5. (a) $\sim p$: Ram is not tall.
Thus, $\sim p \vee q$ means 'Ram is not tall or he is intelligent'.

6. (a)

p	q	$\sim p$	$\sim q$	$p \wedge \sim q$	$(\sim p) \vee (p \wedge \sim q)$	$p \to \sim q$
T	T	F	F	F	F	F
T	F	F	T	T	T	T
F	T	T	F	F	T	T
F	F	T	T	F	T	T

7. (a) $\sim\{q \vee \sim(p \wedge r)\}$
$$\equiv \sim(q) \wedge [\sim\{\sim(p \wedge r)\}] \equiv \sim q \wedge (p \wedge r)$$

8. (b) Let p : Paris is in France,
q : London is in England
The given statement is $p \wedge q$.
Its negation is $\sim(p \wedge q) = \sim p \vee \sim q$
So, the required statement is 'Paris is not in France or London is not in England'.

9. (b) $(p \wedge q) \wedge (q \wedge r)$ is true which means that $p \wedge q$ and $q \wedge r$ are both true.
Therefore, p, q, and r are all true.

10. (a) $(p \wedge \sim r) \to (q \vee r)$ is false.
Thus, $p \wedge \sim r$ is true and $q \vee r$ is false.
Hence, p must be true.

11. (c) $(p \wedge \sim r) \to (\sim p \vee q)$ is false.
Thus, $(p \wedge \sim r)$ is true and $(\sim p \vee q)$ is false.
So, (p is true and $\sim r$ is true)
and ($\sim p$ is false and q is false)
Therefore, p is true, r is false and q is false.

12. (b)

p	q	$\sim q$	$p \wedge \sim q$	$\sim p$	$p \to \sim(p \wedge \sim q)$	$\sim p \vee q$
T	T	F	F	F	T	T
T	F	T	T	F	F	F
F	T	F	F	T	T	T
F	F	T	F	T	T	T

$\therefore p \to \sim(p \wedge \sim q)$ is equivalent to $\sim p \vee q$

13. (d) $p \to (\sim p \vee q)$ is false, then p is true and $\sim p \vee q$ is false
So, p is true and both $\sim p$ and q are false.
Therefore, p is true and q is false.

14. (d)

p	q	$\sim p$	$\sim q$	$\sim p \vee \sim q$	$p \to (\sim p \vee \sim q)$
T	T	F	F	F	F
T	F	F	T	T	T
F	T	T	F	T	T
F	F	T	T	T	T

15. (c)

p	q	$\sim q$	$p \wedge \sim q$	$p \to (p \wedge \sim q)$
T	T	F	F	F
T	F	T	T	T
F	T	F	F	T
F	F	T	F	T

16. (b) To solve this, let's construct a truth table for $p \to q$

p	q	$p \to q$
T	T	T
T	F	F
F	T	T
F	F	T

For (A), statements p and q are as follows,

$p: 3+3=7$

$q: 4+3=8$

Since, p is false, then

$p \rightarrow q$ is true.

Hence, (A) is true.

For (B), statements p and q are as follows,

$p: 5+3=8$

q: Earth is flat

Here, p is true and q is false.

$\therefore$ $p \rightarrow q$ is false.

Hence, (b) is false.

For (c), statements p and q are as follows,

p: Both (A) and (B) are true.

$q: 5+6=17$

Here 'p' is false because (B) is not true, then

$p \rightarrow q$ is true.

Hence, (c) is true.

17. (d) $p \rightarrow q$ is false only when p is true and q is false.

Therefore, $p \rightarrow (q \vee r)$ is false when p is true and $q \vee r$ is false. $q \vee r$ is false only when both q and r are false.

Hence, truth values of p, q and r are respectively, T, F and F.

18. (b)

p	q	$\sim q$	$p \leftrightarrow \sim q$	$\sim p \leftrightarrow \sim q$	$p \leftrightarrow q$
F	F	T	F	T	T
F	T	F	T	F	F
T	F	T	T	F	F
T	T	F	F	T	T

Clearly, equivalent to $p \leftrightarrow q$.

19. (a)

p	q	$p \rightarrow q$	$\sim p$	$\sim p \vee q$	$(p \rightarrow q) \leftrightarrow (\sim p \vee q)$
T	T	T	F	T	T
T	F	F	F	F	T
F	T	T	T	T	T
F	F	T	T	T	T

20. (b) $p \Rightarrow q$, then converse is $q \Rightarrow p$ (q implies p)

21. (a) p : If x is divisible by 4, then x must be divisible by 2.

Then, the converse of p : If x is divisible by 2, then x must be divisible by 4.

22. (d) If p : A number is a prime

q: It is odd

We have, $p \rightarrow q$

The inverse of $p \rightarrow q$ is $\sim p \rightarrow \sim q$

$\sim p \rightarrow \sim q$: If a number is not a prime, then it is not odd.

23. (c) Contrapositive of $p \rightarrow q$ is $\sim q \rightarrow \sim p$

Therefore, contrapositive of $(p \vee q) \rightarrow r$ is

$$\sim r \rightarrow \sim (p \vee q) \equiv \sim r \rightarrow (\sim p \wedge \sim q)$$

24. (c) Contrapositive of $p \rightarrow q$ is $\sim q \rightarrow \sim p$

p : you will work

$\Rightarrow \sim p$: you will not work

q : you will earn money

$\Rightarrow \sim q$: you will not earn money

Then,

$\sim q \rightarrow \sim p$: if you will not earn money, you will not work.

25. (c) Given statement can be written in implication form as 'I am not feeling well'.

$\Rightarrow$ 'I will go to the doctor'.

Contrapositive form

'I will not go to the doctor'.

$\Rightarrow$ 'I am feeling well'.

i.e. 'If I will not go to the doctor, then I am feeling well'.

26. (a) $p \rightarrow q$

Then, contrapositive is $\sim q \rightarrow \sim p$

$\therefore$ If the squares of two numbers are equal, then the numbers are equal.

27. (b) Contrapositive of $p \rightarrow q$ is given by $\sim q \rightarrow \sim p$.

28. (b) Contrapositive statement will be 'If a function f is not continuous at a, then it is not differentiable at a.

29. (d) Contrapositive of $p \rightarrow q$ is $\sim q \rightarrow \sim p$

i.e. Contrapositive of 'If p, then q', is 'If not q, then not p'.

30. (a) Contrapositive statement will be

'For an integer n, 'If n is not odd, then $n^3 - 1$ is not even'.

'For an integer n, 'If n is even, then $n^3 - 1$ is odd'.

31. (a)

p	q	$p \wedge q$	$(p \wedge q) \rightarrow p$
T	T	T	T
T	F	F	T
F	T	F	T
F	F	F	T

Therefore, $(p \wedge q) \rightarrow p$ is tautology.

32. (b) $(p \wedge \sim q) \wedge (\sim p \wedge q)$

$$\equiv (p \wedge \sim p) \wedge (\sim q \wedge q)$$
$$\equiv f \wedge f$$
$$\equiv f$$

Therefore, $(p \wedge \sim q) \wedge (\sim p \wedge q)$ is a contradiction.

33. (c)

p	$\sim p$	$p \to \sim p$	$\sim p \to p$	$(p \to \sim p) \wedge (\sim p \to p)$
T	F	F	T	F
F	T	T	F	F

Clearly, $(p \to \sim p) \wedge (\sim p \to p)$ is contradiction.

34. (a)

p	q	$\sim p$	$\sim q$	$p \wedge \sim q$	$\sim p \vee q$	$(p \wedge \sim q) \wedge (\sim p \vee q)$
T	T	F	F	F	T	F
T	F	F	T	T	F	F
F	T	T	F	F	T	F
F	F	T	T	F	T	F

Clearly, $(p \wedge \sim q) \wedge (\sim p \vee q)$ is a contradiction.

35. (d)

p	q	$\sim q$	$p \wedge \sim q$	$\sim(p \wedge \sim q)$	$p \to \sim(p \wedge \sim q)$
T	T	F	F	T	T
T	F	T	T	F	F
F	T	F	F	T	T
F	F	T	F	T	T

Result is neither tautology nor contradiction.

36. (d) The truth table of $\{p \to (q \to p)\} \to \{p \to (p \vee q)\}$ is

p	q	$p \vee q$	$p \to (p \vee q)$	$q \to p$	$p \to (q \to p)$	$\{p \to (q \to p)\} \to \{p \to (p \vee q)\}$
T	T	T	T	T	T	T
T	F	T	T	T	T	T
F	T	T	T	F	T	T
F	F	F	T	T	T	T

Hence, the statement is tautology.

37. (b) We have,

p	q	$\sim p$	$p \to q$	$\sim p \to q$	$(\sim p \to q) \to q$	$(p \to q) \to \{(\sim p \to q) \to q\}$
T	F	F	F	T	F	T
T	T	F	T	T	T	T
F	F	T	T	F	T	T
F	T	T	T	T	T	T

$\therefore$ It is tautology.

38. (c)

p	q	$p \Rightarrow q$	$\sim p$	$q \Rightarrow \sim p$	$(p \Rightarrow q) \wedge (q \Rightarrow \sim p)$
T	T	T	F	F	F
T	F	F	F	T	F
F	T	T	T	T	T
F	F	T	T	T	T

Clearly, $(p \Rightarrow q) \wedge (q \Rightarrow \sim p)$ is equivalent to $\sim p$.

39. (d) The truth table of both the statements is

p	q	$\sim p$	$\sim q$	$q \vee p$	$p \leftrightarrow \sim q$	(S_1)	$\sim p \leftrightarrow q$	(S_2)
T	T	F	F	T	F	F	F	F
T	F	F	T	T	T	T	T	T
F	T	T	F	T	T	T	T	F
F	F	T	T	F	F	T	F	F

$\therefore S_1$ is not tautology and S_2 is not fallacy.

Hence, both the statements (S_1) and (S_2) are not correct.

40. (b) $\sim \{(\sim(\sim p)) \wedge q\} \equiv \sim p \vee \sim q$

41. (b) $\sim \{p \vee (\sim q)\} \equiv \sim p \wedge \sim(\sim q) \equiv (\sim p) \wedge q$

42. (d) We know that

$$p \to q \equiv \sim p \vee q$$

$$\therefore \quad \sim p \to q \equiv p \vee q$$

$$\therefore \sim(\sim p \to q) \equiv \sim(p \vee q) \equiv \sim p \wedge \sim q$$

43. (a) $\sim[p \vee (\sim p \vee q)] \equiv \sim p \wedge \sim(\sim p \vee q)$

$$\equiv \sim p \wedge (\sim(\sim p) \wedge \sim q)$$

$$\equiv \sim p \wedge (p \wedge \sim q)$$

44. (d) $(p \wedge q) \to (\sim q \vee r) = \sim(p \wedge q) \vee (\sim q \vee r)$

$$= (\sim p \vee \sim q) \vee (\sim q \vee r)$$

$$= (\sim p \vee \sim q \vee r)$$

$\therefore$ $(\sim p \vee \sim q \vee r)$ is false, then $\sim p, \sim q$ and r all these must be false.

$\Rightarrow p$ is true, q is true and r is false.

45. (d) $\sim(p \vee q) \vee (\sim p \wedge q)$

$\Rightarrow (\sim p \wedge \sim q) \vee (\sim p \wedge q)$

$\Rightarrow \quad \sim p \wedge (\sim q \vee q) \Rightarrow \sim p \wedge t \equiv \sim p$

46. (b) Negation of given statement

$$= \sim(p \vee (\sim p \wedge q))$$

$$= \sim p \wedge \sim(\sim p \wedge q) = \sim p \wedge (p \vee \sim q)$$

$$= F \vee (\sim p \wedge \sim q) = \sim p \wedge \sim q$$

47. (a) $(p \wedge \sim q) \vee q \vee (\sim p \wedge q)$

$\Rightarrow \{(p \vee q) \wedge (\sim q \vee q)\} \vee (\sim p \wedge q)$

$\Rightarrow \{(p \vee q) \wedge T\} \vee (\sim p \wedge q)$

$\Rightarrow (p \vee q) \vee (\sim p \wedge q) \Rightarrow \{(p \vee q) \vee \sim p\} \wedge (p \vee q \vee q)$

$\Rightarrow T \wedge (p \vee q) \Rightarrow p \vee q$

48. (a) $p: x \leftrightarrow \sim y = (x \to \sim y) \wedge (\sim y \to x)$

$$= (\sim x \vee \sim y) \wedge (y \vee x)$$

$$= \sim(x \wedge y) \wedge (x \vee y) \quad [\because \sim(x \wedge y) = \sim x \vee \sim y]$$

Negation of p is $\sim p = (x \wedge y) \vee \sim(x \vee y)$

$$= (x \wedge y) \vee (\sim x \wedge \sim y)$$

49. (c) Consider the Boolean expression

$\{(p \wedge q) \vee (p \vee \sim q)\} \wedge (\sim p \wedge \sim q)$

$$= (p \vee \sim q) \wedge (\sim p \wedge \sim q)$$

$$= \{(p \wedge \sim q) \wedge \sim p\} \wedge \{(p \vee \sim q) \wedge \sim q\}$$

$$= \{(p \wedge \sim p) \vee (\sim q \wedge \sim p)\} \wedge \sim q$$

$$= (\sim p \wedge \sim q) \wedge \sim q$$

$$= (\sim p) \wedge (\sim q)$$

CHAPTER 11

STATISTICS

LEVEL 01 \ Starter Level

11.1 Measure of Central Tendency

1 If the height of 5 people are 142 cm, 150 cm, 149 cm, 156 cm and 153 cm, then find the mean height.

2 If a variate x is expressed as a linear function of two variates u and v in the form $x = au + bv$, then find the mean $\bar{x}$ of x in terms of $\bar{u}$ and $\bar{v}$.

3 If the mean of the numbers $27 + x$, $31 + x$, $89 + x$, $107 + x$ and $156 + x$ is 82, then find the mean of $130 + x$, $126 + x$, $68 + x$, $50 + x$ and $1 + x$.

4 The mean of 100 items is 49. It was discovered that three items which should have been 60, 70 and 80, were wrongly read as 40, 20 and 50, respectively. Then, find the correct mean.

5 The mean weight per student in a group of seven students is 55 kg. If the individual weights of six students are 52, 58, 55, 53, 56 and 54, then find the weight of the seventh student.

6 If the arithmetic mean of the numbers $x_1, x_2, x_3, \ldots, x_n$ is $\bar{x}$, then find the arithmetic mean of numbers $ax_1 + b, ax_2 + b, ax_3 + b, \ldots, ax_n + b$, where a and b are two constants.

7 Find the mean of the following distribution

x	4	6	9	10	15
f	5	10	10	7	8

8 Find the mean of the following distribution

Class interval	3-5	5-7	7-9	9-11	11-13
Frequency	5	10	10	7	8

9 Find the weighted mean of the first n natural numbers, whose weights are equal to the squares of the corresponding numbers.

10 Let's consider the data : 56, 67, 54, 34, 78, 43 and 23. What is the median?

11 Let's consider the data : 50, 67, 24, 34, 78 and 43. What is the median?

12 If the median of $\frac{x}{5}, \frac{x}{4}, \frac{x}{3}, \frac{x}{2}$ and $x\,(x > 0)$ is 8, then find the value of x.

13 If a variable takes the discrete values $\alpha + 4, \alpha - \frac{7}{2}, \alpha - \frac{5}{2}, \alpha - 3, \alpha - 2, \alpha + \frac{1}{2}, \alpha - \frac{1}{2}$ and $\alpha\,(\alpha > 0)$, then find the median.

14 Find the median of ${}^{2n}C_0, {}^{2n}C_1, {}^{2n}C_2, {}^{2n}C_3, \ldots, {}^{2n}C_n$, where n is even.

15 Calculate the median for the following data

Marks	0-10	10-20	20-30	30-40	40-50
Number of students	5	15	20	2	8

16 Find the median marks for the following distribution

Classes	0-10	10-20	20-30	30-40	40-50
Frequency	2	12	22	8	6

17 Find the mode of given data

Marks obtained	0-20	20-40	40-60	60-80	80-100
Number of students	5	10	12	6	3

11.2 Range and Mean Deviation

18 Find the range and coefficient of range for the data

Marks	10	20	30	40	50	60
Number of students	5	7	8	7	5	2

19 Find the range and coefficient of range of the following distribution

Marks	10-20	20-30	30-40	40-50
Number of students	5	1	3	10

20 Find the mean deviation about the mean for the following data
12, 3, 18, 17, 4, 9, 17, 19, 20, 15, 8, 17, 2, 3, 16, 11, 3, 1, 0, 5.

21 (i) Find the mean deviation about the mean for the following data
6, 7, 10, 12, 13, 4, 8, 12.

(ii) Find the mean deviation about the mean for the following data

Marks obtained	10-20	20-30	30-40	40-50	50-60	60-70	70-80
Number of students	2	3	8	14	8	3	2

22 Find the mean deviation about the median for the following data
3, 9, 5, 3, 12, 10, 18, 4, 7, 19, 21.

23 Find the mean deviation about median for the following data

Classes	0-10	10-20	20-30	30-40	40-50	50-60
Frequency	6	7	15	16	4	2

11.3 Variance, Standard Deviation and Coefficient of Variation

24 Find the variance of the first n natural numbers.

25 Find the standard deviation of the following data
6, 8, 10, 12, 14, 16, 18, 20, 22, 24.

26 Consider the frequency distribution, where A is a positive integer

x	A	2A	3A	4A	5A	6A
f	2	1	1	1	1	1

If the variance is 160, then evaluate the value of A.

27 Find the Variance of the data 6, 5, 9, 13, 12, 8, and 10.

28 Let a, b, c, d and e be the observations with mean m and standard deviation. Find the standard deviation of the observations $a+k, b+k, c+k, d+k$ and $e+k, k \in R$.

29 Find the variance and standard deviation for the following data

x_i	4	8	11	17	20	24	32
f_i	3	5	9	5	4	3	1

30 Calculate mean, variance and standard deviation for the following distribution

Classes	30-40	40-50	50-60	60-70	70-80	80-90	90-100
Frequency	3	7	12	15	8	3	2

31 The mean of five observations is 4 and their variance is 5.2. If three of these observations are 1, 2 and 6, then find the other two observations.

32 The following values are calculated in respect of heights and weights of the students of a section of class XI.

	Height (in cm)	Weight (In kg)
Mean	162.6	52.36
Variance	127.69	23.1361

Can we say that the weights show greater variation than heights?

33 Coefficient of variation of two distributions are 60 and 70 and their standard deviations are 21 and 16, respectively. What are their arithmetic means?

LEVEL 02 \\ JEE Main Level

Single Choice Correct Questions

11.1 Measure of Central Tendency

1 The AM of the sequence 1, 2, 4, 8, 16, ... 2^n is

(a) $\frac{2^n-1}{n}$ (b) $\frac{2^{n+1}-1}{n+1}$ (c) $\frac{2^n+1}{n}$ (d) $\frac{2^n-1}{n+1}$

2 The mean of a set of numbers is $\bar{x}$. If each number is divided by 3, then the new mean is

(a) $\bar{x}$ (b) $\bar{x}+3$ (c) $3\bar{x}$ (d) $\frac{\bar{x}}{3}$

3 The mean of the data set comprising of 16 observations is 16. If one of the observation valued 16 is deleted and three new observations valued 3, 4 and 5 are added to the data, then the mean of the resultant data is [JEE Main 201

(a) 16.8 (b) 16.0 (c) 15.8 (d) 14.0

4 If the mean of the distribution is 2.6, then the value of y is

Variate x	1	2	3	4	5
Frequency f of x	4	5	y	1	2

(a) 24 (b) 13 (c) 8 (d) 3

5 If the mean of the set of numbers $x_1, x_2, x_3, \ldots, x_n$ is $\bar{x}$, then the mean of the numbers $x_i + 2i$ $1 \le i \le n$ is

(a) $\bar{x} + 2n$ (b) $\bar{x} + n + 1$ (c) $\bar{x} + 2$ (d) $\bar{x} + n$

6 The average of n numbers $x_1, x_2, x_3, \ldots, x_n$ is M. If x_n replaced by x', then new average is

(a) $M - x_n + x'$ (b) $\dfrac{nM - x_n + x'}{n}$ (c) $\dfrac{(x-1)M + x'}{n}$ (d) $\dfrac{M - x_n + x'}{n}$

7 The mean of n items is $\bar{x}$. If the first item is increased by 1, second by 2 and so on, then the new mean is

(a) $\bar{x} + n$ (b) $\bar{x} + \dfrac{n}{2}$ (c) $\bar{x} + \dfrac{n+1}{2}$ (d) None of these

8 The following data give the distribution of heights of students

Height (in cm)	160	150	152	161	156	154	155
Number of students	12	8	4	4	3	3	7

The median of the distribution is

(a) 154 (b) 155 (c) 160 (d) 161

9 If μ is the mean of distribution (y_i, f_i), then $\Sigma f_i (y_i - \mu)$ is equal to

(a) MD (b) SD (c) 0 (d) relative frequency

10 For a normal distribution if the mean is M, mode is M_0 and median is M_e, then

(a) $M > M_e > M_0$ (b) $M < M_e < M_0$
(c) $M = M_e M_0$ (d) $M = M_e = M_0$

11 In a moderately skewed distribution, the values of mean and median are 5 and 6, respectively. The value of mode in such a situation is approximately equal to

(a) 8 (b) 11 (c) 6 (d) None of these

11.2 Range and Mean Deviation (About Mean and Median)

12 The range of the following set of observations 2, 3, 5, 9, 8, 7, 6, 5, 7, 4 and 3 is

(a) 11 (b) 7 (c) 5.5 (d) 6

13 The mean deviation of the data 3, 10, 10, 4, 7, 10 and 5 from the mean is

(a) 2 (b) 2.57 (c) 3 (d) 3.75

14 Runs scored by a batsman in 10 innings are 38, 70, 48, 34, 42, 55, 63, 46, 54 and 44. The mean deviation is

(a) 8.6 (b) 6.4 (c) 10.6 (d) 9.6

15 If the mean deviation about the median of the numbers $a, 2a, \ldots, 50a$ is 50, then $|a|$ equals

(a) 5 (b) 2 (c) 3 (d) 4

16 If the mean deviation of the numbers $1, 1+d, 1+2d, \ldots, 1+100d$ from their mean is 225, then d is equal to

(a) 10.0 (b) 20.0 (c) 10.1 (d) 20.2

17 Following are the marks obtained by 9 students in Mathematics test

50, 69, 20, 33, 53, 39, 40, 65, 59.

The mean deviation from the median is

(a) 9 (b) 10.5 (c) 12.67 (d) 14.76

18 Consider any set of observations $x_1, x_2, x_3, \ldots, x_{101}$. It is given that $x_1 < x_2 < x_3 < \ldots < x_{100} < x_{101}$, then the mean deviation of this set of observations about a point k is minimum when k equals

(a) x_1 (b) x_{51}

(c) $\frac{x_1 + x_2 + \ldots + x_{101}}{101}$ (d) x_{50}

19 If the sum of the diameter of 50 observations from 30 is 50, then the mean of these observation is **[JEE Main 2019]**

(a) 50 (b) 30 (c) 57 (d) 31

11.3 Variance, Standard Deviation and Coefficient of Variations

20 Consider the number 1, 2, 3, 4, 5, 6, 7, 8, 9 and 10. If 1 is added to each number, the variance of the number so obtained is

(a) 6.5 (b) 2.87 (c) 3.87 (d) 8.25

21 The variance of first 50 even natural numbers is

(a) $\frac{833}{4}$ (b) 833 (c) 437 (d) $\frac{437}{4}$

22 If the mean and variance of the following data :

6, 10, 7, 13, a, 12, b, 12

are 9 and $\frac{37}{4}$ respectively, then $(a-b)^2$ is equal to **[JEE Main 2021]**

(a) 24 (b) 12 (c) 32 (d) 16

23 The mean of 100 observations is 50 and their standard deviation is 5. The sum of squares of all the observation is

(a) 50000 (b) 250000

(c) 252500 (d) 255000

24 If a variance of the following frequency distribution

Classes	10-20	20-30	30-40
Frequency	2	x	2

is 50, then x is equal to **[JEE Main 2020]**

(a) 4 (b) 6 (c) 8 (d) 5

25 The following information relates to a sample of size 60, $\Sigma x_i^2 = 18000$, $\Sigma x_i = 960$. The variance is

(a) 6.63 (b) 16 (c) 22 (d) 44

26 The mean of 6 distinct observations is 6.5 and their variance is 10.25. If 4 out of 6 observations are 2, 4, 5 and 7, then the remaining two observations are **[JEE Main 2021]**

(a) 10, 11 (b) 3, 18 (c) 8, 13 (d) 1, 20

27 If a variable x takes values 0, 1, 2, ... n with frequencies proportional to the binomial coefficients ${}^nC_0, {}^nC_1, {}^nC_2, \ldots, {}^nC_n$, then variance of x is

(a) $\frac{n^2 - 1}{12}$ (b) $\frac{n}{2}$ (c) $\frac{n}{4}$ (d) None of these

28 The mean and variance of 20 observations are found to be 10 and 4, respectively. On rechecking, it was found that an observation 9 was incorrect and the correct observation was 11, then the correct variance is **[JEE Main 2020]**

(a) 3.99 (b) 4.01 (c) 4.02 (d) 3.98

29 The mean and variance of 7 observations are 8 and 16, respectively. If five observations are 2, 4, 10, 12 and 14, then the absolute difference of the remaining two observations **[JEE Main 2020]**

(a) 1 (b) 4 (c) 2 (d) 3

30 For the frequency distribution

Variate (x)	x_1	x_2	$x_3, \ldots. x_{25}$
Frequency (f)	f_1	f_2	$f_3, \ldots f_{25}$

where $0 < x_1 < x_2 < x_3 <, \ldots\ldots\ldots\ldots < x_{25} = 10$ and $\sum_{i=1}^{25} f_i > 0$, the standard deviation cannot be **[JEE Main 2020]**

(a) 4 (b) 1 (c) 6 (d) 2

31 Consider the first 10 positive integers. If we multiply each number by -1 and then add 1 to each number, the variance of the numbers so obtained is

(a) 8.25 (b) 6.5 (c) 3.87 (d) 2.87

32 Let in a series of $2n$ observations, half of them are equal to a and remaining half are equal to $-a$. Also, by adding a constant b in each of these observations, the mean and standard deviation of new set become 5 and 20, respectively. Then, the value of $a^2 + b^2$ is equal to **[JEE Main 2021]**

(a) 425 (b) 650 (c) 250 (d) 925

33 For two data sets, each of size 5, the variance are given to be 4 and 5 and the corresponding means are given to be 2 and 4, respectively. The variance of the combined data set is

(a) $\frac{13}{2}$ (b) $\frac{5}{2}$ (c) $\frac{11}{2}$ (d) 6

34 Let $X = \{n \in N : 1 \le n \le 17\}$ and $y = \{ax + b : n \in N \text{ and } a, b, \in R, a > 0\}$. If mean and variance of elements of Y are 17 and 216, respectively, then $a + b$ is equal to **[JEE Main 2020]**

(a) 7 (b) -7 (c) -27 (d) 9

35 Consider three observations a, b and c, such that $b = a + c$. If the standard deviation of $a + 2, b + 2, c + 2$ is d, then which of the following is true? **[JEE Main 2021]**

(a) $b^2 = 3(a^2 + c^2) + 9d^2$ (b) $b^2 = a^2 + c^2 + 3d^2$

(c) $b^2 = 3(a^2 + c^2 + d^2)$ (d) $b^2 = 3(a^2 + c^2) - 9d^2$

36 If $\sum_{i=1}^{9}(x_i - 5) = 9$ and $\sum_{i=1}^{9}(x_i - 5)^2 = 45$, then the standard deviation of the 9 items $x_1, x_2, \ldots, x_9$ is **[JEE Main 2018]**

(a) 3 (b) 9 (c) 4 (d) 2

37 The standard deviation of 25 numbers is 40. If each of the number is increased by 5, then the new standard deviation will be

(a) 40 (b) 45 (c) 35 (d) None of these

38 If the mean and the standard deviation of the data 3, 5, 7, a and b are 5 and 2, respectively, then a and b are the roots of the equation **[JEE Main 2020]**

(a) $x^2 - 10x + 18 = 0$ (b) $2x^2 - 20x + 19 = 0$

(c) $x^2 - 10x + 19 = 0$ (d) $x^2 - 20x + 18 = 0$

39 The mean and standard deviation of 20 observations were calculated as 10 and 2.5 respectively. It was found that by mistake one data value was taken as 25 instead of 35. If α and $\sqrt{\beta}$ are the mean and standard deviation respectively for correct data, then (α, β) is **[JEE Main 2021]**

(a) (11, 26) (b) (10.5, 25) (c) (11, 25) (d) (10.5, 26)

40 The standard deviation of a variate x is σ. The standard deviation of the variate $(ax + b)/c$, where a, b and c are constants, is

(a) $\left(\frac{a}{c}\right)\sigma$ (b) $\left|\frac{a}{c}\right|\sigma$ (c) $\left(\frac{a^2}{c^2}\right)\sigma$ (d) None of these

41 The standard deviation of some temperature data in C is 5. If the data were converted into F, the new variance would be

(a) 81 (b) 57 (c) 36 (d) 25

42 The first of the two samples in a group has 100 items with mean 15 and standard deviation 3. If the whole group has 250 items with mean 15.6 and standard deviation $\sqrt{13.44}$, then the standard deviation of the second sample is **[JEE Main 2021]**

(a) 8 (b) 6 (c) 4 (d) 5

43 Let r be the range and $s^2 = \frac{1}{n-1}\sum_{i=1}^{n}(x_i - \bar{x})^2$ be the standard deviation of 's' a set of observations $x_1, x_2, \ldots, x_n$, then

(a) $s \le r\sqrt{\frac{n}{n-1}}$ (b) $s = r\sqrt{\frac{n}{n-1}}$

(c) $s \ge r\sqrt{\frac{n}{n-1}}$ (d) None of these

44 If the standard deviation of the numbers 2, 3, a and 11 is 3. 5, then which of the following is true?

(a) $3a^2 - 32a + 84 = 0$ (b) $3a^2 - 34a + 91 = 0$

(c) $3a^2 - 23a + 44 = 0$ (d) $3a^2 - 26a + 55 = 0$

45 What is the standard deviation of the following data?

Measurement	0-10	10-20	20-30	30-40
Frequency	1	3	4	2

(a) 81 (b) 7.6 (c) 9 (d) 2.26

46 Let $x : (1 \le i \le 10)$ be ten observations of a random variable X. If $\sum_{i=1}^{10}(x_i - p) = 3$ and $\sum_{i=1}^{10}(x_i - p)^2 = 9$, where $0 \ne p \in R$, then the standard deviation of the observations is **[JEE Main 2020]**

(a) $\sqrt{\frac{3}{5}}$ (b) $\frac{4}{5}$ (c) $\frac{9}{10}$ (d) $\frac{7}{10}$

47 If $\sum_{i=1}^{n}(x_i - a) = n$ and $\sum_{i=1}^{n}(x_i - a)^2 = na, (n, a > 1)$, then the standard deviation of n observations $x_1, x_2, \ldots, x_n$ is **[JEE Main 2020]**

(a) $a - 1$
(b) $n\sqrt{a-1}$
(c) $\sqrt{n(a-1)}$
(d) $\sqrt{a-1}$

48 For $(2n+1)$ observations $x_1, -x_1, x_2, -x_2, \ldots, x_n, -x_n$ and 0, where all x's are distinct, let SD and MD denote the standard deviation and median, respectively. Then, which of the following is always true?

(a) SD < MD
(b) SD > MD
(c) SD = MD
(d) Nothing can be said in general about the relationship between SD and MD

Numerical Value Type Questions

49 The average marks of boys in a class is 52 and that of girls is 42. The average marks of boys and girls combined is 50. The percentage of boys in the class is ....

50 The mean age of 25 teachers in a school is 40 yr. A teacher retires at the age of 60 yr and a new teacher is appointed in his place. If the mean age of the teachers in this school now is 39 yr, then the age (in years) of the newly appointed teacher is .... **[JEE Main 2021]**

51 The median of a set of nine distinct observations is 20.5. If each of the largest four observations of the set is increased by 2, then the median of the new set is ....

52 Consider the following frequency distribution

Class	0-6	6-12	12-18	18-24	24-30
Frequency	a	b	12	9	5

If mean $= \frac{309}{22}$ and median = 14, then the value $(a-b)^2$ is equal to .... **[JEE Main 2021]**

53 Consider the following frequency distribution

Class	10-20	20-30	30-40	40-50	50-60
Frequency	α	110	54	30	β

If the sum of all frequencies is 584 and median is 45, then $|\alpha - \beta|$ is equal to ....

54 If the variance of 10 natural numbers 1, 1, 1,...., 1, k is less than 10, then the maximum possible value of k is .... **[JEE Main 2021]**

55 Consider a set of $3n$ numbers having variance 4. In this set, the mean of first $2n$ numbers is 6 and the mean of the remaining n numbers is 3. A new set is constructed by adding 1 into each of first $2n$ numbers and subtracting 1 from each of the remaining n numbers. If the variance of the new set is k, then $9k$ is equal to ....

56 An online exam is attempted by 50 candidates out of which 20 are boys. The average marks obtained by boys is 12 with a variance 2. The variance of marks obtained by 30 girls is also 2. The average marks of all 50 candidates is 15. If μ is the average marks of girls and σ^2 is the variance of marks of 50 candidates, then $\mu + \sigma^2$ is equal to **[JEE Main 2021]**

57 The standard deviation of a variable x is 10. Then, the standard deviation of $50+5x$ is

58 If the mean and variance of eight numbers 3, 7, 9, 12, 13, 20, x and y be 10 and 25, respectively, then xy is equal to .... **[JEE Main 2020]**

59 If the variance of the terms is an increasing AP, $b_1, b_2, b_3, \ldots, b_{11}$ is 90, then the common difference of this AP is .... **[JEE Main 2020]**

60 The mean of 5 observations is 4.4 and their variance is 8.24. If three of the observations are 1, 2 and 6, then sum of other two observations is ....

61 Let $X_1, X_2, \ldots, X_{18}$ be eighteen observations, such that $\sum_{i=1}^{18}(X_i-\alpha)=36$ and $\sum_{i=1}^{18}(X_i-\beta)^2=90$, where α and β are distinct real numbers. If the standard deviation of these observations is 1, then the value of $|\alpha-\beta|$ is .... **[JEE Main 2021]**

ANSWER KEY

Level 1

1. 150 cm	**2.** $a\bar{u}+b\bar{v}$	**3.** 75	**4.** 50	**5.** 57 kg	**6.** $a\bar{x}+b$
7. 9	**8.** 8.15	**9.** $\frac{3n(n+1)}{2(2n+1)}$	**10.** 54	**11.** 46.5	**12.** $x=24$
13. $\alpha-\frac{5}{4}$	**14.** ${}^{2n}C_{n/2}$	**15.** 22.5	**16.** 25	**17.** 45	**18.** 0.714
19. 0.66	**20.** 6.2	**21.** (i) 2.75	(ii) 10	**22.** 5.27	**23.** 10.16
24. $\frac{n^2-1}{12}$	**25.** 5.74	**26.** $A=7$	**27.** $\frac{52}{7}$	**28.** $\sqrt{\frac{\Sigma x_i^2}{n}-m^2}$	

29. Variance = 45.8 and standard deviation = 6.77

30. Mean = 62, variance = 201 and standard deviation = 14.18

31. $x=4, y=7$ or $x=7, y=4$

32. CV in weight is greater than CV in heights

33. $\bar{x}_1=35$ and $\bar{x}_2=22.85$

Level 2

1. (b)	**2.** (d)	**3.** (d)	**4.** (c)	**5.** (b)	**6.** (b)	**7.** (c)
8. (b)	**9.** (c)	**10.** (d)	**11.** (a)	**12.** (b)	**13.** (b)	**14.** (a)
15. (d)	**16.** (c)	**17.** (c)	**18.** (b)	**19.** (d)	**20.** (d)	**21.** (b)
22. (d)	**23.** (c)	**24.** (a)	**25.** (d)	**26.** (a)	**27.** (c)	**28.** (a)
29. (c)	**30.** (c)	**31.** (a)	**32.** (a)	**33.** (c)	**34.** (b)	**35.** (d)
36. (d)	**37.** (a)	**38.** (c)	**39.** (d)	**40.** (b)	**41.** (a)	**42.** (c)
43. (a)	**44.** (a)	**45.** (c)	**46.** (c)	**47.** (d)	**48.** (b)	**49.** (80)
50. (35)	**51.** (20.5)	**52.** (4)	**53.** (164)	**54.** (11)	**55.** (68)	**56.** (25)
57. (50)	**58.** (54)	**59.** (3)	**60.** (13)	**61.** (4)		

SOLUTIONS

Level 1

1. Mean height,

$$\bar{x} = \frac{(142 + 150 + 149 + 156 + 153)}{5}$$

$$= \frac{750}{5} = 150$$

Mean, $\bar{x} = 150$ cm

Thus, the mean height is 150 cm.

2. We have, $\Sigma x = a\,\Sigma u + b\,\Sigma v$

Therefore,

$$\bar{x} = \frac{1}{n}\Sigma x = a \cdot \frac{1}{n}\Sigma u + b \cdot \frac{1}{n}\Sigma v = a\bar{u} + b\bar{v}$$

3. Given,

$$82 = \frac{(27 + x) + (31 + x) + (89 + x) + (107 + x) + (156 + x)}{5}$$

$\Rightarrow\ 82 \times 5 = 410 + 5x$

$\Rightarrow\ 410 - 410 = 5x$

$\Rightarrow\ x = 0$

Therefore, the required mean is

$$\bar{x} = \frac{(130 + x) + (126 + x) + (68 + x) + (50 + x) + (1 + x)}{5}$$

$$= \frac{375 + 5x}{5}$$

$$= \frac{375 + 0}{5} = \frac{375}{5}$$

$\bar{x} = 75$

4. Sum of 100 items $= 49 \times 100 = 4900$

Sum of items added $= 60 + 70 + 80 = 210$

New sum $= 4900 + 210 - 110 = 5000$

$\therefore$ Correct mean $= \frac{5000}{100} = 50$

5. The total weight of seven students is $55 \times 7 = 385$ kg

The sum of the weights of six students is

$52 + 58 + 55 + 53 + 56 + 54 = 328$ kg

Hence, the weight of the seventh student is

$385 - 328 = 57$ kg

6. Required mean $= \frac{(ax_1 + b) + (ax_2 + b) + \ldots (ax_n + b)}{n}$

$$= \frac{a(x_1 + x_2 + \ldots + x_n)}{n} + b$$

$$= a\bar{x} + b \qquad \left[\because \frac{x_1 + x_2 + \ldots + x_n}{n} = \bar{x}\right]$$

7.

x_i	f_i	$x_i f_i$
4	5	20
6	10	60
9	10	90
10	7	70
15	8	120
	$\Sigma f_i = 40$	$\Sigma x_i f_i = 360$

Mean, $\bar{x} = \frac{\Sigma x_i f_i}{\Sigma f_i} = \frac{360}{40} = 9$

Thus, mean $= 9$

8.

Classes interval	Frequency (f_i)	x_i	$f_i x_i$
3-5	5	4	20
5-7	10	6	60
7-9	10	8	80
9-11	7	10	70
11-13	8	12	96
	$\Sigma f_i = 40$		$\Sigma f_i x_i = 326$

Mean $= \frac{\Sigma f_i x_i}{\Sigma f_i} = \frac{326}{40} = 8.15$

$\therefore$ Mean $= 8.15$

9. Weighted mean $= \frac{1 \cdot 1^2 + 2 \cdot 2^2 + \ldots + n \cdot n^2}{1^2 + 2^2 + \ldots + n^2}$

$$= \frac{\Sigma n^3}{\Sigma n^2}$$

$$= \frac{\frac{n(n+1)}{2} \cdot \frac{n(n+1)}{2}}{\frac{n(n+1)(2n+1)}{6}}$$

$$= \frac{3n(n+1)}{2(2n+1)}$$

10. Arranging in ascending order, we get

23 , 34, 43, 54, 56, 67, 78

Here, $n = 7$ (odd)

So, $\frac{(7+1)}{2} = 4$

$\therefore$ Median = 4th observation

Median = 54

11. Arranging in ascending order, we get

24, 34, 43, 50, 67, 78

Here, $n=6$ (even)

Using the median formula,

$$\text{Median} = \frac{\left(\frac{n}{2}\right)\text{th observation} + \left(\frac{n}{2}+1\right)\text{th observation}}{2}$$

$$\text{Median} = \frac{(\text{3rd observation} + \text{4th observation})}{2}$$

$$= \frac{(43+50)}{2}$$

Median = 46.5

12. In $\frac{x}{5}, \frac{x}{4}, \frac{x}{3}, \frac{x}{2}, x,$

Median = 8 [given]

$\Rightarrow \quad \frac{x}{3} = 8 \quad \left[\because \frac{x}{3} \text{ is the most middle terms}\right]$

$\Rightarrow \quad x = 24$

13. Arrange the data as follows

$\alpha - \frac{7}{2}, \alpha - 3, \alpha - \frac{5}{2}, \alpha - 2, \alpha - \frac{1}{2}, \alpha, \alpha + \frac{1}{2}, \alpha + 4$

Since, total number of terms = 8 (even), then

Median = $\frac{1}{2}$ [value of 4th item + value of 5th item]

$$\therefore \text{ Median} = \frac{\alpha - 2 + \alpha - \frac{1}{2}}{2} = \frac{2\alpha - \frac{5}{2}}{2} = \alpha - \frac{5}{4}$$

14. ${}^{2n}C_0, {}^{2n}C_1, ..., {}^{2n}C_n$ are binomial coefficients which are in odd numbers (because n is even) and middle binomial coefficient is ${}^{2n}C_{n/2}$, which is required median.

15.

Marks	Frequency	Cumulative frequency	Mid-points
0-10	5	5	5
10-20	15	20	15
20-30	20	40	25
30-40	2	42	35
40-50	8	50	45

Thus, $N=50$ and $\frac{N}{2}=25$

So, the median class is 20-30

$l=20, f=20, h=10$ and $C=20$

$$\text{So, median} = l + \left(\frac{\frac{N}{2}-C}{f}\right) \times h = 20 + \frac{(25-20)}{20} \times 10 = 22.5$$

16.

Classes	Number of students	Cumulative frequency
0-10	2	2
10-20	12	14
20-30	22	36
30-40	8	44
40-50	6	50

Here, $N=50$

and $\frac{N}{2} = \frac{50}{2} = 25$

So, median class = (20 – 30),

$l=20, f=22, C=14$ and $h=10$

Using median formula,

$$\text{Median} = l + \left[\frac{\frac{N}{2}-C}{f}\right] \times h$$

$$= 20 + \frac{(25-14)}{22} \times 10 = 20 + \left(\frac{11}{22}\right) \times 10$$

$$= 20 + 5 = 25$$

$\therefore$ Median = 25

17. The highest frequency = 12, so the modal class is 40-60

l = Lower limit of modal class = 40

f_m = Frequency of modal class = 12

f_1 = Frequency of class preceding modal class = 10

f_2 = Frequency of class succeding modal class = 6

h = Class width = 20

Using the mode formula,

$$\text{Mode} = l + \left[\frac{f_m - f_1}{2f_m - f_1 - f_2}\right] \times h$$

$$= 40 + \left[\frac{12-10}{2 \times 12 - 10 - 6}\right] \times 20$$

$$= 40 + \left(\frac{2}{8}\right) \times 20 = 45$$

18. Using limits method,

Range = Highest marks – Lowest marks

$= 60 - 10 = 50$

$$\text{Coefficient of range} = \frac{60-10}{60+10} = \frac{50}{70} = 0.714$$

19. $\therefore$ Range = 50 – 10 = 40

$$\text{Coefficient of range} = \frac{50-10}{50+10} = \frac{40}{60} = \frac{2}{3} = 0.66$$

20. Mean $(\bar{x}) = \frac{1}{20}\sum_{i=1}^{20} x_i = \frac{200}{20} = 10$

The respective absolute values of the deviations from mean i.e. $|x_i - \bar{x}|$ are

2, 7, 8, 7, 6, 1, 7, 9, 10, 5, 2, 7, 8, 7, 6, 1, 7, 9, 10, 5

Therefore, $\sum_{i=1}^{20} |x_i - \bar{x}| = 124$

and MD $(\bar{x}) = \frac{124}{20} = 6.2$

21. (i) Mean, $\bar{x} = \frac{6+7+10+12+13+4+8+12}{8} = \frac{72}{8} = 9$

The deviations of the respective observations from the mean $\bar{x}$ i.e. $x_i - \bar{x}$ are

$6-9, 7-9, 10-9, 12-9, 13-9, 4-9, 8-9, 12-9$ or $-3, -2, 1, 3, 4, -5, -1, 3$

The absolute values of the deviations i.e. $|x_i - \bar{x}|$ are 3, 2, 1, 3, 4, 5, 1, 3

The required mean deviation about the mean is

$$\text{MD}(\bar{x}) = \frac{\sum_{i=1}^{8} |x_i - \bar{x}|}{8}$$

$$= \frac{3+2+1+3+4+5+1+3}{8} = \frac{22}{8} = 2.75$$

(ii)

Marks obtained	Number of students (f_i)	x_i	$f_i x_i$	$\lvert x_i - \bar{x}\rvert$	$f_i\lvert x_i - \bar{x}\rvert$
10-20	2	15	30	30	60
20-30	3	25	75	20	60
30-40	8	35	280	10	80
40-50	14	45	630	0	0
50-60	8	55	440	10	80
60-70	3	65	195	20	60
70-80	2	75	150	30	60
	$\Sigma f_i = 40$		$\Sigma f_i x_i = 1800$		

Here, $N = \sum_{i=1}^{7} f_i = 40$,

$\sum_{i=1}^{7} f_i x_i = 1800$

Therefore, $\bar{x} = \frac{1}{N}\sum_{i=1}^{7} f_i x_i = \frac{1800}{40} = 45$

Now, $\sum_{i=1}^{7} f_i |x_i - \bar{x}| = 400$

and MD $(\bar{x}) = \frac{1}{N}\sum_{i=1}^{7} f_i |x_i - \bar{x}|$

$= \frac{1}{40} \times 400 = 10$

22. Here, the number of observations is 11, which is odd. Arranging the data into ascending order, we have

3, 3, 4, 5, 7, 9, 10, 12, 18, 19, 21

Now, median $= \left(\frac{11+1}{2}\right)$th or 6th observation $= 9$

The absolute values of the respective deviations from the median i.e. $|x_i - M_e|$ are

6, 6, 5, 4, 2, 0, 1, 3, 9, 10, 12

Therefore, $\sum_{i=1}^{11} |x_i - M_e| = 58$

So, mean deviation about median

and $(\text{MD}(M_e)) = \frac{1}{11} \times 58 = 5.27$

23.

Classes	Frequency (f_i)	Cumulative frequency	Mid-points (x_i)	$\lvert x_i - M_e\rvert$	$f_i\lvert x_i - M_e\rvert$
0-10	6	6	5	23	138
10-20	7	13	15	13	91
20-30	15	28	25	3	45
30-40	16	44	35	7	112
40-50	4	48	45	17	68
50-60	2	50	55	27	54
	$\Sigma f_i = 50$				508

The class interval containing $\frac{N}{2}$th or 25th item is 20-30.

Therefore, 20-30 is the median class

We know that

$$\text{Median}(M_e) = l + \left(\frac{\frac{N}{2} - C}{f}\right) \times h$$

Here, $l = 20, C = 13, f = 15, h = 10$ and $N = 50$

Therefore, median $= 20 + \frac{25-13}{15} \times 10 = 20 + 8 = 28$

Thus, mean deviation about median is given by

$$\text{MD}(M_e) = \frac{1}{N}\sum_{i=1}^{6} f_i |x_i - M_e|$$

$$= \frac{1}{50} \times 508 = 10.16$$

24. Variance $= (SD)^2 = \frac{1}{n}\Sigma x^2 - \left(\frac{\Sigma x}{n}\right)^2 \left[\because \bar{x} = \frac{\Sigma x}{n}\right]$

$$= \frac{n(n+1)(2n+1)}{6n} - \left(\frac{n(n+1)}{2n}\right)^2 = \frac{n^2-1}{12}$$

25. The number of observations is $n = 10$

x_i	$d_i = \frac{x_i - 14}{2}$	**Deviations from mean**	$(x_i - \bar{x})^2$
6	−4	−9	81
8	−3	−7	49
10	−2	−5	25
12	−1	−3	9
14	0	−1	1
16	1	1	1
18	2	3	9
20	3	5	25
22	4	7	49
24	5	9	81
	$\Sigma d_i = 5$		$\Sigma(x_i - \bar{x})^2 = 330$

Therefore, mean $(\bar{x})$ = Assumed mean $+ \frac{\sum_{i=1}^{n} d_i}{n} \times h$

$$= 14 + \frac{5}{10} \times 2 = 15$$

and variance $(\sigma^2) = \frac{1}{n}\sum_{i=1}^{10}(x_i - \bar{x})^2 = \frac{1}{10} \times 330 = 33$

Thus, standard deviation $(\sigma) = \sqrt{33} = 5.74$

26.

x	f_i	$f_i x_i$	$f_i x_i^2$
A	2	$2A$	$2A^2$
$2A$	1	$2A$	$4A^2$
$3A$	1	$3A$	$9A^2$
$4A$	1	$4A$	$16A^2$
$5A$	1	$5A$	$25A^2$
$6A$	1	$6A$	$36A^2$
Total	$N = 7$	$\Sigma f_i x_i = 22A$	$\Sigma f_i x_i^2 = 92A^2$

$$\therefore \quad \sigma^2 = \frac{\Sigma f_i x_i^2}{N} - \left(\frac{\Sigma f_i x_i}{N}\right)^2$$

$$\Rightarrow \quad 160 = \frac{92A^2}{7} - \left(\frac{22A}{7}\right)^2$$

$$\Rightarrow \quad 160 = \frac{92A^2}{7} - \frac{484A^2}{49}$$

$$\Rightarrow \quad 160 = (644 - 484)\frac{A^2}{49}$$

$$\Rightarrow \quad 160 = \frac{160A^2}{49}$$

$$\Rightarrow \quad A^2 = 49$$

$$\therefore \quad A = 7$$

27. Given data are 6, 5, 9, 13, 12, 8 and 10

x_i	x_i^2
6	36
5	25
9	81
13	169
12	144
8	64
10	100
$\Sigma x_i = 63$	$\Sigma x_i^2 = 619$

$$\therefore \text{ Variance } (\sigma^2) = \frac{\Sigma x_i^2}{n} - \left(\frac{\Sigma x_i}{n}\right)^2$$

$$= \frac{619}{7} - \left(\frac{63}{7}\right)^2$$

$$= \frac{7 \times 619 - 3969}{49}$$

$$= \frac{4333 - 3969}{49}$$

$$= \frac{364}{49} = \frac{52}{7}$$

28. Given observations are a, b, c, d and e.

$$\text{Mean} = m = \frac{a+b+c+d+e}{5}$$

$$\Sigma x_i = a+b+c+d+e = 5m$$

$$\text{New mean} = \frac{a+k+b+k+c+k+d+k+e+k}{5}$$

$$= \frac{(a+b+c+d+e)+5k}{5} = m+k$$

$$\therefore \quad SD = \sqrt{\frac{\Sigma(x_i^2 + k^2 + 2kx_i)}{n} - (m^2 + k^2 + 2mk)}$$

$$= \sqrt{\frac{\Sigma x_i^2}{n} - m^2 + \frac{2k\,\Sigma x_i}{n} - 2mk}$$

$$= \sqrt{\frac{\Sigma x_i^2}{n} - m^2 + 2km - 2mk} \quad \left[\because \frac{\Sigma x_i}{n} = m\right]$$

$$\therefore \quad SD = \sqrt{\frac{\Sigma x_i^2}{n} - m^2}$$

29.

x_i	f_i	f_ix_i	$x_i - \bar{x}$	$(x_i - \bar{x})^2$	$f_i(x_i - \bar{x})^2$
4	3	12	– 10	100	300
8	5	40	–6	36	180
11	9	99	–3	9	81
17	5	85	3	9	45
20	4	80	6	36	144
24	3	72	10	100	300
32	1	32	18	324	324
	$\Sigma f_i = 30$	$\Sigma f_ix_i = 420$			$\Sigma f_i(x_i - \bar{x})^2 = 1374$

Here, $N = 30, \sum_{i=1}^{7} f_ix_i = 420$

Therefore, $\bar{x} = \dfrac{\sum_{i=1}^{7} f_ix_i}{N} = \dfrac{1}{30} \times 420 = 14$ and $\sum_{i=1}^{7} f_i (x_i - \bar{x})^2 = 1374$

Hence, variance $(\sigma^2) = \dfrac{1}{N} \sum_{i=1}^{7} f_i (x_i - \bar{x})^2 = \dfrac{1}{30} \times 1374 = 45.8$

and standard deviation $(\sigma) = \sqrt{45.8} = 6.77$

30.

Classes	**Frequency** (f_i)	**Mid-points** (x_i)	$y_i = \dfrac{x_i - 65}{10}$	y_i^2	f_iy_i	$f_iy_i^2$
30-40	3	35	– 3	9	– 9	27
40-50	7	45	– 2	4	– 14	28
50-60	12	55	– 1	1	– 12	12
60-70	15	65	0	0	0	0
70-80	8	75	1	1	8	8
80-90	3	85	2	4	6	12
90-100	2	95	3	9	6	18
	$N = 50$				$\Sigma f_iy_i = -15$	$\Sigma f_iy_1^2 = 105$

Therefore, mean $\bar{x} = A + \dfrac{\Sigma f_iy_i}{50} \times h = 65 - \dfrac{15}{50} \times 10 = 62$

$$\text{Variance, } \sigma^2 = \frac{h^2}{N^2} [N \,\Sigma\, f_iy_i^2 - (\Sigma\, f_iy_i)^2]$$

$$= \frac{(10)^2}{(50)^2} [50 \times 105 - (-15)^2]$$

$$= \frac{1}{25} [5250 - 225] = 201$$

and standard deviation $(\sigma) = \sqrt{201} = 14.18$

31. Let the two unknown items be x and y, then

Mean = 4

$$\Rightarrow \quad \frac{1+2+6+x+y}{5} = 4$$

$$\Rightarrow \quad x+y = 11 \quad \text{...(i)}$$

and variance = 5.2

$$\Rightarrow \frac{1^2+2^2+6^2+x^2+y^2}{5} - (\text{mean})^2 = 5.2$$

$$41+x^2+y^2 = 5[5.2+(4)^2]$$

$$41+x^2+y^2 = 106$$

$$x^2+y^2 = 65 \quad \text{...(ii)}$$

Solving Eqs. (i) and (ii) for x and y, we get

$$x=4, y=7 \text{ or } x=7, y=4$$

32. Given, variance of height = 127.69 cm^2

Therefore, standard deviation of height $= \sqrt{127.69}$

$$= 11.3 \text{ cm}$$

Also, variance of weight = 23.1361 kg^2

Therefore, standard deviation of weight

$$= \sqrt{23.1361} = 4.81 \text{ kg}$$

Now, the CV are given by

$$\text{CV in heights} = \frac{\text{Standard deviation}}{\text{Mean}} \times 100$$

$$= \frac{11.3}{162.6} \times 100 = 6.95$$

$$\text{and CV in weights} = \frac{4.81}{52.36} \times 10 = 9.18$$

Clearly, CV in weight is greater than CV in heights. Therefore, we can say that weights show more variability than heights.

33. Given, CV (1st distribution) = 60, $\sigma_1 = 21$

and CV (2nd distribution) = 70, $\sigma_2 = 16$

Let $\bar{x}_1$ and $\bar{x}_2$ be the means of 1st and 2nd distribution, respectively.

$$\text{CV (1st distribution)} = \frac{\sigma_1}{\bar{x}_1} \times 100$$

$$\Rightarrow \quad 60 = \frac{21}{\bar{x}_1} \times 100$$

$$\Rightarrow \quad \bar{x}_1 = \frac{21}{60} \times 100 = 35$$

$$\text{and CV (2nd distribution)} = \frac{\sigma_2}{\bar{x}_2} \times 100$$

$$\Rightarrow \quad 70 = \frac{16}{\bar{x}_2} \times 100$$

$$\Rightarrow \quad \bar{x}_2 = \frac{16}{70} \times 100 = 22.85$$

Level 2

1. (b) The required AM is

$$n = \frac{1+2+2^2+2^3+\ldots+2^n}{n+1}$$

$$= \frac{(2^{n+1}-1)}{(n+1)(2-1)} = \frac{2^{n+1}-1}{n+1}$$

2. (d) Let $x_1, x_2 \ldots x_n$ be n numbers.

$$\text{Then, } \bar{x} = \frac{1}{n}\sum_{i=1}^{n} x_i$$

If each number is divided by 3, then the new mean $\bar{y}$ is given by

$$\bar{y} = \frac{1}{n}\sum_{i=1}^{n}\left(\frac{x_i}{3}\right) = \frac{1}{3}\left(\frac{1}{n}\sum_{i=1}^{n} x_i\right) = \frac{\bar{x}}{3}$$

3. (d) $\dfrac{x_1+x_2+\ldots+x_{16}}{16} = 16$ ...(i)

$$\frac{x_1+x_2+\ldots+x_{16}-16+3+4+5}{18}$$

$$= \frac{16\times 16 - 16+12}{18} = 14 \quad \text{[using Eq. (i)]}$$

4. (c) $\text{Mean} = \dfrac{\sum_{i=1}^{n} f_i x_i}{\sum_{i=1}^{n} f_i}$

$$\therefore \quad 2.6 = \frac{1\times 4+2\times 5+3\times y+4\times 1+5\times 2}{4+5+y+1+2}$$

$$\Rightarrow \quad 31.2+2.6y = 28+3y$$

$$\Rightarrow \quad 0.4y = 3.2 \Rightarrow y = 8$$

5. (b) We know that

$$\bar{x} = \frac{\sum_{i=1}^{n} x_i}{n}$$

$$\Rightarrow \quad \sum_{i=1}^{n} x_i = n\bar{x}$$

$$\therefore \quad \frac{\sum_{i=1}^{n}(x_i+2i)}{n} = \frac{\sum_{i=1}^{n} x_i + 2\sum_{i=1}^{n} i}{n}$$

$$= \frac{n\bar{x}+2(1+2+\ldots n)}{n}$$

$$= \frac{n\bar{x}+2\dfrac{n(n+1)}{2}}{n}$$

$$= \bar{x}+(n+1)$$

6. (b) $M = \frac{x_1 + x_2 + x_3, ..., x_n}{n}$

$\Rightarrow \quad nM = x_1 + x_2 + x_3 + ... + x_{n-1} + x_n$

$\Rightarrow \quad nM - x_n = x_1 + x_2 + x_3 + ... + x_{n-1}$

$\Rightarrow \quad nM - x_n + x' = x_1 + x_2 + x_3 + ... + x_{n-1} + x'$

$\Rightarrow \quad \frac{nM - x_n + x'}{n} = \frac{x_1 + x_2 + x_3 + ... + x_{n-1} + x'}{n}$

$\therefore$ New average $= \frac{nM - x_n + x'}{n}$

7. (c) Let $x_1, x_2, ..., x_n$ be n items. Then, $\bar{x} = \frac{1}{n} \Sigma x_i$.

Let $y_1 = x_1 + 1, y_2 = x_2 + 2, y_3 = x_3 + 3, ..., y_n = x_n + n$.

Then, the mean of the new series is

$$\frac{1}{n} \Sigma (x_i + i) = \frac{1}{n} \Sigma x_i + \frac{1}{n}(1 + 2 + 3 + ... + n)$$

$$= \bar{x} + \frac{1}{n} \frac{n(n+1)}{2} = \bar{x} + \frac{n+1}{2}$$

8. (b) Arranging the data in ascending order of magnitude, we obtain

Height (in cm)	150	152	154	155	156	160	161
Number of Students	8	4	3	7	3	12	4
Cumulative frequency	8	12	15	22	25	37	41

Here, the total number of items is 41 i.e. an odd number.

Hence, the median is $\frac{41 + 1}{2}$ th i.e. 21st item.

From cumulative frequency table, we find that the median,

i.e. 21st item is 155.

(all items from 16th to 22nd are equal, each 155).

9. (c) We have, $\Sigma f_i (y_i - \mu)$

$= \Sigma f_i y_i - \mu \Sigma f_i$

$= \mu \Sigma f_i - \mu \Sigma f_i = 0 \quad \left[\because \mu = \frac{\Sigma f_i y_i}{\Sigma f_i}\right]$

10. (d) The normal distribution is a symmetrical, bell-shaped distribution in which the mean, median and mode are all equal.

11. (a) For a moderately skewed distribution, we have

Mode = 3 Median − 2 Mean

$\Rightarrow$ Mode $= 3(6) - 2(5) = 8$

12. (b) Range $= X_{highest} - X_{lowest} = 9 - 2 = 7$

13. (b) Given observations are 3, 10, 10, 4, 7, 10 and 5

$\therefore \quad \bar{x} = \frac{3 + 10 + 10 + 4 + 7 + 10 + 5}{7} = \frac{49}{7} = 7$

x_i	$d_i = \lvert x_i - \bar{x} \rvert$
3	4
10	3
10	3
4	3
7	0
10	3
5	2
Total	$\Sigma d_i = 18$

Now, MD $= \frac{\Sigma d_i}{N} = \frac{18}{7} = 2.57$

14. (a) Arranging the given data in ascending order, we have

34, 38, 42, 44, 46, 48, 54, 55, 63, 70

Median, $M_e = \frac{46 + 48}{2} = 47$

[$\because n = 10$, median is the mean of 5th and 6th terms]

$\therefore$ Mean deviation $= \frac{\Sigma |x_i - M_e|}{n} = \frac{\Sigma |x_i - 47|}{10}$

$= \frac{13 + 9 + 5 + 3 + 1 + 1 + 7 + 8 + 16 + 23}{10} = 8.6$

15. (d) From the given data,

Median $= \frac{25 + 26}{2} a = 25.5a$

Required mean deviation about median

$\frac{2|0.5 + 1.5 + 2.5 + ... + 24.5|}{50} |a| = 50$

$\Rightarrow \quad |a| = 4$

16. (c) Mean $(\bar{x}) = \frac{\text{Sum of quantities}}{n}$

$= \frac{\frac{n}{2}(a + l)}{n} = \frac{1}{2}[1 + 1 + 100d] = 1 + 50d$

$\text{MD} = \frac{1}{n} \Sigma |x_i - \bar{x}|$

$\Rightarrow 225 = \frac{1}{101}[50d + 49d + 48d + ... d + 0 + d + ... + 50d]$

$= \frac{2d}{101}\left[\frac{50 \times 51}{2}\right]$

$\Rightarrow \quad d = \frac{255 \times 101}{50 \times 51} = 10.1$

17. (c) Marks obtained by 9 students in Mathematics are 50, 69, 20, 33, 53, 39, 40, 65, 59

Let us write the given data in ascending order

20, 33, 39, 40, 50, 53, 59, 65, 69

Here, $n = 9$ (odd)

Median $(M_e) = 5\text{th term} = 50$

x_i	$d_i = \lvert x_i - Me \rvert$
20	30
33	17
39	11
40	10
50	0
53	3
59	9
65	15
69	19
Total = 9	$\Sigma d_i = 114$

$$MD = \frac{114}{9} = 12.67$$

18. (b) Mean deviation is minimum when it is considered about an item equidistant from the beginning and the end, i.e. the median.

In this case, the median is $\frac{101+1}{2}$th i.e. 51st item

i.e. x_{51}.

19. (d) Let the 50 observations are $x_1, x_2, x_3, \ldots, x_{50}$.

Now, deviations of these observations from 30 are $(x_1 - 30), (x_2 - 30), (x_3 - 30), \ldots, (x_{50} - 30)$

$$\because \quad \sum_{i=1}^{50} (x_i - 30) = 50 \qquad \text{(given)}$$

$$\Rightarrow \quad \sum_{i=1}^{50} x_i - (30 \times 50) = 50$$

$$\Rightarrow \quad \sum_{i=1}^{50} x_i = 50(30+1) = 50 \times 31$$

$$\therefore \text{ Mean of 50 observations} = (\bar{x}) = \frac{\sum_{i=1}^{50} x_i}{50} = 31$$

20. (d) Given numbers are 1, 2, 3, 4, 5, 6, 7, 8, 9 and 10.

If 1 is added to each number, then observations will be 2, 3, 4, 5, 6, 7, 8, 9, 10 and 11.

$$\therefore \quad \Sigma x_i = 2 + 3 + 4 + \ldots + 11 \qquad \ldots(i)$$

$$= \frac{10}{2}[2 \times 2 + 9 \times 1] = 5[4+9] = 65$$

and $\Sigma x_i^2 = 2^2 + 3^2 + 4^2 + 5^2 + \ldots + 11^2$

$$= (1^2 + 2^2 + 3^2 + \ldots + 11^2) - (1)^2$$

$$= \frac{11 \times 12 \times 23}{6} - 1 = 505$$

$$\therefore \quad \sigma^2 = \frac{\Sigma x_i^2}{n} - \left(\frac{\Sigma x_i}{n}\right)^2 = \frac{505}{10} - \left(\frac{65}{10}\right)^2 = 8.25$$

21. (b) $\sigma^2 = \left(\frac{\sum x_i^2}{n}\right) - \bar{x}^2$

$$\bar{x} = \frac{\sum_{r=1}^{50} 2r}{50} = 51$$

$$\sigma^2 = \frac{\sum_{r=1}^{50} 4r^2}{50} - (51)^2 = 833$$

22. (d) Given, data = {6, 10, 7, 13, a, 12, b, 12}

Mean = 9, Variance = $\frac{37}{4}$

Now, mean $= \frac{\Sigma x_i}{n} = \frac{6 + 10 + 7 + 13 + a + 12 + b + 12}{8}$

or $\quad 9 = \frac{60 + a + b}{8}$

$\Rightarrow \quad 72 = 60 + a + b$

$\Rightarrow \quad a + b = 12 \qquad \ldots(i)$

Variance $= \frac{\Sigma x_i^2}{n} - \Sigma\left(\frac{x_i}{n}\right)^2$

$\Rightarrow \quad \frac{37}{4} = \frac{6^2 + 10^2 + 7^2 + 13^2 + a^2 + 12^2 + b^2 + 12^2}{8} - 9^2$

$\Rightarrow \quad \frac{37}{4} = \frac{642 + a^2 + b^2}{8} - 81$

$\Rightarrow \quad \frac{37}{4} = \frac{642 + a^2 + b^2 - 648}{8}$

$\Rightarrow \quad 74 = a^2 + b^2 - 6$

$\Rightarrow \quad a^2 + b^2 = 80 \qquad \ldots(ii)$

$\Rightarrow \quad (a+b)^2 = a^2 + b^2 + 2ab$

Putting the values from Eqs. (i) and (ii), we get

$$2ab = (a+b)^2 - (a^2 + b^2)$$

$\Rightarrow \quad 2ab = 12^2 - 80 = 64$

Now, $\quad (a-b)^2 = a^2 + b^2 - 2ab$

$$= 80 - 64 = 16$$

23. (c) Given, $\bar{x} = 50, n = 100$ and $\sigma = 5$

$\Rightarrow \quad 50 = \frac{\Sigma x_i}{100}$

$\Rightarrow \quad \Sigma x_i = 50 \times 100 = 5000$

Now, $\quad \sigma^2 = \frac{\Sigma x_i^2}{n} - (\bar{x})^2$

$\Rightarrow \quad 25 = \frac{\Sigma x_i^2}{100} - (50)^2$

$\Rightarrow \quad 2525 = \frac{\Sigma x_i^2}{100}$

$\therefore \quad \Sigma x_i^2 = 252500$

24. (a)

x_i	15	25	35
f_i	2	x	2

$$\bar{x}=\frac{\Sigma f_i x_i}{\Sigma f_i}=\frac{30+70+25x}{4+x}=\frac{100+25x}{4+x}=\frac{25(4+x)}{4+x}=25$$

$$\sigma^2=\frac{\Sigma f_i x_i^2}{\Sigma f_i}-(\bar{x})^2$$

$$\Rightarrow \quad 50=\frac{450+625x+2450}{4+x}-625$$

$$\Rightarrow \quad 675=\frac{2900+625x}{4+x}$$

$$\Rightarrow \quad 50x=200$$

$$\therefore \quad x=4$$

25. (d) Variance $=\dfrac{\Sigma x_i^2}{n}-\left(\dfrac{\Sigma x_i}{n}\right)^2$

$$=\frac{18000}{60}-\left(\frac{960}{60}\right)^2$$

$$=300-256=44$$

26. (a) Mean = 6.5

$\therefore$ Total $=6\times 6.5=39$

Variance $=\dfrac{\sum x_i^2}{n}-\sum\left(\dfrac{x_i}{n}\right)^2$

$$\Rightarrow 10.25=\frac{\sum x_i^2}{6}-6.5^2$$

$$\therefore \quad \sum x_i^2=6(10.25+42.25)=52.5\times 6$$

$$\sum x_i=39$$

$$\Rightarrow \quad 2+4+5+7+x_1+x_2=39$$

$$\Rightarrow \quad x_1+x_2=21$$

$$\sum x_i^2=52.5\times 6$$

$$4+16+25+49+x_1^2+x_2^2=315$$

$$\Rightarrow \quad x_1^2+x_2^2=221$$

$$\Rightarrow \quad x_1^2+(21-x_1)^2=221$$

$$\Rightarrow \quad 2x_1^2-42x_1+441=221$$

$$\Rightarrow \quad x_1^2-21x_1+110=0$$

$$\Rightarrow \quad (x_1-11)(x_1-10)=0$$

$$\Rightarrow \quad x_1=10 \text{ or } 11$$

27. (c) We have,

$$\bar{x}=\frac{0\,{}^nC_0+1\,{}^nC_1+2\,{}^nC_2+\ldots+n\cdot{}^nC_n}{{}^nC_0+{}^nC_1+{}^nC_2+\ldots+{}^nC_n}=\frac{\sum_{r=0}^{n} r\,{}^nC_r}{\sum_{r=0}^{n}{}^nC_r}$$

$$=\frac{1}{2^n}\sum^{n} r\,\frac{n}{r}\,{}^{n-1}C_{r-1}$$

$$\left[\because \sum_{r=0}^{n}{}^nC_r=2^n,\ {}^nC_r=\frac{n}{r}\,{}^{n-1}C_{r-1}\right]$$

$$=\frac{n}{2^n}\sum_{r=1}^{n}{}^{n-1}C_{r-1}=\frac{n}{2^n}2^{n-1}=\frac{n}{2}$$

$$\left[\because \sum_{r=1}^{n}{}^{n-1}C_{r-1}=2^{n-1}\right]$$

and $\dfrac{1}{N}\sum f_i x_i^2=\dfrac{1}{2^n}\sum r^2\,{}^nC_r$

$$=\frac{1}{2^n}\sum_{r=0}^{n}[r(r-1)+r]\,{}^nC_r$$

$$=\frac{1}{2^n}\left\{\sum_{r=0}^{n} r(r-1)\,{}^nC_r+\sum_{r=0}^{n} r\,{}^nC_r\right\}$$

$$=\frac{1}{2^n}\left\{\sum_{r=1}^{n} r(r-1)\frac{n}{r}\frac{n-1}{r-1}\,{}^{n-2}C_{r-2}+\sum_{r=1}^{n} r\frac{n}{r}\,{}^{n-1}C_{r-1}\right\}$$

$$=\frac{1}{2^n}\{n(n-1)2^{n-2}+n2^{n-1}\}$$

$$=\frac{n(n-1)}{4}+\frac{n}{2}$$

$$\therefore \text{Var}(x)=\frac{1}{N}\sum f_i x_i^2-\bar{x}^2=\frac{n(n-1)}{4}+\frac{n}{2}-\frac{n^2}{4}=\frac{n}{4}$$

28. (a) Let $x_1, x_2, \ldots x_{20}$ be 20 observations, then

$$\text{Mean}=\frac{x_1+x_2+\ldots+x_{20}}{20}=10$$

$$\Rightarrow \quad \frac{\sum_{i=1}^{20}x_i}{20}=10 \qquad \ldots\text{(i)}$$

$$\text{Variance}=\frac{\Sigma x_i^2}{n}-(\bar{x})^2$$

$$\Rightarrow \quad \frac{\Sigma x_i^2}{20}-100=4 \qquad \ldots\text{(ii)}$$

$$\Sigma x_i^2=104\times 20=2080$$

$$\text{Actual mean}=\frac{200-9+11}{20}=\frac{202}{20}$$

$$\text{Variance}=\frac{2080-81+121}{20}-\left(\frac{202}{20}\right)^2$$

$$=\frac{2120}{20}-(10.1)^2$$

$$=106-102.01=3.99$$

29. (c) Let two remaining observations are x_1 and x_2.

So, $\bar{x} = \dfrac{2+4+10+12+14+x_1+x_2}{7} = 8$ [given]

$\Rightarrow \quad x_2 + x_1 = 14$ ...(i)

Now, $\sigma^2 = \dfrac{\Sigma x_i^2}{n} - \left(\dfrac{\Sigma x_i}{n}\right)^2 = 16$ [given]

$\Rightarrow \dfrac{4+16+100+144+196+x_1^2+x_2^2}{7} - 64 = 16$

$\Rightarrow \quad 460 + x_1^2 + x_2^2 = (16+64) \times 7$

$\Rightarrow \quad x_1^2 + x_2^2 = 100$ ...(ii)

From Eqs. (i) and (ii), we get

$\Rightarrow \quad x_1 x_2 = 48$ ...(iii)

$\because \quad (x_1 - x_2)^2 = (x_1 + x_2)^2 - 4x_1x_2 = 196 - 192 = 4$

$\Rightarrow \quad |x_1 - x_2| = 2$

30. (c) If variate varries from a to b, then variance

$$\text{Var}(x) \le \left(\frac{b-a}{2}\right)^2$$

$\Rightarrow \quad \text{Var}(x) < \left(\dfrac{10-0}{2}\right)^2$

$\Rightarrow \quad \text{Var}(x) < 25$

$\Rightarrow$ Standard deviation < 5

It is clear that standard deviation can't be 6.

31. (a) The first 10 positive integers are 1, 2, 3, 4, 5, 6, 7, 8, 9 and 10.

On multiplying each number by – 1, we get

$-1, -2, -3, -4, -5, -6, -7, -8, -9, -10$

On adding 1 to each number, we get

$0, -1, -2, -3, -4, -5, -6, -7, -8, -9$

$\therefore \quad \Sigma x_i = \dfrac{-9 \times 10}{2} = -45$

and $\Sigma x_i^2 = 0^2 + (-1)^2 + (-2)^2 + \ldots + (-9)^2$

$= \dfrac{9 \times 10 \times 19}{6} = 285$

$\therefore \quad \text{SD} = \sqrt{\dfrac{285}{10} - \left(-\dfrac{45}{10}\right)^2} = \sqrt{\dfrac{285}{10} - \dfrac{2025}{100}}$

$= \dfrac{2850 - 2025}{100} = \sqrt{8.25}$

Now, variance $= (\text{SD})^2$

$= (\sqrt{8.25})^2 = 8.25$

32. (a) Let observations are denoted by x_i for $1 \le i < 2n$.

$\therefore \quad \text{Mean}(\bar{x}) = \dfrac{\Sigma x_i}{2n}$

$= \dfrac{(a+a+a+\ldots+a) - (a+a+a+\ldots+a)}{2n}$

$\Rightarrow \quad \bar{x} = 0$

and $\quad \sigma_x^2 = \dfrac{\Sigma x_1^2}{2n} - (\bar{x})^2$

$= \dfrac{\overbrace{a^2 + a^2 + a^2 + \ldots + a^2}^{2n \text{ times}}}{2n} - 0 = a^2$

$\Rightarrow \quad \sigma_x^2 = a^2 \Rightarrow \sigma_x = a$

According to the question, adding a constant b, then new mean $(\bar{y}) = \bar{x} + b = 5$

$\Rightarrow \quad 0 + b = 5$

$\Rightarrow \quad b = 5$

and new SD $\quad (\sigma_y) = \sigma_x$ (no change in SD)

$\Rightarrow \quad \sigma_y = \sigma_x = 20$

$\Rightarrow \quad a = 20$

$\therefore \quad a^2 + b^2 = 400 + 25 = 425$

33. (c) $\sigma_x^2 = 4$

$\sigma_y^2 = 5$

$\bar{x} = 2$

$\bar{y} = 4$

$\dfrac{\Sigma x_i}{5} = 2$

$\Rightarrow \quad \Sigma x_i = 10$ and $\Sigma y_i = 20$

$\sigma_x^2 = \dfrac{1}{5}(\Sigma x_i^2) - (\bar{x})^2, \sigma_y^2 = \dfrac{1}{5} = (\Sigma y_i^2) - 16$

$\Rightarrow \quad \Sigma x_i^2 = 40$

$\Rightarrow \quad \Sigma y_i^2 = 105$

Therefore, variance of the combined data is

$$\sigma^2 = \frac{1}{10}(\Sigma x_i^2 + \Sigma y_i^2) - \left(\frac{\bar{x}+\bar{y}}{2}\right)^2$$

$= \dfrac{1}{10}(40 + 105) - 9$

$= \dfrac{145 - 90}{10} = \dfrac{55}{10} = \dfrac{11}{2}$

34. (b) $\because \bar{x} = \dfrac{1+2+3+\ldots+17}{17} = \dfrac{17 \times 18}{17 \times 2} = 9$

$\therefore \quad \bar{y} = a\bar{x} + b = \dfrac{a(1+2+3+\ldots+17)}{17} + b = 17$

$\Rightarrow \quad \dfrac{a(17 \cdot 18)}{17 \cdot 2} + b = 17$

$\Rightarrow \quad 9a + b = 17$ ...(i)

$\text{Var}(x) = \dfrac{\Sigma x^2}{n} - (\bar{x})^2$

$= \dfrac{1^2 + 2^2 + \ldots + 17^2}{17} - (9)^2$

$$= \frac{17 \cdot 18 \cdot 35}{6 \cdot 17} - (9)^2 = 105 - 81 = 24$$

$$\text{Var}(y) = a^2 \, \text{Var}(x) = a^2 \cdot 24 = 216$$

$$\Rightarrow \quad a^2 = \frac{216}{24} = 9 \Rightarrow a = 3$$

$\therefore$ From Eq. (i),

$$b = 17 - 9a$$
$$= 17 - 27 = -10$$

$$\therefore \quad a + b = 3 + (-10) = -7$$

35. (d) Given, three observations $= a, b, c$

$$b = a + c$$

Standard deviation $= \sigma$

$$\sigma = \sqrt{\frac{\Sigma (x_i - \mu)^2}{N}}$$

When $x_1 = a + 2, x_2 = b + 2, x_3 = c + 2$

$$\mu = \frac{a+2+b+2+c+2}{3} = \frac{2b}{3} + 2$$

$$x_1 - \mu = (a+2) - (2b/3 + 2) = a - 2b/3$$
$$x_2 - \mu = (b+2) - (2b/3 + 2) = b - 2b/3$$
$$x_3 - \mu = (c+2) - (2b/3 + 2) = c - 2b/3$$

$$d = \sqrt{\frac{(a - 2b/3)^2 + (b - 2b/3)^2 + (c - 2b/3)^2}{3}}$$

$$3d^2 = a^2 + b^2 + c^2 + \frac{12b^2}{9} - \frac{4}{3}(ab + b^2 + bc)$$

$$\Rightarrow 3d^2 = a^2 + b^2 + c^2 + \frac{4}{3}b^2 - \frac{4}{3}[b(a+c) + b^2]$$

$$\Rightarrow 3d^2 = a^2 + b^2 + c^2 + \frac{4}{3}b^2 - \frac{4}{3}[b \cdot b + b^2]$$

$$\Rightarrow 3d^2 = a^2 + b^2 + c^2 + \frac{4}{3}b^2 - \frac{4}{3} \cdot 2b^2$$

$$\Rightarrow 3d^2 = a^2 + c^2 - \frac{b^2}{3}$$

$$\Rightarrow b^2 = 3a^2 + 3c^2 - 9d^2$$

36. (d) Let $x_i - 5 = y_i$

$$\therefore \sum_{i=1}^{9} y_i = 9 \text{ and } \sum_{i=1}^{9} y_i^2 = 45$$

So, required standard deviation is

$$\sigma = \sqrt{\left(\frac{\sum_{i=1}^{9} y_i^2}{9}\right) - \left(\frac{\sum_{i=1}^{9} y_i}{9}\right)^2} = \sqrt{\frac{45}{9} - \left(\frac{9}{9}\right)^2} = 2$$

37. (a) If each item of the data is increased or decreased by the same constant, then the standard deviation of the data remains unchanged.

38. (c) Mean $= \dfrac{3+5+7+a+b}{5} = 5$

$$\Rightarrow a + b = 10$$

$$\text{Variance} = \frac{3^2 + 5^2 + 7^2 + a^2 + b^2}{5} - (5)^2 = 4$$

$$\Rightarrow \quad a^2 + b^2 = 62$$

$$\Rightarrow \quad (a+b)^2 - 2ab = 62 \Rightarrow ab = 19$$

Hence, a and b are the roots of equation

$$x^2 - 10x + 19 = 0$$

39. (d) $\quad \bar{x} = \dfrac{\Sigma x_i}{20} = 10$

$$\Rightarrow \quad \Sigma x_i = 200$$

$$\sigma^2 = \frac{\Sigma x_i^2}{20} - 100 = 6.25$$

$$\Sigma x_i^2 = 20 \times 106.25 = 2125$$

Now, replacing 25 with 35 as one data, then

$$\Sigma x_i - 25 + 35 = 210$$
$$\Sigma x_i^2 - 25^2 + 35^2 = 2725$$

$$\text{New mean} = \frac{210}{20} = 10.5 = (\alpha)$$

$$\text{New SD} = \sqrt{\frac{2725}{20} - (10.5)^2}$$

$$\text{SD} = \sqrt{136.25 - 110.25} = \sqrt{26} = \sqrt{\beta}$$

$$(\alpha, \beta) = (10.5, 26)$$

40. (b) Let $y = \dfrac{ax + b}{c}$ i.e. $y = \dfrac{a}{c}x + \dfrac{b}{c}$ i.e. $y = \dfrac{a}{c}x + \dfrac{b}{c}$

$$\therefore \quad \bar{y} = A\bar{x} + B$$

$$\therefore \quad y - \bar{y} = A(x - \bar{x})$$

$$\Rightarrow \quad (y - \bar{y})^2 = A^2 (x - \bar{x})^2$$

$$\Rightarrow \quad \Sigma (y - \bar{y})^2 = A^2 \, \Sigma (x - \bar{x})^2$$

$$\Rightarrow \quad n \cdot \sigma_y^2 = A^2 \cdot n\sigma_x^2$$

$$\Rightarrow \quad \sigma_y^2 = A^2 \sigma_x^2 \Rightarrow \sigma_y = |A| \sigma_x$$

$$\Rightarrow \quad \sigma_y = \left|\frac{a}{c}\right| \sigma_x$$

Thus, new SD $= \left|\dfrac{a}{c}\right| \sigma$

41. (a) Given, $\sigma_c = 5$

Relation between C and F is given by

$$F = \frac{9C}{5} + 32$$

$$\therefore \quad \sigma_F = \frac{9}{5}\sigma_c = \frac{9}{5} \times 5 = 9$$

Here, $\sigma_F^2 = (9)^2 = 81$

42. (c) Combined mean $=15.6$

$$\Rightarrow \quad \frac{100\times 15+150\times \bar{x}_B}{250}=15.6$$

$$\Rightarrow \quad 1500+150\bar{x}_B=3900$$

$$\Rightarrow \quad 150\bar{x}_B=2400$$

$$\Rightarrow \quad \bar{x}_B=\frac{2400}{150}=16$$

Combined SD $=\sqrt{13.44}$

$\Rightarrow$ Combined variance $(\sigma^2)=13.44$

$$\sigma^2=\frac{\sum x_i^2}{n}-(\bar{x})^2$$

$$\Rightarrow \quad 13.44=\frac{\sum x_i^2}{250}-(15.6)^2$$

$$\Rightarrow \quad \sum x_i^2=64200 \quad \text{...(i)}$$

For sample A,

$$9=\frac{\sum (x_i)_A^2}{100}-225$$

$$\Rightarrow \quad \sum (x_i)_A^2=23400$$

Now, $\sum (x_i)_B^2=64200-23400=40800$

$$\text{SD of sample } B=\sqrt{\frac{\sum (x_i)_B^2}{n_B}-(\bar{x}_B)^2}$$

$$=\sqrt{\frac{40800}{150}-256}=4$$

43. (a) We have, $r=\max|x_i-x_j|$

and $s^2=\frac{1}{n-1}\sum_{i=1}^{n}(x_i-\bar{x})^2$

$$\text{Now, } (x_i-\bar{x})^2=\left(x_i-\frac{x_1+x_2+\ldots+x_n}{n}\right)^2$$

$$=\frac{1}{n^2}[(x_i-x_1)+(x_i-x_2)+\ldots+(x_i-x_{i-1})$$

$$+(x_i-x_{i+1})+\ldots+(x_i-x_n)]$$

$$\le\frac{1}{n^2}[(n-1)r]^2 \quad [\because |x_i-x_j|\le r]$$

$$\Rightarrow \quad (x_i-\bar{x})^2\le r^2$$

$$\Rightarrow \quad \sum_{i=1}^{n}(x_i-\bar{x})^2\le nr^2$$

$$\Rightarrow \quad \frac{1}{n-1}\sum_{i=1}^{n}(x_i-\bar{x})^2\le\frac{nr^2}{(n-1)}$$

$$\Rightarrow \quad s^2\le\frac{nr^2}{(n-1)}$$

$$\Rightarrow \quad s\le r\sqrt{\frac{n}{n-1}}$$

44. (a) Standard deviation of numbers 2, 3, a and 11 is 3.5

$$\therefore \quad (3.5)^2=\frac{\Sigma x_i^2}{4}-(\bar{x})^2$$

$$\Rightarrow \quad \left(\frac{7}{2}\right)^2=\frac{4+9+a^2+121}{4}-\left(\frac{2+3+a+11}{4}\right)^2$$

On solving, we get

$$3a^2-32a+84=0$$

45. (c)

Classes	Frequency	x_i	$u_i=\frac{y_i-A}{10}$, $A=25$	f_iu_i	$f_iu_i^2$
0-10	1	5	-2	-2	4
10-20	3	15	-1	-3	3
20-30	4	25	0	0	0
30-40	2	35	1	2	2
Total	10			$\sum f_iu_i=-3$	$\sum f_iu_i^2=9$

Standard deviation

$$\sigma=h\times\sqrt{\frac{\Sigma f_iu_i^2}{N}-\left(\frac{\Sigma f_iu_i}{N}\right)^2}$$

$$=10\times\sqrt{\frac{9}{10}-\left(\frac{3}{10}\right)^2}=10\times\sqrt{\frac{9}{10}-\frac{9}{100}}$$

$$=10\times\frac{9}{10}=9$$

46. (c) $$\text{SD}=\sqrt{\frac{\sum_{i=1}^{10}(x_i-p)^2}{10}-\left[\frac{\sum_{i=1}^{10}(x_i-p)}{10}\right]^2}$$

$$=\sqrt{\frac{9}{10}-\left(\frac{3}{10}\right)^2}=\frac{9}{10}$$

47. (d) $$\text{Standard deviation}=\sqrt{\frac{\sum_{i=1}^{n}(x_i-a)^2}{n}-\left[\frac{\sum_{i-1}^{n}(x_i-a)}{n}\right]^2}$$

$[\because n, a>1]$

$$=\sqrt{\frac{na}{n}-\left(\frac{n}{n}\right)^2}=\sqrt{a-1}$$

48. (b) On arranging the given observations in ascending order, we get $(n+1)$th term would be 0.

$\therefore$ Median of given observations $=(n+1)$th term $=0$

$\therefore$ SD > MD

49. (80) Let the number of boys and girls be x and y, respectively

$$52x + 42y = 50(x + y)$$
$$\Rightarrow \quad 2x = 8y$$
$$\Rightarrow \quad x = 4y$$

So, total number of students in the class $= x + y = 5y$

$\therefore$ Required number of % boys $= \frac{4y}{5y} \times 100\% = 80\%$

50. (35) Given, $n = 25$, $\bar{x} = 40$

$$\therefore \quad \bar{x} = \frac{\sum x}{n}$$
$$\Rightarrow \quad \sum x = 25 \times 40 = 1000$$

New sum $= 1000 - 60 + x$
$= 940 + x$

New average $= 39$

$$\therefore \quad 39 = \frac{940 + x}{25}$$
$$\Rightarrow \quad 975 = 940 + x$$
$$\Rightarrow \quad x = 35$$

51. (20.5) $\because n = 9$,

So, the median term $= \left(\frac{9+1}{2}\right)$th term $=$ 5th term

Now, the last four observations have increased by 2.

The median of 5th observation, which is remain unchanged.

Then will be no changed in median.

52. (4)

Class	Frequency	x_i	$f_i x_i$
0-6	a	3	$3a$
6-12	b	9	$9b$
12-18	12	15	180
18-24	9	21	189
24-30	5	27	135
	$\Sigma n = a + b + 26$		$\Sigma f_i x_i = 3a + 9b + 504$

$$\text{Mean} = \frac{\Sigma f_i x_i}{\Sigma n} = \frac{3a + 9b + 504}{a + b + 26} = \frac{309}{22}$$
$$\Rightarrow \quad 66a + 198b + 11088 = 309a + 309b + 8034$$
$$\Rightarrow \quad 243a + 111b = 3054$$
$$\Rightarrow \quad 81a + 37b = 1018$$
$$\text{Median} = 12 + \frac{\frac{a+b+26}{2} - (a+b)}{12} \times 6 = 14$$
$$\Rightarrow \quad \frac{13}{2} - \left(\frac{a+b}{4}\right) = 2$$
$$\Rightarrow \quad \frac{a+b}{4} = \frac{9}{2}$$
$$\Rightarrow \quad a + b = 18$$

So, $81a + 37b = 1018$
$a + b = 18$
$a = 8$ and $b = 10$

$$\therefore \quad (a - b)^2 = 4$$

53. (164) Sum of frequencies $= 584$

$$\alpha + \beta + 110 + 54 + 30 = 584$$
$$\alpha + \beta = 390$$

Median is at $\frac{584}{2} = 292$th

$\because$ Median $= 45$ (lies in class $40 - 50$)

$$\Rightarrow \quad \alpha + 110 + 54 + 15 = 292$$
$$\Rightarrow \quad \alpha = 113$$
$$\Rightarrow \quad \beta = 390 - 113 = 277$$
$$\Rightarrow \quad \alpha - \beta = 113 - 277 = -164$$
$$\therefore \quad |\alpha - \beta| = 164$$

54. (11) Given, 10 natural numbers $= 1, 1, 1, \ldots 1, k$

According to the question, variance $(\sigma^2) < 10$

$$\because \quad \sigma^2 = \frac{\Sigma x^2}{n} - \left(\frac{\Sigma x}{n}\right)^2$$
$$\sigma^2 = \frac{(9 + k^2)}{10} - \left(\frac{9+k}{10}\right)^2 < 10$$
$$\Rightarrow \quad 10(9 + k^2) - (81 + k^2 + 18k) < 1000$$
$$\Rightarrow \quad 90 + 10k^2 - 81 - k^2 - 18k < 1000$$
$$\Rightarrow \quad 9k^2 - 18k + 9 < 1000$$
$$\Rightarrow \quad (k-1)^2 < \frac{1000}{9}$$
$$\Rightarrow \quad k - 1 < \frac{10\sqrt{10}}{3}$$
$$\text{and} \quad k - 1 < \frac{-10\sqrt{10}}{3}$$
$$\Rightarrow \quad k < \frac{10\sqrt{10}}{3} + 1$$

It is not possible because $k \in N$.

$\therefore$ Maximum possible integral value of k is 11.

55. (68) $x : x_1, x_2, x_3, \ldots x_{n+1}, x_{n+2}, \ldots x_{2n}, x_{2n+1} \ldots x_{3n}$

$$\frac{1}{3n}\sum_{i=1}^{3n} x_i^2 - (\bar{x})^2 = 4$$
$$\left[\text{where, } \bar{x} = \frac{n_1\bar{x}_1 + n_2\bar{x}_2}{n_1 + n_2} = \frac{2n(6) + n(3)}{3n} = 5\right]$$
$$\frac{1}{3n}\sum_{i=1}^{3n} x_i^2 = 25 + 4 = 29$$

$$\Rightarrow \quad \sum_{i=1}^{3n} = 87n$$

$x' : x_1 + 1, x_2 + 1, \ldots, x_{2n} + 1, x_{(2n+1)} - 1 \ldots, x_{3n} - 1$

$$\bar{x}' = \frac{2n(6+1) + n(3-1)}{3n} = \frac{16}{3}$$

$$k = \frac{1}{3n} \sum_{i=1}^{3n} x_i^2 - \frac{256}{9}$$

$$= \frac{1}{3n}\left[\sum_{i=1}^{3n} x_i'^2 + 2\sum_{i=1}^{2n} x_i' - 2\sum_{i=2n+1}^{3n} x_i' + 3n\right] - \frac{256}{9}$$

$$= \frac{1}{3n}[87n + 2(12n) - 2 \cdot 3n + 3n] - \frac{256}{9}$$

$$\Rightarrow \quad k = 36 - \frac{256}{9}$$

$$\Rightarrow \quad 9k = 324 - 256 = 68$$

56. (25) Number of boys $= 20$

Number of girls $= 30$

$$\sigma_B^2 = 2, \bar{X}_B = 12, \sigma_G^2 = 2$$

$$\bar{X}_G = \frac{50 \times 15 - 20 \times 12}{30}$$

$$= \frac{510}{30} = 17 = \mu$$

Variance of 50 candidates

$$\sigma^2 = \frac{20\sigma_B^2 + 30\sigma_G^2}{50} + \frac{20 \cdot 30}{(20+30)^2}(\bar{X}_B - \bar{X}_G)^2$$

$$= \frac{20 \times 2 + 30 \times 2}{50} + \frac{600}{2500} \times 25 = 8$$

$$\therefore \quad \mu + \sigma^2 = 17 + 8 = 25$$

57. (50) $\because \sigma(x) = 10$

$\sigma(50 + 5x) = \sigma(50) + \sigma(5x)$

$= 0 + 5\sigma(x) = 5 \times 10 = 50$

58. (54) Mean, $\bar{x} = \dfrac{3 + 7 + 9 + 12 + 13 + 20 + x + y}{8} = 10$

$\Rightarrow \quad x + y = 16 \quad$ ...(i)

Variance $= \sigma^2 = \dfrac{\Sigma(x_i)^2}{8} - (\bar{x})^2 = 25$

$$\sigma^2 = \frac{9 + 49 + 81 + 144 + 169 + 400 + x^2 + y^2}{8} - 100 = 25$$

$\Rightarrow \quad x^2 + y^2 = 148 \quad$ ...(ii)

From Eq. (i), $\quad (x + y)^2 = (16)^2$

$\Rightarrow \quad x^2 + y^2 + 2xy = 256$

Using Eq. (ii), $148 + 2xy = 256 \Rightarrow xy = 54$

59. (3) Variance $= \dfrac{\sum_{i=1}^{11} b_i^2}{11} - \left[\dfrac{\sum_{i=1}^{11} b_i}{11}\right]^2$

Let common difference of AP be d.

$$\frac{\sum_{r=0}^{10}(b_1 + rd)^2}{11} - \left[\frac{\sum_{r=0}^{10}(b_1 + rd)}{11}\right]^2$$

$$= \frac{11b_1^2 + 2b_1 d\left(\frac{10 \times 11}{2}\right) + d^2\left(\frac{10 \times 11 \times 21}{6}\right)}{11} - \left[\frac{11b_1 + \frac{10 \times 11}{2} d}{11}\right]^2$$

$$= (b_1^2 + 10b_1 d + 35d^2) - (b_1 + 5d)^2 = 10d^2$$

$\because$ Variance $= 90 \quad$ [given]

$\Rightarrow \quad 10d^2 = 90 \Rightarrow d = 3$

60. (13) Let the other two observations be x and y.

Therefore, the series is 1, 2, 6, x and y

Now, mean $(\bar{x}) = 4.4 = \dfrac{1 + 2 + 6 + x + y}{5}$

$\Rightarrow \quad 22 = 9 + x + y \Rightarrow x + y = 13 \quad$ ...(i)

Also, variance $= 8.24 = \dfrac{1}{n}\sum_{i=1}^{5}(x_i - \bar{x})^2$

i.e. $8.24 = \dfrac{1}{5}[(3.4)^2 + (2.4)^2 + (1.6)^2 + x^2 + y^2 - 2 \times 4.4(x + y) + 2 \times (4.4)^2]$

or $41.20 = 11.56 + 5.76 + 2.56 + x^2 + y^2 - 8.8 \times 13 + 38.72$

Therefore, $x^2 + y^2 = 97 \quad$ ...(ii)

But from Eq. (i), we have

$x^2 + y^2 + 2xy = 169 \quad$ ...(iii)

From Eqs. (ii) and (iii), we have

$2xy = 72 \quad$ ...(iv)

Subtracting Eq. (iv) from Eq. (ii), we get

$x^2 + y^2 - 2xy = 97 - 72$ i.e. $(x - y)^2 = 25$

or $\quad x - y = \pm 5 \quad$ ...(v)

So, from Eqs. (i) and (v), we get

$x = 9$ and $y = 4$, when $x - y = 5$

or $\quad x = 4$ and $y = 9$, when $x - y = -5$

Thus, the remaining observations are 4 and 9.

61. (4) Given, $\sum_{i=1}^{18}[x_i - \alpha] = 36$

and $\sum_{i=1}^{18}|x_i - \beta|^2 = 90$

Now, $\sum_{i=1}^{18}(x_i - \alpha) = 36$

$$\sum_{i=1}^{18} x_i - 18\alpha = 36$$

$$\sum_{i=1}^{18} x_i = 36 + 18\alpha \quad \text{...(i)}$$

Again, $\sum_{i=1}^{18} x_i^2 + 18\beta^2 - 2\beta \sum_{i=1}^{18} x_i = 90$

$$\Rightarrow \sum_{i=1}^{18} x_i^2 + 18\beta^2 - 2\beta(36 + 18\alpha) = 90. \quad \text{[using Eq. (i)]}$$

$$\Rightarrow \sum_{i=1}^{18} x_i^2 = 90 - 18\beta^2 + 2\beta(36 + 18\alpha) \quad \text{...(ii)}$$

Given, standard deviation = 1

i.e $\sigma^2 = 1$

$$\Rightarrow \frac{\sum_{i=1}^{18} x_i^2}{18} - \left(\frac{\sum_{i}^{18} x_i}{18}\right)^2 = 1$$

$$\Rightarrow \frac{1}{18}(90 - 18\beta^2 + 36\alpha\beta + 72\beta) - \left(\frac{36 + 18\alpha}{18}\right)^2 = 1$$

$$\Rightarrow 5 - \beta^2 + 2\alpha\beta + 4\beta - (2 + \alpha)^2 = 1$$

$$\Rightarrow 5 - \beta^2 + 2\alpha\beta + 4\beta - 4 - \alpha^2 + (-4\alpha) = 1$$

$$\Rightarrow -\beta^2 + 2\alpha\beta + 4\beta - \alpha^2 - 4\alpha = 0$$

$$\Rightarrow \alpha^2 + \beta^2 + (-2\alpha\beta) - 4\beta + 4\alpha = 0$$

$$\Rightarrow (\alpha - \beta)^2 + 4(\alpha - \beta) = 0$$

$$\Rightarrow (\alpha - \beta)(\alpha - \beta + 4) = 0$$

$$\Rightarrow \alpha - \beta = -4 \quad [\because \alpha \neq \beta]$$

$$|\alpha - \beta| = |-4| = 4$$